International Conference on Applied Sciences (ICAS2016)

IOP Conference Series: Materials Science and Engineering
Volume 163

Hunedoara, Romania
25 – 27 May 2016

Editors:

Ludovic Dan Lemle
Yiwen Jiang

ISBN: 978-1-5108-3526-9
ISSN: 1757-8981

Printed from e-media with permission by:

Curran Associates, Inc.
57 Morehouse Lane
Red Hook, NY 12571

Some format issues inherent in the e-media version may also appear in this print version.

Printed by Curran Associates, Inc. (2017)

For permission requests, please contact the Institute of Physics at the address below.

Institute of Physics
Dirac House, Temple Back
Bristol BS1 6BE UK

Phone: 44 1 17 929 7481
Fax: 44 1 17 920 0979

techtracking@iop.org

Additional copies of this publication are available from:

Curran Associates, Inc.
57 Morehouse Lane
Red Hook, NY 12571 USA
Phone: 845-758-0400
Fax: 845-758-2633
Email: curran@proceedings.com
Web: www.proceedings.com

Table of contents

Volume 163

International Conference on Applied Sciences (ICAS2016)
25–27 May 2016, Hunedoara, Romania

Accepted papers received: 30 November 2016
Published online: 6 January 2017

Preface

Papers

International Conference on Applied Sciences ICAS2016

Hunedoara, Romania, May 25-27, 2016

Organizer:

Universitatea Politehnica Timişoara

Politehnica University of Timişoara

in cooperation with

Military Economics Academy of Wuhan
"Henri Coanda" Air Force Academy of Brasov
Academy of Romanian Scientists
Academy of Technical Sciences of Romania - Timisoara Branch
General Association of Romanian Engineers - Hunedoara Branch

The **International Conference on Applied Sciences ICAS2016** took place in Hunedoara, Romania, in the period May 25–27, 2016 at the Engineering Faculty of Hunedoara, with the aims to serve as a platform for exchange of information between various areas of applied sciences, and to promote the communication between the scientists of different nations, countries and continents.

Topics of the conference covers a comprehensive spectrum of issues from:
>*Fundamental Sciences:* Numerical approximation and analysis, Interdisciplinary applications of mathematics and physics, Modeling and Simulation, and others...
>*Materials Science:* Metallic Materials, Composite Materials, Metal Alloys, Metallurgy, Heat Transfer, Mechanical Engineering, Mechatronics, Reliability, and others...
>*Electrical Engineering:* Circuits and Systems, Signal Processing, Computational Methods in Engineering, Software Engineering, Data Bases, and others...

The conference gathered qualified researchers whose expertise can be used to develop new engineering knowledge that has applicability potential in Industry. The number of participants was almost 190 from 6 countries (China, Romania, Germany, Denmark, Italy and Republic of Serbia, Bosnia and Herzegovina). During the three days of the conference, 3 invited speakers and 81 oral talks were delivered.

Based on the works presented at the conference, 53 selected papers have been included in this volume of *IOP Conference Series: Materials Science and Engineering*. These papers present new researches in the various fields of Materials Engineering, Mechanical Engineering, Computers Engineering, and Electrical Engineering. It's our great pleasure to present this volume of *IOP Conference Series: Materials Science and Engineering* to the scientific community to promote further researches in these areas. We sincerely hope that the papers published in this volume will contribute to the advancement of knowledge in the respective fields.

Special thanks should be directed to the **Organizing Committee** for its tremendous efforts in organizing the conference:

President
Serban Viorel-Aurel, Politehnica University of Timisoara
Honorary President
Harau Carmen, Parliament of Romania - Chamber of Deputies
Vice-President
Panoiu Caius, Politehnica University of Timisoara
Coordinator
Lemle Ludovic Dan, Politehnica University of Timisoara
Members
Abrudean Cristian, Politehnica University of Timisoara
Alexa Vasile, Politehnica University of Timisoara
Dinis Corina, Politehnica University of Timisoara
Ratiu Sorin, Politehnica University of Timisoara
Socalici Ana, Politehnica University of Timisoara
Tirian Ovidiu, Politehnica University of Timisoara
Gherman Lucian, Politehnica University of Timisoara
Secretariat
Berdie Adela, Politehnica University of Timisoara
Serban Sorina, Politehnica University of Timisoara

The members of the **Scientific Committee** who reviewed and selected the papers included in the Proceedings of ICAS2016 are:

Chairmen
Lemle Ludovic Dan, Politehnica University of Timisoara
Jiang Yiwen, Military Economics Academy of Wuhan
Vice-Chairmen
Gherman Laurian, Review of the Air Force Academy
Kiss Imre, Acta Technica Corviniensis
Session Chairmen
Fundamental Sciences Session: Pater Lucian Flavius, Politehnica University of Timisoara
Materials Science Session: Bordeasu Ilare, Politehnica University of Timisoara
Electrical Engineering Session: Deaconu Sorin, Politehnica University of Timisoara
Video-conference Session: Muscalagiu Ionel, Politehnica University of Timisoara
Members
Cioaca Catalin, "Henri Coanda" Air Force Academy of Brasov
Constantinescu George, "Henri Coanda" Air Force Academy of Brasov
Fang Zhengqi, Military Economics Academy of Wuhan
Guo Zhonghou, Military Economics Academy of Wuhan
Heput Teodor, Politehnica University of Timisoara
Lesenciuc Adrian, "Henri Coanda" Air Force Academy of Brasov
Li Xuewu, Military Economics Academy of Wuhan

Luculescu Doru, "Henri Coanda" Air Force Academy of Brasov
Miron Liliana Ecaterina, "Henri Coanda" Air Force Academy of Brasov
Osaci Mihaela, Politehnica University of Timisoara
Panoiu Manuela, Politehnica University of Timisoara
Pinca-Bretotean Camelia, Politehnica University of Timisoara
Popa Gabriel Nicolae, Politehnica University of Timisoara
Vilceanu Lucia, Politehnica University of Timisoara
Zhang Lijun, Military Economics Academy of Wuhan
Zhang Tao, Military Economics Academy of Wuhan

We thank the authors for their contributions and we would also like to express our gratitude to everyone who contributed to the success of this conference.

Ludovic Dan Lemle and *Yiwen Jiang*
Editors of the Proceedings of ICAS2016

Peer review statement

All papers published in this volume of *IOP Conference Series: Materials Science and Engineering* have been peer reviewed through processes administered by the proceedings Editors. Reviews were conducted by expert referees to the professional and scientific standards expected of a proceedings journal published by IOP Publishing.

IOP Conf. Series: Materials Science and Engineering **163** (2017) 012001 doi:10.1088/1757-899X/163/1/012001

FEM modelling of soil behaviour under compressive loads

N Ungureanu[1], V Vlăduţ[2] and S Şt Biriş[1]

[1]Politehnica University of Bucharest, Department of Biotechnical Systems Engineering, 313 Splaiul Independenţei blv., sector 6, 060042, Bucharest, Romania
[2]National Institute of Research-Development for Machines and Installations Designed to Agriculture and Food Industry – INMA Bucharest, 6 Ion Ionescu de la Brad blv., sector 1, 013813, Bucharest, Romania

E-mail: nicoletaung@yahoo.com

Abstract. Artificial compaction is one of the most dangerous forms of degradation of agricultural soil. Recognized as a phenomenon with multiple negative effects in terms of environment and agricultural production, soil compaction is strongly influenced by the size of external load, soil moisture, size and shape of footprint area, soil type and number of passes. Knowledge of soil behavior under compressive loads is important in order to prevent or minimize soil compaction. In this paper were developed, by means of the Finite Element Method, various models of soil behavior during the artificial compaction produced by the wheel of an agricultural trailer. Simulations were performed on two types of soil (cohesive and non-cohesive) with known characteristics. By applying two loads (4.5 kN and 21 kN) in footprints of different sizes, were obtained the models of the distributions of stresses occuring in the two types of soil. Simulation results showed that soil stresses increase with increasing wheel load and vary with soil type.

1. Introduction

In the actual context of continuously increasing world population, it became necessary the development of mechanized agriculture, which requires the use of heavy agricultural machinery, often on soils with high moisture content. This practice increases the risk of degradation of agricultural soil through artificial compaction, a phenomenon with multiple negative effects on the environment and on crop yields. The first paragraph after a heading is not indented.

According to Soil Science Society of America (1996), soil compaction is the process by which soil particles are rearranged so that is reduced the volume of voids, and solid particles are put into contact, thereby contributing to the increase of bulk density [1].

Soil compaction is a reduction of soil volume [2] under the action of compressive loads (external forces forming soil stress) applied on soil surface by the agricultural machinery [3], [4].

Soil compaction can be produced by natural phenomena such as rainfall impact, soaking, freeze-thaw cycles, internal tensions of soil water, rooting system of plants, soil drying. Among the artificial causes, a certain influence has the trampling of grazing animals, but the most important factor is traffic intensity of tractors and agricultural machinery, often performed on inadequate moisture conditions, with high wheel loads and high tire inflation pressures.

From an environmental point of view, soil compaction leads to: erosion, landslides, leakage of pesticides and nutrients intro groundwater, reduced capacity of water infiltration into the soil, ncreased emanations of N_2O, CH_4 and CO_2, forming of ruts. In agronomic terms, in compacted soils were reported negative effects such as: increased penetration resistance, inhibition of root development and plant growth, followed by reduced yield of agricultural crops, increased resistance to plowing and therefore, higher fuel consumption.

The most important factors influencing the artificial compaction of soil are: size of external load (wheel load), type of soil, tire inflation pressure, moisture content, footprint area between the soil and tire, shape of footprint, and the number of passes of agricultural machinery.

Stresses applied by the agricultural machinery are transmitted to different soil depths, through the footprint between the soil and tire, resulting in topsoil compaction and/or subsoil compaction.
Topsoil compaction has a significant effect on crop yield and may last for some years, but is can easily be alleviated by tillage, drying-wetting and freeze-thaw cycles, and by the action of soil biota [5]. Subsoil compaction (occurring on layers of soil below 25 cm depth) is particularly persistent [5] and the compacted layer cannot be removed by conventional tillage [6].

Distribution of stresses into the soil is influenced by factors such as: surface of contact area (footprint area) between the rolling body and soil, wheel load, tire inflation pressure, soil moisture and tire design. According to [7], the vertical stress down to 1 m depth depends both on soil contact stress (contact pressure) and wheel load.

Artificial compaction of soil is very difficult to alleviate and it became of special concern because the weight of tractors and agricultural machinery has increased significantly [8]. Hence, various modeling techniques have been used to predict the response of soil to the traffic of agricultural vehicles. The modeling predicts the distribution of stresses in the soil profile, giving informations about the depth at which the compaction takes place, and is also useful in making recommendations to farmers and designers about which agricultural machinery to use in order to minimize soil compaction [9].

Agricultural soil is not a homogeneous, isotropic and elastic ideal material, and the mathematical modeling of stress distribution is quite difficult. The Finite Element Method (FEM) is a numerical method for obtaining approximate solutions of ordinary and differential equations that describe the distribution of stresses and strains into the soil, being extremely useful for the modeling of this phenomenon [10].

However, the soil can be idealized as an elastic-plastic material, and in FEM modeling it is necessary to define some parameters, such as: Young's modulus and Poisson's ratio, which describe the elasticity, respectively the cohesion and the angle of internal friction, describing soil plasticity.

2. Methods

The aim of this FEM study was to simulate behavior of agricultural soil under the influence of wheel loads applied by an agricultural trailer equipped with tires model Danubiana 11.5/ 80-13.5 profile D179 (width 290 mm, diameter 845 mm).

The distribution of equivalent stress in the soil by FEM was performed in the QUICKFIELD Student program. The analysis was carried out on the 2D geometric model of plane load. A volume of soil with dimensions of 1 m x 1 m x 1 m was considered, and on its surface was applied a pressure plate with the area equal to the area of footprint between the soil and tire. Over the pressure plate was applied a uniformly distributed load (located in the upper corner of the meshed section in Figure 1).

By meshing the physical model of soil through FEM, was obtained the meshed half model presented in Figure 1. To obtain high precision of results, the model was meshed in maximum 250 nodes, with smaller discrete elements in the area where stresses are concentrated.

The simulation was performed for the two types of soil, whose characteristics are presented in Table 1. Besides the characteristics of the two types of soil, other parameters used for the modeling of wheel load effect on the soil were the width of the footprint between the soil and tire, respectively the contact pressure applied on the contact surface between soil and tire, for two wheel loads (4.5 kN and 21 kN). For each wheel load, tire inflation pressure was varied (180 kPa, 240 kPa, and 300 kPa), thereby obtaining different footprint areas. Contact pressures were computed as ratio between wheel load and footprint area. The values of these parameters are presented in Table 2.

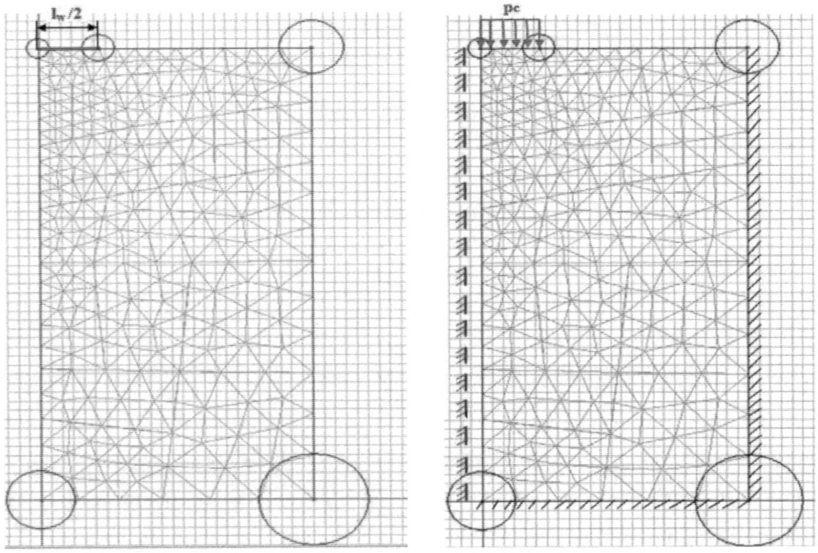

Figure 1. Meshed model of the analyzed soil (half model)

Table 1. Characteristics of cohesive soil [11] and non-cohesive soil [12] used in FEM modeling

Soil characteristic	Cohesive soil (clay)	Non-cohesive soil (sandy-loam)
Young's modulus of elasticity, E [kPa]	3000	13000
Poisson's ratio, ν	0.329	0.3
Cohesion, c [kPa]	18.12	2
Angle of internal friction, φ [°]	30	32
Density, ρ [kg/m^3]	1270	2040
Moisture, w [%]	24	6.5

Table 2. Other parameters used in FEM modeling

Wheel load, Q [kN]	Tire inflation pressure, p_i [kPa]	Footprint area, A [m^2]	Footprint width, l_w [m]	Contact pressure, p_c [kPa]
	180	0.03	0.206	150
4.5	240	0.028	0.190	161
	300	0.023	0.156	196
	180	0.099	0.248	212
21	240	0.077	0.238	273
	300	0.073	0.220	288

For each simulation, footprint width was fixed on the surface of the analyzed soil volume. The force applied to the soil by the agricultural machinery (contact pressure) was considered uniformly distributed in the footprint.

By applying different loads on the meshed model of the agricultural soil, were obtained various distributions of equivalent stresses in the two types of soil, during the artificial compaction caused by the tire of agricultural machinery.

3. Results and discussions

Figures 2 and 3 show the distribution of equivalent stress by the von Mises criterion, occurring in the two types of soil, for a wheel load of 4.5 kN and tire inflation pressure of 180 kPa.

These half-models correspond to contact pressure p_c = 150 kPa, uniformly distributed in a footprint area A= 0.03 m^2 (footprint width l_w = 0.206 m).

In the cohesive soil (Figure 2), the maximum stress of 88.2 kPa was concentrated in the topsoil at a depth of 10-12 cm, and the minimum stress was 1.91 kPa. Referring to Figure 3, it can be seen that in non-cohesive soil (sandy-loam) the highest stress had a value of 90.6 kPa and the minimum stress was 1.88 kPa.

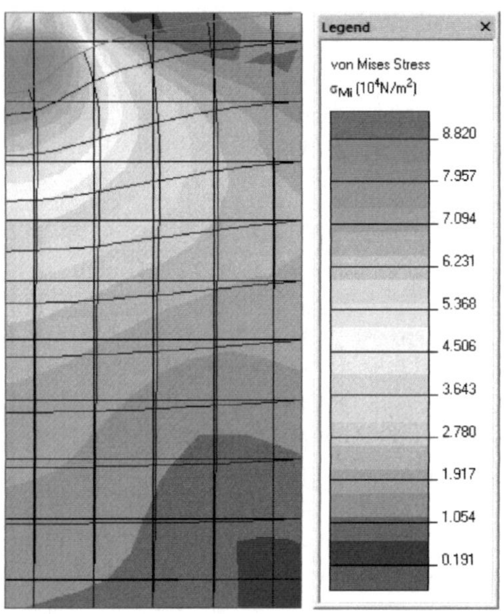

Figure 2. Distribution of equivalent stress in the cohesive soil, for p_c = 150 kPa and l_w = 0.206 m

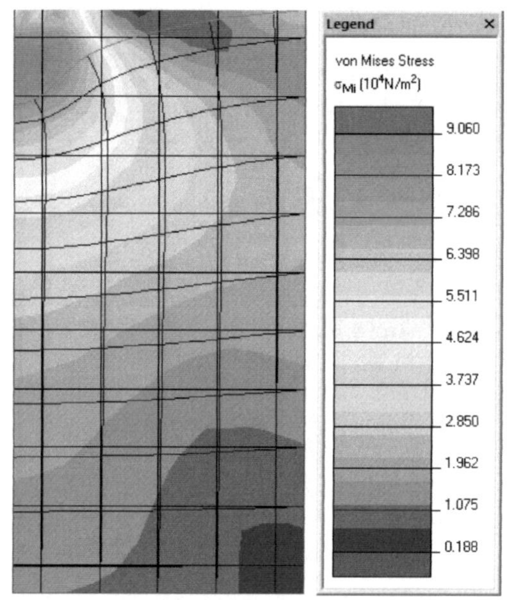

Figure 3. Distribution of equivalent stress in the non-cohesive soil, for p_c = 150 kPa and l_w = 0.206 m

In Figure 4 and Figure 5 are presented the distributions of equivalent stresses by von Mises criterion in the two types of soil, obtained by simulation, considering a wheel load of 4.5 kN at tire inflation pressure 240 kPa. Contact pressure p_c = 161 kPa is considered to be uniformly distributed in a footprint of area A= 0.028 m^2 (with footprint width l_w = 0.190 m). By analyzing the half-models it can be noted that in the cohesive soil (Figure 4), the maximum stress was 97 kPa and the minimum stress was 1.97 kPa. In the non-cohesive soil (Figure 5), the highest stress was 99.81 kPa and the minimum stress was 1.94 kPa.

In Figures 6 and 7 are presented the distributions of equivalent stresses by von Mises criterion, considering a wheel load of 4.5 kN at tire inflation pressure 300 kPa. These half- models correspond to a contact pressure of 196 kPa, uniformly distributed in the footprint with area of 0.023 m^2 (footprint width l_w = 0.156 m).

From the analysis of Figure 6 it can be seen that in the cohesive soil (clayey), the highest stresses were recorded to about 10-12 cm depth under the wheel and they had a value of 96.11 kPa, while the minimum stress was 1.63 kPa.

Figure 7 shows that in the non-cohesive soil (sandy-loam) for a wheel load of 4.5 kN, the highest stress had a value of 98.6 kPa and the minimum stress was about 1.6 kPa, similar to that occuring in the clayey soil.

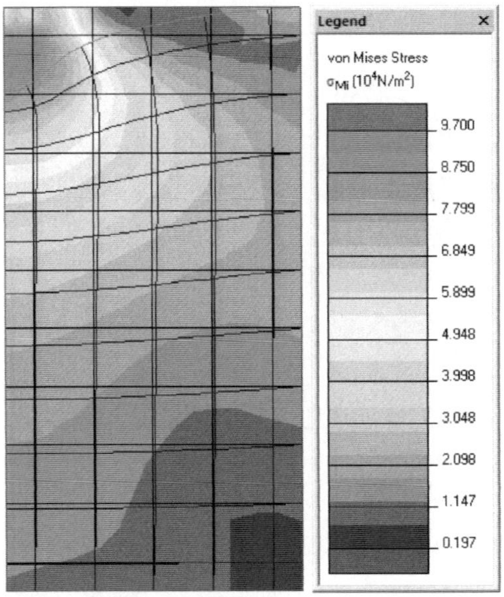

Figure 4. Distribution of equivalent stress in the cohesive soil, for p_c = 161 kPa and l_w = 0.190 m

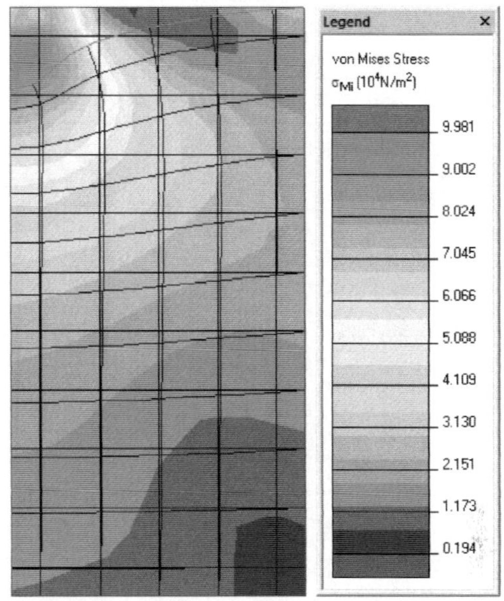

Figure 5. Distribution of equivalent stress in the non-cohesive soil, for p_c = 161 kPa and l_w = 0.190 m

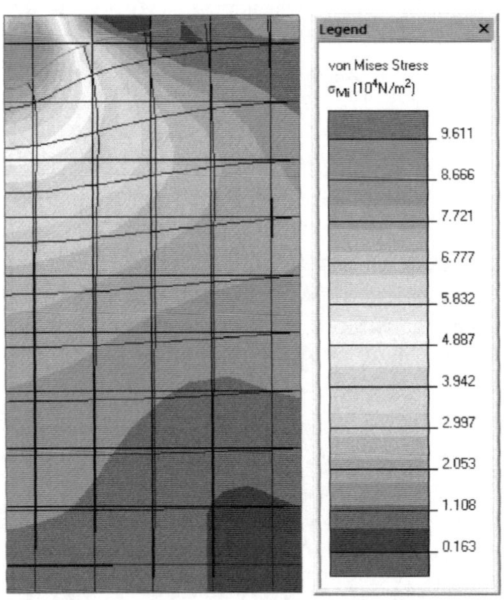

Figure 6. Distribution of equivalent stress in the cohesive soil, for p_c = 196 kPa and l_w = 0.156 m

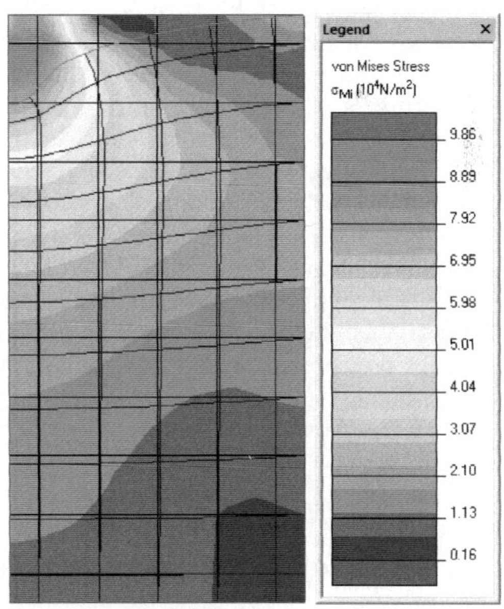

Figure 7. Distribution of equivalent stress in the non-cohesive soil, for p_c = 196 kPa and l_w = 0.156 m

IOP Conf. Series: Materials Science and Engineering **163** (2017) 012001 doi:10.1088/1757-899X/163/1/012001

Table 3 presents some values of equivalent stresses obtained by FEM modeling, in the two analyzed types of soil, from soil surface to 1000 mm depth, considering a wheel load of 4.5 kN.

Table 3. Distribution of equivalent stress for 4.5 kN wheel load, for different footprint widths

Soil depth [mm]	l_w = 206 mm	l_w = 190 mm	l_w = 156 mm
Equivalent stress in the cohesive soil (clayey), σ_{Mi} [N/m^2]			
0	58927	53951	53675
200	72682	70936	62337
400	38739	36911	30942
600	24138	22847	18966
800	17643	16724	13825
1000	12818	11998	9905
Equivalent stress in the non-cohesive soil (sandy-loam), σ_{Mi} [N/m^2]			
0	58928	62596	62214
200	72682	72980	63946
400	38740	38348	32126
600	24138	23777	19734
800	17643	17333	14329
1000	12818	12587	10392

From the analysis of Figures 2-7 and data presented in Table 3 it can be observed that the maximum equivalent stresses are generally concentrated in the layer of soil located between 50 -150 mm depth, in both types of soil. This is due to flow and agglomeration phenomenon of soil particles under the action of wheel load, resulting in a denser layer of soil.

The value of stress that propagates under this layer decreases with soil depth, and at the same soil depth, the value of equivalent stress decreases with decreasing footprint width. Hence, the two types of soil will compact in the arable layer at depths up to 400 mm, where the equivalent stresses exceed 27.5 kPa (as noted in various studies in the literature).

Half-models obtained for a wheel load of 21 kN at different footprint widths are presented next.

Figures 8 and 9 present the distribution of equivalent stress by von Mises criterion in the two types of soil, for a wheel load of 21 kN and tire inflation pressure of 180 kPa. These half-models correspond to a contact pressure p_c = 212 kPa, uniformly distributed in a footprint whose area is 0.099 m^2 (footprint width l_w = 0.248).

Study results showed similar values for the two types of soil. Thus, in the cohesive soil, maximum stress was 133 kPa and minimum stress 3.4 kPa, while in the non-cohesive soil were obtained maximum stresses of 137 kP and minimum stresses of 3.3 kPa.

Figures 10 and 11 show the distribution of equivalent stress by von Mises criterion, obtained by simulation for the two types of soil, considering a wheel load of 21 kN and tire inflation pressure of 240 kPa. These half-models correspond to a contact pressure p_c = 273 kPa, uniformly distributed in a footprint area of 0.077 m^2 (footprint width l_w = 0.238). In the cohesive soil, the highest stress was 167 kPa and the minimum stress was 4.1 kPa, while in non-cohesive soil the highest stress was 172 kPa and the minimum stress was 4 kPa.

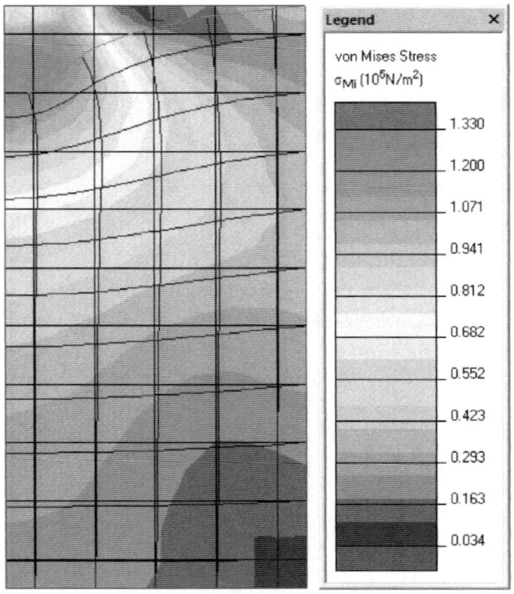

Figure 8. Distribution of equivalent stress in the cohesive soil, for p_c = 212 kPa and l_w = 0.248 m

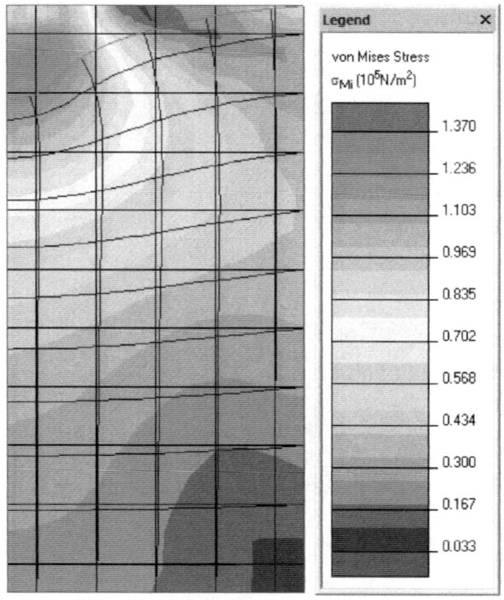

Figure 9. Distribution of equivalent stress in the non-cohesive soil, for p_c = 212 kPa and l_w = 0.248 m

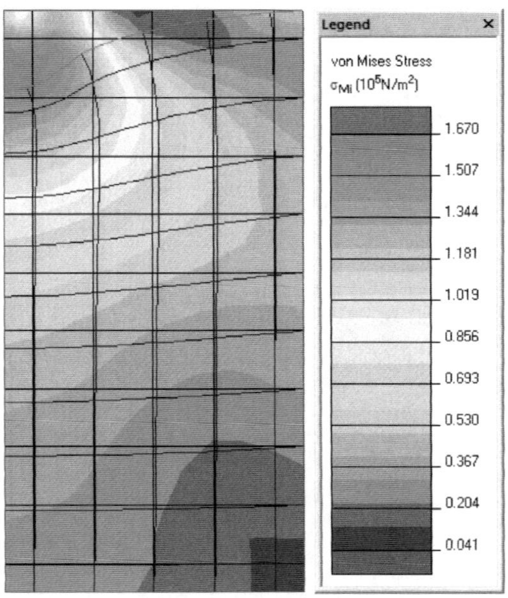

Figure 10. Distribution of equivalent stress in the cohesive soil, for p_c = 273 kPa and l_w = 0.238 m

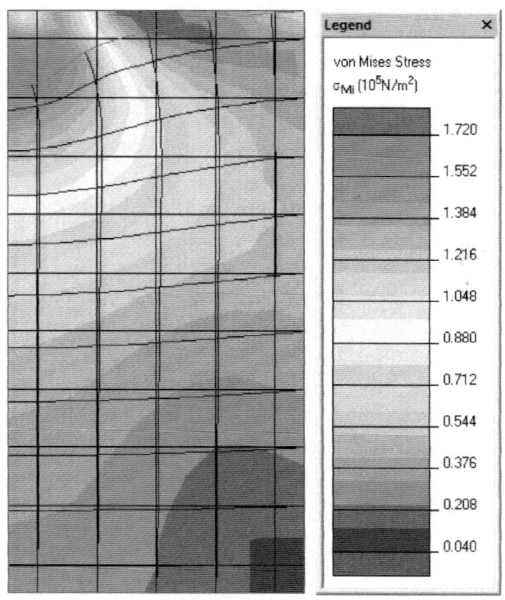

Figure 11. Distribution of equivalent stress in the non-cohesive soil, for p_c = 273 kPa and l_w = 0.238 m

Figures 12 and 13 present the distribution of equivalent stress by von Mises criterion, considering a wheel load of 21 kN and a tire inflation pressure of 300 kPa. These half-models correspond to a contact pressure p_c = 288 kPa, uniformly distributed in a footprint whose area is 0.073 m^2 (footprint width l_w = 0.220). In the cohesive soil, the highest stress of 175 kPa was concentrated in the topsoil, and in the non-cohesive soil the highest stress was 180 kPa. In both types of soil, the minimum stress was 4 kPa.

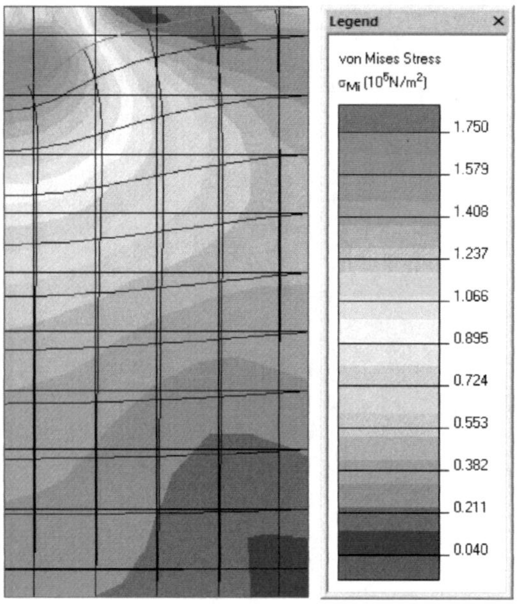

Figure 12. Distribution of equivalent stress in the cohesive soil, for p_c = 288 kPa and l_w = 0.220 m

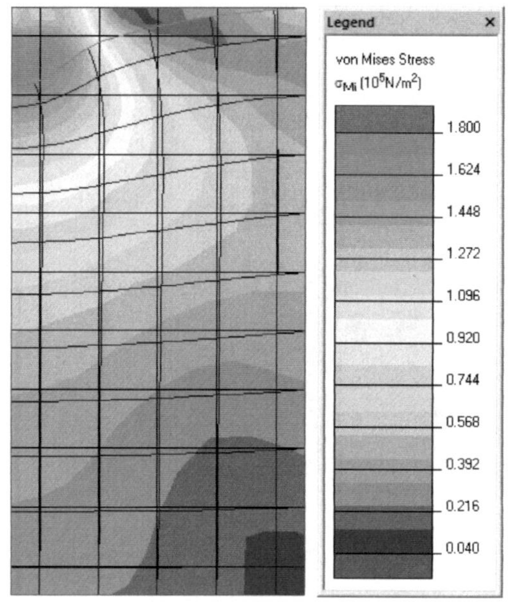

Figure 13. Distribution of equivalent stress in the non-cohesive soil, for p_c = 288 kPa and l_w = 0.220 m

In Table 4 are given some values of equivalent stresses obtained by FEM modeling, in the two analyzed types of soil, from soil surface to 1000 mm depth, considering a wheel load of 21 kN.

Table 4. Distribution of equivalent stress for 21 kN wheel load, for different footprint widths

Soil depth [mm]	l_w = 248 mm	l_w = 238 mm	l_w = 220 mm
Equivalent stress in the cohesive soil (clayey), σ_{Mi} [N/m^2]			
0	71201	104739	95477
200	110757	136984	138082
400	61038	75110	74331
600	38469	47119	46425
800	28384	34714	34144
1000	20414	24956	24524
Equivalent stress in the non-cohesive soil (sandy loam), σ_{Mi} [N/m^2]			
0	82583	117989	110775
200	113940	140968	142135
400	63503	78086	77273
600	40059	49057	48335
800	29417	35977	35386
1000	21413	26176	25724

From the analysis of Figures 8-13 and data presented in Table 4 it can be observed that at a load of 21 kN applied in footprint areas of various sizes, on two types of soil, compaction will occur in the arable layer, but also under this layer, on depths at equivalent stresses in the soil exceed 27.5 kPa, determined empirically in the literature.

Thus, we can recommend the use of agricultural trailers equipped with tires having large width and diameter, because this would increase the footprint in which the loads are applied, and therefore stresses will propagate at shallower depths in the soil, which is much easier to alleviate compared to deep compaction.

FEM analysis was verified by laboratory tests and test results proved that FEM is an accurate method for predicting stress distribution in soil depth.

4. Conclusions

Numerical methods such as the Finite Element Method require the generation of a large number of finite elements to achieve a high accuracy of the results, but they can be successfully applied to simulate the phenomenon of soil compaction.

The results obtained by FEM simulation showed that the values of stresses occurring in the soil increase with increasing the wheel load and they vary depending on soil characteristics.

By comparing the results obtained for the same load, it was found that the clayey soil (cohesive) has the highest predisposition to compaction, while the sandy-loam soil (non-cohesive) is less predisposed to compaction.

In agricultural soil, due to higher tire inflation pressures, smaller footprint areas are formed, the soil deforms more and the stresses are distributed deeper into the soil, requiring remedial works such as deep loosening. At lower tire inflation pressures, the tire deforms more, footprint area is greater, contact pressure is lower, the soil deforms less and stresses are transmitted at shallower depths.

The increase of the footprint area not necessarily leads to lower stress intensity in the soil, but rather to limiting the distribution of high stresses on soil depth, respectively to their expansion in horizons closer to soil surface.

FEM simulation models of the behavior of the two types of soil can be useful to farmers and designers of agricultural machinery, for optimizing wheel loads and tire inflation pressure, in order to reduce the risk of artificial compaction of agricultural soil.

Acknowledgement

This work has been funded by the Ministry of National Education and Research, through the UEFISCDI, within the project entitled ***„Conservative tillage technology"*, contr. 181/2014.**

References

[1] Hamza M A and Anderson W K 2005 Soil compaction in cropping systems. A review of the nature, causes and possible solutions, *Soil & Tillage Research* **82** 121-145

[2] Taghavifar H and Mardani A 2014 Prognostication of vertical stress transmission in soil profile by adaptive neuro-fuzzy inference system based modeling approach, *Measurement* **50** 152-159

[3] Abou-Zeid A, Kushwaha R L and Stilling D S D 2004 Distributed soil displacement associated with surface loading, St. Joseph (MI), *ASAE Paper 031024*

[4] Pytka J and Szymaniak G *Investigations of stress state in soil under tractor tyres*, pp 14-18 (http://www.pan-ol.lublin.pl/wydawnictwa/TMot4/Pytka.pdf)

[5] Schjønning P, van den Akker J J H, Keller T, Greve M H, Lamandé M, Simojoki A, Stettler M, Arvidsson J and Breuning-Madsen H 2016 *Soil compaction*, Chapter **6**, pp 69-78, Soil Threats in Europe - Status, methods, drivers and effects on ecosystem services, Publisher: EU Joint Research Centre

[6] Wolkowski R and Lowery B *Soil compaction: causes, concerns, and cures*, **A3367** (http://www.soils.wisc.edu/extension/pubs/A3367.pdf)

[7] Nankali N, Namjoo M and Maleki M R 2012 Stress analysis of tractor tire interacting with soil using 2D Finite Element Model, *Int J Advanced Design and Manufacturing Technology* **5**(3) 107-111

[8] Rashidi M, Gholami M, Ranjbar I and Abbasi S 2010 Finite Element Modeling of soil sinkage by multiple loadings, *American-Eurasian J. Agric & Environ. Sci* **8**(3) 292-300

[9] Cueto O G, Iglesias Coronel C E, Recarey Morfa C A, Urriolagoitia Sosa G, Hernández Gómez L H, Urriolagoitia Calderón G and Herrera Suárez M 2013 Three dimensional finite element model of soil compaction caused by agricultural tire traffic, *Computers and Electronics in Agriculture* **99** 146-152

[10] Biriş S Şt, Ungureanu N, Vlăduţ V, Voicu Gh, Manea M and Crăciun V 2010 FEM model for determining the influence of tire pressure on artificial soil compaction of agricultural soil, *Bulletin of the Polytechnic Institute of Iaşi-Section Machinery Construction* **LVI(LX)**(4B) 347 – 356

[11] Kushwaha R L and Shen J 1995 Finite element analysis of dynamic interaction between soil and tillage too, *Transaction of ASAE* **37**(5) 1315-1319

[12] Gysi M, Maeder V and Weisskopf P 2001 Pressure distribution underneath tyres of agricultural vehicles, *Transactions of the ASABE* **44**(6) 1385-1389

A new concept for stainless steels ranking upon the resistance to cavitation erosion

I Bordeasu[1], M O Popoviciu[2], L C Salcianu[1], C Ghera[1], L M Micu[3], R Badarau[1], A Iosif[4], L D Pirvulescu[5] and C E Podoleanu[4]

[1]Politehnica University of Timisoara, Faculty of Mechanical Engineering, Mihai Viteazul Street, Nr. 1, 300222 Timisoara, Romania
[2]Academy of Romanian Scientists, Mihai Viteazul Street, Nr. 1, 300222 Timisoara, Romania
[3]Universitatea de Stiinte Agricole si Medicina Veterinara a Banatului „Regele Mihai I Romaniei" din Timisoara, Calea Aradului nr.119, 300645 Timişoara, România
[4]Politehnica University of Timisoara, Faculty of Civil Engineering, Traian Lalescu Street, Nr. 2, 300223 Timisoara, Romania
[5]Casa Auto Timisoara Merceds-benz

E-mail: ilarica59@gmail.com

Abstract. In present, the ranking of materials as their resistance to cavitation erosion is obtained by using laboratory tests finalized with the characteristic curves mean depth erosion against time MDE(t) and mean depth erosion rate against time MDER(t). In some previous papers, Bordeasu and co-workers give procedures to establish exponential equation representing the curves, with minimum scatter of the experimental obtained results. For a given material, both exponential equations MDE(t) and MDER(t) have the same values for the parameters of scale and for the shape one. For the ranking of materials is sometimes important to establish single figure. Till now in Timisoara Polytechnic University Cavitation Laboratory were used three such numbers: the stable value of the curve MDER(t), the resistance to cavitation erosion ($R_{cav} \equiv 1/MDER_{stable}$) and the normalized cavitation resistance R_{ns} which is the rate between v_s = $MDER_{stable}$ for the analyzed material and v_{se}= $MDER_{se}$ the mean depth erosion rate for the steel OH12NDL (R_{ns} = v_s/v_{se}). OH12NDL is a material used for manufacturing the blades of numerous Kaplan turbines in Romania for which both cavitation erosion laboratory tests and field measurements of cavitation erosions are available. In the present paper we recommend a new method for ranking the materials upon cavitation erosion resistance. This method uses the scale and shape parameters of the exponential equations which represents the characteristic cavitation erosion curves. Till now the method was applied only for stainless steels. The experimental results show that the scale parameter represents an excellent method for ranking the stainless steels. In the future this kind of ranking will be tested also for other materials especially for bronzes used for manufacturing ship propellers.

1. Introduction

The concern of the researchers in ranking the materials after the cavitation erosion resistance is a very old one, perhaps the first attempt were made by Parsons in 1883 [1] preoccupied for ship propellers and afterwards by Pilaev and Edel interested in hydraulic turbines [2].

With the increasing of the studies the problem was analyzed by numerous researchers such as Garcia-Hammitt (1960) [3], Thiruvengadam (1963) [4], Heymann (1970) [5], Noskievici (1983) [6], Sakai & Shima (1987) [7], Bordeasu (1997) [8], [9]. It was put into evidence the influence of various factors such as the chemical composition, the mechanical characteristics, the heat treatments and metallographic structure, etc. For the laboratory devices the standardization recommended by ASTM for vibratory equipments [10] was extremely benefic. For a first rapid evaluation of different materials there is necessary a single figure. Till now as such a single figure was chosen from the following parameters: the final value of the lost mass/volume, the maximum value of the erosion rate, the stable value of the mean depth erosion rate or the inverted of the mean depth erosion rate (Rcav≡1/MDER).

It must be noted that every model have its benefits and its limits. That is why the present contribution proposes a new concept of correlating the cavitation erosion resistance with the numerical values of the scale/shape parameters from the equations describing the cavitation erosion characteristic curves MDE(t) and MDER(t). It was observed that for numerous stainless steels tested in our laboratory only the scale parameter has great importance. In consequence we propose the use of this parameter as the most useful single figure method for rapid comparisons.

2. Methods for determining the cavitation erosion resistance

Commonly the cavitation erosion resistance is evaluated in all the important laboratories such as those in Michigan [3], Gdansk [11], Wuxi [8], Timişoara [6], [8], [12-17], respecting the new ASTM G32-2010 Standard [10]. After tests, the results are plotted as cavitation erosion characteristic curves and compared with a standard material for which the behavior both in laboratory and field conditions is known. In the comparisons there are considered one or more of the following parameters: the final slope of the cumulative mean depth erosion (MDE) curves (Fig. 1), the normalized cavitation erosion resistance, the mean depth erosion rate (Fig. 2), the incubation period, the time interval necessary to obtain some penetration depth, some mass loss or some volume loss. Because the deterioration degree of a certain material is strongly influenced by the test device running parameters it was necessary to establish the ASTM G32-2010 Standard for all laboratories. Until now, neither of the parameters indicated by the Standard ASTM G32-2010 is accepted as unique figure for establishing the cavitation erosion resistance [8].

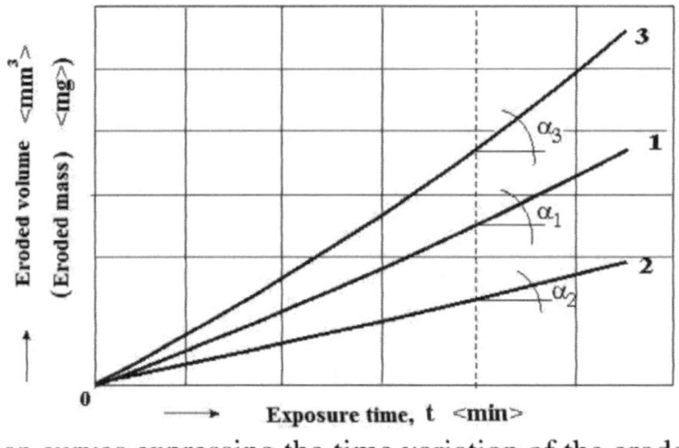

Figure 1. Ideal mean curves expressing the time variation of the eroded mass/volume [2]
1, 2, 3 – Materials with different cavitation erosion behavior

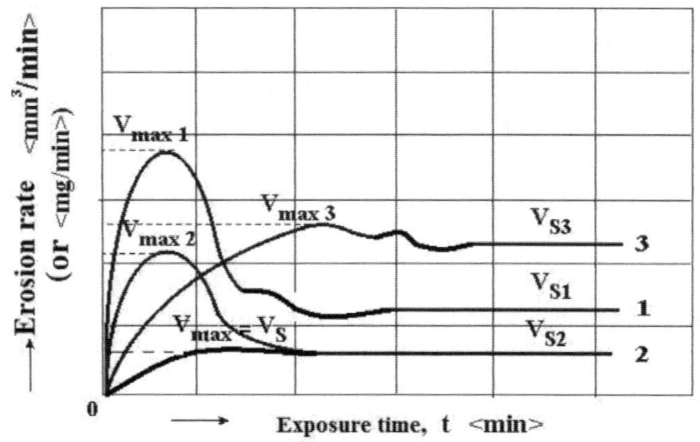

Figure 2. Ideal mean curves expressing the time variation of the erosion rate [2]
(v_{max}- maximum value of the erosion rate; v_s- value of the stable erosion rate)
1, 2,3 – Materials with different cavitation erosion behavior

In Timisoara „Politehnica" University Cavitation Laboratory, taking into account the 70 years of experience in the field as well as the ASTM G32-2010 Standard recommendations, in order to characterize the resistance to cavitation erosions are used three parameters. Two of them represent the characteristic curves of the mean depth erosion MDE(t) and the mean depth erosion rate MDER(t) [3], [12], [15], [17]. Those curves for the tested material are compared with that of a standard material. This method is an excellent criterion to certify the behavior of various stainless steels, marine bronzes with or without hardening heat treatments or even for those pulverized with ceramics powders and afterward treated with laser beams [18]. The third one R_{ns} (normalized resistance) is a dimensionless parameter established by dividing the final rate (the stable value) of the researched material to that of the standard material. As standard material is used the stainless steel OH12NDL of Russian receipt for which we have both laboratory and field results (in Romanian Power Plants) for extremely long periods.

Below are presented the relations established by Bordeasu and co-workers [9], [12], [14], [16] used for plotting the curves MDE(t) and MDER(t) which allow to determine the parameters used for evaluation/comparison of the cavitation erosion resistance:

- for the mean depth erosion

$$\mathbf{MDE\ (t) = A \cdot t \cdot (1\text{-}e^{-B \cdot t})} \quad (1)$$

- for the mean depth erosion rate

$$\mathbf{MDER(t) = A \cdot (1\text{-}e^{-B \cdot t}) + A \cdot B \cdot t \cdot e^{-B \cdot t}} \quad (2)$$

where: ***A*** is a scale parameter statistically established for realizing the approximation curve, with the condition of a minimum scatter for the measured points with regard to the curve;
B is a shape parameter.

The normalized resistance to cavitation erosion R_{ns} can be written as:

$$\mathbf{R_{ns} = v_s/v_{se}} \quad (3)$$

where:

v_s - is a tendency value at which the erosion rate curve remains approximately stable (regardless of the criterion taken in to discussion: the eroded mass, the lost volume or the penetration depth);

v_{se} – is the same value for the standard material (for our laboratory the material employed is the stainless steel OH12NDL).

3. Researched materials. Test devices

The materials used in the present work are cast or rolled stainless steels, annealed, destined for manufacturing hydraulic machineries pieces subjected to intense cavitation. Because those steels are not standard ones, with the exception of the steel OH12NDL, for identification were used the following symbols C**x**Cr**y**Ni**z**, where:

x = 1; 036; 030; represent the carbon content (1- represent 0.1 %, 036 represent 0.036 %, 030- represent 0.030 %);

y = 2, 6, 10, 12, 13, 14, 16, 18, 24 – represent the approximate chromium content in %

z = 05, 2, 4, 6, 8, 10 - represent the approximate nickel content in %.

All the necessary data fore those steels were taken from the following references:

- for steels C1Cr6Ni10, C1Cr10Ni10, C1Cr18Ni10, C1Cr24Ni10, C036Cr13Ni10, C036Cr14Ni10, C036Cr16Ni10, C036Cr18Ni10, OH12NDL [15], [16],
- for steels C1Cr12Ni05, C1Cr12Ni2, C1Cr12Ni6, C1Cr12Ni10, C036Cr12Ni6, C036Cr12Ni8, C036Cr12Ni2, C036Cr12Ni4 [13], [14]
- for steels C030Cr12Ni2, C030Cr12Ni6 [17].

Table 1 gives the ranking after the normalized cavitation resistance presented by Jurchela and Karabenciov [14], [16]. In the same table are given also the structure of constituents determined by using the Schaffler diagram (see Fig.3).

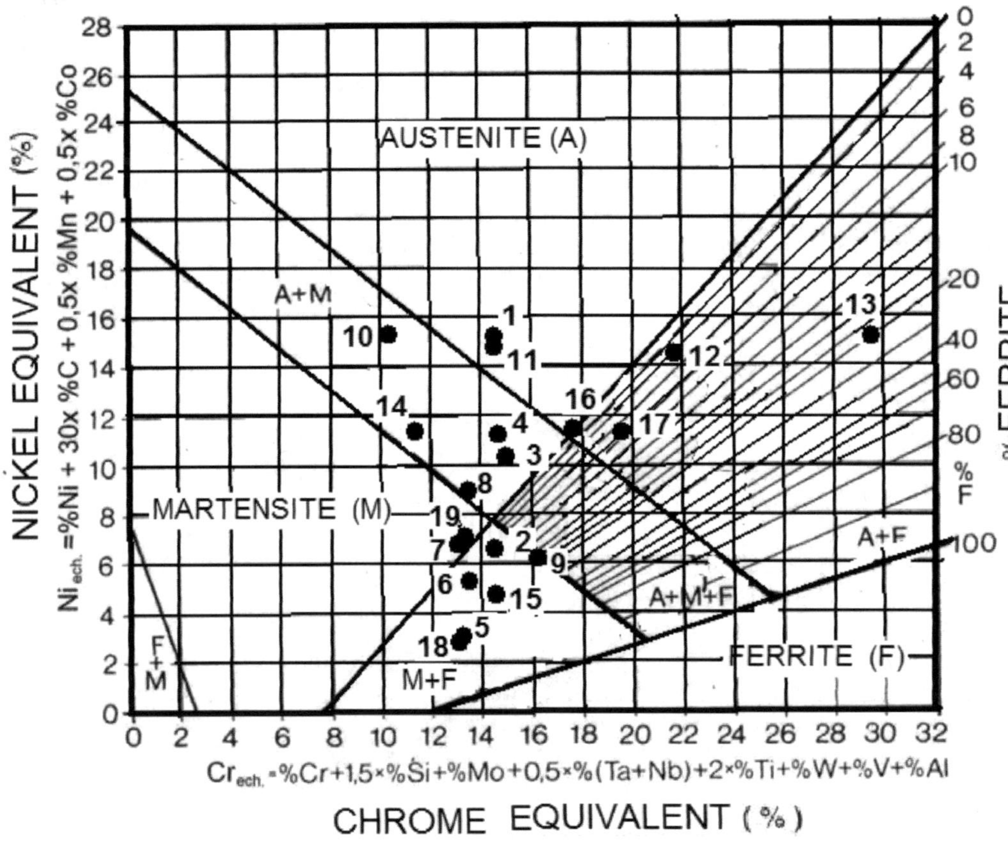

Figure 3. Position of the analyzed stainless steels in Schäffler diagram
1- C1Cr12Ni10, 2- C1Cr12Ni2, 3- C1Cr12Ni6, 4- C036Cr14Ni10, 5-C036Cr12Ni2, 6- C036Cr12Ni4, 7- C036Cr12Ni6, 8- C036Cr12Ni8, 9- OH12NDL, 10- C1Cr6Ni10, 11- C1Cr10Ni10, 12- 1Cr18Ni10, 13- C1Cr24Ni10, 14- C036Cr13Ni10, 15- C1Cr12Ni05, 16- C036Cr16Ni10, 17- C036Cr18Ni10, 18- 030Cr12Ni2, 19- C030Cr12Ni6

Table 1. Ranking of the steels after the normalized cavitation erosion resistance [9], [11]

Cavitation resistance	Material	R_{ns}	Structure
Super resistant, $R_{ns} < 0,2$	C1Cr12Ni6	0,11	60%A+40%M
Excellent $R_{ns} = [0,2 \div 0,4)$	C1Cr6Ni10	0,24	32% M + 68% A
	C1Cr12Ni10	0,29	100 %A
Very good $R_{ns} = [0,4 \div 0,8)$	C1Cr24Ni10	0,41	81% A + 19% F
	C1Cr18Ni10	0,43	98% A + 2% F
	C1Cr10Ni10	0,44	100% A
	C1Cr12Ni2	0,51	90%M+10%F
	C030Cr12Ni6	0.57	100 % M
	C036Cr12Ni8	0,59	90 %M +10%A
	C036Cr12Ni6	0,62	100 %M
	C036Cr13Ni10	0,73	55% M + 45% A
	C036Cr14Ni10	0,76	30% M + 70% A
Good $R_{ns} = [0,8 \div 1,6)$	C1Cr12Ni05	0,88	75% M+ 25 % F
	C036Cr12Ni2	0,88	55 %M+45 %F
	C036Cr12Ni4	0,88	86%M+14 %F
	C036Cr16Ni10	0,92	100% A
	OH12NDL	1,0	84%M+16%F
	C036Cr18Ni10	1,05	93% A + 7% F
	C030Cr12Ni2	1,14	40%M+60%F

All the 19 steels were tested in the Timisoara „Politehnica" University Cavitation Laboratory using a device with piezoelectric crystals. All the testing conditions regarding the procedure and the installation parameters (vibration frequency and amplitude, power of the ultrasonic electric generator, specimen diameter, water temperature) were in conformity with the recommendation of the Standard ASTM G32-2010 and respecting the laboratory custom regarding: the total cavitation exposure, the intermediary measurement intervals, the state of the specimen surface, the preparation of the specimen before testing, preservation of the specimen during pauses between test intervals, the mass loss measurements and the construction of the specific curves [14], [16], [18], [19].

4. Correlation between cavitation erosion resistance and the parameters of characteristic curves

Figure 3 present the diagrams for each cavitation erosion resistance in conformity with the data of Table 1 by using the relation (2). Table 2 presents the values of the parameters A and B as well as the cavitation resistance defined by the relation (4):

$$\mathbf{R_{cav} = 1/MDER_s} \quad (4)$$

where $MDER_s$ is the mean depth erosion rate in the final stages (the constant level of MDER).

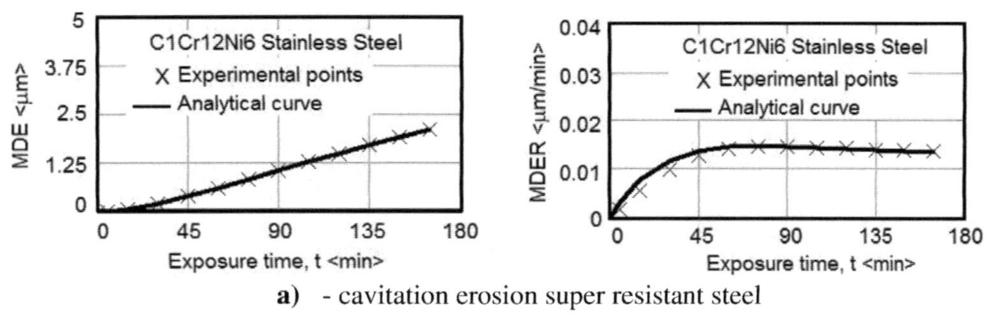

a) - cavitation erosion super resistant steel

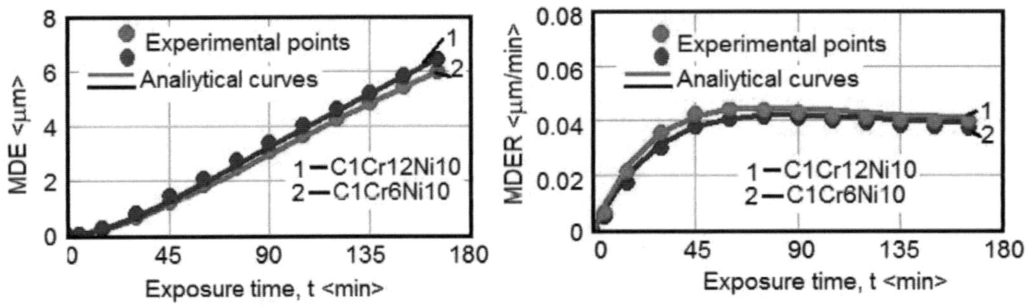

b) - steel with excellent resistance

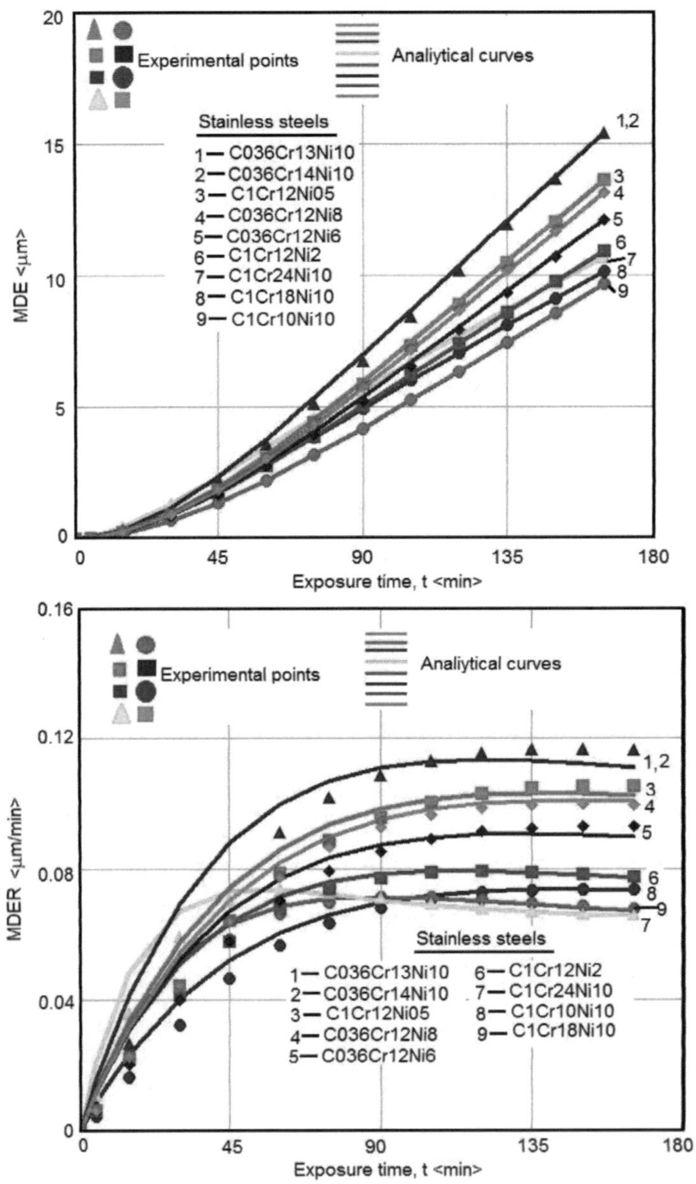

c) - steel with very good resistance

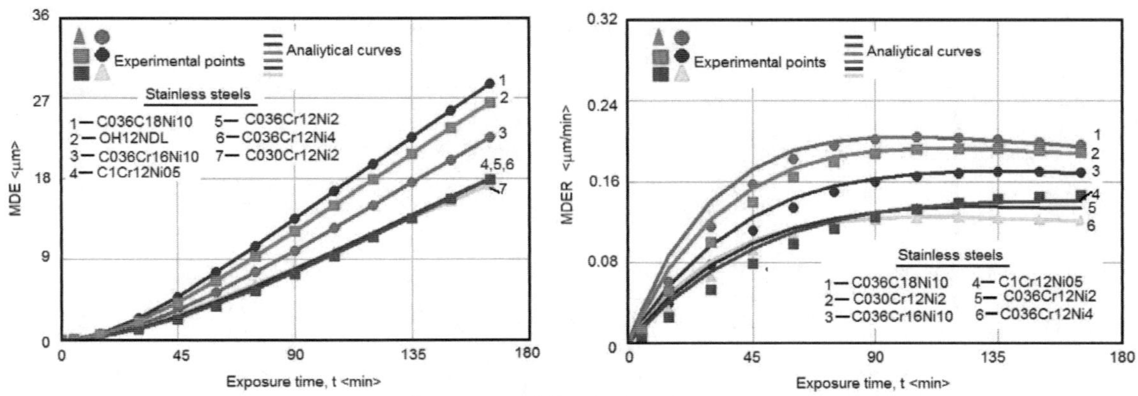

d) - steel with good resistance

Figure 4. Cavitation erosion characteristic curves

a- cavitation erosion super resistant steel; **b**-steel with excellent resistance; **c**- steel with very good resistance; **d**- steel with good resistance

Table 2 presents the values of the scale parameters A and shape parameters B for the mediation curves as well as the cavitation erosion resistance R_{cav}. The values are ranked after the parameter R_{ns} from Table 1.

Table 2. Ranking of steels after cavitation erosion resistance R_{cav} and comparisons with the values of scale and shape parameters

Cavitation erosion	Material	R_{cav}	Parametrul A	Parametrul B
Super resistant	C1Cr12Ni6	78	0.013	0.027
Excellent	C1Cr6Ni10	23	0.040	0.022
	C1Cr12Ni10	22	0.037	0.024
Very good	C1Cr24Ni10	14	0.066	0.025
	C1Cr12Ni2	13	0.070	0.018
	C1Cr18Ni10	13	0.068	0.014
	C1Cr10Ni10	13	0.067	0.013
	C036Cr12Ni6	11	0.080	0.015
	C036Cr14Ni10	10	0.090	0.015
	C036Cr12Ni8	10	0.090	0.013
	C030Cr12Ni6	10	0.086	0.016
	C036Cr13Ni10	9	0.100	0.017
Good	C1Cr12Ni05	8	0.110	0.018
	C036Cr12Ni2	7	0.120	0.014
	C036Cr12Ni4	7	0.120	0.014
	C036Cr16Ni10	6	0.150	0.015
	OH12NDL	5	0.170	0.018
	C036Cr18Ni10	5	0.180	0.020
	C030Cr12Ni2	5	0.191	0.026

In Figure 5 is presented the correlation between the cavitation erosion resistance R_{cav} and the parameter A. It can be seen that this variation is approximate exponential.

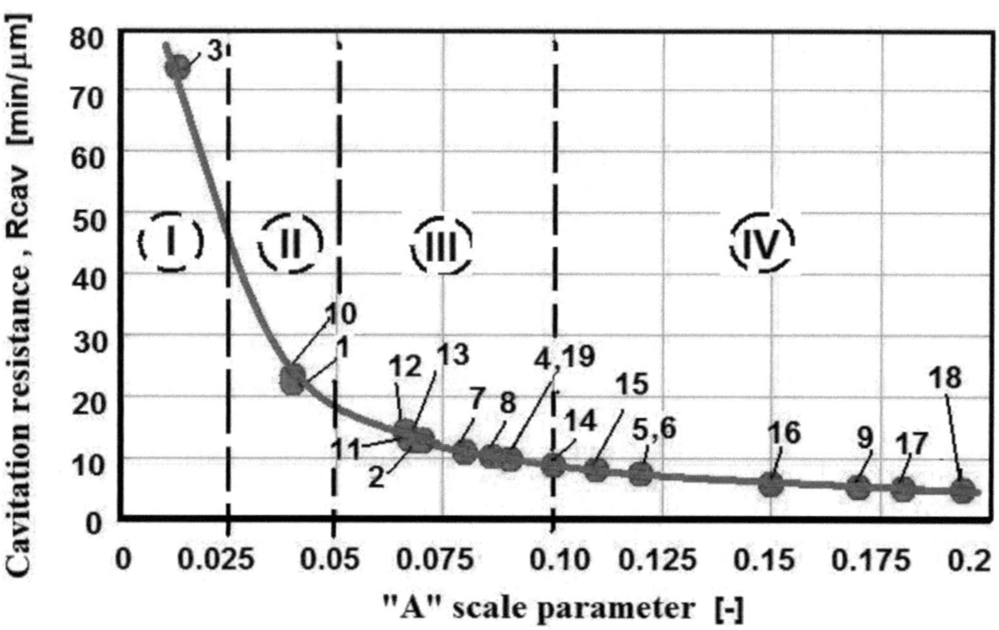

I **Steel with superresistance to cavitation erosion**

II **Steel with excellent resistance to cavitation erosion**

III **Steel with very good resistance to cavitation erosion**

IV **Steel with good resistance to cavitation erosion**

Figure 5. Correlations between cavitation erosion resistance and the scale parameter A
1- C1Cr12Ni10, 2- C1Cr12Ni2, 3- C1Cr12Ni6, 4- C036Cr14Ni10, 5-C036Cr12Ni2, 6- C036Cr12Ni4, 7- C036Cr12Ni6, 8- C036Cr12Ni8, 9- OH12NDL, 10- C1Cr6Ni10, 11- C1Cr10Ni10, 12- 1Cr18Ni10, 13- C1Cr24Ni10, 14- C036Cr13Ni10, 15- C1Cr12Ni05, 16- C036Cr16Ni10, 17- C036Cr18Ni10, 18- 030Cr12Ni2, 19- C030Cr12Ni6

Taking into account the ranking in Tables 1 and 2 it is observed that the same ranking can be obtained also by using the parameter A, respecting the same principle (doubling the interval limits at passing form one class to another). In consequence, the resistance classes for the cavitation erosion of stainless steels, for the Cavitation Laboratory of Timisoara Polytechnic University have the following limits:

- super resistant for $A < 0.025$
- excellent resistance for $A \in [0.025\text{-}0.05)$
- very good resistance for $A \in [0.05\text{-}0.1)$
- good resistance for $A \in [0.1\text{-}0.2)$

This variation manner shows that the value of parameter A is dependent on the material characteristics, at least for stainless steels. The researches must be extended also to other materials.
The dependence on the cavitation intensity is difficult to be appreciated. In consequence the comparisons with good standard materials used long time in industrial facilities (especially power plants) remains a necessity. If in the future will appear great differences between the laboratory result and the industrial behavior this is a signal that the laboratory devices must be modified to have the same type of cavitation as those encountered in the industrial installations.

Figure 6 shows that the shape parameter B of the relations (1) and (2) do not have significant influences upon the cavitation resistance and can not be used for rankings. Regardless of the resistance class, for the stainless steels its value is comprised in the interval (0.012-0.028) and no correlation between R_{cav} and B can be established.

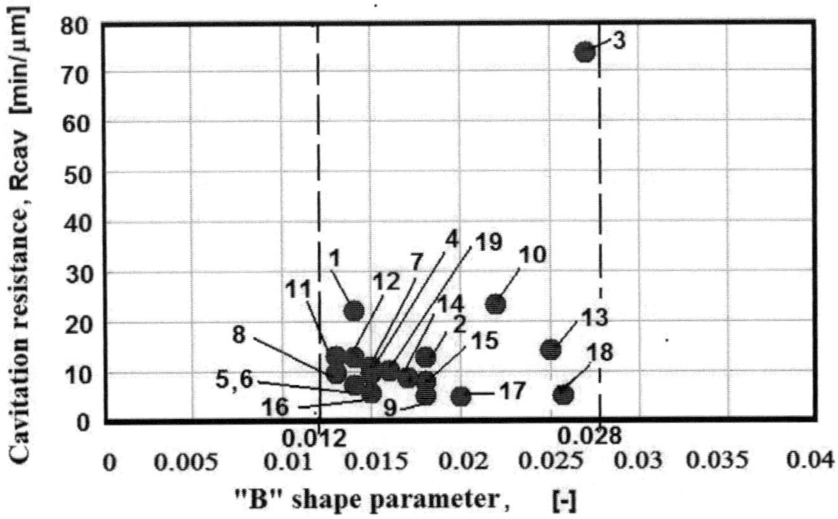

Figure 6. Correlations between cavitation erosion resistance and the shape parameter B
1- C1Cr12Ni10, 2- C1Cr12Ni2, 3- C1Cr12Ni6, 4- C036Cr14Ni10, 5-C036Cr12Ni2, 6- C036Cr12Ni4, 7- C036Cr12Ni6, 8- C036Cr12Ni8, 9- OH12NDL, 10- C1Cr6Ni10, 11- C1Cr10Ni10, 12- C1Cr18Ni10, 13- C1Cr24Ni10, 14- C036Cr13Ni10, 15- C1Cr12Ni05, 16- C036Cr16Ni10, 17- C036Cr18Ni10, 18- C030Cr12Ni2, 19- C030Cr12Ni6

On the other hand, analyzing the shape of the curves MDER(t) from Figure 4 it resulted, that regardless the resistance class, the value of this parameter depends on the curve shape (with a pronounced maxim, with an extended maxim or with the steady value equal with the maxim one). It is possible that the value of this parameter to be dependent on the surface quality at the beginning of the cavitation and not to the mechanical characteristics or the intensity of the hydrodynamic phenomena.

This is consistent with values of B equal for steels placed in different classes (for example the value of B is approximate the same for the super resistant steels - C1Cr12Ni6 - and those with good resistance - C1Cr24Ni10 and C030Cr12Ni2 -).

5. Conclusions

1. The scale parameter A, from the equations (1) and (2) is used for representing the cavitation erosion characteristic curves but it can be chosen as a representative numeric value for the cavitation erosion resistance, in vibratory devices, playing the same role as the normalized resistance R_{ns}.
2. The reference steels (for example OH12NDL) are materials subjected to cavitation both in laboratory devices and in field condition (for example the cavitation erosion resistance of OH12NDL was examined during 40 years in the Iron Gates Power Plant). The comparison with such a steel give great assurances regarding the behavior of the new material, chosen in laboratory conditions. The comparisons are possible also by using the values of the parameter “A”.
3. The scale parameter “A” can be used also for establishing the classes of behavior to cavitation erosion.
4. The shape parameter “B” from relations (1) and (2) being is determined by the state of the specimen subjected to cavitation which dictates the behavior in the first moment of exposure, can not be used as cavitation erosion ranking criterion. The influence of this criterion there must be undertaken in future analyses, on large classes of materials (carbon steels, stainless steels, bronzes etc.), to establish the manner in which their values are influenced by the nature of the material.

5. The use of the scale parameter must be considered as a very good single figure for comparisons of cavitation erosion behavior but, for exhaustive analyses, must not exclude the use of the characteristic curves, especially MDE(t).

References

[1] Manzana M E 2012 *Studii şi cercetări experimentale privind modificările structurale produse prin cavitaţie-eroziune în diferite materiale metalice*, Politehnica University, Bucuresti, Romania, Doctoral Thesis

[2] Pîlaev N I and Edel I U 1974 *Kavitatia v ghidroturbinah*, Masinostroenie, Leningrad, Russia

[3] Garcia R, Hammitt F G and Nystrom R E 1960 *Corelation of cavitation damage with other material and fluid properties*, Erosion by Cavitation or Impingement, ASTM, STP 408 Atlantic City

[4] Thiruvegadam A and Prieser H S 1964 On testing materials for cavitation damage resistence, *Journal of Ship Research* 39-56

[5] Heymann F J 1970 Toward Quantitative Prediction of Liquid Impact Erosion, *ASTM STP* **474** 212

[6] Noskievic J 1983 *The extend mathematical model of cavitation and erosion wear*, Proceedings of the Sixth International Conference on Erosion by Liquid and Solid Impact, Cambridge, UK, 5-8 September

[7] Sakai I and Shima A 1987 *On a New Representative Equation for Cavitation Damage Resistance of materials*, Report No. 385,Tokyo

[8] Bordeasu I 2006 *Cavitation Erosion of Materials*, Polytechnic Publishing House, Timisoara, Romania (in Romanian)

[9] Bordeasu I, Popoviciu M O, Balasoiu V and Patrascoiu C 2004 An Analytical Model for the Cavitation Erosion Characteristic Curves, *Scientific Bulletin "Politehnica" University of Timişoara, Transaction of Mechanics* **49**(63) 253-258

[10] *** 2010 *Standard method of vibratory cavitation erosion test*, ASTM, Standard G32-10

[11] Steller J K 1992 *International cavitation erosion test – summary of results*, IMechE Cavitation Conference, Cambridge, UK, 9-11 December, pp 95-102

[12] Bordeasu I, Popoviciu M O, Micu L M, Salcianu L C and Bordeasu C-tin 2014 Cavitation Erosion Researches upon Two Ampco Bronzes, *Machine Design* **6**(3) 97-102

[13] Jurchela A D, Bordeasu I, Mitelea I and Karabenciov A 2012 *Considerations on the Effects of Carbon Content on the Cavitation Erosion Resistance of Stainless Steels with Controled Content of Chromium and Carbon*, 21st International Conference on Metallurgy and Materials, Brno, Czech Republic, 23-25 May

[14] Jurchela A D 2012 *Cercetări asupra eroziunii produse prin cavitatie vibratorie la otelurile inoxidabile cu continut constant în crom şi variabil de nichel*, University Politehnica Timisoara, Romania, Doctoral Thesis

[15] Bordeasu I, Popoviciu M O, Mitelea I, Ghiban B, Ghiban N, Sava M, Duma S T and Badarau R 2014 Correlations between mechanical properties and cavitation erosion resistance for stainless steels with 12% Chromium and variable contents of Nickel, *IOP Conf. Ser.: Mater. Sci. Eng.* **57** 012006

[16] Karabenciov A 2011 *Cercetări asupra eroziunii produse prin cavitatie vibratorie la otelurile inoxidabile cu continut constant în nichel si variabil de crom*, University Politehnica Timisoara, Romania, Doctoral Thesis

[17] Micu L M, Bordeasu I, Popoviciu M O, Oanca V O, Salcianu L C, Ghera C and Anton I 2015 Researches upon Cavitation Erosion Behavior of Some Stainless Steels with Different Structures, *Hidraulica* **4** 46-54

[18] Oancă O 2014 *Tehnici de optimizare a rezistenei la eroziunea prin cavitaie a unor aliaje CuAlNiFeMn destinate executiei elicelor navale*, University Politehnica Timisoara, Romania, Doctoral Thesis

[19] Bordeasu I, Popoviciu M O, Mitelea I, Balasoiu V, Ghiban B and Tucu D 2007 Chemical and mechanical aspects of the cavitation phenomena, *Revista de Chimie* **58**(12) 1300-1304

Experiments on composites meant for making brake pads for the rolling stock

A Socalici[1], L Pascu[2], E Ardelean[1] and V Putan[1]

[1]Engineering and Management Department, Polytechnic University of Timisoara, 5 Revolutiei street, Hunedoara, 331128, Romania
[2]Technical College of Railway Transportation "Anghel Saligny", 137 Nationala Street Simeria, 335900, Romania

E-mail: virginia.socalici@fih.upt.ro

Abstract. For noise rail decrease and observance of the limits in UE rolling stock must be equipped with silent brake blocks of composite material to replace the conventional cast-iron blocks. For testing composite materials test pieces have been produced according to the characteristics of the experimental installation. The paper presents the laboratory experimentation phase on obtaining composite materials for brake blocks and wear behavior of the experimental test pieces.

1. Introduction

The railway is considered to be an ecological means of transport, but its contribution to noise pollution is significant. This is a threat to human health and its reduction is needed both for comfort reasons and in order to dim out its impact upon human health. Experts have identified the noise of freight carriages as being the most important source of railway noise [1], [2].

In some European regions there is significant opposition to rail noise from public opinion, demanding political initiatives on its decrease. If no remedial actions are taken, this might lead to restrictions on rail traffic along the most important European rail corridors, especially on freight trains, and the resulted activity gaps could have adverse effects on European economies. European legislation [2], [3] imposes limits for both new rolling stock as well as the reconditioned and put into circulation in the European area. In this way, equipping the rolling stock with noise brake blocks is required, which will significantly reduce noise emissions.

Brake blocks from composite materials are used to replace the conventional cast-iron blocks, which are considered an important source of noise. Experts [4], [5] recommend to give priority to measures at source (vehicles and tracks) because they offer a better" cost / benefit" rate. Accoustic barriers could incontestable be, an efficient element part of noise reduction programmes, where necessary for example in dense urban areas. If barriers are coupled with measures at source, the length or height of barriers can be reduced, leading to significant cost savings.

In order to solve the problems at source and achieve interoperability along railway lines, in December 2005 the Commission adopted technical specifications for interoperability relating to rail noise (STI "noise") [3], [5] introducing limits for rolling stock used in European Union. These limits apply to new and renewed rolling stock including for freight wagons, which must be equipped with noise brake blocks (of composite material), which reduce noise emission.

At present brakes are equipped with two types of K-blocks and LL type [6], [7]. The composite material K is recommended for new vehicles or wagons in circulation when it has the funds necessary to change the brake system. Also is noticed an series of advantages in case of using composite materials LL type, due to low expenses with the adjustment to brake blocks assembly, lower the weight of the shoe, the coefficient of friction similar to that of cast iron blocks.

Evaluating the quality of blocks in operation is made based of the criteria [8-11]: road the braking stability, wear the smallest specific capacity as high thermal load, susceptibility to breakage as small as and low tendency to formation of sparks.

From the analysis of the characteristics of the thermo-physical characteristics of the materials from which are manufactured the brake shoes (cast iron or composite materials) is observed behavior superior to composite materials characterized by obtaining lighter products, which work quite well at temperatures sometimes high during the braking process.

In the case of using the composite material the brake shoes is noted that [7], [8], [11]:

- Small adjustment expenses in order to mount brake blocks (for LL-blocks);
- Lower weight of the shoe;
- Coefficient of friction similar to that of the conventional;
- Values similar to wheel wear;
- Noise reduction by approx. 50%;
- Decrease costs by approx. 30% the braking system maintenance;
- Higher wear resistance of approx. 5 times;
- Does not give off toxic dusts braking.

2. The Experimental Part

Objectives of the experimental part (laboratory experiments): manufacturing the composite materials for rolling stock brake pads, characterization of the samples obtained and interpretation of the results.

For the preliminary tests of the composite materials the following have been established:

- The composition of the experimental recipes;
- The technology of sample manufacturing;
- The assessment of the tribological behavior of the samples;
- The analysis of the results.

For the laboratory the following test rods have been established 10 test rods made of composite material. For the composite material, 10 experimental test rods were made, using the following components: 20-38% novolac, 6-10% hexamethylenetetramine, 0-5% sulphur, 10-12% carbon fiber, 0-10% graphite, 0-28% aluminum, 0-28% brass and 0-30% rubber.

Composition of recipes is shown in Table 1.

Table 1. Components recipes

Component recipes, %	Recipes									
	1	2	3	4	5	6	7	8	9	10
Novolac	22	20	22	24	38	28	30	20	25	36
Hexamethylenetetramine	6	6	6	6	10	8	6	6	8	10
Sulphur	4	4	5	5	0	0	0	5	0	0
Carbon fibre	10	10	10	10	12	10	10	10	10	12
Graphite	0	11	0	0	8	10	0	0	10	10
Aluminum powder	16	0	0	0	0	20	0	15	23	12
Aluminum chips	0	13	15	15	12	0	14	0	0	0
Brass powder	20	0	0	0	20	24	14	14	24	20
Brass chips	0	6	22	28	0	0	0	10	0	0
Rubber	22	30	20	12	0	0	26	20	0	0

The (cylinder - type) test rods have a diameter of ø14 mm and a height of about 30 mm. The composite sample is presented in Figure 1.

Figure 1. Composite sample

The resulting test rods were submitted to a wear test on an installation of testing abrasive wear with abrasive disc (Figure 2).

Figure 2. Installation of testing abrasive wear with abrasive disc

The trials were performed according to STAS 9639 – 81 [12], [13]. HE 16 polishing silicon carbide paper was used as abrasive, which is supplied as ø310 mm diameter discs. The length of the wear spiral trajectory results from the combination of the two movements (rotation and advance) and has the value of 70 m during 7,08 minutes [14], [15]. The assessment of the behavior of the test rods has been done by the gravimetric method (mass loss).

3. Results and Discussions

The variation of the mass wear of the experimental samples is given in Figure 3.

As a result, the mass wear was determined at various time intervals. The variation of the test rod mass with time for the sample under analysis is given in Figure 4. Processing the resulting data in Excel leads to the correlations for the parameters in question, correlations that are expressed both graphically and analytically and shown in Figure 5-9.

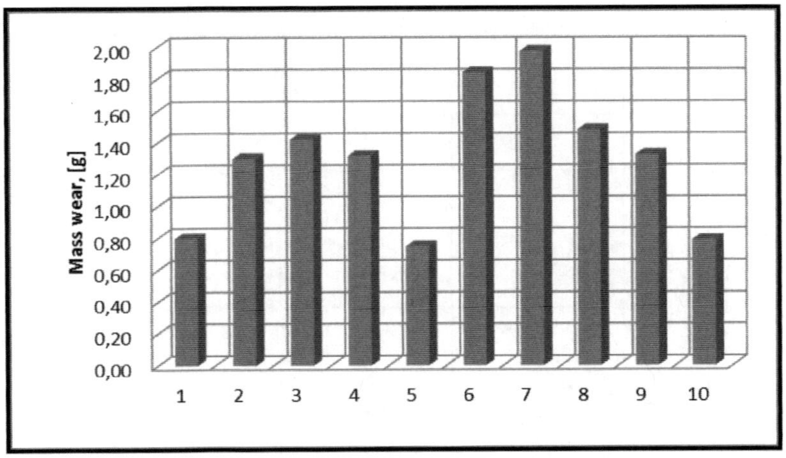

Figure 3. Wear mass experimental samples

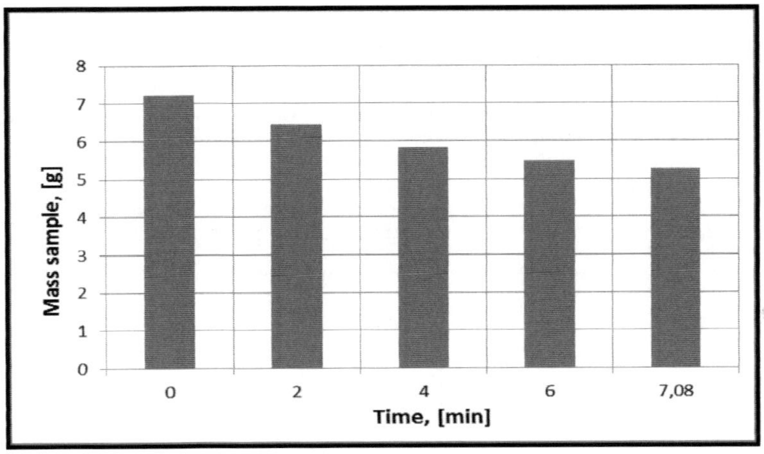

Figure 4. Sample mass variation function of time

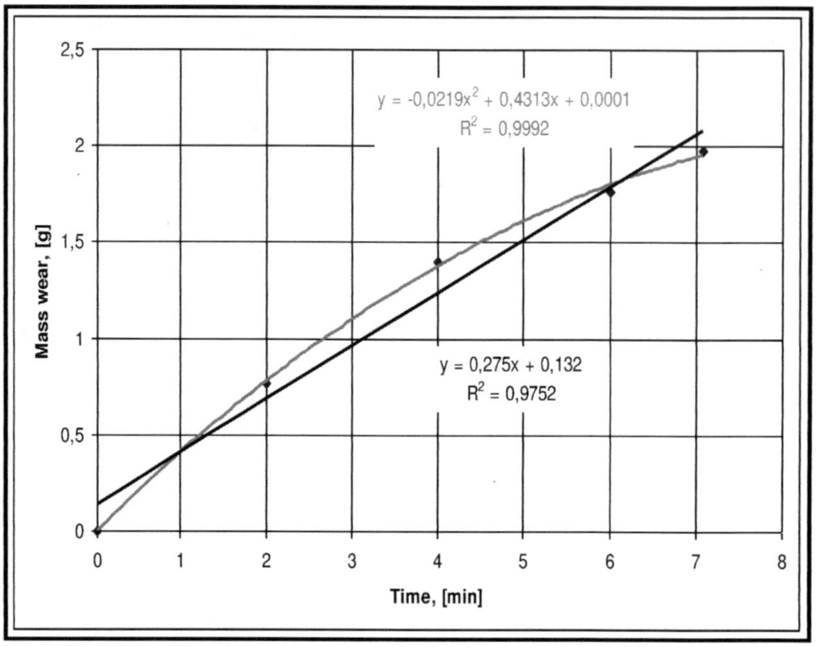

Figure 5. The variation in time of the mass wear for sample 7

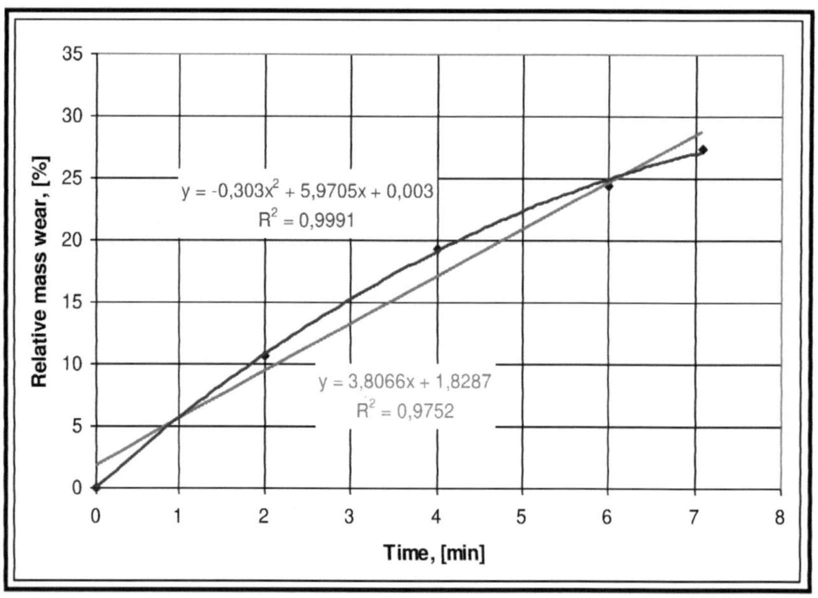

Figure 6. The variation in time of the relative mass wear for sample 7

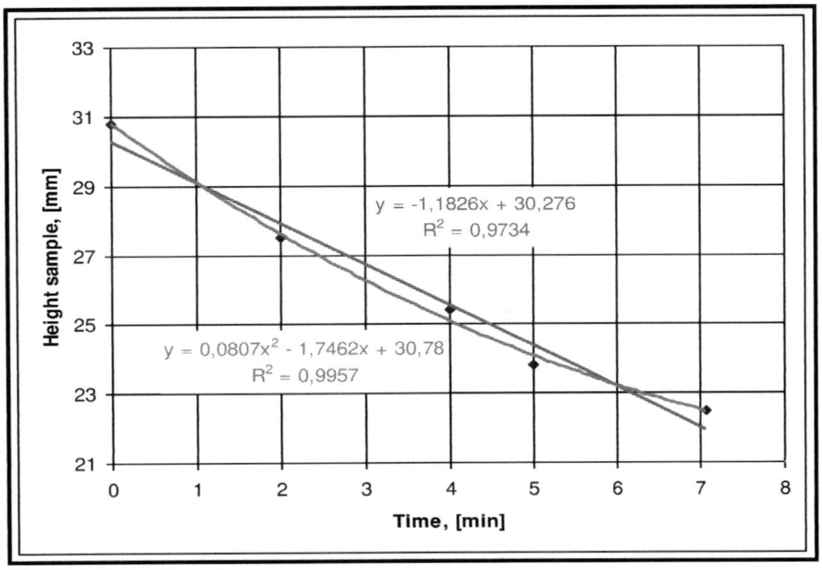

Figure 7. The variation in time of the height for sample 7

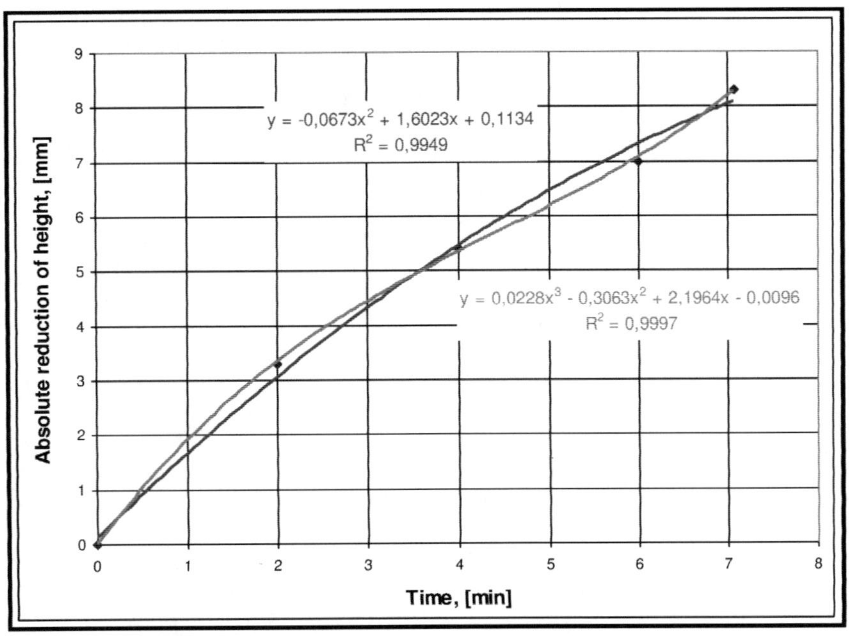

Figure 8. Absolute reduction of height for sample 7

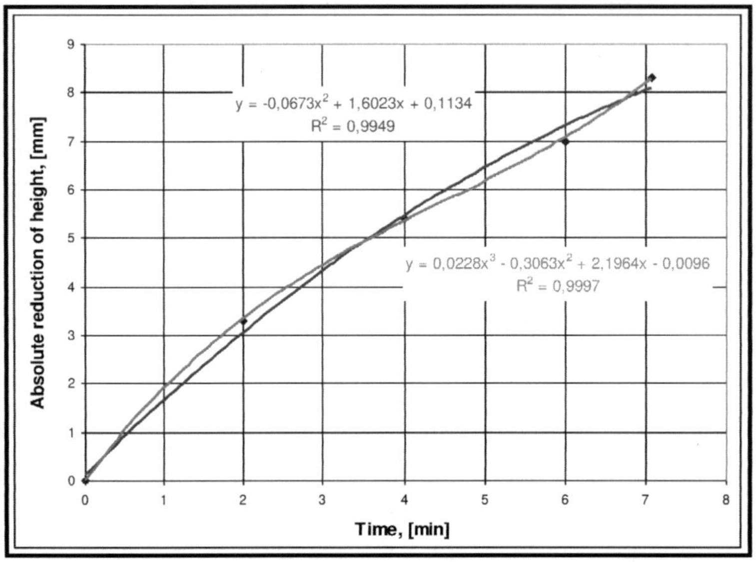

Figure 9. Relative reduction of height for sample 7

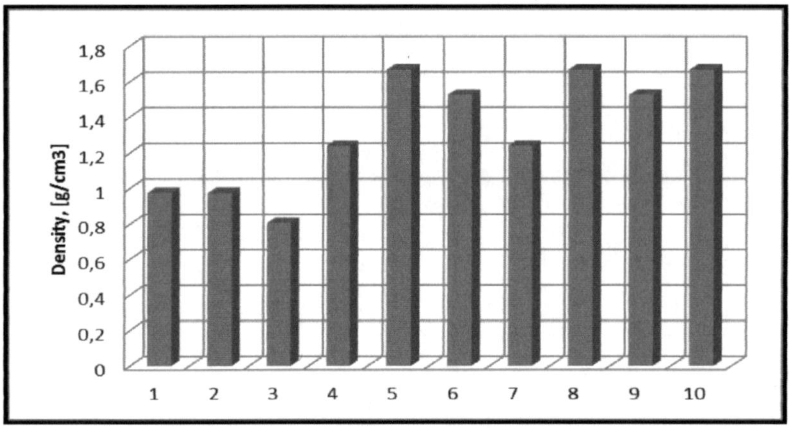

Figure 10. Sample mass variation function of time

The analysis of the data shows that the lowest mass wear of the composite material is that of samples rod 1, 5 and 10. Experimental samples rod 7 and 8 showed the highest wear.

4. Conclusions

Through the research and experiments in laboratory phase was aimed at obtaining of composite materials for the manufacture of brake shoes with characteristics superior to those already in operation.

The analysis of the experiments and their results leads to the following conclusions:

- The test rod with the highest resistance to abrasive wear is 1, 5 and 10;
- The resistance to abrasive wear is satisfactory in case of test rods 2, 3, 4, 8 and 9, due to the fact that the manufacturing technology is the same and the structure of the components varies within narrow limits.

In comparison with iron phosphorous (conventional material for brake shoes) test pieces experimental of composite admit a larger wear mass but have a similar behavior to composite materials used in practice. You also need to take into account the behavior of friction and high temperature considering operational use of brake blocks. Considering the results, the best behavior to wear of the experimental composite materials, was the one of test rods produced according to recipe 5. Experiments will continue in order to finalize the recipe and the manufacturing technology, in order to obtain the best values for the indices under consideration.

References

[1] Directive 2002/49/CE, JO L 189, 18.7.2002, p 12
[2] EU Commission Decision 2006/66/EC concerning the technical specification for interoperability relatingto the 'material stock – noiseof the conventional rail system
[3] http://www.uic.asso.fr/download.php/environnement/reductionbruitinfra_en.pdf. (UIC: Noise reduction European railway infrastructure. Assessment Report 2007)
[4] Oertli J 2013 *Railway Noise Control In Europe*: *Current Status*, Proceedings of the 11th International Workshop on Railway Noise, Uddevalla, Sweden, September 9–13, pp 1-6
[5] Commission of European Communities 2008 Measures to reduce rail noise from the existing fleet, COM 432, Bruxelles
[6] UIC SET 07/2012 Design rules of composite brake blocks (K)
[7] UIC SET 07/2013 Rules for use of composite brake blocks (LL)
[8] Josan A and Pinca Bretotean C 2015 Using special additions to preparation of the moulding mixture for casting steel parts of drive wheel type, *IOP Conf. Ser.: Mater. Sci. Eng.* **85** 012017
[9] Georgescu C, Deleanu L and Botan M 2014 Dry sliding of composites with PBT matrix and micro glass beads on steel, *Industrial Lubrication and Tribology* **66**(3) 424-433
[10] Botan M, Dumitru D, Pirvu C and Deleanu L 2015 Influence of interface quality on stress and strain distribution in a micro cell of a composite, *Revista de Materiale Plastice* **52**(1) 20-23
[11] Ardelean E, Ardelean M, Josan A and Pinca-Bretotean C 2014 Contribution on the influence of steel ladle processing (LF) upon the nitrogen removal rate, *IOP Conf. Ser.: Mater. Sci. Eng.* **57** 012001
[12] STAS 9639 – 81, Testing of metals. Trying to wear on rotating disc sanding paper
[13] Nicoara I 2006 *Reliability and terotehnica*, Politehnica Publishing House, Timisoara, Romania
[14] Pascu L 2015 *Researches on Improving the Quality of Brake Shoes Meant for Use with the Rolling Stock*, University Politehnica of Timisoara, Romania, Doctoral Thesis
[15] Socalici A, Pascu L, Ardelean E and Popa E 2015 Researches on obtaining composite material stobe used in manufacturing brake pads, *IOP Conf. Ser.: Mater. Sci. Eng.* **106** 012015

IOP Conf. Series: Materials Science and Engineering **163** (2017) 012004 doi:10.1088/1757-899X/163/1/012004

EAF optimal managing elements

A Ioana, N Constantin and E C Dragna

University Politehnica of Bucharest, Engineering and Management of Metallic Materials Obtaining Department, Bucharest, Romania

E-mail: claudiadragna90@gmail.com

Abstract. Electric Arc Furnaces (EAF) is an important and complex aggregate. We present elements of EAF operating efficiency. The reliability of the EAF is mainly determined by electric circuit reliability and especially of the transformator of the furnace. This due to the fact that, in an electric steel plant, the objective „24 tapping/day" involves. Due to the complexity of the process, the EAF operation needs a computer usage and, in his componence, it must have two independent calculation units (UC1) and (UC2). Based on these two input sets (Σi_1) and ((Σi_2), the calculation unit (UC1) builds the general operation procedure based on mathematical methods. For that purpose, there are used the results of the 5 mathematical methods: the mathematical model to write-off the function objective (M.F.O.); the mathematical model of calculating the charge (M.C.C.); the mathematical model of conducting the effective melt (M.C.M.); the mathematical model of reheating the charge (M.R.C.); the mathematical model of blasting the reactive dusts (M.B.R.D.).

1. Introduction

The changing of the electric energy into heat takes place mostly inside of the Electric Arc Furnace (EAF), where the temperature goes above 2500^{0}C. The heat transfer from the electric arc to the charge in the furnace it is done through conduction and radiation [1], [2].

One of the main advantages of the EAF is also the possibility of using scrap iron as a main charge to obtain the steel. This allowed the EAF development and affirmation as a steel melting assembly [3].

The main parts of the direct action EAF are:
- refractory masonry;
- steel construction;
- electrical equipment.

The refractory masonry of the direct action EAF consists of: the hearth furnace masonry, the side walls masonry and the roof. In Figure 1 we present the scheme of EAF shell with a capacity of 10t [5], [6].

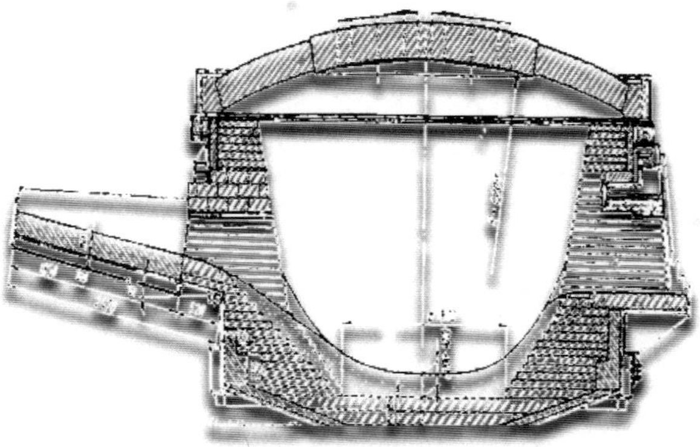

Figure 1. The EAF masonry (10t capacity)

Metallic construction of direct action EAF consists of:
- the furnace bowl (borders the bedstone - which can have a cone head or spherical shape)
- the furnace work door;
- the molten steel outlet and spout;
- vault ring;
- the tilting mechanism;
- the supporting - coupling and handling electrodes devices.

The EAF tilting mechanism has a great importance for its efficiency. It must comply with the following tasks [4]:

- to ensure the fine tilting of the furnace (without any bumps) with 40...50° towards the outlet spout and with 10...20° towards the work furnace door;
- to allow changing the tilting speed according to necessities (emergencies); the tilting speed is 1,5°/s for small furnaces and 0,4...0,8°/s for big furnaces;
- to be located so that it will not be damaged in case of hearth boring.

For this purpose the tilting mechanism is to be located to the bottom of the furnace and it consists of two runners (sledge) which the furnace is built on. The tilting force applies in a point between the two sledge runners (for small furnaces) or it applies to the two runners (for large furnaces) and it is realized hydraulic (with water or oil) or electro mechanic [7], [8].

2. Elements of EAF Optimal Managing

The work power (Pt) at EAF is calculated with the ratio (1):

$$P_t = \frac{W}{t \cdot \eta \cdot \cos\varphi} \quad \text{[KVA]} \tag{1}$$

where :
W – is the total electrical consumption in the melting time period (kWh);
t – is the melting time (h);
η – is the efficiency (0.8...0.9);
cosφ – is the power factor (about 0.85).

The electrodes serve to conduct the current from cathead to the metal charge and to ensure the electric arcs forming. The EAF most often uses the graphite electrodes.

The diameter of the electrode (d) is calculated with the ratio (2):

$$d = \sqrt{\frac{4 \cdot I}{\pi \cdot \Delta}} \quad \text{[cm]} \tag{2}$$

where:

I - is the phase current, [A];
Δ - is the current density [A/cm^2]

In Table 1 are synthetically presented the main constructive and functional characteristics of EAF.

Table 1. Constructive and functional characteristics of an EAF

Nr Crt	EAF type Parameters	DSN-3	DSP-6	DSP-12	DSP-25	DSP-50
1	Charge capacity [t]	3	6	12	25	50
2	Working Power [kVA]	2000	4000	8000	12500	25000
3	The specific melting electrical consumption [kWh/t]	500	500	470	460	440
4	The graphite electrode diameter [mm]	200	300	350	400	500
5	The inner diameter of the coat [mm]	2764	3190 ... 3500	3760 ... 4260	4450 ... 4950	5800 ... 6050
6	The bath diameter at the limit level [mm]	-	2230	2740	3540	4500
7	The depth of the bath at the limit level, in mm	400	425	555	775	890
8	The height of the melting space from the border to the vault [mm]	1050	1110	1365	1500	1950
9	The work door size [mm]	650X500	750X500	980X680	1000X800	1200X570
	The angles:					
10	-the rotation angle of the vault to the spout [degrees]	-	61	70	75	70
11	-the tilting angle of the furnace to the spout [degrees]	-	40	45	45	45
12	- the tilting angle of the furnace to the working door [degrees]	-	15	15	15	15
13	The steel construction mass, in t	35.5	45	80	140	235
	The power in kW:	-		-	-	-
14	- the lifting mechanism of the furnace working door		1,0			
15	- The oil pump	-	10X4	10X4	10X4	10X4
16	- The vault rotation mechanism	-	-	-	3.2	3.5

The reliability of the EAF is mainly determined by its electric circuit reliability and especially of the transformator of the furnace. This due to the fact that, in an electric steel plant, the objective „24 tapping/day" involves:

- 150.000 adjusting charge manoeuvre/year;
- 50.000 closing – opening;

And therefore:

- 45% of the incidents destroy the transformer coil;
- 40% of the incidents affect the short power supply of the furnace;
- 15% of the incidents are related to the voltage adjustment.

In Figure 2 is presented the basic diagram for computer assisted EAF operation.

Due to the complexity of the process, the EAF operation needs a computer usage and, in his componence, it must have two independent calculation units (UC1) and (UC2). Furthermore, considering that the most EAF's in the country don't have AMCR (automatic model charge reheating) to work in a continuous mode and in real time, the operator presence is requested.

For the proper operation of an EAF it is required to ensure two inputs sets for the calculation unit (UC1):

- Σi_1 – inputs obtained from the outputs of the process Σy_1, possible to be quantified through direct measurements - ensured by the measuring elements (ΣEM).
- Σi_2 – inputs ensured by the operator (these are the values that cannot be measured continuously and in real time).

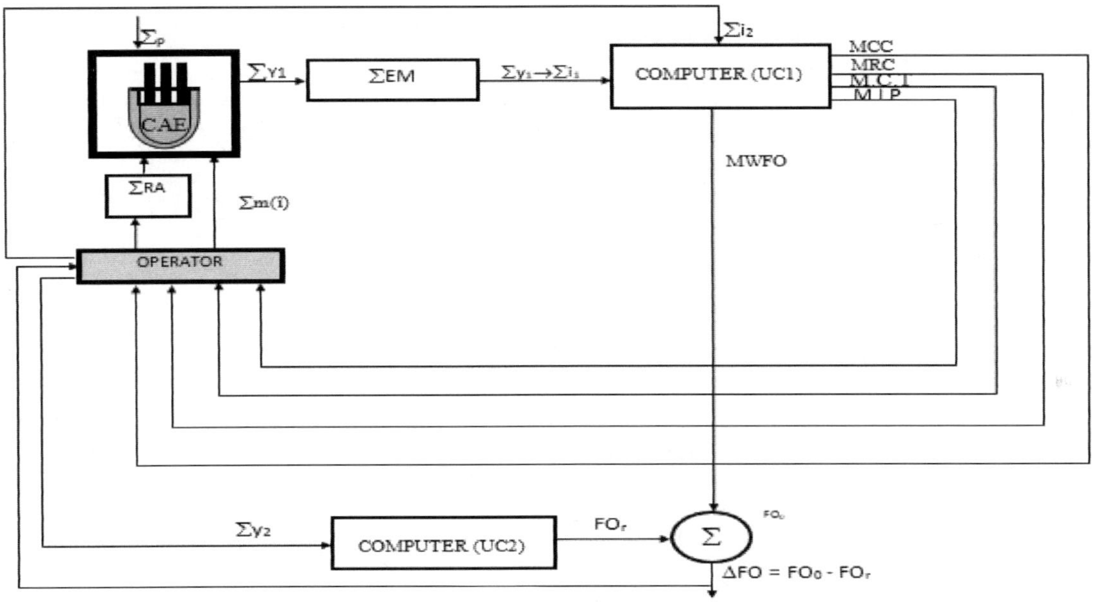

Figure 2. Basic diagram for computer assisted EAF operation

Based on these two input sets (Σi_1) and (Σi_2), the calculation unit (UC1) builds the general operation procedure based on mathematical methods. For that purpose, there are used the results of the 5 mathematical methods:

- The mathematical model to write-off the function objective (MWFO);
- The mathematical model of calculating the charge (MCC);
- The mathematical model of conducting the effective melt (MCT);
- The mathematical model of reheating the charge (MRC);
- The mathematical model of blasting the reactive dusts (MIP).

3. Results and conclusions

The operator takes over the general operating procedure parameters of EAF and, for some categories, it launches tasks to the line of automated regulators (ΣRA) which ensures the guidance of the whole process, and for other parameters it acts directly in the process through the execution extensions as it is the case for ensuring the right dosage of the charge when manual operating.

Simultaneously, the calculation unit (UC1) elaborates the prescript level of the function objective (FO_0).

When the process is done (in our case it is considered the elaboration ending), the operator takes over all the outputs of the process (Σy_2) which couldn't be ensured through the continuous measurements and in real time. These data are provided to a second calculation unit (UC2) which, along with the calculation unit (UC1), elaborates the accomplished level of the function objective (FO_r).

According to the level of this deflection value, the operator decides to modify the general procedure of operating the EAF and through these two calculating units, he elaborates a new operating procedure.

References

[1] Ioana A, Semenescu A, Marcu D, Pollifroni M and Březinová M 2015 *Some Aspects about Product Management of Electric Arc Furnace Elements*, *Applied Mechanics and Materials* **809/810** 1319-1324

[2] Ioana A 2013 Metallurgy's Impact on Public Health, *Review of Research and Social Intervention* **43** 169-179

[3] Ioana A, Semenescu A, Marcu D, Pollifroni M and Březinová M 2015 Quality and Environmental Evaluation Methods for Metallic Materials Industry, *Applied Mechanics and Materials* **809/810** 1600-1605

[4] Ioana A, Constantin N and Moldovan P 2015 Constructive and functional modernization of EAF, *IOP Conf. Ser.: Mater. Sci. Eng.* **85** 012014

[5] Ioana A, Constantin N and Moldovan P 2015 About EAF and environment, *IOP Conf. Ser.: Mater. Sci. Eng.* **85** 012015

[6] Ioana A and Nicolae A 2002 *Optimal Managing of Electric Arc Furnaces*, Fair Partners Publishing, Bucharest, Romania

[7] Odhano S A, Bojoi R, Boglietti A, Rosu S G and Griva G 2015 Maximum Efficiency per Torque Direct Flux Vector Control of Induction Motor Drives, *IEEE Transactions on Industry Applications* **51**(6) 4415-4424

[8] Dragna E C, Ioana A, Nicolae C and Pollifroni M 2016 *About optimal management procedures of EAF*, 11th European Electric Steelmaking Conference & Expo EEC 013, Venice, Italy, May 25-27

Elements of EAF automation processes

A Ioana, N Constantin and E C Dragna

University Politehnica of Bucharest, Engineering and Management of Metallic Materials Obtaining Department, Bucharest, Romania

E-mail: claudiadragna90@gmail.com

Abstract. Our article presents elements of Electric Arc Furnace (EAF) automation. So, we present and analyze detailed two automation schemes: the scheme of electrical EAF automation system; the scheme of thermic EAF automation system. The application results of these scheme of automation consists in: the sensitive reduction of specific consummation of electrical energy of Electric Arc Furnace, increasing the productivity of Electric Arc Furnace, increase the quality of the developed steel, increasing the durability of the building elements of Electric Arc Furnace.

1. Introduction

Considering the great complexity of the specific procedures for steelmaking in an Electric Arc Furnace (EAF), the complex operation of this technological aggregate implies the systematically covering of the following steps [1], [2]:

- Quantifying and maintaining a certain prescript technological state (state of inertia) for the aggregate – achievable through conventional usual automation;
- Complex automation of EAF, which relies on the controlling of the processes for obtaining the maximum criteria function (objective) FO - prescript according to the mathematical model.

The conventional automation of EAF mainly targets [3], [4]:

- the automation of the electric regimen;
- the automation of the thermic regimen.

The characteristics of the performant level furnaces from the world ferrous metallurgy is the electro-hydraulic positioning and actioning system, at 15-20 MPa pressures. The type of electromechanical actioning and positioning of the EAF electrodes, used at the furnaces in our country, which had developed optimal operating procedures validation experiments (furnace 9 – OE 2 – COS Târgovişte and furnaces 1 and 2 – OE – SIDERCA Călăraşi) is based on an inductance coupling and electromagnetic brake. The electro-hydraulic adjustment system of the EAF electrodes is much more performant. In the Table 1 there are presented, by comparison, the performances of two actioning and positioning systems of the EAF electrodes [5-7].

Table 1. The performances of the actioning and positioning systems of EAF' electrodes

Nr. crt.	Actioning and positioning of the EAF electrodes system type	System performance	
		Speed transit of the electrodes (mm/s)	Reaction time (s)
1.	Electro mechanic	70	0,2
2.	Electro hydraulic	90	max 0,01

2. Elements of EAF Automation

The main objective of the EAF automation electric regime consists of adjusting the electric power absorbed from the system on each phase (associated with each electrode). For this purpose, the algebraic composition through algebraic summing (SA) of the signals offered by the current transformer (TRI) and voltage transformer (TRU) constitute inputs for the driver of the action motor (DMA).

The adjustment of the electrodes of EAF, which has as a criterion to maintain constant of the impedance of the electric arc (Za = ct), is ensured by the action motor (MA) through the position of the electrodes adjustment device (DRPE).

There can also be used position regulators of the electrodes based on the following adjustment algorithms:

- the adjustment at a constant voltage of the electric arc (Ua = ct);
- the adjustment at a constant current in the electric arc (Ia = ct).

In Figure 1 we shown the basic diagram of automation (adjustment) of the EAF electrical regimen.

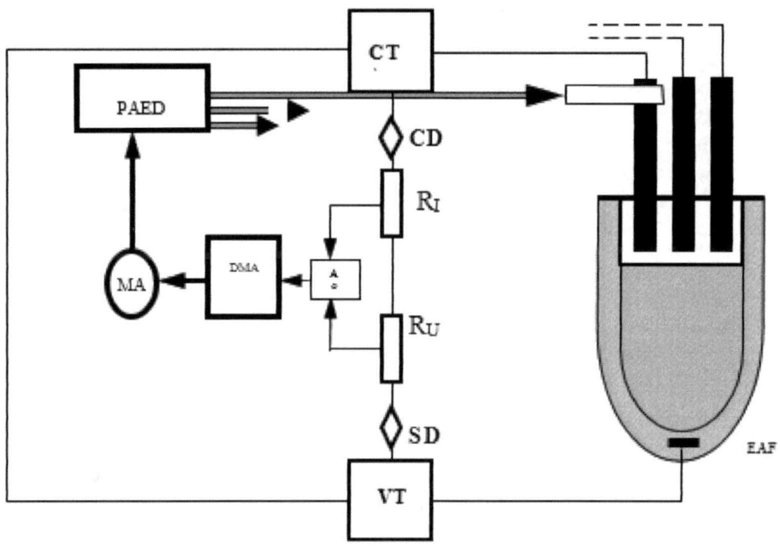

Figure 1. The basic diagram of automation of the EAF electrical regimen

EAF – Electric Arc Furnace; CT – Current Transformer; VT – Voltage Transformer; CD – Command Device; SD – Separation Device; R_I, R_U – Additional Resistances; AS – Algebraic Sumator; DMA – Driver of the Motor Action; MA – Motor Action; DRPE – Position Adjustment of the Electrodes Device

The analysis of the opportunity of replacing the electro mechanic system of actioning and positioning the electrodes with an electro hydraulic one, for a 100t EAF, reveals the high level of investment and determines the obtaining of a pay off time for about 2-3 years by a level of steel production of minimum 250 000 t/year.

The automation of thermic regime of EAF is based on the basic diagram shown in Figure 2. The thermic regime of EAF is defined by a lot of technological parameters, of which has been chosen as main parameters:

- The temperature of the shell masonry of EAF, t_{ZP};
- The temperature of the hearth furnace masonry of EAF, t_{ZV};
- The temperature of the molten steel, t_o.

All of these basic parameters - technological and functional - can be determined through direct thermo-technological measurements.

The adjustment loop of the EAF thermic regime has, as a central element, the temperature regulator (RT) which, based on the input signals offered by measurements of the base parameters (t_{ZP}, t_{ZV}, t_o), launches switching actions of the furnace transformer on a new voltage bench – through the voltage bench control unit (BCTU) – and/or respectively modifying (including limitation) of absorbed electrical power – through the power regulator (RP).

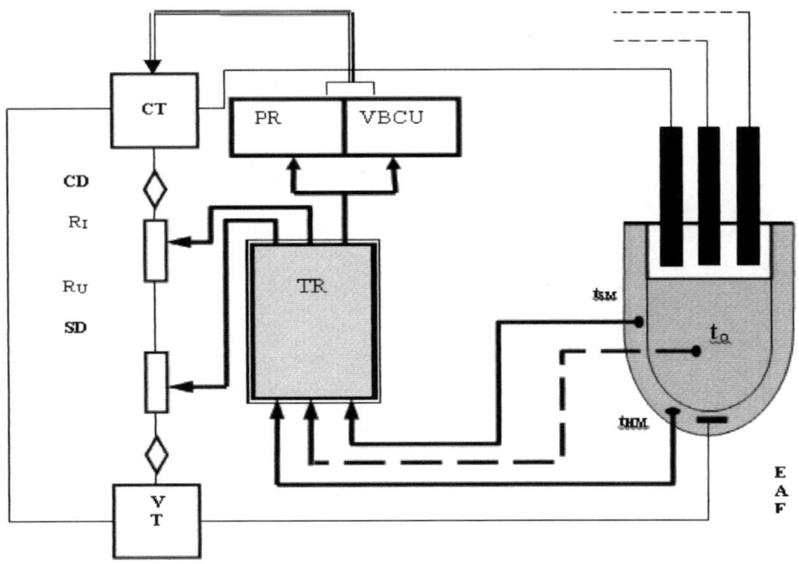

Figure 2. The basic diagram of the EAF thermal condition automation

EAF – Electric Arc Furnace; CT – Current Transformer; VT – Voltage Transformer; CD – Command Devices; SD – Separation Devices; R_I, R_U – additional resistances; t_{SM} – Temperature of the Shell Masonry of EAF; t_{HM} – Temperature of the EAF Hearth Masonry; t_o – Temperature of the Molten Steel; TR – Temperature Regulator; VBCU – Voltage Bench Control Unit; PR – Power Regulator

3. Conclusions

The two schemes of EAF process automation, presented and analyzed in our article are particularly important for increasing the efficiency of production of steel.

The electrical automation system of EAF is particularly important on one hand, on reducing specific consumption of electricity of EAF, and on the other hand, on increasing productivity of EAF.

Automating the thermal regime of the EAF has a great importance not only on the quality of produced steel, but on the sustainability of the parts of EAF.

References

[1] Ioana A and Nicolae A 2002 *Optimal Managing of Electric Arc Furnaces*, Fair Partners Publishing, Bucharest, Romania

[2] Ioana A, Constantin N and Moldovan P 2015 *Constructive and functional modernization of EAF, IOP Conf. Ser.: Mater. Sci. Eng.* **85** 012014

[3] Ioana A 2013 Metallurgy's Impact on Public Health, *Review of Research and Social Intervention* **43** 169-179

[4] Ioana A, Semenescu A, Marcu D, Pollifroni M and Březinová M 2015 Some Aspects about

Product Management of Electric Arc Furnace Elements, *Applied Mechanics and Materials* **809/810** 1319-1324

[5] Ioana A, Semenescu A, Marcu D, Pollifroni M and Březinová M 2015 Quality and Environmental Evaluation Methods for Metallic Materials Industry, *Applied Mechanics and Materials* **809/810** 1600-1605

[6] Ioana A, Constantin N and Moldovan P 2015 About EAF and environment, *IOP Conf. Ser.: Mater. Sci. Eng.* **85** 012015

[7] Odhano S A, Bojoi R, Boglietti A, Rosu S G and Griva G 2015 Maximum Efficiency per Torque Direct Flux Vector Control of Induction Motor Drives, *IEEE Transactions on Industry Applications* **51**(6) 4415-4424

IOP Conf. Series: Materials Science and Engineering **163** (2017) 012006 doi:10.1088/1757-899X/163/1/012006

FEM analysis of an single stator dual PM rotors axial synchronous machine

L N Tutelea[1], S I Deaconu[2] and G N Popa[2]

[1]Politehnica University of Timisoara, Electrical Engineering Department, Vasile Parvan str., no. 1-2, 300223 Timisoara, Romania
[2]Politehnica University of Timisoara, Department of Electrical Engineering and Industrial Informatics, 5 Revolution Street, Hunedoara, 331128, Romania

E-mail: sorin.deaconu@fih.upt.ro

Abstract. The actual e - continuously variable transmission (e-CVT) solution for the parallel Hybrid Electric Vehicle (HEV) requires two electric machines, two inverters, and a planetary gear. A distinct electric generator and a propulsion electric motor, both with full power converters, are typical for a series HEV.
In an effort to simplify the planetary-geared e-CVT for the parallel HEV or the series HEV we hereby propose to replace the basically two electric machines and their two power converters by a single, axial-air-gap, electric machine central stator, fed from a single PWM converter with dual frequency voltage output and two independent PM rotors. The proposed topologies, the magneto-motive force analysis and quasi 3D-FEM analysis are the core of the paper.

1. Introduction

The (AFPM) machine (the permanent magnet machine with an axial flux), which is known also called the disc-type machine, as well, is an attractive substitute because of its shape resembling a pancake, solid(tight) compact constructions and high power density. AFPM motors are particularly especially advisable (applicable) in case of the electrical vehicles, ventilators, valve supervision, machine tools, robots and industrial [1], [2].

Axial flux machinery has been a part of the specialized literature since the early 70's, followed by the motors based on axial flux induction a couple of years later [3-6]. At the moment, direct drive appliances which are based upon actuators or low-speed generators with increased torques have drawn attention once again to the axial flux machines, particularly to the more capable ones when it comes to increased torque density and to the efficient ones (such as the PM variety) [3], [7-11].

However, every time a couple of design goals are met, the AFPM synchronous machines are a good option to consider. More exactly, it's a common fact that the number pole pairs have to be comfortably high [3], [12].

It is also a common occurrence that the fractional slot windings can be manufactured in robust outlines. This is the case when the windings extensions are not overlapped and the stator teeth are surrounded by single coil windings. Fractional slot robust windings provide consistent assets to the end user. Actually, their usage grants a physical phase separation between the phases and of the magnetic circuits of the phases, resulting in a reduced exposure to issues between the phases and also decreasing the reciprocal inductance during the phases [3], [13].

The common material used for the back cores of the AFPM synchronous machines is the laminated steel. However, Soft Magnetic Composites (SMC) are heavily considered because of their match with increased frequencies and their reduced weight [3].

The coveted increased achievements may be obtained by the usage of PM machines and different supervision ways, one of which the direct torque control (DTC) or other increased performance vector management methods stand out [14].

This study nominates a brand new dual PM rotor, SM run with basically a single stator (with a Gramme-ring or dual winding, placed on either extremities of the stator) and dual rotor equipped with distinct leg calculations and increased winding elements, in order to decrease the size, the mass and the expenditure for both series and parallel planetary-gear HEVs. The paper is organized as follows: Section 2: constructive elements, Section 3: concentrated winding axial PM synchronous machine, Section 4: analytical quasi – 3D modeling, Section 5: FEM analysis, Section 6: conclusions.

2. Constructive elements

In the first figure the scheme of the machine seen from the front and from the side is presented.

The single stator dual PM rotors axial synchronous machine with the centred stator cluster (1) with two three-phase windings (2) found in free sockets fixed rigid in the casing (3), coming with two extremity covers (4), (5) in which the two ball bearing backings(6), (7), radial and axial, are put. The ball bearings create the possibility that the two shafts (8), (9) will pivot each on its own, each shaft having a solid steel platter (on which the permanent magnet extremities are arranged in a circular and symmetric fashion) placed towards the stator. The other end of the shaft is is added into a half-coupling linked to the thermal engine (10), respectively to the gears placed against the drive wheels (11).

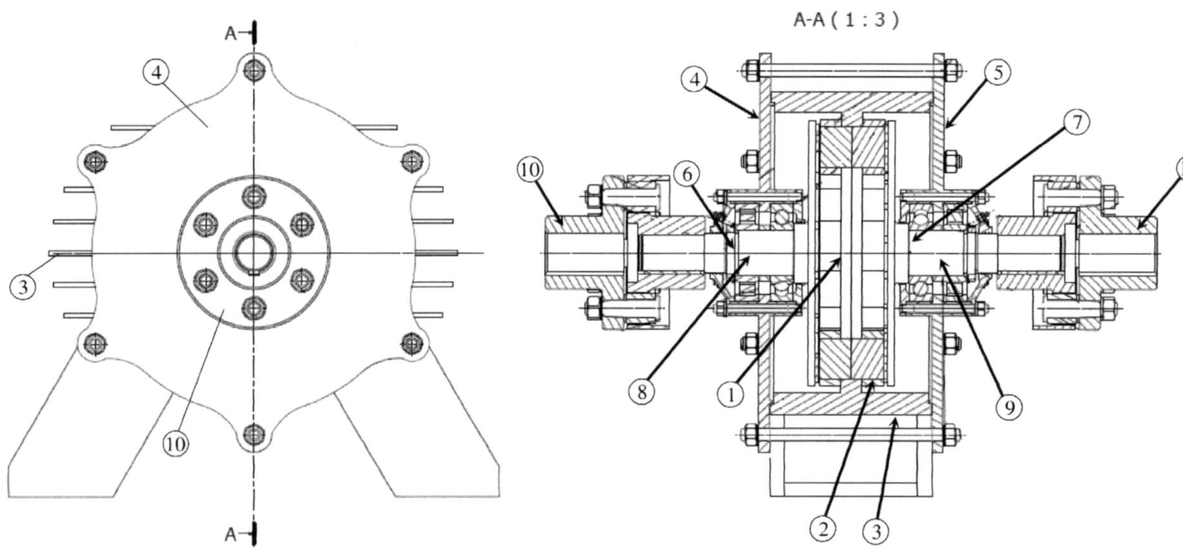

Figure 1. Front view and longitudinal section through the dual PM rotors machine

For the stator magnetic core soft magnetic composite (SMC) is suggested, but a rolle-laminated magnetic core works just as fine. The same explanation is true in case of the rotors magnetic core.

A surface-mounted PM design has been taken into account for the rotors, 14 poles for rotor 1 and 10 poles for rotor 2 (Figure 2).

The material of the permanent magnets is the Neodymium Iron Boron (Nd-Fe-B) type MPN 40SH, with flux density remanence B_r = 1,13T and cohercitive field intensity $H_c = 860 \cdot 10^3\ A/m$.

IOP Conf. Series: Materials Science and Engineering **163** (2017) 012006 doi:10.1088/1757-899X/163/1/012006

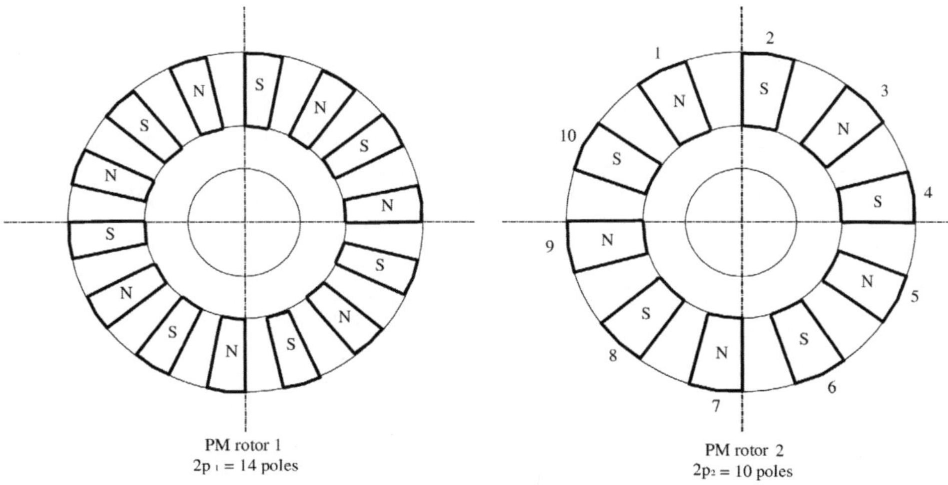

a) b)

Figure 2. Rotor 1 with 14 PM poles (a) and rotor 2 with 10 PM poles (b)

3. Concentrated winding axial PM synchronous machine

3.1. Armature Windings Design

An AFPM machine can compete with the traditional radial-flux machines in terms of torque and power density if the number of pole-pairs is sufficiently high and the ratio axial length/outer diameter is low [1], [3], [12].

The various combinations of slots and poles which allow the realization of balanced winding can be determined by the relation:

$$\frac{N_{st}}{[GCD(N_{st}, 2p)]} = 3k \quad , \tag{1}$$

where N_{st} is the number of slots, p the number of pole pairs and k an integer number (GCD: Greatest Common Divisor). As is well known, the number of slots per pole and per phase of an electrical machine is equal to:

$$q = \frac{N_{st}}{2pm} \quad , \tag{2}$$

where m is number of phases.

Fixing N_{st}, if the machine has a high number of pole pairs, q decreases.

In concentrated non-overlap windings, there is only one coil in a coil group or coil phase belt (z = 1). The distribution factor for the fundamental space harmonic is calculated by [1]:

$$k_{d1} = \frac{sin\left(\frac{\pi}{2m}\right)}{z \cdot sin\left(\frac{\pi}{2mz}\right)} \quad , \tag{3}$$

and the pitch factor for the fundamental space harmonic is [1]:

$$k_{p1} = sin\left(\frac{\pi}{2} \cdot \frac{y}{\tau}\right) \quad , \tag{4}$$

where y is the pitch of one coil and τ the pole pitch. The fundamental winding factor:

$$k_{w1} = k_{d1} \cdot k_{p1} \quad , \tag{5}$$

For our case $k_{w1} = 0.966$ for both windings. (N_{st}=12, $p_1 = 7$, $p_2 = 5$).

Figure 3 and Figure 4 shows the AFPM windings layout.

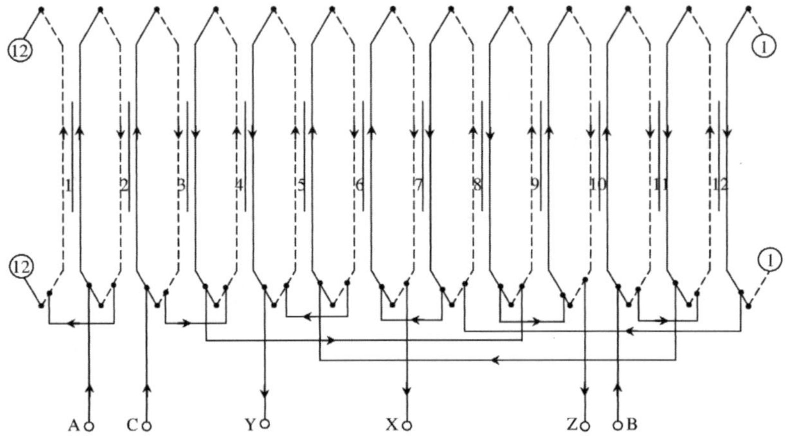

Figure 3. AFPM winding arrangement (N_{st}=12, p=7, m=3)

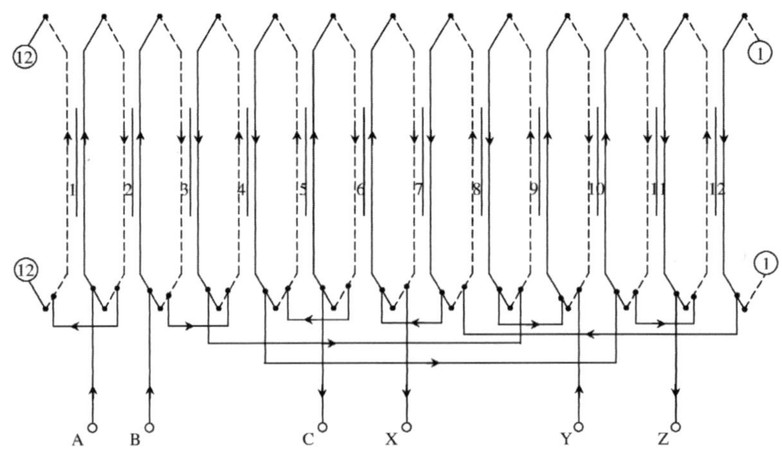

Figure 4. AFPM winding arrangement (N_{st}=12, p=5, m=3)

3.1. MMF Analysis

The parameter τ is the average pole pitch.

$$\tau = \frac{\pi D}{2p}, \tag{6}$$

where D is the average diameter

$$D = \frac{D_{out} + D_{in}}{2}, \tag{7}$$

with D_{out} the outer and D_{in} the inner diameter of machine.

The MMFs F_{ad} and F_{aq} in the d-axis and q-axis are [1]:

$$F_{ad} = \frac{m_1\sqrt{2}}{\pi} \cdot \frac{N_1 \cdot k_{w1}}{p} \cdot I_{ad}, \tag{8}$$

$$F_{aq} = \frac{m_1\sqrt{2}}{\pi} \cdot \frac{N_1 \cdot k_{w1}}{p} \cdot I_{aq}, \tag{9}$$

where I_{ad} and I_{aq} are the d and q axis stator (armature) currents respectively.

Let us consider the case where the concentrated winding PM machine is fed by sinusoidal currents, with a maximum current I in phase "A" (i_A=I and i_B=i_C=-I/2).

In order to investigate the spatial distribution of the armature MMF, the surface mounted PMs in the rotor are supposed to be discarded. Giving the fact that each phase is made up of N coils, and accounting for the notations considered in Figure 5, the MMF F_{A1} produced by one coil (A^+, A^-) is expressed as [15]:

$$F_{A1}(\theta)=\begin{cases}(2-\beta)\cdot\dfrac{N\cdot I}{8}, & -\dfrac{\delta}{2}<\theta<\dfrac{\delta}{2}\\ -\beta\cdot\dfrac{N\cdot I}{8}, & elsewhere\end{cases}, \tag{10}$$

where β is the coil pitch ratio, such that

$$\beta=\frac{\delta}{\tau_{st}}, \tag{11}$$

where τ_{st} is the armature pole pitch. Taking into account the hypothesis which considers that the magnetic circuit is not saturated, one can apply the superposition theorem in order to obtain the magneto-motive force (MMF) $F_A(\theta)$ of phase "A" as follows:

$$F_A(\theta)=F_{A1}(\theta)+F_{A2}(\theta)+F_{A3}(\theta)+F_{A4}(\theta), \tag{12}$$

with

$$\begin{cases}F_{A2}(\theta)=-F_{A1}\left(\theta-\dfrac{\pi}{6}\right)\\ F_{A3}(\theta)=-F_{A1}(\theta-\pi)\\ F_{A4}(\theta)=F_{A1}\left(\theta-\dfrac{7\pi}{6}\right)\end{cases}, \tag{13}$$

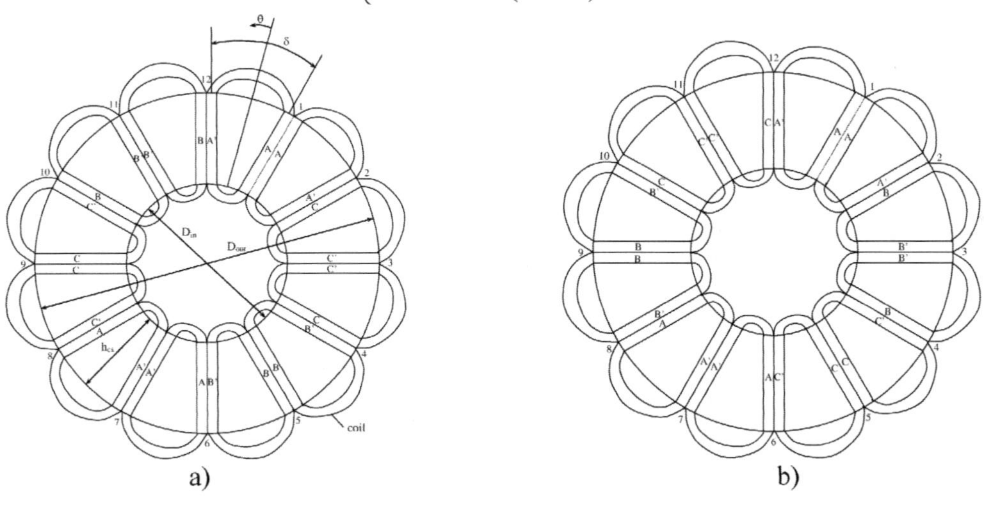

Figure 5. Two-dimensional view of the proposed three phase concentrated winding surface-mounted PM machine with N_{st}=12 slots. a) p=7 and b) p=5 pole pairs

Now let us consider the spatial distribution of the MMFs produced by phases "B" and "C", witch are deduced from the one produced by phase "A" as follows:

$$\begin{cases}F_B(\theta)=-\dfrac{1}{2}F_A\left(\theta+\dfrac{2\pi}{3}\right)\\ F_C(\theta)=-\dfrac{1}{2}F_A\left(\theta-\dfrac{2\pi}{3}\right)\end{cases}, \tag{14}$$

Then, the superposition of F_A, F_B and F_C has let to the spatial profile of the resulting armature MMFs F(θ).

4. Analytical Quasi-3D Modelling

Concerning the quasi-3D modeling the axial-flux PM machine may be considered to be composed of several axial-flux machines with differential radial length. The overall performance of an axial-flux machine is obtained by summing the performance of individual machines and by neglecting the possible flux flow in the depth direction of the machine [14]. This approach allows the consideration of different magnet shapes and variations of the teeth width in the direction of the machine radius. The method of transforming the 3D geometry of an axial-flux machine to a corresponding 2D model, to be used in quasi-3D computation, is illustrated in Figure 6.

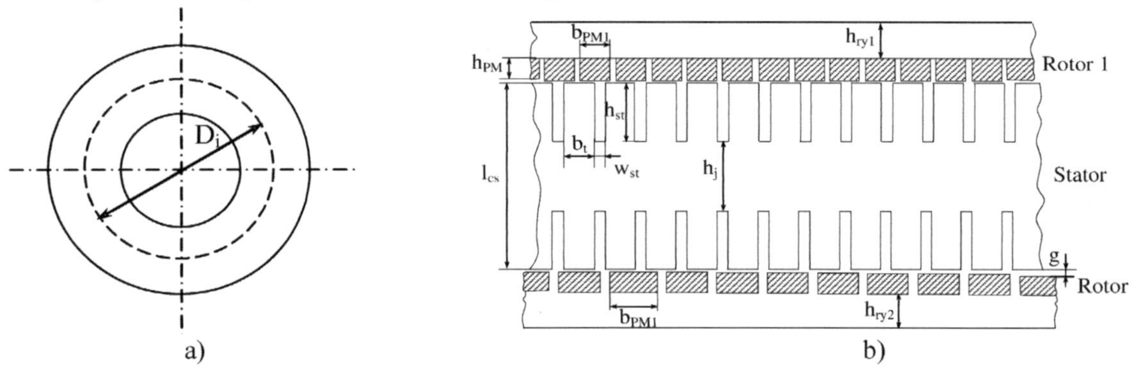

Figure 6. The selected i computation plan. a) selected i plan and b) 2D model of plan

For the quasi-3D computation, the average diameter D_i of a particular computation plan i, starting from the outer diameter of the machine is given by the equation [14]:

$$D_i = D_{out} - j \cdot \frac{h_{cs}}{N_p} , \tag{15}$$

where N_p is the number of computation planes used in a computation and h_{cs} is the length of the stator stack. The parameter j is defined as

$$j = 2i - 1 , \tag{16}$$

where index i goes from 1 to N_p. The length of the stator stack h_{cs} is defined as:

$$h_{cs} = \frac{D_{out} - D_{in}}{2} . \tag{17}$$

The pole pitch for each computation plane is given by the equation:

$$\tau_i = \frac{\pi D_i}{2p} . \tag{18}$$

Based on the set of equations (15)-(18) it is possible to divide the machine into a certain amount of computation planes. The number of computation plans needed for computation depends on the purpose (precision) of the computation.

5. Quasi – 3D FEM analysis

When applied to electrical machines, the described problem is usually reduced to cover only one pole or one pole pair with the help of boundary and symmetry conditions in order to reduce the computation time. In addition, it is often sufficient to reduce the problem to a 2D plane and use the finite element analysis (FEA). A finite element program enabling the modelling of linear movement is recommended [14], [16].

The main parameters of the model are given in Table 1.

The windings are in star connections and thus the current shares among phases are $I_A = I$, $I_B = I_C = -\frac{I}{2}$.

The computation results are calculated with the Simpson formula:

$$C = \left(C_{out} + 4C_{average} + C_{in}\right) \cdot \frac{1}{6}, \tag{19}$$

where C are result from computation and C_{out}, $C_{average}$ and C_{in} are the results from computation plane at D_{out}, D and D_{in}.

Table 1. Parameters and machine dimensions

Frequency (f)	50 Hz
Number of poles rotor 1	14
Number of poles rotor 2	10
Current density (J_s)	5.83 A/mm^2
Air-gap length (g)	2 mm
Outer diameter (D_{out})	260 mm
Inner diameter (D_{in})	156 mm
Pole-arc-ratio rotor 1 (α_{i1})	1
Pole-arc-ratio rotor 2 (α_{i2})	0.6944
Number of stator slots (N_{st})	12
Slot depth (h_{st})	21 mm
Axial length of stator core (h_{cs})	72 mm
Axial length of rotor 1 core (h_{ry1})	15 mm
Axial length of rotor 2 core (h_{ry2})	17.5 mm
Magnet axial length (h_{PM})	5 mm
PM rotor 1 width (b_{pm1})	38.6 mm
PM rotor 2 width (b_{pm2})	54.1 mm
Permanent magnet material	Ne Fe B

Figure 7 shows the finite element mesh for initial position. A background object that fills all free space between material objects is added. When the interaction forces are computed using the Maxwell tensor, it is better to design the integration path through finite elements that have all nodes in a linear medium without currents, in order to reduce the integration error. This condition could be achieved if the mesh has at least four layers in the air-gap. In order to enforce four mesh layers in the air-gap, we can define two extra regions along the air-gap with air features [16].

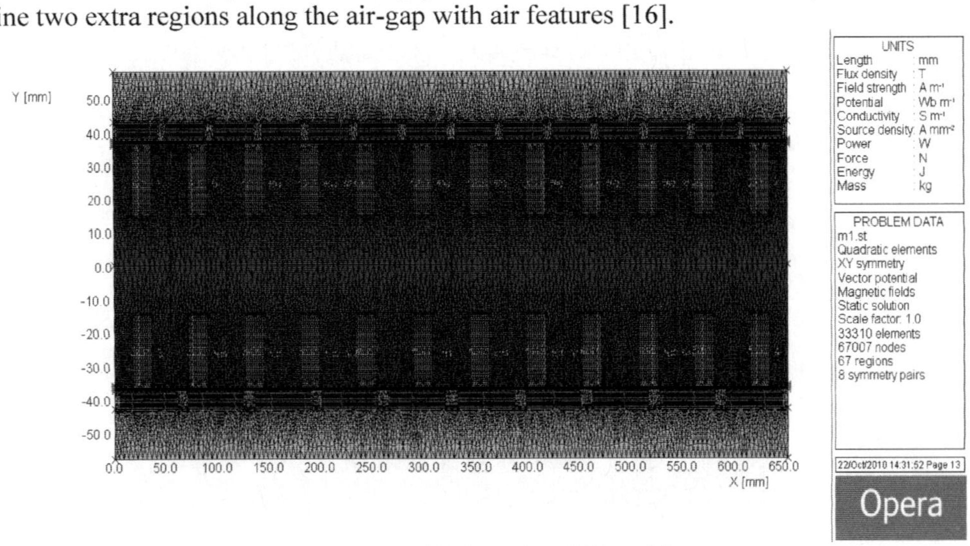

Figure 7. AFPM mesh – full problem

IOP Conf. Series: Materials Science and Engineering **163** (2017) 012006 doi:10.1088/1757-899X/163/1/012006

In our case, such a region was defined around the PM rotor 1 and PM rotor 2, placed in such a manner that it does not reach this region when it is moved. Figure 8 presents mesh details. The mesh has 67 regions, 8 symmetry pairs, 68 341 nodes and 33 958 elements.

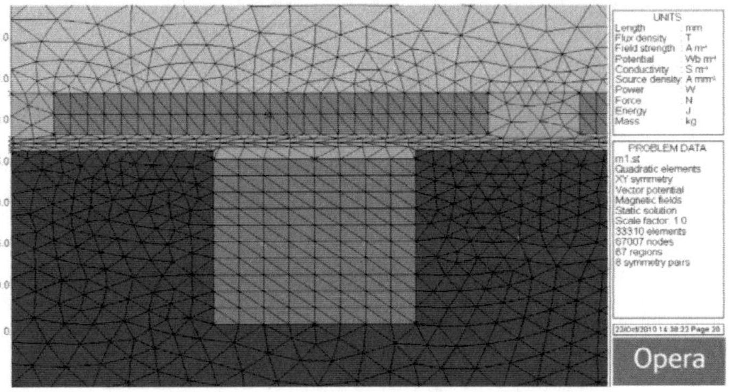

Figure 8. AFPM mesh – details

The FEM setup and magnetic field lines for initial position, are show for full load, I=720 Aturns in Figure 9.

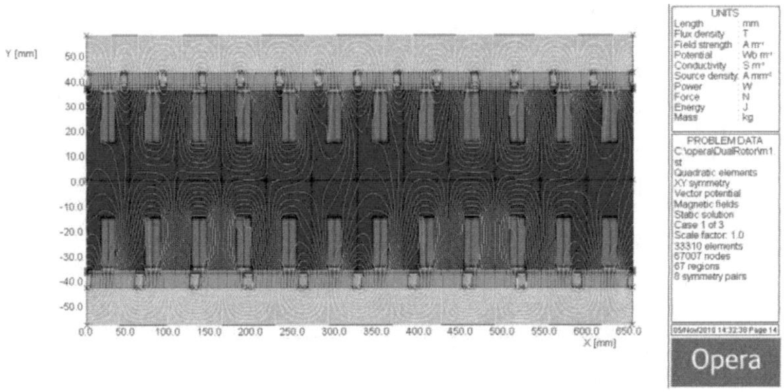

Figure 9. Magnetic field lines at I = 720 Aturns, and initial position of rotor 1

The total distribution of the flux density for initial position is presented in Figure 10 and the air-gaps flux density for rotor 1 and rotor 2 are presented in Figure 11 and Figure 12.

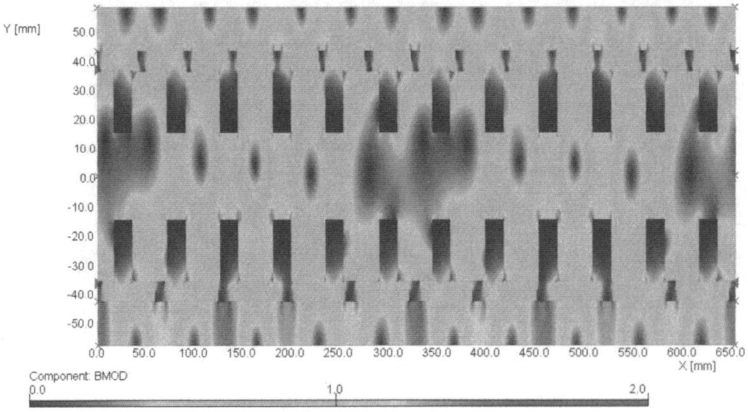

Figure 10. The total distribution of the flux density at full load (I=720 Aturns), and initial position of rotor 1

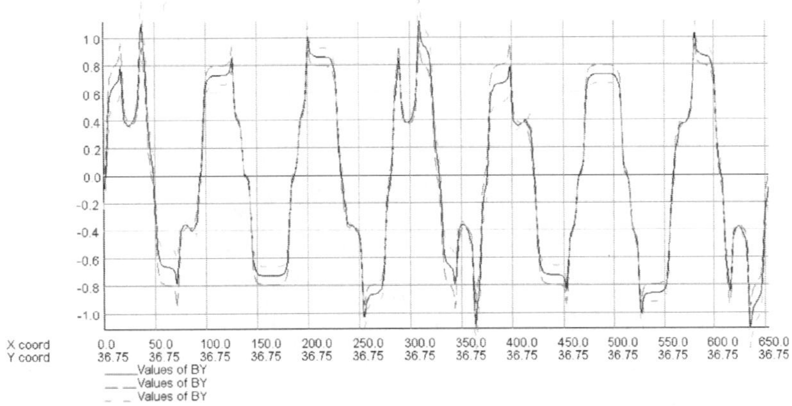

Figure 11. The air-gap 1 flux density for I = 0 (red), I=720 Aturns (black) and I = 1440 A (green)

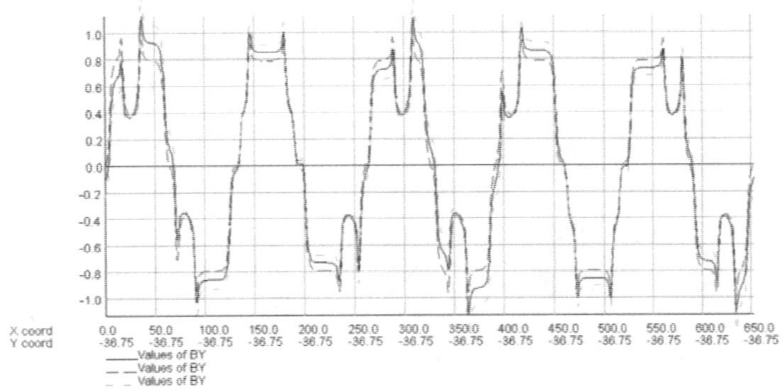

Figure 12. The air-gap 2 flux density for I = 0 (red), I=720 Aturns (black) and I = 1440 Aturns (green)

In Figure 13 is presented the magnetic induction on a pair of poles in idle run for the airgap 1, respectively 2, obtained by FEM-3D [2]. Comparing with the results presented in Figure 11 and Figure 12, one can notice that both the curves' shape and the induction values mostly coincide (3% and 3,5% maximum difference for $p_1=7$ and $p_2=5$ respectively).

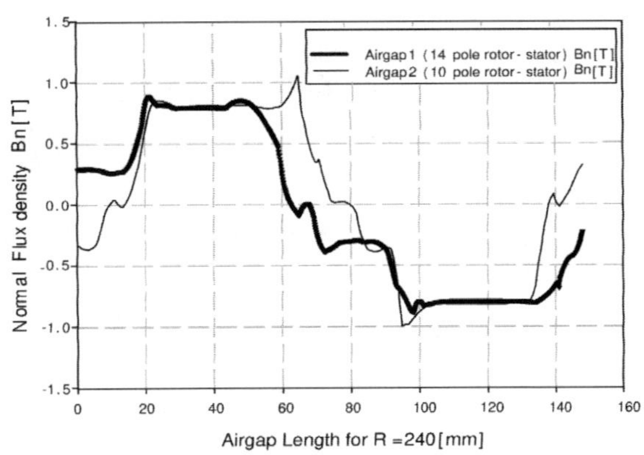

Figure 13. No load air-gap flux density variation from 3D-FEM

In quasi 3D-analysis the rotating of rotor 1 is converted in translation between rotor 1 and stator. Figure 14 presents the position of rotor 1 for maximum torque, and in Figure 15 a mesh detail for this position. The real scale between the machine's radial and axial dimensions are presented in Figure 16 (with magnetic field lines produced by the PMs – I=0).

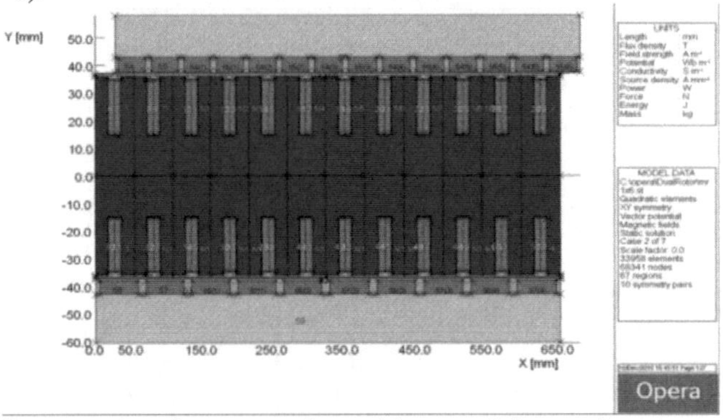

Figure 14. Computational average plan with rotor 1 in maximum torque position

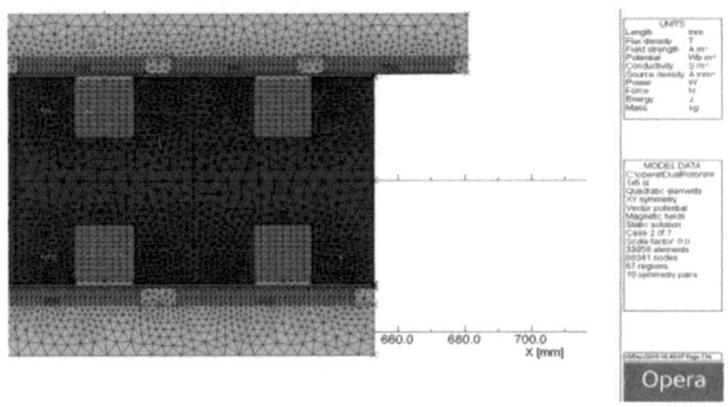

Figure 15. Mesh details at maximum torque position of rotor 1

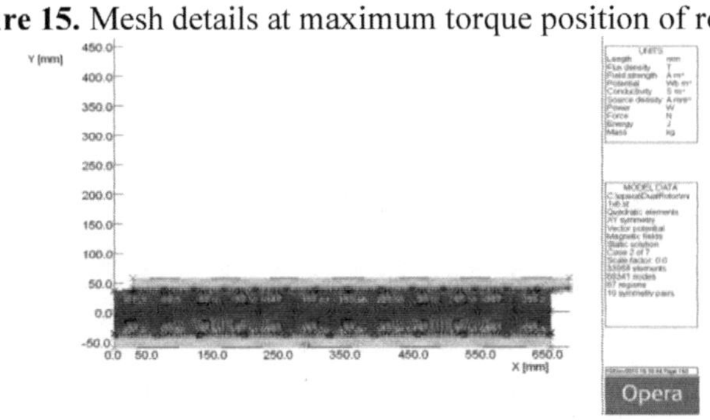

Figure 16. Real scale of the machine. Magnetic field lines with rotor 1 in maximum torque position. I=0

The computation was made with seven factor scale current and τ displacement of rotor 1 in ten intermediary positions. The results are for phase a1 flux (Figure 17), for PM flux in rotor 1 and rotor 2 (considered stationary) (Figure 18), for tangential and axial forces for rotor 1 in Figure 19 and Figure 20. The inductance variation is very low with position (Figure 21) and the torque (Figure 22) is comparable with the value computed by 3D FEM analysis (Figure 23) (3,7% maximum difference) [2]. The cogging torque is very low (I=0 Figure 22).

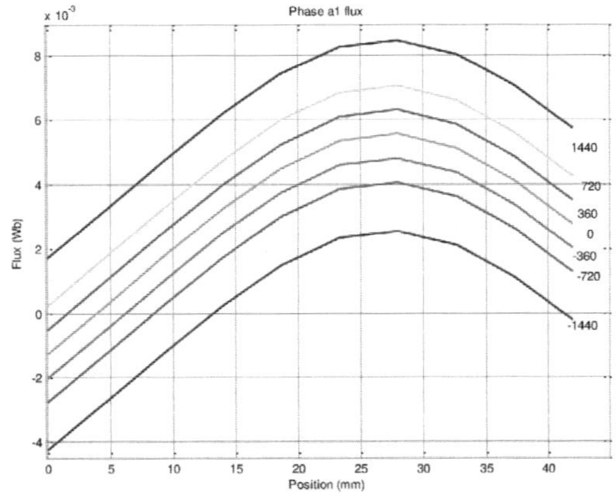

Figure 17. Phase a1 flux at seven factor scale current

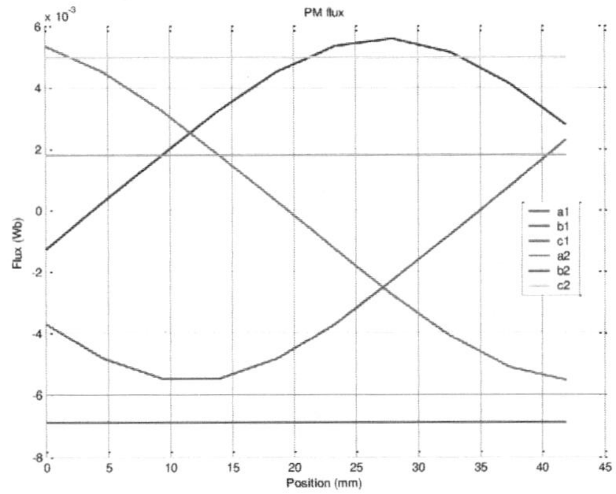

Figure 18. PM flux in rotor 1 and rotor 2

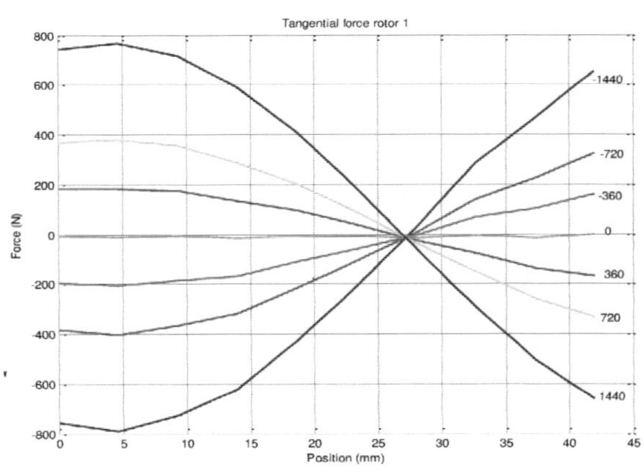

Figure 19 Tangential force rotor 1 for seven factor scale current

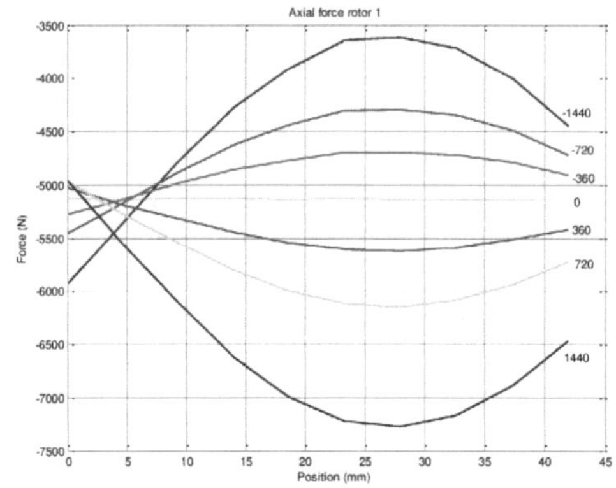

Figure 20. Axial force rotor 1 for seven factor scale current

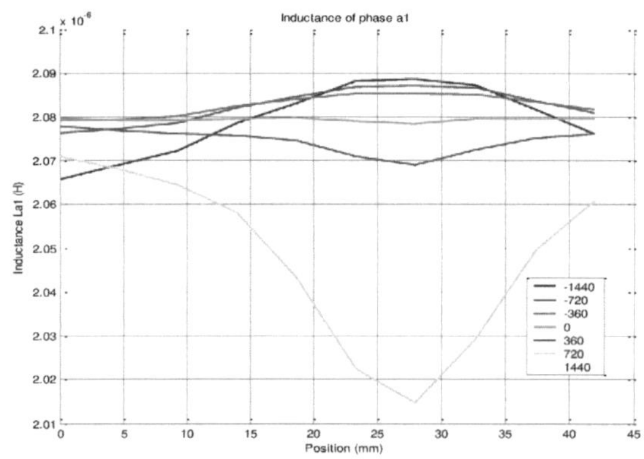

Figure 21. Inductance per phase at seven factor scale current

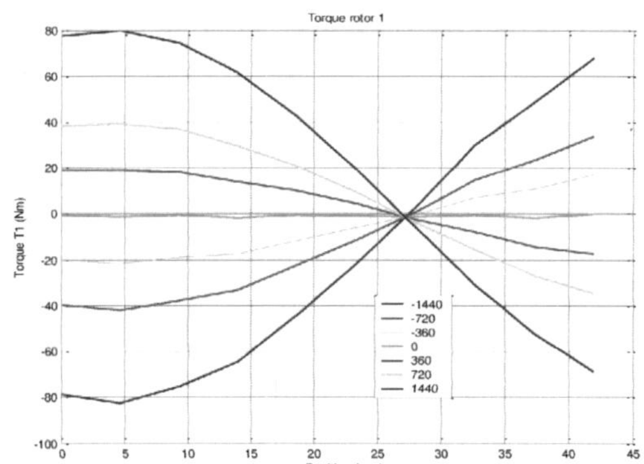

Figure 22. Torque of rotor 1 at seven factor scale current

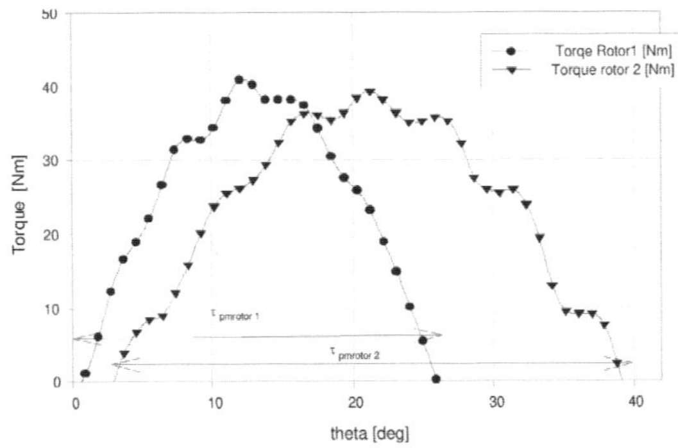

Figure 23. 3D Analysis torque for the two rotors

Comparing the results obtained by the quasi-3D method with 3D-FEM (2,1% mean square error for torque calculation), is noticed that both the torque and the magnetic induction in the air gap have closed variation forms and values. The results of the quasi-3D analysis were obtained for three calculation plans at diameters D_{in}, D and D_{out}. For a higher precision can be taken into account additional plans. The calculation time for 3D-FEM analysis is much bigger than for quasi-3D. From here, it results the advantage of using this type of analysis.

6. Conclusions

Using an analytical design method or the 2D finite element analysis (FEA) only for the average radius on the machine does not generally yield sufficiently accurate computation results. With the 3D FEA, it is possible to take into consideration the actual 3D structure of the machine, but performing the computations is too often time-consuming, especially if the objective is to achieve a preliminary design of the machine.

The basic idea of the hereby proposed design method is to firstly subdivide the axial flux machine into a sufficient amount of independent computation plans, next perform the required 2D computations on each plane, and finally compose the overall performance of the machine from the computation results obtained for each design plan.

References

[1] Gieras J F, Wang R J and Kamper M J 2008 *Axial Flux Permanent Brushless Machines*, Second Edition, Springer Science, Boston, MS USA

[2] Boldea I, Topor M, Marignetti F, Deaconu S I and Tutelea L N 2010 *A Novel Single Stator Dual PM Rotor, Synchronous Machine: topology, circuit model, controlled dynamics simulation and 3D FEM Analysis of Torque Production*, 12th International Conference on Optimization of Electrical and Electronic Equipment OPTIM 2010, Brasov, Romania, May 20-22, pp 343-351

[3] Marignetti F, Colli V D, Di Stefano R and Cavagnino A 2007 D*esign Issues of a Fractional – Slot Windings Axial Flux PM Machine with Soft Magnetic Compound Stator*, IECON 2007, Taipei, Taiwan, November 5-8, pp 187-192

[4] Campbell P 1974 Principles of a permanent magnet axial-field DC machine, *Proceedings IEE* **121**(12) 1489-1494

[5] Campbell P 1975 The magnetic circuit of an axial field DC electrical machine, *IEEE Transactions on Magnetics* **11**(5) 1541-1543

[6] Leung W 1978 *Sandwich synchronous machines and DC machines*, patent specifications, London, UK

[7] Profumo F, Zhang Z and Tenconi A 1997 Axial flux machine drives a new viable solution for electric cars, *IEEE Transaction on Industrial Electronics* **44**(1) 39-45

[8] Eastham J F, Profumo F, Tenconi A, Hill-Cottingham R J, Coles P C and Gianolio G 2002 Novel Axial flux Machine for aircraft drive: design and modeling, *IEEE Transactions on*

Magnetics **38**(5) 3003-3005

[9] Brooking P and Bumby J R 2002 A*n integrated engine-generator set with power electronic interface for hybrid electric vehicle applications*, International Conference on Power Electronics, Machines and Drives, June 4-7, pp 153-158

[10] Cavagnino A, Lazzari M, Profumo F and Tenconi A 2002 A comparison between the axial flux and the radial flux structures for PM synchronous motors, *IEEE Transactions on Industrial Applications* **38**(6) 1517-1524

[11] Sitapati K and Krishnan R 2001 Performance Comparisons of Radial and Axial Field, permanent-Magnet, Brushless Machines, *IEEE Transactions on Industrial Applications* **37**(5) 1219-1226

[12] Marignetti F and Scarano M 1999 *An Axial-flux PM Motor Wheel*, Electromotion'99, Patras, Greece, pp 1-6

[13] Cros J and Viarouge Ph 2002 Synthesis of High performance PM Motors with Concentrated Windings, *IEEE Transactions on Energy Conversion* **17**(2) 248-253

[14] Parviainen A 2005 *Design of AFPM low-speed Machines and Performance Comparison between Radial-Flux and Axial-Flux Machines*, Lappeenranta University of Technology, Finland, Doctoral Thesis

[15] Hamadou G B, Masmoudi A, Abdennadher I and Masmoudi A 2009 Design of a Single-Stator Dual-Rotor Permanent-Magnet Machine, *IEEE Transactions on Magnetics* **45**(1) 127-132

[16] Boldea I and Tutelea L N 2009 *Electric Machines. Steady State, Transients and Design with MATLAB*, CRC Press

IOP Conf. Series: Materials Science and Engineering **163** (2017) 012007 doi:10.1088/1757-899X/163/1/012007

The modified nodal analysis method applied to the modeling of the thermal circuit of an asynchronous machine

O Nedelcu[1], C I Salisteanu[1], F Popa[1], B Salisteanu[2], C V Oprescu[2] and V Dogaru[3]

[1]Valahia University of Targoviste, Department of Electronics, Telecommunications and Energy Engineering, Targoviste, Romania
[2]Valahia University of Targoviste, PhD Student, Targoviste, Romania
[3]Valahia University of Targoviste, Doctoral School, Targoviste, Romania

E-mail: otilia.nedelcu@valahia.ro

Abstract. The complexity of electrical circuits or of equivalent thermal circuits that were considered to be analyzed and solved requires taking into account the method that is used for their solving. Choosing the method of solving determines the amount of calculation necessary for applying one of the methods.
The heating and ventilation systems of electrical machines that have to be modeled result in complex equivalent electrical circuits of large dimensions, which requires the use of the most efficient methods of solving them.
The purpose of the thermal calculation of electrical machines is to establish the heating, the overruns of temperatures or over-temperatures in some parts of the machine compared to the temperature of the ambient, in a given operating mode of the machine.
The paper presents the application of the modified nodal analysis method for the modeling of the thermal circuit of an asynchronous machine.

1. Introduction

By modeling the heating and ventilation systems of the electrical machines, equivalent complex electrical circuits that have large dimensions are achieved and they require a complex analysis. The analysis of these circuits, in stationary and dynamic regime, requires the use of effective methods of solving, [1]. Due to the simplicity and flexibility of the modified nodal analysis method, it is one of the most frequently used as it formulates equations for electrical circuits and your computer simulates, [2].

Generally, circuit analysis in a dynamic regime is reduced to the analysis of a string of resistive circuits, by replacing coils and condensers with discrete patterns resistive of circuit, associated to an implicit numerical algorithm of numerical integration. Developing methods for analyzing nonlinear resistive electrical circuits, as effective as possible, has a special importance, resulting from substituting dynamic circuit elements with discrete resistive models associated with an implicit algorithm of integration.

The two theorems of Kirchhoff, together with the constitutive equations of the sides, are the base of nodal analysis, [3], [4]. Currents on certain sides cannot always be expressed according to the parameters that are characteristic of the respective sides and the potential nodes, situation that does not allow the use of a classical nodal analysis method, for this reason its extension was appealed, to yield the modified nodal analysis method.

By doing the analogy between a network of cooling ventilation and a non-linear resistive circuit it can be determined the flow values of the cooling fluid of an electrical machine.

2. Description of modified nodal method

The modified nodal analysis method is a generalization of the classical nodal method, which allows the analyzed circuit to contain circuit elements that are incompatible with the classical method.

The structure of circuits to which they apply the nodal method consists in the following circuit elements: resistors, linear capacitors and non-coupled magnetically coils; nonlinear resistors commanded in voltage; nonlinear capacitors; independent sources of current; independent sources of power that do not form by themselves a side.

The difference between the classical nodal method and the modified nodal method is that in the modified nodal method currents of the circuit elements that are non-compatible with the classical nodal method (such as currents of non-linear coils commanded in current and currents of linear coils magnetically coupled) appear additionally, as independent variables. Linear coils that are non-coupled do not introduce additional variables, [4-7].

For the modified nodal analysis method, the reference is considered the potential of one of the nodes of the circuit for which it will be deemed $\mathbf{V} = 0$, this node of reference in the matrix of incidence A in the node which is missing.

It is applied the Kirchhoff's first theorem to the other nodes, i.e. $(n-1)$ nodes, writing the currents of sides depending on the potentials at nodes. The currents of sides, having null resistance, cannot be expressed by themselves depending on the potentials at nodes, but shall be considered as independent variables found in the vector $\mathbf{V}_{(n-1),1}$, together with potentials at nodes.

Depending on potentials, all nodes and tensions on the sides will be expressed as shown in relation (1):

$$\begin{bmatrix} U_P \\ U_E \\ U_J \end{bmatrix} = \begin{bmatrix} A_P^t \\ A_E^t \\ A_J^t \end{bmatrix} \cdot \mathbf{V} = \begin{bmatrix} A_P^t \cdot \mathbf{V} \\ A_E^t \cdot \mathbf{V} \\ A_J^t \cdot \mathbf{V} \end{bmatrix}. \tag{1}$$

The first Kirchhoff theorem shall be written as in relation (2):

$$A_P I_P + A_E I_E + A_J I_J = \mathbf{0}_{(n-1),1}. \tag{2}$$

The tensions of the passive sides will determine the currents of the passive sides, as shown in (3):

$$I_p = G \cdot U_P = G \cdot A_P^t \cdot \mathbf{V}. \tag{3}$$

After they determined the currents of the passive sides, the expression obtained will replace the first Kirchhoff theorem, and then the characteristic equations of sides with null resistance will be added, affording a corresponding matrix system for the modified nodal method, relations (4) - (6):

$$A_P \cdot G \cdot A_P^t + A_E I_E = -A_J J, \tag{4}$$

$$U_E = A_E^t \cdot V; \quad U_E = -E, \tag{5}$$

$$\begin{cases} A_P \cdot G \cdot A_P^t \cdot V + A_E \cdot I_E = -A_J \cdot J \\ A_E^t \cdot V = -E \end{cases}. \tag{6}$$

IOP Conf. Series: Materials Science and Engineering **163** (2017) 012007 doi:10.1088/1757-899X/163/1/012007

3. Applying the modified nodal method in modeling an asynchronous machine with rotor in cage

In Figure 1 it is presented a section of an asynchronous machine with a rotor in a cage, in which case the rotor is the thermal source and calculations must be made for the thermal resistance's corresponding heat transmission, developed in the rotor, outwards, [8].

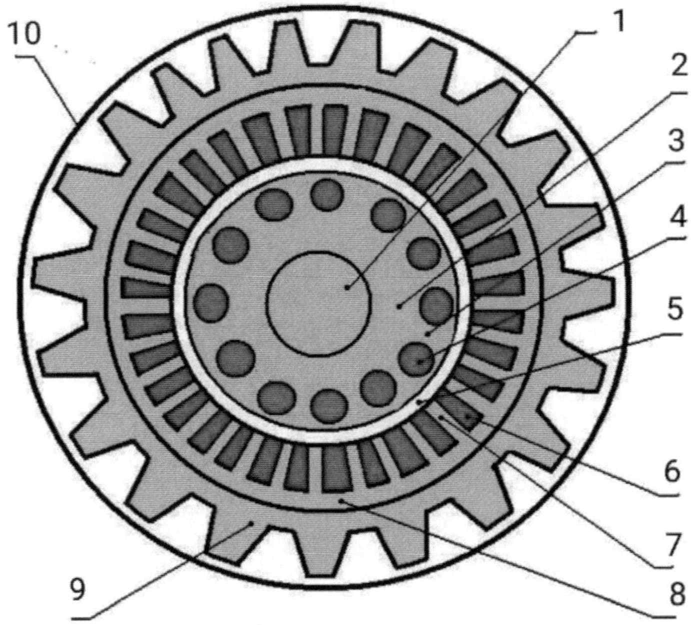

Figure 1. Asynchronous machine with rotor in cage
1 –ax; 2 –rotor yoke; 3 –rotor notch; 4 –rotor windings;
5 –air gap; 6 –stator notch; 7 –stator tooth; 8 –stator yoke;
9 –radiator; 10 –housing

Figure 2 shows the equivalent circuit diagram with heating systems (in stationary regime). Using the program PDCEN – Program of Rotting Decompose of Electrical Circuit after the Nodes, it will decompose „after the nodes" the circuit from Figure 2, [9].

The program, [10], is written in C++ and implemented on a microcomputer type PC compatible with IBM, used for the numerical integration of an ordinary differential equation generalized with the first order regressive method. Dynamic circuit elements are replaced with discreet resistive circuits associated with these numerical methods, and for the analysis of electrical circuits the program used the modified nodal method, [10], [11].

Running the program the results presented in tables 1 and 2 were obtained.

The sub-circuits structure is presented in Table 1.

Table 1. Sub-circuits structure

SUBCIRCUITS=8.
SUBCIRCUIT 1 of size 4
4 5 6 7
END
SUBCIRCUIT 2 of size 1
2
END

SUBCIRCUIT 3 of size 4
9 29 30 31
END
SUBCIRCUIT 4 of size 4
11 13 14 15
END
SUBCIRCUIT 5 of size 1
25
END
SUBCIRCUIT 6 of size 3
21 22 23
END
SUBCIRCUIT 7 of size 2
27 28
END
SUBCIRCUIT 8 of size 4
16 17 18 19
END

The nodes of interconnection are: {1,3,8,10,12,20,24,26,32}.
Table 2 shows the distribution of 9 interconnection nodes to the 8 sub-circuits obtained.

Table 2. Distribution the 9 interconnection nodes to the 8 sub-circuits obtained

9 INTERCONNECTION NODES
INTERCONNECTION NODE 32 IS ADJACENT TO:
CLUSTER 8 OF SIZE 4 (5 CONNECTIONS)
CLUSTER 7 OF SIZE 2 (2 CONNECTIONS)
CLUSTER 6 OF SIZE 3 (2 CONNECTIONS)
CLUSTER 5 OF SIZE 1 (1 CONNECTIONS)
CLUSTER 4 OF SIZE 4 (3 CONNECTIONS)
CLUSTER 3 OF SIZE 4 (2 CONNECTIONS)
BEST CLUSTER FOR NODE 32: 8 OF SIZE 4 (5 CONNECTIONS)
INTERCONNECTION NODE 12 IS ADJACENT TO:
CLUSTER 8 OF SIZE 4 (1 CONNECTIONS)
CLUSTER 4 OF SIZE 4 (1 CONNECTIONS)
BEST CLUSTER FOR NODE 12: 8 OF SIZE 4 (1 CONNECTIONS)
INTERCONNECTION NODE 20 IS ADJACENT TO:
CLUSTER 8 OF SIZE 4 (1 CONNECTIONS)
CLUSTER 6 OF SIZE 3 (1 CONNECTIONS)
CLUSTER 3 OF SIZE 4 (1 CONNECTIONS)
BEST CLUSTER FOR NODE 20: 6 OF SIZE 3 (1 CONNECTIONS)
INTERCONNECTION NODE 26 IS ADJACENT TO:
CLUSTER 7 OF SIZE 2 (1 CONNECTIONS)
CLUSTER 4 OF SIZE 4 (1 CONNECTIONS)
BEST CLUSTER FOR NODE 26: 7 OF SIZE 2 (1 CONNECTIONS)
INTERCONNECTION NODE 3 IS ADJACENT TO:
CLUSTER 6 OF SIZE 3 (1 CONNECTIONS)
CLUSTER 2 OF SIZE 1 (1 CONNECTIONS)
CLUSTER 1 OF SIZE 4 (1 CONNECTIONS)
BEST CLUSTER FOR NODE 3: 2 OF SIZE 1 (1 CONNECTIONS)
INTERCONNECTION NODE 24 IS ADJACENT TO:
CLUSTER 5 OF SIZE 1 (1 CONNECTIONS)
CLUSTER 4 OF SIZE 4 (1 CONNECTIONS)
BEST CLUSTER FOR NODE 24: 5 OF SIZE 1 (1 CONNECTIONS)

INTERCONNECTION NODE 10 IS ADJACENT TO: CLUSTER 4 OF SIZE 4 (2 CONNECTIONS) CLUSTER 1 OF SIZE 4 (1 CONNECTIONS) BEST CLUSTER FOR NODE 10: 4 OF SIZE 4 (2 CONNECTIONS)
INTERCONNECTION NODE 8 IS ADJACENT TO: CLUSTER 3 OF SIZE 4 (1 CONNECTIONS) CLUSTER 1 OF SIZE 4 (1 CONNECTIONS) BEST CLUSTER FOR NODE 8: 3 OF SIZE 4 (1 CONNECTIONS)
FINAL NUMBER OF INTERCONNECTION NODES: 9; 0 COULD BE MOVED DIRECTLY TO CLUSTERS END_INTERCONN CLUSTERS=8
CLUSTERS CLUSTER 1 OF SIZE 4 4 5 6 7 END
CLUSTER 2 OF SIZE 1 2 END
CLUSTER 3 OF SIZE 4 9 29 30 31 END
CLUSTER 4 OF SIZE 4 11 13 14 15 END
CLUSTER 5 OF SIZE 1 25 END
CLUSTER 6 OF SIZE 3 21 22 23 END
CLUSTER 7 OF SIZE 2 27 28 END
CLUSTER 8 OF SIZE 4 16 17 18 19 END
END_CLUSTERS
RENUM
4 5 6 7 2 9 29 30 31 11 13 14 15 25 21

```
22
23
27
28
16
17
18
19
1
3
8
10
12
20
24
26
32
END_RENUM
```

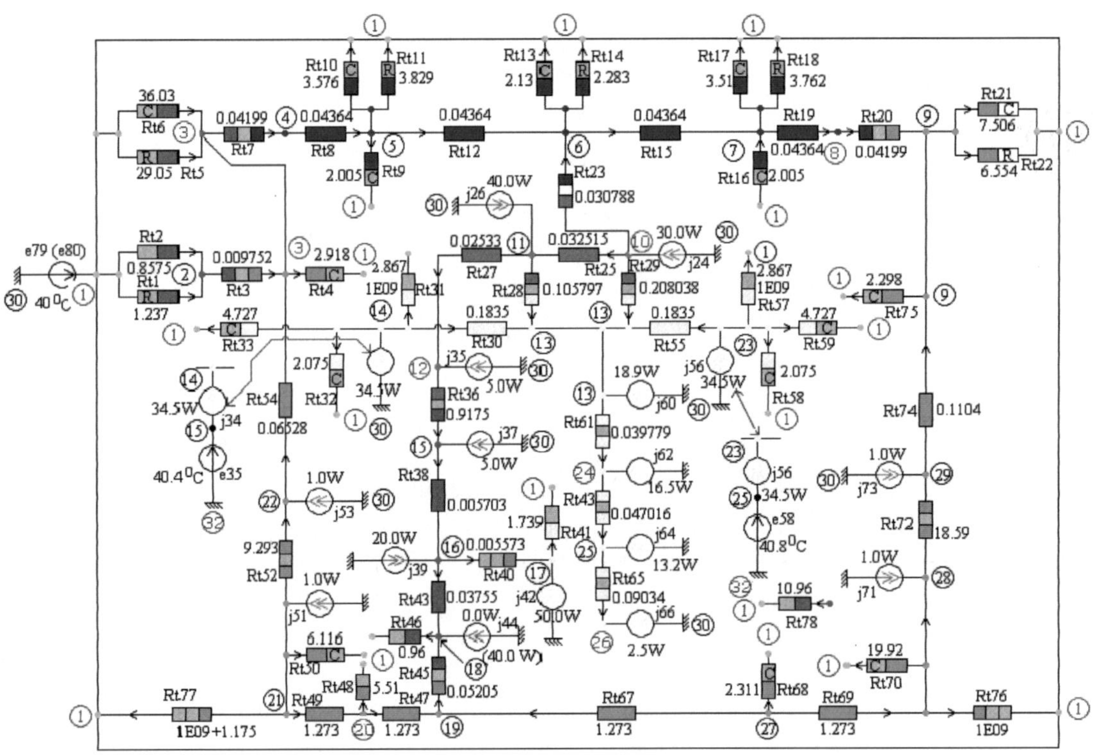

Figure 2. The decomposition of thermal circuit of the asyncronous machine in subcircuits

After the decomposition circuit, the nodal matrix has the bordered block diagonal structure as in Figure 3.

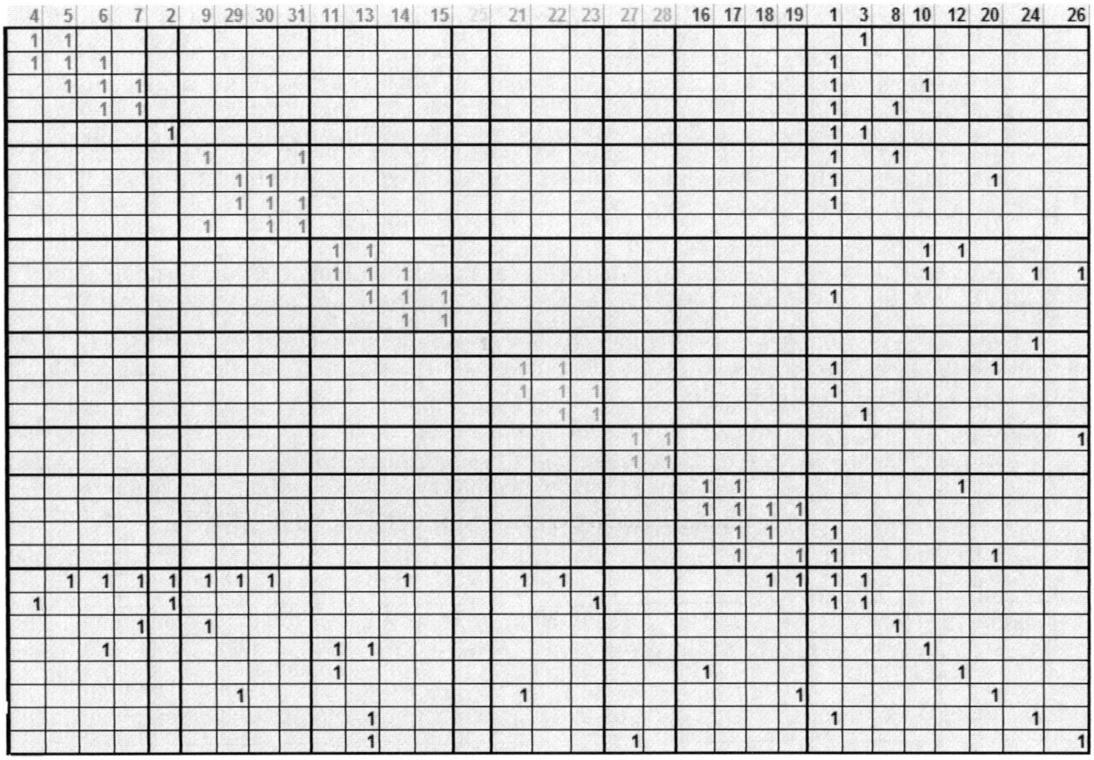

Figure 3. Nodal matrix structure after decomposition circuit

4. Conclusions

The modified nodal method does not require restrictions of the circuit structure to be analyzed, [12].

For the analysis, in a stationary regime and also in a dynamic regime, of the equivalent complex electrical circuits with large dimensions which are obtained from modeling heating and ventilation systems of the electrical machines, the modified nodal analysis method is used.

The bordering nodal matrix consists of rows and columns corresponding with potential interconnection nodes. In the case of the modified nodal method, the bordering is formed with rows and columns which correspond to the nodes of interconnection and the command currents of interconnection.

The purpose of a thermal calculation for electrical machines is to establish heating, overruns of temperatures or over-temperature in some parts of the machine compared to the ambient temperature, at a given operating mode of the machine.

References

[1] Hănţilă F 1979 A Method for Solving Nonlinear Resistive Networks, *Roum. Sci. Techn. Électrotechn. et Énerg.* **24**(2) 21-30

[2] Iordache M and Dumitriu L 2004 *Computer-aided analysis of non-linear analogue circuits* (in Romanian: *Analiza asistată de calculator a circuitelor analogice neliniare*), Politehnica Press Publishing House, Bucharest, Romania

[3] Iordache M, Dumitriu L and Matei I 1999 *Symbolic analysis based on Modified NOdal Method, User's Manual* (in Romanian: *ASINOM – Analiza SImbolică bazată pe Metoda Nodală Modificată, Manual de utilizare*), Department of Electrical Engineering, University Politehnica of Bucharest

[4] Schwarz A E 1987 *Computer-aided design of microelectronic circuits and systems*, Academic Press, London, UK

[5] Chua L O, Desoer C A and Kuh E S 1987 *Linear and nonlinear circuits*, McGraw-Hill, N.Z.

[6] Topan D and Mandache L 2002 *Methods of analysis in complex electric circuits* (in Romanian: *Metode de analiză în circuite electrice complexe*), Universitaria Publishing House, Craiova, Romania

[7] Iordache M 2007 *Modified Nodal Method* (in Romanian: *Metoda Nodală Modificată*), Simpozionul National de Electrotehnica Teoretica SNET'07, Bucharest, Romania, October 12-14, pp 10 - 20

[8] Magureanu R and Vasile N 1990 *Brushless synchrounous servomotors* (in Romanian: *Servomotoare fara perii tip sincron*), Technical Publishing House, Bucharest, Romania

[9] Iordache M and Perpelea M 1995 *Computer-aided analysis of large complex linar and non-linear electric circuits* (in Romanian: *Analiza asistată de calculator a circuitelor electrice şi electronice neliniare complexe de mari dimensiuni*), Didactic and Pedagogical Publishing House, Bucharest, Romania

[10] Dumitriu L and Iordache M 1998 *Modern theory of electric circuits- Vol. I: Theoretical background, Applications, Algorithms and Computational softwares* (in Romanian: *Teoria modernă a circuitelor electrice – Vol. I: Fundamentare teoretică, Aplicaţii, Algoritmi şi Programe de calcul*), All Educational S.A. Publishing House, Bucharest, Romania

[11] McCalla W J 1988 *Fundamentals of computer-aided circuit simulation*, Kluwer Academic Publishers, Boston, USA

[12] Nedelcu O, Iordache M and Enescu D 2006 *An efficient method for computing of the electric machine ventilation*, The Sixth WESC, Torino, Italy, July 10-12, p 251

Specific loss power in superparamagnetic hyperthermia: nanofluid versus composite

M Osaci[1] and M Cacciola[2]

[1]Politehnica University of Timisoara, Department of Electrical Engineering and Industrial Informatics, 5 Revolutiei street, 331128 Hunedoara, Romania
[2]„Mediterranea" University of Reggio Calabria, DICEAM, Italy

E-mail: mihaela.osaci@upt.ro

Abstract. Currently, the magnetic hyperthermia induced by nanoparticles is of great interest in biomedical applications. In the literature, we can find a lot of models for magnetic hyperthermia, but many of them do not give importance to a significant detail, such as the geometry of nanoparticle positions in the system. Usually, a nanofluid is treated by considering random positions of the nanoparticles, geometry that is actually characteristic to the composite nanoparticles. To assess the error which is frequently made, in this paper we propose a comparative analysis between the specific loss power (SLP) in case of a nanofluid and the SLP in case of a composite with magnetic nanoparticles. We are going to use a superparamagnetic hyperthermia model based on the improved model for calculating the Néel relaxation time in a magnetic field oblique to the nanoparticle magnetic anisotropy axes, and on the improved theoretical model LRT (linear response theory) for SLP. To generate the nanoparticle geometry in the system, we are going to apply a Monte Carlo method to a nanofluid, by minimising the interaction potentials in liquid medium and, for a composite environment, a method for generating random positions of the nanoparticles in a given volume.

1. Introduction

The magnetic nanoparticles in suspension in a liquid medium (nanofluid), subjected to a high frequency alternating magnetic field, can generate heat through magnetic losses. This local heat can lead to the destruction of cancer cells. Depending on the increasingly higher nanoparticle diameters, there are, according to the literature, three mechanisms for generating heat by magnetic hyperthermia: susceptibility loss, hysteresis loss and viscous heating (magnetic stirring loss). The specific loss power (SLP) is defined as electromagnetic power lost per magnetic material mass unit and is expressed in watts per kilogram. The mostly used particles in magnetic hyperthermia are the magnetite ones, because of their biocompatibility and low toxicity. The susceptibility loss occurs in superparamagnetic particles. The agglomeration and sedimentation tendency increases considerably with the transition from small nanoparticles to stable single-domain ferromagnetic particles [1], in which case the mechanism of generating hysteresis heat is predominant. Unfortunately, while the particle size is increasing, the agglomeration and sedimentation tendency determines the clogging of blood vessels, situation which is therapeutically unacceptable. The heat generation by magnetic stirring loss occurs when the particles are large enough. Most of these particles are not aligned to the applied field axis, and therefore the particle stirring leads to viscous heating [2]. This stirring can add heat to the system.

These mechanisms for generating heat by magnetic hyperthermia are generally quite complex, because they involve a number of parameters, such as: external magnetic field strength and frequency, particle size distribution, magnetic anisotropy constants distribution, particle concentration, colloid properties, etc.

Recent modelling studies have investigated the main mechanisms for generating heat by magnetic hyperthermia, i.e. susceptibilities loss and hysteresis loss, without taking into account the particle positions [3-7].

Other current models take also into account the particle positions, such as the particle concentration. Some of these models are studying the influence of anisotropy axes orientation (with respect to the external magnetic field) on the properties of magnetic hyperthermia, when the particles have random positions in a given volume, as in the case of particles spread in non-magnetic solid matrix [8-10], others take into account the particle organisation (as in liquid matrix), but under the influence of a generic repulsive dimensionless force [11].

In a liquid medium, the particle organisation is clearly different than in a solid medium, and depends on the attractive and repulsive interaction forces exerted on the particle. Some recent papers showed experimentally a tendency to agglomeration, even in the absence of an external magnetic field [12], [13]. This trend, which can determines the clogging of the blood vessels through which the hyperthermia agent is circulating (see [1]), can also increase the influence of magnetic dipolar interactions on the heat generation process. If we apply a nanoparticle positioning like in solid matrix for magnetic hyperthermia therapy, it means to further use an approach beyond reality.

This paper adopts a model for superparamagnetic hyperthermia with spherical nanoparticles. Using this model, we can study how the nanoparticle organisation in non-magnetic liquid or solid matrix influences the SLP. Since the magnetic hyperthermia agent is a nanofluid reaching the affected zone through the bloodstream, it is shown the fact that using in studies a random-type organisation in a given volume, specific to the solid matrix, is not the best choice.

2. Nanoparticle organisation in non-magnetic liquid or solid matrix

In our model, we define a system of N spherical nanoparticles. Each nanoparticle i has a magnetic moment $\bar{\mu}_i$ and a radius r_i. Initially, the orientation of the magnetic moments is random.

In case of solid matrix, the possible positions are: simple-centred cubic lattice, base-centred cubic lattice, and face-centred cubic lattice. Among all the possible positions, the particles have been located in randomly selected suitable locations, according to the ratio between the volumes of the considered nanoparticles and the whole volume of the simulation box. The periodic boundary conditions have been applied on all sides of the simulation box. In Figure 1, we show such arrangement for 500 nanoparticles in a face-centred cubic lattice, where the volume fraction of nanoparticles is 0.05.

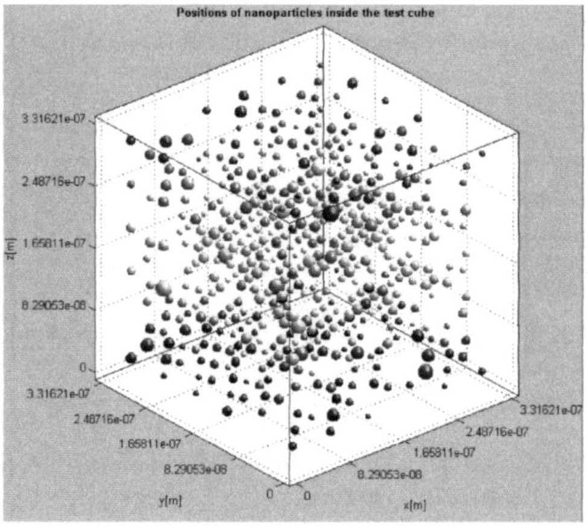

Figure 1. Nanoparticles randomly positioned in a face-centred cubic grid

The nanoparticles dispersed in a carrier fluid form superparamagnetic colloids or nanofluids. The nanofluids are basically classified into two groups: ionic nanofluids and surfactant nanofluids, depending on whether the colloidal stability is provided by electrostatic or steric repulsive interactions, respectively. In the recent years, a new class of ferrofluids, which are ultrastable and often biocompatible, have been developed using a combination of both stabilisation methods [14].

We further describe the positioning in case of liquid matrix. The nanoparticles are coated with a surfactant layer whose thickness is δ, and the number of molecules adsorbed on each nanoparticle surface is ζ. In case of electrosteric stabilisation, the nanoparticle surface charge is q. The 3D arrangement of the particles was initially made randomly in grid, and then we fixed the nanoparticle positions in a simple-centred cubic lattice, a body-centred cubic lattice, and a face-centred cubic lattice.

The stability of a magnetic colloidal suspension results from the equilibrium between the attractive and repulsive forces. Theoretically, two types of forces can contribute to the interparticle potential in the system: van der Waals force and electrostatic repulsive force. The Van der Waals forces induce strong short-range isotropic attractions [15-16]. The electrostatic repulsive forces [15], [17-19], can be partially screened by adding salt to the suspension. The steric repulsion forces [20], [21] have to be taken into account for coated nanoparticles. Finally, the magnetic dipolar forces [22] between two particles must be also added in case of magnetic suspensions.

The local magnetic field $\vec{H}_i$ on each nanoparticle results from a vectorial sum of the applied external magnetic field ($\vec{H}_{ext}$) and the internal dipolar magnetic field $\vec{H}_{id}$:

$$\vec{H}_i = \vec{H}_{ext} + \vec{H}_{id} \tag{1}$$

In colloid, the nanoparticles form aggregations leading to an increase in the local concentration of nanoparticles, and so the effect of dipolar magnetic interaction can be significant.

In simulating the positioning of the nanoparticle and the orientation of their magnetic moments, we neglect the internal dynamics of the magnetic moments and consider only their external dynamics caused by the nanoparticle rotation. We considered the initial size of the cubic simulation box (i.e. the size calculated with the algorithm for random 3D layout in grid) in a simple-centred cubic lattice, body-centred cubic lattice, and face-centred cubic lattice.

For positioning and orientation by minimising the potential energy of the system, we are going to use a Monte Carlo method applying the Metropolis Hastings algorithm – the so called *"off-lattice"* model [23].

The method used in our paper is based on changing the nanoparticle position and orientation of its magnetic moment by minimising the assembly of potentials exercised between the concerned nanoparticle and the other nanoparticles of the colloid. The *"assembly of potentials"* means: the van der Waals potential, the potential of double layer electrostatic interaction and/or the steric stabilisation potential and magnetostatic potential:

$$V_i = \sum_{j\neq i} V_{vdw,ij} + \sum_{j\neq i} V_{el,DL,ij} + \sum_{j\neq i} V_{steric,ij} + \sum_{j\neq i} V_{d,ij} \tag{2}$$

The dipolar magnetic potential is calculated by direct summation. To equilibrate the system by minimising the potential energy, we use 5000 MCS moves.

In Figure 2, we show the simulation result for 500 spherical magnetite nanoparticles whose volume fraction is 0.08, dispersed in water and positioned in a self-organised system, the intensity of external magnetic field being 15kA/m.

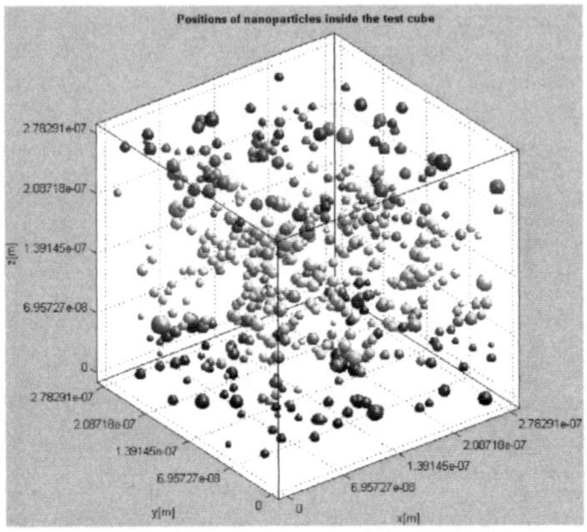

Figure 2. Nanoparticles positioned in a self-organised system

One can easily see a clear tendency of nanoparticles towards self-organisation. The nanoparticles can form chains and other forms of clusters, which can cause a local increase in the nanoparticle volume fraction, influencing therefore the magnetic behaviour of the system.

3. Numerical calculation of the specific loss power

The specific loss power (SLP) is defined as the electromagnetic power lost per nanofluid mass unit, and it is expressed in watt per kilogram. For SLP calculation, within the range of external magnetic field amplitudes suitable for medical applications [16], we can use the linear response theory (LRT) [24]. In this theory, the volumetric power dissipation is expressed as:

$$P = \pi\mu_0\chi_0 H_{0,ext}^2 \nu \frac{2\pi\nu\tau}{1+(2\pi\nu\tau)^2} \tag{3}$$

where ν and $H_{0,ext}$ are the frequency and magnitude of the applied field, μ_0 is the magnetic permeability of free space, χ_0 is the equilibrium susceptibility and τ is the magnetic relaxation time. The equation (3), in its original form, applies to nanoparticles monodispersed in sizes, which do not interact magnetically dipolarly, and for which χ_0 is constant [24].

We are going to use a generalised variant of the equation (3) for spherical nanoparticles, polydispersed in sizes, which interact magnetically dipolarly, showing distribution of the magnetic anisotropy constants [24]. Moreover, compared to [25], we will take into account the influence of directions of easy magnetisation axes on the equilibrium susceptibility [26-29]. The power dissipation on nanoparticle is volumetrically calculated, after which we find the average.
For dispersion in solid matrix:

$$p = \pi\mu_0 H_{0,ext}^2 \nu \frac{1}{N}\sum_{i=1}^{N} \frac{2\pi\nu\tau_N^i \chi_{0i}}{1+\left(2\pi\nu\tau_N^i\right)^2} \tag{4}$$

where τ^i_N stands for *"Néel relaxation time"*. The Néel relaxation process is the solid state mechanism occurring inside the nanoparticle, which corresponds to the magnetic moment rotation between two equilibrium positions, when considering the nanoparticle as fixed in space.

For dispersion in liquid matrix:

$$p = \pi\mu_0 H_{0,ext}^2 \nu \frac{1}{N}\sum_{i=1}^{N} \frac{2\pi\nu\tau_{eff}^i \chi_{0i}}{1+\left(2\pi\nu\tau_{eff}^i\right)^2} \tag{5}$$

Using the effective relaxation time for the particle *i* (τ^i_{eff}) and the equilibrium susceptibility for the particle *i* (χ_{0i}), the effective relaxation time can be described as follows [30]:

$$\frac{1}{\tau_{eff}^i} = \frac{1}{\tau_N^i} + \frac{1}{\tau_B^i} \tag{6}$$

where τ^i_B is the Brownian relaxation time for the particle *i*. The Brownian process represents the nanoparticle rotation in the fluid environment. For spherical particles, the Brownian relaxation time is usually described by:

$$\tau_B^i = \frac{3 v_{iH} \eta}{k_B T} \tag{7}$$

where k_B is the Boltzmann constant, T is the temperature, v_{iH} is the hydrodynamic volume of the particle i, and η is the coefficient of dynamic viscosity.

The relaxation mechanism which dominates the magnetic behaviour of the colloidal suspensions is determined by the nanoparticle properties. This two relaxation processes are treated separately.

For χ_{0i}, we use the expression [26-29]:

$$\chi_{0i} = 3\varepsilon_i \chi_L^i \tag{8}$$

$$\varepsilon_i = \frac{\left(\frac{\sigma_i}{3.4}\right)^{1.47} \cos^2\psi_i + \frac{1}{3}}{\left(\frac{\sigma_i}{3.4}\right)^{1.47} + 1} \tag{9}$$

$$\text{where } \sigma_i = \frac{K_{eff,i} v_i}{k_B T} \tag{10}$$

and Ψ_i is the angle between the easy magnetisation axis and the direction of the applied field (the field direction is the same with the measurement direction). $K_{eff,i}$ is effective magnetic anisotropy constant of of the i-th nanoparticle and v_i is magnetic core volume of the i-th nanoparticle.

$$\chi_L^i = \frac{f_m M_S}{H_{0,ext}} \left(\coth(\xi_i) - \frac{1}{\xi_i} \right) \tag{11}$$

$$\text{where } \xi_i = \frac{\mu_0 M_s v_i H_{0,ext}}{k_B T} \tag{12}$$

and f_m is magnetic volume fraction of the nanoparticles. f_m is correlated with volume fraction of nanoparticles by the relation [30]:

$$f_m = \frac{f}{\left(1 + \frac{2\delta}{d_m - 2\delta}\right)} \tag{13}$$

In relation (13) d_m is the average diameter of the nanoparticles.

Let M_s be the spontaneous magnetisation, k_B the Boltzmann constant and T the temperature. For solid matrix:

$$SLP = \frac{p}{f\rho} = 3\pi\mu_0 H_{0,ext} \nu \frac{M_s}{\rho N} \sum_{i=1}^{N} \varepsilon_i \left(\coth(\xi_i) - \frac{1}{\xi_i} \right) \frac{2\pi\nu\tau_N^i}{1 + \left(2\pi\nu\tau_N^i\right)^2} \tag{14}$$

For the nanofluid:

$$SLP = \frac{p}{f\rho} = 3\pi\mu_0 H_{0,ext} \nu \frac{M_s}{\rho N} \sum_{i=1}^{N} \varepsilon_i \left(\coth(\xi_i) - \frac{1}{\xi_i} \right) \frac{2\pi\nu\tau_{eff}^i}{1 + \left(2\pi\nu\tau_{eff}^i\right)^2} \tag{15}$$

where ρ is the density of the nanoparticle material.

The numerical calculation of Néel relaxation time is made taking into account the fact that the dipolar magnetic field on the nanoparticle and the external magnetic field create together a resultant magnetic field $\vec{H}_i$ that does not generally act along the direction of the nanoparticle easy magnetisation axis. If denoting with Ψ_i the angle between $\vec{H}_i$ and the easy anisotropy axis of the *i*-th nanoparticle, we obtain:

$$h_i = \frac{\mu_0 M_s H_i}{2K_i^{eff}} \tag{16}$$

where K_i^{eff} is the effective magnetic anisotropy constant of the *i*-th nanoparticle.

The free diffusion time of magnetisation $\tau^i{}_{0N}$ for low damping is [31] :

$$\tau^i{}_{0N} = \frac{v_i M_s}{2\gamma\alpha k_B T} \tag{17}$$

where α is the damping constant and γ is the gyromagnetic ratio. For most systems of ferro- and ferrimagnetic nanoparticles, the values of the magnetic damping constant α are close to the lower limit (α<<1) [32-34].

The nanoparticle relaxation in oblique magnetic fields was studied by Coffey and other colleagues [35]. This calculation shows the relaxation time dependence on the magnetic damping constant (α). In these conditions, the relaxation time relationship for an oblique magnetic field is [35]:

$$\tau^i_{N(oblique\_field)} = \frac{4\pi\tau^i{}_{0N}\left(S_{i1}^{-1} + S_{i2}^{-1}\right)}{\left(\sqrt{c_{i1}^{(1)} c_{i2}^{(1)}} \cdot e^{-\Delta E_{i12}} + \sqrt{c_{i1}^{(2)} c_{i2}^{(2)}} \cdot e^{-\Delta E_{i21}}\right)} \tag{18}$$

where ΔE_{i12} and ΔE_{i21} are the normalised energy barriers for magnetic moment re-orientations.

In equation (18):

$$c_{i1}^{(p)} = 2\sigma_i(\cos 2\theta_{ip} + h_i \cos(\theta_{ip} - \psi_i)),\ c_{i2}^{(p)} = 2\sigma_i(\cos^2\theta_{ip} + h_i \cos(\theta_{ip} - \Psi_i)),\ \text{with } p = 1,2 \quad (19)$$

where θ_{ip} are the solutions of the transcendental equation $\sin 2\theta_i = 2h_i \sin(\Psi_i - \theta_i)$.
If $h_i < h_{ic}(\Psi_i) < 1$:

$$\cos\theta_{i1,2} = \pm 1 \mp \frac{h_i^2}{2}\sin^2\psi_i + h_i^3 \sin^2\psi_i \cos\psi_i \mp \frac{h_i^4}{16}(13 + 11\cos 2\psi_i)\sin^2\psi_i + \\ + \frac{h_i^5}{2}(3 + \cos 2\psi_i)\sin^2\psi_i \cos\psi_i \mp \frac{h_i^6}{64}(183 + 156\cos 2\psi_i - 19\cos 4\psi_i)\sin^2\psi_i + \ldots \quad (20)$$

$$\Delta E_{i12} = \sigma_i[1 - 2h_i(\sin\Psi_i - \cos\Psi_i) + h_i^2 + \frac{h_i^3}{2}\sin 2\Psi_i(\cos\Psi_i - \sin\Psi_i) + \frac{h_i^4}{2}\sin^2 2\Psi_i + \\ + \frac{h_i^5}{32}\sin 2\Psi_i(7\cos\Psi_i - 3\cos 3\Psi_i - 7\sin\Psi_i - 3\sin 3\Psi_i) + \frac{h_i^6}{2}\sin^2 2\Psi_i + \ldots] \quad (21)$$

$$\Delta E_{i21} = \sigma_i[1 - 2h_i(\sin\Psi_i + \cos\Psi_i) + h_i^2 + \frac{h_i^3}{2}\sin 2\Psi_i(\cos\Psi_i + \sin\Psi_i) + \frac{h_i^4}{2}\sin^2 2\Psi_i + \\ + \frac{h_i^5}{32}\sin 2\Psi_i(7\cos\Psi_i - 3\cos 3\Psi_i + 7\sin\Psi_i + 3\sin 3\Psi_i) + \frac{h_i^6}{2}\sin^2 2\Psi_i + \ldots] \quad (22)$$

$$S_{i1,2} = \sigma_i\sqrt{h_i \sin\Psi_i}\,[16 - \frac{104}{3}h_i \sin\Psi_i + h^2(1 - 21\cos 2\Psi_i) + \frac{h_i^3}{2}\sin\Psi_i(45 + 51\cos 2\Psi_i) + \ldots] \pm \\ \pm 2\pi\sigma_i h_i^2 \sin 2\Psi_i(4 - 3h_i \sin\Psi_i - 2h_i^2 \sin^2\Psi_i + \ldots) \quad (23)$$

In addition to our model presented in [36], we considered the coated nanoparticles, size and uniaxial magnetic anisotropy constants distribution and we incorporates the dependence on the distributions of particle size, anisotropy values, and anisotropy easy axis directions in the equilibrium susceptibility calculs.

4. Results and discussions

We chose a system consisting of 500 spherical magnetite nanoparticles with spontaneous magnetisation $4.46 \cdot 10^5$ A/m, whose sizes and effective magnetic anisotropy constants had lognormal distribution, were dispersed in water (dynamic viscosity: $8.9 \cdot 10^{-4}$ Pa·s), and the relative electrical permittivity was 78.5.

For this study, we considered the case of a colloid with electrosteric stabilisation. The parameters of the system were: the average diameter of the nanoparticles d_m = 10 nm and d_m = 15 nm with the standard deviation parameter $0.25 \cdot d_m$, the effective magnetic anisotropy constant $K_{eff} = 2.5 \cdot 10^4$ J/m^3 with the standard deviation parameter $0.1 \cdot K_{eff}$, the coating thickness of 2 nm, the number of polymer molecules adsorbed on the nanoparticle surface $\zeta = 10^{-18}$ molecules/m^2, the surface charge $1.6 \cdot 10^{-16}$ C, the concentration of ions in solution 10^{26} ions/m^3, the Hamaker constant for magnetite in water $A = 39 \cdot 10^{-20}$ J and the temperature T = 298.15 K. We varied the nanoparticle volume fraction from 0.02 to 0.24.

In the first stage, we calculated SLP using the equation (14) for solid matrix and the equation (15) for liquid matrix (Figures 3 and 4).

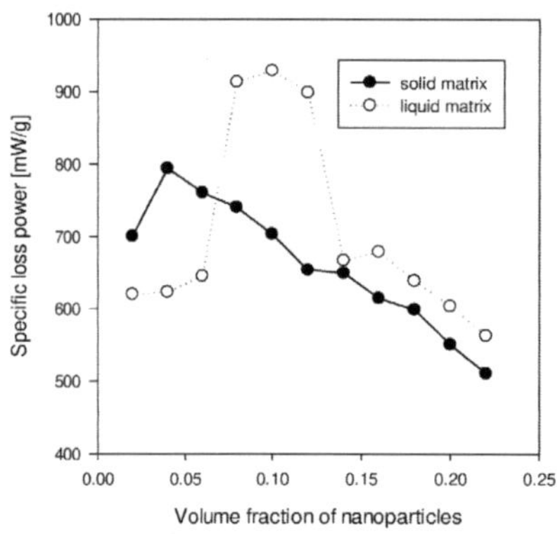

Figure 3. SLP versus volume fraction of nanoparticles (the average diameter of the nanoparticles 10 nm)

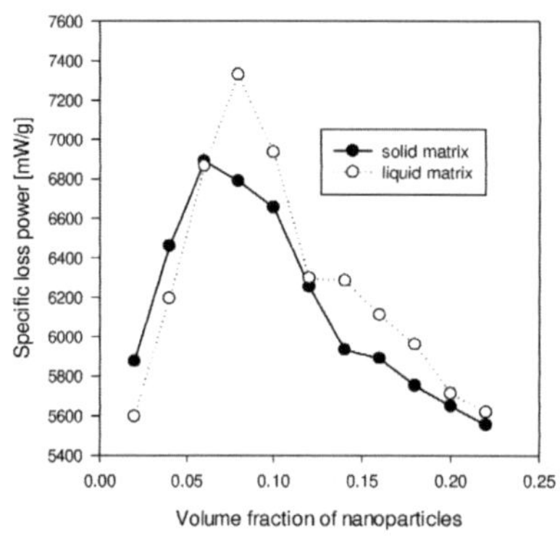

Figure 4. SLP versus volume fraction of nanoparticles (the average diameter of the nanoparticles 15 nm)

In the second stage of numerical experiment, we calculated the SLP value using the effective relaxation time, i.e. for the nanofluid, but with random positioning (as in case of solid matrix), and then with the particle positioning like in liquid matrix (Figures 5 and 6).

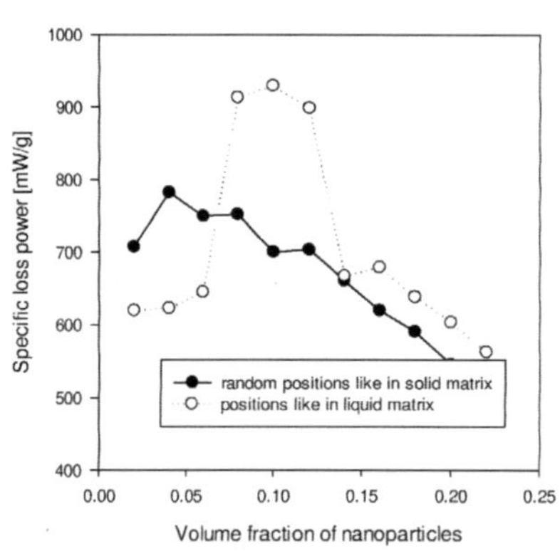

Figure 5. SLP versus volume fraction of nanoparticles for a nanofluid (the average diameter of the nanoparticles 10 nm)

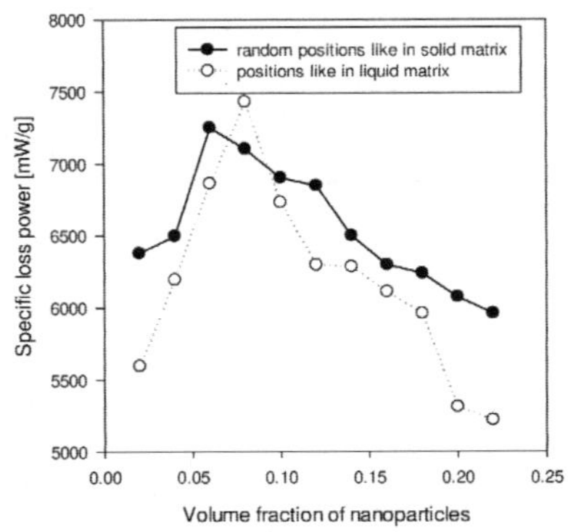

Figure 6. SLP versus volume fraction of nanoparticles for a nanofluid (the average diameter of the nanoparticles 15 nm)

Regardless of the average diameter, it can be seen that either in case of solid matrix or liquid matrix stands out a value of the concentration for which SLP is maximal. Similar results with regard to maximum heating at a specific volume fraction were obtained in other studies current experimental [37]. In case of a liquid matrix, the maximum is moving towards higher concentrations of nanoparticles and has higher values than for solid matrix. It can be seen that, at large and very low concentrations, there is no difference between the values of SLP, but the differences are significant at low and medium concentrations. It can be seen that, for nanoparticles with 15 nm average diameter, there is no so much difference between the values of SLP, but the differences are significant, for nanoparticles with 10 nm average diameter, at low and medium concentrations.

5. Conclusions

In this paper, we wanted to point out that, in magnetic hyperthermia modelling, the nanoparticle positioning geometry in a nanofluid, aspect neglected in most models, is especially important for low and medium concentrations of nanoparticles, case used in the therapeutic practice.

References

[1] Dutz S and Hergt R 2014 Magnetic particle hyperthermia: a promising tumour therapy? *Nanotechnology* **25** 452001-452029

[2] Vallejo-Fernandez G, Whear O, Roca A G, Hussain S, Timmis J, Patel V and O'Grady K 2013 Mechanisms of hyperthermia in magnetic nanoparticles, *J. Phys. D: Appl. Phys.* **46** 312001

[3] Salunkhe A B, Khot V M and Pawar S H 2014 Magnetic Hyperthermia with Magnetic Nanoparticles: A Status Review, *Current Topics in Medical Chemistry* **14**(5) 572-594

[4] Deatsch A E and Evans B A 2014 Heating efficiency in magnetic nanoparticle hyperthermia, *Journal of Magnetism and Magnetic Materials* **354** 163–172

[5] Ota S, Yamada T and Takemura Y 2015 Magnetisation Reversal and Specific Loss Power of Magnetic Nanoparticles in Cellular Environment Evaluated by AC Hysteresis Measurement, *Journal of Nanomaterials* 2015 1-8

[6] Reeves D B and Weaver J B 2014 Approaches for modelling magnetic nanoparticle dynamics, *Critical Reviews Biomedical Engineering* **42**(1) 85-93

[7] Caizer C 2014 Computational study on superparamagnetic hyperthermia with biocompatible SPIONs to destroy the cancer cells, *Journal of Physics: Conference Series* **521** 012015

[8] Conde-Leborán I, Serantes D and Baldomir D 2015 Orientation of the magnetisation easy axes of interacting nanoparticles: Influence on the hyperthermia properties, *Journal of Magnetism and Magnetic Materials* **380**(15) 321–324

[9] Munoz-Menendez C, Conde-Leboran I, Baldomir D, Chubykalo-Fesenko O and Serantes D 2015 The role of size polydispersity in magnetic fluid hyperthermia: average versus local infra/over-heating effects, *Phys. Chem. Phys.* **17** 27812-27820

[10] Conde-Leboran I, Baldomir D, Martinez-Boubeta C, Chubykalo-Fesenko O, Del Puerto Morales M, Salas G, Cabrera D, Camarero J, Teran F J and Serantes D 2015 *A* Single Picture Explains Diversity of Hyperthermia Response of Magnetic Nanoparticles, *J. Phys. Chem. C* **119**(27) 15698–15706

[11] Ruta S and Chantrell R 2015 Unified model of hyperthermia via hysteresis heating in systems of interacting magnetic nanoparticles, *Scientific Reports* **5** 1-7

[12] Sanz B, Calatayud M P, Cassinelli N, Ibarra M R and Goya G F 2015, Long-Term Stability and Reproducibility of Magnetic Colloids are Key Issues for Steady Values of Specific Power Absorption Over Time, *European Journal of Inorganic Chemistry* **27** 4524-4531

[13] Etheridge M L, Hurley K R, Zhang J, Jeon S, Ring H L, Hogan C, Haynes C L, Garwood M and Bischof J C 2014 Accounting for biological aggregation in heating and imaging of magnetic nanoparticles, *Technology* **2**(3) 1-15

[14] Heinrich D, Goñi A R and Thomsen C, 2007, Dynamics of magnetic-field-induced clustering in ionic ferrofluids from Raman scattering, *The Journal of Chemical Physics* **126** 124701

[15] Lebovka N I 2014 Aggregation of Charged Colloidal Particles, *Adv. Polym. Sci.* **255** 57-96

[16] Chuan Lim E W and Feng R 2012 Agglomeration of magnetic nanoparticles, *The Journal of Chemical Physics* **136** 124109

[17] Prakash S and Yeom J 2014 *Nanofluidics and Microfluidics: Systems and Applications*, Elsevier Oxford, UK

[18] Chang H, Chen X Q, Jwo C S and Chen S L 2009 Electrostatic and Steric Stabilisation of CuO Nanofluid Prepared by Vacuum Arc Spray Nanofluid Synthesis System (ASNSS), *Materials Transactions* **50**(8) 2098-2103

[19] Hunter J R 1981 *Zeta Potential in Colloid Science: Principles and Applications*, Academic Press Inc., San Diego, USA

[20] Scherer C and Figueiredo Neto A M 2005 Ferrofluids: Properties and Applications, *Brazilian Journal of Physics* **35**(3A) 1678-4448

[21] Napper D H 1983 *Polymeric Stabilisation of Colloidal Dispersions*, Academic Press, Inc., New York, USA

[22] Haase C and Nowak U 2012 Role of dipole-dipole interactions for hyperthermia heating of magnetic nanoparticles ensembles, *Physical Review B* **85**(4) 045435

[23] Landaun D P and Binder K 2003 *Simulations in Statistical Physics*, Cambridge University Press, Cambridge, UK

[24] Rosensweig R E 2002 Heating magnetic fluid with alternating magnetic field, *Journal of Magnetism and Magnetic Materials* **252** 370–374

[25] Osaci M and Cacciola M 2015 Studies about the influence of self-organisation of colloidal magnetic nanoparticles on the magnetic Néel relaxation time, *Beilstein Journal of Nanotechnology* **6** 2173-2182

[26] Gittleman J I, Abeles B and Bozowski S 1974 Superparamagnetism and relaxation effects in granular Ni-SiO_2 and Ni-Al_2O_3 films, *Phys. Rev. B* **9** 3891

[27] Carrey J, Mehdaoui B and Respaud M 2011 Simple models for dynamic hysteresis loop calculations of magnetic single-domain nanoparticles: Application to magnetic hyperthermia optimization, *J. Appl. Phys.* **109** 083921

[28] Bean C P and Livingston J D 1959 Superparamagnetism, *Journal of Applied Physics*, Supplement to **30**(4) S120–S129

[29] Sánchez F H, Mendoza Zélis P, Arciniegas M L, Pasquevich G A and Fernández van Raap M B, 2015, *Dipolar interaction and demagnetizing effects in magnetic nanoparticle dispersions: introducing the Mean Field Interacting Superparamagnet Model (MFISP Model)*, source arXiv:1507.05192 [cond-mat.mes-hall]

[30] Rosensweig R E 1985 *Ferrohydrodynamics*, Cambridge University Press, Cambridge, UK

[31] Coffey W T, Gregg P J and Kalmzkov Y P 1992 *On the theory of Debye and Néel relaxation of single domain ferromagnetic particles*, I Prigogine, S A Rice editors, Advances in Chemical Physics, John Wiley & Sons, Inc., Hoboken, NJ, USA

[32] Branquinho L C, Carrião M S, Costa A S, Zufelato N, Sousa M H, Miotto R, Ivkov R and Bakuzis A F 2013 Effect of magnetic dipolar interactions on nanoparticle heating efficiency: Implications for cancer hyperthermia, *Scientific Reports* **3** 3637

[33] Leliaert J, Coene A, Crevecoeur G and Dupré L 2014 Regarding the Neel relaxation time constant in magnetorelaxometry, *Journal of Applied Physics* **116**(16) 163914

[34] Mălăescu I 1998 *Ferrofluids in radio-frequency field,* Mirton Publishing House, Timişoara, Romania

[35] Coffey W T and Kalmykov Y P 2012 Thermal fluctuations of magnetic nanoparticles: fifty years after Brown, Applied Physics Reviews, Section of the *Journal of Applied Physics* **112** 121301

[36] Osaci M and Cacciola M 2015 An addapted Coffey model for studying susceptibility losses in interacting magnetic nanoparticles, *Beilstein Journal of Nanotechnology* **6** 2173-2182

[37] Coral D F, Zélis P M, Marciello M, Morales M P, Craievich A, Sánchez F H and Fernández van Raap M B 2016 Effect of Nanoclustering and Dipolar Interactions in Heat Generation for Magnetic Hyperthermia, *Langmuir* **32** (5) 1201–1213

IOP Conf. Series: Materials Science and Engineering **163** (2017) 012009 doi:10.1088/1757-899X/163/1/012009

Overview of the wind energy market and renewable energy policy in Romania

C P Chioncel[1], G O Tirian[2], N Gillich[1], C Hatiegan[1] and E Spunei[1]

[1]Eftimie Murgu University of Resita, Electrical Engineering and Informatics Department, P-ta Tr. Vuia, no. 1-4, 320085 Resita, Romania
[2]Politehnica University of Timisoara, Department of Electrical Engineering and Industrial Informatics, 5 Revolution Street, Hunedoara, 331128, Romania

E-mail: c.chioncel@uem.ro

Abstract. The modern, developed society becomes aware of the necessity to conserve and protect the environment, increasing the gained benefits from a rational use of the natural resources. The pollution and the limitation of the fossil fuels, associated with the political situation worldwide that affects direct the energy strategies, have opened opportunities in the area of operation renewable energy sources. The development of the exploitation of renewable energy sources is directly linked to the energy politic, which, in terms of Romania, has the focus to integrate into the European Union energy strategy. The year 2014 brought in Romania many legislative changes to the renewable support scheme, that proves, once again, the legislative unpredictability and limitations introduced by the legislator "during the game" that overthrew all economic profitability calculation of the existent and planned investments in this sector. The actual stage of the wind energy across Europe and the particular situation in Romania are highlighted; also a 2020 forecast for Romania tries to evaluate the perspective for the wind, and general, renewable energy market. The actual Romanian renewable energy support scheme, mainly regulated by "Law 220/2008" ends December 2016. The so-called "ready to build" projects especially wind- or hydropower, can't be finalized until this deadline, being unable to qualify to the existing, mainly to inoperable, support scheme. Another legislation that has to clarify how investments in renewable energy will be supported is still not in place, blocking any project development, implementation and economical benefit of the producer. The paper presents in this respect an updated overview of the Romanian renewable energy sector and its perspective.

1. Introduction

The total installed wind power decreased at the global level in 2013, compared with previous years, being for the first time in more than 20 years that this sector has experienced a downtrend. Despite this, the total worldwide installed wind power capacity raised to 318.105 MW at the end of 2012, establishing itself as one of the main sources of electricity generation [1].

In this context, we can indicate countries and cities that have proposed to achieve, or already are, a 100% renewable energy in their energy production: Iceland, Cook Islands, Scotland, Denmark, a.o [2].

With more than 1.000 MW installed wind capacity, Romania is among the twenty-four countries worldwide beyond this limit. Closest to the milestone of 100.000 MW installed wind capacity is China [3], which will attain it at the end of 2014 [1], [2].

An overview of the growth of the global cumulative installed wind capacity is shown by the chart reflecting this evolution for the period 1996 – 2015, Figure 1 [2]. The global capacity installed in wind farms at the end of 2013, placed China in the first position with 91.412 MW, followed by USA, with 61.091 MW and in third position, the first European country, Germany, with 34.250 MW.

In other classification, those of installed wind power capacity during 2013, places Romania in „top 10" worldwide, at the 10's position, with 695 MW. According to the Romanian wind energy association (RWEA), Romania has now a total number of 75 wind farms with 1200 installed wind turbines, totaling 3.008 MW, state 1 April 2016, of the nearly 5.000 MW renewable energies on the Romanian market [4].

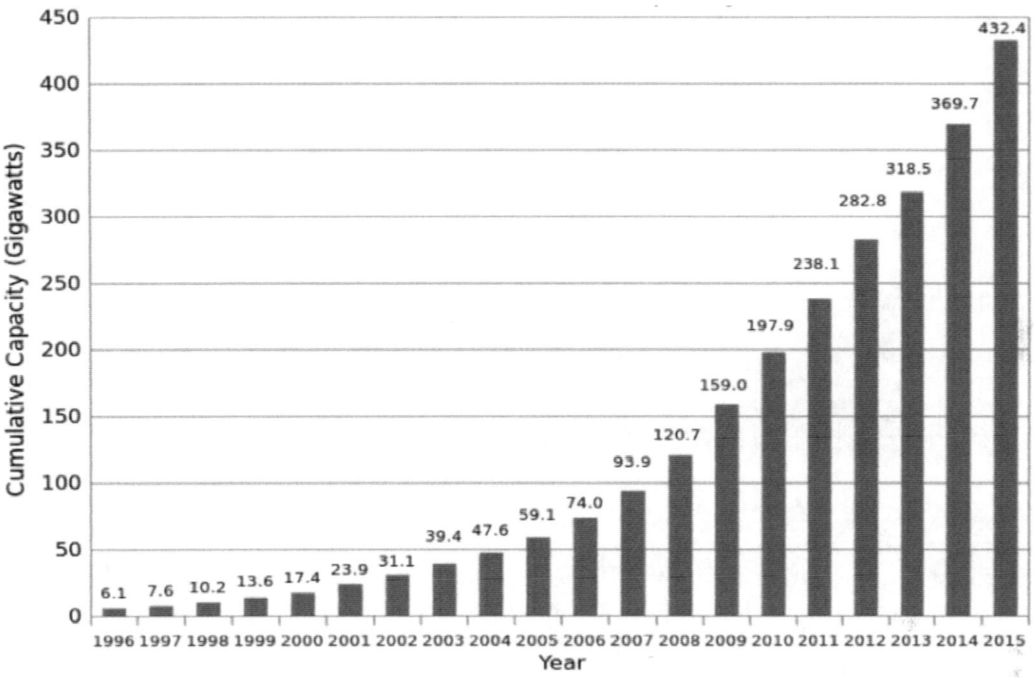

Figure 1. Global installed wind power in the period 1996 – 2015 [12]

2. Wind energy in context of economic downturn

The regulatory and political uncertainty across Europe since 2011 began to influence direct existing wind power installations as well as planned and „ready to build" wind power plants starting with 2013. So, the boom of build and installed wind turbines in Romania, 2013, have been authorized and financed based on the prior legislation changes that affect today the wind energy sector, especially in Eastern Europe.

The pressure on the renewable energy sector, especial on the wind farms with their higher installed capacity, will act across Europe and especially in Romania, to a falling and almost stagnant installation of new capacities. Many projects, as well as operational capacities in the renewable energy sector, are put up for sale by smaller companies in the hope of escape from bankruptcy. Other companies take into account the losses made with the project development (grid connection, wind measurements and monitoring [5], [6] building permits, environmental studies, zoning classification and land designation, securing land through land purchase and / or rental, legal advice, payment of various fees, a.o.), amounts that can easily reach the order of several million euro / project, knowing that, in the today legal framework, building and operating a wind farm in Romania would have meant even higher losses, which could have caused instability in the mother companies, specifically in small and medium sized companies. The companies that have built on the Romanian market, in specially wind power plants and big photovoltaic plants, are big players, as ENEL Green Power, CEZ, EON

owning also parts of the electric energy distribution networks (ENEL Distributie, EON or CEZ), which are able to overcome easier the market blockade of green certificates trading and maintaining a low level of the renewable energy quota.
Based on the GWEC estimation [2], a positive trend in the wind energy market is expected, especially in the regions of Asia, Europe, and North America, Figure 2 [7].

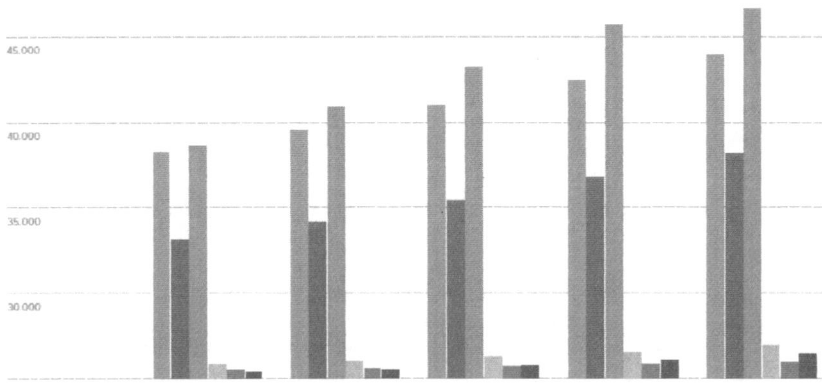

	2014	2015	2016	2017	2018
Europe	132,5	145,5	159,5	174,5	189,5
North America	80,8	91,8	104,3	117,8	131,8
Asia	136,4	158,9	182,4	206,9	232,9
Latin America	7,8	9,8	12,0	15,0	19,0
Pacific	4,7	5,7	6,7	7,9	8,9
Middle East and Africa	3,3	4,8	7,3	10,3	14,3

Figure 2. Cumulative forecast by region, 2014 - 2018

3. Wind energy in Europe

Considering that the value of 100 GW of total installed wind power was reached in 2012, in 2013 a shift could be finding in the interest of the investor in wind power in Europe. At the end of 2013, begin of 2014, Germany has strengthened their position in Europe achieving the highest installed capacity in wind farms across Europe, total 34.3 GW.

Across Europe, in 2015, a total amount of 12.800 MW has been installed in wind power, accounting 15.6% from the total installed power, overtaken hydro as the third largest source of power generation in the EU [8], [9].

The importance of the wind energy compared with other energy sources is highlighted in the next figure, Figure 3 [8], [9].

The installed wind power in Europe at the end of 2015, is highlighted in Figure 4.

At European level, the annual installed wind power since 2000, is revealed in Figure 5 [8], [9]. As shown, the best year with new installed wind power capacities was the last year, 2015, with 12.800 MW.

A significant growth of the offshore wind power installation can be noticed: towards 2014, in 2015 those capacities have more than doubled, reaching now 24% of the annual EU wind energy market, from 13% in 2014, Figure 6 [8].

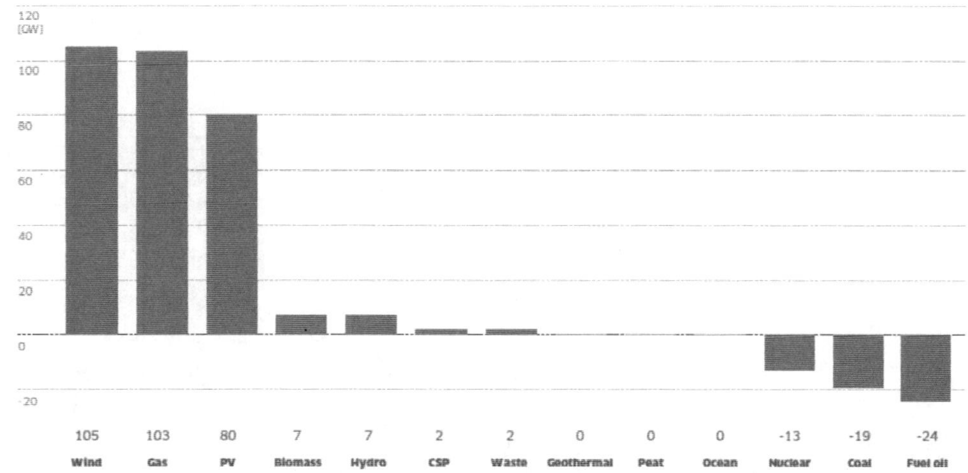

Figure 3. Europe: importance gain of wind energy with other energy sources

Figure 4. Europe: installed wind power by end of 2015

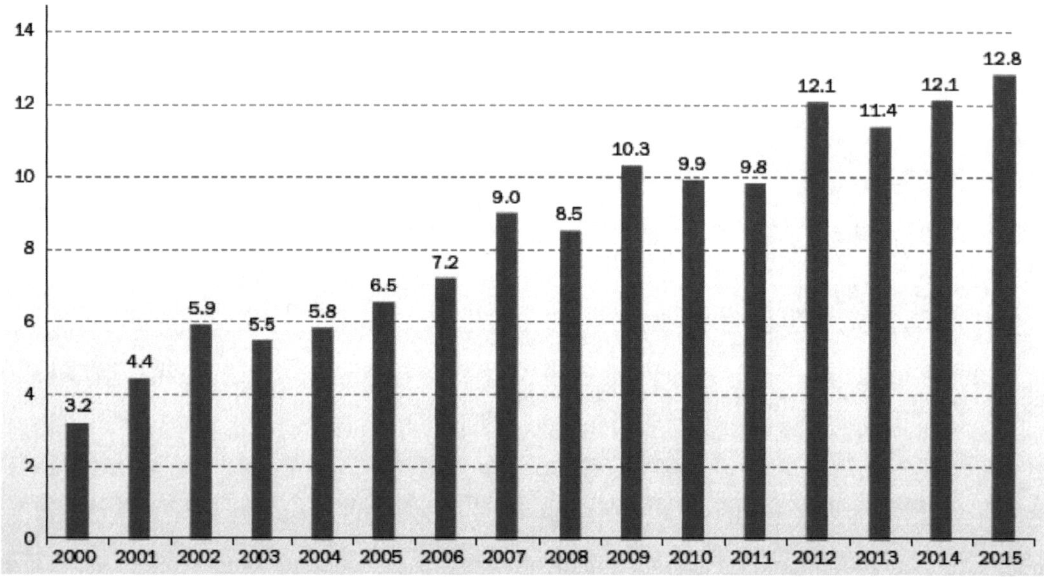

Figure 5. Annual new wind power capacities in EU (GW)

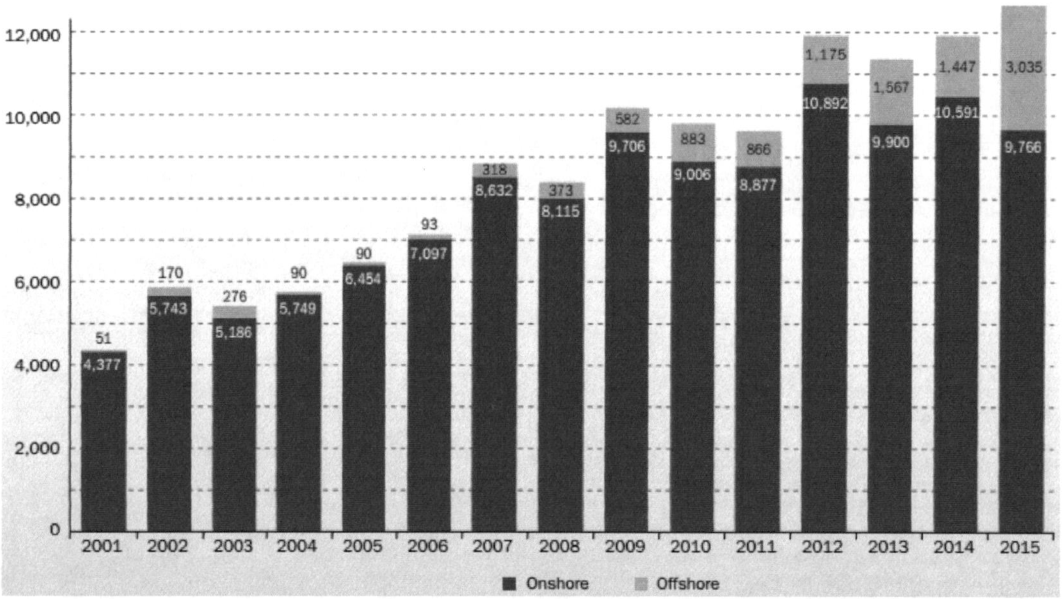

Figure 6. Overview of annual offshore and onshore installations (MW) [16]

4. Wind energy in Romania

Even with the major legislative unpredictability, generated by changes of the law 22 from 2008 through repeated government decisions, culminating with the violation of the green energy quota in the last years – being much lower to the prescribed value in law 220 / 2008, as well as the more restrictive secondary legislation, 3.008 MW installed wind power were registered in Romania, compared to the 2.599 MW at end of 2013. The increase with almost 13.6% - 354 MW compared to 2013, is due to the projects that made the financing closing, ordered the wind turbines and start construction work before the occurred legislation changes. Besides the classic developers risk in the wind power plant development [6], [10], the Romanian market raises with legislative instability.

This rapid growth of the Romanian wind energy market has exceeded the expectations of the Energy Regulatory Authority (ANRE) [11] and those predicted in the National Renewable Energy Action Plan (NREAP), as highlighted in Figure 7 [7], [4].

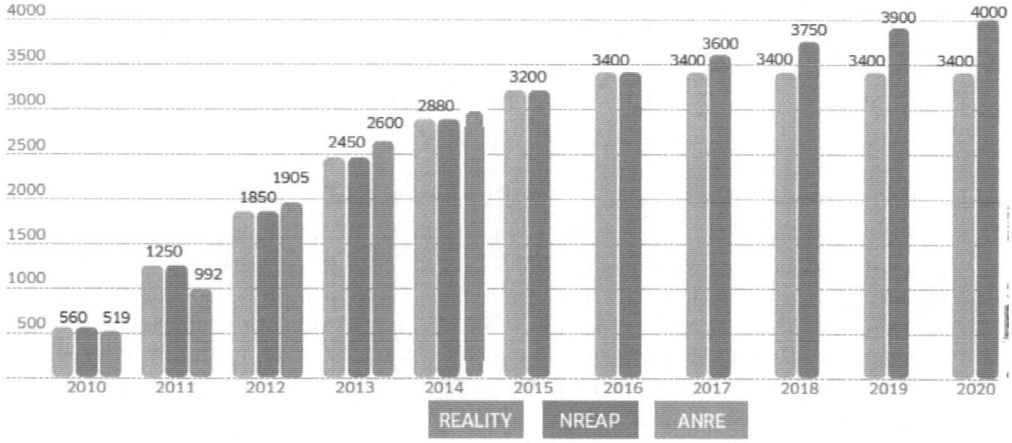

Figure 7. Estimates and achievements in the Romania wind power market

According to Transelectrica, the Romania wind power plant's injected 2014 in the grid 4.7235 TWh, compared to 4.70TWh in 2013, considering that the total energy consumption fell by 6%. The electricity produced by the wind farms in 2015 covered 11.03% of gross Romania electricity consumption [12].

Based on the last statistic dates, Romania has a number of 75 wind farms, with installed capacities between 0.008 MW and 600 MW, with a total number of about 1.200 wind turbines. A Romanian wind park has in average 40 MW, which is much above the European average. This value could be achieved due to a low population density in Dobrogea, the main windy region of Romania, with over 85% of the installed wind power capacity.

The Romanian wind market is dominated by „big players", which own also, in the different region, the distribution network, as ENEL or CEZ. Other multinational companies are EDPR, Verbund or Monsson Group, that registers around 80% of the total installed wind capacity. The figure of investments made in the Romania wind energy totals EUR 5 billion.

The biggest problem that investors that already installed their wind parks or those that are now remaining in expectation have, is the unstable and thereby unpredictable legislation, with numerous inefficient – as the normal functioning of the market for green certificates, which brought mistrust of the investors, abolishes the initial financial calculations and makes it impossible to finance other investments in this area.

Despite the increase in efficiency of the wind power plants, through control techniques [13-18] or the possibility of increasing their gauge – nacelle height, diameter of the rotor blades and the installed power of a generator, a good and courageous environmental [19] policy and economic policy [1], [7], [20] is needed at European Union level, followed by their takeover through the member states.

5. Conclusion

In this condition, shortly, Romania has a negative premier: the first investor that announced the disassembly of his wind park, with a capacity of 27 MW in the Dobrogea region.

In the absence of new legislation [8], the existing support scheme is applicable just to the projects commissioned by 31 December 2016, a serious and constructive approach from the decision makers to this industry will make Romania impossible to achieve 2020 the quota of 20% of the total gross energy consumption for renewable energy, foreseen by the European Union, and on the other hand would undermine investor confidence.

References

[1] Böhringer C and Löschel A 2006 Promoting renewable energy in Europe: A hybrid computable general equilibrium approach, *The Energy Journal*, *Special issue* **27**(2) 135-150

[2] www.gwec.net (Global Wind Energy Council)

[3] Li J F, Shi J L and Wang Z Y 2007 New progress of renewable energy policy in Europe and its reference for China, *Renewable Energy Resources* 3: 000

[4] www.rwea.ro (Romania Wind Energy Association)

[5] Chioncel C P, Gillich N, Chioncel P and Gillich G R 2007 *CMS solutions in monitoring and real time data transfer of photovoltaic plants*, XVth International Symposium on Electrical Apparatus and Technologies SIELA 2007, Plovdiv, Bulgaria, May 31-June 1, pp 14-17

[6] Chioncel C P, Chioncel P and Gillich N 2001 Overview of Classic and Modern Wind Measurement Techniques, Basis of Wind Project Development, *Analele Universității "Eftimie Murgu" Reșița, Fasciula de Inginerie* **XVIII**(3) 73-80

[7] ***TPA Horwath and Schönherr 2014 *Wind Energy and other renewable energy sources in Romania*, 4th Edition

[8] www.ewea.org (European Wind Energy Association)

[9] www.windeurope.org

[10] Prostean G, Vasar C, Prostean O and Vartosu A 2015 Risk analysis for renewable energy projects due to constraints arising, *IOP Conf. Ser.: Mater. Sci. Eng.* **106** 012017

[11] www.anre.ro (Romanian Regulatory Authority for Energy)

[12] www.transelectrica.ro (Romanian Transmission and System Operator)

[13] Chioncel C, Chioncel P and Gillich N 2008 *Scalar control structure of an asyncronous motor at maximum torque*, 19th International Symposium of the Danube-Adria-Association-for-Automation-and-Manufacturing, Trnava, Slovakia, Ocober 22-25, pp 233-234

[14] Chioncel C P, Babescu M, Chioncel P, Gillich N and Gillich G R 2007 *Speed control method for asynchronous motor*, 18th International Symposium of the Danube-Adria-Association-for-Automation-and-Manufacturing, Zadar, Croatia, October 24-27, pp 137-138

[15] Chioncel C P, Babescu M and Chioncel P 2011 *Optimization of the Wind Systems, Based on the Maximum Wind Energy*, International Conference on Optimization of the Robots and Manipulators (OPTIROB 2011), Sinaia, Romania, May 26-28, pp 978-981

[16] Chioncel C P, Erdodi G M, Petrescu D I, Spunei E and Gillich N 2015 *Control of wind power systems imposing the current in the intermediate circuit of the converter at variable wind speed*, 9th International Symposium on Advanced Topics in Electrical Engineering (ATEE), University Politehnica Bucharest, Bucharest, Romania, May 07-09, pp 923-928

[17] Chioncel C P, Tirian G O, Gillich N and Raduca E 2016 Vector control structure of an asynchronous motor at maximum torque, *IOP Conf. Ser.: Mater. Sci. Eng.* **106** 012005

[18] Chioncel C P, Bereteu L, Petrescu D I and Babescu M 2014 *Determination of reference mechanical angular speed for wind power systems*, 6th International Workshop on Soft Computing Application (SOFA 2014), Timisoara, Romania, July 24-26, pp 1213-1222

[19] Lehmann P, et al. 2012 Carbon Lock-Out: Advancing Renewable Energy Policy in Europe, *Energies* **5**(2) 323-354

[20] Klessmann C, Held A, Rathmann M and Ragwitz M 2011 Status and perspectives of renewable energy policy and deployment in the European Union—What is needed to reach the 2020 targets?, *Energy policy* **39**(12) 7637-7657

Water hammer effect in the spiral case and penstock of Francis turbines

D Pepa[1], C Ursoniu[1], G R Gillich[2] and C V Campian[2]

[1]UCM Resita, Department of Mechanical Engineering, Golului str., no.1, 320053, Resita, Romania
[2]"Eftimie Murgu" University of Resita, Department of Engineering and Management, Piata Traian Vuia str., no. 1-4, 320085, Resita, Romania

E-mail: pepadamaschin58@yahoo.com

Abstract. Sudden pressure increases in the penstock or spiral case of a hydraulic turbine are the effect of sudden flow variation that occur during transient processes of type opening / closing or load rejection of the hydro unit. The consequence of the pressure rise in the spiral case and penstock is the water hammer phenomenon, whose effects can be devastating in some cases, up to breaking pipes and calamities produced in the area. This paper aims to analyze the method of calculation of the maximum pressure values that might occur in load rejection situations to a hydraulic turbine, in spiral case and in penstock, conditioned by the limiting of the values of the over speed and measures of limiting the increase in pressure in conjunction with limiting the increase in speed in these specific processes. As an example, we studied and analyzed the situation of a hydroelectric power plant equipped with a 7.8 MW Francis turbine without pressure regulator and the inflow surge. The results of analytical calculation overlaid on the experimental measurements performed during the performance tests of the hydro unit lead to the conclusion that the calculation algorithm proposed has been chosen correctly and the 2-stage closing law of the wicket gate promoted in this case is effective in such situations.

1. Introduction

The impermanent regime of fluid dynamics is a frequent event, met during operation of hydraulic installations which are carrying fluids under pressure [1], [2], [3], [4]. It appears whenever the dynamic regime is varying very fast, meaning whenever changes occur in the flow limiting conditions, such as opening or closing a valve, stopping or turning on pumps, rejecting air, water pipe breaking, so on [5],[6].

Such an impermanent regime can introduce important mechanical stresses in the system. The motion of the fluid in the pipes may cause an over-pressure, capable to emulate several or tens of times the operating pressure. Also, an important depression can be expected. In both cases partial damages or even the total destruction of such an installation can occur [7], [8], [9].

The motion of the fluids in the pipes in impermanent regime can be of a slow or fast variation, depending on the relative speed of changing the limit conditions of the system [10], [11].

The more the variations are slower in time, the more the wave character is reduced. The movement becomes increasingly more oscillating in mass and compressibility and plays a less important role. To the limit, for very slow movements varying in time, the compressibility can be neglected and it can be reached the simplified model of the incompressible fluid which can however be used only in this case, of the very slowly varying in time movements.

The quick variation impermanent movement phenomena will be referred to as "water hammer effect" and the slow variation will be called "oscillation in mass movements". Parts, fittings or installations that are required in order to control the water hammer phenomena will be called "protection devices" [12].

The effect of increasing the pressure in a pipe or chamber due to rapid flow changes is called in technical terms "water hammer". Water hammer represents a very destructive force that can appear in any pumping system where the rate of flow is varying abruptly in operation from different reasons. Most of the engineers recognize the existence of water hammer effect, but few realize its destructive impact. In fact, too much time and many expenses have been spent repairing pipelines and damaged pumps and valves because of water hammer. Thus, it is essential for an engineer to know exactly when to expect water hammer occurrences, how to estimate the maximum pressure rise, and if possible, how to provide a solution in order to reduce the maximum pressure rise up to a safe limit.

The water hammer phenomena have a wave characteristics. Any change of limiting conditions in any point of the pipe, is causing local variations in flow and pressure, changes which are transmitted from point to point, with a finite speed called propagation speed, due to the elasticity of the fluid and of the material of the pipe, forming plane waves of flow and pressure.

In water hammer case, the propagation speed practically merges with celerity, namely the propagation speed in the fluid at rest [16]. Of course, there are many methods of calculating the maximum values of pressure attainable in transitory processes [13], [14], [15], as well as we can find many solutions for limiting the effects of water hammer. In all these cases, the solution is to be determined, from case to case, according to the technical characteristics of the hydraulic system in which the phenomenon may occur.

Since we cannot establish a clear demarcation between the fast and slow water motion (pressure variation) and we cannot decide for sure whether the simplified model can be applied, it is recommended that in all cases to apply only the precise model of the compressible fluid.

This paper proposes to take the fast varying phenomena only, meaning the water hammer effects that appear in spiral case of the Francis turbines or in the penstocks from the hydro arrangements without surge tank.

2. Analytical approach to the phenomenon

This paper shows the case of impermanent phenomena that occur during transitory processes of the stopping with load rejection of a hydro power plant equipped with one Francis turbine powered by a single penstock with a diameter of 1.2 m. The water intake of the penstock is placed into a polder made on the derivation from a river with almost constant flow. The arrangement is placed on the Tana River in Kenya.

2.1. Initial data

Technical characteristics of the hydro unit (head, flow, power) are mentioned in Table 1. Supplementary necessary information to operate the calculation [17]:

- Hydro unit runaway speed: $n_{rs} = 610.9$ rpm
- Generator inertia momentum: $GD^2_{generator} = 150$ tm^2
- Turbine inertia momentum: $GD^2_{turbine} = 7.3$ tm^2
- The function $f = f(\sigma)$
- The experienced function $T_a = f(Tw)$
- Maximum over speed: 150% at nominal speed
- Maximum overpressure: 130% at nominal head

Table 1. Technical data for the Francis turbine

	Q (m^3/s)	H_{netto} (m)	P (Kw)	n(rpm)
1a	16.5	52.45	7913	333.33
1b	13	54.05	6315	333.33
1c	9	55.39	4019	333.33
1d	7	55.89	2865	333.33

2.2. Calculation Hypothesis

In the present calculation there were assumed the following hypothesis [17]:

- Correctly operation of the speed governor which leads to perform the closing time which was taken into consideration;
- Linear closing lure of the wicket gate;
- Hydraulic hammer;
- $g = 9.777$ m/s^2 (value for Equator area)

2.3. Calculation algorithm

The first step of calculation is determining LV_{tot} according to the equation (1):

$$L_{vtot}=\Sigma L_{vad}+\Sigma L_{vcs}+\Sigma L_{vst} \tag{1}$$

It is taken into account the length and the speed that are specific to the 2 segments of the hydraulic circuit, namely: penstock and spiral case.
The calculation algorithm determines the following variables:
a) ρ – Allievi constant according to the equation:

$$\rho=\frac{a\cdot V_m}{2gH_n} \tag{2}$$

where:

Vm = the medium water speed (m/s)
a = speed of water hammer pressure wave (1425 m/s according to the nomogram for the situation of metallic pipes and in the case in which the ratio D/wall width = 10…15)
g = acceleration of gravity (m/s^2)
Hn = nominal head (m)

b) $T_{w\,(sec)}$ - inertia time constant of the water masses in the penstock according to:

$$T_W=\frac{\Sigma L_{Vtot}}{gH_n} \tag{3}$$

c) σ = coefficient of overpressure in the penstock (no dimensional) according to

$$\sigma=\frac{T_w}{T_i} \tag{4}$$

while T_i is the closing time of wicket gate (s).
The closing time is chosen according the imposed limit for speed rise in many variants according to the results shown in Table 1.2
d) T_r - reflection time of the pressure wave is calculated according to the equation:

$$T_r=\frac{2\cdot L_{ad}}{a} \tag{5}$$

e) In the first part of the wicket gate closing, the pressure rise coefficient ζ_1, is calculated with the equation:

$$\zeta_1 = \frac{2 \cdot \sigma}{1 + \rho\tau_0 - \sigma} \tag{6}$$

where $t_0 = 1$ for load rejection from the rated power.

f) After a number of phases had been passed, the limit of the pressure rise coefficient is almost reaching the limit value ζ_m and further remain constant. The constant ζm is derived as:

$$\zeta_m = \frac{\sigma}{2}\left(\sqrt{\sigma^2 + 4} + \sigma\right) \tag{7}$$

In case of a sudden closing of the wicket gate after a load rejection from the rated power, the pressure rise coefficient on the installation is divided between the construction elements as follows:

g) For a pressure rise in the penstock the rise coefficient is calculated according to the equation:

$$\zeta_{ad} = \frac{\Sigma L_{Vad}}{\Sigma L_{Vtot}} \cdot \zeta_m \tag{8}$$

h) For a pressure rise in the spiral case the rise coefficient is calculated with the equation:

$$\zeta_{cs} = \frac{\Sigma L_{Vad} + \Sigma L_{Vcs}}{\Sigma L_{Vtot}} \zeta_m \tag{9}$$

i) The effective pressure rise in the penstock ΔH_{ad} is calculated with the equation:

$$\Delta H_{ad} = \zeta_{ad} \cdot H_n \tag{10}$$

j) The effective pressure rise in the spiral case is calculated from the equation:

$$\Delta H_{cs} = \zeta_{cs} \cdot H_n \tag{11}$$

k) The pressures in the penstock and in the spiral case are:

$$H_n + \Delta H_{ad} = H_n(1 + \zeta_{ad}) \tag{12}$$

$$H_n + \Delta H_{cs} = H_n(1 + \zeta_{cs}) \tag{13}$$

The permissible limit values of over pressure in the penstock and spiral case are closely linked to the limit values allowed for hydro unit over pressure from technical and economic reasons. In such a situation, treatment of water hammer phenomenon cannot be done without taking into consideration the packaging of the machine in the worst case of load rejection, usually from the maximum value of power that can be given by the power plant.

The over speed is calculated by equation:

$$\mathbf{\Delta n} = -1 + \sqrt{\frac{364 \cdot P \cdot T_{ia} \cdot f}{GD^2 \cdot n^2} + 1} \tag{14}$$

where $T_{ia\ (sec)}$ - closing time of the wicket gate up to the no load position and it will be calculated with the equation:

$$T_{ia} = 0.9 \cdot T_i \tag{15}$$

- P - turbine output (Kw)
- f - corrective coefficient of speed rise
- GD^2 - fly wheel moment of the generator and turbine (tm^2)
- n – hydro unit speed (rpm)

2.4. Establishing the proper solution for avoiding overpressure and over speed

According to Table 2, results:

- It can be inferred it is an indirectly water hammer, namely the head maximum rise appear at the end of wicket gate closing curve ($\zeta 1 < \zeta m$ and $Tr < Ti$)
- For a closing time of 7 seconds, the head maximum rise in the penstock is of 12.5 m and in the spiral case is of 15.26 m
- For the same closing time $T_i = 7$ s the hydro unit speed maximum rise is of approx. 53% (see Figure 1- law no. 1)
- The highest value of the head and speed rise are reached when the unit is loaded at rated power (P = 7913 Kw).

Because the over speed value goes beyond the admissible value (50%) it is adopted the solution of closing the wicket gate in two steps (see Figure 1 – law no. 2), as follows:

- First step of closing has influence upon the over speed value, it is active up to 30% opening in a time T1i of 5 seconds;
- The second step of closing goes from 30% up to the end of the stroke in a time of T2i = 12 seconds.

Taking into consideration the input data and the calculation hypothesis mentioned above, further in the following tables are centralized the values of the calculated parameters with the equations already presented in Table 3.

In such a situation there are guaranteed the following:

- The maximum value of the over speed will not go over 50 %;
- The maximum value of overpressure will not go over 30 %;

Table 2. Value calculated for closing law no. 1

Parameter	MU				
P	(Kw)	7913	6315	4019	2865
Hn	(m)	52.45	54.05	55.39	55.83
Q	(m^3/s)	16.5	13	9	7.00
L_{ad}	(m)	238.46	238.46	238.46	238.46
L_{Vad}	(m^2/s)	891.39	594.26	486.21	378.15
L_{cs}	(m)	10.565	10.565	10.565	10.565
L_{Vcs}	(m^2/s)	65.79	43.58	35.2	21.9
L_{ST}	(m)	54.52	54.52	54.52	54.52
L_{VST}	(m^2/s)	29.56	20.31	15.21	12.95
L_{Total}	(m)	303.545	303.545	303.545	303.545
L_{Vtotal}	(m^2/s)	986.75	680.37	567.22	412.1
V_m	(m/s)	2.99	1.98	1.63	1.302
A	(m/s)	850	850	850	850
P	-	2.47	1.57	1,31	1.01
Tr	(s)	0.71	0.71	0.71	0.71
T_W	(s)	1.92	1.27	1.04	0.75
T_i	(s)	7	7	7	7
Σ	-	0.27	0.18	0.148	0.107

ζ_1	-	0.16	0.123	0.136	0.112
ζ_m	-	0.30	0.159	0.159	0.126
$H_{ad/Hn}$	(m)	65.24	62.38	62.9	61.6
$H_{cs/Hn}$	(m)	67.71	63.23	63.65	62.66
GD^2	(tm^2)	157.3	157.3	157.3	157.3
Tia	(s)	6.3	6.3	6.3	6.3
F	-	1.33	1.28	1.25	1.18
Δn/n	(%)	53	36	28.2	20

Table 3 Values calculated for closing law no. 2

Parameter	MU				
P	(Kw)	7913	6315	4019	2865
Hn	(m)	52.45	54.05	55.39	55.83
Q	(m^3/s)	16.5	13	9	7.00
L_{ad}	(m)	238.46	238.46	238.46	238.46
L_{Vad}	(m^2/s)	891.39	594.26	486.21	378.15
L_{cs}	(m)	10.565	10.565	10.565	10.565
L_{Vcs}	(m^2/s)	65.79	43.58	35.2	21.9
L_{ST}	(m)	54.52	54.52	54.52	54.52
L_{VST}	(m^2/s)	29.56	20.31	15.21	12.95
L_{Total}	(m)	303.545	303.545	303.545	303.545
L_{Vtotal}	(m^2/s)	986.75	680.37	567.22	412.1
V_m	(m/s)	2.99	1.98	1.63	1.302
A	(m/s)	850	850	850	850
P	-	2.47	1.57	1,31	1.01
Tr	(s)	0.71	0.71	0.71	0.71
T_W	(s)	1.92	1.27	1.04	0.75
T_{itotal}	(s)	17	17	17	17
T_{1i}	(s)	5	5	5	5
T_{2i}	(s)	12	12	12	12
Σ	-	0.11	0.07	0.06	0.04
ζ_1	-	0.16	0.123	0.136	0.112
ζ_m	-	0.11	0.07	0.06	0.04
$H_{ad/Hn}$	(m)	10.6	6.7	5.8	3.9
$H_{cs/Hn}$	(m)	11	7	6	4
GD^2	(tm^2)	157.3	157.3	157.3	157.3
-	-	-	-	-	-
ia	s)	.5	.5	.5	.5
-	-	-	-	-	-
		.33	.28	.25	.18
-	-	-	-	-	-
n/n	%)	0.9	7.3	1.2	4.7

The closing diagram of the wicket gate is given in Figure 1.

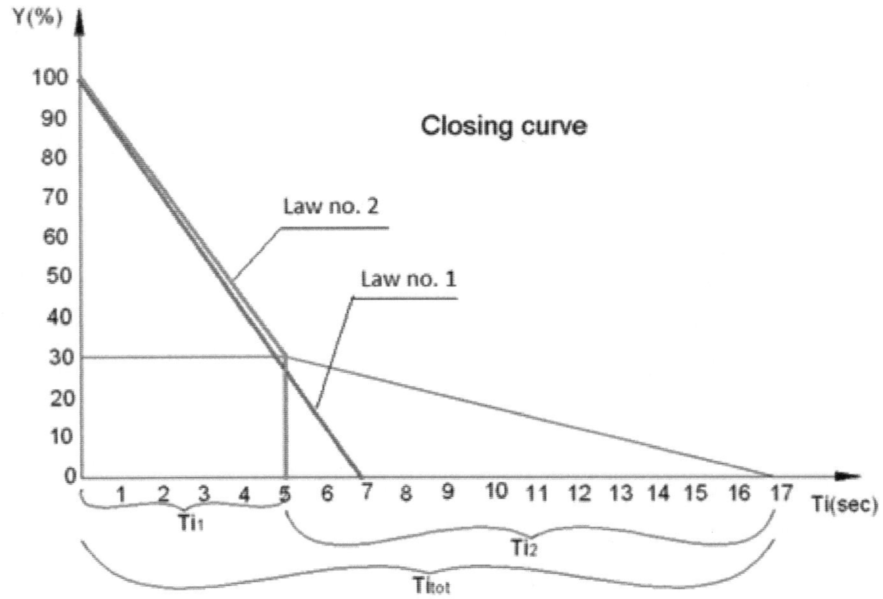

Figure 1. Closing laws

2.5. The technical solution applied to avoid water hammer effect

In order to avoid the phenomenon of water hammer, in case in which the hydro power arrangement has no surge and no pressure regulator fitted to the inlet of the spiral case, given that the automatic speed governor is of the type digital and electro-hydraulic, it was used an electro-hydraulic system installed on the closing pipe and controlled by the same PLC that implemented in digital equipment.

The system consists of several hydraulic elements from Bosch, adequate for necessary flow used for closing the wicket gate in the calculated time.

The reason for which it was proposed this solution of 2-stage closing is as follows:

- The first part of the closing to be as fast as possible in order to avoid overpressure, having in attention that over speed has the biggest gradient immediately after the disconnecting of the main circuit breaker (the 5 s time corresponds to the rise limitation of the speed up to 150% from the rated speed).
- The second part of the closing should be as smooth as possible, having in attention that over speed is not increasing anymore and we have in intention to avoid water hammer (the total time of 17 s is fitted for an overpressure of 130% from the maximum static head).

The inflection point in the closing law of the wicket gate is an adjustable variable by application software in this case, depending on the current position of the wicket gate, but it can be realized through a mechanical cam, adjustable as position, tightly connected of the servomotor piston of the wicket gate. Of course, we can use as input data too, the instantaneous value of the speed, but in such a case should be adopted a different software implemented algorithm or a different hydraulic actuation system corresponding to the proper law that intended to be implemented.

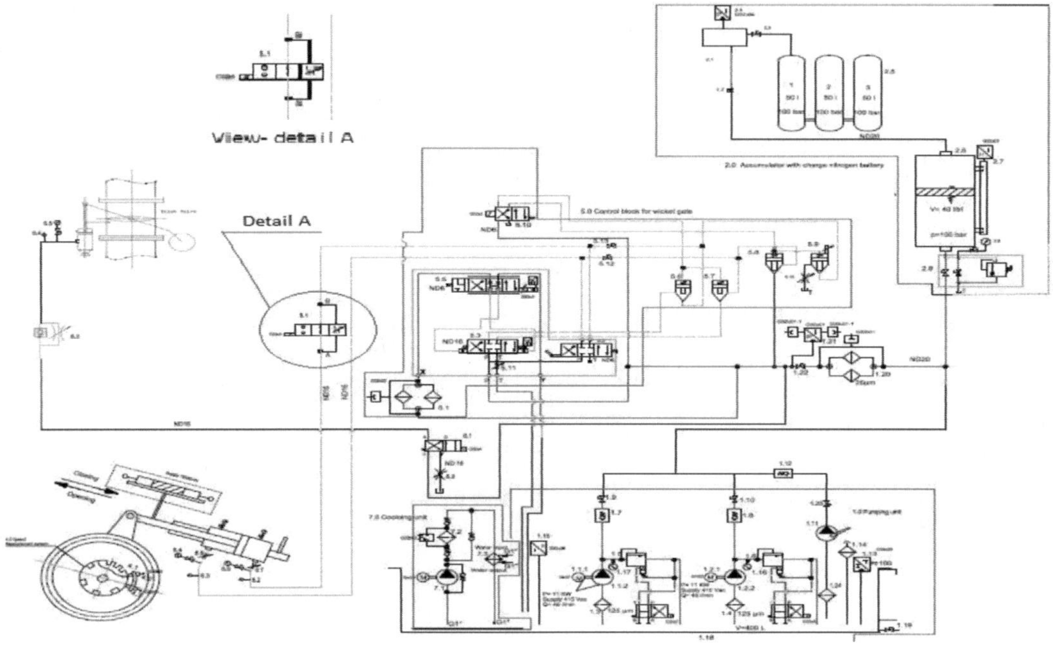

Figure 2. Hydraulic scheme of governing system

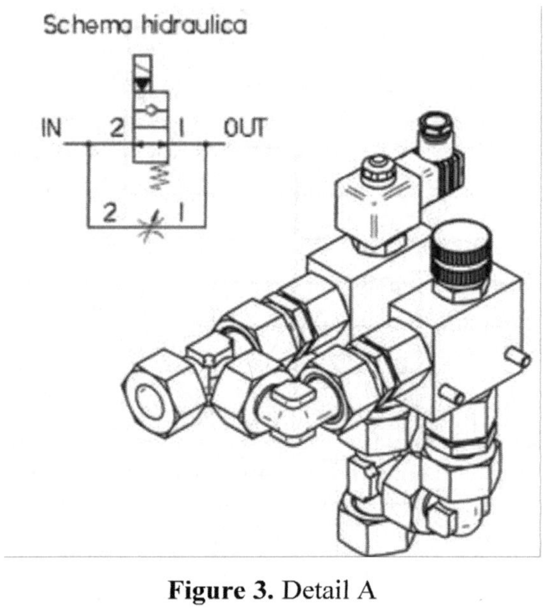

Figure 3. Detail A

3. Experimental measurements (records)

Checking the analytical calculation results was made in site during commissioning tests.

The present diaphragm on the hydraulic control circuit of the wicket gate has a 2 mm hole. The closing time for the wicket gate (Figure 3) was recorded as almost 17 s (the same value as it was calculated in the design). The changing point of the two slopes is around 35% of total opening. So, for the first slope (from 100% to 35%) we have around 5 s and for the second slope (from 35% to 0%) we have around 12 s (according to designed value).

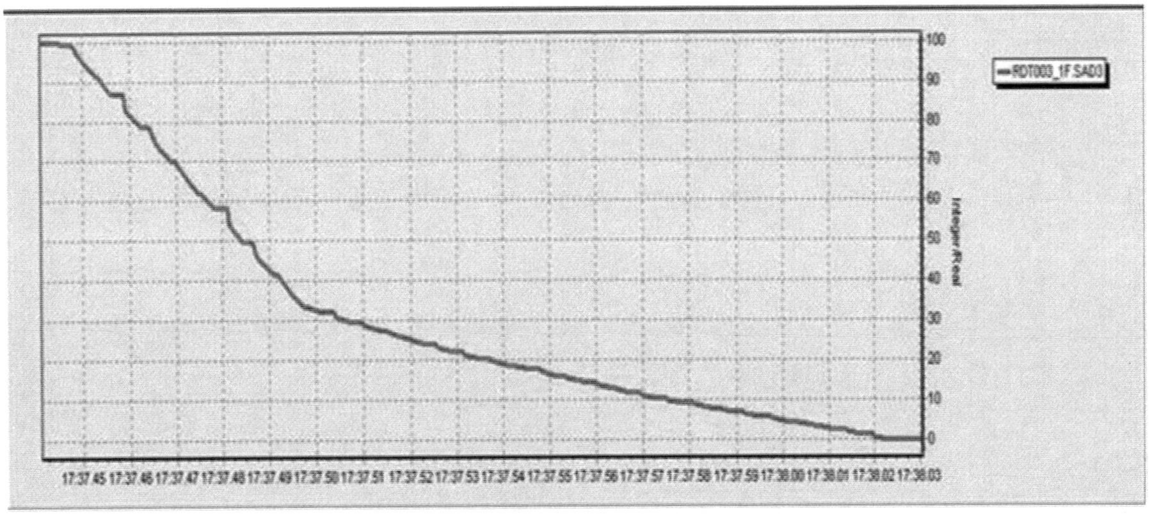

Figure 4. Record of closing time of the wicket gate

We recorded exactly the total stroke of the wicket gate servomotor. It is 260 mm for 100% of opening. The wicket gate opening temporized time (increasing/decreasing from keys) from 0% to 100% is around 50 sec. (0.01%/5.6 ms scan time). It was done the test of closing the wicket gate through de-energizing the safety valve by losing power supply to the solenoid valves hydraulic block.

There were performed the load rejection tests for 75% and maximum power obtained. Following we will present some conclusions resulted from the records. The over speed and overpressure values which were recorded are according to the closing law. This closing law is performing the most optimal situation regarding the rise of the speed and of the pressure in the spiral case. The maximum power can be reached in the range 75-78% opening of the wicket gate (see records). The maximum power obtained was 7.3 MW, with the line voltage of approx. 10.5 kV, the water pressure in the spiral case of 5.3 bar.

3.1. Recording for STOPPING LOAD REJECTION at 75% of nominal power

The values resulted from the records for over-speed and overpressure are:

- Over-speed: 126 % from nominal speed
- Pressure rise in the spiral case: 118.8 from static head
- Wicket gate closing law: Inflection point is set at 32% and the total closing time of the wicket gate is 8 s, suitable to the opening on which the load rejection occurred.)

IOP Conf. Series: Materials Science and Engineering **163** (2017) 012010 doi:10.1088/1757-899X/163/1/012010

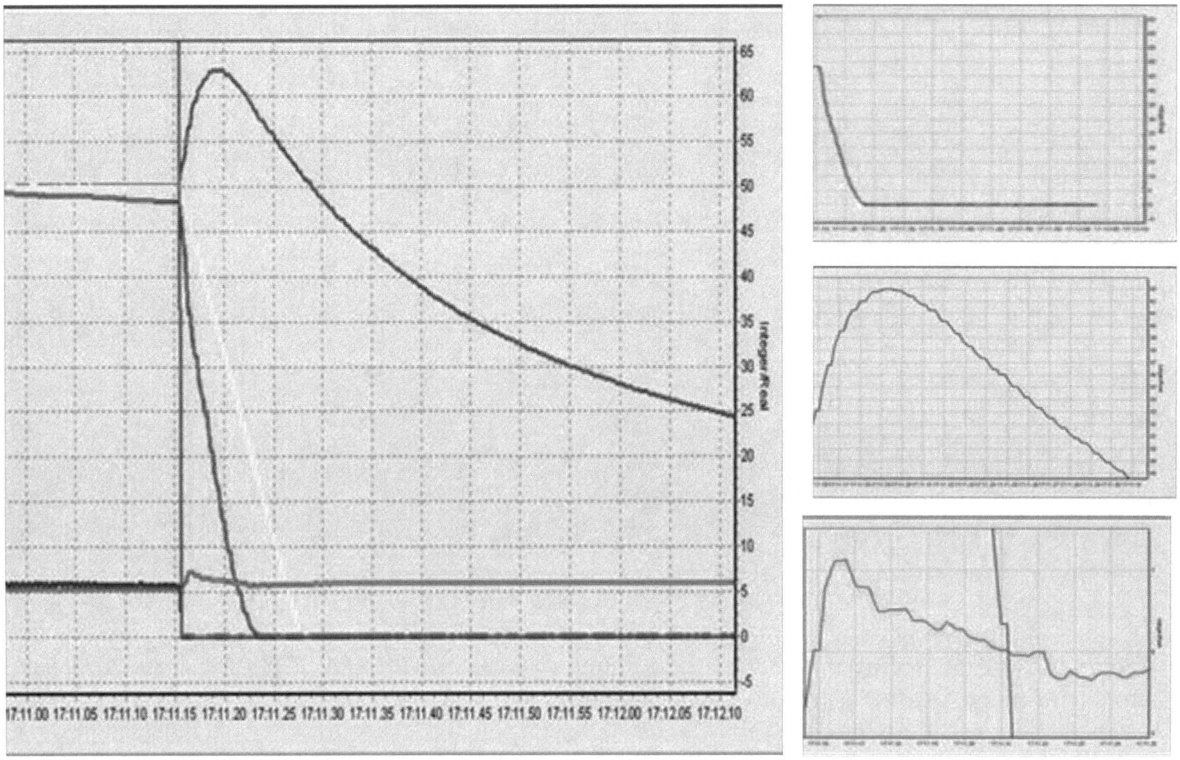

Figure 5. Stopping Load rejection from 75% of nominal power (left image); Over-speed: 126 % (right image middle); Pressure rise in the spiral case: from 5.3 bar to approx. 6.3 bar (right image down); Wicket gate law (right image up)

3.2. Recording for STOPPING LOAD REJECTION AT MAXIMUM POWER

The values resulted from the records for over-speed and overpressure are:

- Over-speed: 142.6% from nominal speed (This value is obtained in case of remaining in no load regime, and it is not under guarantee)
- Pressure rise in the spiral case: 129.2 % from static head
- Wicket gate law: Inflection point is set at 32% and the total closing time of the wicket gate is 9.5 seconds, suitable to the opening on which the load rejection occurred.)

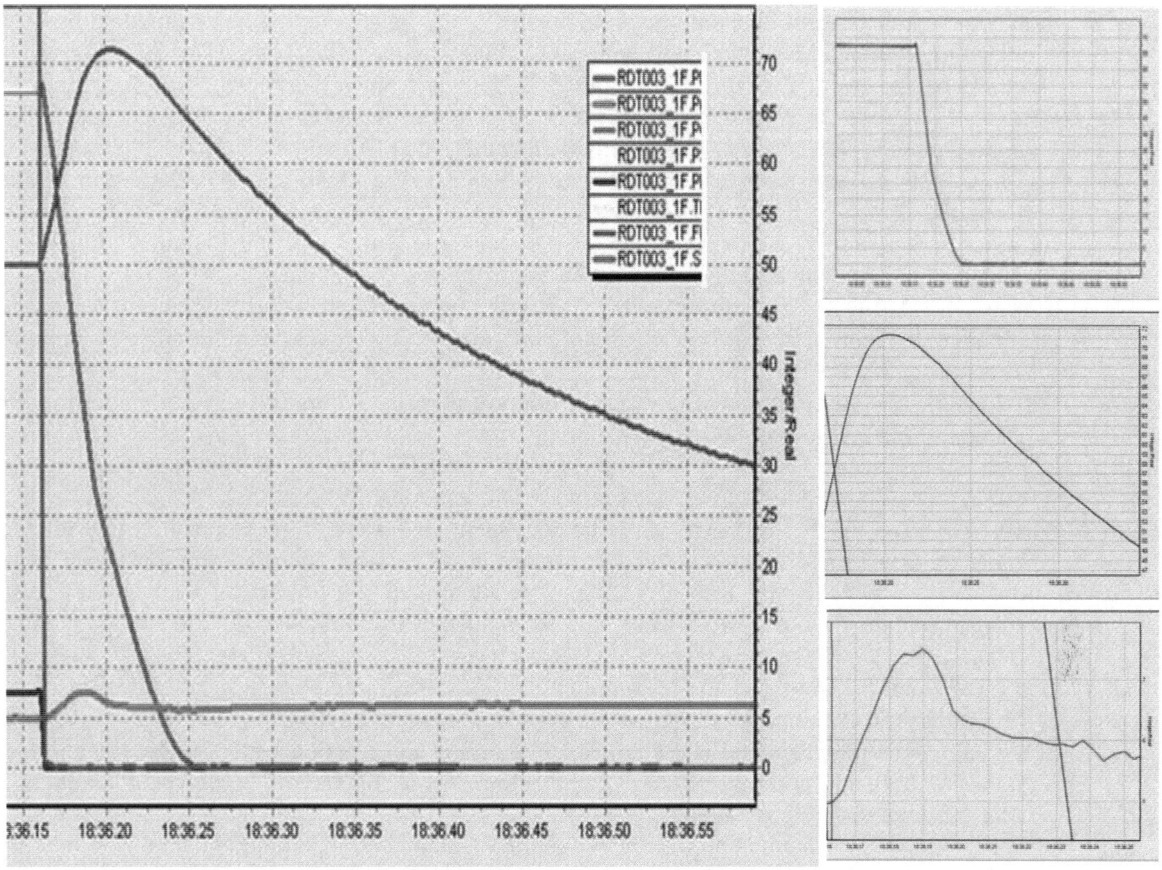

Figure 6. Stopping Load rejection from maximum power (left image); Over-speed: 140.9 % (right image middle); Pressure rise in the spiral case: from 4.95 bar to approx. 6.4 bar (right image down); Wicket gate law(right image up)

4. Conclusion

Following the completion of the hydraulic 2-stage closing system it was harmonized the allowable values for pressure rise or speed rise to load rejection from nominal power, transient phenomenon that occurs frequently in hydropower units operation.

The calculus algorithm is our own and uses equations taken by the studied bibliography.

As seen from the analysis of recordings that made, the values of interest, namely over speed and over pressure so as the closing time of the wicket gate did not exceed the values which calculated.

In those circumstances the penstock and the whole arrangement can operate safely even after a simplified calculation, if calculation assumptions are chosen appropriately.

Graphs realized to the commissioning tests of the hydro unit shows the 2-stage closing law of the wicket gate and the operation times realized by adjusting the holes of the diaphragms from the return pipe are very close to analytical calculation results.

Acknowledgments
The work has been funded by the Sectoral Operational Programme Human Resources Development 2007-2013 of the Ministry of European Funds through the Financial Agreement POSDRU/159/1.5/S/132395.

References

[1] ***2009 *Calculul și combaterea loviturii de berbec la conductele pentru transportul apei,*

Universitatea Tehnică de Constructii Bucureşti
[2] Arsenie D I, Popescu M, Omer I and Niţescu C 2000 Lovitura de berbec în cazul conductei compuse, *Analele Universităţii "OVIDIUS" Constanţa, Seria Construcţii* 47-54
[3] Bergant A, Simpson A R and Tijsseling A S 2006 Water hammer with column separation: A historical review, *Journal of fluid and structures* **22** 135-171
[4] Martin C S 2014 *Water hammer*, School of Civil Engineering - Institute of Technology, Atlanta, Georgia, USA
[5] Hemmat H S 1972 On the elastic behavior of the pipe wall for water hammer applications, *Nuclear engineering and design* **21**(1) 85-94
[6] Choon T W, Aik L K, Aik L E and Hin T T 2012 Investigation of water hammer effect through pipeline system, *International Journal on Advance Science Engineering Information Technology* **2**(3) 246-251
[7] Paffel K 2011 Water hammer: the number one problem in a stem system US Department of energy, *Plant engineering* (October)
[8] Kiyama A, Tagawa Y, Ando K and Kameda M 2016 Effects of water hammer and cavitation on jet formation in a test tube, *Journal of Fluid Mechanics* **787** 224-236
[9] Subani N and Amin N 2015 Analysis of water hammer with different closing valve laws on transient flow of hydrogen-Natural gas mixture, *Abstract and Applied Analysis* **2015** 510675
[10] Koohzadi E, Borghei S and Kabiri A 2016 Water hammer in a horizontal rectangular conduit containing air water two phase slug flow, *Journal of Hydraulic Engineering* **142**(3) 04015059
[11] Wang L, Wang F J, Zou Z C and Li N X 2015 Effects of air vessel on water hammer in a high-head pumping station, *IOP Conf. Ser.: Mater. Sci. Eng.* **52** 072010
[12] Sukamta S 2015 Experimental Investigation of slugging as initiating water hammer phenomena through indirect contacts steam condensing in a horizontal pipe heat exchanger, *International Journal of Technology* **6**(6)
[13] Morozov A A 1958 *Turbinoe oborudovanie ghidroelectrostantii*, Leningrad, Russia
[14] Morozov A A 1952 *Utilisation of water energy*, Moskva, Russia
[15] Krivcenko M 1960 *Ghidromehaniceskie perehodie protesi v ghidroenergheti ceskih ustanovkah,* Moskva, Russia
[16] Florea M 2000 Despre aproximaţii în calculul loviturii de berbec, *Analele Universităţii "OVIDIUS" Constanta, Seria Construcţii* 87-94
[17] ***2015 *Calculul garanţiilor de reglaj la CHE Gâlceag,* UCM Resita

Experimental research on the behavior of the pneumatic transport of fine-grained iron

V Andrei, M Hritac, N Constantin and C Dobrescu

University Politehnica of Bucharest, Department of Engineering and Management of Metallic Materials Elaboration, 313 Splaiul Independentei, Bucharest, 060042, Romania

E-mail: victortandrei11@yahoo.com

Abstract. Mixed injection of fine–grained iron ore and pulverized coal in the furnace, involves determining the behavior of these materials during pneumatic transport in a dense state through the pipe and setting possibilities for adjusting the flow rate of material transported with the corresponding values of the process. Parameters of the pneumatic transport were determined for the main types of iron ore and chalk used in Arcelor Mittal Galati. Outside the intended purpose of injecting iron ore and flux, it was considered also the experimental check of the possibility for injecting ilmenite in the furnace for crucible protection purpose. The possibility of injecting cinder mill into the furnace was also considered. Injecting cinder could be taken into account for the recycling of ferrous waste in the furnace, also as additive for intensifying the combustion process around the tuyeres.

1. Introduction

All experimental research had the following deployment: preparing the research materials, mounting the installation, checking the installation for starting operations, achieving actual experiments, data processing.

During the experiments the following materials were considered:

- ore from India
- concentrate from Brazil
- ore from Australia (Robe River)
- 80% ore from India and 20% dolomite
- ilmenite concentrate
- rolling mill cinder

The ore from India and Robe River, the concentrate from Brazil and the dolomite are the materials that are being presently used in Arcelor Mittal Galati.

The ilmenite concentrate was obtained from Preparation Facility in Merisani, Arges County, as part of Rare Metals Autonomous Regia Bucharest. The rolling mill cinder came from the Pipe Factory in Roman.

The chemical composition of the used materials is presented in Figure 1.

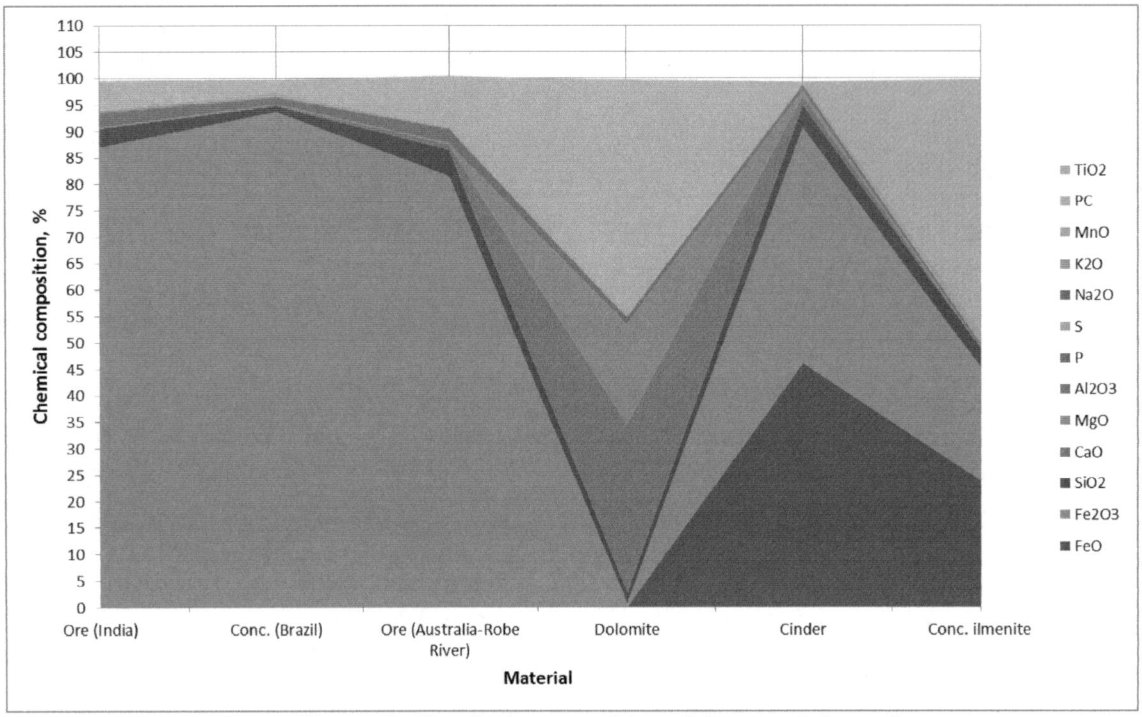

Figure 1. Chemical composition of the materials used

In terms of physical characteristics, the corresponding requirements for experimental materials required previous preparation [1], [2], [3]. The preparation process covered the following operations: grinding in order to bring the grains to a 1 mm dimension; grain classification for removing the grains larger than 1 mm and drying to eliminate the humidity.

It is worth mentioning that from a chemical point of view all materials are suitable for use in the furnace [4].

The grain distribution and the density of the materials are presented in Figure 2 and Table 1.

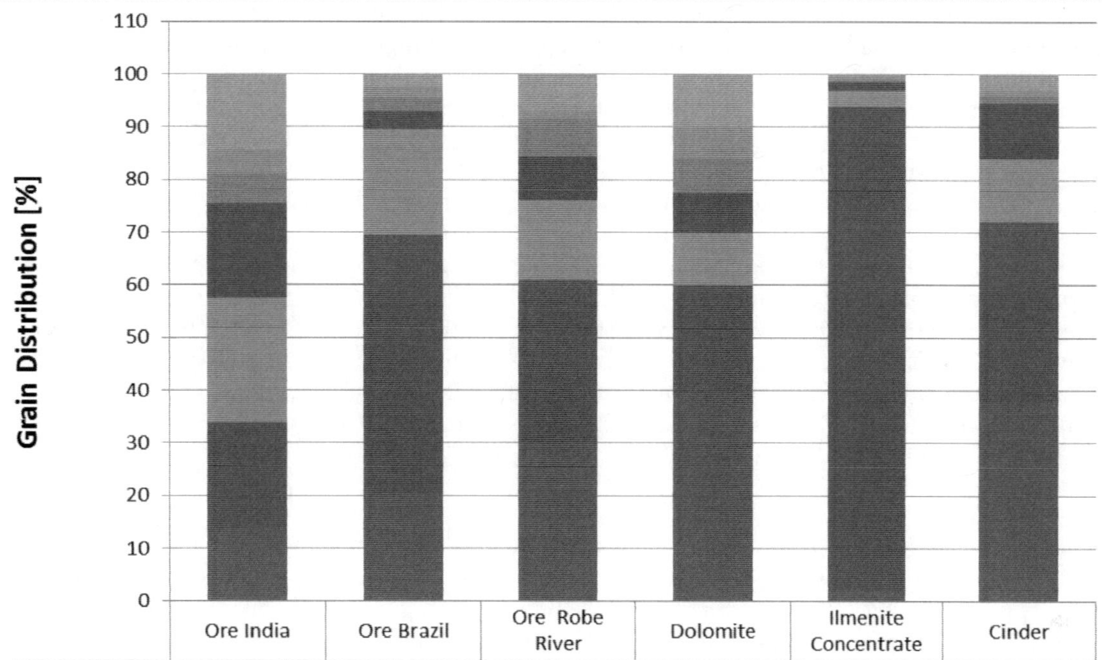

	Ore India	Ore Brazil	Ore Robe River	Dolomite	Ilmenite Concentrate	Cinder
< 0.063	14.5	2.5	7	10	0.5	3
0.071 - 0.063	4.5	2	1.5	6	0.5	1
0.09 - 0.071	5.5	2.5	7	6.5	0.5	1.5
0.1 - 0.09	18	3.5	8.5	7.5	1.5	10.5
0.2 - 0.1	23.5	20	15	10	3	12
0.5 - 0.2	20	49	31	20.5	92	34
1.0 - 0.5	14	20.5	30	39.5	2	38

Figure 2. Grain characteristics of the used materials

Table 1. The density of used materials

Material	Density kg/m^3	Bulk density kg/m^3		Gap fractions	
		uncompact	compacted	uncompact	compacted
Ore India	3504	1780	2136	0.492	0.390
Concentrate Brazil	4625	2381	3030	0.485	0.345
Ore Robe River	3688	1923	2222	0.479	0.397
Ore Venezuela	3892	2020	2424	0.481	0.377
Dolomite	2901	1299	1786	0.552	0.384
Ilmenite concentrate	4660	2160	-	0.536	-
Cinder	3640	1800	-	0.505	-

2. Experimental approach

2.1. Experimental facility

The facility consists of the following component parts: compressed air supply circuit, powder distributor, a circuit for pneumatic transport of the material.

The operating principle was the following: the compressed air brought in the distributor through the supply chain was streamlining the material and created a certain pressure due to which the fluid flows through the pipeline to the tuyeres. The flow rate is determined by the difference between the pressure and the resistance on the transport path of the streamlined mixture.

2.2. Experimental procedure

Prior to the experiment the preparation steps were established: checking the outside of the facility for any faults, placing the pipes for transport of the compressed air and of the streamlined mixture at ground level, checking if the valves are closed, connecting the pipe for the compressed air to the compressor, checking the pressure in the compressed air circuit by turning the valve on and reading the manometer at the reducers, purging the installation for 10 minutes with compressed air in order to eliminate the condensate or other impurities that could disturb the experiment (for this the valves are opened at the entrance to the distributor and at the entrance in the transport circuit).

In case of a required TiO_2 of 10-20 kg/t iron results a quantity of TiO_2 of 57-114 kg/h.g.v [5], [6], [7], [8].

2.3. Experimental results

Two groups of experiments were made:

- Group A: with material evacuated directly from the distributor through the connector, Φ_{int}=6 mm.

A set of tests were performed during the A experiments, A_0, with material evacuated through the pipe with overflow, Φ_{int} = 15 mm, using Robe River ore to verify the ability of the device to evacuate the overflow.

- Group B: with pneumatic transport and evacuation through a rubber pipe, Φ_{int}=8 mm and 12 m length, in horizontal position and with the evacuation end lifted up to approximately 1.6 m to simulate the placement on the furnace platform.

All experiments were done with pressures in the distributor of 0.2 up to 4 bar with evacuation at atmospheric pressure. During the experiments it was noticed that all materials studied had good behavior and were transported pneumatically even at low pressure of 0.2 bar in the distributor.
In the situation of material evacuation directly from the distributor with small pressures, up to 1 bar, the streamlined mixture flowed still, being maintained in a laminar domain; at higher pressures, over 1.5 bar, the mixture jet becomes increasingly strong and the flow had a turbulent character with intense dust releases.

In the case of material transport through the pipe, in general, the flow of streamlined mixture, material / air, is stable. A flow with fewer beats and dusty exhaust is observed at pressures above 1.5 bar for ore from India, from Robe River and the mixture 80% ore from India and 20% dolomite, with a grain showing a high content of disperse and ultrafine fractions below 0.063 mm. It was noticed that the flow was resumed easily after the temporary interruption of the transport process.

Throughout the experiments no blocking phenomena occurred besides accidental situations because of moisture or contaminants with the size greater than 1 mm entered in the distributor. Based on experimental data obtained, mass flow of material discharged at different pressures and the parameters of pneumatic transport through the pipeline were calculated, also actual transmission velocities and the velocities at the exit of the pipe of the mixture and the mass loading coefficients m.

Figure 3 and Figure 4 present the mass flow at different pressures. It was observed also the equipment ability to operate in a wide range of flow rates and its sensitivity to changes in pressure.

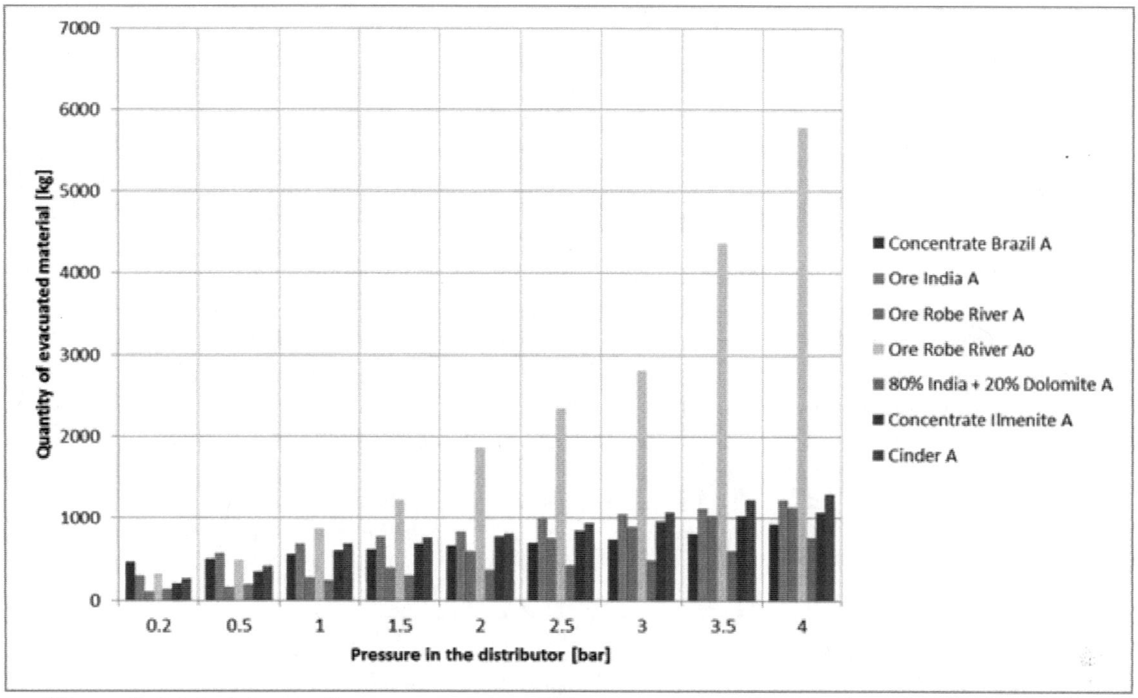

Figure 3. Material mass flow evacuated at different pressures directly from the distributor through the connection Φ 6 mm (A) and through the overflow pipe Φ 15 mm (A_0)

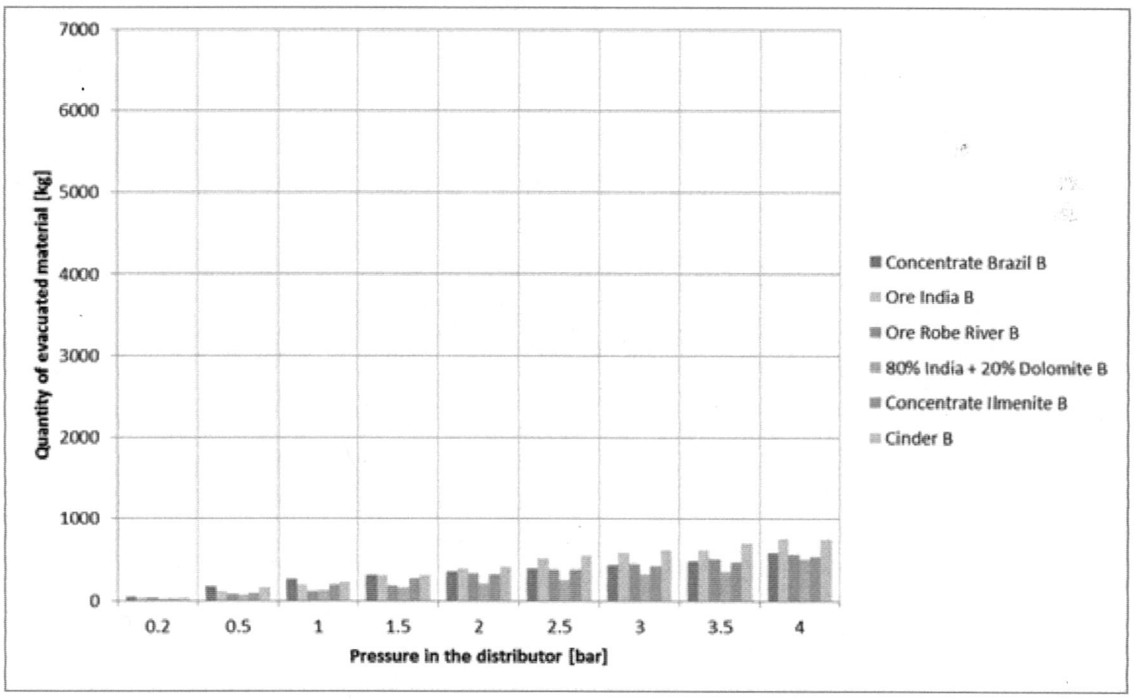

Figure 4. Material mass flow evacuated at different pressures through the rubber pipe Φ = 8 mm x 12m (B)

The mass flow rates obtained by evacuating Robe River ore directly from the distributor through the pipe with overflow $\Phi_{int.}$ = 15 mm at pressures between 0.2 - 4 bar is between 324 - 5784 kg / h. By fitting a bottleneck connection $\Phi_{int.}$ = 6 mm it was obtained a mass flow reduction by 70-80%. Minimum flows are between 108 kg / h (ore from Robe River) and 468 kg / h (ore from Brazil) and the maximum between 768 kg / h (80% ore India and 20% dolomite) and 1296 kg / h (cinder).

In case of pneumatic transport by pipeline F_{int} 8 mm x 12 m due to pressure loss a reduction in the minimum flow to between 24-48 kg / h and a maximum of between 504-756 kg / h is observed. Flow ranges of 45-340 kg / h can be achieved at low pressure in the distributor at maximum 2.4 bar, except mixture 80% India ore and 20% dolomite requiring max pressure 3.2 bar.

The possibility of injecting the ilmenite in the ranges 125-260 kg/h for pressures in the distributor of 0.6-1.4 bar. These results show that the resistance on the transport path is lower and it is possible to accomplish the pneumatic transport in the flow ranges at a pressure in the distributor of 0-4 bar. Adjusting the flows at a value required by the process could be done by choosing the working overpressure in the distributor. Tables 2-8 transport process parameters for all materials studied.

Table 2. Transport process parameters for each material

Material	Evacuated flow, [kg/h]	Overpressure distributor, [bar]	Maximum required air flow, [Nm^3/h]	Maximum transport velocity, [m/s]	Maximum pipe output velocity, [m/s]	Maximum mass loading coefficient, [kg mat/ kg air]
Concentrate Brazil	-	0.2 - 1.8	4.6	10	25.8	57
Ore India	45-340	0.2 - 1.65	4.5	10	25.4	58
Ore Robe River	-	0.2 - 2.2	4.8	10	27	55
80% ore India + 20% dolomite	-	0.2 - 3.2	5.4	10	30.4	49
Ilmenite	125-260	0.6 -1.4	2.7-4.2	10	23.5	63
Cinder	36-744	0.2 - 4	5.7	10	32.6	101

Table 3. Pneumatic transport parameters through the pipeline of the concentrate from Brazil

No.	Pressure in the distributor, [bar]	Air flow, [m^3/h]	Mass flow evacuated material, [kg/h]	Transport velocity [m/s]	Pipeline output velocity, [m/s]	Mass loading coefficient, μ [kg mat./kg air]
1.	0.2	1.80	48	8.3	9.96	20
2.	0.5	2.60	168	9.57	14.36	51
3.	1.0	3.60	264	10.06	20.11	57
4.	1.5	4.20	312	9.48	23.71	57
5.	2.0	4.80	360	8.99	26.97	58
6.	2.5	5.20	396	8.32	29.12	59
7.	3.0	5.30	432	7.46	29.83	63
8.	3.5	5.50	480	6.9	30.98	68
9.	4.0	5.70	588	6.4	32.2	80

Table 4. Pneumatic transport parameters through the pipeline of the ore from India

No	Pressure in the distributor, [bar]	Air flow, [m^3/h]	Mass flow ejected material [kg/h]	Transport velocity, [m/s]	Pipeline output velocity, [m/s]	Mass loading coefficient, μ [kg mat./kg air]
1	0.2	1.8	36	8.3	9.95	15
2	0.5	2.66	120	9.56	14.35	36
3	1	3.6	192	10.05	20.1	42
4	1.5	4.2	312	9.53	23.83	57
5	2	4.8	396	9.06	27.17	64
6	2.5	5.2	516	8.42	29.45	77
7	3	5.3	588	7.6	30.24	86
8	3.5	5.5	624	6.98	31.4	88
9	4	5.7	756	6.54	32.7	103

Table 5. Pneumatic transport parameters through the pipeline of the ore from Robe River

No.	Pressure in the distributor, [bar]	Air flow, [m^3/h]	Mass flow ejected material [kg/h]	Transport velocity, [m/s]	Pipeline output velocity, [m/s]	Mass loading coefficient, μ [kg mat./kg air]
1	0.2	1.8	36	8.29	9.95	16
2	0.5	2.6	84	9.52	14.28	25
3	1	3.6	120	9.98	19.97	26
4	1.5	4.2	180	9.4	23.6	33
5	2	4.8	336	9	27.04	54
6	2.5	5.2	384	8.35	29.22	57
7	3	5.3	444	7.5	29.98	65
8	3.5	5.5	504	6.93	31.2	71
9	4	5.7	564	6.47	32.37	77

Table 6. Pneumatic transport parameters through the pipeline of the mixture 80% ore India + 20% dolomite

No.	Pressure in the distributor, [bar]	Air flow, [m^3/h]	Mass flow ejected material [kg/h]	Transport velocity, [m/s]	Pipeline output velocity, [m/s]	Mass loading coefficient, μ [kg mat./kg air]
1	0.2	1.8	24	8.28	9.93	10
2	0.5	2.6	72	9.52	14.27	22
3	1	3.6	132	10	20	29
4	1.5	4.2	156	9.4	23.6	29
5	2	4.8	204	8.9	26.9	33
6	2.5	5.2	252	8.3	29.06	38
7	3	5.3	324	7.46	29.83	47
8	3.5	5.5	360	6.9	31	51
9	4	5.7	504	6.47	32.3	69

Table 7. Pneumatic transport parameters through the pipeline of the ilmenite concentrate

No.	Pressure in the distributor, [bar]	Air flow, [m^3/h]	Mass flow ejected material [kg/h]	Transport velocity, [m/s]	Pipeline output velocity, [m/s]	Mass loading coefficient, μ [kg mat./kg air]
1	0.2	1.8	24	8.27	9.93	10
2	0.5	2.6	96	9.51	14.27	29
3	1	3.6	192	10.01	20.03	42
4	1.5	4.2	276	9.47	23.66	51
5	2	4.8	324	8.98	26.93	52
6	2.5	5.2	384	8.31	29.1	57
7	3	5.3	420	7.45	29.8	62
8	3.5	5.5	468	6.88	30.97	66
9	4	5.7	540	6.43	32.16	74

Table 8. Pneumatic transport parameters through the pipeline of cinder

No.	Pressure in the distributor, [bar]	Air flow, [m^3/h]	Mass flow ejected material [kg/h]	Transport velocity, [m/s]	Pipeline output velocity, [m/s]	Mass loading coefficient, μ [kg mat./kg air]
1	0.2	1.8	36	8.29	9.95	16
2	0.5	2.6	156	9.6	14.39	47
3	1	3.6	228	10.07	20.14	50
4	1.5	4.2	312	9.5	23.8	57
5	2	4.8	408	9.05	27.16	66
6	2.5	5.2	552	8.42	29.48	83
7	3	5.3	624	7.56	30.26	91
8	3.5	5.5	696	6.99	31.47	98
9	4	5.7	744	6.53	32.65	101

By analyzing the process parameters for each material from the Tables 3-8, it was observed, in general, that they are maintained in the transport ranges in a dense state ($v_{transport} \leq 10$ m/s, m>20). The maximum air flow that was necessary for a productivity of the installation of 340 kg/h is, in general, 4.2-4.8 Nm^3/h, exception is the mixture 80% ore India and 20% dolomite, which requires a flow of approximately 5.4 Nm^3/h. The velocities at the outlet of the transport pipe (injection velocity) are similar with the ones at the outlet of the transport pipe for pulverized coal (18-25 m/s). This is a necessary condition for not disturbing the injection of the pulverized coal in the furnace.

The results presented in Tables 3-8, regarding cinder, like the observation made during the experiments demonstrated its capacity for being transported pneumatic in dense state. At overpressures of 0-4 bar, in the distributor, flows of 36-744 kg/h can be injected in the furnace. The checks that were done on the installation after assembly have shown that the installation has the capacity of operating in safety condition at pressures of maximum 7 bar.

Experimental results with evacuation through the pipe with overflow $\Phi_{int} = 15$ mm, indicated that at pressures of 0-4 bar in the distributor and 1 bar at the evacuation, the evacuation capacity of the

IOP Conf. Series: Materials Science and Engineering **163** (2017) 012011 doi:10.1088/1757-899X/163/1/012011

dispensing device with overflow is situated in the ranges 324-5784 kg/h. The evacuation capacity with connection $\Phi_{int} = 6$ mm is reduced by 70-80%, the mass flow was 108-1296 kg/h.

The experimental results for the injection process simulation show that the studied materials had a good behavior and can be pneumatic transported even at low pressures of 0.2 bar in the distributor and 1 bar at evacuation, the evacuation capacity through the rubber pipe Φ_{int} = 8mm x 12 m is situated between the ranges 24-756 kg/h. this demonstrates the applicability of the process when injecting ore and flux in the furnace with the flows 45-340 kg/h, also when injecting ilmenite with the flows 125-260 kg/h.

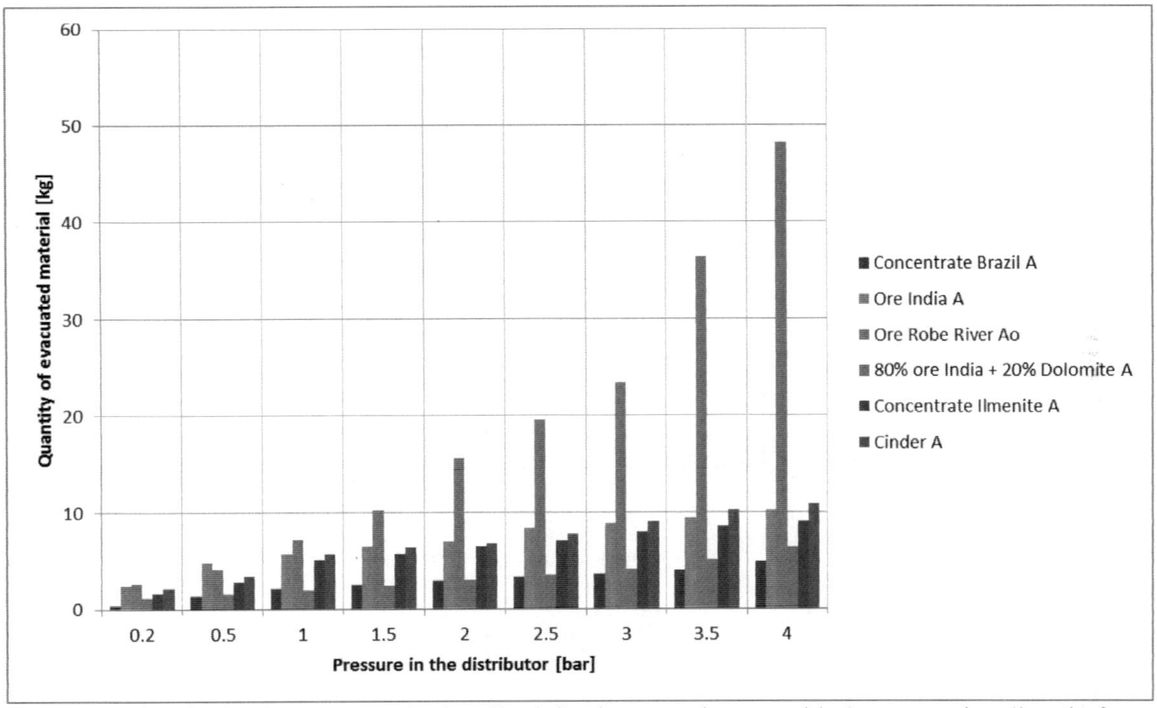

Figure 5. The obtained parameters in the pilot injection experiments with the evacuation directly from the distributor through the connection $\Phi = 6$ mm (A) and through the overflow pipe $\Phi = 15$ mm (A_0)

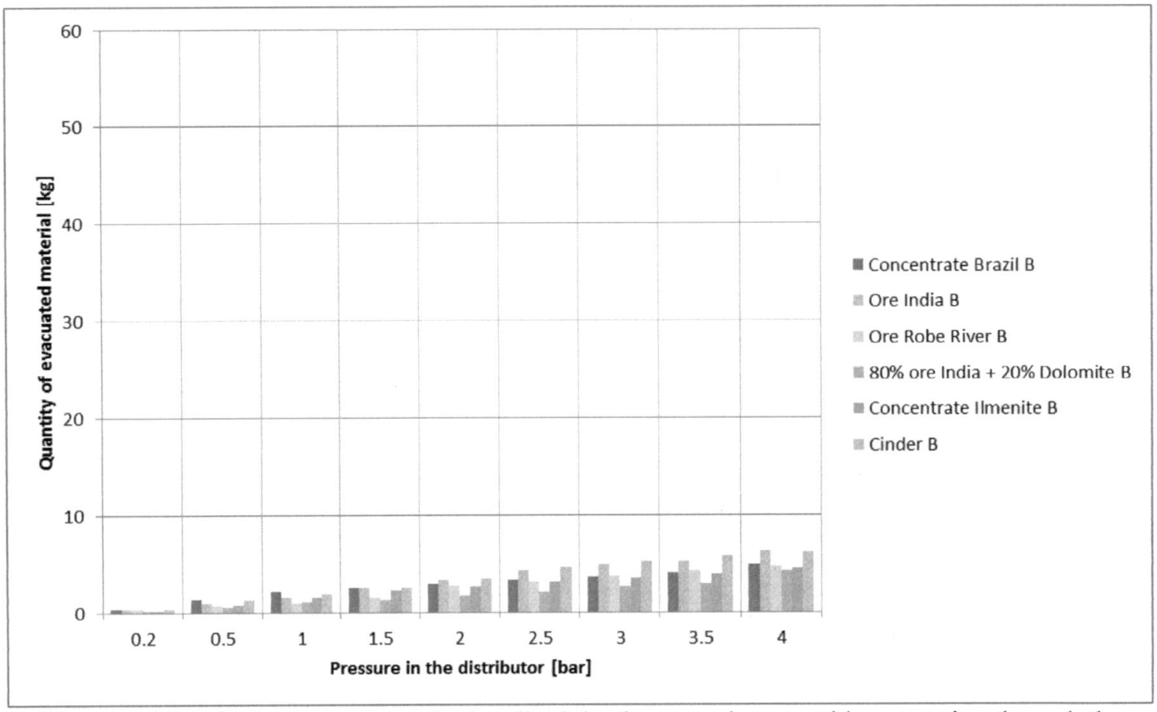

Figure 6. The obtained parameters in the pilot injection experiments with evacuation through the rubber pipe Φ = 8 mm x 12 m

3. Conclusions

The presented results demonstrate the installations capacity to operate in a regime of transport in a dense state ($v_{transport} \leq 10$ m/s, $m>20$).

Pneumatic transport of the materials in the flow ranges 45-340 kg/h and 125-260 kg/h is accomplished at low overpressures in the distributor which indicates that the resistance on the transport path is small.

The velocity at the outlet transport pipe and the mass loading coefficient is situated in the injection parameter ranges of injecting pulverized coal in the furnace (y_i = 18-25 m/s, m = 40 – 50).

Experimental results also show the cinders capacity of being transported pneumatic in a dense state and as a result we can consider recycling it in the furnace by injection through the tuyeres. The experiments have also allowed us to establish a correlation between the evacuated material mass flow and the pressure in the distributor. On this basis, in the injection technology is now possible to determine a working overpressure in the distributor and adjusting the flow at a value required by the process.

References

[1] Constantin N and Cătană C 1997 *Recherches experimentales sur les caracteristique qualitatives des materiaux utilisees a la production des fontes de premiere fusion,* International Symposium - Jubilee 35 years of Pitesti University, Pitesti, Romania, November 21-22, pp 53 – 57

[2] Heput T, Socalici A and Constantin N 2011 *Valorificarea deseurilor feroase marunte si pulverulente*, Editura PolitehnicaTimisoara, Romania

[3] Burlacu J C and Constantin N 2012 Experimental research on pneumatic conveying reaction to iron ore fine-grained, *Metalurgia International* **17**(4) 184-188

[4] Oancea V, Mihaiu A C, Dobrescu C and Constantin N 2011 Using pulverized coal as an alternative energotehnological source to produce pig-iron for steelmaking, *Journal of Sustainable Energy* **II**(1) 62-65

[5] Dobrovici D, Constantin N, Şoit I and Geantă V 1990 Informaţii asupra utilizării în încărcătura furnalului a materialelor greu fuzibile în scopul protejării creuzetului şi a vetrei, Second edition, *Metalurgia Revue* **9-12** 449-452

[6] Constantin N, Geantă V and Cătană C 1997 *La production des fontes speciales dans haut fourneau en utlisant des minerais de titane,* International Symposium - Jubilee 35 years of Pitesti University, Pitesti, Romania, November 21-22, pp 58-62

[7] Branescu E, Blajan A O and Constantin N 2015 Experimental research on the characteristics of softening and melting of iron ores as significant factor of influence on gas permeability of blast furnace charge, *IOP Conf. Ser.: Mater. Sci. Eng.* **85** 012001

[8] Comsa M, Buzduga M, Constantin N and Heput T 2007 *Analysis methods in ordin to fully characterise molding powders*, AMCSIT -Excellence research as a way to E.R.A., Brasov, Romania

IOP Conf. Series: Materials Science and Engineering **163** (2017) 012012 doi:10.1088/1757-899X/163/1/012012

Change management methodologies trained for automotive infotainment projects

G Prostean, S Volker and A Hutanu

Politehnica University of Timisoara, Faculty of Management in Production and Transportation, Remus str., no. 14, 300191 Timisoara, Romania

E-mail: andreihutanu@yahoo.com

Abstract. An Automotive Electronic Control Units (ECU) development project embedded within a car Environment is constantly under attack of a continuous flow of modifications of specifications throughout the life cycle. Root causes for those modifications are for instance simply software or hardware implementation errors or requirement changes to satisfy the forth-coming demands of the market to ensure the later commercial success. It is unavoidable that from the very beginning until the end of the project "requirement changes" will "expose" the agreed objectives defined by contract specifications, which are product features, budget, schedule and quality. The key discussions will focus upon an automotive radio-navigation (infotainment) unit, which challenges aftermarket devises such as smart phones. This competition stresses especially current used automotive development processes, which are fit into a 4 Year car development (introduction) cycle against a one-year update cycle of a smart phone. The research will focus the investigation of possible impacts of changes during all phases of the project: the Concept-Validation, Development and Debugging-Phase. Building a thorough understanding of prospective threats is of paramount importance in order to establish the adequate project management process to handle requirement changes. Personal automotive development experiences and Literature review of change- and configuration management software development methodologies led the authors to new conceptual models, which integrates into the structure of traditional development models used in automotive projects, more concretely of radio-navigation projects.

1. Introduction

Evaluating the development of an Electronic Control Units (ECU) in the specific area of automotive engineering and focusing for example on todays Electronic-Infotainment-System, which is divided into several Network-Community partners such as the Main-Unit, display, telephone and several comfort and driver assistance units, as shown in Figure 1. Especially complex are Network-Partners or Content-Provider, which are connected via Internet (e.g. navigation-information or Plug & Charge of electro-mobility Systems). Each ECU will be assigned and development by a specialized automotive supplier. It is common practice, not only in the automotive business to develop complex projects with partners "to respond quickly to the changing need of customers" [1] or market requirements and concentrate on the core business competency [1].

The objective of this paper is to examine the question how to handle automotive requirement changes in general and taking into account the challenging aftermarket driven infotainment development life cycle, the side effects of multi-partner-relationships, and most important to suggest such an automotive specific Change-Management (CM) process. This paper contributes to several papers and books [1-5] dealing already with CM processes, but commonly neglecting the consideration of the faced risk of requirement changes. and human resource management. Thus, the 2nd focus is deliver a Risk-Assessment-Model embedded within such a CM construction, which is the key to meet the adjusted objectives to be ready as expected. The research will aim to drive a simple to use guided-tour to judge the risk, which is carried by each single change and as well as reviewing the total risk at a particular milestone. Another key issue to discuss is how to handle the impact of a must-have RfC, which appears at the final stage (i.e. Preparation of Production) of the project. Literature research of agile software development methodologies have led the authors to new conceptual supportive models, which integrates into the structure of the investigated automotive specific change management process. The graphical visualization of the RfC inflow with respect to the development phase will give a description of the critical issues the project is facing and possible remedies to sustain control of the project. The result of the analysis will provide instruments to judge the Risk and control the inflow of RfC's to meet the agreed project objectives.

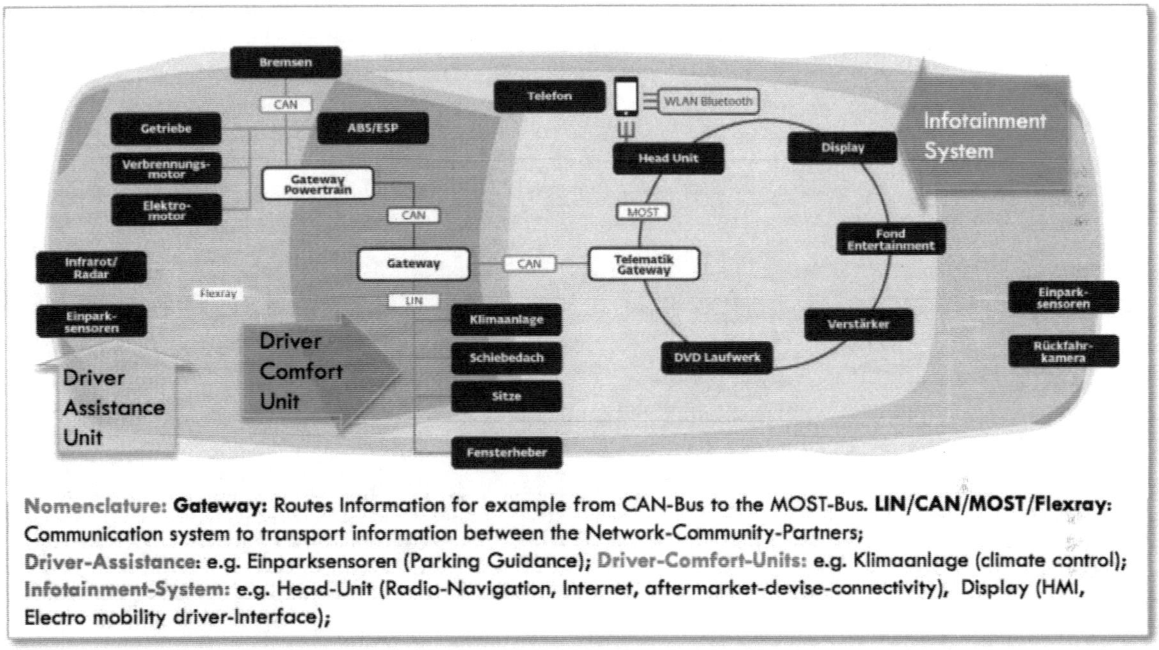

Figure 1. ECU Network Community [6]

2. Research Environment

Automotive systems are broken down into specialized project areas. Considering for example todays Electronic-Infotainment-Systems (p.r.t. Figure 1), which are divided into several Network-Community partners. Each ECU will be assigned and development by a specialized suppliers [1] section "Requirements for collaborative, multi-company ECM" [1]. The research project assumes that that all project partner have proven their potential to meet the project objectives which is part of an initial project launch supplier assessment, proving their company Management- and "domain knowledge in your own field" [7]. The supplier selection is part of the Supplier agreement management; but only the 1st tear supplier (i.e. navigation system developer) will be reviewed. The 2nd tear supplier (i.e. CPU supplier) might be assessed as well but is within the contractual responsibility of the 1st tier supplier. The innovation target will a risk, but must be also considered during supplier selection [7]. Supplier management is most certainly important [7] and will be part the Risk-Management investigation.

All modifications of the requirement specification must be evaluated regarding all possible impacts, s as development schedule, budget and especially possible side effects, including impacts to of the network community partners. Another key issue to discuss is how to handle the impact of a must-have RfC, which appears at the final stage (i.e. Preparation of Production) of the project.

3. Project life cycle of Automotive Projects

At this point it is not necessary to describe the complete automotive product development phases of a car manufacturer, but the key milestones of an ECU development. The titles and descriptions of the project milestones vary with the OEM (Original Equipment Manufacturer), understanding the story and the to-do's behind the milestones is of utmost significant to appreciate the variations of possible negative effects of RfC's and their dependencies, i.e. at which HW/SW release timing of the modification appears in the project schedule. If for example, the implementation of an RfC is release 1st time just after the debugging phase (see Figure 2) and the verification required intensive test drives, it might be hardly possible to solve possible negative side effect or bugs before the 0-bug sample.

The following description will only briefly outline the process steps required until finally the ECU (or Use Case) will meet the customer; in addition explain the terminologies used the following discussion. The original OEM milestone expressions and schedules are not used due to arrangements of confidentiality and especially to prevent that the research's findings are bound to be valid only for a particular OEM.

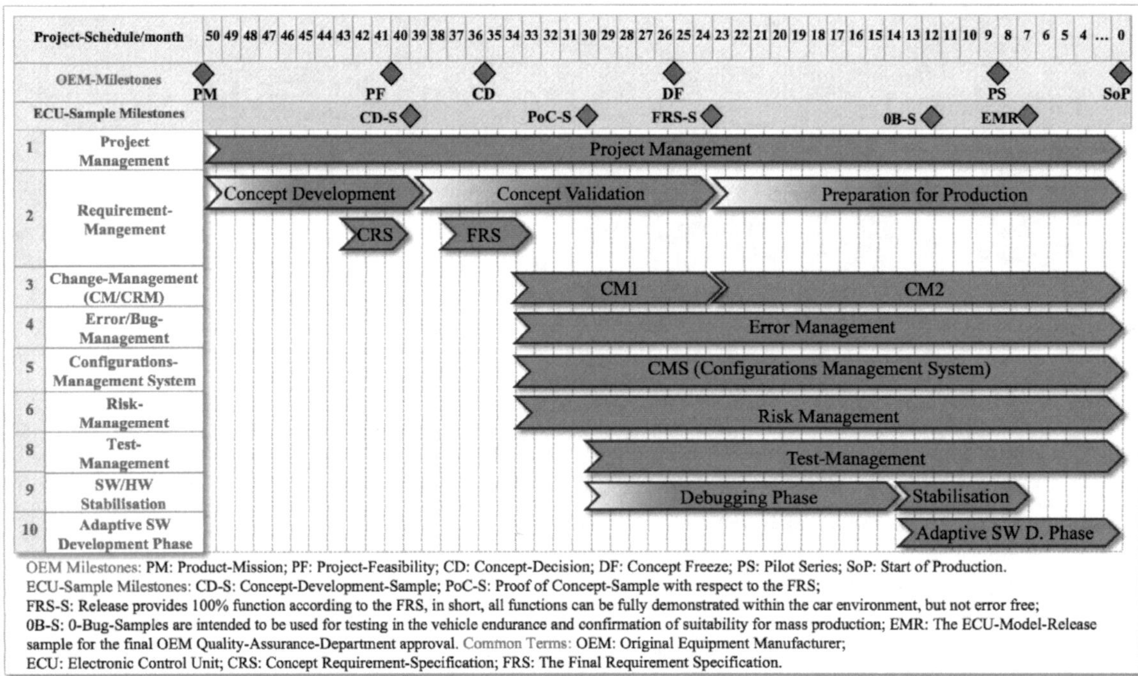

Figure 2. Automotive Project Milestone-Plan with Process-Introduction and Deliverables (adapted after [8], [9])

Figure 2 shows the development life cycle of automotive projects can be divided into three main parts such as Concept-Development, Concept-Validation and finally Preparation-of-Production.

3.1. Concept-Development

The start of the Concept-Development phase is marked with the Project-Mission (PM) milestone and ends with the Concept-Development-Sample (CD-S), [8]. The goal of this phase is to develop upon a Requirements-Brief, stating for example the market positioning, technology and finances the Concept-

Requirement-Specification (CRS). The final review of the Concept-Requirement-Specification will be based firstly upon a feasibility study (PF) and finally lead to a Concept-Development-Sample which could be a first ECU sample or just a computer simulation.

3.2. Concept Validation

The Concept-Validation phase lies in between the Concept-Development-Sample (CD-S) and the Final-Requirement-Specification-Sample (FRS-S), or more precisely the concept sample will reach within this period 100 % functionally. The task of the FRS-Sample is to show all (final customer) features in the car according to the Final-Requirement-Specification (FRS). The FRS is the baseline for the nomination of the supplier of the ECU. All contractual agreements such as functional content, schedule, finances and quality targets and so on are fixed within this document. In fact, the FRS is the INPUT of the supplier mission or contract given to several prospective suppliers for quotation and the FRS-Sample the OUTPUT.

The Concept-Validation phase has two noticeable deliveries, the Proof-of-Concept-Sample (PoC-S) and as already explained the Requirement-Specification-Sample (FRS-S). The question of this paper is how to deal with possible changes of specification, which could not be seen while compiling the FRS, but sprung into life after the commercial agreement with the supplier, and cannot be ignored for reasons stated already in the paper. The worst-case scenario is when the Requirement-Specification-Sample (FRS-S) milestone must be shifted to realize a redesigned of the Final-Requirement-Specification. Automotive projects have shown that even adequate requirement documentation reviews cannot prevent the appearances of specification limitations. Reviews are most certainly useful, but the dynamic behaviour of a car network and market demand changes are not all foreseeable. This paper must therefore define the period to allow FRS changes and the research must clearly identify the side effect.

3.3. Preparation of Production

The start of the Preparation-of-Production phase is marked by the acceptance of the OEM of the Requirement-Specification-Sample (FRS-S). The initial aim of this period is to locate all hardware, software and system problems and meet finally FRS defined quality targets. Thus, found problems require to be handled with according a to be defined process and solved until the 0-Bug-Sample (0B-S).

4. Automotive Tailored Change Management (CM) including Configuration Management.

"Is there an ideal requirements engineering process? No- processes must be tailored to organizational needs" [3]. The following definitions and responsibilities have to be clarified to setup an automotive specific change management process shown in Figure 3a/b. The outlined process does not show all required "RfC-questionnaire" formal or technical reviews for the sake of simplicity. However, the essential task of the shown CM process is to establish and sustain a relationship with the stakeholders regarding the status and access of the RfC's which respect to the documentation database (CMDB), [3] and current service transition (see Figure 3a/3b).

4.1. Definition of the Input parameters or RfC-input:

a) New requirements due to must have changes to satisfy the strong demands of market to ensure the later commercial success. b) Secondly, requirement changes due to insufficient or faulty function description of the FRS. c) Finally, software or hardware implementation errors.

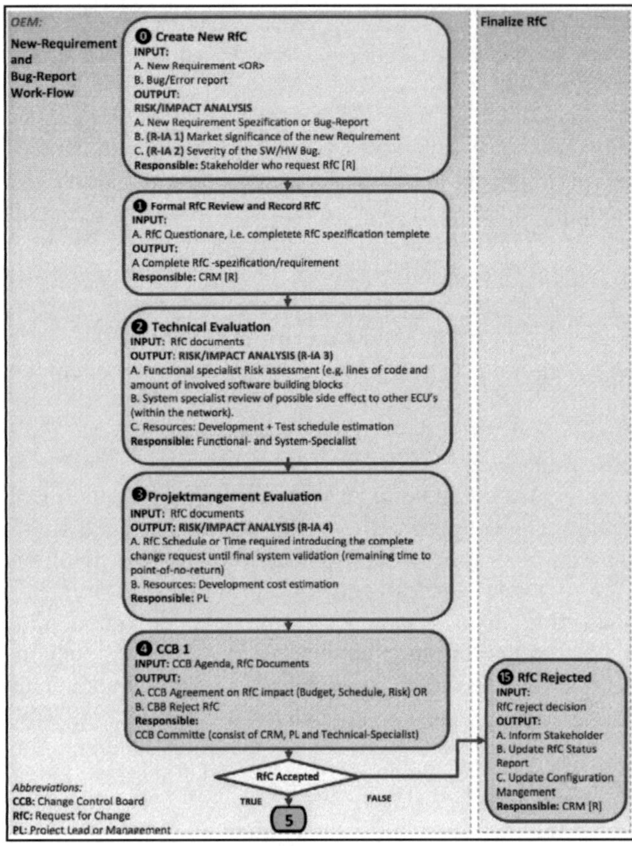

Figure 3a. Automotive Change Management (CM) Work Flow [11]

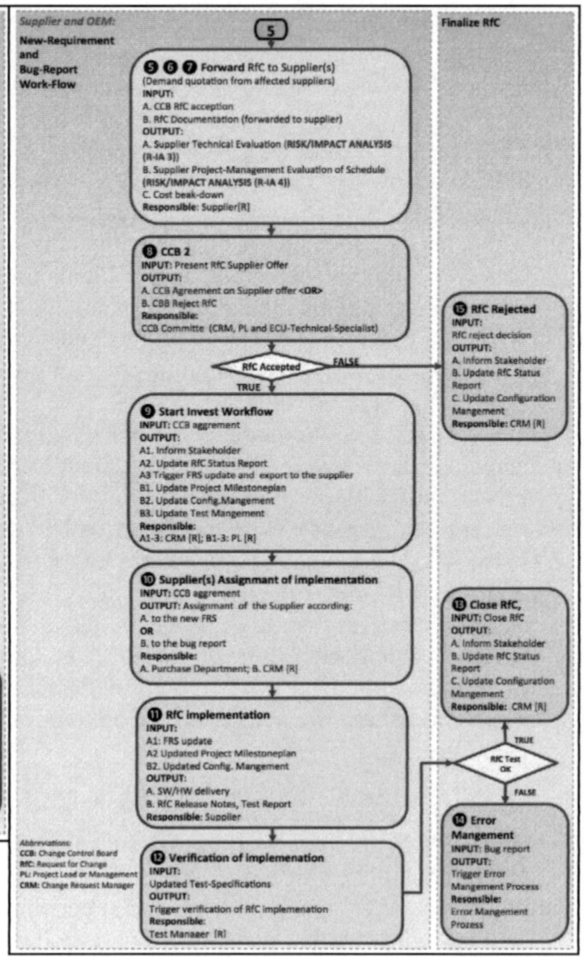

Figure 3b. Automotive Change Management (CM) Work Flow [11]

4.2. Start of the Change-Management (CM) process

From Figure 2 can be seen that it is suggested to start with the CM process at the start of the CM1 or alternatively at the CM2 phase. This discussion sets the CM start point at CM1, which is directly after the Final Requirement-Specification (FRS), since the FRS is used as the requirement base line to start with the development. In addition, the Configuration Management System (CMS), [3] will run simultaneously with the CM (see Figure 3c) to record all relevant data (CI), [3]. The benefit to start at CM1 phase is simply that all discussion, test result finding are feedback into the FRS. Secondly and may be more important is introduction of risk or impact analysis (see chapter 5). In addition, the consequence of neglecting to finalize the required change on contractual basis is the commercial impact review or influence is neglected. Thus the discussion might be concluded with a solution, which is not cost effective and due to budget limitations impossible to implement. Lessons-learned workshops of previous automotive projects state clearly that CM1 should be the trigger for the CM Work Flow in order to document all agreed specification changes. Without this process all other activities such as deriving a test specification based upon an invalid FRS is pointless. In addition, the CM Workflow must insure that all possible stakeholders or other departments are included in the RfC review process to insure not to neglect possible side effects.

4.3. Change Management (CM) Work Flow

The presented CM Work Flow in Figure 3 follows the concept of combining the requirement changes and bug report within the same CM work flow. Consequently the formal treatment of RfC for both cases must be harmonized in the way that the tooling (GUI) or RfC questionnaire using the same parameter such as priorities.

4.4. CM Work Flow transitions ('0'-15' see Figure 3: CM Work Flow)

In order to prevent or at least reduce delays and effort spent waiting for decisions from the Change Control Board (CCB), [2] or quotations from suppliers the transition timings should be defined.

4.5. RfC Status Report

The Status report will summarize the following parameters: a) RfC-in: New RfC; b) RfC-out: RfC is closed if the SW or HW change is delivered, implemented and in car successfully tested; c) RfC-Rejected; Rejected CR's will be added to the RfC-out count; d) RfC-active/open CM process step ('2'-'12'): The number of RfC-active/open corresponds with all RfC (RfC-in – RfC-out), which are still active; e) RfC-new workflow process step ('0'-'1'): The RfC will be taken for the RfC Status after the review of the CRM.

In addition, "the change record holds the full history of the change, incorporating information from the RFC and subsequently recording agreed parameters such as priority and authorization, implementation and review information." [5].

4.6. Automotive - Configuration Management System (CMS)

The Configuration Management System (CMS) is the backbone of Change-Management and "should prevents changes from being made to IT infrastructure or service configuration baseline without valid authorization." [2]. The CMS Workflow and CI (configuration item [3] or entities [4]) dataflow is shown in figure 3c. The CMS applied for automotive system should at least provide the following services:

a) A definition of the configuration database (CMBD), which relevant CI's to be recorded [4]

b) "Support for CIs of varying complexity, e.g. entire systems, releases, single hardware items, software modules" [2]

c) "Hierarchic and networked relationships between CIs; by holding information on the relationships between CIs, Configuration Management tools facilitate the impact assessment of RfCs" [2]

d) "Integration of problem management data within the CMS, or at least an interface from the Configuration Management System to any separate problem management databases that may exist" [2]

e) "Ease of interrogation of the CMS and good reporting facilities, including trend analysis (e.g. the ability to identify the number of RFCs affecting particular CIs)" [2]

f) "Ease of reporting of the CI inventory so as to facilitate configuration audits" [2]

g) "Flexible reporting tools to facilitate impact analyses" [2]

h) "Assign people to each role and maintain the assignment in the service (CMS)" [2], [4]

The RfC authorization should be recorded within the Configuration Management Database (CMDB) at least by the time that the change is implemented. The Software "Integration- and Delivery-process" should only be started, if all Configuration-Items (CI) of the CMDB are reviewed and approved by all significant stakeholders. The Integration-Call (stakeholder meeting, see Figure 3c) can give a final valid authorization for implementation if all stakeholders are present.

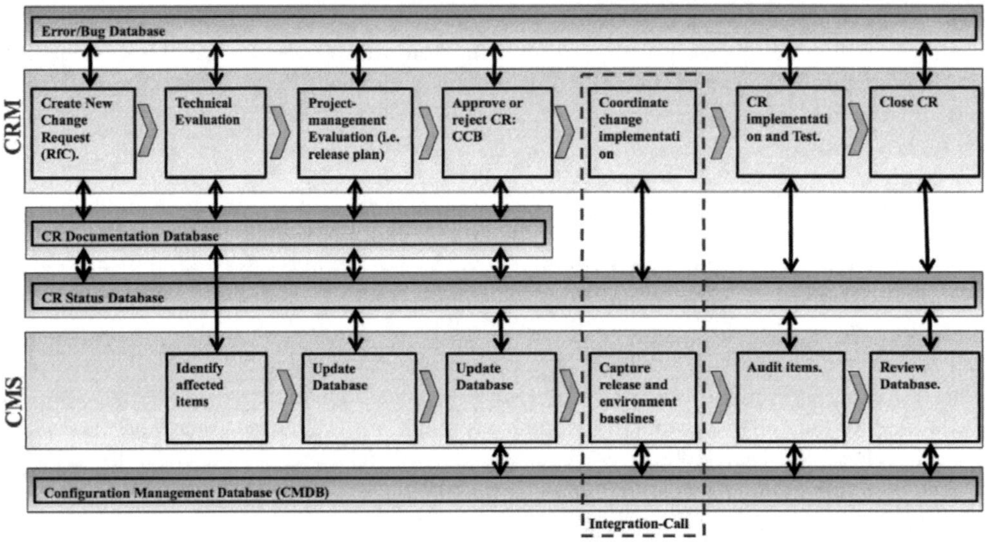

Figure 3c. Automotive Configuration Management System (CMS)

5. Implementation the of Change Management Process in combination with Risk-Management Assessment Simulations

The demand is how to evaluate the risk or impact of a change request (RfC) to the project and to enable the project management to decide a) to implement c) postpone to a later release or c) reject the deviation of the initial requirement specification. The following discussion will assume the CM Process is implemented as discussed.

5.1. Practical analysis. Total of valid RfC in-flow w.r.t. the project schedule (time)

For the analysis of the effects of a continuous in-flow of RfC's (New-Bug/-Req.) throughout the development life cycle until SoP the previous discussion has recognized:

a) The start-point of the Change-Management (CM) process should be CM1, which is the milestone of the freeze of the contract specification (FRS).

b) As also previously mentioned, this FRS is after nomination of the supplier the base line of the requirement (and CMDB) to be developed and handed over to the supplier via the purchase department.

c) The end-point of the analysis (of this paper) will be the SoP.

Utilizing the explained Automotive development key milestones (FRS-S, PoC-S, FRS-S, 0B-S and EMR) and the RfC-in and consequently the RfC-out flow the graphs shown in figure 4 (RfC-in/-out/-active) can be drawn along the milestone FRS to EMR or SoP. The trend of simulated RfC-in and RfC-out flow corresponds to lessons-learned of previous projects [8].

The visualization of the in- and out-flow of RfC w.r.t. time shown in Figure-4 is already common practice [8]. From the shape of the simulated RfC in-/out-flow graph shown in figure 4 (RfC-in/-out/-active) can be deduced:

a) The amount of active RfC shows the maturity of the ECU or of the concept and final specification. Bearing in mind the agreed expectations (fixed by contract) of the ECU quality defined and scheduled by the key milestones (FRS-S, PoC-S, FRS-S, 0B-S and EMR). If the quality goals are not meet countermeasure might be necessary. The countermeasures depend on the root cause of the RfC, therefore a precise remedy will depend upon the analysis result of the technical specialists.

b) The FRS-S milestone marks the 100% feature complete ECU maturity expectation. Moving towards the 0B-S (0-bug Sample) milestone the RfC inflow should decrease as a statement of functional stability. The expectation is that the RfC inflow and the RfC total will be 'zero' at the 0B-S milestones. It is an unfortunate fact of life that software and hardware is never error free and any bug solution might inflict another or several bugs into the system. Therefore it is important to introduce a SW/HW stabilization phase (just before the 0B-S - until the EMR-milestone) [8]. The described procedure only cultivates the RfC active curve into a fever (quality) chart of the project or a particular milestone. The graphs give no risk assessment support.

5.2. Risk-Management Assessment via RfC-Forecast Model

It is common practice at the beginning of a project to describe within the Project-Brief the required Project-Infrastructure, see "Infrastructure Management Process" [10]. The Project-Brief describes the demanded "resources for the achievement of project objectives" [10]. In order to estimate the required project infrastructure such as human resources or budget, similar projects can be used as a baseline and compared to be project at hand. Thus, data or the amount of requirement changes and bug report experienced in on-going or closed project can be use to derive a graph to be used as a Forecast or expected RfC inflow. In fact this data can be used to establish and reason the Project-Briefs statement of required resources. Figure 5 shows such estimated RfC-Forecast graph.

Reviewing the RfC-Forecast and the RfC-in graph with respect to the given milestone plan the following observations could be made, bearing in mind the maturity expectations of the ECU key milestones. If the inflow excites the forecast, countermeasures might to be necessary to return to the planned inflow (forecast).

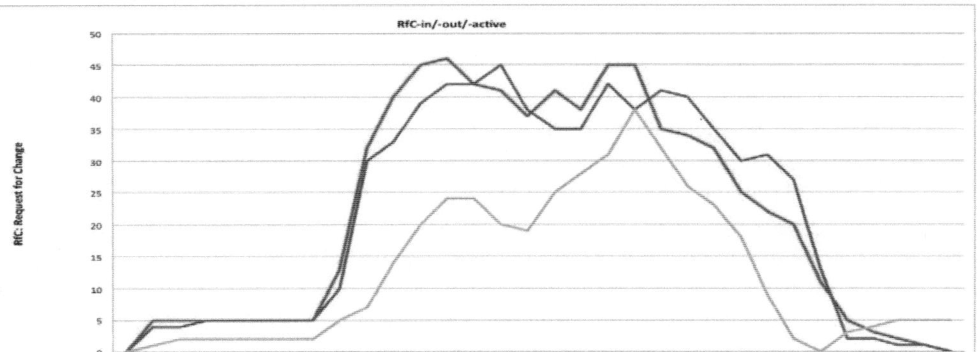

	-36	-35	-34	-33	-32	-31	-30	-29	-28	-27	-26	-25	-24	-23
	FRS								PoC-S					
RfC-in	0	5	5	5	5	5	5	5	13	32	40	45	46	42
RfC-out	0	4	4	5	5	5	5	5	10	30	33	39	42	42
RfC-active/open	0	1	2	2	2	2	2	2	5	7	14	20	24	24

	-22	-21	-20	-19	-18	-17	-16	-15	-14	-13	-12	-11	-10	-9	-8	-7	-6	-5
	FRS-S												OB-S					EMR
RfC-in	41	37	41	38	45	45	35	34	32	25	22	20	11	5	3	2	1	0
RfC-out	45	38	35	35	42	38	41	40	35	30	31	27	13	2	2	1	1	0
RfC-active/open	20	19	25	28	31	38	32	26	23	18	9	2	0	3	4	5	5	5

Nomenclature: RfC: Request for Change; RfC-in: Total inflow of new RfC; RfC-out: Total implemented or rejected RfC's; RfC-active/open: Total amount of RfC-in - RfC-out.

Figure 4. RfC-in/-out/-active [8], [11]

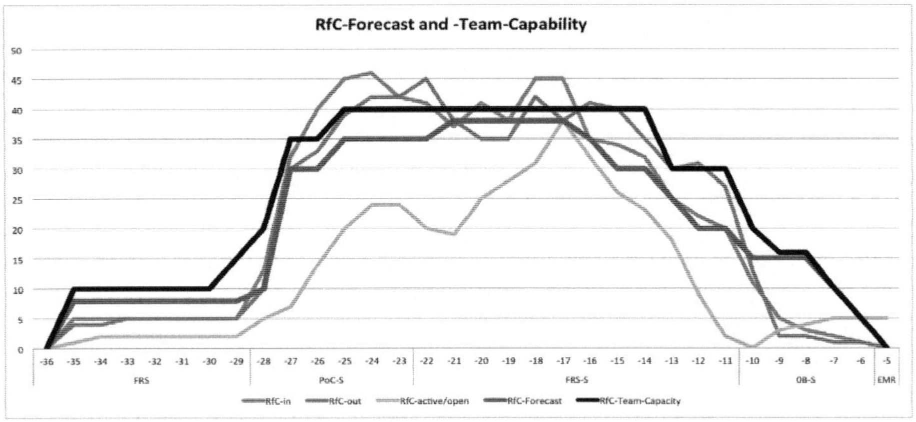

Figure 5. RfC-Forecast Model [8], [11]

5.3. Risk-Management Assessment via RfC-Team-Capacity Model

The shape of the simulated RfC in-/out-flow graph shown in figure 4 (RfC-in/-out/-active) can be observed reviewing the data of several representative automotive projects [8], as stated in the previous chapter. Utilizing the data of such projects it is possible to construct the forecast model, most likely RfC inflow. Ideally the shape of the forecast curve must not be adjusted over the life cycle of the project, which is also a weakness of the Forecast-Model. The Forecast-Model take not into account the typical alterations of e.g. human resources and skills available. In order to improve the forecast model a "Human Resource" graph could be added to show the maximum team capacity to deal with the project requirements, i.e. "Human Resource Management Process" [10].

Plainly the graph must show the maximum human resources and skills available over time. Thus, the resulting team capacity must be able to produce the required RfC outflow to meet the project milestones. Introducing the described risk management control mechanism into figure 4 will conclude to figure 5 (RfC Forecast and Team Capacity). The data for forecast curve is idealized and leads to the shown curve. The data for the Team Capacity were derived as follows: During the early and the final phase of the project the Team-Capacity was set to equal or slightly larger values as the RfC-Forecast. During project phases were the probability growths that the RfC-inflow increases as described in chapter 'Timeline and Constraints of valid RfC inflow' the value of the Team-Capacity was set to at least 10% higher than the value given by the forecast as a save guard. The resulting data should be reused in the project brief, which "assures the providing of a supply of skilled and experienced personnel qualified to perform life cycle processes to achieve organization, project and customer objectives." [10].

5.4. Risk-Management or Control of RfC inflow via Priority and Impact-Analysis Model

The Team-Capacity-Model curve states the maximum inflow of new RfC's, which the total available project members at a specific time can handle. All previous discussed simulations only view amount of new changes requests, but not the impact of the bug or new requirement regarding: a) Market significance or consequences of introducing the feature at a later release; b) Severity of the SW/HW Bug to the customer or vehicle; c) Functional specialist Risk assessment (e.g. lines of code, amount of involved software building blocks); d) Project-Management Evaluation of Schedule impact, such as Time required introducing the complete change request until final system validation; e) Remaining time to until SW Freeze or SW stabilization; f) It is important to investigate the impact of an RfC to possible involved Network-Partners (see Figure 1) as discussed in the previous chapters; g) Implementation cost.

The question is therefore how to improve the discussed models to take the above stated possible arguments in to account, adding the impact of each argument to each single RfC. This Impact-Analysis (IA) is identifying or estimating the potential consequences and modifications to realize the change request.

Table 1 shows an example of Impact-Analysis (IA) items (R-IA 1 to R-IA-6) and the corresponding assessments focus for each impact analysis item; whereas each items links to a rating or score (IA Score):

a) R-IA 1, R-IA 2 and R-IA 6 (IA: 4) gives the project management team a risk evaluation if the RfC is not implemented, with respect to market issues and to customer complaints. If the IA results into the highest impact sore, there is are very limited choices of not implementing the RfC as reasoned and shown in Table 1. The Change Management (CM) Work Flow (see Figure 3a.) suggests how to systemize or schedule the impact analysis R-IA 1 and R-IA 2. Responsible for the scoring is the stakeholder who initiates the RfC.

b) R-IA 3 and R-IA 4 gives the project management a risk evaluation if the RfC is implemented, with respect to engineering or SW/HW implementation effort, including involving of network partner. The CM Work Flow (see Figure 3a.) suggests how to systemize or schedule the impact analysis R-IA 3 and R-IA 4. Responsible for the scoring is the technical group that evaluates the RfC from the OEM point of view (see CW workflow step '2' and '3'). The scoring is might be adjusted after the suppliers review of the RfC via the supplier technical specialist (see Figure 3b, CM process step '5')

c) R-IA 6 states the implementation cost. It is of vital importance that a new requirement is only realized after commercial agreement; otherwise the HW/SW implementation work might be stopped or even dismantled. Loss of budget and resources are the consequence. The implementation costs are part of the CMS/CI tracking.

Table 1. Impact-Analysis (IA) Items and Assessments [11]

Risk (R)	IA Focus	IA Priority: Assessments Description	IA Score
R-IA 1	Market significance of the new Requirement	1. Immediate: *Meets legislative requirements"* [2]. 2. High: "*Responds to short term market opportunities or public requirements; Supports new business initiatives that will increase company market position"* [2]. 3. Medium: "*Maintains business viability; Supports planned business initiatives"* [2]. 4. Low: "*Improvements in usability of a service"* [2]	1. IA: 4 2. IA: 3 3. IA: 2 4. IA: 1
R-IA 2	Severity of the SW/HW Bug	1. Immediate: Car Immobility expected (low battery), *Putting life at risk* [2]. 2. High: Customer complaint to be expected; *Severely affecting some key users, or impacting on a large number of users* [2]. 3. Medium: "*Severely affecting some key users, or impacting on a large number of users* [2]. 4. Low: "*No severe impact, but rectification cannot be deferred until the next scheduled release or upgrade"* [2].	1. IA: 4 2. IA: 3 3. IA: 2 4. IA: 1
R-IA 3	Technical Evaluation	1. Immediate: High and Network-Partner are involved 2. High: High SW/HW implementation risk, without Network-Partners involvement or else jump to the next risk level. 3. Medium: Medium SW/HW implementation risk, without Network-Partners involvement or else jump to the next risk level. 4. Low: Low SW/HW implementation risk, without Network-Partners involvement or else jump to the next risk level.	1. IA: 4 2. IA: 3 3. IA: 2 4. IA: 1
R-IA 4	Project-Management Evaluation of Schedule impact*	1. Immediate: RfC implementation during and after the stabilization phase. 2. High: RfC implementation during the Preparation of Production phase, but before the SW/HW stabilization phase. 3. Medium: RfC implementation during the Concept Validation-Development phase 4. Low: RfC implementation during the Concept-Development phase *RfC SW/HW realization, test and final system implementation.	1. IA: 4 2. IA: 3 3. IA: 2 4. IA: 1
R-IA-5	Network-Partner are involved	1. RfC Propagation: RfC on hold until Risk Analysis or Risk Impact of Network-Partner available. Critical Path Analysis: Worst Case Scenario! 2. No Network-Partner is involved	1. IA: 1 2. IA: 0
R-IA 6	Budget management	1. Required budget not available (no commercial agreement)! Consequently, the RfC is rejected, i.e. not realized. 2. Implementation cost evaluation necessary. Risk: Los of time due to evaluation and commercial agreement. 3. No Implementation cost, only control for CMS/CI tracking.	1. IA: 4 2. IA: 1 3. IA: 0

5.1. Combination of RfC-Team-Capacity Model and Impact-Analysis Model

RfC priorities [2] as explained for R-IA 1 and R-IA 2 are partially known in the IT. The impact priorities listed for R-IA 2 are common practice in the automotive developments. R-IA 1 and R-IA 2 states the severity of the problem to be solved if no solutions is found or implemented.

Figure 6 shows for each new RfC the impact evaluation result according to table 1, utilizing the R-IA 3 and R-IA 4 priorities. R-IA 3 and R-IA 4 are considering the impact from the SW/HW/System-implementation threats and schedule perspective inflicted to the team. The advantage of preassigned priorities to each RfC are as follows:

a) The project development team can operate the implementation and test order in accordance to the priorities, i.e. 1st the RfC with highest IA score.

b) For the configuration management the top score (immediate) RfC can influence the decision to release a SW or postpone a SW/HW release until this particular problem is solved, depending as well on the issue. However, the pre-marked top problem is marked to be especially treated with to priority and easily traceable.

c) From the shape of the simulated Team-Capacity graph shown in Figure 6 (RfC-in/-out/-active) in combination with the total impact analysis can be observed that if the maximum calculated impact comes near by or even crosses the Team-Capacity border appropriate management activities neat to be in place. "Resources utilization against capacity" [2] needs to be considered. Figure 6 shows that for month '-26' until '-17' the resources will be sufficient to release all RfC's.

Consequently, several obvious management decisions are possible:

a) The RfC' marked with the Low priority need to be postponed or re-evaluated to reject candidates.

b) The team capacity needs to be upgraded until the month '17' (see Figure 6).

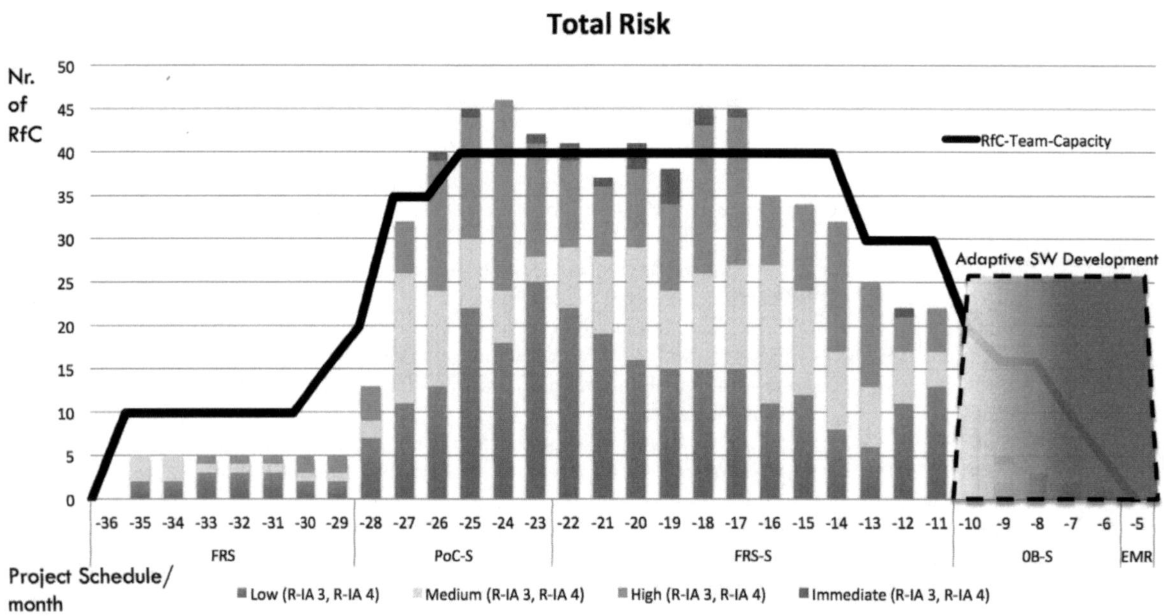

Figure 6. Total (Implementation) Risk-Analysis and Team-Capacity (adapted after [8], [11])

6. Conclusions

6.1. Configuration Management System (CMS) and Change Management (CM)

The discussion suggests that the 'Change Management' (CM) process" in automotive ECU development projects starts directly after the freeze of Final Requirement-Specification (FRS), which is the basis of the contract between OEM and supplier. The real benefit of the CM process is to collect the data describing the changes request and guide them through all the CM transitions steps necessary to prepare the information required to take the adequate decisions with the best possible confidence and benefit for the business.

The CMS and CM process secures the effective handling of RfC's from the initial need to change a requirement until the delivery and test of the adopted functionality. Summarizing the services of the described CM workflow as shown in figure 3a-c will lead to the following conclusion: The CM Process is a service, which collects all the predefined information at defined transition steps, as well as triggering supportive processes such as requirement-, configuration-, risk-management (impact analysis) and CCB (Change-Control-Board). The specification of the data required depends on the specific field of development. The RfC information (CI) required and collected by the CMS service should be defined by the (technical and project lead) project team. Involving the team in defining the CM process will serve to define the best practice relegated RfC questionnaire and getting familiar with the goals and purpose of the service.

The CMS/CM service requires to be maintained via a change request manager, which is required for both parties, for the OEM and the supplier to lead all stakeholders along a defined CM steps, providing data or information at the right time to the matching RfC stakeholder(s).

6.2. Risk Management

The shown Risk-Management Model depends completely on the quality of information (i.e. CI) gathered by the CM service. The Forecast- and Team-Capacity processes depending also on the data gathered of previous projects and on the team member transferring the data to the new project (-brief). The Impact-Analysis Model depends especially on the engineering expertise of the stakeholders judging and deciding on the IA score. Clearly, this is the weakness of all risk models discussed, for example the impact score decision might not be reproducible for any specific RfC. The suggested remedy for the IA scoring of 'R-IA 3 Technical Evaluation' and 'R-IA 4 Evaluation of Schedule impact' is the review of the scoring (see Figure 3b). However, the marking of 'R-IA 1 - Market significance' and 'R-IA 2 - Severity of the Bug' are depending on personal experience and possibly fondness on a use-case or customer feature.

However the Risk-Management Models are still valid supporting tools for the project-lead and especially for the CCB to judge and to decide to implement or reject the RfC. The Team-Capacity Model is particularly important at the end of the project timeline, since misjudgments of RfC inflow rates are more difficult to compensate. In addition, the graph of figure 5/6 could help at this point to argue against reduction of resources if the total amount of new of RfC requires the opposite.

6.3. Boundaries of the discussed models

Considering the following Use-Case of new requirements (RfC), with an Impact-Analysis (IA) assessment results (according to Table 1):

a) R-IA '1' (Market significance of the new Requirement) with IA: 3 (High: Responds to short term market opportunities) and

b) R-IA '4' (Project-Management Evaluation of Schedule impact) with IA: 3 (Immediate: RfC implementation during and after the stabilization phase).

If the CCB decides to pro-actively to integrate the RfC, due to the must have change to satisfy the strong demand of market (i.e. to ensure commercial success), the models can support with CM or Risk analysis processes, but not with the successful integration and final test which fits into the former explained automotive project life cycle concept of to meet the defined quality targets which can only be confirmed with long-term endures-test. Thus, during the stabilization phase the integration of a RfC require a revised software design and test process with will minimize the risk of lost development an endures-test time.

IOP Conf. Series: Materials Science and Engineering **163** (2017) 012012 doi:10.1088/1757-899X/163/1/012012

6.4. Further research (Adaptive Software Development Phase)

The key issue of infotainment systems is that there are embedded within the 4 years car development cycle against a one-year update cycle of an aftermarket devises such as a smart phones. Unfortunately the smart phones are already part the infotainment system. From personal automotive project experiences, the discussed CRM methods are useful for requirement changes, which appear before stabilization phase or 0-Bug sample (see Figure 2). The discussed methods are beyond those milestones still valid to give support to judge the risk, but this does not help the team to deliverer the most wanted feature (RfC) after the planned 0-bug sample; and to be integrated for SoP. Further research is necessary to establish "Adaptive Software Development" methods at the beginning of the automotive development life cycle.

References

[1] K. Rouibaha and K. R. Caskey 2003 Change management in concurrent engineering from a parameter perspective, *Computers in Industry* **50** 15-34

[2] ITIL3 *Service Transition* 2007 ISBN 978 0 11 331048 7

[3] ITIL® *glossary and abbreviations* English 2011 http://www.itilofficialsite.com/InternationalActivities/ITILGlossaries_2.aspx, ITIL, www.itil-officialsite.com/InternationalActivities/TranslatedGlossaries.aspx, ITIL®, Last access, 12.03.2014

[4] Sommerville I 2004 *Software Engineering*, 6th Edition, Addison-Wesley

[5] Kotonya G and Sommerville I 1998 *Requirements Engineering - Processes and Techniques*, First Edition, Wiley Publishing, Chichester, UK

[6] https://www.telemotive.de/fileadmin/user_upload/content/Grafiken/Engineering/Bussysteme.pn g, last access 06.06.2016

[7] Ebert C 2007 *Optimizing Supplier Management in Global Software Engineering*, IEEE Computer Society Press

[8] ***Primary Source: Automotive OEM. The OEM Terminology, graphics and data are altered; the original OEM Company-Name, milestone expressions and schedules are not used due to strict arrangements of confidentiality. The examples are based on personal experience

[9] Volker S, Prostean G and Hutanu A 2016 Research of Automotive Change Management and Supportive Risk Management, *Soft Computing Application* **357** 1097-1108

[10] ***INTERNATIONAL STANDARD 2008 *Systems and software engineering - Software life cycle processes*; ISO/IEC 12207; IEEE Std 12207-2008; Second edition 2008-02-01

[11] Volker S and Prostean G 2016 Research of Automotive Change Management and Combined Risk-management Models, *Procedia - Social and Behavioral Sciences* **221** 395–404

IOP Conf. Series: Materials Science and Engineering **163** (2017) 012013 doi:10.1088/1757-899X/163/1/012013

Impact of agile methodologies on team capacity in automotive radio-navigation projects

G Prostean, A Hutanu and S Volker

Politehnica University of Timisoara, Faculty of Management in Production and Transportation, Remus str., no. 14, 300191 Timisoara, Romania

E-mail: andreihutanu@yahoo.com

Abstract. The development processes used in automotive radio-navigation projects are constantly under adaption pressure. While the software development models are based on automotive production processes, the integration of peripheral components into an automotive system will trigger a high number of requirement modifications. The use of traditional development models in automotive industry will bring team's development capacity to its boundaries. The root cause lays in the inflexibility of actual processes and their adaption limits. This paper addresses a new project management approach for the development of radio-navigation projects. The understanding of weaknesses of current used models helped us in development and integration of agile methodologies in traditional development model structure. In the first part we focus on the change management methods to reduce request for change inflow. Established change management risk analysis processes enables the project management to judge the impact of a requirement change and also gives time to the project to implement some changes. However, in big automotive radio-navigation projects the saved time is not enough to implement the large amount of changes, which are submitted to the project. In the second phase of this paper we focus on increasing team capacity by integrating at critical project phases agile methodologies into the used traditional model. The overall objective of this paper is to prove the need of process adaption in order to solve project team capacity bottlenecks.

1. Introduction

The development of automotive radio-navigation projects became even more complex because of the impact of peripheral components development processes. The integration of new communication protocols, which are used for new developed smartphones or IT components, into automotive project raised the necessity to harmonize the processes used in the development of different system components area. As the project has a strict deadline and the success factor depends also on the number of resources involved, not only the harmonization of processes will decide the success of projects. In order to achieve project success by implementing the critical factors like time, cost, scope and innovation, the project leaders have to be careful on the influence of changes on project requirements. Applying changes on requirement will bring confusion and delays in such complex projects.

The objective of this paper is to analyse potential process adaption and their influence on team capacity while requirements changes in a fast manor. The requirement changes in projects using traditional development models endangers project objective. Current literature review shows that the main objective of study is the adaption of development processes used, without taking into consideration the team capacity factor. By assuring that the initial team capacity stays under the planned level, the financial factor will not overtake the initial planning. On the other side, by not fulfilling project capacity, initial planning will lead to the necessity of cutting project scope or opening an update-project.

2. Development models applied in complex automotive systems

Different automotive projects components derived from various areas implies the integration of the different development models used. While the V-cycle development model is used in the development of traditional in car ECUs (Electronic Control Unit), the peripheral components like servers, smartphone integration are developed using agile methodologies. The main differences between the two main development methodologies are presented in Table 1.

Table 1. Difference between traditional development model and agile methodologies (adapted after [1])

	Traditional	**Agile**
Fundamental Assumptions	Systems are fully specifiable, predictable, and can be built through meticulous and extensive planning. In automotive projects the planning of full implementation has its roots in production processes.	High-quality, adaptive software can be developed by small teams using the principles of continuous design improvement and testing based on rapid feedback and change.
Control	Process centric	People centric
Management Style	Command-and-control. Every sub-area of one automotive project has a leader whose responsibility is to plan, command and track the project.	Leadership-and-collaboration
Knowledge Management	Explicit	Tacit
Role Assignment	Individual—favours specialization	Self-organizing teams—encourages role interchangeability
Communication	Formal	Informal
Customer's Role	Important, still in automotive projects some decisions are taken even if customer does not agree to them.	Critical
Project Cycle	Guided by tasks or activities	Guided by product features
Development Model	Life cycle model (Waterfall, Spiral, or some variation). In automotive projects most often the V-cycle model is used.	The evolutionary-delivery model
Desired Organizational Form/Structure	Mechanistic (bureaucratic with high formalization)	Organic (flexible and participative encouraging cooperative social action)
Technology	No restriction. Particular to radio-navigation system projects, the object-oriented technology is used.	Favours object-oriented technology
Acceptance of changes	*Rigid, requirement changes are hard to implement*	*Flexible, easy to implement changed or new requirements*

The use of traditional development models in automotive projects is based mainly by historical car manufacturing reasons. The automotive software projects appeared on a later time than production processes. The inexistence of software processes pushed the project leaders to use production processes, which are based mainly on the V-cycle model.

In comparison with automotive traditional development models, the automotive peripheral components are developed using agile methodologies. According to literature the most popular agile methodologies used are:

- SCRUM;
- Extreme Programming;
- Crystal;
- ASD (Adaptive Software Development).

3. Change management methods applied in automotive projects

During our research we found several papers regarding change management. There are some opinions on requirement changes during project development. For example Muhammad Nabeel Mirza et al. [2] said that a requirements should not change during project development, while De Bakker et al. [3] affirmed that initial requirements will certainly change. From our practice experience and also according to Milosevic D. and Patanakul P. [4], citing Eisenhardt [5], said that a fast developing environment abounds in fast requirements changes.

The main reason for requirement changes are:

- Changing technologies during project development. "The life cycle of products does not respect anymore the profile of the classic curve, this curve turned in the profile of a saw tooth. Before ending the phase of introducing the product on the market, the product is considered already outdated" [6];
- Unclear specification [7];
- Incomplete specification [7];
- "New requirements due to must have changes to satisfy the (strong) demand of market to ensure the later commercial success" [8];
- Improvement for robust design. [8];
- Project has a big temporal pressure [9];
- Project client does not want to inform all project details in the early phases of the project [9];
- Developing an innovative and complex system for the first time [9].

While changes will certainly occur during project development and organization of changes needs to be done, the practice taught us that change management processes need to be used. It has to be distinguished at this point between change management - and error tracking processes as "the tracking of errors is a separate process, with the objective" [10] "to ensure that all discovered problems are identified, analysed, managed and controlled to resolution" [9]. The purpose of the change request management process is to ensure that change requests are managed, tracked and controlled" [9]:

- "Managing the approved changes when and as they occur, by regulating the flow of requested changes" [11];
- "Controlling and updating the scope, cost, budget, schedule and quality requirements based upon approved changes, by coordinating changes across the entire project" [12].

The decision to implement a change request is taken mostly depending on the following factors [7]:

- "Market significance";
- "Number of lines of code which has to be modified";
- "Number of modules which have to be adapted in order to implement the change";
- "Time required for complete change request until final system validation".

Beside controlling and documenting changes, the change management process has the role to minimize project risk. Taking into consideration the tight project schedule of automotive project, the later the changes appears into the project, the greater will be the implementation risk. Particular to the automotive industry, the objective of change management process "is to establish and sustain a relationship with the stakeholders, to support all with up-to-date information such as the RfC (Request for Change) transition status (i.e. reject, planned, implemented) and impacts to the initial agreed objectives, which are features, budget, schedule and quality" [8].

Volker et al. [12] described the change management process applied in automotive projects according to Figure 2. The same way the effects of changed requirements have been described by Prostean and Volker [8]. By changing requirements the whole project structure is "under attack".

Nomenclature: RfC: Request for Change; RfC-in: Total inflow of new RfC; RfC-out: Total implemented or rejected RfC's; RfC-active/open: Total amount of RfC-in - RfC-out.

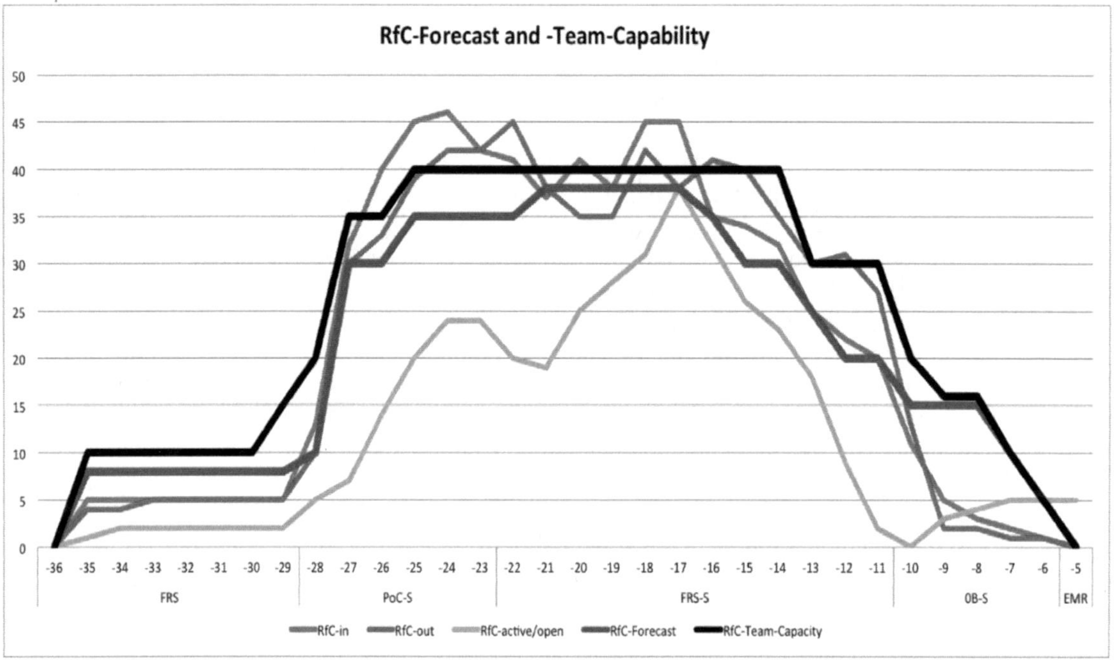

Figure 1. RfC-in/-out/-active [8]

First of all, the project timeline is endangered and automatically the risk of overtaking project budget increases with every changed requirement. In the practice of automotive projects not holding project milestone can have an immense impact on one company's business. The risk grows even more by delaying the market entry of the product which can lead to the loss of market shares, as the product does not fulfil the innovation characteristic. For this reason before starting the change management process, the project responsible have to estimate the number of change requests which can be implemented by current project team and structure (Figure 1).

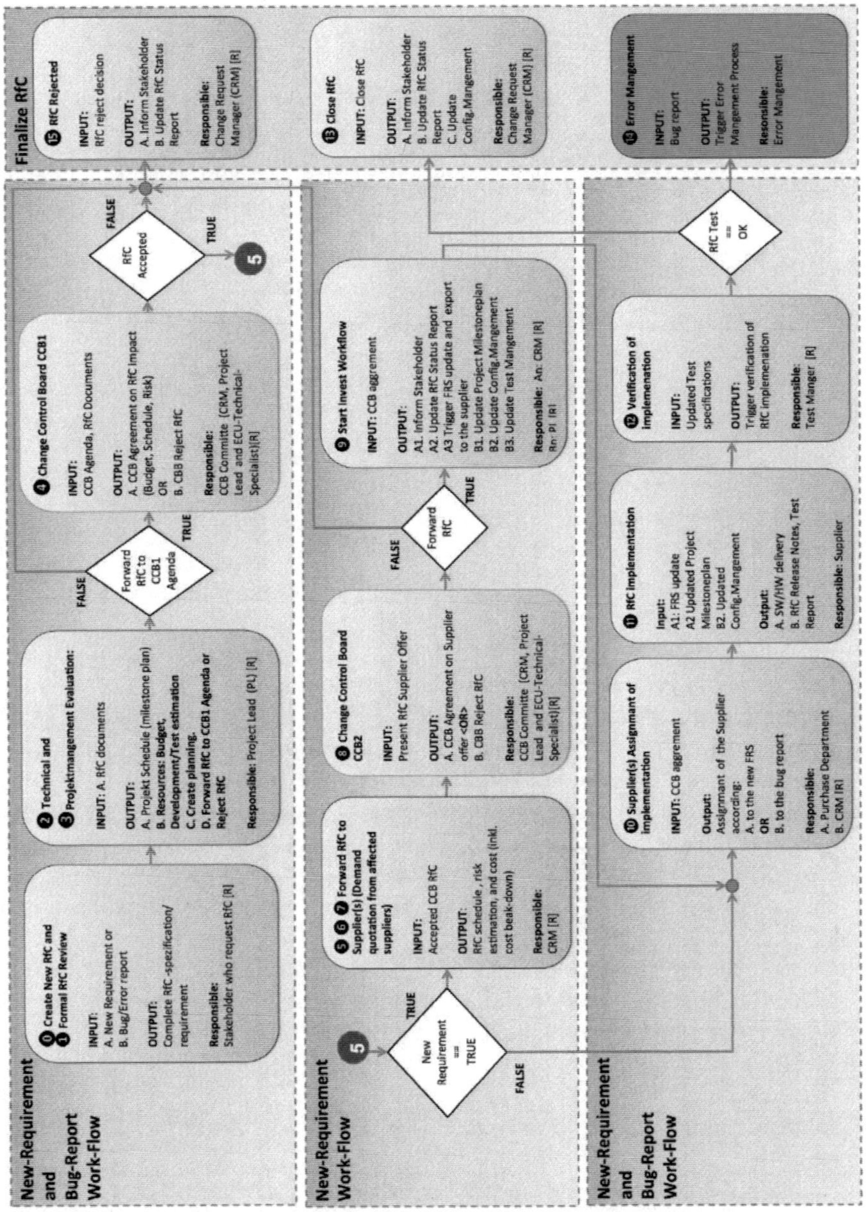

Figure 2. Automotive Change Management Process [12]

As team capacity is essential in reaching project goals, the authors tried to find solutions to protect the team capacity workload. The authors observed in their practical experience that a project cannot be successful if the team capacity is overwhelmed for a long period of time. Figure 3 presents the impact of RfCs (Request for Change) on team capacity during project development.

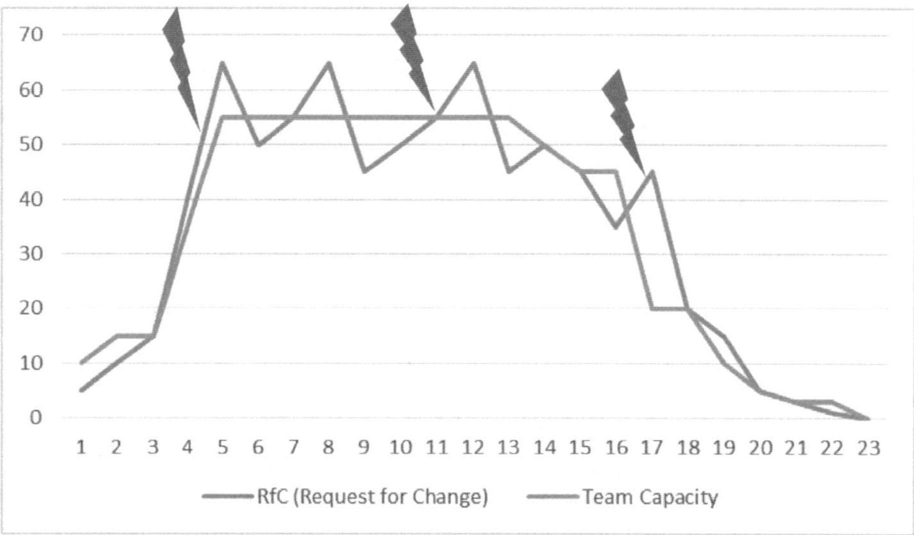

Figure 3. Team Capacity compared to RfC-input (adapted after [8])

4. Improvement of team capacity by applying agile methodologies

While the most automotive project are struggling keeping initial project planning, we identified a way to reduce team capacity workload by keeping the same amount of requirements. As automotive products have to preserve fix delivery dates to production, the first condition is to keep the actual structure of models used. As the greatest challenge is the fast evolution of requirements which tend to become obsolete [13], [14] the authors recognized the solution with respect to the project constraints:

- Keep the structure of traditional v-cycle model;
- Other methods like e.g. Goldratts Critical Chain [15] does not bring the required improvement.

Based on the V-cycle model the authors changed the essential phases of the traditional model with agile methodologies. The objective is to reduce the implementation of unnecessary software. The phases changed in the traditional V-cycle development model are:

- Requirement specification;
- System specification;
- Architectural design;
- Implementation.

The agile methodologies used in the new development model are:

- ASD (Adaptive Software Development), the collaboration phase;
- SCRUM.

The main advantages of using agile methodologies in the phase of requirement-, system specification and architectural design are:

- All requirements are matched with customer wish;
- The architecture is build change tolerant;
- All requirements are implemented following a priority list agreed with the customer;
- Better collaboration between project teams and between customer and project supplier;
- Better customer monitoring on current project status by participation on the SCRUM sprint reviews [16];
- Better possibility to predict and prioritize which requirements have to be implemented first.

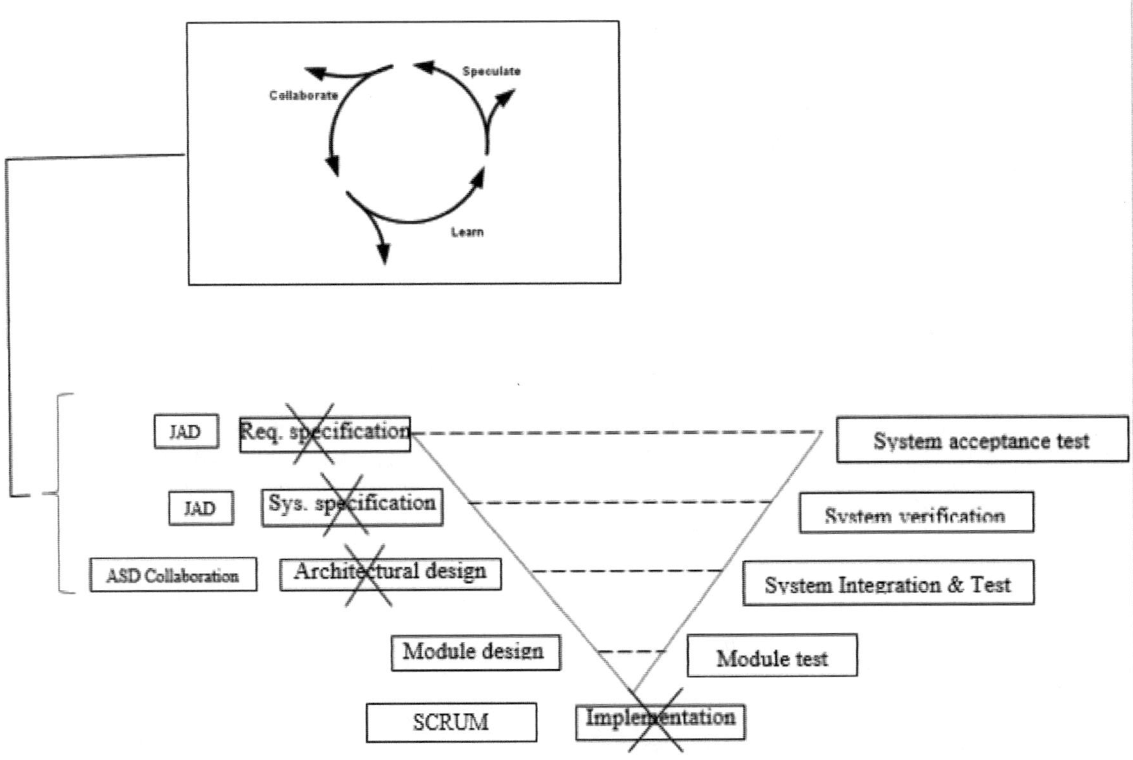

Figure 4. V-cycle – Agile development model

5. Discussion on integration of agile methodologies into traditional development models

The new resulted development model brings several advantages with it. Still the new development model has to prove his usefulness by removing the disadvantage of traditional models and solving the defined problem. In this particular case, the issue followed was the protection of team capacity while the project is confronted with a lot of requirement changes. The way of verifying models effectiveness is to compare the performance with regard to team capacity of the v-cycle model and the integrated agile model.

We decided to use this two models in two different projects and analyze the impact of changed requirements on team capacity. The first validation phase of the model was the evaluation of the first two specification phases (Figure 5). We compared every sub-phase and noticed that the project running with the integrated agile methodology managed to save 20.7% of time. By involving the client in the phase of specification, the project team saved time in the sub-phase of requirement analyzation. It is a known practice method that while using agile methodology the requirement phase is accompanied by supplier and customer. The savings arise mostly from the sub-phase of requirement analysis, requirement review and requirement adaption as a result of requirement review.

Our result was confirmed also by literature review, where savings have been measured between 15% [17] and 75% [18].

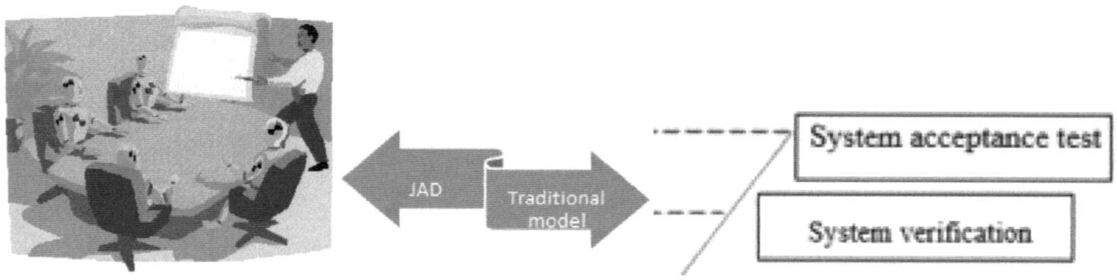

Figure 5. Comparison of requirement specification phases (only picture from [20])

In the next phase we compared the phases of architectural design (Figure 6). Even if according to our measurements applying agile methodologies did not save time, we faced the advantage of building a change tolerant architecture which allows future requirement changes.

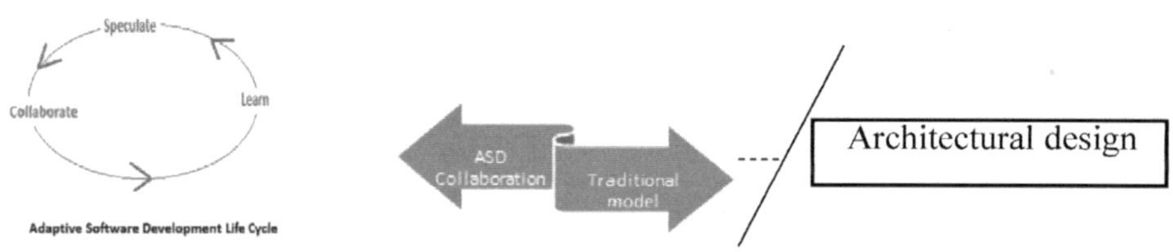

Figure 6. Comparison of architectural design phase (only ASD circle from [19])

Most important and time consuming phase in the development of project is the implementation phase. While being in the construction phase of the project, the implementation order of requirements in V-cycle model projects where not determined by strategic grow of the project. On the other hand, while using the SCRUM methodology the prediction of risk of change played a major role for the future development of the project (Figure 7). By implementing at first the requirements with a low risk of change, the amount of scrap software was reduced to the minimum. Another reason for time saving was the agreed requirement implementation with customer and less task interruption. This kind of approach allowed project team to estimate an improvement of capacity by 20%. As in automotive projects the time allocated for the implementation phase is fixed, the time saving is used for the implementation of changed specification.

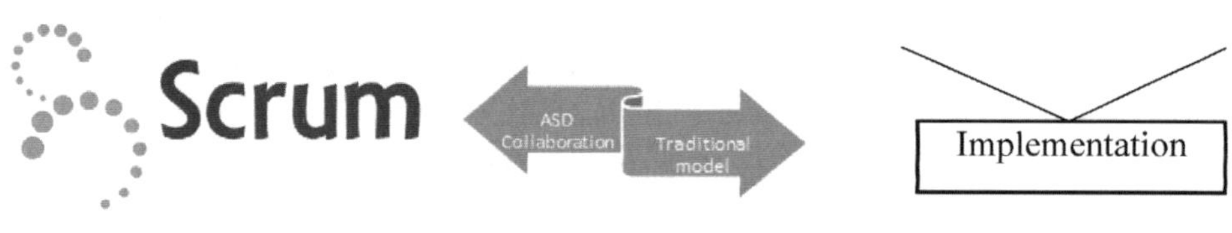

Figure 7. Comparison of implementation phase [21]

6. Conclusion

Our practical experience in project management using traditional development models and expertize on critical issues led us to this study. We observed in the past few years a continuous change of automotive projects environment and integration of even more peripheral components with different lifecycles and development models used. Additionally these differences bring with them a continually increasing amount of requirement changes in automotive projects. Car manufacturer projects are not intended to support a lot of requirement changes, not only due to the high quality or lifetime endurance standards and expectations of customers. Every requirement change brings a high pressure on team capacity, this being the reason for changing the project approach and integrating agile methodologies into automotive projects. Main advantages of using agile methodologies are:

- Project time savings;
- Financial savings due to time savings;
- Team capacity improvements and better project disposition due to precise requirements and less scrap software implementation.

Further work should handle all sub-phases of the traditional v-cycle model and try to establish and integrate new methodologies in order to achieve specific project objectives.

References

[1] Nerur S, Mahapatra R and Mangalaraj G 2005 Challenges of Migrating to Agile Methodologies, *Magazine Communications of the ACM - Adaptive complex enterprises* **48**(5) 72-78
[2] Mirza M N, Pourzolfaghar Z and Shahnazari M 2013 Significance of Scope în Project Success, *Procedia Technology* **9** 722-729
[3] de Bakker K, Boonstra A and Wortmann H 2010 Does risk management contribute to IT project success? A meta-analysis of empirical evidence, *International Journal of Project Management* **28**(5) 493–503
[4] Milosevic D and Patanakul P 2005 Standardized project management may increase development projects success, *International Journal of Project Management* **23**(3) 181-192
[5] Eisenhardt K 1989 Making fast strategic decisions în high-velocity environments, *Acad. Manage J* **32**(2) 543–576
[6] Prostean G 2001 *Management Prin Proiecte,* Editura Orizonturi Universitare, Timisoara, Romania
[7] Hutanu A, Prostean G, Mnerie D and Badea A 2016 *Opportunity Analysis of Change Requests in Automotive Projects*, Soft Computing Application, Springer
[8] Prostean G and Volker S 2015 *Research of Automotive Change Management and supportive Risk-Management*, 13th International Symposium in Management SIM2015, Timisoara, Romania, October 9-10
[9] Hoermann K, Mueller M, Dittmann L and Zimmer J 2008 *Automotive Spice in Practice*, dpunkt.verlag GmbH
[10] Hutanu A, Prostean G, Mnerie D and Badea A 2013 Research of Requirements Constraints in Automotive Projects, *Procedia Technology* **9** 750-756
[11] ***2005 Project Management Body of Knowledge *TenStep Student Handbook*
[12] Volker S, Prostean G and Hutanu A 2016 *Research of Automotive Change Management and Supportive Risk Management*, Soft Computing Application, Springer
[13] DeBellis M and Haapala C 1995 User-centric software engineering, *IEEE Expert* **10**(1) 34–41
[14] Hood C, Wiedemann S, Fichtinger S and Pautz U 2008 *Requirements Management The Interface Between Requirements Development and All Other Systems Engineering Processes*, Springer, Berlin, Germany
[15] Goldratt E M 1997 *Critical Chain*, The North River Press Publishing Corporation, Great Barrington
[16] http://www.scrumguides.org/docs/scrumguide/v1/scrum-guide-us.pdf. Last accesed on 19th of June, 2016

[17] http://www.w3computing.com/systemsanalysis/joint-application-design, last accesed on 25.12.15
[18] http://www.lifecyclestep.com/browse/411.1.4JADSessions.htm, last accesed on 25.12.15
[19] http://wiki.expertiza.ncsu.edu/images/thumb/0/01/ASD_lifecycle.png/400px-ASD_lifecycle.png
[20] http://www.swengcon.com/meeting.gif
[21] http://ric.mclaughlin.today/assets/themes/ricify/images/scrum.png

IOP Conf. Series: Materials Science and Engineering **163** (2017) 012014 doi:10.1088/1757-899X/163/1/012014

Tribological properties of nonasbestos brake pad material by using coconut fiber

A L Craciun[1], C Pinca-Bretotean[1], D Utu[2] and A Josan[1]

[1]Politehnica University of Timisoara, Engineering and Management Department, Revolutiei Street, no. 5, 331128 Timisoara, Romania
[2]Politehnica University of Timisoara, Materials and Manufacturing Department, Mihai Viteazu str., no. 1, Timisoara, 3300222, Romania

E-mail: camelia.bretotean@fih.upt.ro

Abstract. In automotive industry, the brake system is influenced by a large number of variables including geometry of components, materials of brakes, components interaction and many operating condition. Organic fiber reinforced metallic friction composites are increasingly being used in automotive brake shoes, disc and pads, linings, blocks, clutch facings, primarily because of awareness of health hazards of asbestos. Current trend in the research field of automotive industry is to utilization of different wastes as a source of raw materials for composite materials. This will provide more economical benefit and also environmental preservation by utilize the waste of natural fibre In this paper it has performed a tribological study to determine the characteristics of the friction product by using coconut natural fibred reinforced in aluminium composite. In this sense, two different laboratory formulation were prepared with 5% and 10% coconut fibre and other constitutes like binder, friction modifiers, abrasive material and solid lubrificant using powder mettallurgy. These dnew materials for brake pads are tested for tribological behaviour in a standard pin on disc tribometer. To know the wear behavior of composite materials will determine the parameters that characterize there tribological properties.

1. Introduction

The most important element of braking system is friction material, which is expected to continue its functioning, reliably and efficiently for a prolonged time, in adverse operating conditions, [1]. Nowadays, factors such as road development and road traffic demand more efficient braking system which requires improved brake friction materials, [1].

In automotive industry, the brake system is influenced by a large number of variables including geometry of components, materials of brakes, components interaction and many operating condition. Friction behavior is the most critical factor in brake system design and performance. In specialty literature have been reported a great number of ingredients for being used to tailor the friction composites. These are classified into four major categories: binder, structural modifiers, friction modifiers and fillers. They contribute to controlling friction and wear performance. Nonasbestos organic fibre reinforced metallic friction composites are increasingly being used in automotive brake shoes, disc and pads, linings, blocks, clutch facings, and so forth, primarily because of awareness of health hazards of asbestos, [1].

The use of asbestos fibre as reinforcement in the friction materials started at the beginning of the 20th century. After that, this material became popular all over the world. Later on, medical reports regarding asbestos exposure to humans proved that asbestos is carcinogenic and can cause deadly diseases. This motivated researchers all over the world to find safe substitute for asbestos fibers. In paper [2], the authors list the following replacement materials for asbestos: calcium silicate, vermiculite (hydrated calcium aluminum silicate), aluminum silicate, basalt fiber, blast furnace slag, ceramic fiber, polyacrylonitrile, polyester, chopped glass fiber and aramid fibers.These components are used solely or in cocktail in friction formulations and research is continuing for more improved these materials, [3–5]. Current trend in the research field of automotive industry is to utilization of different wastes as a source of raw materials for composite development, [6].This will provide more economical benefit and also environmental preservation by utilize the waste of natural fibre, [7]. In this paper it has performed a tribological study to determine the characteristics of the friction product by using coconut natural fibred reinforced in aluminium composite.

2. Making samples for tribological experiments

In order to conduct this study were achieved iron discs and pin composite for disc - pad brake assembly. Iron discs play the role of brake discs of all real vehicle braking system and the pin`s play brake pad role. Tribological tests will be performed on a pin on disc tribometer, under dry friction.

2.1. The iron disc samples product

The iron discs sample were produced in "Molten Metal Laboratory" placed in the Faculty of Engineering Hunedoara. For this, it is use an oven of 10 kg capacity shown in Figure 1.

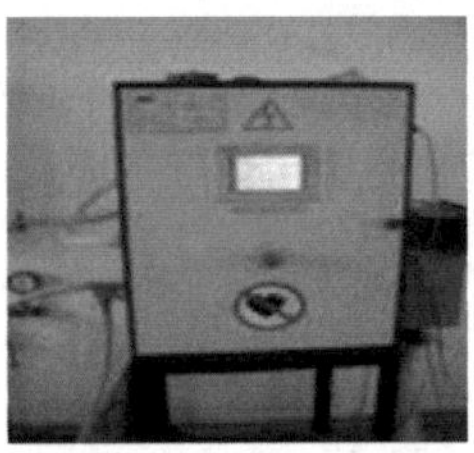

Figure 1. Oven for produced the disc sample

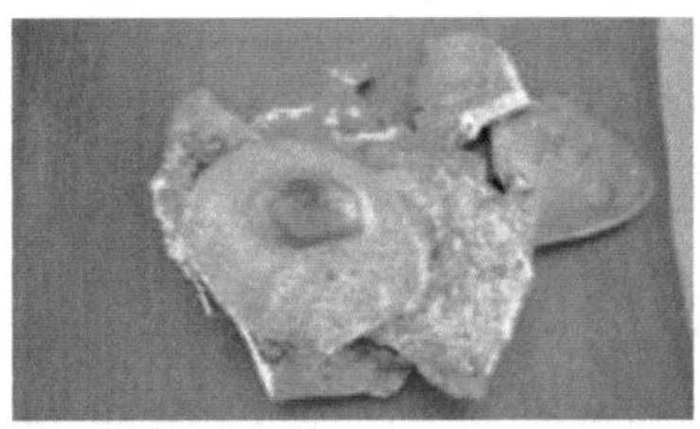

Figure 2. Metal load placed in the oven

Figure 3. Metal batch out of the oven

In the oven was placed brake discs, out of use and other constituents, like: 35 g coke, 25g FeSi and 15g FeMn, Figure 2. Melting duration was 90 min and after that the metal bath is discharge into a shape, Figure 3.

Figure 4. Cylindrical shape

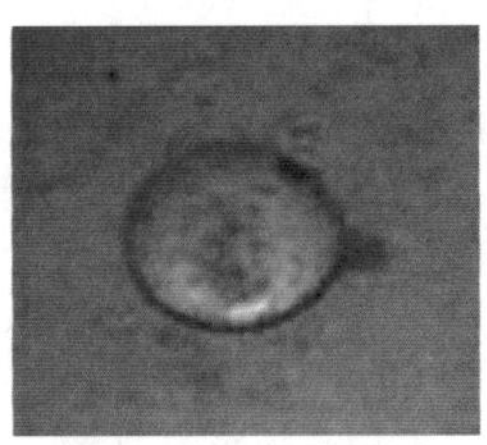

Figure 5. Iron disc sample results

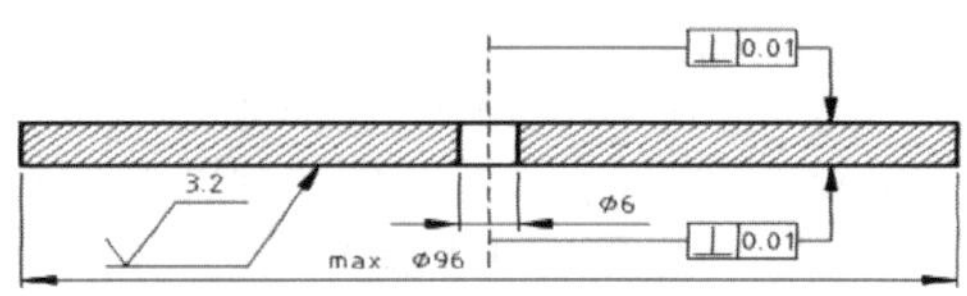

Figure 6. Disc sample size

The metal bath was dispensed into a cylindrical shape, shown in Figure 4 and results a disc sample which is shows in Figure 5. The iron disc sample size shows in Figure 6.

2.2. The composite pin product

In order to achieve the tribological experiments were performed composite pin`s. These samples together with disc samples will be mounted in the pin on disc tribometer. The composite pin will take cylindrical form. In this sense, it was designed and executed a mold shown in Figure 7. The pin`s size are presented in Figure 8.

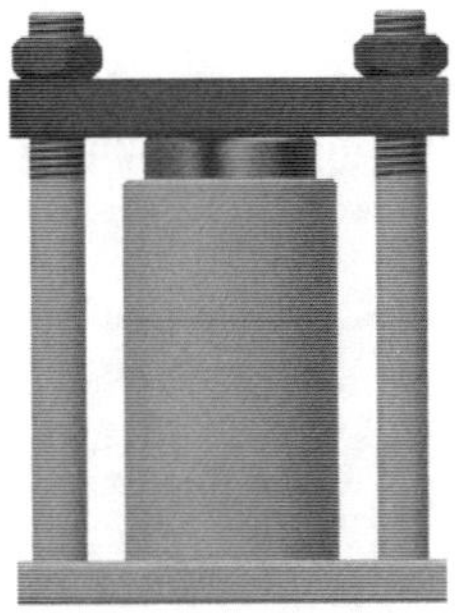

Figure 7. Mold for making the pin`s

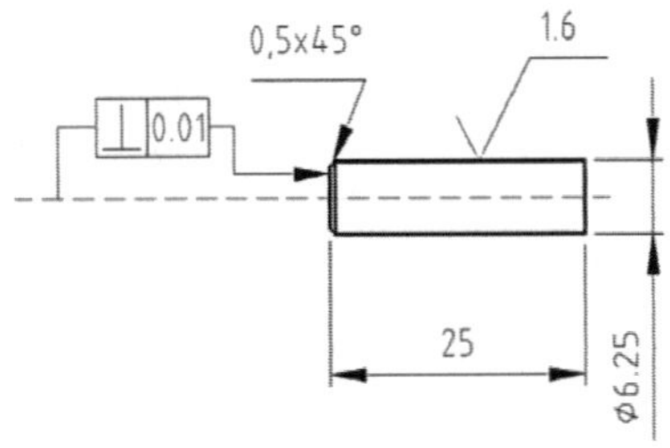

Figure 8. Composite pin size

Figure 9. The finished composite pin

The recipes of composites containing seven ingredients which maintaining five ingredients (around 75%) constant and varying two ingredients, aluminium and coconut fibre (around 25 %) in complementary manner as shown in Table 1.

Table 1. The recipes used in the product of the pin composites

Patern composite	Aluminium (%)	Graphite (%)	Zirconia oxide (%)	Silicon carbide (%)	Titanium oxide (%)	Phenolic resin (%)	Coconut fibre (%)
Composite1	20	10	2	10	13	40	5
Composite 2	15	10	2	10	13	40	10

The ingredients were mixed in a drum mixing machine to ensure the macroscopic homogeneity using a speed of 200 rot/min for 5 minutes. It was found that the mixing sequence and time of mixing of ingredients represents factors which influencing the fabrication of composites. If mixing time is low, proper homogeneity cannot be achieved. If it is too high, it does not improve the homogeneity further. The mixture was then placed into a mold and compress under a force of 10 N for 3 min. The mold with the sample was inserted into an oven and heat at 200^0 C. The time for keeping the sample in the oven was 150 min, after which cooling was carried out in air for 12 hours. Extracting the samples from the mold was done without sticking problems. Then the surfaces of the pin were polished to obtained a smooth surface. The finished pin is shown in Figure 9.

3.Tribological test by pin on disc tribometer

In order to determine the tribological characteristics, the iron disc and composite pin were mounted in pin on disc tribometer, Figure 10. Tribological experiments consists on pressing the composite pin on the surface of a rotating iron disc. Disk speed depends on sliding speed and range of work. The time of each test depends on the speed of sliding.

Table 2. Testing parameters

Iron disc speed (rot min^{-1})	Range of work (mm)	Sliding speed (m s^{-1})	Angular speed (rad s^{-1})
1500	25	3.92	157

Table 2 shows the values of parameters used in the tribological experiments.

For each pin composite were performed tests for four different lengths. Trial time for each experiment was determined by the sliding speed and the length of work.The experimental measurements are carried out with two forces of 5 N and 10 N. The weight and thickness of the samples were noted before and after the friction test. These are used to calculate the total wear of each sample. A thermography camera Thermo CAM Quick View was used to record the temperature of the contact interface during the test and readings were recorded every second. Captured images can provide information about the evolution of the temperature of the parts in contact. Figure 11 shows the disc – pin assembly and Figure 12 shows the auxiliary support for mounted the composite pin.

Figure 10. Pin on disc tribometer

Figure 11. Disc-pin assembly

Figure 12. Pin suport

4. Discusion and results

4.1. Characterization of iron disc sample

The chemical composition of cast iron used in this study is shown in Table 3. Thus, the carbon equivalent (CE) and the degree of saturation of carbon (SC) according to [8], have been determined by the relations (1) and (2). The calculated values for the two parameters are shows in Table 3.

$$CE = C + 0{,}3(Si + P) - 0{,}03Mn + 0{,}4S + 0{,}07Ni + 0{,}05Cr + 0{,}074Cu + 0{,}25Al \quad (1)$$

$$SC = C/[4{,}26 - 0{,}3(Si + P)] \quad (2)$$

Table 3. The chemical composition, carbon equivalent and the degree of saturation of carbon of iron disc sample used in the study

Chemical composition (%)												
C	Mn	Si	S	P	Cr	Ni	Co	Ti	V	CE	SC	HV_{10}
2.82	0.81	1.96	0.093	0.250	0.07	0.07	0.09	0.02	0.009	3.50	0.78	550

The low saturation level of carbon (SC) increase the proportion of perlite, the only constituent that can provide high strength for cast iron, [9]. The chemical composition of iron must ensure a small amount of graphite which is why it contains a low carbon. After modification, the iron structure is perlitic, because of the effect of the graphitized, [9].

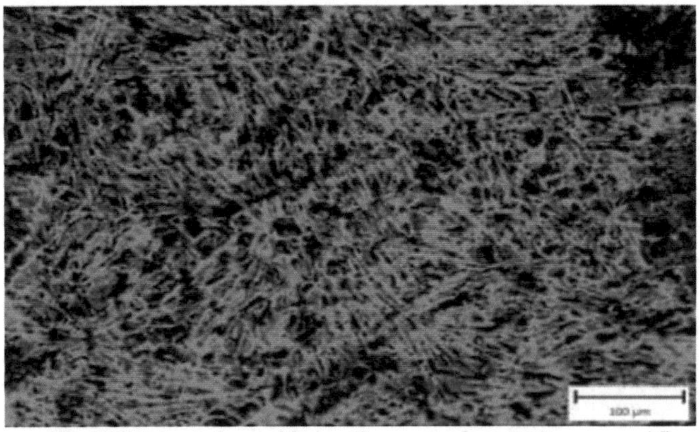

Figure 13. SEM imagine of iron disc sample - 100 x magnification

The microstructures of the produced disc samples were investigated using light microscope with magnification of 100x and is shown in Figure 13. The microstructure shows mainly type D flake graphite in a matrix of ferrite-perlite base, [8-10]. Graphite is smaller due to the phosphorus content (0.25%), [10].

4.2. Characterization of composite pin

A density measurement test has been carried out on a laboratory scale to examine the density of the new material obtained. The true density was determined by weighing each sample on a digital weighing machine and calculated their volume. The results shown in Table 4.

Porosity has an important role in automotive brake pad materials, [11]. The function of porosity is to absorb energy and heat, which is very important for effectiveness of the brake system. Theoretically, lower porosity conduct to higher friction coefficient and wear rate due to higher contact areas between the surfaces in contact. The samples was cut to a dimension of 25x25x7 mm. The structure should be preserved during sample grinding and polishing, Figure 14. Then, the samples were left for 24 hours in a desiccator at room temperature, Figure15. For the porosity test, the samples were weighed, placed in a container with oil and keep at 90^{0}C for 8 hours, Figure16. After that, the samples were left in an oil container for 12 hours, until the oil cools to the room temperature. Finally, the sample was rolled on a piece of cloth for remove oil from the sample. The sample was weighed again and the value is presented in Table 4.

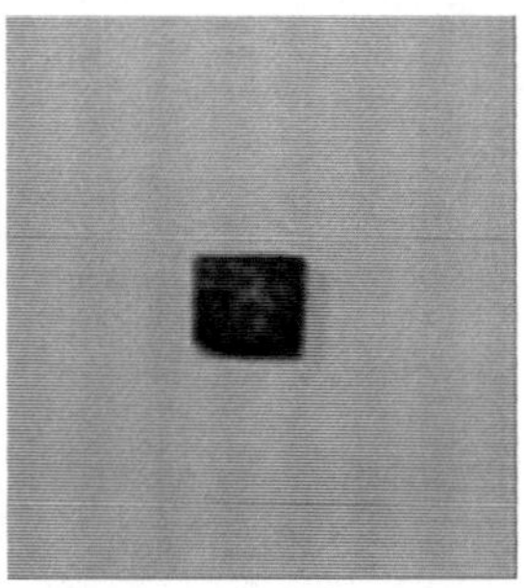

Figure 14. The samples for the porosity test

Figure 15. The sample placed in a desiccator

Figure 16. Sample placed in a container with oil

The compressive strength test was done using a universal testing machine type Zwich Roell Z005. The sample has the dimensions 25x25x7mm and was subjected to compressive force, loaded continuously until failure occurred. The load at which failure occurred was then recorded. Table 4 shows the results obtained after the experimental determinations for both composites.

Table 4. Dimensional characteristics, density, porosity and mechanical properties of composite pin

Height (mm)	Diameter (mm)	Density (g cm^{-3})		Porosity (%)		Compressive strength (MPa)	
		5 % coconut	10 % coconut	5 % coconut	10 % coconut	5 % coconut	10 % coconut
21.4	14	1.82	1.34	11.31	13.12	405.11	398.77

It is seen from Table 4 that the density of the 10 % coconut fibre composite shows lower than 5% coconut fibre.

The brake pads must have a certain porosity to minimize the effect of water and oil on the friction coefficient. The paper [12] found that increasing porosity by more than 10% could reduce the brake noise. From the porosity results as shown in Table 4 it can be seen that two brake pad formulations shows percentage more than 10 %.

The compressive strength value for 5 % coconut fibre composite shows higher compared with 10% coconut fibre composite.

4.3. Tribological results

Tribological test has the next objectives: influence of material factors, parameters influence of the operating mode (pressing force, sliding speed, time trial, the regime of friction) on the tribological characteristics of the sample tested, process analysis that occur in the superficial layers of the composite material testing. In this regard we will notice: the evolution of a parameter of wear and the evolution of the temperature in the contact area [13]. Finding the behavior of materials samples is by gravimetric method. During experiments disc sample and composite pin was weighted. For the iron disc will determine the weight loss and weight loss/route for each forces test. The results are presented in Table 5.

Table 5. Experimental results for the cast disc sample

Test force (N)	Initial mass (g)	Final mass (g)	Total road length (km)	Weight loss (g)	Weight loss / route (g km^{-1})
5	866.7119	866.5577	4	0.1542	0.03855
10	866.5577	866.4145	4	0.1432	0.03580

Table 5 shows that weight loss and weight loss/route for iron disc sample decreases with increasing workload. Table 6 shows test parameters and temperature values on the contact area.

Table 6. Experimental results for the composite pin

Test force (N)	Specific pressure (MPa)	Temperature (^{0}C)		Length test (m)	Trial time $t = L v^{-1}$(min)
		5% coconut	10 % coconut		
		40.1	60.4	500	2.12
5	0.032	43.8	74.3	1000	4.25
		44.0	74.8	1500	6.37
		32.5	61.1	2000	8.50
		47.3	74.8	500	2.12
10	0.064	55.2	75.8	1000	4.25
		56.1	61.1	1500	6.37
		56.0	60.4	2000	8.50

Also, were determined the variation in height of composite pin according to the trial time, with test force of 5N, Figure 17 and with force of 10 N, Figure 18. It can see that height loss is less for composite 2, with 10 % coconut fiber.

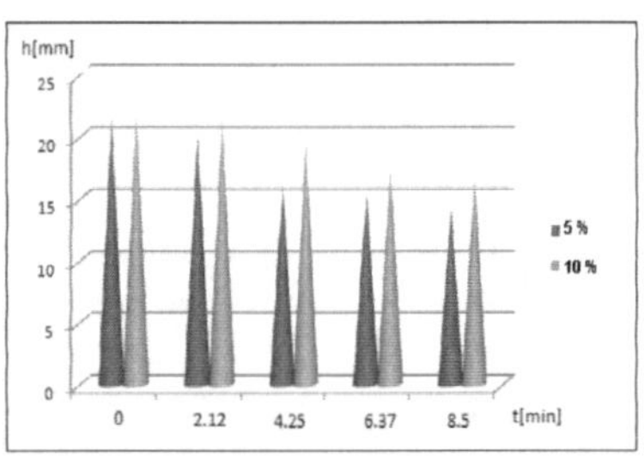

Figure 17. Variation in height of pin according to the trial time, F = 5N

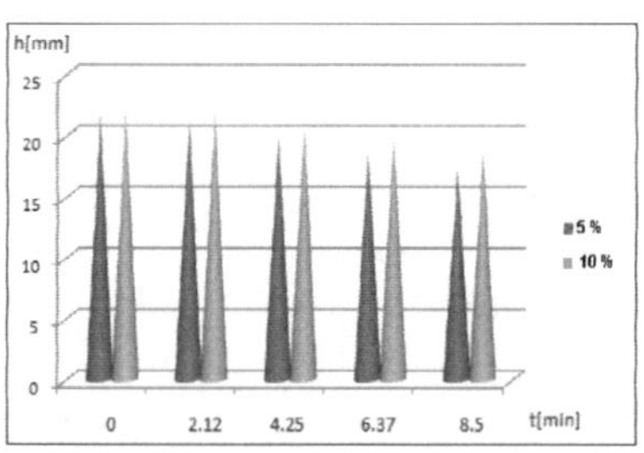

Figure 18. Variation in height of pin according to the trial time, F = 10 N

Figure 19 and Figure 20 shown the loos mass of composite pin according to the trial time for both force applied. It is observed that the mass loss increases with the trial time. For composite with 10% coconut fiber, the mass loss is higher.

Figure 21 and Figure 22 presents the linear wear test parameters with sliding distance using the normal force of 5N, respectively 10 N. The best wear reaction has the composite pin with 10% coconut fibre.

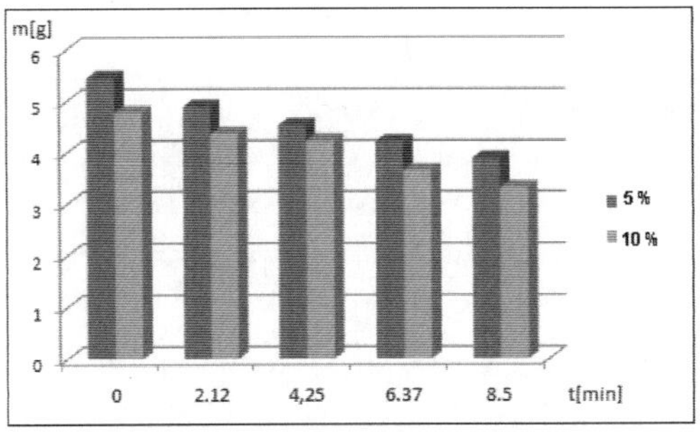

Figure 19. Weight loos of composite pin according to the trial time, F = 5N

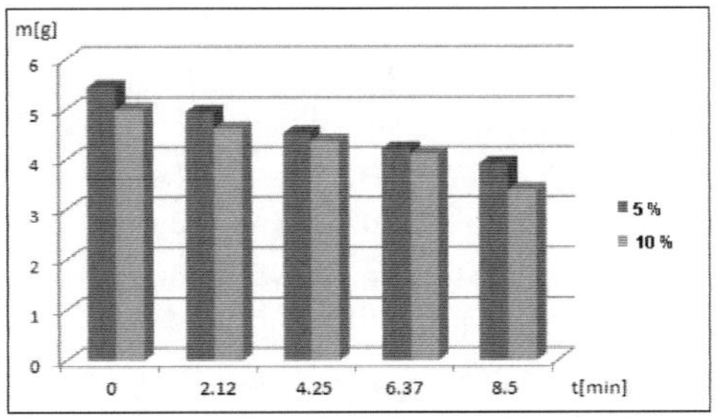

Figure 20. Weight loos mass of composite pin according to the trial time, F = 10N

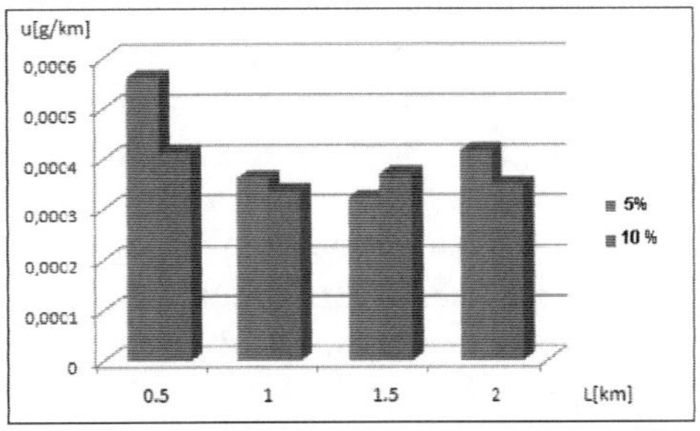

Figure 21. Linear wear test parameters with sliding distance using F = 5N

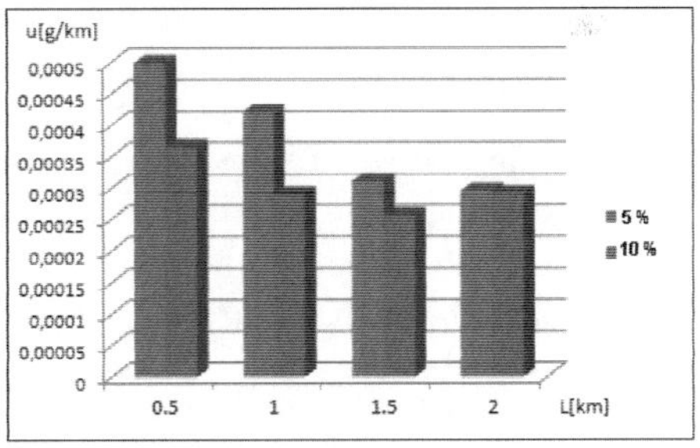

Figure 22. Linear wear test parameters with sliding distance using F = 10N

Figure 23 and Figure 24 shows the temperature in the contact area for composite pin with 5 % coconut fiber at the end of wear experiments with 5 N force, respectively 10 N.

Figure 25 and Figure 26 shows the temperature in the contact area for pin with 10 % coconut fiber at the end of wear experiments with 5 N force, respectively 10 N.

In the contact zones the temperature rises rapidly in the first part of the test, this period finishing asperities being treated with the composite structure.

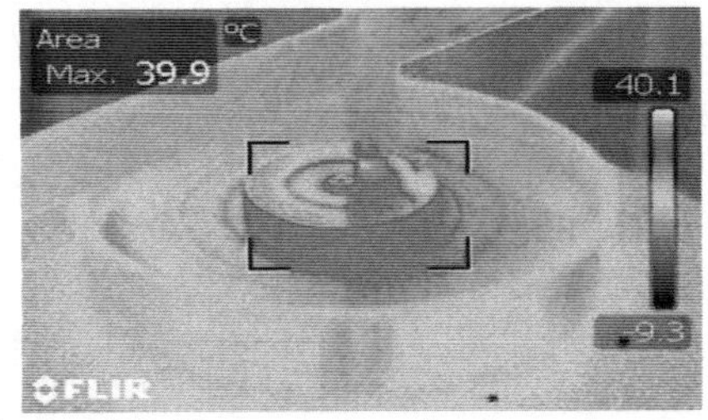

Figure 23. Temperature in the contact area at the end of wear experiments with 5 N force, for pin with 5 % coconut fiber

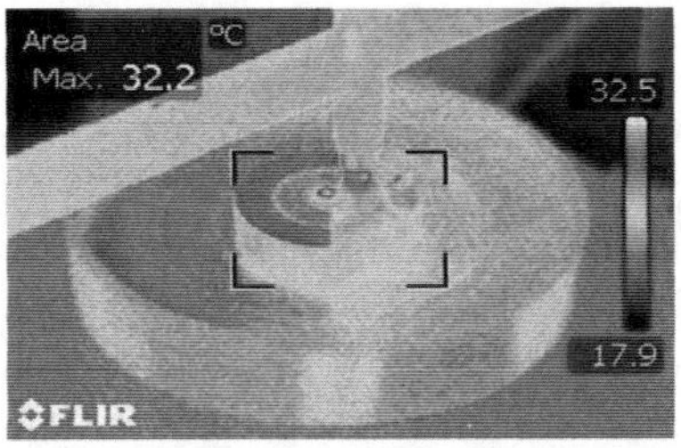

Figure 24. Temperature in the contact area at the end of wear experiments with 10 N force, for pin with 5 % coconut fiber

Figure 25. Temperature in the contact area at the end of wear experiments with 5 N force, for pin with 10 % coconut fiber

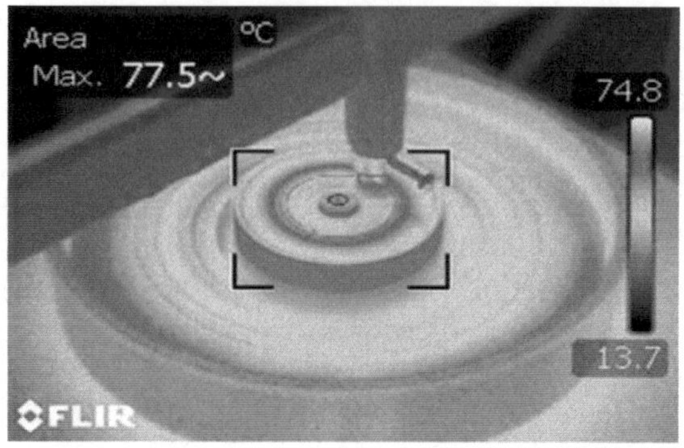

Figure 26. Temperature in the contact area at the end of wear experiments with 10 N force, for pin with 10 % coconut fiber

Between 500-1000 m, temperature rises less and in the end of the tests, the temperature drops. At the end of the test period, the iron disc has bigger temperature than pin composite which shows a better distribution of temperature on the composite material.

5.Conclusions

From the results and discussion of this study it can be observed:

- the cooling rate influences the mechanical properties of cast iron, increases resistance and hardness of cast iron, due to the increase in the percentage of pearlite structure and finishing of graphite separation. It must thus preheating the mold before casting, in order to reduce the cooling rate and preventing the occurrence of interdendritic graphite;
- the fabrication of the composite material depends on mixing sequences and time of mixing;
- following the study you can observe that the iron disc wears faster than the composite material and wear resistance does not depend greatly on the amount of metal parts by mass of the sample;
- iron disc wears faster than the composite material, which justifies the importance of tribological studies;
- the study led that natural coconut fibre can be used for the automotive brake pads;
- it can be concluded that both composite showed almost similar properties, hence coir could be a candidate fibre for the mass-scale fabrication of asbestos-free brake pad without any harmful effect.

References

[1] Vijay R, Jees J M, Saibalaji M A and Thiyagarajan V 2013 Optimization of Tribological Properties of Nonasbestos Brake Pad Material by Using Steel Wool, *Advances in Tribology* **2013** 165859

[2] Nicholson G 1995 *Facts About Friction* P&W Price Enterprises, Inc., Croydon, PA, pp 11

[3] Gopal P, Dharani L R and Blum F D 1995 Load, speed and temperature sensitivities of a carbon-fiber-reinforced phenolic friction material, *Wear* **181–183**(2) 913–921

[4] Kato T and Magario A 1964 Wear of aramid fiber reinforced brake pads: the role of aramid fibers, *Tribology Transactions* **37**(3) 559–565

[5] Österle W, Griepentrog M, Gross T and Urban I 2001 Chemical and microstructural changes induced by friction and wear of brakes, *Wear* **250-251**(2) 1469–1476

[6] Leman Z, Sapuan S M, Saifol S M, Maleque A M and Ahmad M M 2008 Mosture absorption behavior of sugar palm fibre reinforced apoxy composites, *International Journal of Materials and Design* **29** (8) 1666-1670

[7] Maleque A M, Atiqua A, Talib R J and Zahurin H 2012 New natural fibre reinforced aluminium

composite for automotive brake pad, *International Journal of Mechanical Engineering* **7**(2) 166-170

[8] Stefanescu C and Cazacu I 1981 *Implementing technologies molded parts,* Tehnica Publisher, Bucuresti, Romania

[9] Sofroni I *Develop and casting alloys* 1975 Didactic and Pedagogic Publisher, Bucuresti, Romania

[10] Ripoşan I and Chisamera M 1981 *Technology development and pig iron*, Didactic and Pedagogic Publisher, Bucuresti

[11] Mackin T J 2002 Thermal cracking in disc brakes, *Engineering Failure Analysis* **9**(1) 63-67

[12] Sasaki Y 1995 *Development Philosophy of Friction Materials for Automobile Disc Brakes*, The 8th International Pacific Conference on Automobile Engineering. Society of Automobile Engineers of Japan, pp 407- 412

[13] Pascu L V 2014 *Research on improving the quality of brake shoes intended for rolling stock*, University Politehnica Timisoara, Romania, Doctoral Thesis

IOP Conf. Series: Materials Science and Engineering **163** (2017) 012015 doi:10.1088/1757-899X/163/1/012015

Research on the reduction of the hydrogen content in steels treated in vacuum

M Magaon, S Şerban and T Heput

Politehnica University of Timisoara, Engineering and Management Department, 5 Revolution Street, Hunedoara, 331128, Romania

E-mail: miruna_magaon@yahoo.com

Abstract. This paper discusses the results of the research on the influence of the vacuuming technological factors on the hydrogen content in the steel produced in an electric furnace type EBT, treated outside the furnice in Ladle-Furnace (LF) and Vacuum Degassing (VD) facilities, and then continuously casted as semi round products designed for pipes manufacturing. Vacumming main parameters were followed such as: the total duration of vacuming, the duration of advanced vacuuming, the pressure in the vacuum facility, the vacumming temperature drop during the treatment (considered independent parameters) and the amount of hydrogen removed from the metal bath during the vacuuming. The obtained data was processed in the MATLAB calculation program, using three types of equations, the results obtained being presented both in analytical and graphical form. The technological analysis of the correlation equations and the graphical representations allowed to establish the variation limits of the independent parameters in order to advancely reduce the steel hydrogen content.

1. Introduction

After the elaboration, casting and lamination processes, impurities remain in the steel structure such as hydrogen, nitrogen and oxygen, in addition to non-metallic inclusions and gases. It is well known that the presence of these gases has negative influence on the quality of steel, which is why their content should be reduced to values as low as possible (particularly hydrogen and oxygen).

Regarding hydrogen, it represents an impurity for steel products, its negative influence is manifested especially in steel because [1-4]:

- It's one of the causes of sulphides in ingots and calmed steel castings;
- It contributes to the occurrence of the defect called "flakes" (very fine small cracks disposed as a rosette) in steels alloyed with chromium and nickel, which substantially reduces fatigue strength of steel parts;
- Decreases the elasticity and toughness of the steel;
- Affects the electrical and magnetic properties of steel.

To produce steel with low hydrogen content, a series of measures must be taken on the manufacturing flow such as [3], [5], [6]:

- the cargo used must be of good quality, without moisture, oils, calcinated additions, freshly burnt limestone, etc;
- the use of chromium ferroalloys, molybdenum, calcinated nickel;
- the bubbling of liquid steel with inert gases (usually argon);
- the treatment of steel under vacuum;

- the adequate heating of the casting machine (the ladle and the distributor);

2. Presentation of the experimental and data processing framework

In the research conducted, the possibilities to reduce the hydrogen content from steels designed for pipes manufacturing were investigated.

The researches were carried out at an electrical steelwork, equipped with an electric arc furnace type EBT (Electric Bottom Tapping) capacity 100t, LF (Ladle-Furnace) facility, VD (Vacuum Degassing) facility and a 5 wires continuous casting (ITC) facility. On the flow presented a total of 30 steel batches were followed, OLT45 type steel, circular rolled with a diameter of Ø180mm.

The 30 batches were monitored throughout the technological flow, stage by stage, as follows:

- Elaboration in the furnace: loads of good quality, without scrap with oil, moisture, rust under about 2%, freshly burnt limestone, iron ore (calcined when appropriate), adequately heated casting ladle;
- LF and VD treating facility: respecting the limits of variation of technological parameters (heating, bubbling), additions without moisture (for slag formation and alloying), mixing chemical and thermal by bubbling with argon followed by treatment in the vacuum facility without VD heat input, in the casting process the ladle and the distributor were heated adequately;
- The continuous casting compliance parameters: the casting speed corelated with the casting temperature, flow rate and pressure of cooling water in each cooling zone, the lubricating powders without moisture, the proper functioning of MTC.

The following parameters were:

- the hydrogen content before (initially) and after (final) vacumming, $[H]_i$ [ppm], respectively $[H]_f$ [ppm], which are considered dependent parameters;
- the total duration of vacuuming, $D_{t.v.}$ [min], independent parameter;
- the duration of advanced vacuuming, $D_{v.a.}$ [min], independent parameter;
- the pressure in the vacuum facility, $P_{V.D.}$ [mBar], independent parameter;
- the steel temperature at the entry and exit of the vacuum facility, T_i şi T_f, based on which is calculated the reduction of temperature, $\Delta[T] = [T]_i - [T]_f$ [^{0}C], independent parameter;

By processing the obtained data in the MATLAB calculation program, multiple correlation equations have resulted between the independent and dependent parameters, the results being presented both in analytical and graphical form. The hydrogen removal process is analyzed based on the obtained relations.

In order to obtain the correlation in MATLAB program I used three different equations type z - dependent parameter, x, y - independent parameters:

EQUATION 1:
z=a(1)+a(2).*x+a(3).*x.^2+a(4).*x.^3+a(5).*y+a(6).*y.^2+a(7).*y.^3+a(8).*y.^4+a(9).*y.^5

EQUATION 2: z=a(1)+a(2).*log(x)+a(3).*log(x).^2+a(4).*log(x).^3+a(5)./y+a(6)./(y.^2)+a(7)./(y.^3)+a(8)./(y.^4)+a(9)./(y.^5)

EQUATION 3: z=a(1).*x.^2+a(2).*y.^2+a(3).*x.*y+a(4).*x+a(5).*y+a(6)

We have to mention that all correlations are significant both in terms of values for correlation coefficients, respectively in mathematical terms as well as for technological direction.

The results of data processing are presented in the next figures.

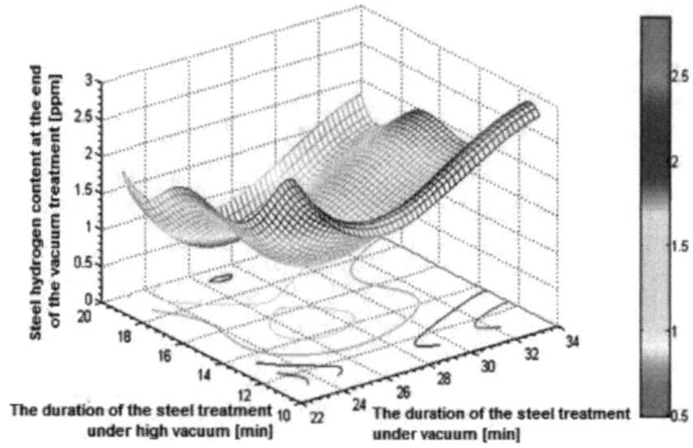

(a)

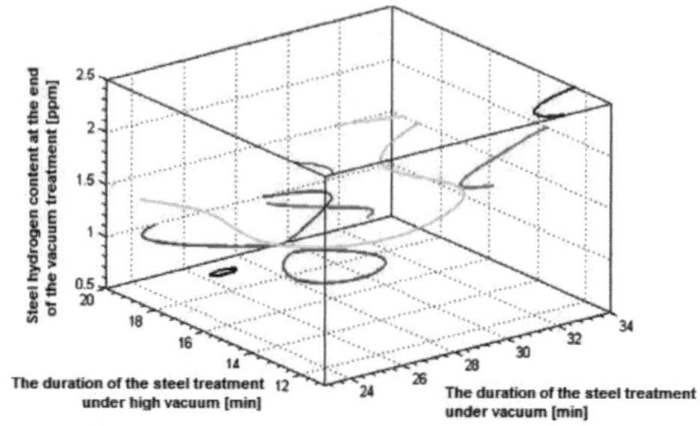

(c)

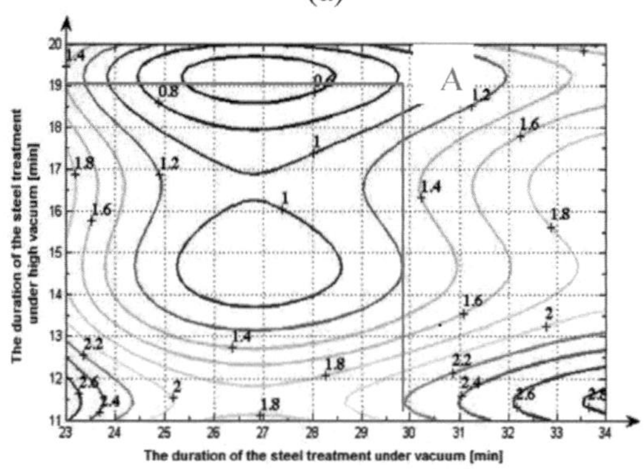

(b)

Figure 1. Correlation [H] =f(Dv.., Dv.a.)
Equation 1
z = a(1)+a(2).*x+a(3).*x.^2+a(4).*x.^3+a(5).*y+a(6).*y.^2+a(7).*y.^3+a(8).*y.^4+a(9).*y.^5
D = 50143.37411; H = -2291.32815;
R2 = 0.69283; r2= 0.68779
a1=-1145.66407: a2=-10.94201;
a3=0.361301; a4=-0.003909;
a5=426.291125; a6=-57.08538:2
a7=3.77343; a8=-0.12323; a9=0.00159

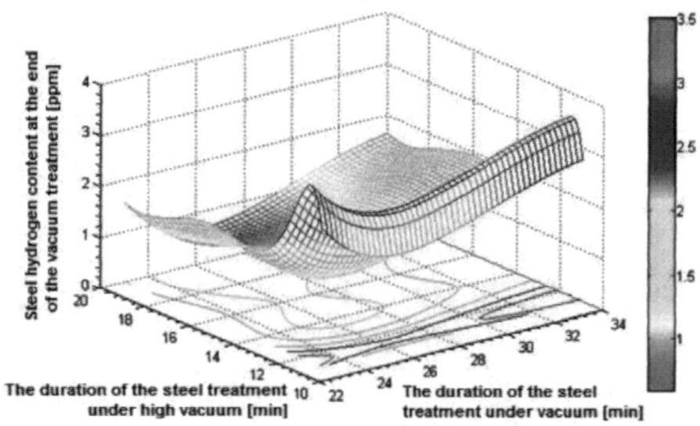

(a)

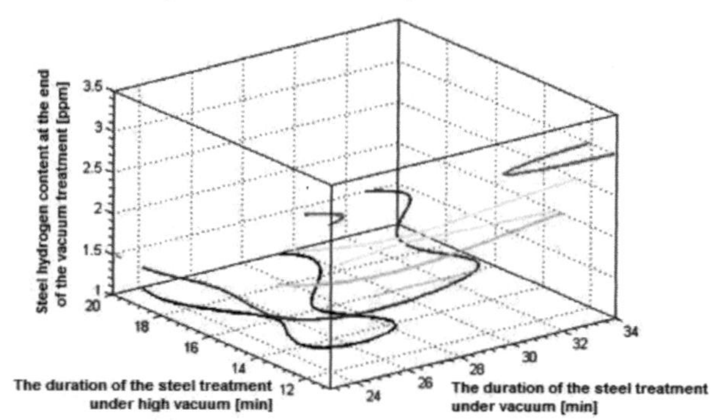

(c)

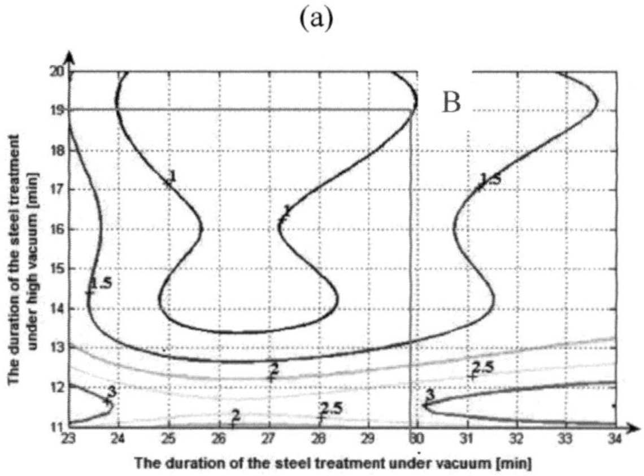

(b)

Figure 2. Correlation [H] =f(Dv., Dv.a.)
Equation 2
D= -53361734.97141;
H= 9129.39647;
R2= 0.45837; r2= 0.67703;
a1=4564.69823;
a2=-2883.35866; a3=845.627604;
a4=-82.52299; a5=-99622.18974;
a6=3037813.21626; a7=-45789786.20388;
a8=341126005.604936; a9=-1004504954.21221

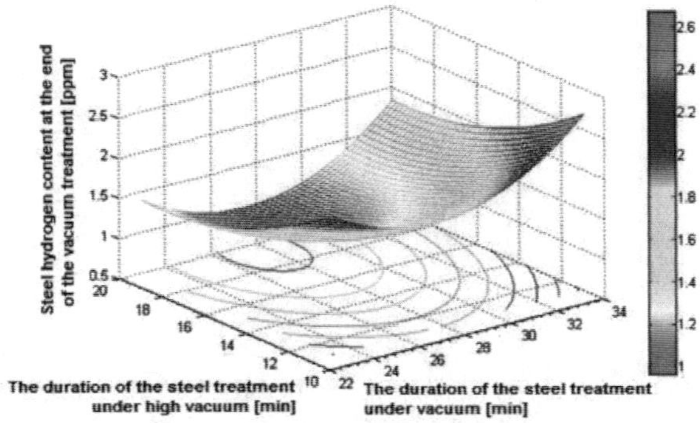

(a)

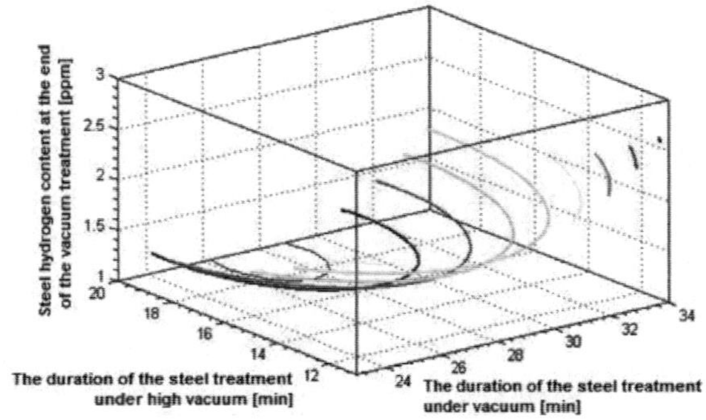

(c)

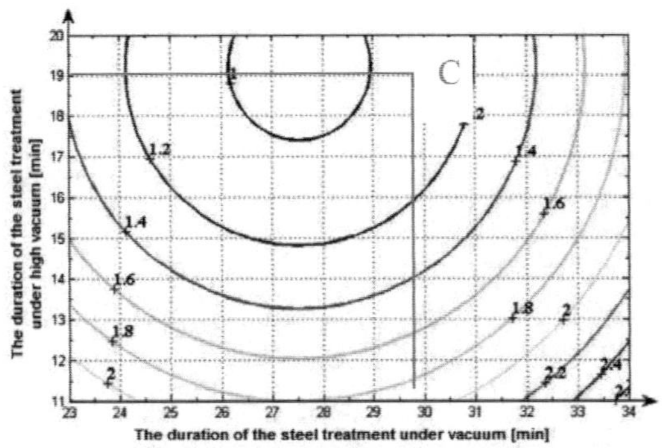

(b)

Figure 3. Correlation [H] =f(Dv., Dv.a.)
Equation 3
R2 = 0,59876
r2= 0.58246

a1=0.02052; a2=0.01238;
a3=-0.000285 a4=-1.12525;
a5=-0.46862 a6=20.96577

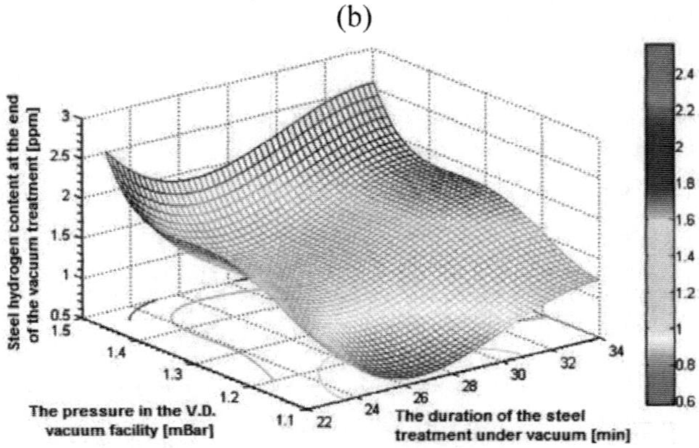

(a)

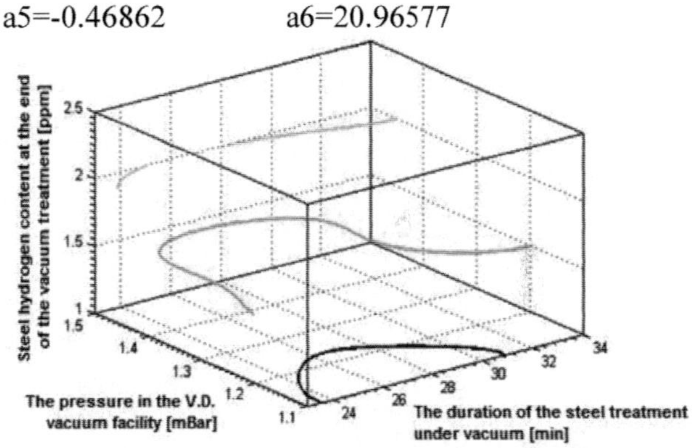

(c)

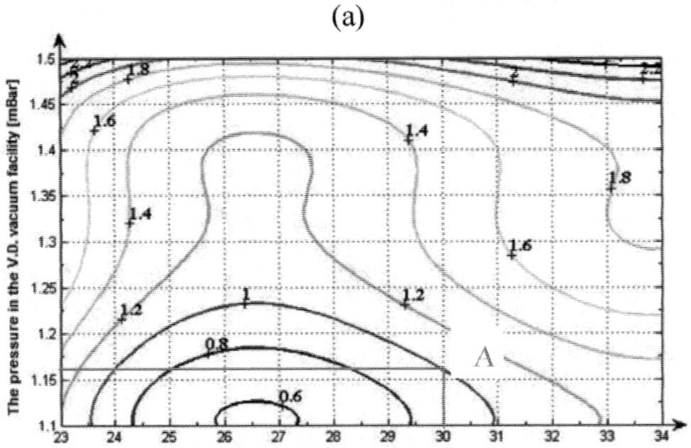

(b)

Figure 4. Correlation [H] =f(Dv., Pv.)
Equation 1
D= -21402.968209; H= 1183.96614;
R2= 0.59917; r2= 0.77406;
a1=591.98307;
a2=-9.03863; a3=0.303679;
a4=-0.003351; a5=-1006.38884;
a6=0; a7=1270.083880;
a8=-1001.29237; a9=235.80519;

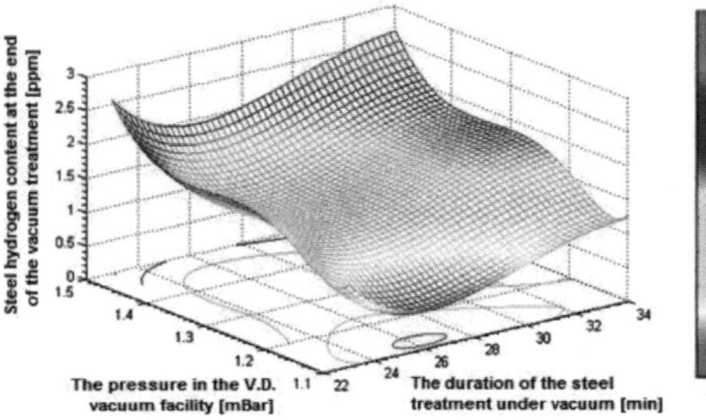

(a)

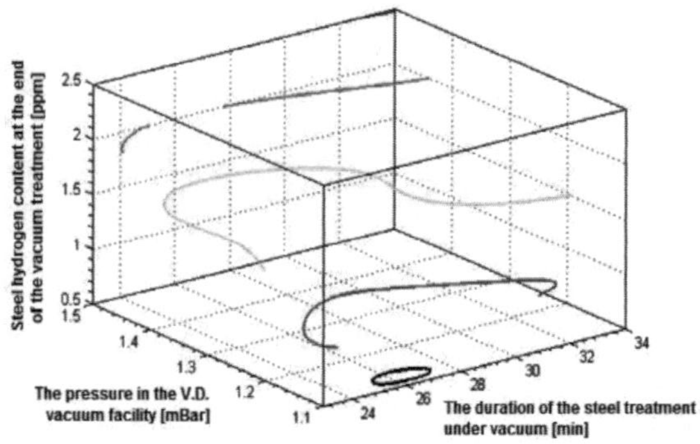

(c)

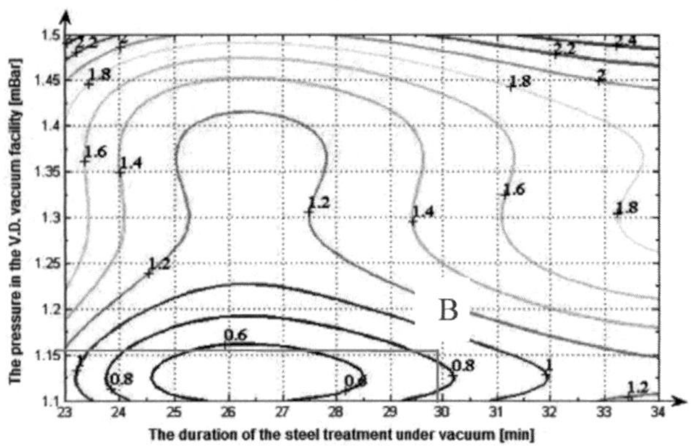

(b)

Figure 5. Correlation [H] =f(Dv.., Pv.)
Equation 2
D= -34820868.25972; H= 7849.62416;
R2= 0.59881; r2= 0.77382;
a1=3924.81208;
a2=-2191.59653; a3=643.77787;
a4=-62.92921; a5=-6095.10921;
a6=7714.16717: a7=0;
a8=-6119.69409; a9=3066.89523;

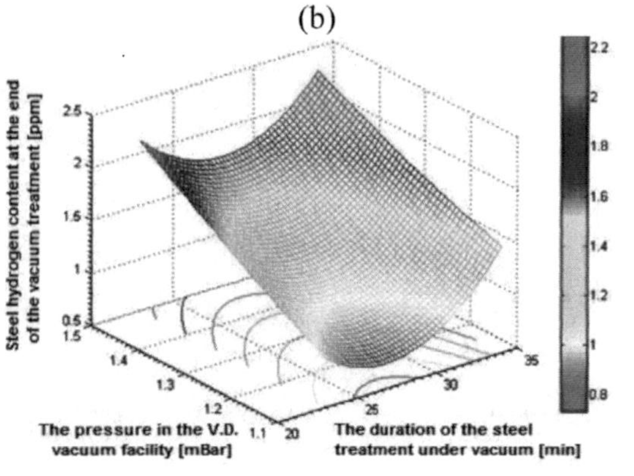

(a)

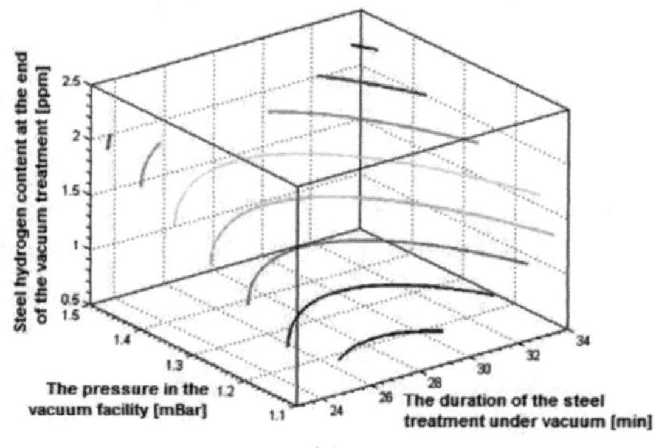

(c)

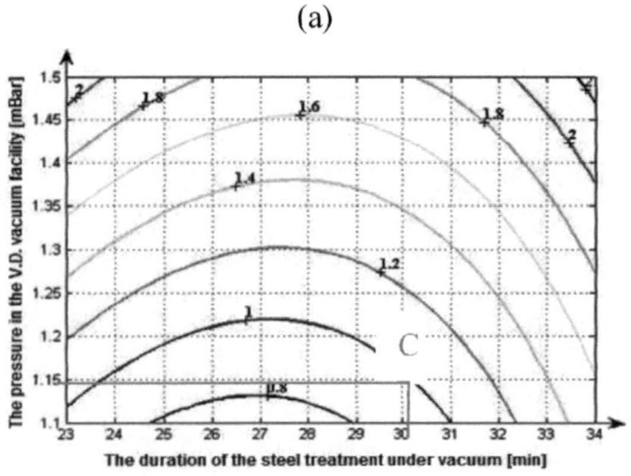

(b)

Figure 6. Correlation [H] =f(Dv., Pv.)
Equation 3
D= 0.054954; H= 0.030394,
R2 = 0,82365 r2= 0.72162;
a1=0.0151972; a2=1.04289;
a3=-0.09187; a4=-0.714028;
a5=2.28989; a6=7.88193;

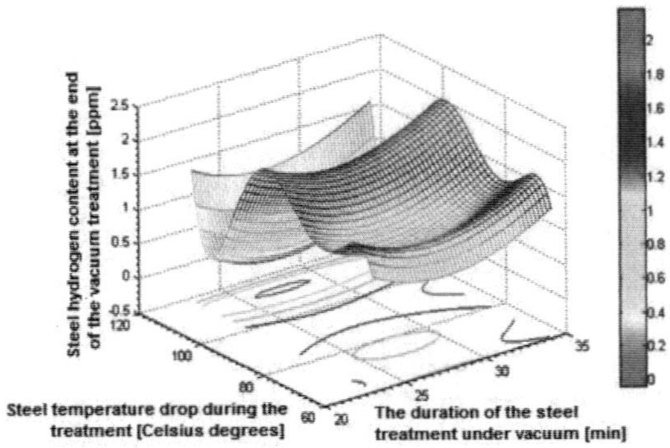

(a)

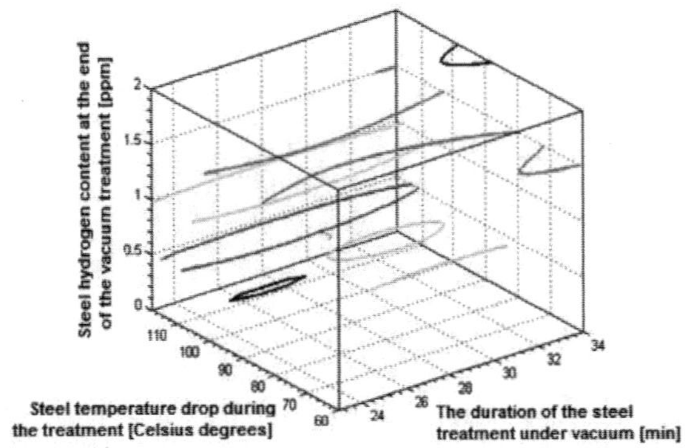

(c)

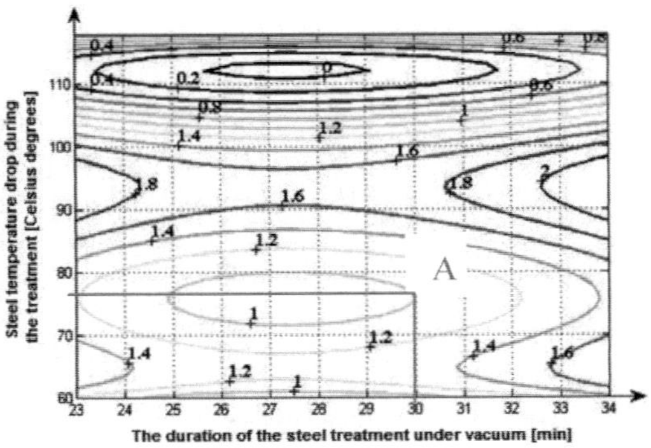

(b)

Figure 7. Correlation [H] =f(Dv., D.T.)
Equation 1
D= 7791.336304; H= -2473.444308;
R2= 0.554809; r2= 0.74485;
a1=-1236.72215; a2=-1.57499;
a3=0.043454; a4=-0.000357;
a5=76.78162; a6=-1.852201;
a7=0.022024; a8=-0.000129; a9=0;

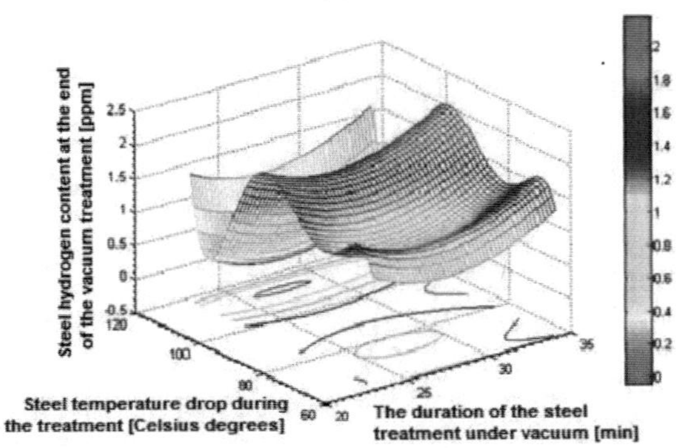

(a)

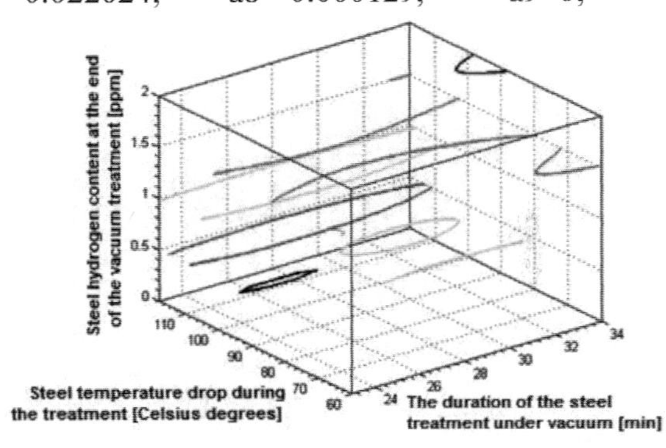

(c)

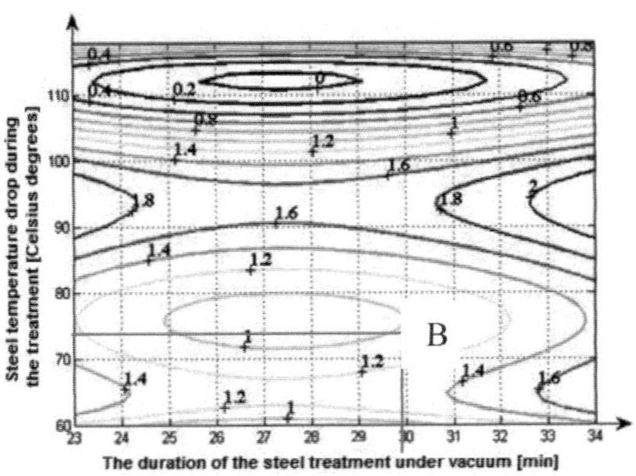

(b)

Figure 8. Correlation [H] =f(Dv.., D.T.)
Equation 2
D=-2931199.4192; H=1729.6901
R2=0.422522; r2=0.650017;
a1=864.84508;
a2=-831.15991; a3=236.43427;
a4=-22.33452; a5=34960.30999;
a6=-4190887.606; a7=219058469.7883;
a8=-4223734695.9615; a9=0;

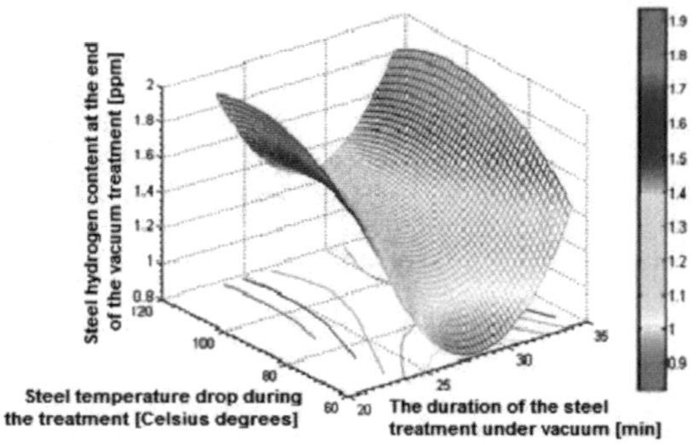

(a)

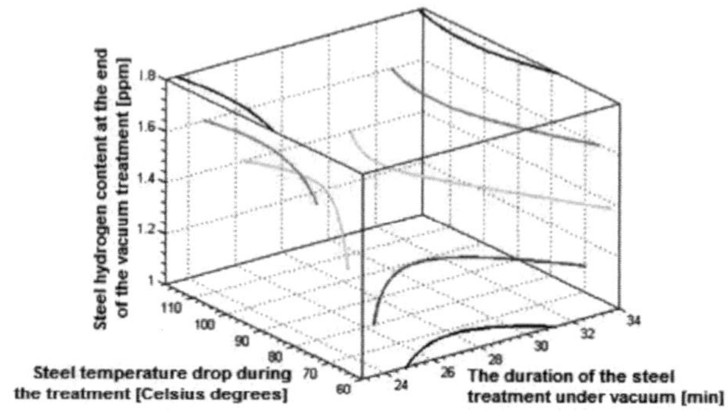

(c)

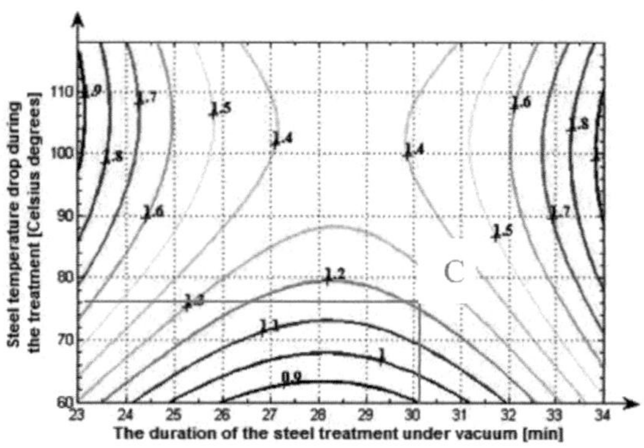

(b)

Figure 9. Correlation [H] =f(Dv., D.T.)
Equation 3

D= -2.16e-05;	H= 0.03710;
R2 = 0,8235;	r2= 0.560541;
a1=0.01855;	a2=-0.0002895;
a3=-0.000382;	a4=-1.01734;
a5=0.07069;	a6=12.20237;

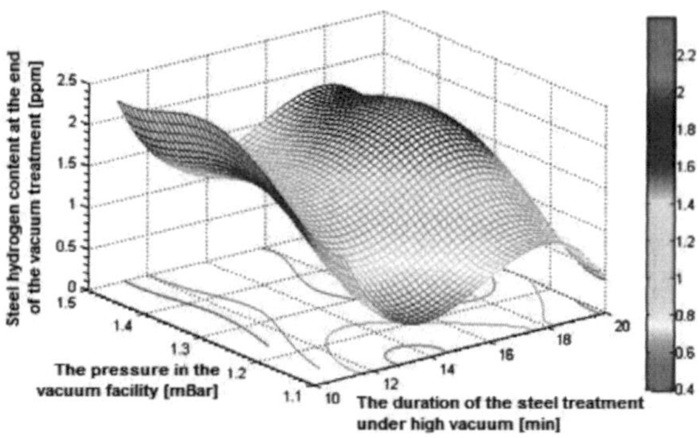

(a)

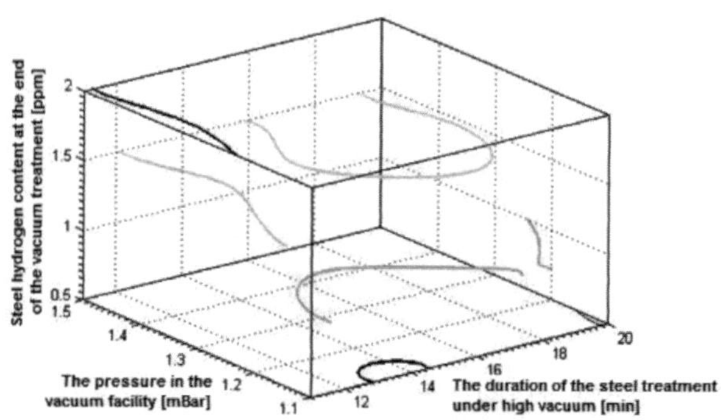

(c)

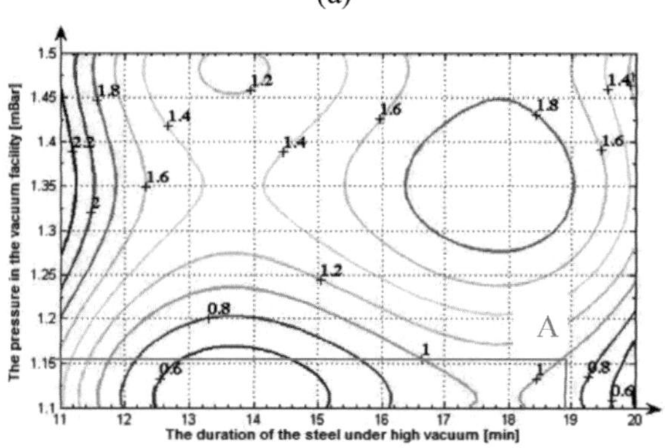

(b)

Figure 10. Correlation [H] =f(Dv.a.., Pv.)
Equation 1

D= -25804.29908;	H= 1100.9862;
R2= 0.52185;	r2= 0.72239;
a1=550.4931;	
a2=-11.71846;	a3=0.75717;
a4=-0.01601;	a5=-958.55512;
a6=0;	a7=1145.16551;
a8=-875.04711;	a9=199.2611;

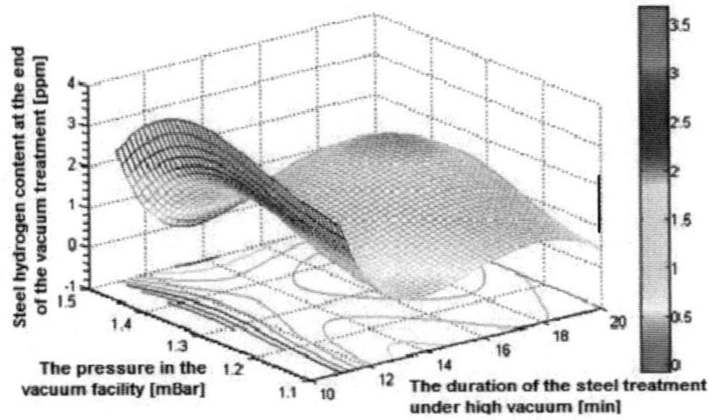

(a)

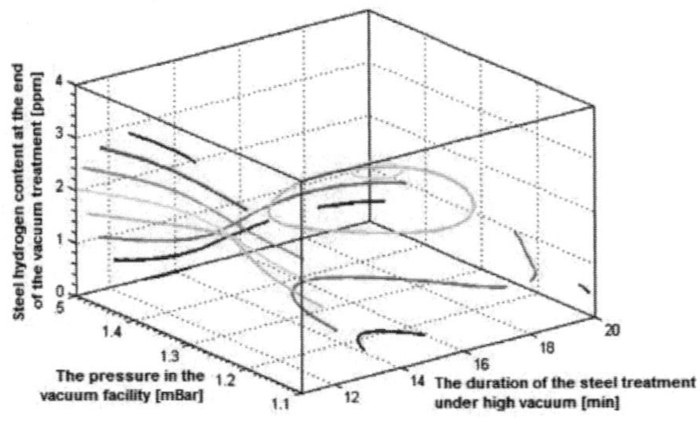

(c)

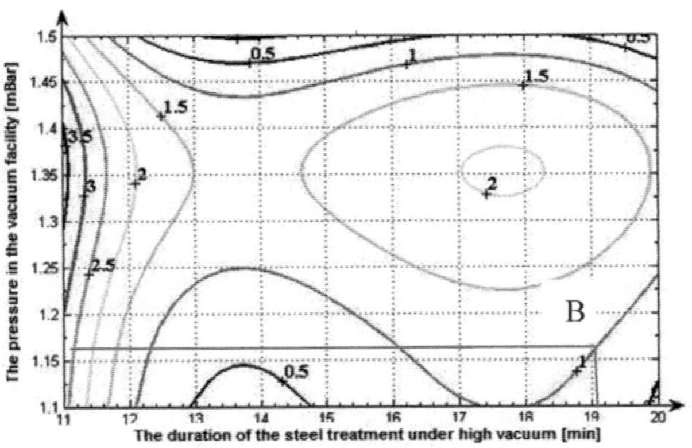

(b)

Figure 11. Correlation [H] =f(Dv.a.., Pv.)
Equation 2
D= -4286170.14509, H= 1105.60627,
R2= 0,78475 r2= 0.72186,
a1=552.80313,
a2=-1753.21723, a3=639.87045,
a4=-77.6772, a5=4189.74343,
a6=-5008.31045, a7=0,
a8=3534.2579, a9=-1671.75313,

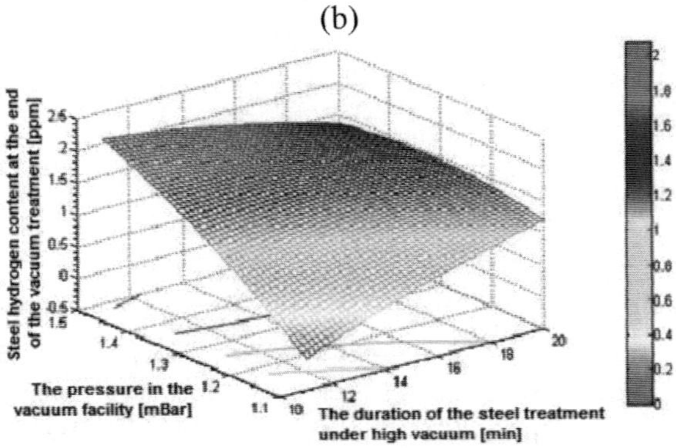

(a)

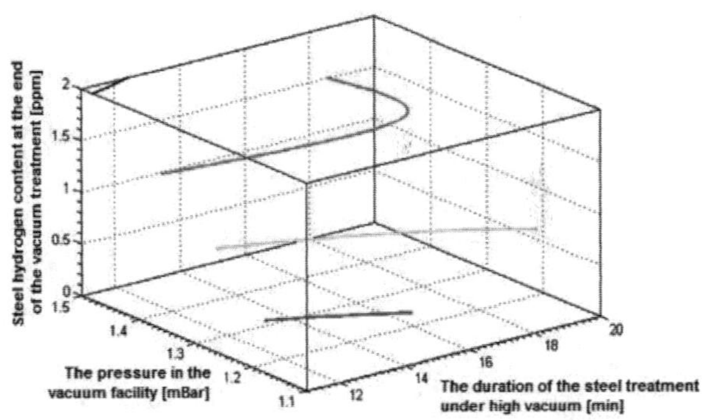

(c)

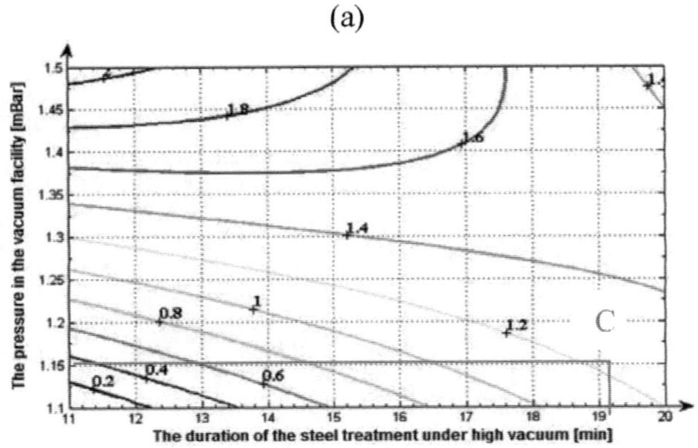

(b)

Figure 12. Correlation [H] =f(Dv.a.., Pv.)
Equation 3
D= -0.219832; H= -0.008069;
R2=0,74832; r2= 0.65253;
a1=-0.004034; a2=-4.26805;
a3=-0.53732; a4=0.85042;
a5=22.19735; a6=-21.62451;

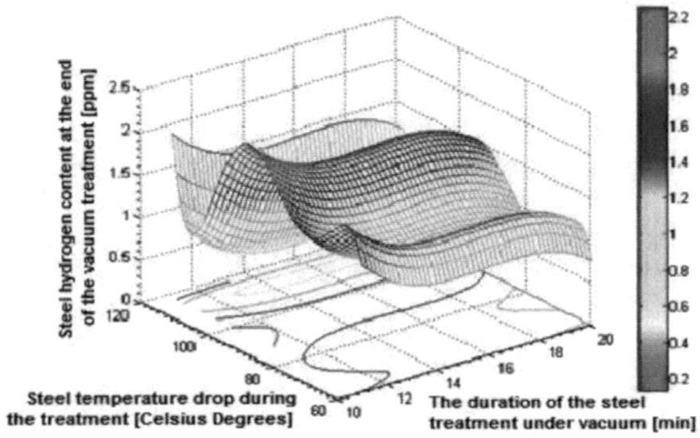

(a)

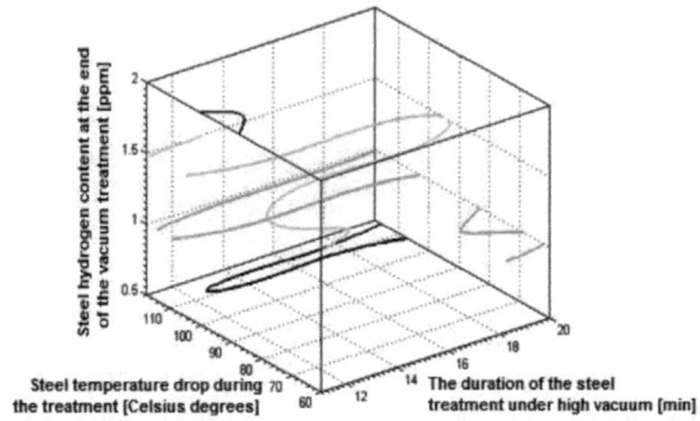

(c)

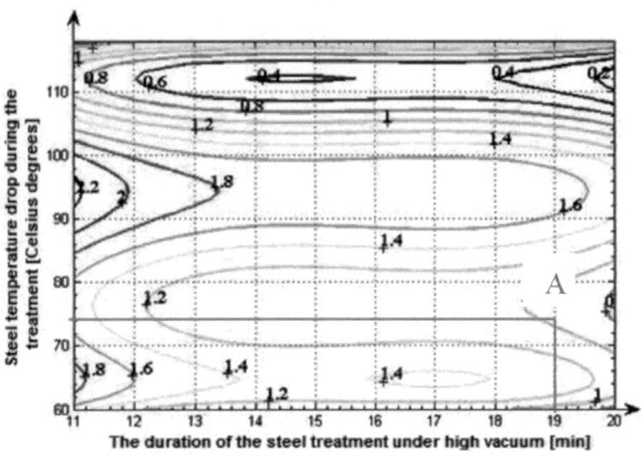

(b)

Figure 13. Correlation [H] =f(Dv.a.., D.T.)
Equation 1

D= 17730.5808013; H= -2354.230908;
R2= 0.53083; r2= 0.72858;
a1=-1177.11545;
a2=-3.76569; a3=0.239211;
a4=-0.005036; a5=73.00332;
a6=-1.754725; a7=0.020788
a8=-0.0001213; a9=0,0;

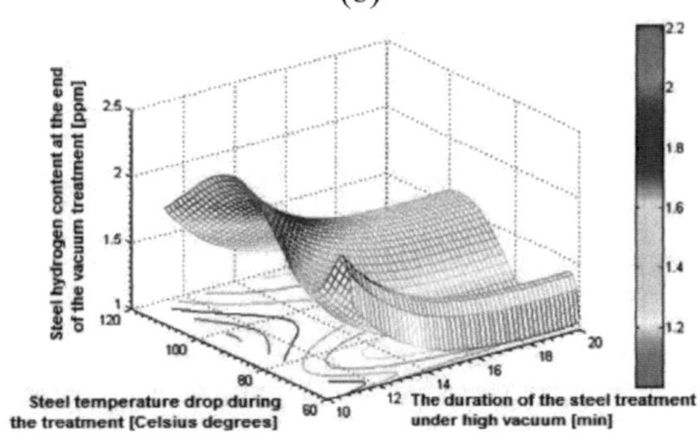

(a)

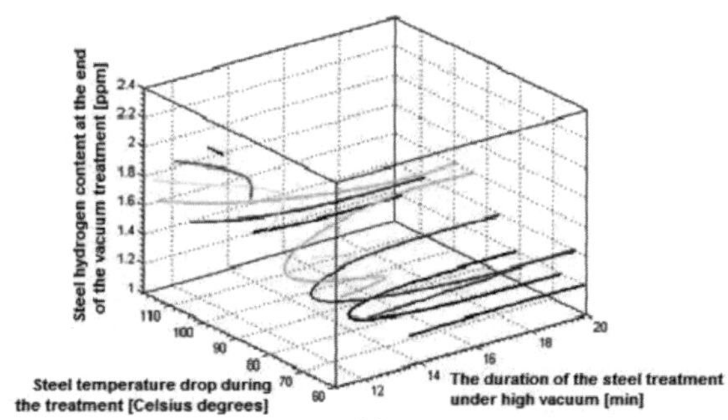

(c)

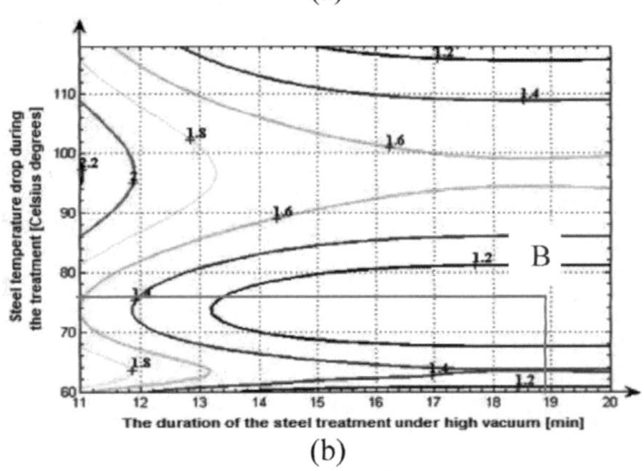

(b)

Figure 14. Correlation [H] =f(Dv.a.., D.T.)
Equation 2

D= -96419.89852; H= 652.58094;
R2= 0.43949; r2= 0.66294;
a1=326.29047; a2=-73.42107;
a3=24.36214; a4=-2.69198;
a5=-117291.97043; a6=21298262.972151
a7=-1882891276.0254; a8=81181908340.2696
a9=-1368431890047.13

IOP Conf. Series: Materials Science and Engineering **163** (2017) 012015 doi:10.1088/1757-899X/163/1/012015

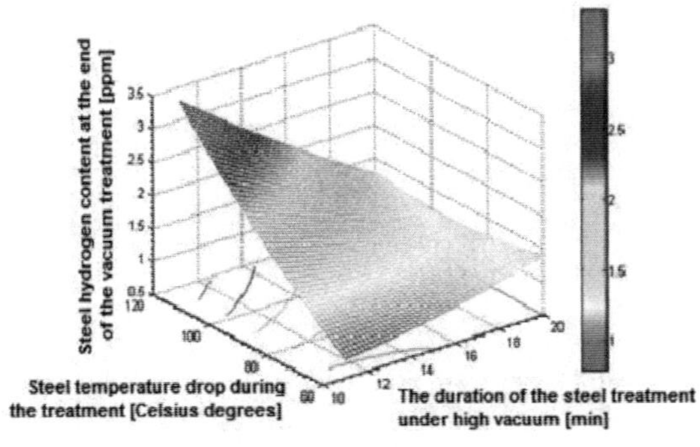

(a)

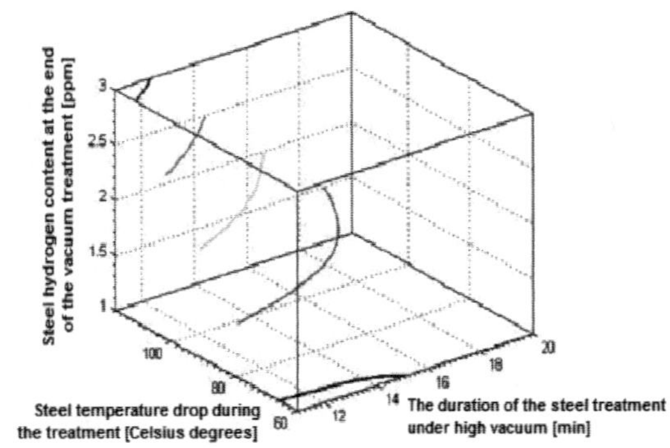

(c)

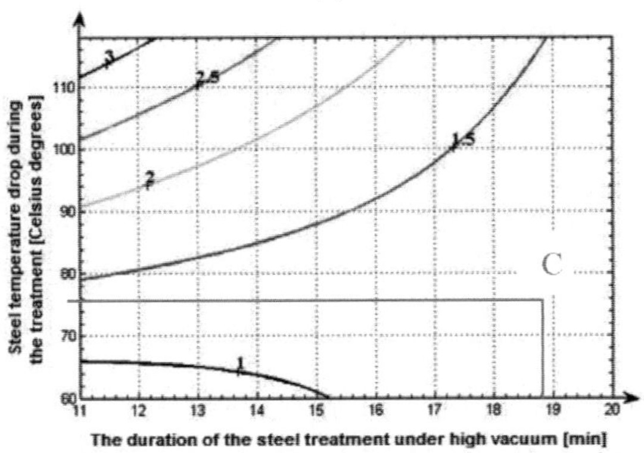

(b)

Figure 15. Correlation [H] =f(Dv.a.., D.T.)
Equation 3
D= -2.35114133680876e-05
H= 0.006804;
R2=0,76341 r2= 0.58542;
a1=0.003402; a2=0.000174;
a3=-0.00508; a4=0.266044;
a5=0.06884; a6=-3.94954;

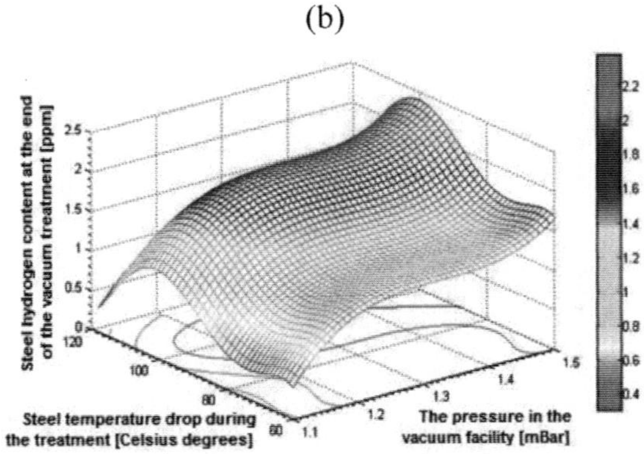

(a)

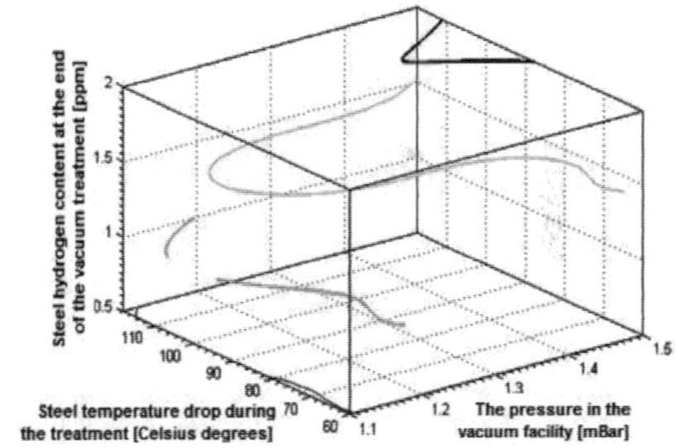

(c)

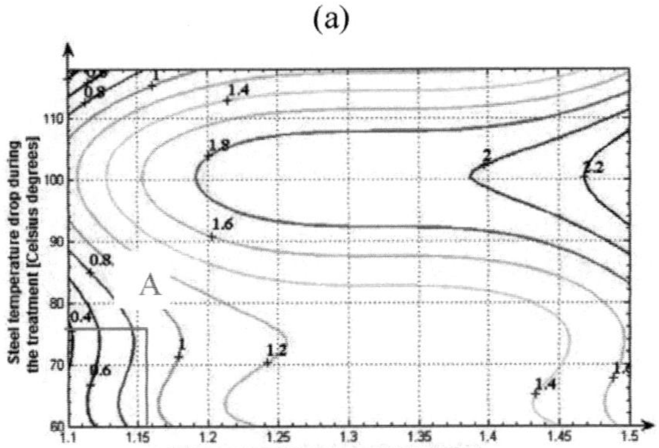

(b)

Figure 16. Correlation [H] =f(Pv..., D.T.)
Equation 1
D= -789108.88797; H= -894.018024;
R2= 0.823008; r2= 0.907197;
a1=-447.009012;
a2=392.34923; a3=-295.92939;
a4=74.40428; a5=16.630155;
a6=-0.39454; a7=0.00458;
a8=-0,00002; a9=0.

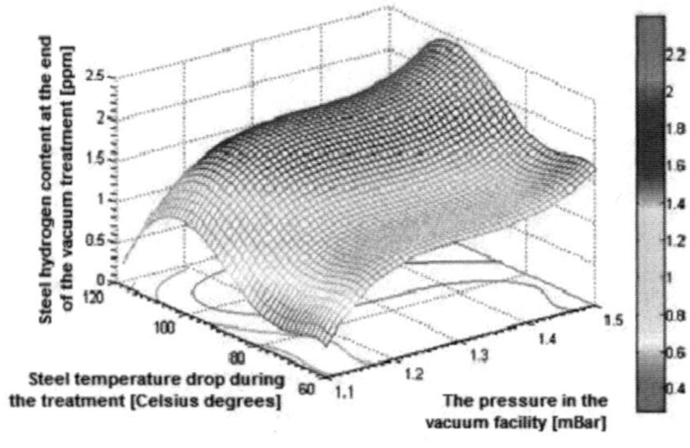

(a)

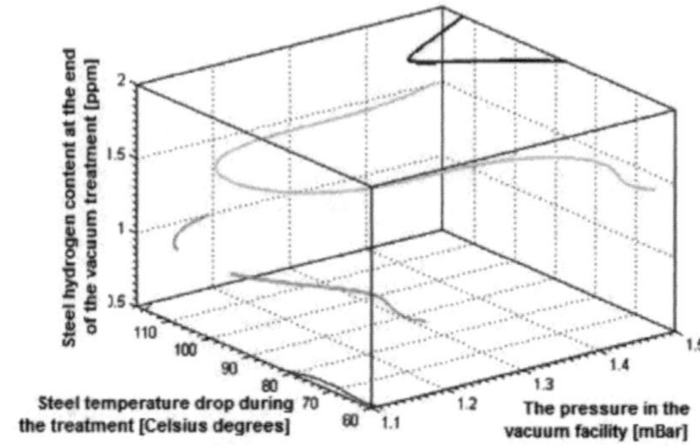

(c)

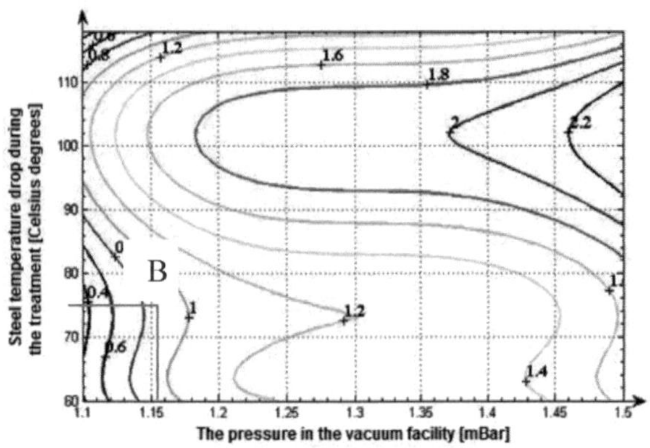

(b)

Figure 17. Correlation [H] =f(Pv.., D.T.)
Equation 2

D= -59073.73998;	H= -610.50587;
R2= 0.82664;	r2= 0.90919;
a1=-305.2523;	a2=34.78972;
a3=-128.82187;	a4=159.586535;
a5=113196.732;	a6=-16493575.611;
a7=11767;	a8=-41196729582.5623
a9=567011467997.132	

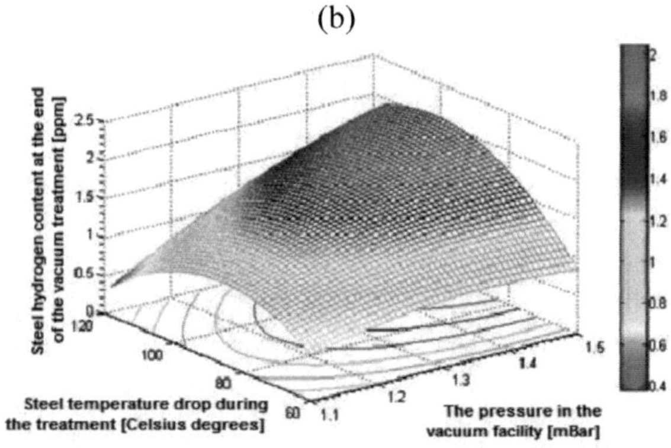

(a)

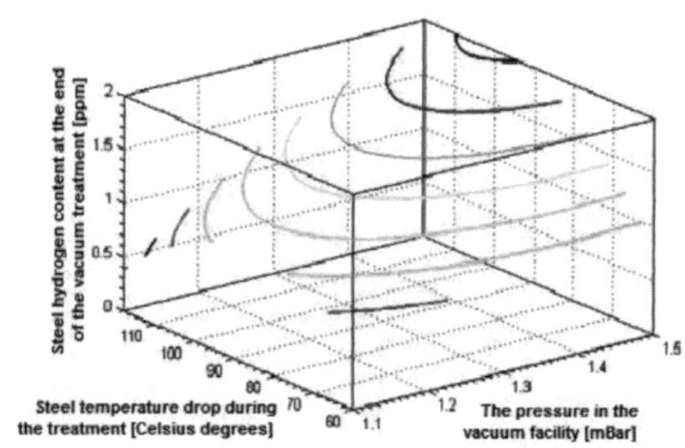

(c)

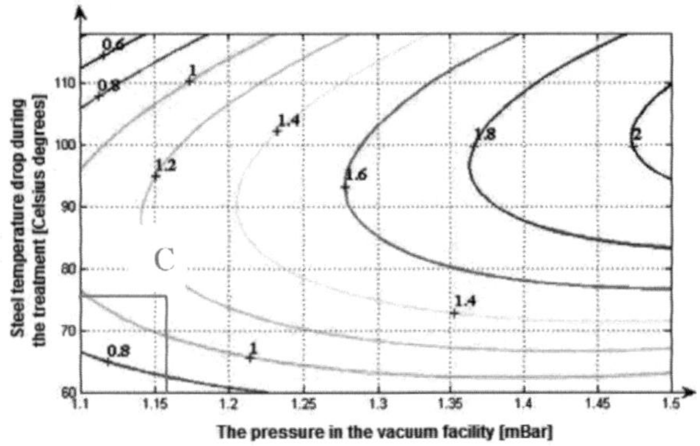

(b)

Figure 18. Correlation [H] =f(Pv.., D.T.)
Equation 3

D=0.00704;	H=-7.350204;
R2 =0,78946,	r2=0.75639;
a1=-3.675102;	a2=-0.000679;
a3=0.054216;	a4=6.8915;
a5=0.057437;	a6=-7.11338;

3. The technological analysis of the obtained results

The analysis of the graphical representations shows that the vacuuming parameters have significant influence on the final content of hydrogen and at the end of processing in the vacuuming facility.

It must also be emphasized that there is a very good correspondence between the values obtained for the dependent parameters at the same values of the independent parameters after these three equations (identical sometimes), which is a confirmation of the validity of the results obtained.

Within the comparative technological analysis for the 4 independent parameters the following values were chosen:

the total duration of vacuuming: D_{tv} = 30 min.;

the duration of advanced vacuuming: $D_{v.a.}$ = 19 min.;

the pressure in the vacuum facility: P_{VD} =1,16 mBar;

the reduction of the temperature: ΔT = 76 ^{0}C.

Consequently the technological analysis is presented for the 6 double correlations groups, each group comprising correlations after the three types of equations.

Correlation 1: [H] = f(D_{tv}, D_{va}) presented in figures 1, 2 and 3 with the spatial coordinate points A(30; 19; 0,9); B(30; 19; 1,0) and respectively C (30; 19; 1,1). The difference of 0,2 ppm [H] is admissible given the max. of 1,1 ppm [H]. Very good results are obtained for D_{tv} = 25-31 min and D_{va} = 15-20 min, resulting $[H]_f$ under 1,4 ppm.

Correlation 2: [H] = f(D_{tv}, P_{VD}) presented in figures 4, 5 and 6 with the spatial coordinate points A(30;1,16; 1,0); B(30; 1,16; 0,85) and respectively C(30;1,16; 1,0). The difference of 0,15 ppm [H] is admissible given the max. of 1,1 ppm [H]. Very good results are obtained for D_{tv} = 25-31 min and P_{VD} under 1,35 mBar , resulting $[H]_f$ under 1,4 ppm.

Correlation 3: [H] = f(D_{tv}, ΔT) presented in figures 7, 8 and 9 with the spatial coordinate points A(30; 76; 1,0); B(30; 76; 1,0) and respectively C(30; 76; 1,2). The difference of 0,20 ppm [H] is admissible given the max. de 1,2 ppm [H]. Very good results are obtained for D_{tv} = 25-31 min and ΔT = 60 – 100min, resulting $[H]_f$ under 1,4 ppm.

Correlation 4: [H] =f(D_{va}, P_{VD}) presented in figures 10, 11 and 12 with the spatial coordinate points A(19;1,16; 0,95); B(19; 1,16. 1,1) and respectively C(19; 1,16; 1,15). The difference of 0,20 ppm [H] is admissible given the max. of 1,15 ppm [H]. Very good results are obtained for D_{va} 15-20 min and P_{VD} under 1,35 mBar, resulting $[H]_f$ under 1,4 ppm.

Correlation 5: [H] = f(D_{va}, ΔT) presented in figures 13, 14 and 15 with the spatial coordinate points A(19;76; 1,0); B(19; 76. 1,0) and respectively C(19; 76; 1,25). The difference of 0,25 ppm [H] is admissible given the max. of 1,25 ppm [H], respectively equal values in points A and B. Very good results are obtained for D_{va} =15-20 min and ΔT = 60 – 100 ^{0}C, resulting $[H]_f$ under 1,4 ppm,

Correlation 6: [H] = f(P_{VD},ΔT) presented in figures 16, 17 and 18 with the spatial coordinate points A(1,16;76; 0,75); B(1,16;76;0,95) and respectively C(1,16; 76; 1,1). The difference of 0,1 ppm [H] is admissible given the max. of 1,1 ppm [H]. Very good results are obtained for P_{VD} under 1,35 mBar and ΔT = 60 – 100 ^{0}C, resulting $[H]_f$ under 1,4ppm.

For all presented correlations can be choosed values for the independent parameters, respectively variation interval, so that for the final content of hydrogen (dependent parameter) to be obtained a value close to the desired one, or even the desired one.

4. Conclusions

By processing the data in the MATLAB calculation program, after these three types of equations and the technological analysis of the obtained results, one can conclude the following:

- the steel treatment in the vacuum facility without heat input determines a value for the hydrogen content in the liquid steel at the exit of the VD vacuum facility under 2,0 ppm (an extent of at least 90% of batches);
- the steel treatment in LF facility ensures, on the one hand, a good thermal and chemical homogeneity and, on the other hand, a refinery concerning the contents of gases and nonmetallic inclusions;

- as a result of the treatment in the LF facility a temperature drop can be determined during the treatment under vacuum up to 110 ^{0}C (less often over 90^{0}C), which can ensure a value for the total duration of vacuuming up to 35 minutes and for the duration of advanced vacuuming (under 1,5 mBar) of 20 minutes;

- for ensuring the proper hydrogen content in the steel the following independent parameter variation limits are recommended:

  -the total duration of the treatment under vacuum: 25-31 minutes;
  -the duration of advanced vacuuming treatment: 15 20minute;
  -the pressure in the vacuum facility: under 1,5 mBar;
  -the temperature drop during the treatment: $\Delta T = 60 - 100\ ^0C$

References

[1] Nica G, Socalici A, Ardelean E and Hepuţ T 2003 *Technologies to improve the quality of steel*, Mirton Publishing House, Timişoara, Romania

[2] Ştefănoiu R 2004 *Research on the refining process of steels, inert gases,* University Politehnica Bucharest, Romania, Doctoral Thesis

[3] Drăgoi F 2012 *Research on reducing the gas content of the steel treated on the technological and EBT- TC,* University Politehnica Timisoara, Romania, Doctoral Thesis

[4] Puţan A 2013 *Research on steel refining elaborated on the flow: electric arc furnace-ladle furnace-continuous casting*, University Politehnica Timisoara, Romania, Doctoral Thesis

[5] Ştefănoiu R and Geantă V 2005 The injection systems of inert gases used in the secondary metallurgy, *Metalurgia* **7** 37-46

[6] Geantă V 2003 *Processes and technologies for refining steel*, PRINTECH Publishing House, Bucharest, Romania

Research on fire hose couplings damages

C Babut, N Ungureanu and M Ungureanu

Technical University of Cluj Napoca, North University Center of Baia Mare, Faculty of Engineering, 62A, Dr. Victor Babes St., 430083 Baia Mare, Romania

E-mail: nicolae.ungureanu@cunbm.utcluj.ro

Abstract. This paper presents a part of a research on the damaging of fire hose coupling. These are important components of the hydraulic systems used by firefighters to deliver one or more suppression agents into the fire. The operating regime parameters, working environment and general functioning conditions during fire suppression operations may lead to malfunctions and/or damaging of the couplings.

1. Introduction

1.1. General presentation and background

Fire suppression remains one of the basic missions of Romanian firefighter units. To accomplish this task, firefighters use a wide range of equipment, designed to deliver suppression agents into the fire.

Water and foam are the two most used substances for fire dousing. [1] These are delivered through a hydraulic chain that starts with the fire truck pumps, continues with the fire hoses and ends with a nozzle.

Fire hoses are connected to the fire truck or to the hydrants with couplings. [2] When needed, one or more sections can be extended with couplings, which makes these components very important parts of fire control operations. Damaging of the couplings leads to serious delays of the fire extinguishing actions and might have serious effects on the rescue operations.

1.1. Fire hose couplings

Couplings' main role is to connect flat hoses in order to ensure the normal flow of fire suppressing substances, mostly water and foam, during firefighting operations. Romanian firefighters are using the so called" Storz" type couplings, that allow connection by interlocking lugs (or hooks) and flanges then rotating these with a quarter of a turn. "Storz" type couplings can be classified, according to usage, as hose couplings, adaptors and blank caps. [3] The characteristics of delivery and suction couplings given by Romanian standards are shown in Table 1.

The Storz couplings used by Romanian firefighters are fabricated of die cast (pressure cast) aluminum, more specifically- AlSi5Cu1 alloy. It's chemical composition, according to [4-6] is given in Table 2. The range of mechanical properties of AlSi5Cu1 alloy, according to [4], [6], [7] is shown in Table 3.

Table 1. Delivery and suction couplings characteristics according to Romanian standards [2]

Type	A	B	C	D
Diameter d [mm]	96	62	43	18
Mass [kg]	2,1	1,67	0,86	0,24

Table 2. Chemical composition of AlSi5Cu1 alloy [4], [6], [7]

Alloy	Chemical composition [%]						Impurities [%]		
	Cu	Si	Mg	Mn	Al	Fe	Zn	Pb	Ni
ATP AlSi5Cu1 Mg SR EN 1706:2010	1-1,5	5-6	0,3-0,6	0,2-0,5	rest	0,8	0,5	0,2	0,3

Table 3. Range of mechanical properties of AlSi5Cu1 alloy, according to [4], [6], [7]

Mechanical properties					
Rp02 [MPa]	*Rm* [MPa]	*A5* [%]	*Rsm* [MPa]	*Rsm/ Rp02*	*Kic* [MPa]
140 - 210	230 - 280	1 - 3	240 - 318	1,51 - 1,71	22,05 – 26,50

where:

- *Rp02* represent yield strength;
- *Rm* -tensile strength;
- *A5*-elongation;
- *Rsm*-sharp notch tensile strength;
- *Kic*-plain strain fracture toughness

These couplings are used to connect hoses to pumps, to fire hydrants and hoses to hoses. Its basic design is shown in Figure 1, an exploded view in Figure 2, and connected couplings in Figure 3 and Figure 4.

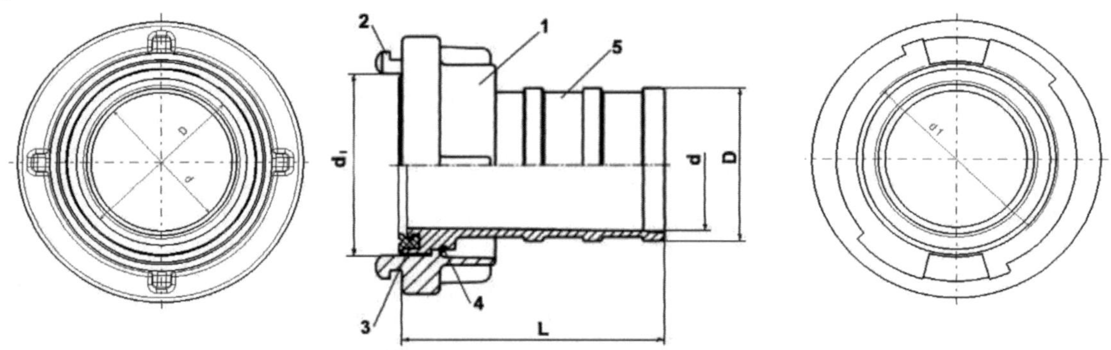

Figure 1. Storz type fire hose coupling basic design
(1 body; 2 hooking lugs; 3 sealing ring/gasket; 4 safety ring for pipe's position locking; 5 tail)

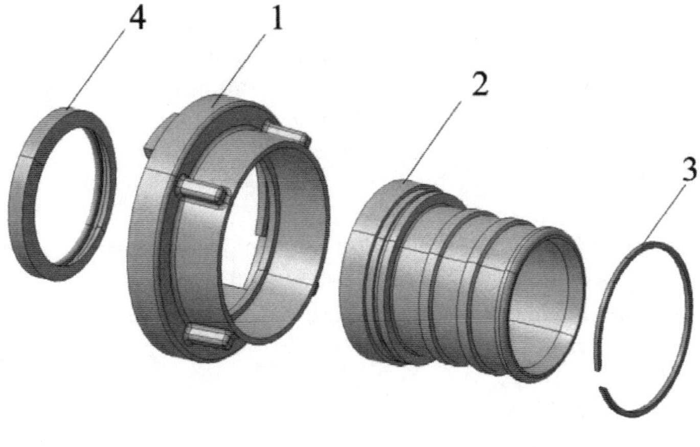

Figure 2. Fire hose couplings- exploded view
(1 body with hooking lugs; 2 tail piece;3 safety ring for pipe's position locking; 4 sealing ring)

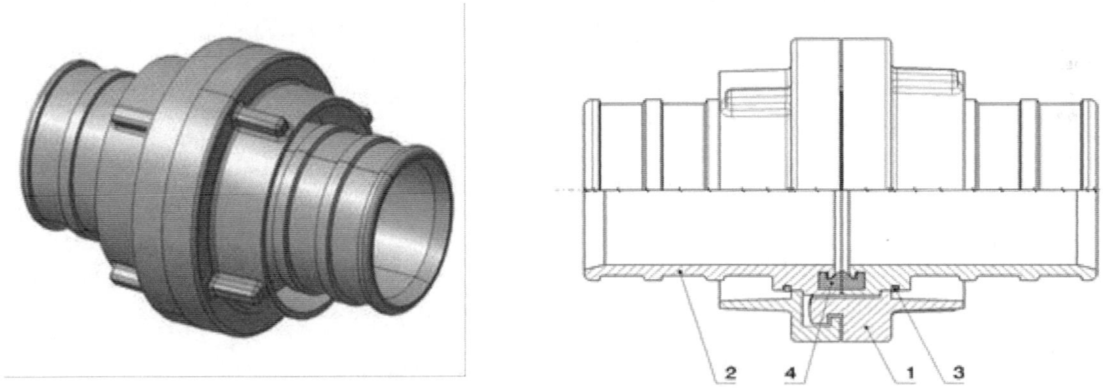

Figure 3. Connected couplings in operating position

Figure 4. Connected couplings

2. Analysis of Potential Failures. Analysis of couplings operating regime

While in use, fire hose couplings are subjected to important strains that could lead to a wide range of damages and even loss of functional properties, i.e. failure. Considering one of the main fire engines in use in Romania, the APCA R 12215 type pumper, the operating parameters for the couplings would include:

- pump pressure in the 7.8 - 22.55 [bar] range;
- flows in the 1500 to 5000 [l/min] range;
- pump rotational speeds from 2200 to 2900 [rpm].

Working and environment conditions of the firefighter units during fire suppression operations also imply [8]:

- mechanical shocks due to human manipulation;
- hydraulic shocks ("water rams");
- high level of vibrations
- accidental exposure to high temperatures;
- incorrect connection of the couplings;
- sand, dirt, vegetation, various type debris;
- wide range of chemicals, petroleum fuels and oils;

Analyzing the operating parameters, the couplings characteristics and the specialty literature [9-12] we found that the following damages might occur during fire suppression operations. These and their effects were identified as process related potential failures:

- accidental exposure to high temperatures and incorrect assembly could lead to breaking of the safety ring- non critical failure - shown in Figure 5

Figure 5. Breaking of the safety ring

- contact with chemicals, petroleum fuels and oils and vibrations could lead to wear of the sealing gasket- important but non critical failure- shown in Figure 6.

Figure 6. Wear of the sealing gasket

- Incorrect connecting, vibrations plus penetration of sand and/or other abrasive materials from the environment in the connected area could lead to leakage and fretting wear marks on the flanges- important but non critical failure- shown in Figure 7.

Figure 7. Wear marks on the flanges

- Mechanical shocks that occur during firefighting operations, due to incorrect manipulation and operating environment such as concrete floors around and inside buildings could lead to the cracking of the couplings' body- important, potentially critical failure-shown in Figure 8.

Figure 8. Cracking of the couplings body ("the Storz head")

- Hydraulic shocks ("water ram"), mechanical shocks due to incorrect manipulation of the equipment, vibrations that generate fretting fatigue could lead to the breaking of the lugs - critical failure-shown in Figure 9.

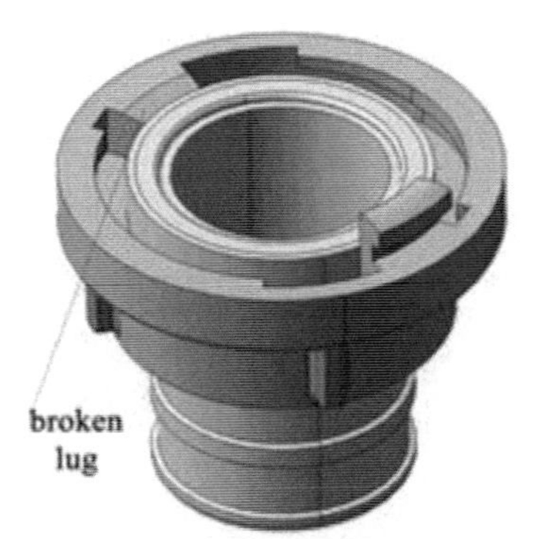

Figure 9. Breaking of the lugs

This leads to the basic failure mode and effects diagram for potential coupling damages during firefighting operations shown in Figure 10.

IOP Conf. Series: Materials Science and Engineering **163** (2017) 012016 doi:10.1088/1757-899X/163/1/012016

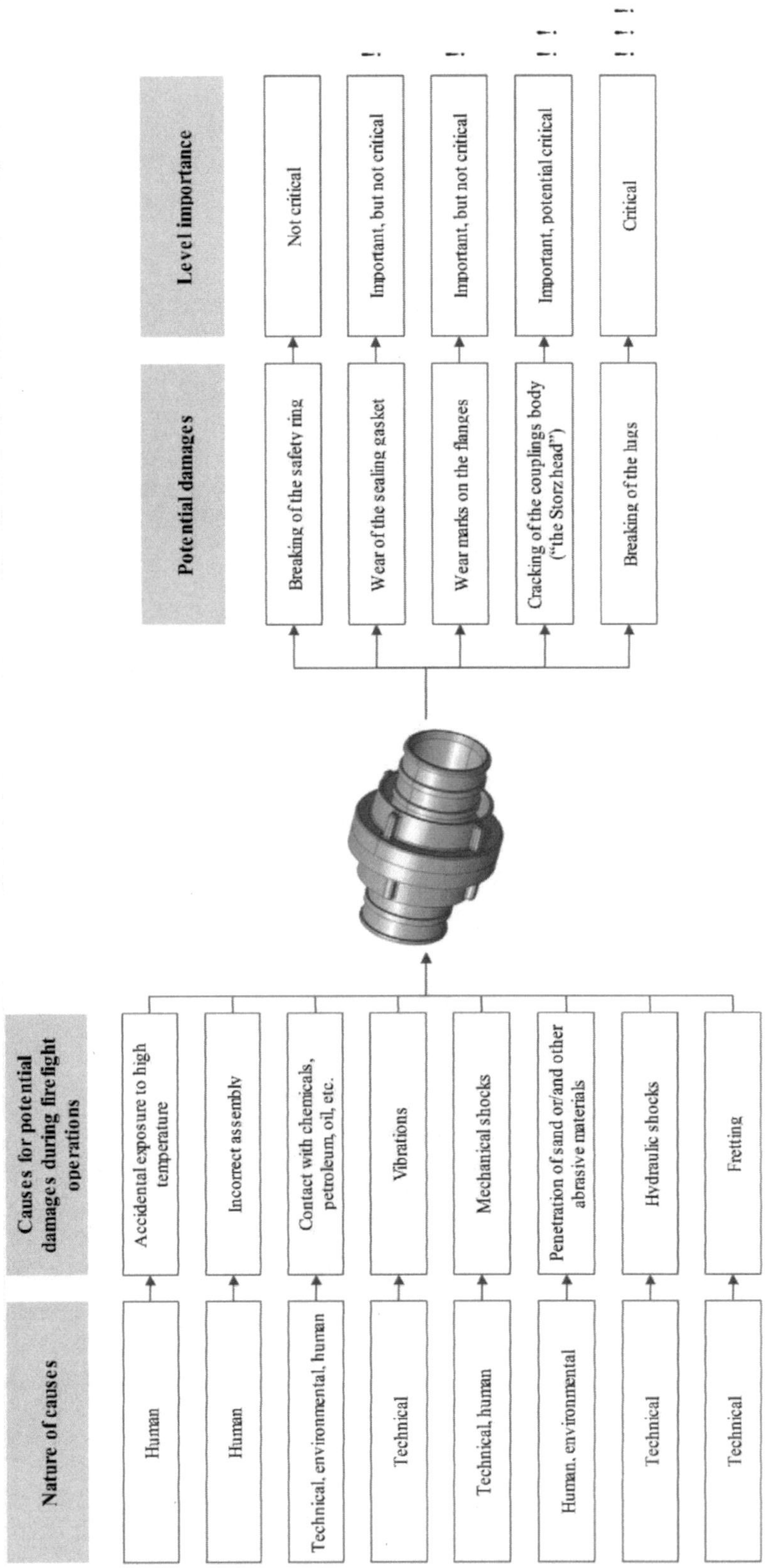

Figure 10. Basic failure mode and effects diagram for potential coupling damages

3. Conclusions

This paper presents a part of a theoretical research on the potential damages of the fire hose couplings during firefighting operations. It is shown that the different damage types could occur due to the operating conditions, that were identified and used for potential failure prediction.

Technical, environmental and human manipulation related parameters could lead to a series of failures that were evaluated from non-critical to critical, with total loss of function of the analyzed fire hose connectors.

This study was conducted based on theoretical research on firefighting operations and the direct practical experience pf the main author in this field. The present study is a starting point for an ensuing research, both theoretical and experimental.

From a theoretical point of view, the effects of hydraulic shocks ("water rams") will be investigated, as well as the influence of the fretting phenomenon- the fretting wear. The following study will investigate the effects of the vibrations generated by the flow of the suppression agents through the components of the hydraulic chain.

From an experimental point of view, the next study will investigate the damages that occur during and/or related to the fire suppression operations. The results will be compared with the predictions, identify the matches and differences, if there are any, and investigate the parameters of the damages that occur in real firefighting situations.

The results of this study will be subsequently used to compile a handbook of guidelines, based on Figure 10, that will address every type of damage identified. The handbook is intended to provide recommendations for the available maintenance and reconditioning methods, that would ensure the life extension and reuse of these parts.

Acknowledgements

The experimental researches are the outcome of the researches done by the Laboratory of the Basis of Experimental Research of the Technical University of Cluj-Napoca, North University Center of Baia Mare, Faculty of Engineering.

References

[1] Bălulescu P and Crăciun I 1993 *Agenda pompierului*, Editura Tehnică, Bucureşti, Romania
[2] Calotă S 2009 *Manualul Pompierului*, Editura Imprimeriei de Vest, Oradea, Romania
[3] ***1992 *Manual pentru cunoasterea accesoriilor, utilajelor şi autospecialelor de stingere a incendiilor*, Editura Ministerului de Interne
[4] ***SR EN 1706:2010
[5] Niţu M I 2010 *Cercetări privind îmbunătăţirea unor proprietăţi fizico-mecanice la aliajele de aluminiu turnabile prin intermediul tratamentelor termice*, Universitatea "Transilvania" Braşov, Doctoral Thesis
[6] Rajic Z, Torkar M et al. 2014 Properties of AlSi5Cu1Mg modified with Sb, Sr and Na, *Materials and Technology Journal* **48**(6) 991–996
[7] Wierzbińska M and Sieniawski J 2006 Effect of morphology of eutectic silicon crystals on mechanical properties and cleavage fracture toughness of AlSi5Cu1 alloy, *Journal of Achievements in Materials and Manufacturing Engineering* **14**(1-2) 31-36
[8] Popescu G et al. 2009 Măsuri generale/specifice pentru controlul riscurilor de deteriorare a furtunurilor plate destinate stingerii incendiilor, *Buletinul Pompierilor* **2**
[9] Ungureanu N S 2003 *Fiabilitate şi diagnoză*, Editura Risoprint, Cluj Napoca, Romania
[10] ***1992 *Friction, Lubrication, and Wear Technology*, ASM Handbook Volume 18
[11] Fenner A J and Field J E 1958 La Fatigue dans les Conditions de Frottement, *Rev. Métall.* **55**
[12] Goto H, Ashida M and Endo K 1987 The Influence of Oxygen and Water Vapour on the Friction and Wear of an Aluminum Alloy under Fretting Conditions, *Wear* **116** 141-155

Collaborative decision-making on wind power projects based on AHP method

A Badea[1], G Proştean[1], M Tămăşilă[1] and A Vârtosu[2]

[1]Politehnica University of Timisoara, Management Department, 14 Remus Street, Timisoara, Romania
[2]Politehnica University of Timisoara, Measurement and Optical Electronics Departament, Vasile Parvan Street, no. 1-2, Timisoara, Romania

E-mail: andra@publicist.com

Abstract. The complexity of projects implementation in Renewable Energy Sources (RES) requires finding collaborative alliances between suppliers and project developers in RES.
Links activities in supply chain in RES, respectively, transportation of heavy components, processing orders to purchase quality raw materials, storage and materials handling, packaging, and other complex activities requiring a logistics system collaboratively to be permanently dimensioned properly selected and monitored. Requirements imposed by stringency of wind power energy projects implementation inevitably involves constraints in infrastructure, implementation and logistics. Thus, following an extensive research in RES project, to eliminate these constraints were identified alternative collaboration to provide feasible solutions on different levels of performance.
The paper presents a critical analysis of different collaboration alternatives in supply chain for RES projects, selecting the ones most suitable for particular situations by using decision-making method Analytic Hierarchy Process (AHP).
The role of AHP method was to formulate a decision model by which can be establish the collaboration alternative choice through mathematical calculation to reduce the impact created by constraints encountered.
The solution provided through AHP provides a framework for detecting optimal alternative collaboration between suppliers and project developers in RES and avoids some breaks in the chain by resizing safety buffers for leveling orders in RES projects.

1. Introduction

Given that presently discussions about dramatic decline in fossil fuel prices are still debated, the benefits provided by renewable energy development has already proven to be a very good solution to the actual problem of global energy. The development of renewable energy sector has the tendency to maximize the benefits that nature has to offer, more than that, wind project developers being aware of the long-term benefits propose major investments in developing countries even more than investments in already industrialized countries [1]. Development projects that support renewable energy production have a significant potential which contributes to a sustainable development in green energy and aims to improve and implement new methods which streamline the implementation process of RES projects [2]. Accomplishing all EU directives through Directive 2001/77/EC, RES projects implementation in Romania has rose by 2.9 percent last year, or 124 MW, to 4.400 MW, according to Transelectrica [3].

IOP Conf. Series: Materials Science and Engineering **163** (2017) 012017 doi:10.1088/1757-899X/163/1/012017

The increase of installed capacity in renewable energy is due to the fact that Romania has complex areas of landforms, which allows it to exploit wind, solar and biomass energy. A particular situation respectively real areas isolation of many human communities, are deprived of RES benefits due to the landform conditions. Such locations are relatively numerous in Romania, with wide range of applications: dwellings, isolated agriculture farms, sheepfolds, touristic chalets, stations broadcast relay etc. In these situations, economic efficiency is the determining factor that lead to the assessment of the opportunity for this kind of projects.

This research presents a RES exploitation project in isolated regime made by a team composed by key representative from a company, direct interested in research theme, and the authors. In this project the research team identifies through long debates and multi-criteria decision an efficient and economical way to choose a wind turbine power and an optimal assembly method in isolated regime. These decisions were based on project activities which involved research, development and implementation correlated with some disciplines that include a wide range of specialties such as: Mechanical structures and wind turbines (MSWT); Electrical machines (EM); Power electronics (PE); Automatic controlling strategies (ACS); Project management and administrative problems (PM).

1.1. Wind power projects implementation

RES exploitation project in isolated regime covers a wide range of activities, this requiring a dispersed set of skills. A major challenge for these kind of project empowered wind power project developers to identify and surpass the obstacles that they encounter in their way, when implementing project specifications. These obstacles were present for the entire project life cycle, namely research, development and implementation.

Activities within the project:

Research: the tower, the nacelle, and the rotor blades.

Development: communication sessions between the parties that made the project: communication with funders, communication with local authorities (to obtain approvals and permits), communication with suppliers

Implementation of a windmill: preparation of the site (roads cut, land graded and leveled), a concrete foundation is made with the underground cables, the tower is assembled on site and erect with a special crane, the fiberglass nacelle is assembled with inner workings off site (main drive shaft, gearbox, and blade pitch and yaw controls), the utility box for each wind turbine and the electrical communication system are installed simultaneously with the placement of the nacelle and blades.

2. Specific collaboration in sharing risks in such projects

The research team identify constraints in RES project in isolated regime and established a complex and suitable decision model which underpin the decision selection of wind power turbine and the way of assembly.

2.1. Bottlenecks in wind power supply chain

Most common situations encountered by developers in wind power project were:

- Developing a project plan, which includes defining project goals, how task and objectives will be achieved.

-What resources are need, associating budgets and timelines for completion, implementing the project plan, and ensure that the plan is being managed according to the project.

-The process of obtaining necessary permits and approvals is laborious and time-consuming and construction works sometimes can be atypical, hampered by placement location without power supply and water. Weather conditions also can influence the development work in the field.

-The analysis and research of the cost reduction possibilities, through diversification and simplification of the equipping, multiple usages of the components.

-The necessary financial funds, various reorganizations and restructurings are necessary to complete the project in time and achieve the desired goals.

- It can be difficult finding companies that have the ability to achieve some different materials elements of high precision.
-Transportation of heavy components, processing orders to purchase quality raw materials, storage and materials handling, packaging.
-Bottlenecks in infrastructure, implementation and logistics (Figure 1).

2.2. Problem statement

The first part of the project was very complex because of the region isolation special conditions so finding suitable sites and transportation routes for turbine installation it was a challenge for project developers.

Another aspect was the deficiency quality of rural roads that are designed for low traffic or light vehicles. New roads construction on slopes to access grows, lead to erosion causing degradation of the landscape location. Higher components transportation that need to be assembly at the farm site implies a higher cost.

A possible solution to obtain minimal costs transportation implies that the assembly process of the wind turbine should integrate more possible operations at the component manufacturing [4].

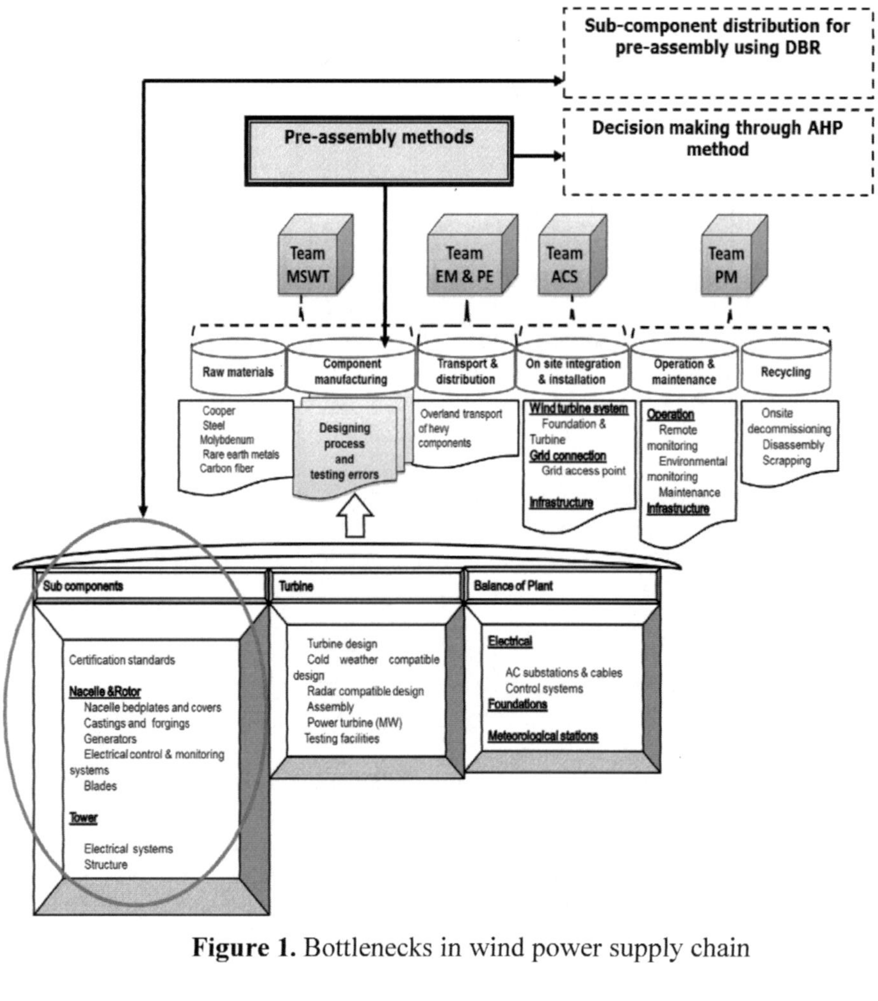

Figure 1. Bottlenecks in wind power supply chain

2.3. Optimal collaboration alternatives

In this paper there are presented several pre-assembly methods with different characteristics that are technically adequate to achieve the economic objective for such projects identified above. The methods are classified according to the number of lifts required for each turbine [5].

Method 1 -low-power wind turbines: less than 12m blades diameter rated power of less than 1.0 MW and turbine pillar 7 m height. Hub and all three blades are assembled at the farm site. Remaining sub-assemblies (lower tower and transition piece, lower and upper tower sections, upper tower and nacelle) are done at the farm site. Two tower sections and the nacelle are transported separately. 4 lifts (operation) are required for each turbine during loading and during installation.

Method 2- medium power wind turbines: less than 45m blades diameter rated power between 1.0 MW and 2.0 MW turbine pillar 9 m height. Two tower sections are assembled at the farm site, also the nacelle and hub are assembled together all the remaining components are transported separately to farm site. For each turbine, 5 lifts (operation) are required during loading and during installation. During installation, first the tower is assembled to the transition piece, then the nacelle and hub sub-assembly is attached to the tower, finally three blades are lifted and assembled to the hub.

Method 3 -high power wind turbines: 46m blades diameter exceeding rated power over 2 .0 MW-2.5 MW turbine pillar 12 m height. The nacelle and hub are assembled together at the farm site all the other components are transported to the farm site separately. 6 lifts are needed for each turbine for loading and for installation. During installation, first the lower tower is assembled to the transition piece, then two tower sections, after that the nacelle and hub sub-assembly is attached to the upper tower; finally, three blades are bolted to the hub one by one.

3. Methodology

Finding suitable sites and transportation routes for wind farms is a complex decision-making problem, involving several, sometimes conflicting, criteria and multiple objectives. Apart from the manufacturing turbines cost, energy cost is significantly affected by transportation and installation costs operations of wind turbines and turbine components maintenance operations.

Through optimum selection of decision variables, such as turbine installation method and rated power output of each turbine, cost of transportation and installation operations can be minimized. Pre-assembly at the farm site area is another controlling variable of transportation and installation cost. More pre-assembly done results in lower number of lifts and assembly.

3.1. Decision making methods

In order to have a clear classification of the presented pre-assembling methods we identify base on our RES project very important criteria that need to be assess when pre-assembly components must be achieved at the component manufacturing.

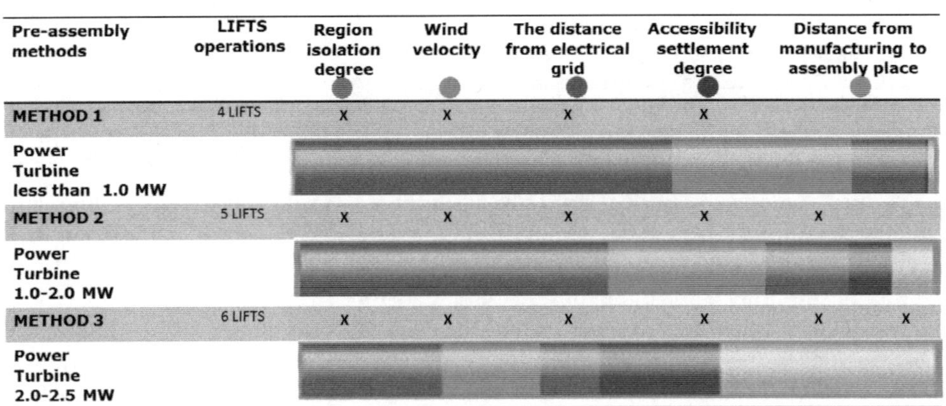

Figure 2. Wind turbine pre-assembling methods

The criteria presented are: region isolation degree, wind velocity, distance from electrical grid, accessibility settlement degree and distance from manufacturing to assembly place. To rank this classifications, the research team used a software online application (Priority Estimation Tool-AHP) [6] (Figure 2).

To achieve optimal selection a MULTI-CRITERIA DECISION MAKING (MCDM) methods provide a logical framework to investigate, analyze, and solve such problems:

- provide an effective structure in decision-making process;
- shows objectives which decision maker has identified;
- measurable evaluation criteria of objectives are established;
- provides several ways of aggregating data concerning criteria for obtaining indicators (scores) of alternatives performance;
- helps to maintain decision makers thinking models by deriving the relative weight of each component;
- it is assigned a numerical value to each alternative of the problem;
- through mathematical calculation is chosen the optimal alternative.

3.2. Analytic Hierarchy Process Method

The algorithm AHP is done in six stages:

1. Hierarchical scheme composition of the problem that need to be analyzed (Figure 3). In this phase it is presented fundamental purpose of the problem to be analyzed. Decision criteria are identified and, if necessary, separating them into sub-criteria. Depending on the preferences of the analyst alternatives are presented and ranked by the decision;
2. Establish relative weights by comparing them in combinations of two. The relative weights are based on a numerical scale from 1 to 9 through Saaty scale followed by subjective evaluation of the decision-maker comparisons are made in pairs. By comparison is obtained the degrees of importance of a criterion to each other (Table 1) [7];
3. The relative prioritization of the sub criteria is obtained by comparing sub-criteria two by two depending on the decision criterion, in order to rank them;

Table 1. Relative importance through Saaty scale [7]

Numerical value	Verbal meaning for risk factor evaluation	Verbal meaning for alternative evaluation
1	Equally important	Equally preferred
2	Equally to moderately more important	Equally to moderately preferred
3	Moderately more important	Moderately preferred
4	Moderately to strongly more important	Moderately to strongly preferred
5	Strongly more important	Strongly preferred
6	Strongly to very strongly more important	Strongly to very strongly preferred
7	Very strongly more important	Very strongly preferred
8	Very strongly to extremely more important	Very strongly to extremely preferred
9	Extremely more important	Extremely preferred

4. Pairwise comparison is made of decision criteria (criteria and sub-criteria) gives them a global priority. Global priority for each decision criterion is determined by the result of multiplying each priority vector for each criterion on the second level of the decisional tree with relative prioritization of sub-criteria from the third level of decisional tree;
5. Performance matrix is achieved by calculating the relative prioritization of the alternative for each sub-criterion. In this step is necessary to assess alternatives relative weights from fourth level of the hierarchy, the process is similar to that has been used in criteria and sub-criteria group for the purpose decision-making. This process is executed as the previous step, but the

difference does that this step is performed assessment of each alternative beside each sub-criterion;

6. The final decision is obtained by choosing alternative priority with maximum points and argumentation of the obtained solution.

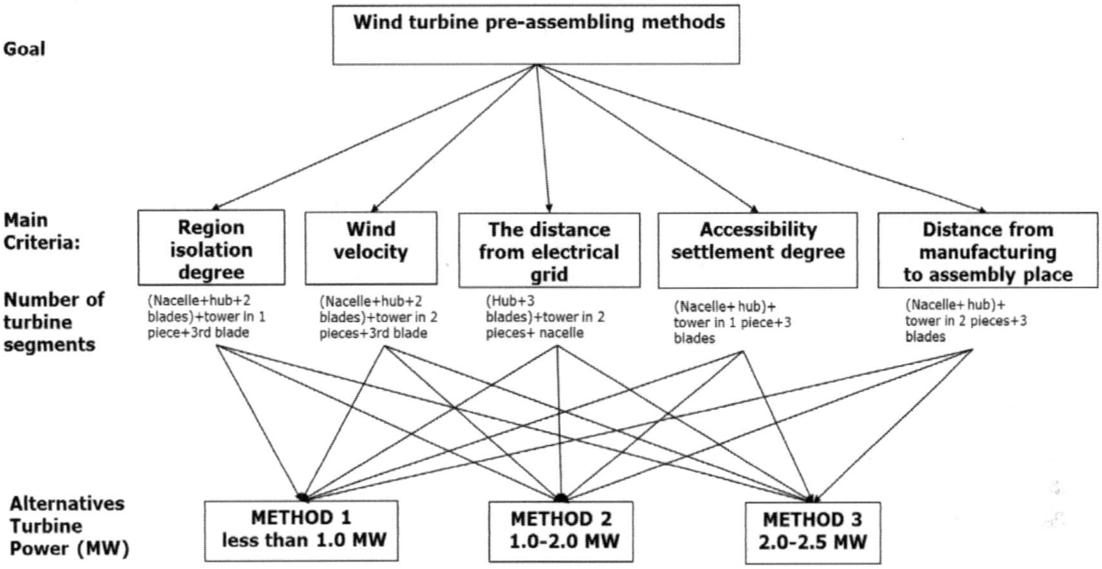

Figure 3. Decision tree through AHP method

Utilization of AHP method involves the decision tree decomposition into criteria and alternatives (Figure 4). AHP method designated as region isolation degree, wind velocity and the distance from electrical grid that they were the main important criteria with higher rank to emerge from the analysis decision. For wind turbine pre-assembling methods, they were chosen turbines with different powers, namely, less than 1.0 MW, 1.0-2.0 MW and 2.0-2.5 MW.

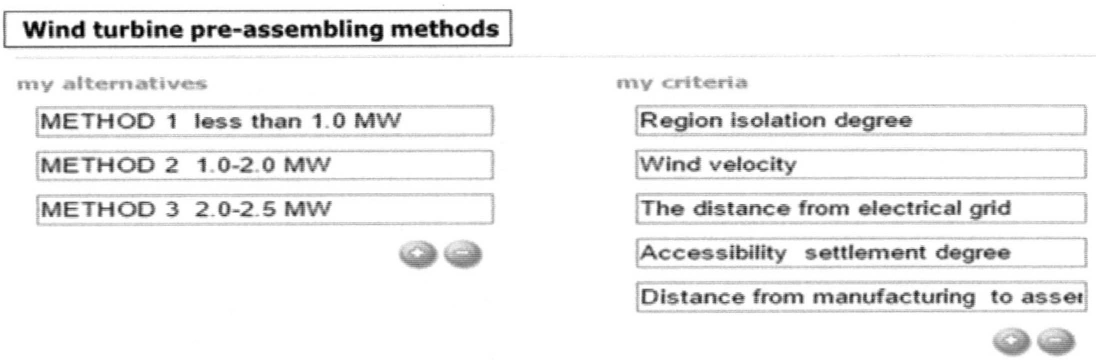

Figure 4. Decision tree decomposition into criteria and alternatives

Applying the method AHP, has required to use Method 1, namely the importance given to each criterion and by comparison matrices, revealed a very high value on **region isolation degree** (0,5271). In this scenario the isolated region characteristics they had relative disadvantages because of the extremely weather conditions and landform, which made it the wind turbine difficult to implement. The number of panels where smaller, and the size of wind generators were also reduced, namely a theft implies a critical impact. In this case it is knowing that Method 2 and Method 3 involve a greater risk for pre-assembling of wind power turbine in such conditions. Inspection of the local areas around the target site and many visits to select the site was made by the research team to estimate wind power potential, namely wind velocity measurements. Analysis of daily and annual **wind velocity** (0,2641) measuring system was delivered and in-house tested together with data processing system and wireless communication. Also the **distance from electrical grid** (0,1122) has implied a system which is not vulnerable to surges from lightning, solar flares, earthquakes, ice storms, wind. This technique provides delivering renewable energy over long distances (Figure 4).

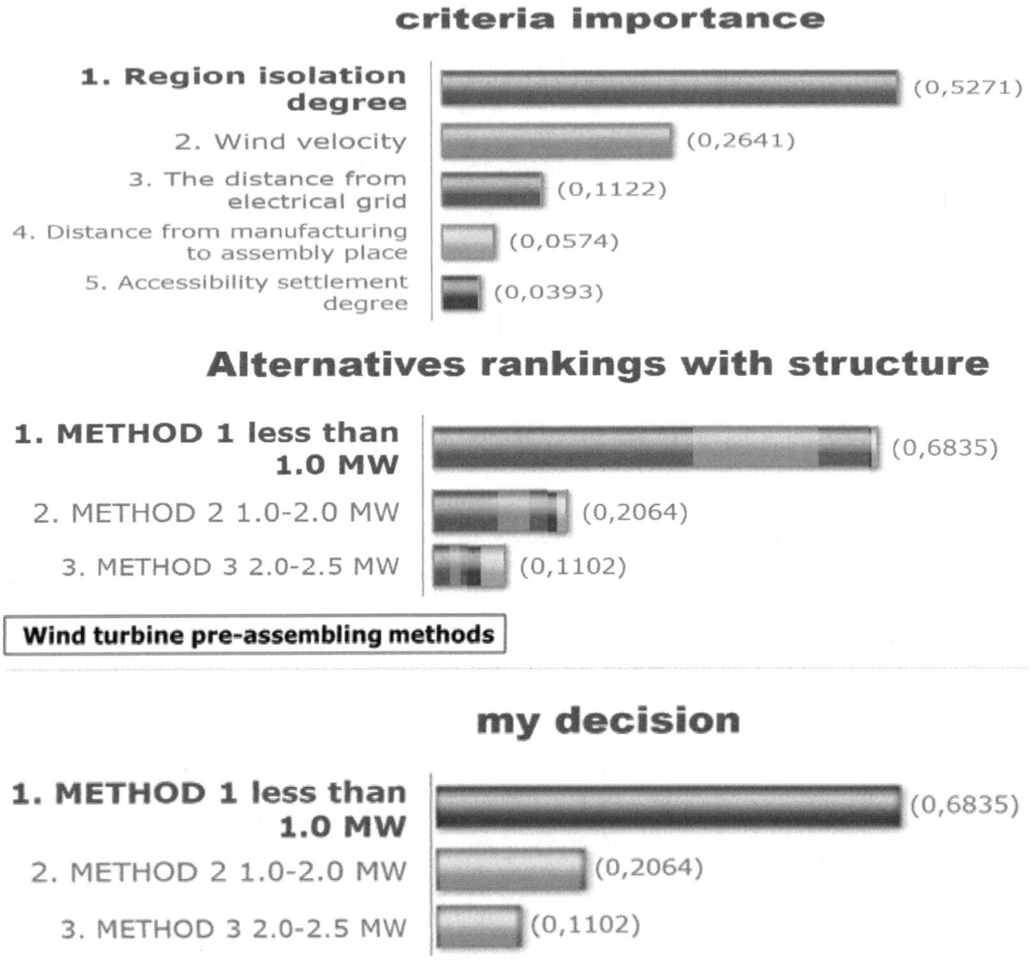

Figure 5. Final results presentation

4. Conclusions and future work

The solution provided through AHP, Method 1 provides a framework for detecting the optimal alternative and avoid less costs in the wind supply chain by choosing the right method of pre-assembling components and obtain minimal transportation cost to install the wind turbine in special conditions. Using AHP algorithm offers both advantages and disadvantages as its defining by the elements it can be decide which combination is suitable to elaborate a wind turbines economical installation plan in isolated regions.

AHP method also highlights the ability to obtain correct weights of each criterion in the case of the present study, these being developed on the economic point of view. This method allows alternatives and criteria analysis to be upgraded during the RES project development so that all beneficial factors to be considered and lead to the fulfillment of the ultimate goal.

RES projects in isolated regime are a compromise solution, but implementation on such projects provides local energy needs in isolated locations. Even in this case these represent an important way for generating green energy which in a visible manner sustain the strategy to achieve Romania's mandatory targets within the EU strategy 20/20.

References

[1] ***REN21 2016 *Renewables 2016 Global Status Report,* (Paris: REN21 Secretariat), ISBN 978-3-9818107-0-7

[2] Zamfir A I 2011 Management of renewable energy and regional development: European experiences and steps forward, *Theoretical and Empirical Researches in Urban Management* **6**(3) 35-42

[3] ***Planul de Dezvoltare a RET perioada 2016- 2025 http://www.transelectrica.ro/documents/10179/25146/Planul+de+dezvoltare+a+RET+2016-2025.pdf/0d1d070f-75c5-43d6-8833-6563dd9e7c58

[4] Faiz T I 2014 *Minimization of Transportation, Installation and Maintenance Operations Costs for Offshore Wind Turbines,* B.Sc.Eng., Bangladesh University of Engineering and Technology

[5] Kaiser M J and Snyder B 2010 *Offshore Wind Energy Installation and Decommissioning Cost Estimation on the U.S. Outer Continental Shelf,* Energy Research Group, LLC, Baton Rouge, LA, USA

[6] Siraj S, Mikhailov L and Keane J A 2015 PriEsT: an interactive decision support tool to estimate priorities from pairwise comparison judgments, *International Transactions in Operational Research* **22** 217–235

[7] Saaty T L 1980 *The Analytic Hierarchy Process,* McGraw-Hill Book Co., New York, USA

IOP Conf. Series: Materials Science and Engineering **163** (2017) 012018 doi:10.1088/1757-899X/163/1/012018

Using the heat flow plate method for determining thermal conductivity of building materials

M Flori, V Puţan and L Vîlceanu

Politehnica University of Timisoara, Engineering and Management Department, Vasile Parvan str., no. 1-2, 300223 Timisoara, Romania

E-mail: flori.mihaela@upt.ro

Abstract. The heat flow plate method is used to determine thermal conductivity of a building material sample made of Rohacell (insulating foam). Experimental technique consists in placing the sample with a reference material on top (polystyrene sample) in a calorimetric chamber and heating from underside. Considering that the heat flux which passes through the two layers is constant and knowing thermal conductivity of the reference material, the sample thermal conductivity is determined. The temperature difference between the two opposite sample's sides is recorded only when the steady state is achieved (constant heat flux).

1. Introduction

The phenomenon by which heat is transported through a material, when a temperature difference is applied between two opposite sides, is called thermal conduction. Thermal conductivity is the property which indicates the ability of materials to conduct heat [1]. Knowing material thermal properties is important in many applications especially in thermal insulation where the heat exchange with surroundings is a particularly important issue [2].

Since conductive heat transfer phenomenon takes place in a thermal field with temperature differences, of interest is how thermal conductivity varies with temperature. So, for materials majority was established the following dependence of thermal conductivity with temperature [1]:

$$\lambda = \lambda_0[1 + C_1(T - T_0)] \quad W/(m \cdot K) \tag{1}$$

where: λ is thermal conductivity at temperature T, λ_0 is thermal conductivity at reference temperature T_0, C_1 is a constant characteristic of each material.

Thermal conductivity values for some classes of substances (gases, liquids and solids) at moderate pressure and temperature are given in Figure 1 [1].

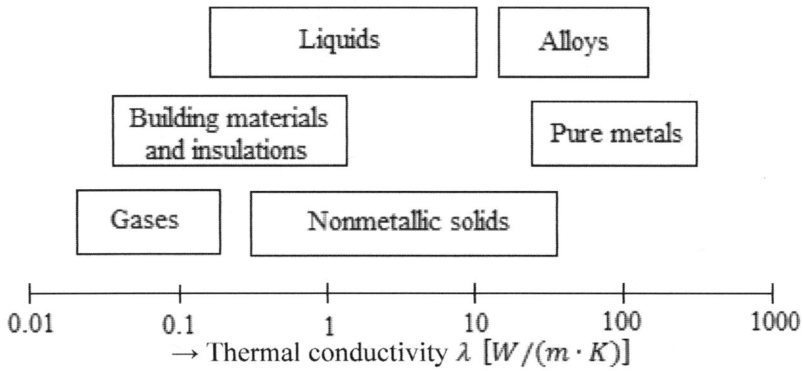

Figure 1. Thermal conductivity values for some categories of substances [1]

For building materials and insulations, λ has values in the range $(0.02 \div 3\ W/m \cdot K)$ and depends on material structure, porosity and humidity. These values increase with increasing temperature or with increasing density [1].

If the heat transfer process is stationary (steady-state condition), the heat flux which passes through each *n* layers of a flat wall is considered identical (Figure 2) [1]:

$$\dot{Q}_1 = \dot{Q}_2 = \cdots = \dot{Q}_n = \dot{Q}\ [W] \tag{2}$$

where: subscript *1…n* denotes layer number.

So, according to Fourier's Law of Heat Conduction, the heat flux passing through the first layer of the wall is [1], [2]:

$$\dot{Q}_1 = \frac{\lambda_1}{d_1} A(T_1 - T_2)\ \ [W] \tag{3}$$

where: $\lambda_1\ [w/(m \cdot K)]$ is thermal conductivity; $d_1[m]$ is the wall thickness; $A[m^2]$ is the wall area normal to the heat flux direction; T_1 and T_2 in [K] are the temperatures on two sides of the layer.

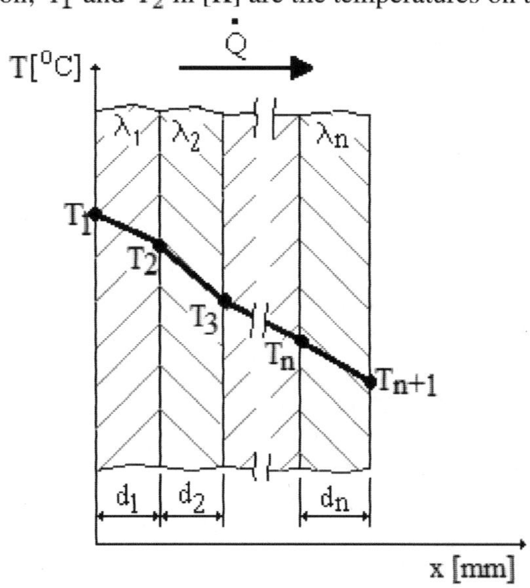

Figure 2. Temperature distribution in a flat wall with *n* layers crossed by a constant heat flux $\dot{Q}$ (steady state condition) [1]

Temperature difference in the first layer of d_1 thickness may be written as:

$$T_1 - T_2 = \frac{d_1}{\lambda_1 \cdot A} \dot{Q}_1 \tag{4}$$

in similar way, for the other layers results:

$$T_2 - T_3 = \frac{d_2}{\lambda_2 \cdot A} \dot{Q}_2 \quad (5)$$

....

$$T_n - T_{n+1} = \frac{d_n}{\lambda_n \cdot A} \dot{Q}_n \quad (6)$$

By adding equations (4) to (6), and taking into account equation (2) results the heat flow through the wall:

$$\dot{Q} = \frac{T_1 - T_{n+1}}{\sum_{i=1}^{n} \frac{d_i}{\lambda_i \cdot A}} \ [W] \quad (7)$$

If the wall has one layer, relation (3) may be used for the heat flux determination.

Several experimental methods have been developed for thermal conductivity measurement, e.g.: steady-state techniques (heat flow plate [4-6], guarded hot plate [6], [7] and guarded hot box [8], [9]) or transient techniques (line heat source, plane heat source and disc heat source [6]).

The steady-state techniques require experimental conditions in which the temperature gradient does not change with time (i.e. constant heat flux as shown above in Figure 2), while the transient ones permits determinations when material's temperature varies in time.

This study is focused on the steady-state techniques which consists in measuring the temperature difference between two opposite sides of sample with certain thickness, knowing the heat flux value which passes through it. Therefore, thermal conductivity in a single layer wall may be deduced from relation 3 which represents the Fourier's Law of Heat Conduction for steady-state conditions [1], [2]:

$$\lambda = \dot{Q} \cdot d / A \cdot \Delta T \ [W/m \cdot K] \quad (8)$$

where: $\Delta T = T_1 - T_2 \ [K]$ is the temperature difference between wall's two sides.

By the heat flow plate method it may be determined thermal conductivity of an unknown material plate (λ_x) with the aid of a reference material plate ("heat flow plate") with known thermal conductivity (λ_R) [2], [3].

The plates are placed one on top of another and crossed from underside by a constant heat flux (steady state condition):

$$\dot{Q}_x = \dot{Q}_R \rightarrow \lambda_x \cdot \Delta T_x \frac{A_x}{d_x} = \lambda_R \cdot \Delta T_R \frac{A_R}{d_R} \quad (9)$$

where subscript x denotes "unknown" and R denotes "reference".

As the two plates have same size ($A_x = A_R$ *and* $d_x = d_R$), when the reference plate is placed on the bottom, results [1], [2]:

$$\lambda_x = \lambda_R \cdot \frac{\Delta T_R}{\Delta T_x} \ [W/m \cdot K] \quad (10)$$

and when the reference plate is on top:

$$\lambda_x = \lambda_R \cdot \frac{\Delta T_x}{\Delta T_R} \ [W/m \cdot K] \quad (11)$$

where: $\Delta T_R = (T_u - T_m)$ and $\Delta T_x = (T_m - T_o)$ are the temperature differences, with T_u – temperature on the underside (heated side); T_m – temperature in the middle, between the plates; T_o – temperature on the outside.

The aim of this paper is determining the thermal conductivity of Rohacell foam sample using the heat flow plate method, with polystyrene sample as the reference material. Two series of experiments were done: the first one using single layer wall of polystyrene, and the second one using a two-layer wall made of Rohacell foam (inside) and polystyrene (outside). So, the first experiment aims determining the thermal conductivity of the reference sample (polystyrene) by means of which the thermal conductivity of Rohacell sample is determined through the second experiment.

2. Experimental details

2.1. The experimental setup

Figure 3 shows a photograph of the experimental ensemble containing: transformer 2-12 V (1), calorimetric chamber with insulated housing (2), digital thermometer with 4 inputs (3) and laptop for data acquisition (4).

The calorimeter housing (2) is insulated but permits through several channels in the exterior walls, insertion of the thermocouple sensors. Inside it is placed a heater having a $80 \times 100\ mm$ surface, working by Joule effect. The heat flux magnitude may be adjusted by modifying the applied current and tension with the transformer (1). So, the heat flux is proportional with the electric power consumed by the heater and may be determined with relation:

$$\dot{Q} = U \cdot I\ [W] \tag{12}$$

where: $U[V]$ is tension and $I[A]$ is current applied to the heater.

For the first experiment, using a single layer wall of polystyrene, the parameters were: $U_1 = 10\ V$ and $I_1 = 1.44\ A$, giving a heat flux of $\dot{Q}_1 = 14.4\ W$. For the second experiment, using a two-layered wall of polystyrene and Rohacell foam, the parameters were: $U_2 = 6\ V$ and $I_2 = 0.87\ A$, giving a heat flux of $\dot{Q}_2 = 5.22\ W$.

A digital temperature indicator with 4 inputs (3) permits simultaneous displaying of temperature values which were recorded with the aid of a data acquisition system (4).

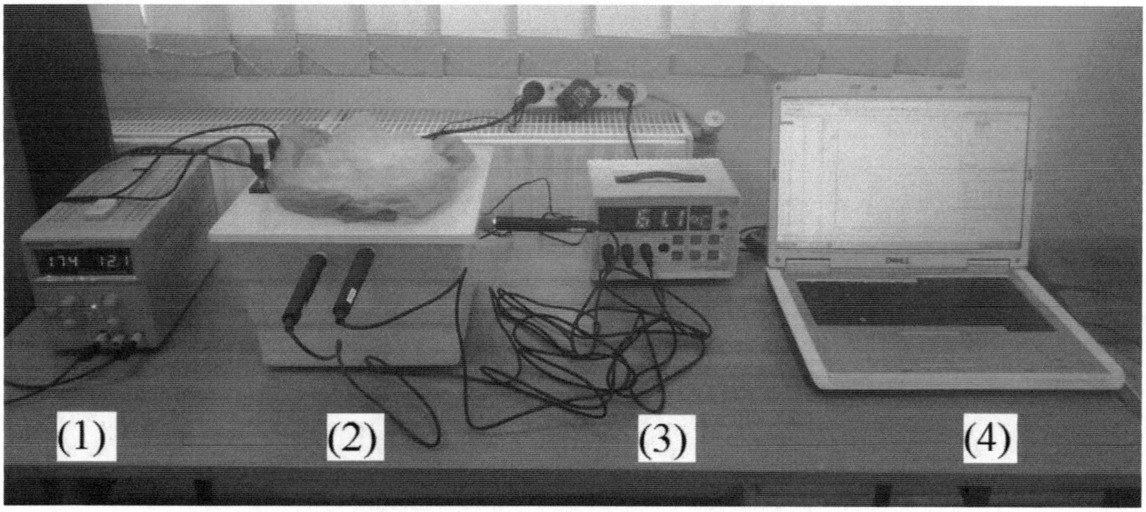

Figure 3. Photograph of experimental setup:
(1) Transformer; (2) Calorimetric chamber; (3) Digital temperature indicator; (4) Laptop

2.2. Temperature measurements

Thermocouples arrangement through the calorimetric chamber walls is shown in Figure 4, a) and b).

Three NiCr-Ni thermocouple sensors were used for temperature measurements as follows: $\mathbf{T_u}$ – on the underside, toward the interior of the calorimetric chamber (heated side); $\mathbf{T_m}$ – in the middle, between the two samples; $\mathbf{T_o}$ – outside sample surface, toward the exterior of the calorimetric chamber (Figure 4). When single sample was used, T_u represented the inside air temperature and T_m, underside sample temperature.

All samples have square surface of $150 \times 150\ mm$ and $10\ mm$ thickness and fit perfectly inside the calorimetric chamber in horizontal position (see Figure 4, b). Also, on both sides each sample has semi-circular grooves of $2\ mm$ diameter for inserting the thermocouples, so when a multilayered structure is formed, a perfect contact is permitted between samples (see Figure 4, b).

As the heat source is positioned under the samples, the heat flux is transmitted from bottom to top (see Figure 4, a). Also the temperature difference over the samples cross-sections is needed to be kept constant during measurements (steady state conditions), and so a constant outside temperature was maintained with the aid of ice, placed in a plastic bag covering the entire sample surface (Figure 3).

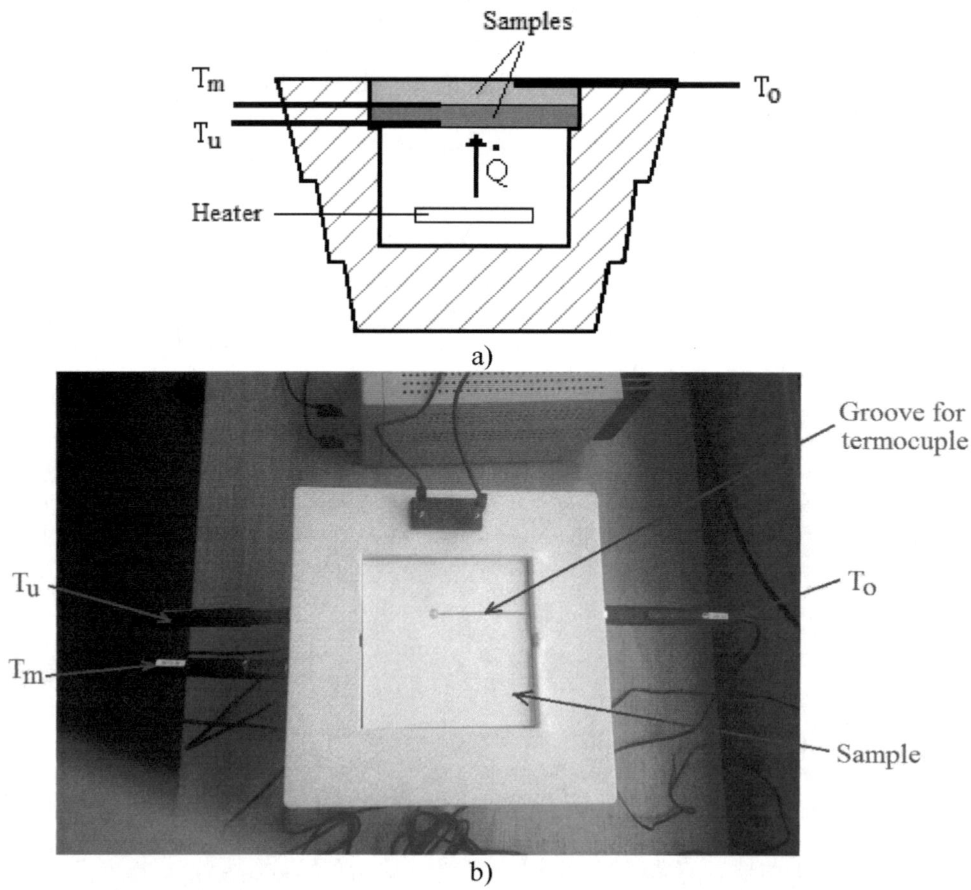

Figure 4. Thermocouples arrangement in the calorimetric chamber: (a) schematic representation, side view and (b) photograph, top view

Temperature values were recorded only when the heating conditions achieved steady state. The procedure was as follows: the samples were positioned in the calorimetric chamber and the heat source was turned on. Further, both underside (T_u) and outside (T_o) temperatures were observed until the higher one started to rise and the lower one was maintained constant. It was recommended that the outside temperature (toward ice) T_o is kept constant in the range $-2°C$ and $+4°C$ [2].

3. Results and discussion

Temperature variations on both sides of polystyrene sample versus time, measured during the first experiment are presented in figure 5.

As one may observe, temperature T_o on the upper side of sample (outside, toward ice) was maintained in the range $+4.3°C$ and $+4.6°C$, so an average of $+4.45°C$ was considered for subsequent calculations. Temperature T_m on the underside of sample (toward the heater) rises from $19.5°C$ up to $34.2°C$ after heating during 50 minutes, when it's maintained constant. Also, temperature inside the calorimetric chamber T_u rises continuously from 35.5°C up to about 84.5°C after 50 minutes.

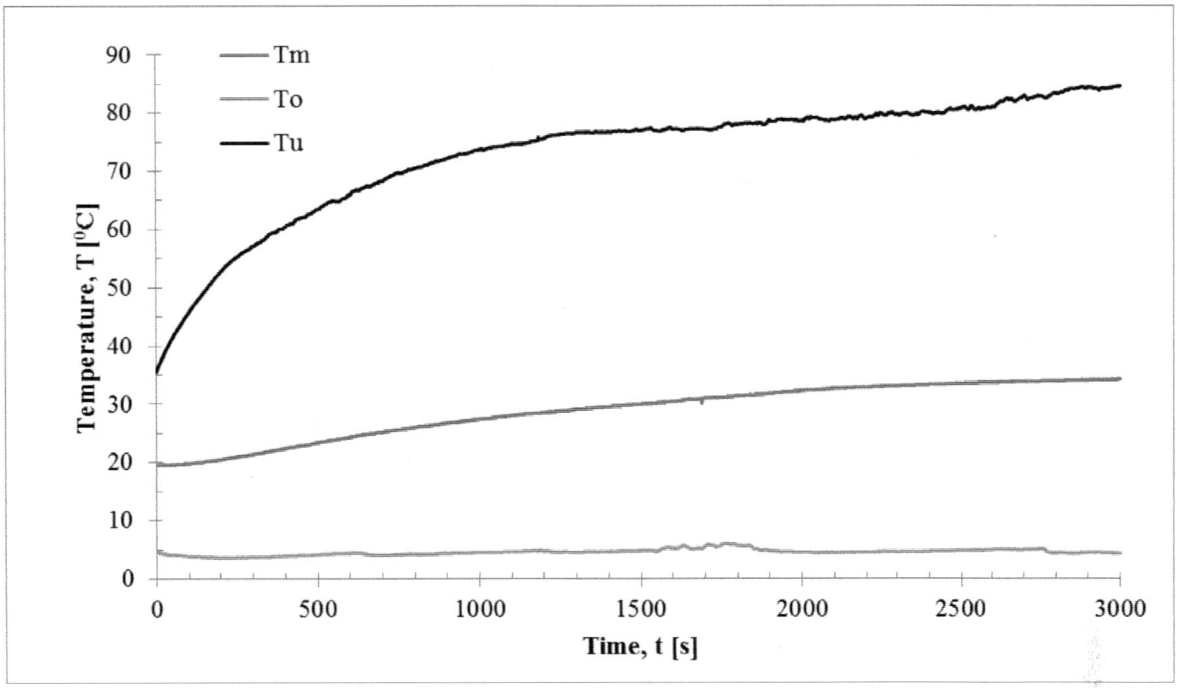

Figure 5. Temperature variation vs. time of polystyrene sample

For calculating the thermal conductivity of polystyrene sample is used formula (8) with the following quantities: $\dot{Q}_1 = 14.4\ W$, $d = 0.01\ m$, $A = 0.0225\ m^2$ and $\Delta T = T_m - T_o = 34.2 - 4.45 = 29.75°C$.

So, the calculated thermal conductivity of the reference material is: $\lambda_R = 0.2151\ \frac{W}{m \cdot K}$. This value is slightly bigger than that of $\lambda_R = 0.16 \div 0.18\ \frac{W}{m \cdot K}$ indicated by the polystyrene sample producer [2]. Also a bigger value is expected since the heat flux crossing the sample cross-section is not equal to the electric power consumption due to heat losses [2].

Temperature measurements during the second experiment when the polystyrene sample was placed on top of Rohacell sample are presented in Figure 6. Temperature T_o on the upper side of sample (outside, toward ice) was maintained in the range $+0.2°C$ and $+1.3°C$, so the average is of $+0.75°C$. After heating during 50 minutes the temperature between the samples is constant and is of $T_m = 10.7°C$, while temperature underside is $T_u = 41.7°C$.

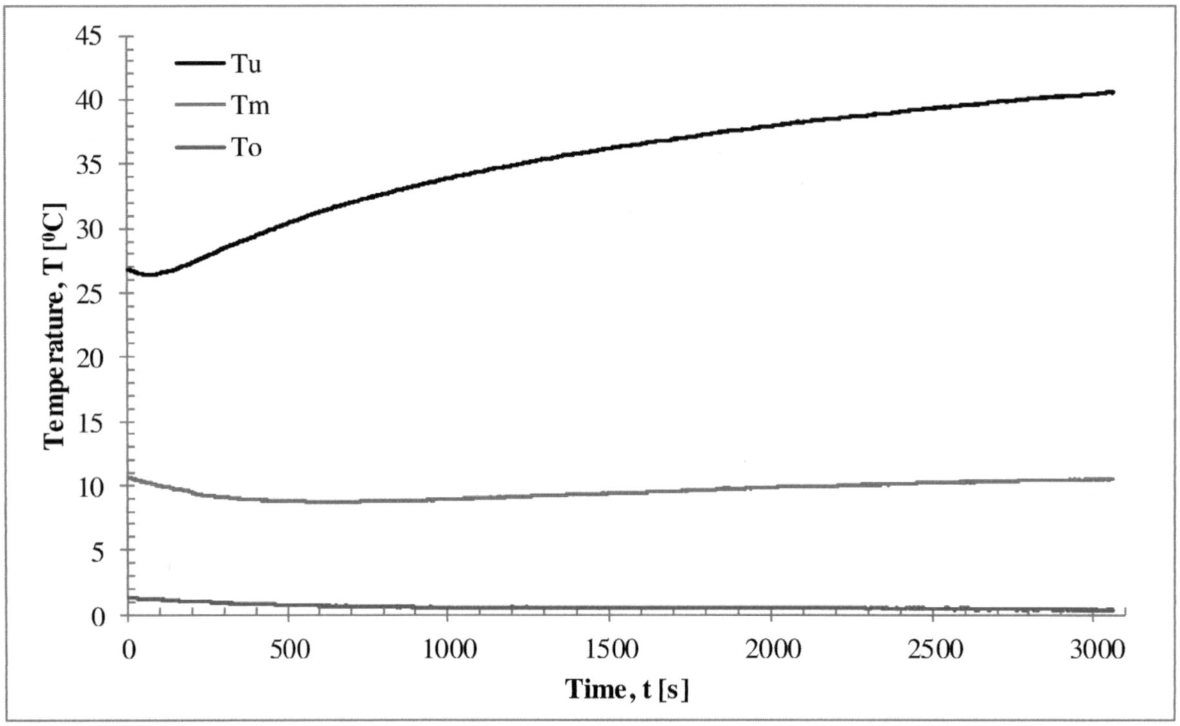

Figure 6. Temperature variation vs. time of polystirene and Rohacell samples

Thermal conductivity of Rohacell sample calculated with the formula (4) is: $\lambda_x = 0.2151\frac{10.7-0.75}{41.7-10.7} = 0.0690\ W/m\cdot K$. As attended and explained above, obtained value is slightly bigger than that recommended by the producer, i.e. $\lambda_x = 0.02 \div 0.05\ W/m\cdot K$ [2].

4. Conclusions

In order to determine thermal conductivity of Rohacell (insulating foam) sample by heat flow plate method, the polystyrene reference sample thermal conductivity was first established. So, from Fourier law of heat conduction thermal conductivity was determined knowing the heat flux which passes through the sample of certain dimensions and the experimentally determined temperature difference between the two opposite sides of the sample. Next, considering a two layered sample (polystyrene sample placed on the Rohacell one) and a constant heat flux which passes through both layers, thermal conductivity of Rohacell sample was determined in function of previous calculated polystyrene thermal conductivity (reference material). A small difference between determined values and the producer recommended ones is probably due to heat losses during experiments.

References

[1] Janna W S 2000 *Engineering heat transfer, Second edition*, CRC Press LLC, US

[2] ***Leybold Didactic, *Physics Experiments Catalogue*, http://www.ld-didactic.de/en.html

[3] Yesilata B and Turgut P 2007 A simple dynamic measurement technique for comparing thermal insulation performances of anisotropic building materials, *Energy and Buildings* **39** 1027-1034

[4] Mahanta N K and Abramson A R 2010 The dual-mode heat flow meter technique: A versatile method for characterizing thermal conductivity, *International Journal of Heat and Mass Transfer* **53** 5581–5586

[5] Hostler S R, Abramson A R, Gawryla M D, Bandi S A and Schiraldi D A 2009 Thermal conductivity of a clay-based aerogel, *International Journal of Heat and Mass Transfer* **52** 665–669

[6] Latif E, Pruteanu M, Rhydwen G R, Wijeyesekera D C, Tucker S, Ciupala M A and Newport D 2011 *Thermal Conductivity of Building Materials: An overview of its determination*, 6th Annual Conference on Advances in Computing and Technology (AC&T), University of East London, London, UK, January, pp 15-22

[7] Kobari T, Okajima J, Komiya A and Maruyama S 2015 Development of guarded hot plate apparatus utilizing Peltier module for precise thermal conductivity measurement of insulation materials, *International Journal of Heat and Mass Transfer* **91** 1157–1166

[8] Buratti C, Belloni E, Lunghi L, Borri A, Castori G and Corradi M 2016 Mechanical characterization and thermal conductivity measurements using of a new 'small hot-box' apparatus: innovative insulating reinforced coatings analysis, *Journal of Building Engineering* **7** 63–70

[9] Asdrubali F and Baldinelli G 2011 Thermal transmittance measurements with the hot box method: Calibration, experimental procedures, and uncertainty analyses of three different approaches, *Energy and Buildings* **43** 1618–1626

International Conference on Applied Sciences (ICAS2016) IOP Publishing
IOP Conf. Series: Materials Science and Engineering **163** (2017) 012019 doi:10.1088/1757-899X/163/1/012019

Dynamic simulation of road vehicle door window regulator mechanism of cross arm type

I Zs Miklos, C Miklos and C Alic

Politehnica University of Timisoara, Department of Engineering and Management, 5 Revolution Street, Hunedoara, 331128, Romania

E-mail: imre.miklos@fih.upt.ro

Abstract. The paper presents issues related to the dynamic simulation of a motor-drive operating mechanism of cross arm type, for the manipulation of road vehicle door windows, using Autodesk Inventor Professional software. The dynamic simulation of the mechanism involves a 3D modelling, kinematic coupling, drive motion parameters and external loads, as well as the graphically view of the kinematic and kinetostatic results for the various elements and kinematic couplings of the mechanism, under real operating conditions. Also, based on the results, the analysis of the mechanism components has been carried out using the finite element method.

1. Introduction

With the development of the car making industry, in addition to an increase in car technical performance, as well as the development of engines & transmissions and improvement of car bodywork aerodynamics, electronic systems (on-board computer) or passive safety features, the automobile manufacturers insist increasingly on the driver and passenger comfort, cabin ergonomics, and on the idea that, in particular the driver but also the passengers, must manipulate the various parts of vehicle systems with minimal effort. This can be achieved through the "mechanisation" of these systems, as well as by placing the actuation commands within reach of the driver. One of these systems is the window drive mechanism [1].

A study method, described below, is the dynamic simulation, which is a component of the computer-aided design, applied to assemblies for showing the movement of the moving parts, taking into account their mass. In this respect, the movable elements of the assembly that need to be defined are the kinematic couplings of various classes and the internal and external loads. By these settings, the assembly becomes a mechanism that can be studied using the graphical or analytical methods provided by the theory of mechanisms and machines. [2]

Given the widespread deployment of computer in design, the analytical analysis methods based on mathematical algorithms are mainly used, because they ensure maximum precision of the results. The analysis can be performed either using specialised software implemented on the platform of a programming language, or high-performance software (MultiBody Systems - MBS), based on the theory of multi-body mechanical systems.

In the case of MBS applications, the equations of motion are self-formulated by the software, based on the geometric model for elasticity of the mechanism, taking into account the restrictions on element movements [3]. Also, true virtual prototypes can be created using the computer in view of obtaining products that meet the high functional requirements imposed by the market.

This means that we can obtain the faithful modelling of the system components, as well as the operating conditions thereof, which enables the rapid testing of numerous geometric-constructive variants for optimising the mechanism. This removes much of the experimental testing stage, which is an expensive and time consuming process [4].

2. Dynamic simulation of the car window regulator mechanism

2.1. Considerations on Autodesk Inventor application (MBS)

The dynamic simulation of the road vehicle door window regulator mechanism was carried out, based on the principle of MultiBody Systems, using the Autodesk Inventor Professional application.

The Autodesk Inventor products provide professional grade engineering solutions for 3D mechanical design, simulation, tooling creation, and design communication which helps create advantage of a Digital Prototyping workflow to make great products, cost-effectively, in less time. Autodesk Inventor software is the foundation of the Autodesk solution for Digital Prototyping. The Inventor model is an accurate 3D digital prototype that enables you to validate the form, fit, and function of a design as you work, minimizing the need to test the design with physical prototypes. By enabling to design, visualise, and simulate your products digitally, Inventor software helps you connect more effectively, reduce errors, and deliver great product designs faster. [5]

Inventor software provides powerful parametric modelling, robust direct editing tools and advanced surfacing modelling capabilities with T-splines. You can automate the advanced geometry creation of intelligent components, such as plastic parts, steel frames, rotating machinery, tube and pipe runs, and electrical cable and wire harnesses. Inventor software also helps reduce the geometry burden so it can rapidly build and refine digital prototypes that validate design functions and help minimize manufacturing costs. Traditionally, validating the operating characteristics of a design before it was built meant hiring expensive specialists. Inventor products include easy-to-use and tightly integrated part and assembly-level motion simulation and stress analysis functionality. By simulating stress, deflection, and motion, you can optimize and validate the design under real-world conditions, before the product or part is ever built. [6]

In this regard, the analysis and simulation of the studied mechanism by using the multibody system method by means of Autodesk Inventor, involves the following stages:

- Modelling of the mechanism as a multibody system (modelling the components and defining the assembly constraints);
- Defining the kinematic couplings of the mechanism, either through self-formulating or by individual definition;
- Defining the mass properties of the kinematic elements, i.e. the force or momentum-type loads;
- Carrying out the actual simulation, graphical display of results and their interpretation;
- Analysis of mechanism elements using the finite element method;
- Graphical display of results and their interpretation after the analysis by finite element method.

2.2. 3D modelling of the mechanism used to manipulate the car window

The 3D modelling of road vehicle door window regulator mechanism was carried out by individual modelling of its kinematic elements, followed by defining the assembly constraints. The model of the mechanism can be seen in Figure 1 [7].

It is very important that the assembly constraints, as geometrical relationship among the kinematic elements, to be properly defined, because when determining the necessary parameters for dynamic simulation, they can be auto-converted into kinematic couplings of various classes. In the Figures 2.a, 2.b, 2.c, 2.d, we can see details of the application of the constraints imposed to the various elements of the mechanism, as follows:

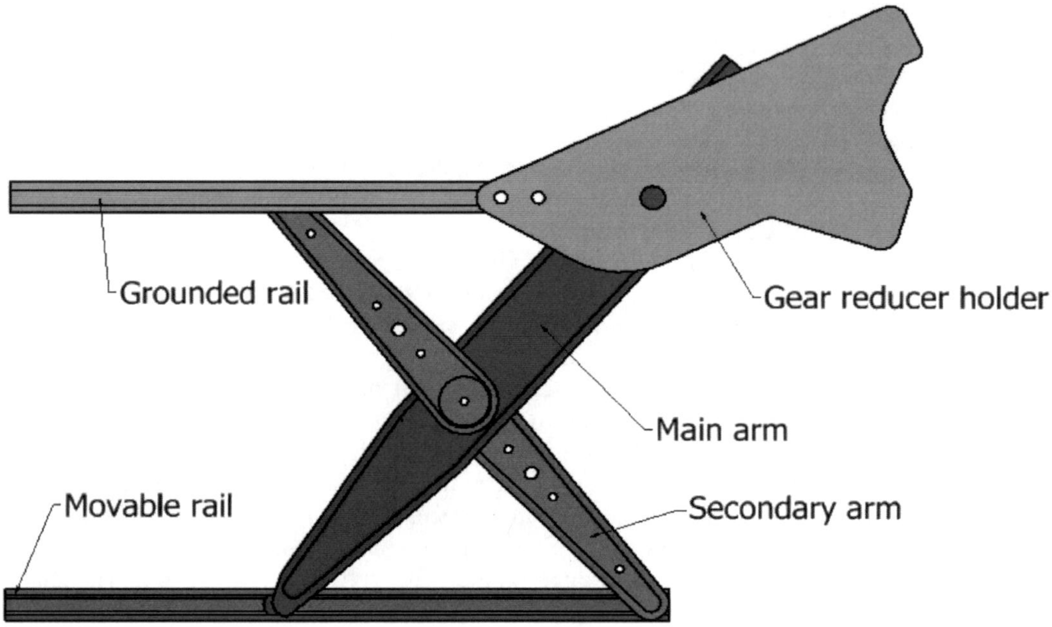

Figure 1. 3D model of the mechanism

a. gear reducer holder – fixed rail

b. main arm – movable rail

c. secondary arm – fixed rail

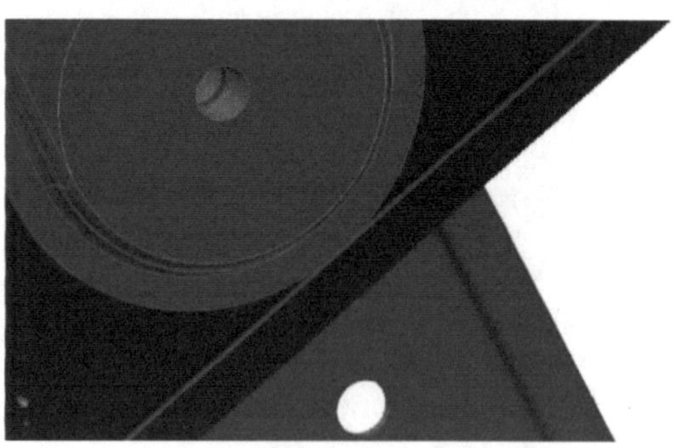

d. main arm – secondary arm

Figure 2. Application of assembly constraints

2.3. Analysis and dynamic simulation of the window regulator mechanism

The simulation of road vehicle door window regulator mechanism was carried out on the basis of the following parameters:

- Power transmitted: P = 0.3 kW
- Rotational speed of the driving element: n = 1.6 rot/min
- Up/down stroke – 510 mm
- Stroke length for lifting / lowering the window – t = 10s
- Payload (weight of the window) – F = 80 N

As stated in the preceding paragraph, the assembly constraints among the elements of the system can be auto-converted into kinematic couplings (Figure 3). In the case of the interface shown in Figure 3, we chose to carry out the analysis through the method of finite elements, using the AIP Stress Analysis module.

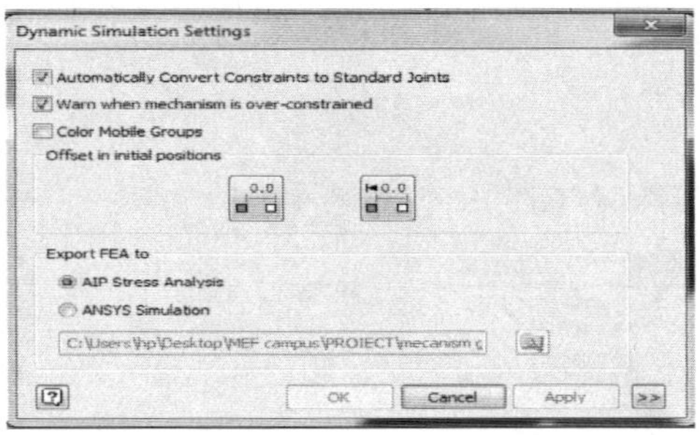

Figure 3. Self-conversion of constraints

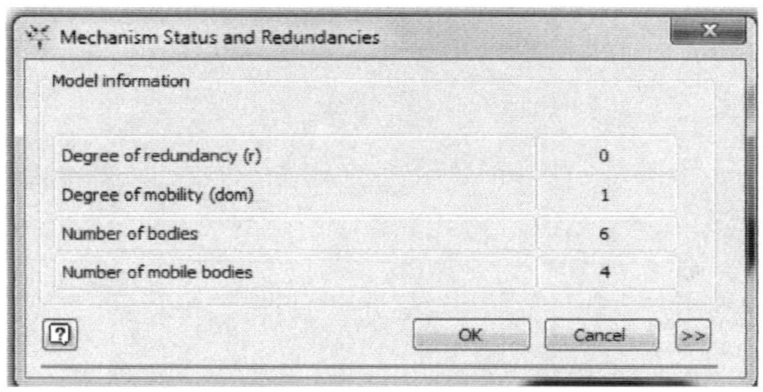

Figure 4. Structure of the mechanism

The articulated, cross arm type mechanism taken into account (Figure 1), parametrically modelled using Autodesk Inventor, by properly applying the assembly constraints, includes a kinematic chain which transmits the motion of the main arm to the movable rail, the geometric constraints being realised by rotational and translational couplings.

In these circumstances, the check of topology model leads obviously to a kinematic system without redundant joints (super type constraints), with r = 0 and the number of degrees of mobility dom = 1 (Figure 4). The indicated number of kinematic elements is 6, but it should be noted that the fixed element consists of two elements: the gear reducer holder and the fixed rail, stiffened one to another, and the secondary arm consists also of two components stiffened one to another, by properly applying the assembly constraints. Thus, the number of movable kinematic elements is equal to 3.

To ensure that the assembly constraints among the elements of the mechanism have been correctly self-converted into kinematic couplings, there is the possibility of viewing and redefining them. Thus, Figure 5.a shows the rotational coupling between the fixed element and the main arm, and Figure 5.b shows the translational coupling between the main arm and the movable rail.

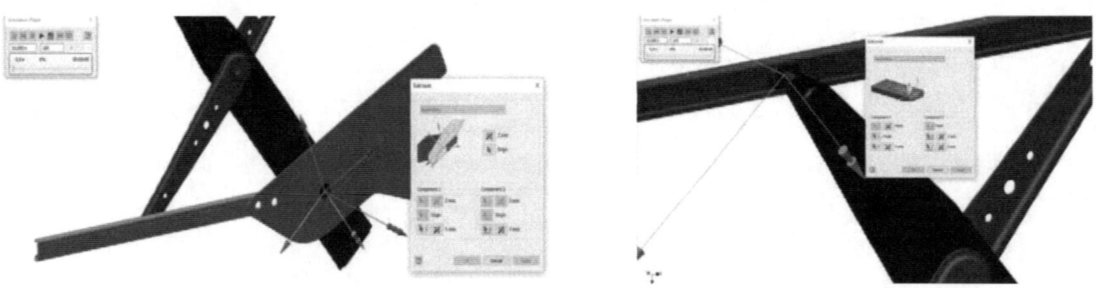

a. Fixed element – Main arm **b.** Main arm – Movable rail

Figure 5. Redefining of the kinematic couplings

The driving motion has been defined in the kinematic coupling, based on the initial data, by introducing the value of angular velocity (9.7 degrees/sec). In the same kinematic coupling, we defined the torque value, known in the theory of mechanisms as *momentum balancing*.

For the mechanism to be regarded as a multibody system, as has been stated in the first paragraph, besides defining the kinematic couplings of the components, it is necessary to define the external loads, i.e. we must take into account the mass of the elements.

The external loads are given, on the one hand, by the weight of the manipulated window (they have been defined by distribution, as two concentrated forces acting on the movable rail, which actually serves as window holder - Figure 6) and, on the other hand, by the tare weight of the kinematic elements of the mechanism. The masses of the elements are determined automatically, by means of the volume or density of the materials selected accordingly. The tare weight results from defining the value and direction of the gravitational acceleration, based on the orientation of the reference system associated to the 3D model of the mechanism. The load pattern can be visualized in Figure 6.

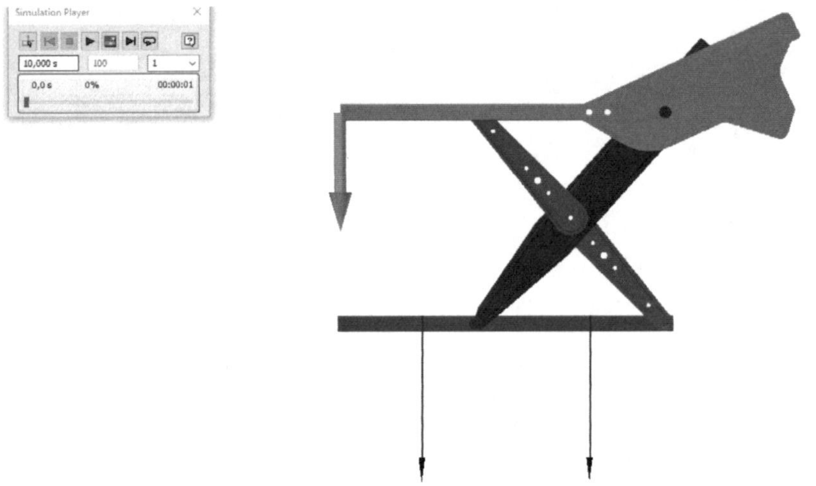

Figure 6. Load pattern of the mechanism

After determining the loading scheme, i.e. defining the motion parameters, the actual simulation of the mechanism has been carried out. As simulation parameter, we settled 100 time steps. Therefore, the results can be seen every 0.1 seconds throughout the kinematic cycle. As results, we can obtain numerical values or graphical variations of the various kinematic and kinetostatic parameters afferent to the couplings and kinematic elements.

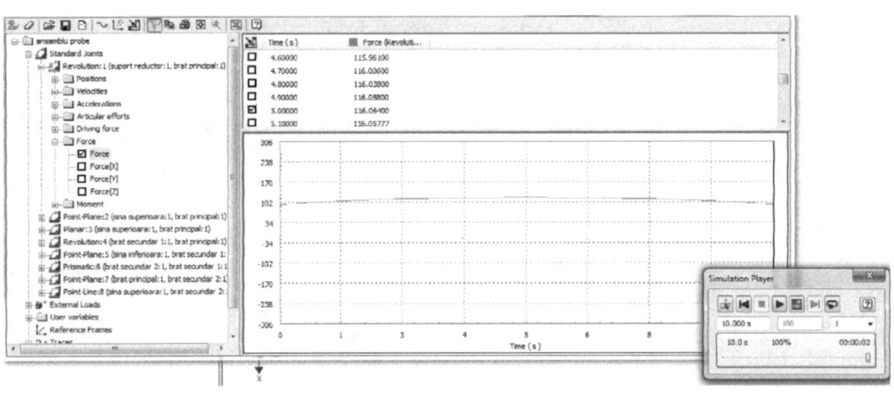

Figure 7. The reaction force in the kinematic driving coupling

Figure 7 shows the variation of reaction force in the kinematic driving coupling, between the fixed element (gear reducer holder) and the main arm of the mechanism, throughout the kinematic cycle. After viewing the result, it was found that the maximum reaction force, 116.06 N, was placed in the mid of kinematic cycle, i.e. at the time step t = 5 sec. It is important to specify this identified time step because it will be taken into account along with the external loads defined and the internal loads resulted from the finite element analysis of the mechanism elements.

3. Analysis of the main arm using the finite element method

The finite element analysis is a method to study the state of stress and deformation in the cases of bearing structures with very complex geometry, for which the calculation is made easier if the whole is divided into simpler areas.

The finite element method is a mathematical method of numerical integration of partial differential equations in variational form. We mean that all the calculation problems of the elastic structures will be reduced to a system of partial differential equations, which usually cannot be solved analytically. This inconvenient has been eliminated by using digital computers which, using dedicated software applications, provide rapid means of making a large volume of calculations involved in the finite element analysis, actually making this method to be easily applicable. [8]

For the road vehicle door window regulator mechanism, i.e. for its components, an analysis based on the finite element method is required to determine the state of stresses and deformations in real operating conditions, in order to assist the sizing process.

Basically, the finite element analysis assumes the completion of specific stages, such as: building the geometric model, discretisation, selection of materials, settling boundary conditions and external loads. But, since the finite element analysis of the studied mechanism was performed based on the results obtained from the dynamic simulation, i.e. maximum reaction force in the kinematic coupling between the fixed element and the main arm, at the time step t = 5 sec, it is sufficient that the kinematic element examined (i.e. the main arm) to be isolated from the rest of the mechanism by specifying its movable couplings with the adjacent elements (Figure 8a). Following this operation, all the external loads defined and the internal loads resulted, as well as the boundary conditions resulted from defining the kinematic couplings, will be taken over automatically, the loading diagram of the mechanism being shown in Figure 8b.

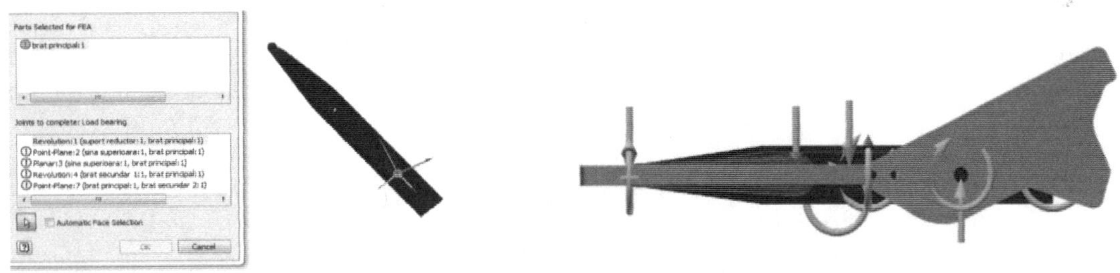

a. Isolation of the main arm – Boundary conditions

b. External and internal loads

Figure 8. Establishing of the loading diagram

As material for the studied element, but also for the other components, we opted for a general purpose steel grade, whose specific properties are shown in Table 1.

Table 1. Properties of the chosen material

Mass Density (g/ cm^3)	Yield Strength (MPa)	Ultimate Tensile Strength (MPa)	Young's Modulus (GPa)	Poisson's Ratio	Shear Modulus (GPa)
7.85	207	345	210	0.3	80.7692

Following the actual analysis, the obtained results are visualised and interpreted, i.e. the Von Mises stress state (Figure 9.a) and the deformation state (Figure 9.b) in the main arm.

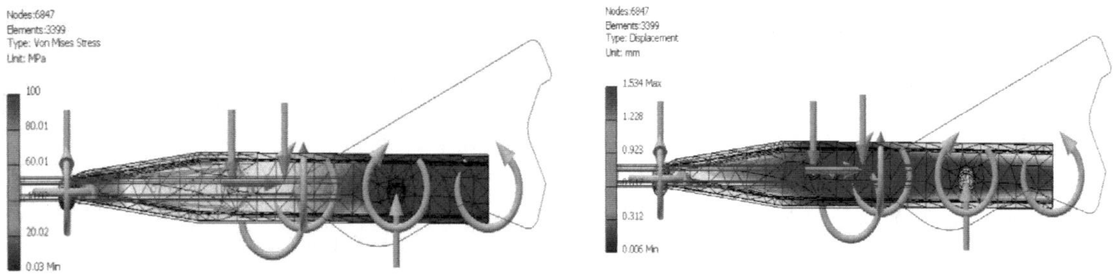

a. Von Mises stress state **b.** Deformation state

Figure 9. The results of the finite element analysis

It appears that the maximum stress value is 100 MPa, found in the area of movable connection between the main arm with the movable rail, and the maximum deformation is 1.53 mm, both results being able to be considered satisfactory in terms of resistance calculation.

Conclusion

The dynamic simulation is a component of the computer-aided design and a method of analysing a product under real operating conditions. It can be likened to a virtual prototyping based on a 3D model, but without the expenses required to build a real prototype.

Following the dynamic simulation of the road vehicle door window regulator mechanism, it was found that the maximum reaction force in the kinematic coupling is between the fixed element and the main arm, in the mid of the kinematic cycle, and the maximum stress in the main arm is in the area of connection with the movable rail.

References

[1] Miklos I Zs, Miklos C and Alic C 2014 Drive mechanism analysis of door windows to road vehicles, *Acta Technica Corviniensis – Bulletin of Engineering* **4** 61-64

[2] Stăncescu C, Manolache D, Pârvu C, Ghionea I and Tarbă C 2012 *Proiectare asistată cu Autodesk Inventor*, Fast, Bucureşti, Romania

[3] Rusu-Anghel S, Lihaciu I C and Rusu-Anghel N 2015 State feedback control of a robotic arm, *Annals of Faculty Engineering Hunedoara – International Journal of Engineering* **13**(1) 75-82

[4] Alexandru C 2006 Aspecte privind analiza mecanismelor ştergătoarelor de parbriz considerate sisteme multicorp, *Annals of the University of Oradea, Fascicle of Management and Technological Engineering* **V**(XV) 592-593

[5] * * * www.manandmachine.ro

[6] Dolga L, Revencu M, Maci C A and Giuchici M 2004 *Parametric and feature based modelling with applications in Catia and Inventor*, Politehnica, Timişoara, Romania

[7] Miklos I Zs, Miklos C and Alic C 2015 Articulated bar mechanism design for manipulating the doors windows of road vehicles, *Annals of Faculty Engineering Hunedoara - International Journal of Engineering* **13**(3) 159-163

[8] Spyrakos C Ch 1994 *Finite element modeling in engineering practice*, West Virginia University Press, Morgantown, USA

IOP Conf. Series: Materials Science and Engineering **163** (2017) 012020 doi:10.1088/1757-899X/163/1/012020

Using the anti-adherence paints to manufacturing of the moulds intended for iron castings

A Josan and C Pinca Bretotean

Politehnica University of Timisoara, Department of Engineering and Management, 5 Revolution Street, Hunedoara, 331128, Romania

E-mail: ana.josan@fih.upt.ro

Abstract. The paper presents the critical technology for obtaining of the lamellar graphite grey cast iron castings (Supporting roll type). Are presented the stages of achievement of the mould and the elaboration technology of the liquid alloy. A view to improving the qualitative characteristics of the castings and reducing the defects due to achievement of moulds it is necessary to use the antiadherence paints for moulds and cores. From the point of view of the cost the antiadherence paints belong to the expensive material category. But these expenses are done with their acquisition are amortized into account of improving the commercial aspect of the castings and shortening of the cleaning-finishing operations. Due to increase the resistance of the moulds and cores which they apply it is possible to decrease the machining allowance.

1. Introduction

In industrial practice of Romanian foundries, a particulary important role has it establishing of moulding-casting technology of the castings. Depending on the alloy that is pouring is necessary to into account certain technological aspects. Thus [1]:

- The mould must be capable to take over all the mechanical, thermal and chemical stresses to which it is subjected to the liquid alloy;
- The casting must have a commercial aspects, must be appropriate in terms of quality and advantageous in terms of price.

In these conditions the Romanian foundries places great importance on the elaboration of the liquid alloy technology, of moulding-casting technology and the materials used to achievement the moulds and cores so that the castings to obtain without defects.

Therefor the elaboration process must be carried out under normal conditions and the alloy elaborated complies with the conditions required by standard [1-4]. From the point of view of moulding-casting technology, achievement the moulds and cores must be carried out correctly using the quality moulding materials. For eliminating defects and rejects is necessary to use the refractory antiadherence paints for both moulds and cores.

In this context, the suppliers of technological materials for foundry provide an extensive range of foundry paints capable of covering all arreas of application. At present there are a great variety of products in this category with properties corresponding to technological stresses, quality requirements and customer requirements [5].

The correlation of this aspects and their implementation in practice lead to savings in Romanian companies.

IOP Conf. Series: Materials Science and Engineering **163** (2017) 012020 doi:10.1088/1757-899X/163/1/012020

2. The analysis of elaborating technology of the grey cast iron

For pointing out the aspects presented above is analyzed the moulding-casting technology for castings type "Supporting Roll". This casting is a component of the hoist gearing of an industrial furnace door. This piece was chosen for study because it was registered a high percentage of defects for castings.

The piece "Supporting Roll" is cast of unalloyed grey cast iron with lamellar graphite, EN-GJL-350, according to the standard [1], [2]. There are analyzed 10 charges. For these, are presented the chemical compositions (Table 1), the carbon equivalent (CE) and the saturation degree of carbon (SC). CE and SC are determined using to the following relations [1]:

$$CE = C + 0.3(Si + P) - 0.03\,Mn + 0.4\,S + 0.07\,Ni + 0.05\,Cr + 0.074\,Cu + 0.25\,Al \quad (1)$$

$$Sc = C / [4.26 - 0.3(Si + P)] \quad (2)$$

Table 1. The chemical composition of the charges studied

No.	Chemical composition, %					CE	SC
crt.	C	Si	Mn	P	S		
1	3.20	2.20	0.60	0.161	0.062	3.81	0.76
2	3.69	2.94	0.84	0.159	0.034	3.90	1.04
3	3.90	2.10	0.82	0.183	0.046	4.41	1.02
4	3.80	2.53	0.91	0.162	0.027	4.41	1.02
5	3.80	3.51	0.88	0.157	0.065	4.41	1.02
6	3.73	3.00	0.90	0.158	0.082	3.94	0.87
7	4.25	3.06	0.70	0.159	0.085	4.54	1.06
8	4.19	3.45	0.80	0.167	0.074	5.22	1.26
9	4.20	2.03	0.82	0.170	0.034	5.05	1.20
10	3.97	2.46	0.68	0.163	0.051	4.83	1.14

Using the data registered in the Table 2, was performed the critical analysis of the chemical composition of the 10 charges studied. Thus, conclusions were as follows [2], [4], [6], [7]:

➢ ***The influence of the carbon and silicon content***

- the carbon and the silicon are the graphitizer elements with the biggest importance their variation determining the greatest changes in the structure of cast irons; in these irons the two main elements varies usually within the limits: C = 2.7...3.8% and Si = 0.8...2.5%; from the point of view of the two elements the cast irons are placed on the Fe-C diagram through the carbon equivalent CE (Figure 1) for the most part in the hypereutectic domain.

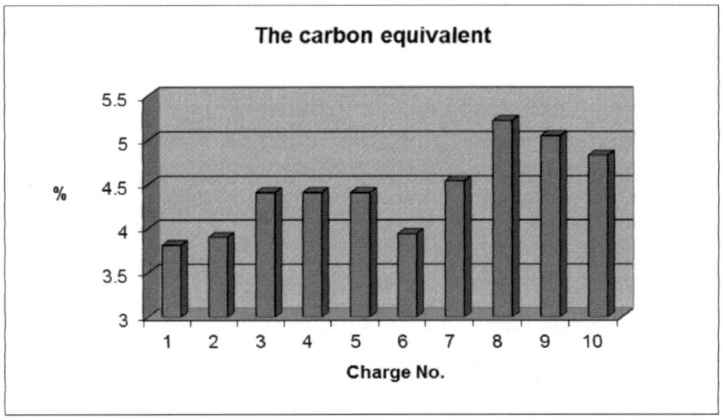

Figure 1. The variation of the carbon equivalent

- The diagram of Figure 2 indicated that the carbon content is greater; with the increase of carbon content a decrease is recorded both tensile strenght and hardness.

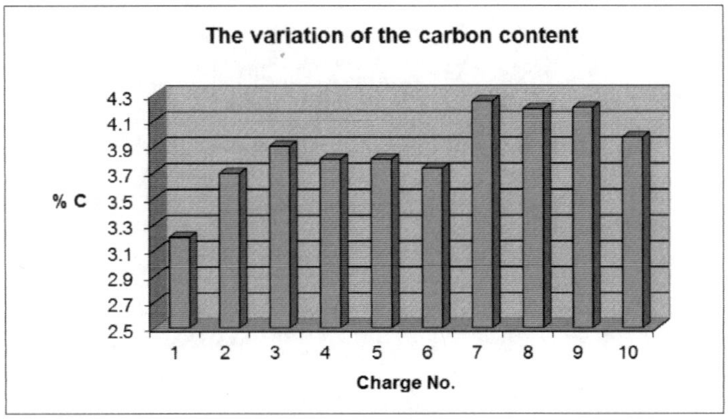

Figure 2. The variation of the carbon content

- To increase the silicon content is recorded a decrease of the hardness and tensile strenght; however, at values more than 3% Si the hardness of iron increases although the structure remains practically the same (ferrite + graphite) (Figure 3).

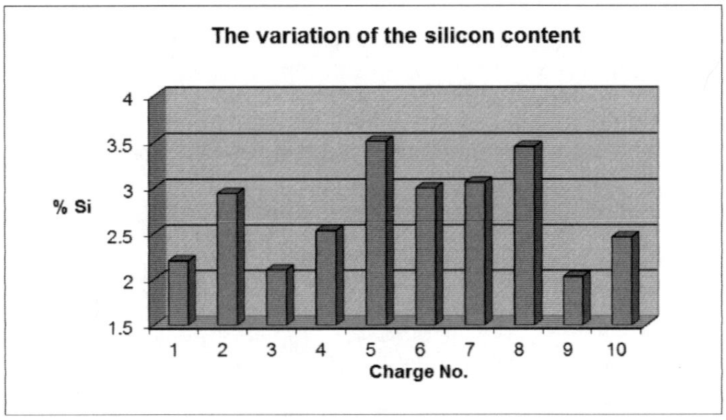

Figure 3. The variation of the silicon content

-The influence of the two elements (Si, C) ist o be regarded and according to the ratio of their concentrations. Thus, the increase of the carbon and silicon contents (respectively the increase of SC) leads to increasing the quantity of graphite and the ferrite in the structure and thereby to the decrease of mechanical properties (R and HB). The variation of the saturation degree of carbon is presented in Figure 4.

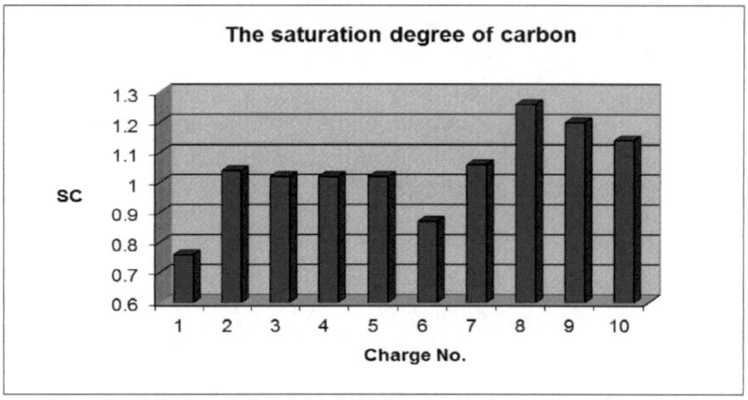

Figure 4. The variation of the saturation degree of carbon

➢ ***The influence of the manganese and sulphur content***

- In grey cast irons, usually the mangan content varies within 0.3...1.5%, and the sulphur content within 0.03…0.15%; of the diagrams in Figure 5 and Figure 6 is observed that the two elements comply with these limits at the charges analyzed;

- The influence of the manganese and sulphur content on the structure and properties of grey irons depend largely on the ratio of their values; to the grey irons with low content of manganese sulphur occurs for the most part in the form of eutectic or sulphides rich in iron (Fe, Mn)S and to the irons with high manganese content - in the form of sulfides rich in manganese (Mn, Fe).

-In the grey cast irons manganese content at values over 0.3% manages to neutralize the sulphur action.

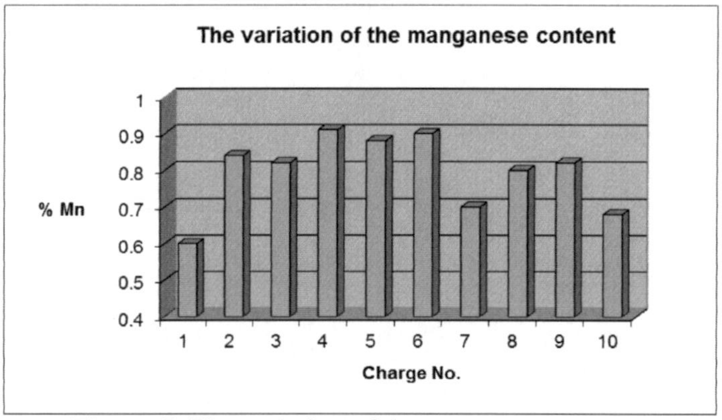

Figure 5. The variation of the manganese content

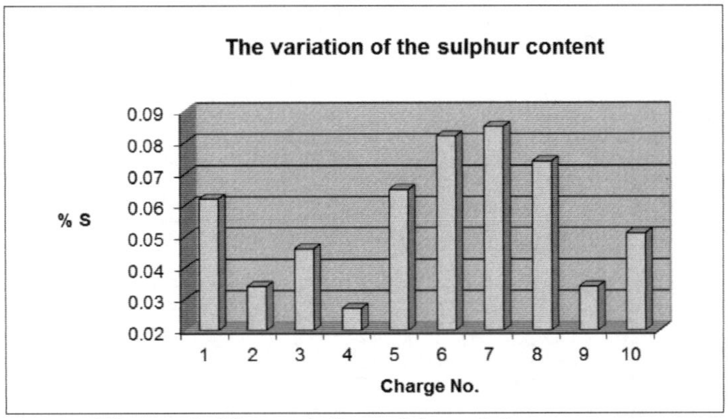

Figure 6. The variation of the sulphur content

➢ ***The influence of the phosphor content***

- In the grey cast irons the phosphor content varies between 0.1…1.5% not exceending 0.3% in all cases (Figure 7); this element favors the process of graphitization (as silicon) an dis almost entirely in the form of eutectic phosphorous (Fe-C-P);

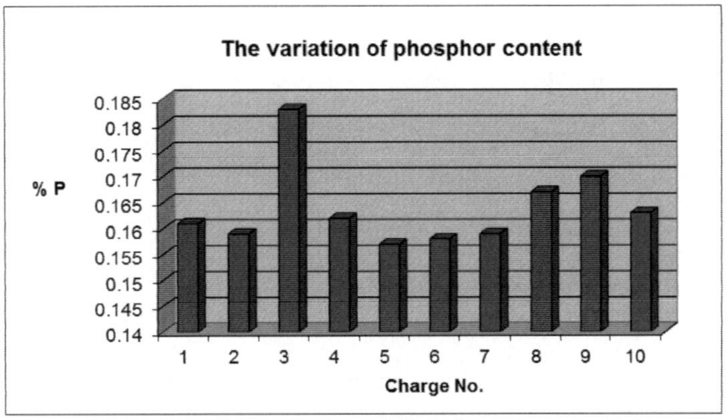

Figure 7. The variation of the phosphor content

- The phosphor content affects the mechanical properties of iron through structural changes which it produces, respectively: graphite finishing, increasind the proportion of pearlite, formation of the eutectic phosphorous (a constituent with high hardness: 500…600HB and fragile: R = 30...70 N/mm^2, which has great influence on the properties of cast iron).

Following this analysis resulted that elaboration is carried out in normal conditions, without any deviations too great of the chemical composition from the standards. The variations in chemical compositions of the charges are insignificant in the technological point of view and, generally, comply with the requirements of the beneficiary.

2. The analysis of moulding technology for obtaining the parts of type *Supporting roll*

Due to the defects registered in industrial practice was analyzed the moulding-casting technology of this type of casting: it started from the finished piece design (Figure 8), was making the technological drawing and the wooden pattern (Figure 9) and then they were achieved and assembled the two mould-part for casting of the liquid alloy [1], [3], [8-10].

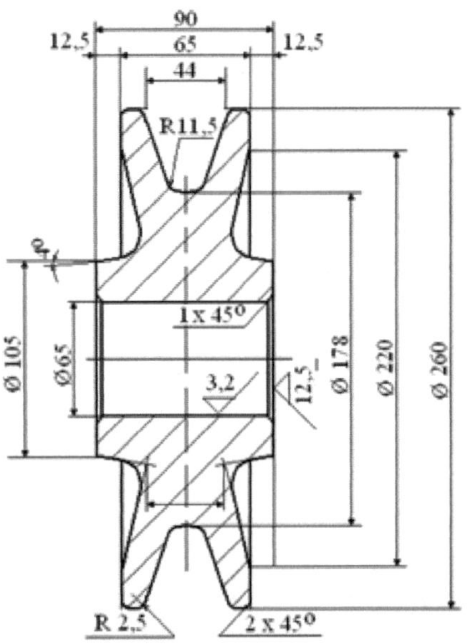

Figure 8. The drawing finished casting (*Supporting roll*)

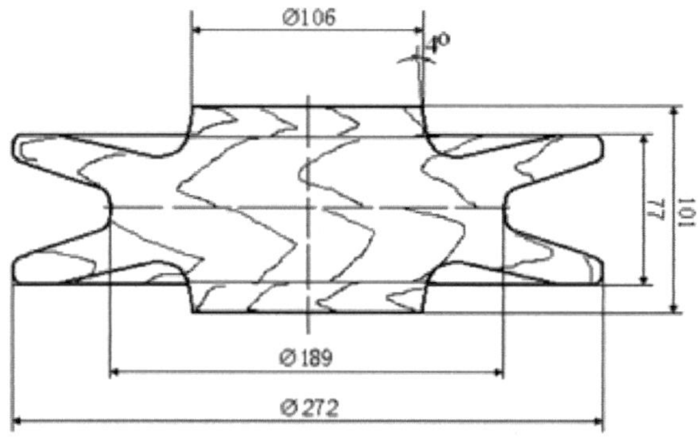

Figure 9. The wooden pattern

The most frequent defects for this type casting were adherences. *Adherences (D 221)* is defects that appear on the casting surfaces due to the following causes [11], [12]:

- The improper achievement of the moulds;
- The use of improper moulding sands;
- Non-use of refractory paints for the mould and core surfaces;
- The refractory quality is improper.

A method of prevention of appearance of the registered defects to moulding is the use of the antiadherence paints for moulds and cores [9].

3. The analysis of moulding technology for obtaining the parts of type *Supporting roll*

The refractory antiadherence paints used in foundry are coating materials that are applied to the surface of the moulds and cores. Their role is to smooth the porous surface of moulds and cores in order to ultimately obtain castings with surfaces as clean, aesthetic and without defects. The basic components of foundry paints are refractory materials finely ground. The refractory paint layer acts as insulation protecting the moulds and cores against the themal and/or chemical action of the molten metal [5].

With time, the foundry paints were transformed from simple materials (mono- or two- component) in complex products that contain material systems with synergistic action. Following the development and continuous change of the binders system, the foundry paints go through significant changes so that they correspond to the new stresses. This is achieved by using of new refractory materials and testing of new products in the laboratory and industrial scale [5].

The main criteria for choosing the sorts of the antiadherece paints are the cast-alloy type and the castings weight. The paints can be used to protect the moulds and cores achieved by various technological methods of moulding. But, it must be provided that the moulds and cores are dry after painting [13].

➢ **For mould:**

For casting studied (*Supporting roll*) is recommended the use of antiaherence paint that requires drying, KGS-type. For this type of paint are specified the following characteristics [13]:

- The refractory material is composed of quartz and graphit
- Humidity of powder – max. 4%

- Density – 1380-1430 kg/m3
- Viscozity – max. 18 s
- Stability – max. 97%
- Film resistance to abrasion– max. 2 kg/mm
- are delivered in the form of powder compositions which contain all the necessary components; to be use in foundry, the powdery paints are dissolved in water up to obtain the necessary consistency

The main advantage of powders consists in the possibility of long-term storage of them. The duration guaranteed by the manufacturer is 12 months, but industrial practice shows that the powders do not lose the properties even longer storage durations [13].

➢ **For core:**

For painting by immersion of the core is recommended to use the paint KGS-1-type due to the following advantages [13]:

- viscozity - 14 -15 s
- Thixotropic properties
- Ensure the formation of a homogeneous film on the surface on the core

It is necessary [13]:

- In the process of painting by immersion the cores are maintained into the paint maximum 5 s in order to prevent the excessive increase of their humidity.
- The drying of the core takes place at temperatures of 150-200°C for not destroy the binder of paint (by burning).

To increase the effectiveness of the drying process and prevent potential overheating, the drying chambers must by provided with fans (recirculation and exhaust air) [13].

5. Conclusions

Preventing the formation of adherences and obtaining of plane surfaces are achieved by the use the refractory paints. The refractory paints are applied to the surface of the moulds and cores, in order to achieve a refractory layer to the contact surface of the mould with the liquid metal. They must be fluid and dense to form a continuous layer, thinly and evenly on the surface of the mould and core.

Regarding of the cost, the antiadherence paints belong to the expensive material category. These expenses are done with their acquisition are amortized taking into account the following advantages:

- Improving the commercial aspect of castings;
- Shortening the cleaning-finishing operations for castings;
- Decreasing the machining allowance due to increasing resistance to surface of the moulds and cores that apply;
- The possibility of long storage in safe conditions;
- Quality / price ratio for this castings is a sure guarantee of obtaining a important economic effect to users.

References

[1] Stefanescu C and Cazacu I 1980 *Technologies of realization of the parts by casting*, Technical Publishing, Bucharest, Romania
[2] ***The Standard *SR EN 1561:2012 - Founding. Grey cast iron*
[3] Josan A 2002 *The technology of moulding and casting alloys*, Politehnica Publishing, Timisoara
[4] L. Sofroni 1975 *The elaboration and pouring of alloys*, Publishing House Didactic and Pedagogical, Bucharest, Romania
[5] ***http://www.huettenes-albertus.ro/produse/vopsele_refractare/index.html
[6] Josan A, Pinca-Bretotean C and Putan V 2015 The critical analysis of austenitic manganese steel T130Mn135 used for castings in the mining industry, *IOP Conf. Ser.: Mater. Sci. Eng.* **106** 012021
[7] Socalici A, Pascu L, Popa E and Hepuţ T 2015 The influence of the cast iron structure upon the

hardness of brake shoes meant for the rolling sock, *IOP Conf. Ser.: Mater. Sci. Eng.* **85** 012026
[8] Josan A, Pinca Bretotean C, Ardelean E and Ardelean M 2015 Using lateral cores to casting of carbon steel parts of drive wheel type in a metallurgical enterprise, *IOP Conf. Ser.: Mater. Sci. Eng.* **85** 012016
[9] Josan Aand Pinca Bretotean C 2015 Using special additions to preparation of the moulding mixture for casting steel parts of drive wheel type, *IOP Conf. Ser.: Mater. Sci. Eng.* **85** 012017
[10] Josan A 2016 Impoving the quality of castings by optimization of the moulding-casting technlogy, *Acta Technica Corviniensis – Buletin of Engineering* **IX**(4) 75-78
[11] Josan A 2010 *The analysis of casting defects recorded in the metallurgical enterprises*, International Symposium on Advanced Engineering & Applied Management - 40th Anniversary in Higher Education (1970-2010), Hunedoara, Romania, November 4-5, CD
[12] ***1977 *International Atlas of casting defects*, Technical Publishing, Bucharest
[13] Davidov N I 2004 Refractory daies for moulds and cores, *Romanian Foundry Journal* **3/4** 17 (translation Phd.Eng. Constantin Cosneanu)

IOP Conf. Series: Materials Science and Engineering **163** (2017) 012021 doi:10.1088/1757-899X/163/1/012021

Experiments assigned to determine the acceleration of 8000kN shear laboratory model elements

A Budiul Berghian, T Vasiu and C Abrudean

[1]Politehnica University of Timisoara, Department of Engineering and Management, 5 Revolution Street, Hunedoara, 331128, Romania

E-mail: adina.budiul@upt.ro

Abstract. In this paper presents an experimental kinetics study by measuring accelerations using a bi-axial accelerometer constructed in the basis of a miniature integrated circuit, included in the class of micro-electrical and mechanical systems - *MMA6261Q* on the *experimental installation* reduced to the 1:5 dividing rule by comparison with the shear existent in exploitation, conceived and projected at the Faculty of Engineering in Hunedoara.

1. Description of the laboratory model of scissors 8000 kN

The laboratory model of scissors dimensional 8000kN is reduced to 1:5 scale, to real scissors. The model is powered by an AC motor with 0.8 kW drive power and a speed of 985 rev / min (Figure 1).

2. The experimental installation for measuring acceleration

To perform the experiments was used the following equipment:
- Biaxial accelerometer constructed in the basis of MMA6261Q circuit;
- Type data acquisition board NI PCI-6024;
- Computing system IBM-compatible PC.

2.1. Biaxial accelerometer built on circuit MMA6261Q

In order to build the accelerometer, it has developed a printed circuit on which components were installed, MMA6261Q chip resistors for adjusting the frequency and duty cycle filtering capacitors as shown in Figure 2.

For this type of accelerometer was provided out for analog, power accelerometer was conducted from a source DC 3.6V.

For fixing quick and easy, on the opposite side of the integrated circuit was placed a magnet. Before the establishment of this type of fixing were made experiments to reveal any influences which may be caused by the behavior of the magnetic field in the operation of the accelerometer. Did not were highlighted such interference, therefore it was considered that this type of fixing is the most suitable in this case[1], [2].

In order to do not achieve measurement errors, it was ensured couple against rotation around its axis during operation.

IOP Conf. Series: Materials Science and Engineering **163** (2017) 012021 doi:10.1088/1757-899X/163/1/012021

Figure 1. The laboratory model of scissors 8000 kN

Figure 2. Performance accelerometer assembly

2.2. Plate of data acquisitions of type NI PCI-6024E

For collecting and transmitting the quantities recorded accelerometer to the computer, using a data acquisition plate specializes in digital computer interfacing with external circuits.

Acquisition plate NI PCI-6024 are installed in an IBM-PC compatible computer and the acquisition system is shown in Figure 3 [3].

Acquisition plate was made an "array" of contacts for the connections where the 8 channels of purchase.

Figure 3. The data acquisition system

Accelerometer allows simultaneous measurement of components after the two-axis acceleration of the reference system, data acquisition was done in analog format.

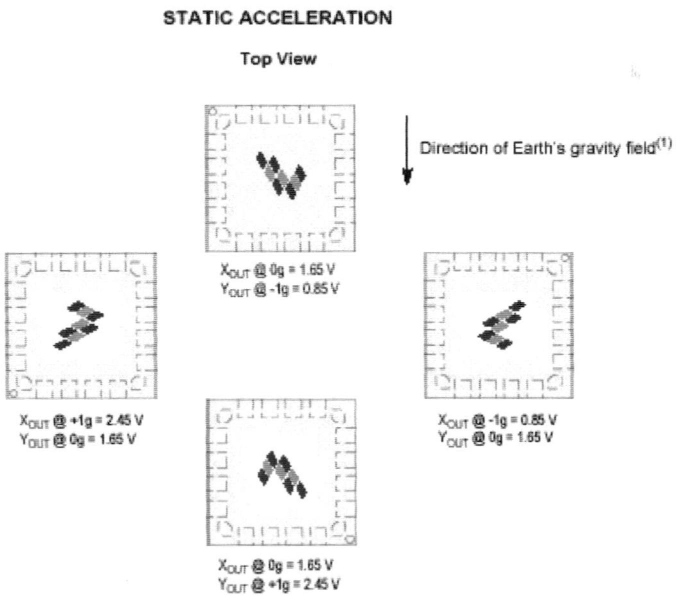

Figure 4. Accelerometer calibration mode

IOP Conf. Series: Materials Science and Engineering **163** (2017) 012021 doi:10.1088/1757-899X/163/1/012021

Calibration of the accelerometer was made by placing it in four positions so that the acceleration of gravity to be the x and y axes in either direction (Figure 4), after which the resulting voltage measured using acquisition program. Linear transformations that convert voltages in acceleration values are obtained simply having two reference points: the measured voltages to -g and g.

A general formula for deducting a linear transformation a(u) having two values A_1 and A_2 corresponding to u_1 respectively u_2 is:

$$a(u) = a_1 + (a_2 - a_1)\frac{u - u_1}{u_2 - u_1} \tag{1}$$

In this case: $a_2 = g$ şi $a_1 = -g$, therefore:

$$a(u) = -g + 2g\frac{u - u_1}{u_2 - u_1} = 2g\left(\frac{u - u_1}{u_2 - u_1} - \frac{1}{2}\right) \tag{2}$$

When has been experimentation, for purposes of convenience has taken positive x direction opposite to the calibration, so it is necessary to change the sign once more in the program. C syntax conversion functions are:

- Ox-axis function has the expression:

```
double scx(double x) {
  return -2 * 9.81 * ((x - 1.112362) / (2.530534 - 1.112362) - 0.5);
}
```

- Oy-axis function has the expression:

```
double scy(double y) {
  return 2 * 9.81 * ((y - 0.78675) / (2.296962 - 0.78675) - 0.5);
}
```

To acquisition of data provided by the accelerometer has been designed and wrote a program in C ++.

3. Measurements and results obtained using accelerometer-based on integrated circuit MMA6261Q

Experiments were performed on experimental stand on the scale of 1: 5 to 8000kN scissors, Figure 1.

For each kinematic coupling were measured components of accelerations after the two axes of the reference system. The measurements were performed for several cycles kinematics, and in Figures 5-8. are shown these graphics dependencies for one cycle cinematic.

Coupling E has a translational movement on OY.

4. Conclusion

From the many methods of measuring the acceleration, the use of an accelerometer-based integrated circuit technology built using micromachining of silicon (silicon micromachining) is an effective and relatively simple to achieve. This type of sensor is used in robotics, the technique for monitoring vibration of any kind, inertial navigation systems etc; accelerometer built on integrated circuit MMA6261Q accelerometer is a biaxial low-g (for small accelerations) can measure both dynamic acceleration and static (eg gravitational acceleration).

From the data obtained by measuring the stand is established that there is interference (vibration, variations in acceleration due to sudden shocks by turning off certain components and others start twisting torque, etc.) but analyzing the results obtained by analytical calculation and simulation it can be concluded that the use of the accelerometer that is appropriate experimental means[4].

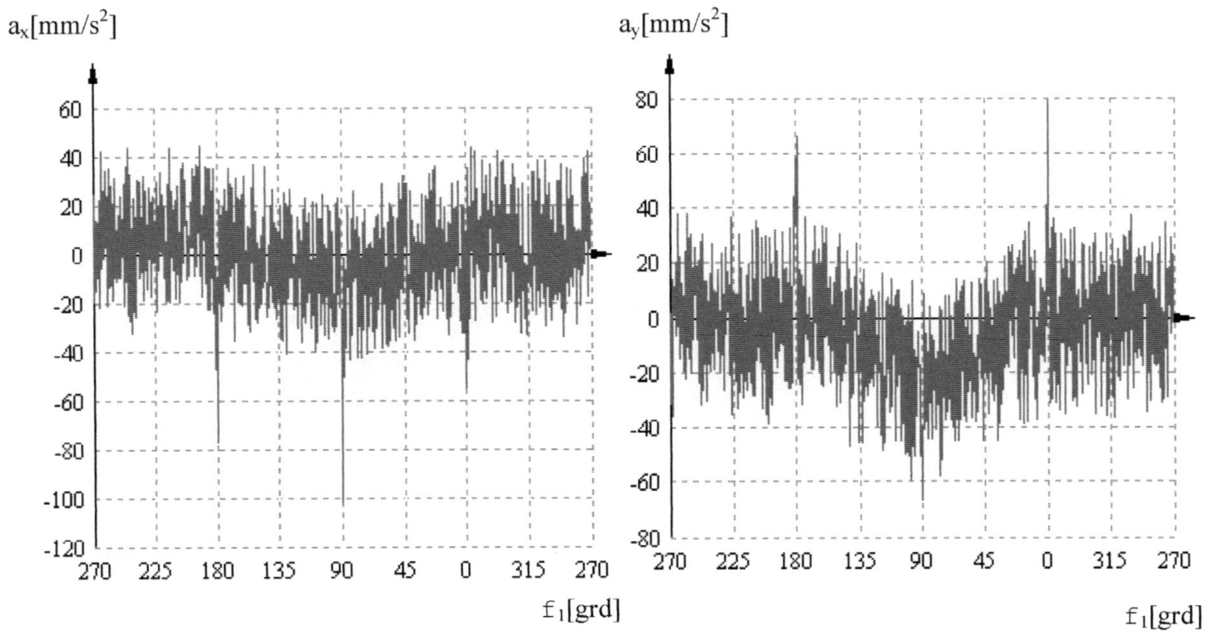

Figure 5. The variation of acceleration components of the measurement approach for the coupling C

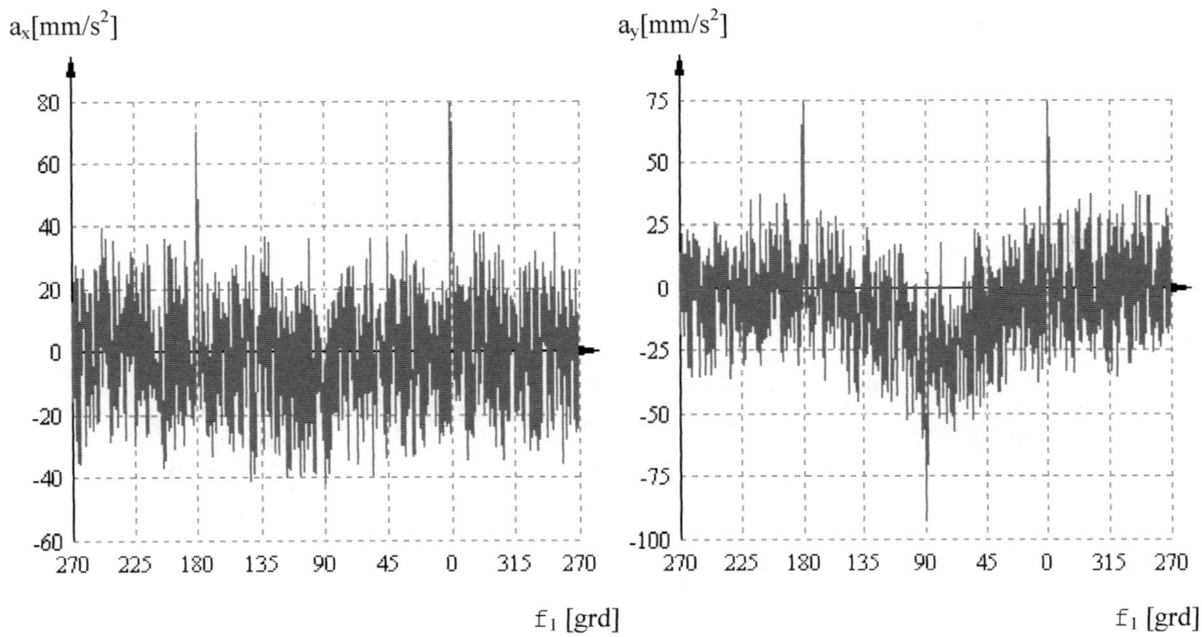

Figure 6. The variation of acceleration components of the measurement approach for the coupling D

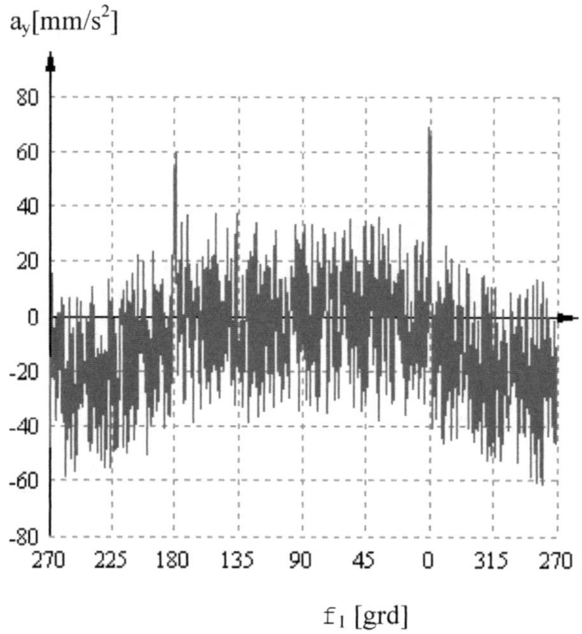

Figure 7. The variation of the acceleration component on OY obtained by measuring for the coupling E

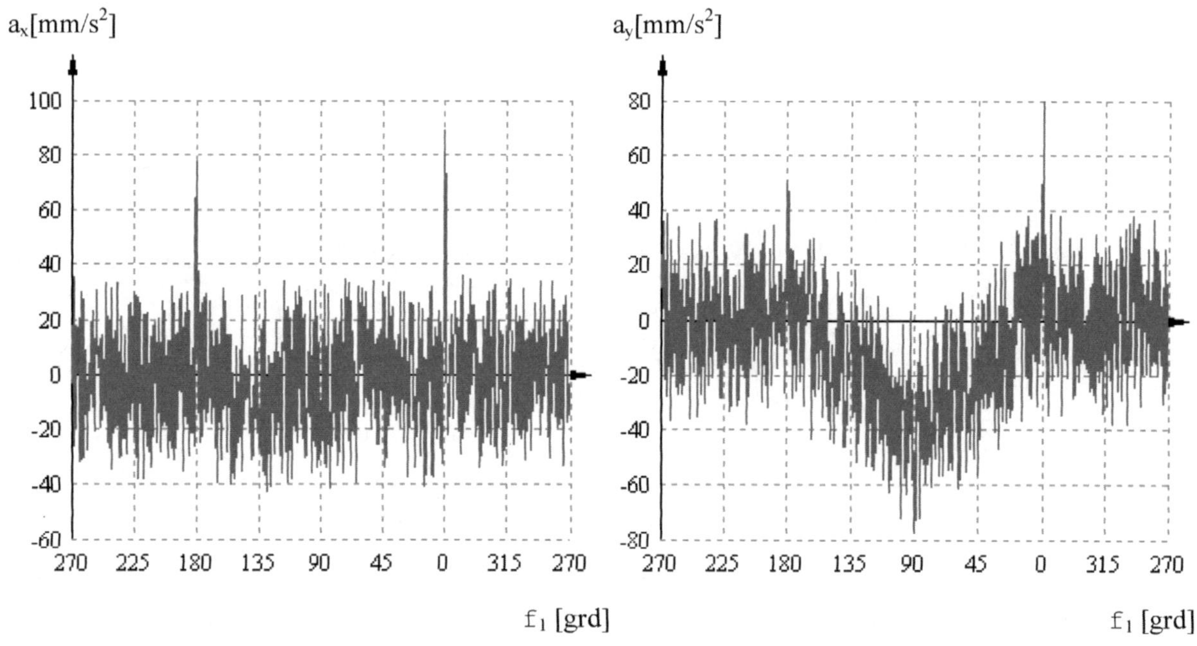

Figure 8. The variation of acceleration components of the measurement approach in point F

References

[1] Viksten F 2001 *On the use of an accelerometer for identification of a flexible manipulator,* Linkoping University, Sweden, Master thesis project in Automatic Control at the Department of Electrical Engineering Reg. Nr: LiTH-ISY-EX-3028
[2] ****±1,5g Dual Axis Micromachined Accelerometer,* www.freescale.com
[3] ***www.ni.com *http://sine.ni.com/nips/cds/view/p/lang/en/nid/10968*
[4] Budiul Berghian A, Vasiu T, Abrudean C and Stoica D 2013 Mathematical model to determine the acceleration to the elements of the laboratory model to the 8000kN shears, *AIP Conf. Proc.* **1558** 1639-1642

Assessment of small hydropower potential for the Olanesti River using advanced software techniques

A Moldoveanu[1], A Galie[1], M Moldoveanu[1], F Popa[2], E Tica[3] and B Popa[3]

[1]National Institute of Hydrology and Water Management, Water Management Department, 97 Bucuresti - Ploiesti, Bucharest, 013686, Romania
[2]ISPH Project Development, 293 Calea Vitan, Bucharest, Romania
[3]University Politehnica of Bucharest, Department of Hydraulics, Hydraulic Machineries and Environmental Engineering, 313 Splaiul Independentei, Bucharest, 060042, Romania

E-mail: moldoveanu.alexandru@hidro.ro

Abstract. The assessment of small hydropower potential for rivers is essential for the Renewable Energy Sources Directive implementation, in order to identify and analyze opportunities for new small hydropower developments. As the Water Framework Directive requirement is the non-deterioration of the status of the river water bodies, the aspects regarding the consistency of a flow regime downstream hydropower plant and the environmental objectives are important. The paper presents a case study, the Olanesti River (Romania), using software techniques dedicated to the assessment of small hydropower potential. The hydropower potential of the river was assessed considering the heads based on the Digital Terrain Model and the mean flow. In addition, the environmental flow was considered for the installed capacity assessment, in order to cope with the Water Framework Directive requirements. The harnessed hydropower potential, the optimal distance between the water intake and power plant location, their optimum positions along the river sector, installed capacity and electricity production, calculated at an average lifetime, are the outcomes of the software. The applicability of the tool might be extended on a regional scale in order to support the decision making authorities, taking into account the increasing demand for energy as well as the environmental issues.

1. Introduction

One of the major challenges for the European Union refers to how we can ensure energy security with competitive and *"clean"* energy taking into account the increasing global energy demand and the uncertain future regarding the access to the energy resources.

The vision of European energy policy corresponds to the concept of *"sustainable development"* and refers to the following:

- consumer access to energy at affordable and stable prices;
- sustainable development of production, transport and consumption of energy;
- security of energy supply;
- reducing of emissions of greenhouse gases.

The Directive 2001/77/EC of the European Parliament and Council on the promotion of electricity produced from renewable energy sources on the single energy market was the first concrete action of the European Union in order to achieve the objectives committed by the ratification of the Kyoto Protocol on reducing the greenhouse gas emissions.

The main directions of action of the Directive 2001/77/EC were:

- increasing the use of renewable energy within the electricity production,
- setting a target on the electricity consumption produced from renewable energy sources,
- appropriate procedures for financing the investments in renewable energy sector,
- simplification of administrative procedures on the projects implementation that harness the renewable energy sources.

In 2009, the European Commission issues the new Directive 2009/28/CE on the promotion of the use of energy from renewable sources, amended by Directive 2013/18/EU, which main objective establish that for each Member State, 20% from the total internal gross consumption of energy must be produced from renewable energy sources until 2020 [1].

The main objective must be transposed in the individual objectives for each Member State, taking into consideration the national particularities: energetic potentials, the existing level of the energy production from the renewable energy sources and the energetic mix (the RES-E rate from the total internal energy production of each Member State: hydropower, solar, wind, thermal etc.).

The national objective for Romania, according to Directive 2009/28/CE, is the following: "24% from the total internal gross consumption of energy must be produced from renewable energy sources until 2020".

In order to implement the RES-e Directive 28/2009/EC on promotion of the use of energy from Renewable Energy Sources, the need to evaluate the hydropower potential of water courses in Romania is highlighted.

The facts that: the hydropower potential is a renewable resource, the small hydropower plants (SHP) locations generate a low environmental impact (at local scale), and the depletion of conventional energy sources at an accelerated rate, are the main prerequisites for the promotion of the SHP development.

In order to implement the RES-E Directive 28/2009/EC on promotion of the use of energy from Renewable Energy Sources one must take into account the provisions of the Water Framework Directive 60/2000/EC. In this context, a hydrological regime consistent with the achievement of the environmental objectives of the Water Framework Directive (ecological flow) should be considered downstream hydropower plant.

2. Materials and methods

The paper presents a case study, the Olanesti River (Romania), using Vapidro-Aste software technique dedicated to the assessment of small hydropower potential of a river sector, identifies the potential locations for new small hydropower plants developments and takes into consideration the environmental flow necessary for ensuring the aquatic ecosystems protection.

The VAPIDRO-ASTE software has been produced by RSE S.p.A in the frame of the SEE HYDROPOWER project, financed by the South East Europe Program and the Research Fund for the Italian Electrical System [2].

2.1. The ecological flow determination

The method to determine the ecological flow (environmental issue) was elaborated by the National Institute of Hydrology and Water Management (NIHWM) in 2015 in order to support the implementation of the Water Framework Directive 2000/60/EC in Romania and to fulfill its requirements related to water quality and quantity necessary to the aquatic ecosystem protection [3].

The Olanesti River (Figure 1) is located in the south-central part of Romania within the Olt River Basin.

The river sector analyzed to assess the harnessed hydropower potential of the Olanesti River, about 30 km length, is located upstream Vladesti Reservoir up to the river springs.

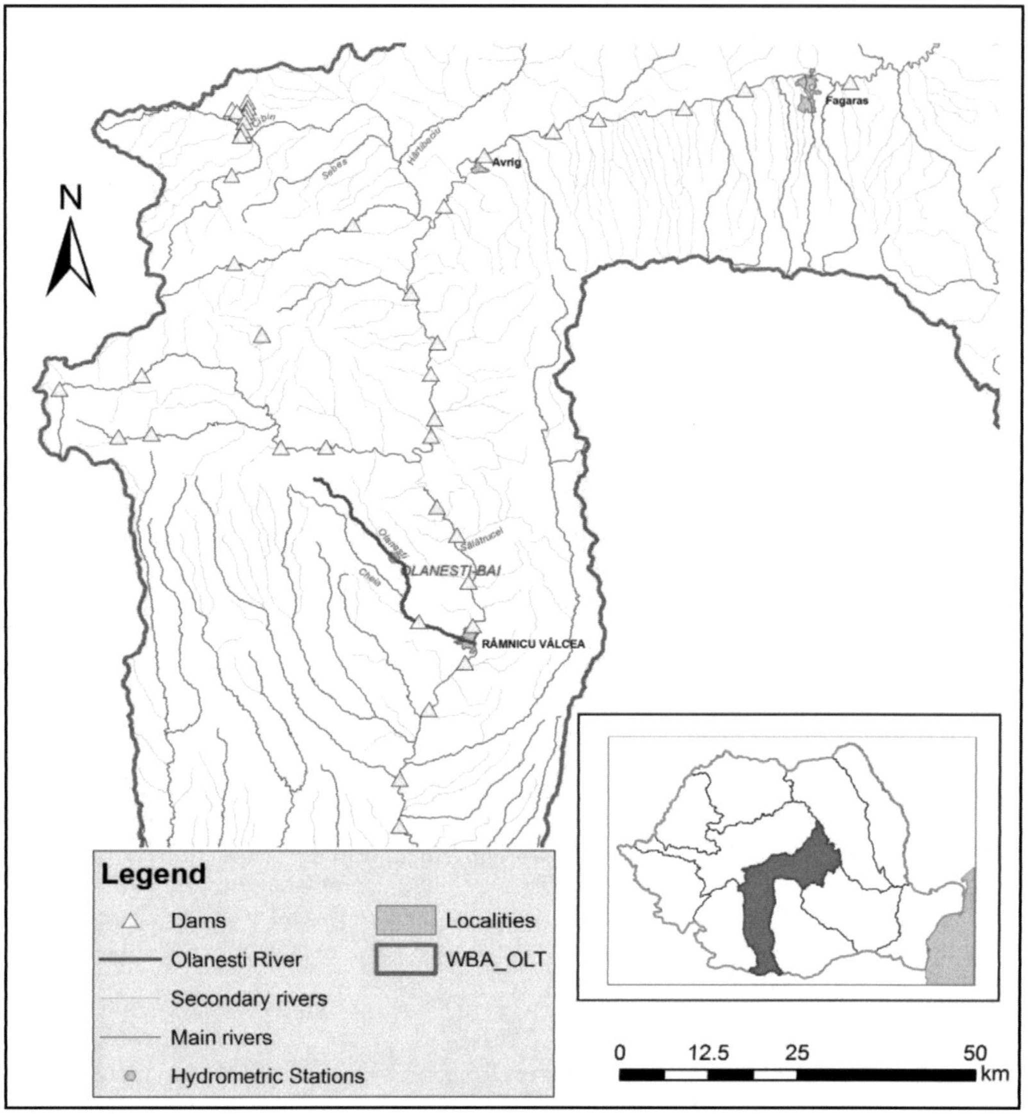

Figure 1. Olanesti River – tributary of the Olt River

The approach of the method to determine the ecological flow is the following:

- the need for a full range of natural variability in hydrological regime for aquatic ecosystem,
- mean flow provides habitat for fish fauna,
- more water during the periods of reproducing fish.

The ecological flow is dynamic having monthly variability, one value for each month it means 12 different values/year (Figure 2). The ecological flow takes into account the natural variability of the flow regime.

The main condition of this method is that the twelve monthly values for each year must be higher than the salubrious flow ($Q_{95\%}$ - minimum monthly mean flow with 95% probability of occurrence).

In this case study on the Olanesti River, the values used for the ecological flow are the following:

- the ecological flow for dry season (Q_{e1}= 0.26 m^3/s), which was computed as a mean of the 6 monthly values for dry season;

- the ecological flow for wet season (Q_{e2}= 0.32 m^3/s), which was computed as a mean of the 6 monthly values for wet season.

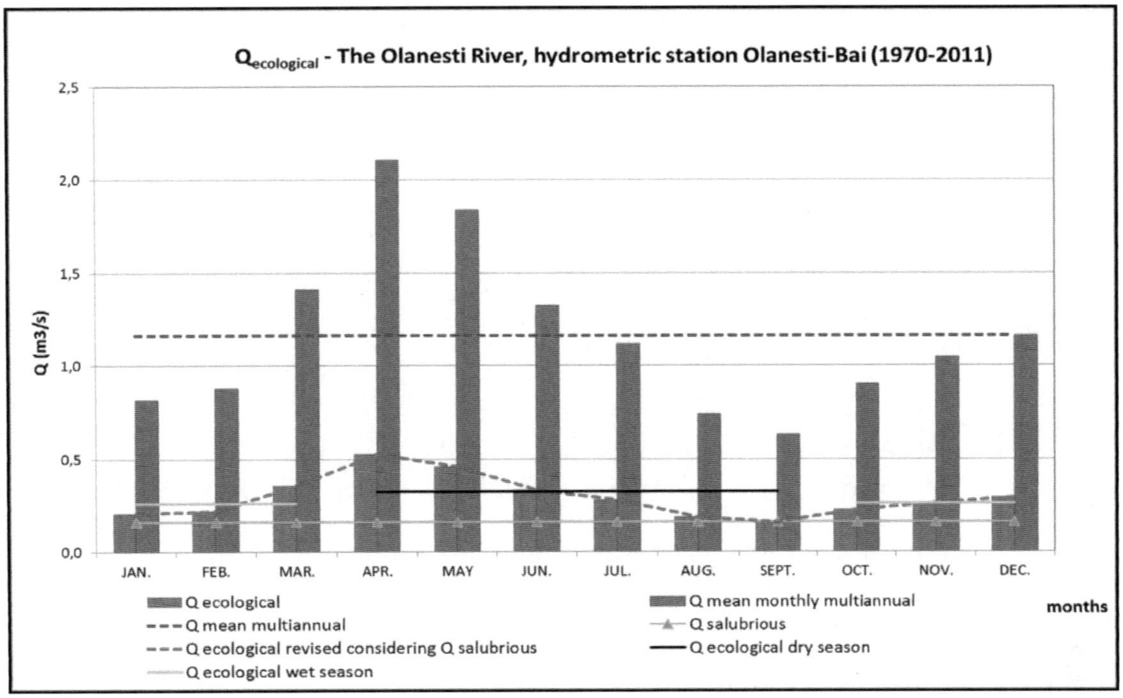

Figure 2. The ecological flow at the Olanesti hydrometric station located on the Olanesti River

2.2. The technical and economical data

The technical and economical data used by Vapidro-Aste software as input data are the following:

- Digital Terrain Model (with a 30 m grid cell);
- the mean flow for the 1970-2011 period at the Olanesti-Bai Hydrometric station both for the dry season (Q_1= 1.292 cm/s) and the wet season (Q_2= 1.035 cm/s);
- the ecological flow for the dry season (Q_{e1}= 0.26 cm/s) and the wet season (Q_{e2}= 0.32 cm/s);
- the hydropower plant's lifetime – 30 years;
- the total cost of the water works – 5000 Euro/kW of installed capacity [4];
- the selling price of energy – 0.0425 Euro/kWh [5];
- the selling price of the Green Certificates (GC) – 40 Euro/GC [6] and there are issued two green certificates (GC) for each MWh produced for the first 15 years of functioning according to the present law.

The economical data (the interest rate, the price of the built water works on km) and the financial data (the price of the GC, of the produced electricity, of the mechanical and electrical equipments, maintenance cost etc.) that were used as input data were obtained from the websites of the Romanian competent authorities in the specific fields of activity: ISPH Project Development [4], Romanian gas and electricity market operator [5], Romanian Energy Regulatory Authority [6]; and transformations in the measurement units and RON (Romanian coin) to Euro conversions were performed according to the Vapidro-Aste requirements.

2.3. Presentation of the Vapidro-Aste software main steps performed for the study case

The Vapidro-Aste software menu allows the user to select the appropriate method for calculus and perform the simulation in order to assess the hydropower potential of the river sector. The selected method takes into account the national regulations, climatic conditions (dry and wet seasons) and environmental issues (ecological flow).

In the present study case, the selected method from the software menu uses two multiannual mean flows for each section of calculus (hydrometric station), one value for dry season and one value for wet season, the corresponding values for the ecological flow to dry and wet season and presents the construction of a small hydropower with derivation (Figure 3).

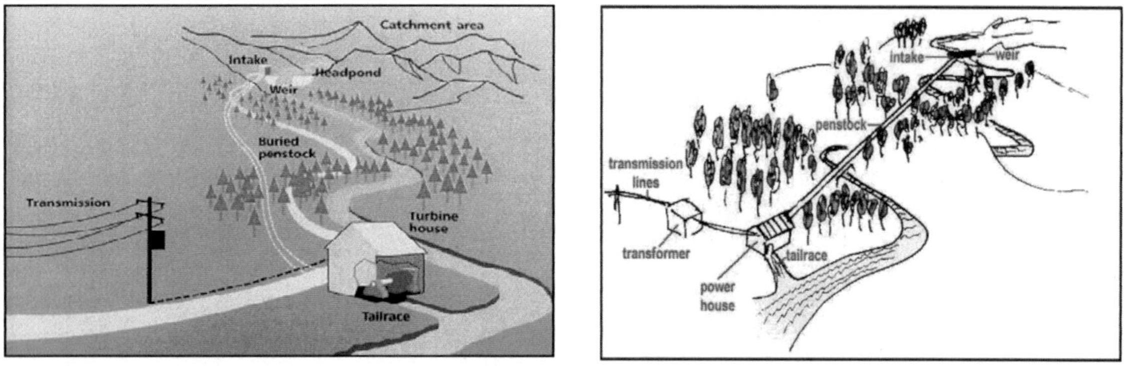

Figure 3. Small hydropower plant (SHP) with derivation [7]

First step after selecting the appropriate method is the ecological flow determination and entering the GIS data required by the software and then the Vapidro-Aste calculates automatically the river network associated to the interesting area.

For the next steps, the user selects the river sector where to calculate the potential hydropower production, inputs the hydrological data, the technical and economical data required by the software and in the end the model is able to determine the optimal hydropower exploitation scheme considering the financial parameters.

Therefore, the results obtained using Vapidro-Aste software consist of GIS maps, tables and graphics that can be exported in different formats for the interpretation and provide the final conclusions.

3. Results

The case study in the paper presents a simulation of the hydropower potential assessment of the Olanesti River, L=41 km according to the Cadastral Water Atlas of Romania [8], for the river sector from the upstream of the Vladesti Reservoir to the Olanesti River springs.

The Olanesti River Basin (S = 237 km^2) is located in the south-central part of Romania within the Olt River Basin, in the North-West of the Oltenia region within the Căpăţânii Mountains from the Oltenia Sub-Carpathians.

The analyzed river sector is 30 km long and the calculus is performed for the hydrological data registered at the Olanesti-Bai hydrometric station within the 1970 – 2011 period.

The results obtained using the Vapidro-Aste software for the assessment of small hydropower potential for the Olanesti River are the following:

- the installed capacity of the small hydropower plant is 978 kW;
- the electricity production computed on an yearly average functioning duration of 3900 hours/year is 3,818 MWh/year;
- the average lifetime of the small hydropower plant (SHP) is 30 years;
- GIS maps and graphical representation of:
  - the installed capacity variation on the analyzed river sector (Figures 4 - 5);
  - the electricity production variation on the analyzed river sector, computed on the average lifetime of 30 years (Figures 6 - 7);
  - the identification of the optimal locations for the water intake and the power plant (Figures 8-9).

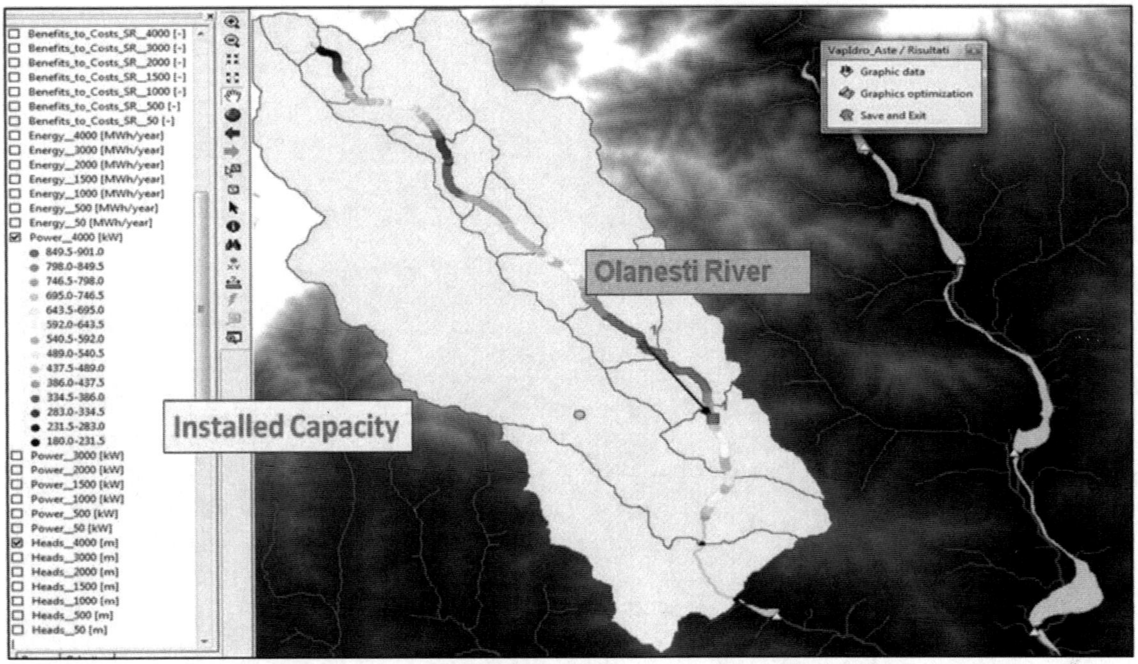

Figure 4. The GIS map of the installed capacity variation on the Olanesti River for the analyzed river sector

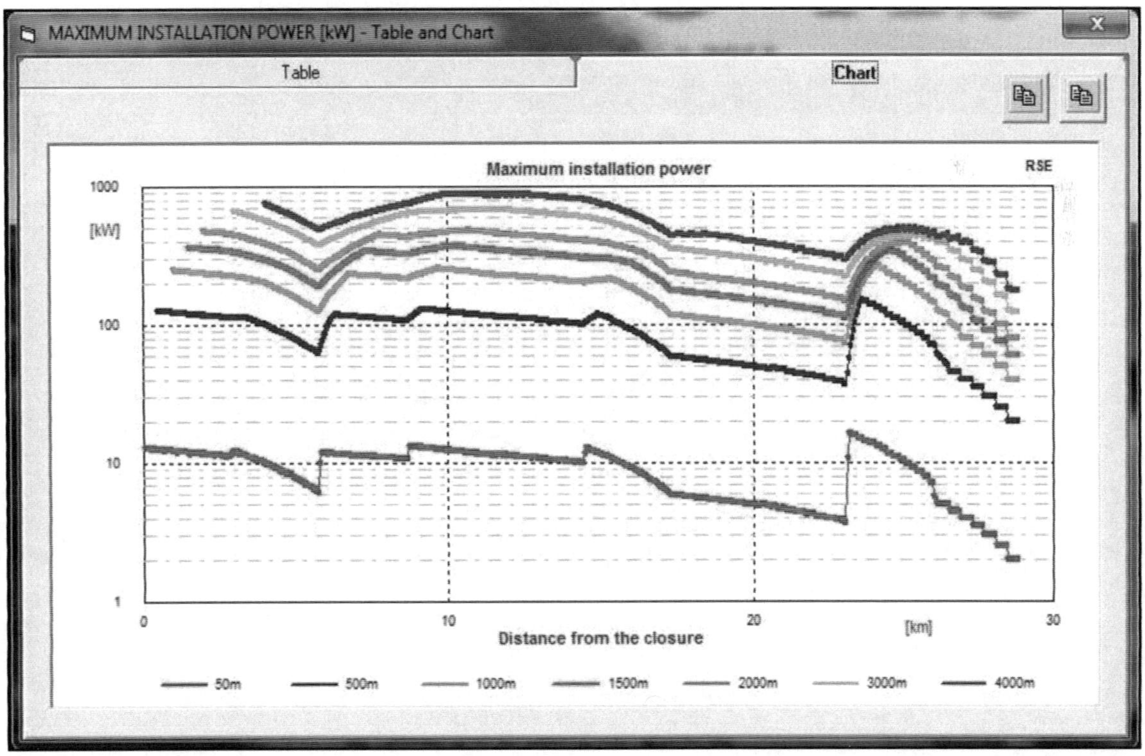

Figure 5. The graphical representation of the installed capacity variation on the Olanesti River for the analyzed river sector

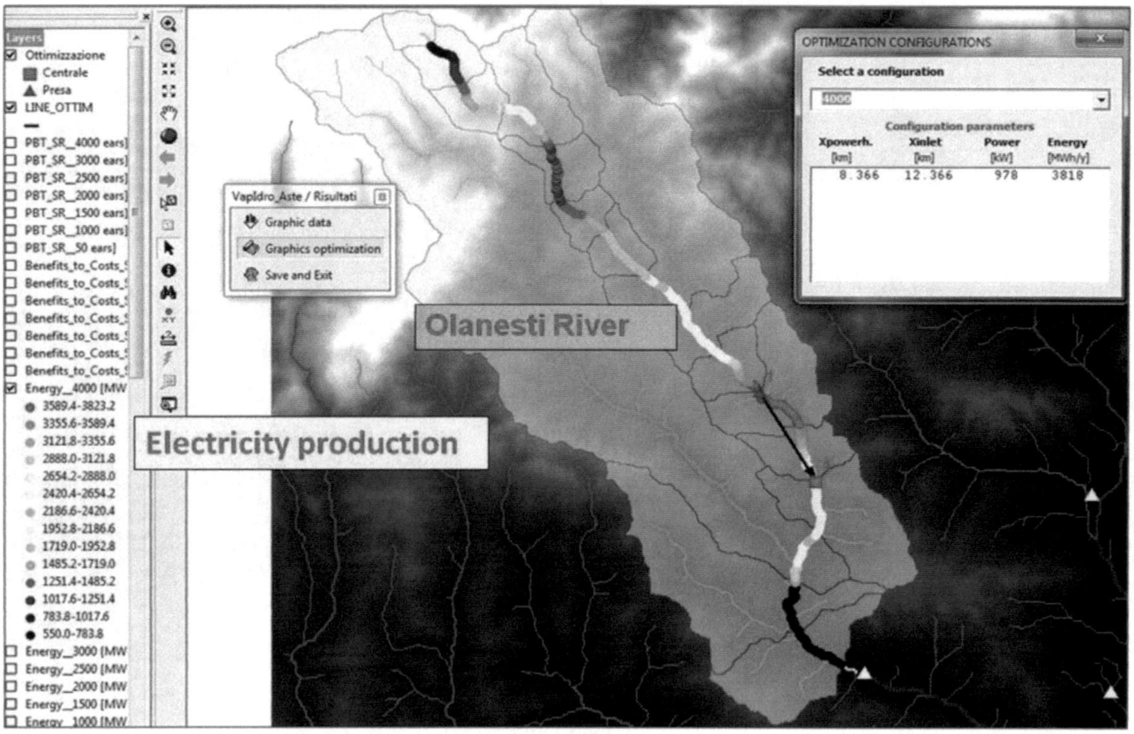

Figure 6. The GIS map of the electricity production variation on the Olanesti River for the analyzed river sector, computed on the average lifetime of 30 years

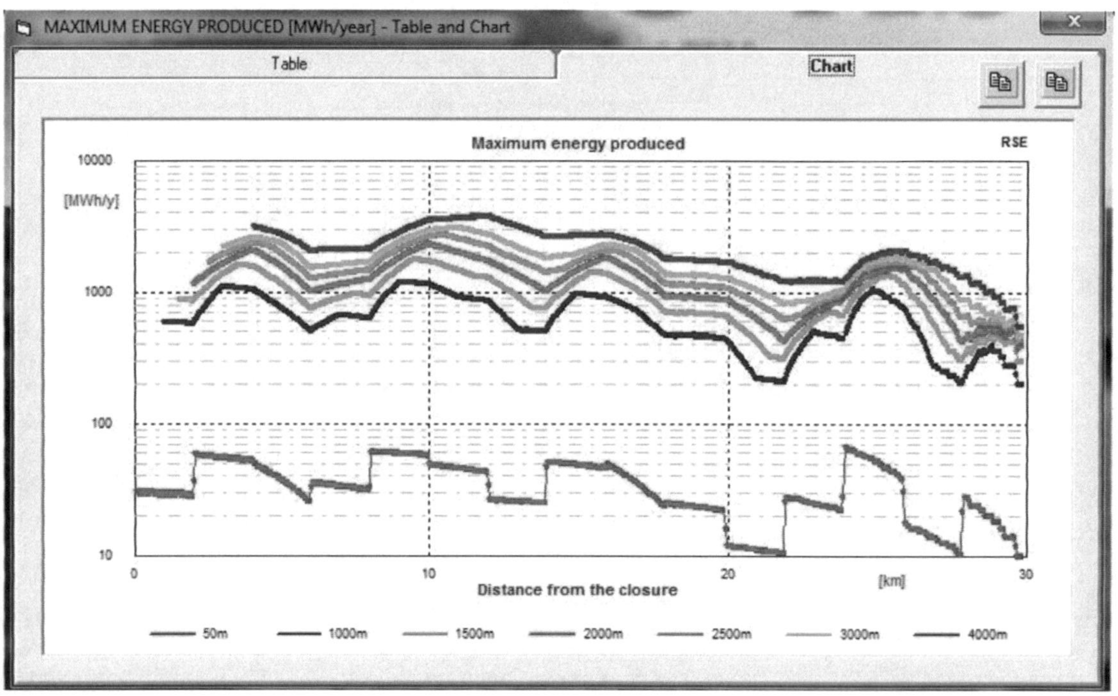

Figure 7. The graphical representation of the electricity production variation on the Olanesti River for the analyzed river sector, computed on the average lifetime of 30 years

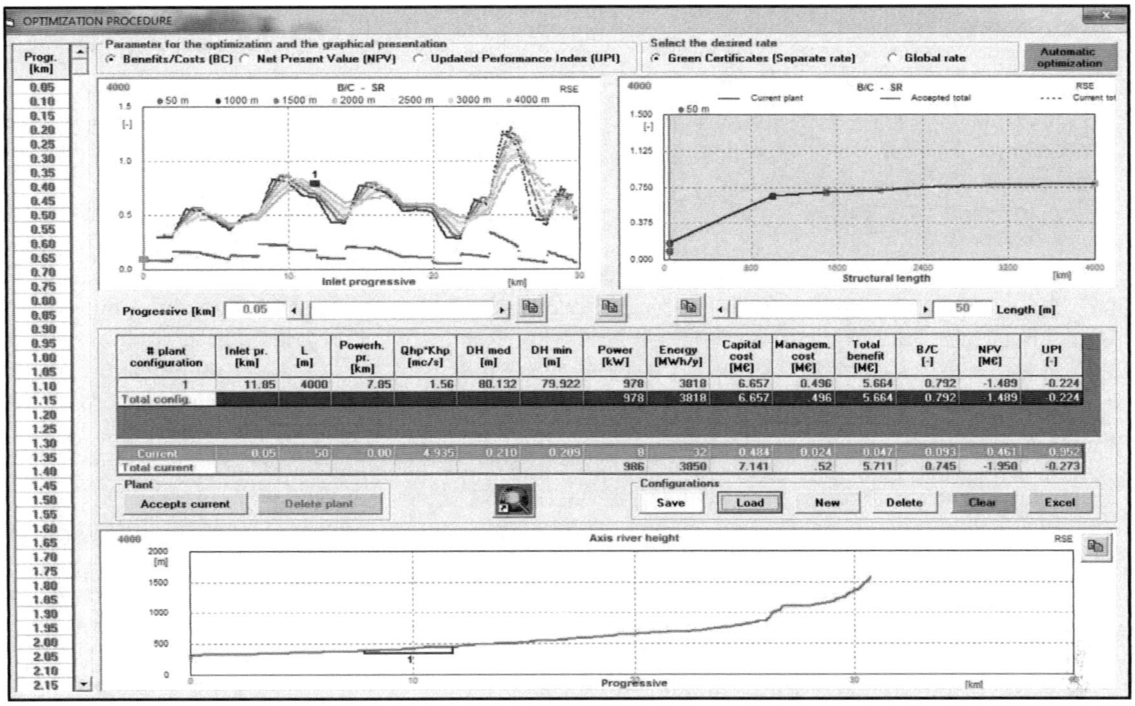

Figure 8. The Vapidro-Aste optimization procedure on the analyzed river sector for the identification of the optimal locations of the water intake and the power plant based on installed capacity

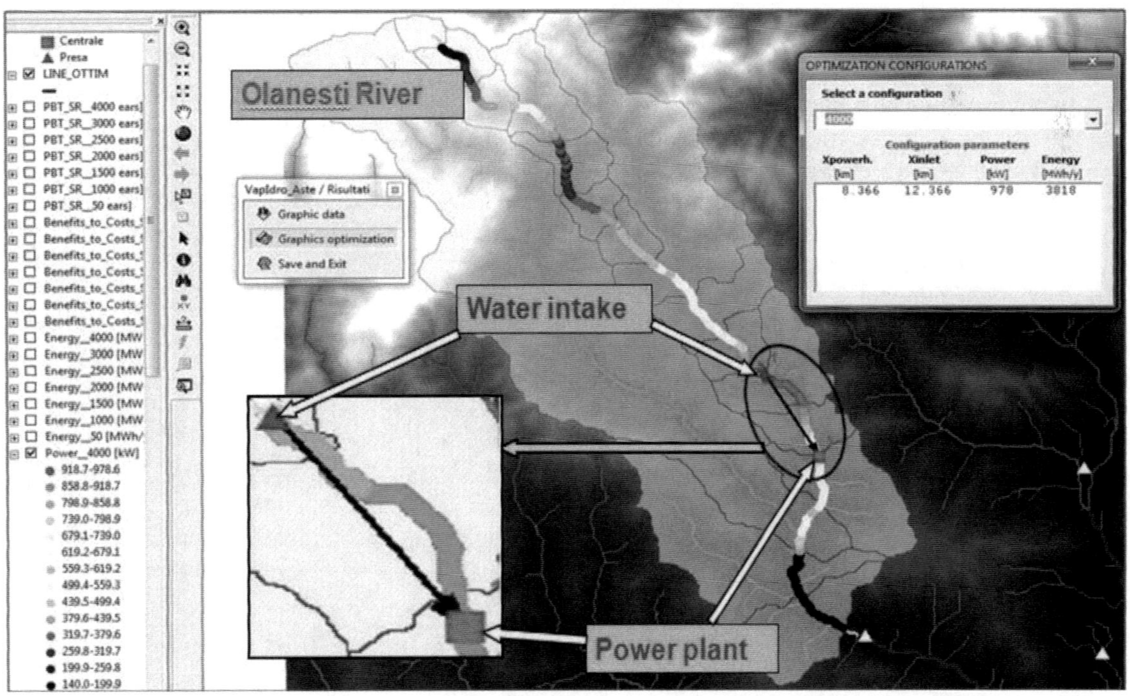

Figure 9. The identification of the optimal locations for the water intake and the power plant based on installed capacity on the Olanesti River for the analyzed river sector

IOP Conf. Series: Materials Science and Engineering **163** (2017) 012022 doi:10.1088/1757-899X/163/1/012022

4. Conclusions

The hydropower potential on the Olanesti River for the analyzed river sector of about 30 km length was assessed considering the heads based on the Digital Terrain Model and the multiannual mean flow for the dry and the wet seasons. In addition, the environmental flow for the dry and the wet seasons was considered for the installed capacity assessment, in order to cope with the Water Framework Directive 60/2000/EC requirements.

The harnessed hydropower potential, the optimal distance between the water intake and power plant location, their optimum positions along the river sector, installed capacity and electricity production, calculated at an average lifetime, the cost-benefit analysis and the payback time are the outcomes of the Vapidro-Aste software and represent important specifications for potential investors.

In the presented case study, the results obtained with Vapidro-Aste revealed the maximum installed power on the analyzed river sector is 978 kW which corresponds to a distance of 4000 meters from the water intake to the power plant and for a mean head of 60-70 meters. The mean electricity production is about 3818 MWh/year and is computed for an operation time of the hydropower plant of 3900 hours/year.

In order to respect the RES-e Directive 28/2009/EC and the Water Framework Directive 60/2000/EC provisions, a periodical hydropower assessment of the rivers which includes the ecological issues is required.

The periodical assessment of the river's hydropower potential is needed also due to the technological progress, modifications of the river flows, changes of the economical conditions, discovery of other alternative energy sources etc.

The applicability of the software might be extended on a regional scale in order to support the decision making authorities, taking into account the increasing demand for energy as well as the environmental issues.

References

[1] ***Directive 2009/28/CE on the promotion of the use of energy from renewable sources, amended by Directive 2013/18/EU

[2] www.seehydropower.eu/, accesed on 10 May 2016

[3] ***Water Framework Directive 2000/60/EC with further modifications

[4] www.isph.ro, accesed on 12 May 2016

[5] www.opcom.ro, accesed on 12 May 2016

[6] www.anre.ro, accesed on 11 May 2016

[7] Peviani M, Alterach J and Danelli A 2011 *SEE HYDROPOWER Project, targeted to improve water resource management for a growing renewable energy production*, HYDRO 2011, Prague, Czech Republic, 17-19 October

[8] ***1992 Cadastral Water Atlas of Romania

IOP Conf. Series: Materials Science and Engineering **163** (2017) 012023 doi:10.1088/1757-899X/163/1/012023

The experimental determination of the luminous flux emitted by a few types of lighting sources

E Spunei, I Piroi and C P Chioncel

Eftimie Murgu University of Resita, Department of Electrical Engineering and Informatic, Traian Vuia str., no. 1-4, 320085 Resita, Romania

E-mail: e.spunei@uem.ro

Abstract. This paper aims to determine the luminous flux of several types of power sources used for the artificial lighting inside. The luminous flux was determined using a photometer integrator (lumenmeter Ulbricht) and a luxmeter from the laboratory of Electrical and Lighting University "Eftimie Murgu" Resita. Based on a reference source, the constant of the integrator photometer k was determined. The paper presents the measuring results of the luminous flux on several lighting sources: with incandescent, fluorescent, and LED. It has been found that the amount of luminous flux in the nominal data written in the box of lighting sources, does not correspond to the actual measured value. Using the photometric sphere, the actual light output from various sources were it is not known, can be determined.

1. Introduction

To ensure a certain visual comfort, lighting sources must emit a certain luminous flux. When acquiring a specific source, its nominal data are marked on the box, including the light output.

From tests conducted on various sources with the same technical features, has been found that for the acquired light source, the luminous flux value marked on the box do not agree with the measured value.

There are situations where the value of the emitted flux by a certain source of lighting is not part of the nominal data and is necessary to be determined. This determination is necessary in order to establish the aging degree of the source and whether it should be replaced. This is done by measuring the flux values issued by it in different time intervals. Keeping in function a lighting source that is no longer capable to emit a corresponding visual luminous flux, causes discomfort and unnecessary electricity consumption.

Therefore it follows that, for certain applications, is necessary to determine the actual value of the light output of the light sources. To help increase the lighting energy efficiency, it is useful, from time to time, to measure the luminous flux emitted by different sources, to be replaced in time.

2. Luminous flux, lighting, luminous efficacy

The luminous flux of a source is defined as the radiant power of that source evaluated after the produced light feeling. The light feeling is dependent on the radiation spectrum wavelength λ according to the relative visibility curve (Figure 1). The sensitivity is maxim ($V_\lambda = 1$), - for a normal eye – at a radiation with the wavelength 0.555 μm (green) [1-3].

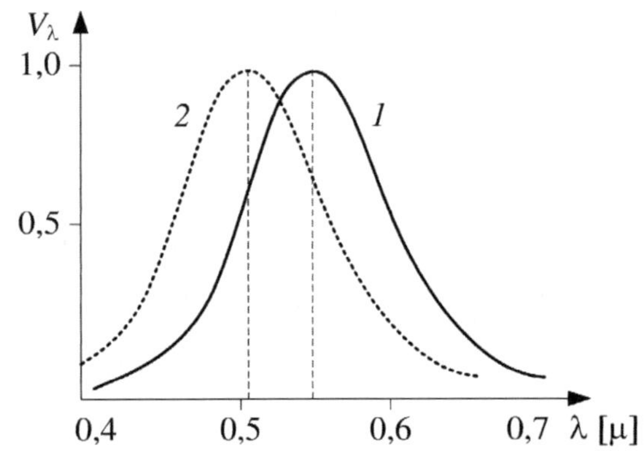

Figure 1. V_λ=f(λ) curve for diurnal vision – 1; night vision – 2

The luminous flux is dependent radiant power (energy flow Φ_e), through the relative visibility curve of the normal eye. The monochromatic light flow Φ_λ is defined by the relationship:

$$\Phi_\lambda = P_\lambda \cdot V_\lambda \tag{1}$$

where: P_λ is the power required for the wavelength λ, and V_λ is the relative visibility of the eye for the wavelength λ. The relative eye visibility is defined as the power corresponding to the exciting agent for the 0,555 μm wavelength and the required power for the λ wavelength, to produce the same lighting feeling:

$$V_\lambda = \frac{P_{0,555}}{P_\lambda} \tag{2}$$

The measurement unit of the luminous flux is a radiated watt per 0,555 μm wavelength, called watt light [Wl]. In the international system (IS) the measuring unit for flux is lumen [lm], one watt light is equal to 683 lm.

The value of the luminous flux can be measured using the photometer integrator – Ulbricht lumen meter, which consists of a empty metal or fiberglass sphere, with a diameter of 0,5÷4 m, painted in interior with matt, very fine grain white color, that produces a multiple diffuse reflectance. The white color used for painting the photometer is characterized by a certain reflection factor *r*.

The illumination of a surface at a certain point of it, is the luminous flux dΦ ratio received by an elementary surface surrounding the point and the area of that surface dS:

$$E = \frac{d\Phi}{dS} \tag{3}$$

The measurement unit of illumination in the international system is lux [lx], defined as illumination of an area that receives a luminous flux of one lumen, uniformly distributed, over a surface of $1m^2$.

To light level is measured using an luxmeter. This is based on the action of the light on a photosensitive element.

The luminous efficacy of a light source is defined as the ratio of luminous flux Φ emitted by the source and the power P consumed by the source:

$$e = \frac{\Phi}{P} = 683 \cdot \eta_l \left[\frac{lm}{W}\right] \tag{4}$$

where η_l is the efficiency of the light source.

The luminous efficacy is the size that establishes a relationship between bright-technical characteristics of the source and electric power consumed by it. This is very important in choosing the sources types in lighting design and analyze their effectiveness.

3. Case Study

To measure the light levels and to determin the luminous flux, we used a benchmark source, were the issued flux was known, a luxmeter and a lumenmeter, those being in the Laboratory of Electrical installation and Lighting.

In Figure 2 we presented the digital luxmetrul, from UNITEST, which uses a silicon photodiode as sensor.

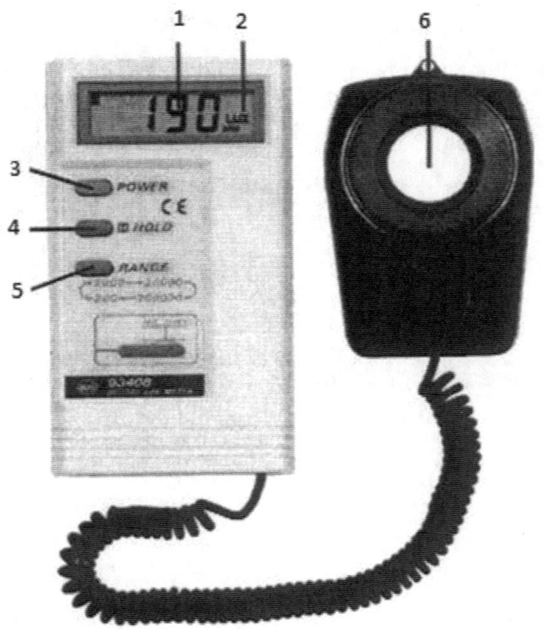

Figure 2. Digital Luxmetru Unitest

It is composed of:

- 1 - LCD display;
- 2 - display measuring range;
- 3 – start button;
- 4 - data memory button;
- 5 - choice button;
- 6 - sensor (photodiode with silicon).

The digital used lux meter has four measuring ranges, ie 0-20 lx, 0-200 lx, 0-2.000 lx and 0-20.000 lx, the corresponding errors to the four areas being 0,01 lx, 0,1 lx, 1 lx and 10 lx.

To determine the luminous flux emitted by different sources, the Ulbricht lumen meter, shown in Figure 3, was used. His property is a uniform light scattering or diffusion effect. The incident light rays on any point on the inner surface of the sphere are evenly distributed at all points of the sphere, by multiple reflections with the purpose to create an equal light flow over the inner surface of the sphere.

The effects of the original light direction are minimized. The integrating photometer is generally used as an light source and a detector to measure optical power.

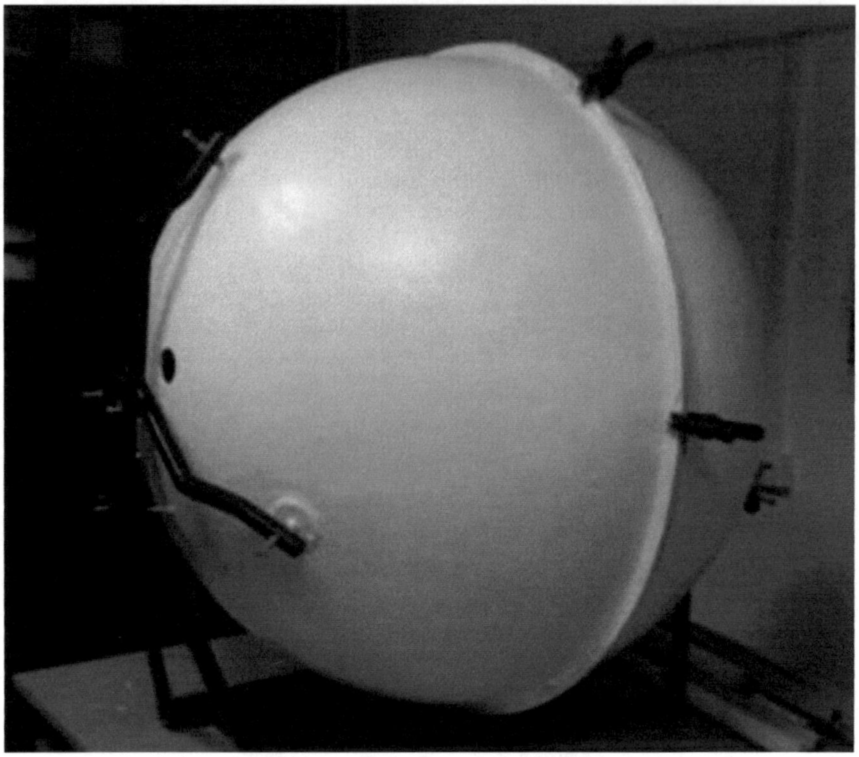

Figure 3. Integrator photometer

The integrator photometer has in its composition:

- glass fiber sphere, with a diameter of 1 m (R = 0.5 m), with one fixed and one mobile hemisphere;
- fastening bracket of the sphere;
- power supply system of the light sources which are to be tested. The system allows the supply of 220V AC , 24V DC voltage and 12V DC voltage;
- Closure gasketed system and mounting gripper;
- fastening bracket of the light source that has to be tested;
- A screen to block the reception by luxmeter of the direct wave emitted by the light source.

The integrator photometer has a window where the photosensitive cell of the luxmeter, that measures the light caused by multiple reflections (infinite) of light rays (Figure 4), is entered.

To hamper the light rays fall directly on the photometer window, different types of screens (Figure 4) for each of the types of tested light sources (Figure 5), were made.

To determine the k constant of the integrator photometer and the reflectance factor r of the paint, an benchmark source was used, where the the value of light output was known. The benchmark source was a halogen lamp whose flux value was 204 lm. Measuring with the luxmeter the brightness from the window, it was found that it had the value of 295 lx.

Knowing the relationship between the measured illumination in the window and the flux of the benchmark source [1-3]:

$$E = k \cdot \Phi \ [\mathrm{lx}] \quad (5)$$

the value of the photometer integrator constant k was determined k= 1,44 lx/lm.

Figure 4. Illumination measurement

Figure 5. Made screens

Figure 6. Tested sources

Knowing the value of this constant, the valueof the reflection factor *r* of the white paint integrator photometer color was determined too:

$$k = \frac{r}{1-r} \cdot \frac{1}{4 \cdot \pi \cdot R^2} \quad [\text{lx/lm}] \tag{6}$$

whence:

$$r = \frac{4 \cdot k \cdot \pi \cdot R^2}{1 + 4 \cdot k \cdot \pi \cdot R^2} \cong 0{,}82 \tag{7}$$

Several types of lighting sources have been tested, some whoere the luminous flux was known from the nominal data box, others where the nominal flow was not known. The types of tested sources and the obtained or calculated values are shown in Table 1.

Table 1. Tested sources, measured and determined values

Current number	Source type	Power [W]	Nominal luminous flux Φ_N [lm]	Measured lighting [lx]	Calculated luminous flux Φ_c [lm]	Real light effectiveness e [lm/W]
1	Halogen	23	204	295	204,86	8,907
2	Incandescence	100	-	1940	1347,22	13,47
3	LED 1	18	1750	2200	1527,77	84,88
4	LED 2	5	400	548	380,55	76,11
5	LED 3	5	400	558	387,5	77,5
6	LED 4	5	400	562	390,27	78,05
7	LED 5	7	700	915	635,41	90,77
8	Auto 24 V	5	-	856	594,44	118,88
9	Fluorescent	18	1040	1280	888,88	49,38

The calculated luminous flux was determined taking into account the value of k constant of the integrator photometer and the real luminous efficacy was calculated with the relationship:

$$e = \frac{\phi_c}{P} \quad [lm/W] \tag{8}$$

To determine whether the value of the luminous flux posted on the box with the nominal data of the sources is equal to the real (determined by measurements) we purchased and tested 25 LED sources with the same power and the same rated light output.

Table 2 presents the measured and calculated values of the parameters of interest. The tested sources were LED type, with 7 W power and luminous flux of 700 lm, resulting a luminous efficacy of 100 lm / W.

Using efficient lighting spurces in the optimal design process of ligthing systems leads to efficient use of electricity by reducing power source at the same luminous flux [4].

Taken in consideration the LEDs characteristics an analysis of harmonics introduced by them is required [5].

4. Conclusion

Interpreting the results of measurements and calculations we found that:

- the luminous efficacy is totally unacceptable for incandescent source, being 5.94 times lower than the acceptable value. This low luminous efficacy lead to the elimination of these sources from the manufacturing process and trading;
- the luminous efficacy is almost of 80 lm / W (an acceptable) for the LED 2, LED 3 and LED 4 sources, for LED 1 source the luminous efficacy exceeds the acceptable value;
- the halogen source is exhausted and must be replaced, due to extremely low luminous efficiency;

Table 2. Tested sources, measured and determined values

Current number	Nominal luminous flux Φ_N [lm]	Measured ligthing [lx]	Calculated luminous flux Φ_c [lm]	Real light effectiveness *e* [lm/W]
1	700	868	602,78	86,11
2	700	917	636,81	90,97
3	700	896	622,22	88,89
4	700	893	620,14	88,59
5	700	887	615,97	88,00
6	700	924	641,67	91,67
7	700	908	630,56	90,08
8	700	915	635,42	90,77
9	700	905	628,47	89,78
10	700	916	636,11	90,87
11	700	912	633,33	90,48
12	700	924	641,67	91,67
13	700	922	640,28	91,47
14	700	928	644,44	92,06
15	700	873	606,25	86,61
16	700	889	617,36	88,19
17	700	895	621,53	88,79
18	700	912	633,33	90,48
19	700	909	631,25	90,18
20	700	890	618,06	88,29
21	700	910	631,94	90,28
22	700	892	619,44	88,49
23	700	910	631,94	90,28
24	700	906	629,17	89,88
25	700	904	627,78	89,68

- for any light source, the output light written in the nominal data on the products box was not corresponding;
- the luminous efficacy for the 25 LED 7W sources tested in the laboratory, is 89.7 lm / W, which represents 89.7% of efficiency resulting from the use of the written information on the products box.

By measurements using lumen meter and luxmeter we can determine the status of the light sources and the decommissioning of those who are no longer adequate. The measurement results can be used in optimizing the design [6], [7] and the use of indoor and outdoor lighting installations.

The results obtained for the 25 LED sources lead to the idea of passing a law througth tha the producing and importing companies are required to verify the accuracy of the written data on the consumer packaging lighting sources.

References

[1] Piroi I 2009 *Instalații electrice și de iluminat*, Editura Eftimie Murgu, Reșița, Romania
[2] Mogoreanu N 2013 *Iluminatul electric*, Editura Lumina, Chișinău, Moldavia
[3] Bianchi G, Mira N, Moroldo D, Gheorghescu A and Moroldo H 2001 *Sisteme de iluminat interior și exterior*, Ediția a-III-a revizuită, Editura Matrixrom, Bucuresti, Romania
[4] Pentiuc R D, Vlad V, Lucache D D and Pavel S 2014 *Street Lighting Power Quality*, 8th International Conference And Exposition On Electrical And Power Engineering EPE 2014, Iași, Romania, October 16-18, pp 1107-1110
[5] Rata G, Rata M and Prodan C 2014 *Analysis of the Deforming Regime Generated by Different*

IOP Conf. Series: Materials Science and Engineering **163** (2017) 012023 doi:10.1088/1757-899X/163/1/012023

Light Sources, using Reconfigurable System – CompactRIO, 8th International Conference And Exposition On Electrical And Power Engineering EPE 2014, Iaşi, Romania, October 16-18, pp 748-751
[6] Spunei E, Piroi I and Piroi F 2014 Optimizing Street Lighting Systems Designs, *Analele Universității Eftimie Murgu Reșița, Fascicola de Inginerie* **XXI**(3) 257-268
[7] Spunei E, Piroi I and Piroi F 2014 Notes On Led Insallation In Street Illumination, *Analele Universității Eftimie Murgu Reșița, Fascicola de Inginerie* **XXI**(3) 269-280

IOP Conf. Series: Materials Science and Engineering **163** (2017) 012024 doi:10.1088/1757-899X/163/1/012024

Preliminary experimental research for silver recovery from radiographic films

L R Cânda and E Ardelean

Politehnica University of Timisoara, Engineering and Management Department, Revolutiei Str., no. 5, 331128 Hunedoara, Romania

E-mail: letitia.canda@fih.upt.ro

Abstract. Global demand for silver remains steadily to about 1,000 million ounces (28349500 kg), of which around 600 million ounces (17009700 kg) are used in industrial applications. Extraction of silver from the ore is expensive and harmful to the environment and low efficiency. X-ray films represent an important worldwide consumer as research on recovery of silver from exposed radiographic films must be oriented to achieve a maximum recovery and a high purity silver, with methods through the by-products will be less polluting for the environment. The paper presents some laboratory tests referring to the recovery of silver from radiographic films by leaching with sodium hydroxide. Two series of experiments were performed with different amounts of used X-ray film.

1. Introduction

Silver is a rare earth metal, seen as the symbol Ag, after the Latin word Argentum, which occurs naturally in the environment only as a soft metal, coloured "silver", there are no artificial sources for it. Silver comes from a combination of mining silver, gold, copper, lead, and zinc ore. It also appears as a white powder (silver nitrate and silver chloride), or dark grey to black compounds (sulphide, silver oxide and silver). These compounds are hazardous to the environment. Photography is a sector representing an important source of silver released into the environment.

The mines that produce silver and other metals are another source [1]. Silver is a very inactive metal. It does not reacts with oxygen in air under normal conditions. However, the sulphur compounds react slowly with air. The product of this reaction is silver sulphide (Ag_2S), a black compound. Silver do not react with water, acids or many other compounds and it burns only as a silver powder.

Silver has been used due to its functionality and beauty. Silver is a material used both in many industrial sectors and also as a precious metal. As an industrial metal it has thousands of uses due to its remarkable qualities. Silver has the highest electrical and thermal conductivity compared to any other element. Silver bullion stocks dropped steadily, while investments in silver are growing, which is reflected in its relatively constant price. Production will be increased to meet demand. In the last 10 years the price of silver was doubled, currently hovering at about $ 15 an ounce, but in the same period it also had prices near $ 50 per ounce [2], as you can see in Figure 1.

The ores that provide silver are limited, so silver mining cannot sustain the consumption. Moreover, the extraction of silver from scrap by products and consumer goods is important to put back into markets that rely on silver [3].

IOP Conf. Series: Materials Science and Engineering **163** (2017) 012024 doi:10.1088/1757-899X/163/1/012024

Figure 1. Silver price in the last 10 years [2]

2. Silver recovery methods from X-ray films

The value of silver hasn't always been compared to that of gold. However, considering the countless uses of silver, gold just does not stack up against silver in terms of functionality. Silver is used in four major areas: industrial, photography, jewellery and manufacture coins. Each year, the photographic sector is allocated about 45% of silver for radiography, which is completely eliminated after use. Despite the fact that many technologies have tried to recover the silver content from this sector, they have not met the environmental demands of less than 5 ppm silver in effluents. In the latest years, to recover silver from the waste generated by the photographic sector are used electrochemical and chemical methods. The ion exchange and electrolytic plating silver units rank as the most acknowledged methods used in most cases. Both methods have been described extensively in the scientific literature, which indicates that electrochemical processes are best suited due to their low operating cost and high recovery rates.

Several authors have done research on the recovery of silver present in effluents generated by the photographic industry. However, information concerning solid waste is not conclusive. Fundamental studies on the subject showed that solutions such as nitric acid, cyanide, ammonia and potassium borohydride are appropriate agents to the dissolution of silver that is found in solid waste from the photographic industry (radiographic films). Besides, it has been shown that it is convenient to use substrates such as stainless steel, titanium, silver, glassy carbon as a working electrode to perform the deposition of silver [4-6]. Moreover, fundamental studies of chemical species in nitrate solid bath generated by waste from radiographic and photographic industry showed that the predominant species in the nitrate is Ag^+ and hydrogen evolution does not mingle with its deposit [4].

Silver recovery by chemical leaching depends on the heating of the film in the solution of oxalic acid ($H_2C_2O_4$), nitric acid (HNO_3), sodium hydroxide (NaOH), at boiling temperature for the separation of the inorganic component from the polymer layer. The development of X-ray film leads to waste from the fixer and from the aqueous solutions containing from 1000 to 10,000 mg Ag/L and respectively 50-200 mg Ag/L silver in form of thiosulfate complex [7].

Regarding the data found in specialized literature [8], the following conclusions can be drawn:

a) The easiest way for silver recovery is by treatment with nitric acid. It is an inexpensive reagent, the process is simple and with low consumption;

b) The efficiency by using nitric acid is superior comparing with the using of oxalic acid and sodium hydroxide;

c) Silver recovery rate depends on the density of pulp;

d) The presence of NO_3^- ions excess (> 30g/L of nitric acid in the solution) had no effect upon improving the silver recovery process;

e) Optimum leaching time was 15 minutes;

f) There is a strong link between the density of pulp, acid concentration and the leaching time.

In electrolysis a current is applied directly between the two electrodes. The main electrolytic equipment are: one having a rotating movement of the cathode in the solution and the other in which the solution is pumped around an immovable cathode. The silver thus obtained has generally a higher purity than 98%, according to several studies. I spite of that, this method is used only for solutions with high concentrations and it is not capable to reduce silver at a concentration of less than 100mg/L, considering that the maximum allowed concentration for environmental is 5mg Ag/L [9]. The raw metal to be refined serves as the anode and the pure metal is deposited on the cathode. A significant experiment for the recovery is by dissolving silver compounds with concentrated nitric acid. It has been used electrolysis with an internal and external steel electrode and a copper electrode at the anode. For extracting silver, enzyme-based methods could be a promising solution since the silver linked gelatine in the emulsion layer may be released using proteolytic enzymes. Thus, research has been directed at the enzymatic hydrolysis of gelatine with Conidiobolus Coronatus alkaline protease. After the whole process, the gelatine layer was completely removed, leaving the clean polyester film and silver was recovered by hydrolysis, and thus can be reused both the polyester film and the silver [10].

Worldwide, approximately 57% of the silver present in discarded products is recovered. Concerning the most commonly used metals, silver has the highest percentage of recovery, but large quantities of it are still lost in the different emissions to the environment [11].

Radiographic films consist of a transparent, flexible, blue-tinted base coated on one or both sides with an emulsion, in layers about 0.0005 inch thick – Figure 2.

X-ray film must have different characteristics from photographic film because X-ray exposure is different from light exposure, and it must be manufactured with consistent uniformity and quality as well. Emulsion is a material in which X-ray or light photons from screen interact and transfer information. It consists of gelatine containing microscopic radiation sensitive silver halide crystals, 90-99% silver bromide and 1-10% silver iodide. Furthermore, it also contains several additives like wetting agents, hardener or antifoggants. Gelatine keeps silver halide grains dispersed, prevents from clumping and allows the ingress of processing solutions without affecting the strength or the efficiency, should be pure and porous so that it transmits light and allows chemicals to penetrate to the crystals of silver halide during this procedure. Radiographs are the result of a complex process, as well as the production of emulsion.

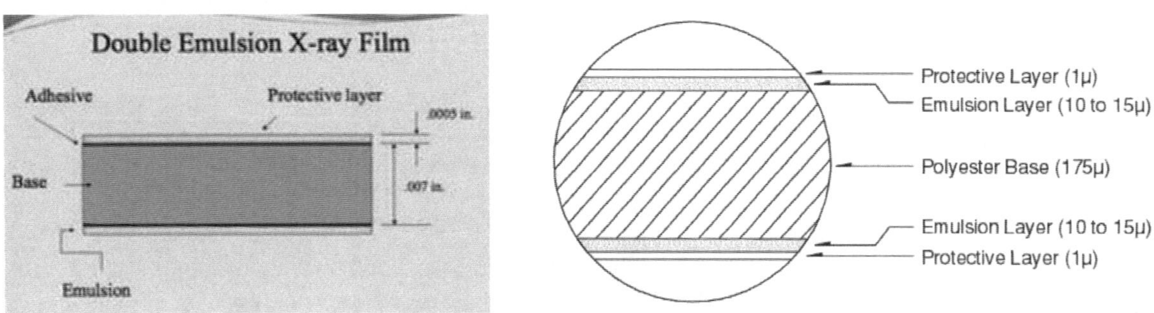

Figure 2. Radiographic films structure [12]

X-rays strike the grains of the silver halide in the emulsion, so some of the Br^- ions are released and captured by the Ag^+ ions. This modification cannot be observed by common, average physical methods, therefore it is defined as a "latent image". In the presence of the developer, the reduction process is improved and the black, metallic silver, suspended in the gelatine layer, creates an image on both sides of the radiographic film. Radiographic film processing implies complex methods that must comprise certain rules related to concentration, temperature, time and physical transformations. To obtain high standard X-ray films, a quality control and a superior level of consistency is demanded [13].

Precious metals' price is less affected by the economic destabilization determined by financial crises. This fact can be proved by price-performance ratio reached by the companies involved in the extraction of metals as a raw material. Sustainable development and increasingly severe legislation regarding the environment are making the precious metals recovery, a secure path in conserving natural resources and reducing waste without high costs. Increasing efficiency of recovery methods opens up new opportunities in research field, due to the world demands, decreasing natural resources and human need of development [14].

3. Laboratory experiments on recovery of silver from used X-ray

On our own research it was used the silver leaching from X-ray films with NaOH [15]. The exposed X-ray films were provided from Municipal Hospital " Alexandru Grigorescu" Hunedoara. There were conducted two sets of experiments on different amounts of radiographic film.

For the first series of experiments, X-ray film was cut into relatively small pieces, washed with distilled water, alcohol and then dried in an oven at 50°C. The dried film was weighed and placed in a beaker in which was placed a NaOH solution of 1.5M concentration (Figure 3). It was used 100ml of 1.5M NaOH for 50 g of X-ray film. The container was put on the stove at 70°C for the leaching of silver layer. The process was conducted for 1h, with the extraction of the pieces of film, this time with a blue colour (Figure 4).

Figure 3. Radiographic film preparation (washing, drying, treatment with NaOH 1.5M)

Figure 4. The leaching, the extraction of the clean film, the evaporation of the liquid phase

The resulting product was again heated on a hotplate and mixed vigorously until complete evaporation of NaOH; the resulting solid residue was washed with distilled water and then filtered (Figure 5), issues during the experiments being listed below.

For the second series of experiments it was taken a larger amount of developed radiographic film (150g) with 300ml leached with 1.5M NaOH. For this group of experiments the film was cut into larger pieces, washed in distilled water and natural dried in air (24h) on a cotton surface (Figure 6). Thus the drying part in oven was avoided. The same steps as in the first case have been complied for the other parts of the experiments.

After the drying process of residues, they were analyzed to the stereomicroscope from Metallurgy Laboratory of Physics where it was found the presence of silver grains, and soda crystals which were formed on them – Figure 7.

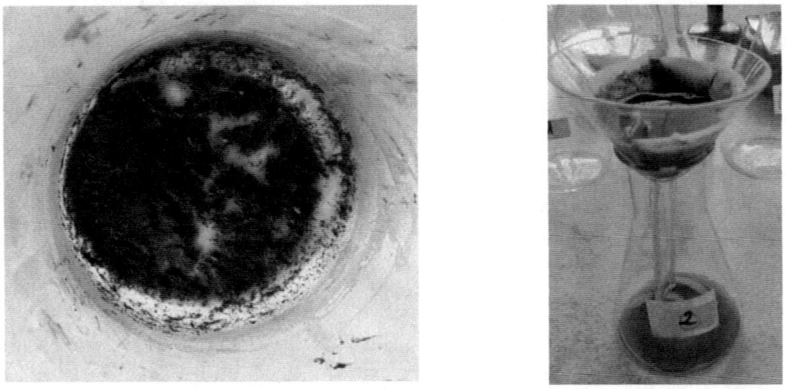

Figure 5. The resulted product, the filtration of the suspension

Figure 6. X-ray preparation

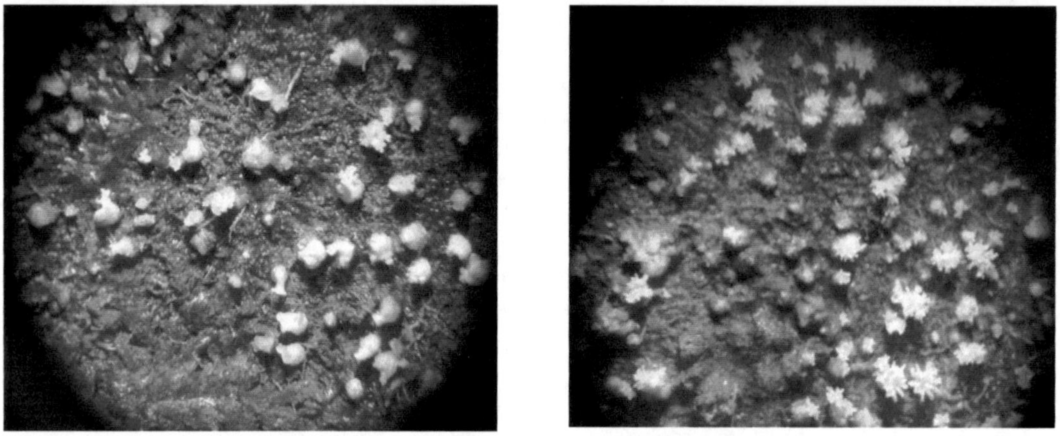

Figure 7. Stereomicroscope view (x20)

4. Conclusions

After the preliminary experiments carried out for the purpose of recovery of silver by leaching with NaOH from radiographic films the following conclusions were revealed:

- The process comprises a series of steps that must be followed so that it can be recovered at least part of silver deposited onto radiographic film.
- The relatively long duration of the process, which would have in industrial practice would have a rather low economic efficiency.
- Improving the washing of the resulting residue with a larger amount of distilled water until no soda crystals will be found beside the recovered silver.

However, the relatively low results obtained by this method leads to the opportunity to apply a more efficient method for the recovery of silver from X-ray films.

Future research will be directed to radiographs leaching with nitric and oxalic acid by determining the optimum parameters of leaching. Subsequently, the purification of silver, thus obtained, will be made either by electrolysis or by other methods, less harmful to the environment. In present circumstances, the research for silver recovery is a promising solution for the vast applications of this noble metal.

In PET-film recycling technologies a clear hierarchy is also required. The most important criteria for the recovery of PET-scrap are the degree of "purity" after silver recovery process and the economics of recovery means. Direct reuse in extrusion, for the cleanest PET grade, is the most economical process. The less clean PET samples could be reused after a partial degradation, as by glycolysis, but with higher costs. Film waste represents more contaminated PET. Therefore it must be degraded into the starting monomers, separated then and re-polymerized at the end of the process, which involves further costs. At industrial scale, methanolysis process is more appropriate to be used, while hydrolysis processes, need improving in the future. The only solution for heavily contaminated PET-shreds is they should be incinerated or brought to a landfill. The structure of PET-films must be modify as that they become biodegradable, so the process can be accelerated. Another line of research will focus on recovering Polyethylene terephthalate (PET) after removing the covering film emulsion. [16]

Given the growing number of medical tests worldwide, due to the population growth and the expansion of this type of waste, research on more efficient methods of silver recovery, as well as the PET recovery are appropriate to minimize environmental impact.

References

[1] ***ATSDR, http://www.atsdr.cdc.gov/, The Agency for Toxic Substances and Disease Registry (ATSDR), 1990

[2] http://silverprice.org/silver-price-history.html

[3] http://web.archive.org/web/20120210133952/http://www.silver-refiner.com/silver-information.html

[4] Ramírez P A, Reyes V E and Veloz M A 2011 Silver recovery from radiographic films using an electrochemical reactor, *International Journal Electrochem. Sci* **6** 6151-6164

[5] Zhouxiang H, Jianyiang W, Ma Z and Jifan H 2008 A method to recovery silver from X-ray films with spent fixing bath, *Hydrometallurgy* **92** (3-4) 148-151

[6] Syed S, Suresha S, Sharma L M and Syed A A 2002 Clean technology for the recovery of silver from processed radiographic films, *Hydrometallurgy* **63**(3) 277-280

[7] Arslan V, Ucurum M, Vapur H and Bayat O 2011 Recovery of silver from waste radiographic films by chemical leaching, *Asian Journal of Chemistry* **23**(1) 67-70

[8] Masebinu S O and Muzenda E 2014 *Rewiew of silver recovery technics from radiographic effluents and X-ray film waste*, Proceeding of the World Congress on Engeneering and Computer Science 2014, Vol II, WCECS 2014, San Francisco, USA, October 22-24, view on-line

[9] Yazici E Y, Bas A D and Deveci H 2010 *Extraction of Silver from Waste X-ray Films by Thiosulphate Leaching*, Proceedings of the XIIth International Mineral Processing Symposium, Cappadocia-Nevşehir, Turkey

[10] Shankar S, More S V and Seeta Laxman R 2010 Recovery of silver from X-ray film by alkaline protease from Conidiobolus Coronatus, *Kathmandu University Journal of Science, Engeneering and Technology* **6**(I) 60-69

[11] González P, Recio F J, Ribera D, González O, DaSilva P, Herrasti P and Avila-Rodriguez M 2011 *Silver Recovery from Acidic Solutions by Formation of Nanoparticles and Submicroparticles of Ag on Microfiltration Membranes,* in Mass Transfer – Advanced Aspects, Nakajima H Ed., InTech, Rijeka, Croatia, pp 439-460

[12] http://armymedical.tpub.com/MD0954/Composition-of-X-Ray-X-Ray-Film-Processing-8.htm

[13] https://www.google.ro/search?q=chemical+composition+x+ray+film+images&client=firefox-b&biw=1067&bih=719&tbm=isch&tbo=u&source=univ&sa=X&ved=0ahUKEwiCocOKgbbNAhUOlxQKHYuZA8cQsAQIGQ#imgdii=C-eewAXmWBDJ0M%3A%3BC-eewAXmWBDJ0M%3A%3Bp5siiVKvm2VN2M%3A&imgrc=C-eewAXmWBDJ0M%3A

[14] Canda L, Heput T and Ardelean E 2016 Methods for recovering precious metals from industrial waste, *IOP Conf. Ser.: Mater. Sci. Eng.* **106** 12020

[15] https://www.researchgate.net/publication/238770816_A_Novel_Silver_Recovery_Method_from_Waste_Photographic_Films_with_NaOH_Stripping

[16] http://www.ewp.rpi.edu/hartford/~ernesto/F2012/EP/MaterialsforStudents/Patel/DeWinter-Poly_ethylene_terephthalate__Film_Recycling.pdf

IOP Conf. Series: Materials Science and Engineering **163** (2017) 012025 doi:10.1088/1757-899X/163/1/012025

The analysis of the steel deoxidation process in a vacuum installation

E Ardelean[1], A Socalici[1], M Ardelean[1], S Şerban[1] and M Vătăşescu[2]

[1]Politehnica University of Timisoara, Engineering and Management Department, Revolutiei Str., no. 5, 331128 Hunedoara, Romania
[2]SC Saint Michele SRL, G. Enescu Str. no.1 bis, 331125 Hunedoara, Romania

E-mail: erika.ardelean@fih.upt.ro

Abstract. The oxygen is an element which has adversely affect to the steel quality because it causes embrittlement at high temperature, evidenced by the increased susceptibility on overheating and the appearance of cracks during deformation plastic, respectively by determining the appearance of the gas holes during castings solidification of products. As a result from a technological point of view, in industrial practice shall be adopted a series of measures that are intended to ensure a low oxygen content (recommended below 40ppm). The paper presents a few practical data taken from the industry in relation to the manner in which it is influenced the content of oxygen according with the vacuum parameters and the steel temperature. The resulting correlations, presented in both form graphical and analytical, are a real benefit to the industry, so can be determined a parameter depending on the other two.

1. Introduction

Deoxidation is the metallurgical process for removing oxygen from molten steel. The oxygen solubility in the molten steel decreases with decreasing of temperature, when the excess of oxygen are separated from solution in the form of oxide products - oxide phase, which will be found on the edge of the crystalline grains in the solidified steel [1], [2]. These compounds can form eutectic phases that can determine embrittlement of steel (cannot be plastically deformed). At medium oxygen contents, the steel presents lower mechanical characteristics (resilience), lower corrosion resistance and greater tendency to aging [1-4].

Steel deoxidation process is done through the following methods: precipitation (by using ferroalloys), diffusion (by using reducing or synthetic slag) - in the furnace [5] or in the secondary treatment aggregate [6], [7], and using a vacuum [8]. By using vacuum steel deoxidation, can be obtained grade steels with high purity.

Steel vacuum treatment aims to reduce the oxygen, hydrogen and nitrogen content in liquid steel, decreasing the content of nonmetallic inclusions as a result of accelerating of the settling process, respectively chemical and thermal homogenization of steel. All these metallurgical effects lead to improved physic-mechanical properties of the steel [9].

The vaccum deoxidation can be made on the folowing methods:

- Ferrous oxide dissociation through reaction:

$$[FeO] = [Fe] + \frac{1}{2}\{O_2\} \quad (1)$$

- Volatilization of oxides dissolved in the steel:

$$[MeO] \rightarrow \{MeO\} \quad (2)$$

- With carbon:

$$[C] + [O] = \{CO\} \tag{3}$$

The first two ways have not a practical importance for steelmaking, instead the third (known as self-deoxidation reaction with carbon) have a special importance because, through this, the oxygen dissolved in the steel bath it is transferred, in the gas phase. Carbon vacuum deoxidation is possible by the fact that the decreasing of the pressure increases the intensity of the carbon deoxidation (decreases the content of oxygen in equilibrium with the carbon) [1], [2]

The equilibrium constant of this chemical reaction is expressed as follows:

$$K_c = \frac{p_{CO}}{a_C \cdot a_O} = \frac{p_{CO}}{f_C \cdot f_O \cdot [\%C] \cdot [\%O]} \tag{4}$$

where: p_{CO} - partial pressure of carbon monoxide in the atmosphere;

a_C and a_O - activities of carbon and oxygen in liquid steel;

f_C and f_O – activities coefficients of carbon and oxygen in liquid steel;

[%C] and [%O] – content of carbon and oxygen in liquid steel [1], [2]

At the 1600°C steel temperature and 1 atm - partial pressure of carbon monoxide [10], the equation (4) became:

$$[\%C] \cdot [\%O] = 0.0025 \tag{5}$$

2. Experimental data's

To analyse how the deoxidation process in a vacuum installation is realized, were analysed a total of 17 charges of steel S 355 JO brand (according to EN 10025). It is noted that the charges were prepared in an electric arc furnace (EBT type), treated in the secondary treatment of LF. From there, the ladle with liquid steel was introduced into vacuum installation (VD) and finally, the steel was continuously cast – as is presented in Figure 1.

Figure 1. The main equipment's used in steel works

Were selected for analysis and further correlations the following parameters: the oxygen content, the duration of vacuum respective high vacuum treatment, steel temperature at the entering the VD facility and temperature drop during the treatment and, not least, the argon flow rate, used in the liquid steel stirring.

3. Results analysis

After data's processing, have been obtained correlations expressed by polynomial functions of 3 and 4 degrees, respectively exponential and logarithmic; but given the closer value for correlation coefficients and simplest form of correlations represented by polynomial functions of second degree, only these were presented. The correlation equations that define the upper and lower variation domain for each correlation are presented. The correlations presented are representative in terms of the values of correlation coefficients, and in the technological sense, all is presented in graphical and analytical form [11].

Regarding the duration of steel vacuum treatment (Figure 2) and high vacuum treatment (Figure 3), is established that with the increasing of these times, it ensures a decrease in oxygen content [1], [2], [12].

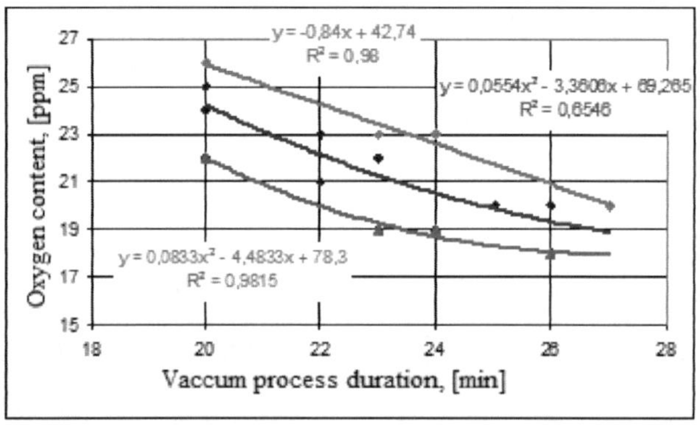

Figure 2. Correlation between oxygen content and the vacuum process duration

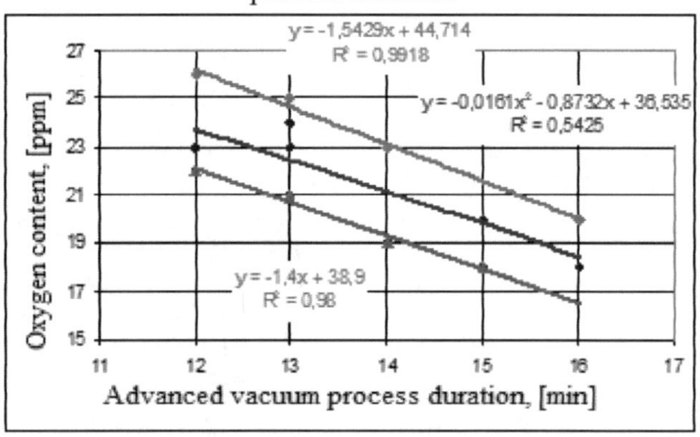

Figure 3. Correlation between oxygen content and the advanced vacuum process duration

A vacuum time treatment varies between 20 to 27 minutes, and the oxygen content of 26-18 ppm. Duration of high vacuum treatment was within 12-16 minutes. Desirable as the duration for the two phases of vacuum cycle should be as close as possible to the maximum duration, but it should be kept in mind that the vacuum system is without any heat input, which limits the duration of treatment (the temperature should not fall below a certain value). Mentioned that by using the superior quality refractory materials for ladle (resistant to steel and slag corrosion activity and advanced thermal insulation capacity) allows increase the vacuum duration from 15 min to 26 min.

The temperature of the steel at the entrance in the plant facility, is situated at the upper limit of the technical variation domain, steel is more fluid, so it provides good conditions for mixing by argon bubbling of the metal bath and the possibility of a larger decreasing of temperature during the vacuum treatment, respectively an increasing of the time of this operation. As a result of these presented data, is confirmed that with increasing values for steel temperature at entrance in the vacuum facility, the steel oxygen content is decreasing (Figure 4) and also with higher values of temperatures drop, the steel oxygen content is decreasing (Figure 5).

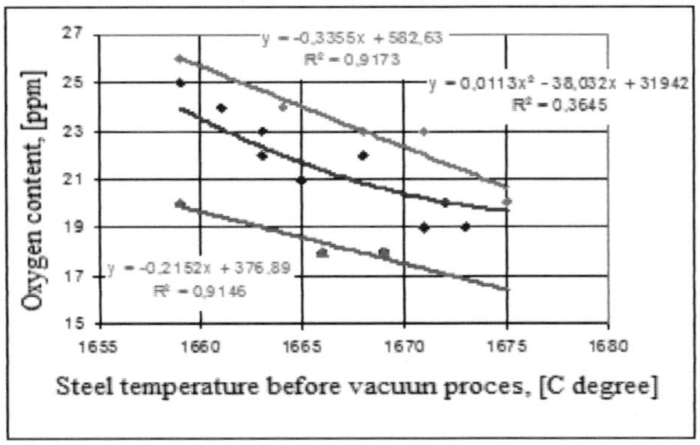

Figure 4. Correlation between oxygen content and the steel temperature before vacuum process

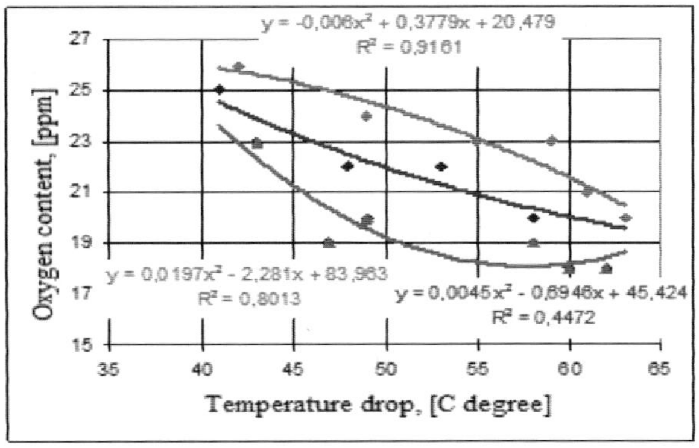

Figure 5. Correlation between oxygen content and the temperature drop

A particular importance for the steel treatment technologies in the ladle ("Ladle Metallurgy") has the process of chemical and thermal mixing of the metal bath, which is achieved in most cases by of inert gas stirring (usually argon, less by electromagnetic stirring) [13]. From the correlation presented in Figure 6 results that with the increasing of argon flow rate it is made a reduction in oxygen content, which is explained by increasing technological surface interphase steel-gas, so the surface diffusion of steel-argon bubbles. The minimum value of the oxygen content from correlation is by (5.40; 18.70) and for correlation that representing the lower limit of the variability, the minimum point has coordinates (3.29; 16.01); therefore can be considered that the argon flow rate should not be increased (there is a risk of slag entrainment in the steel, on the one hand, and on the other hand, an increasing of ladle refractory erosion). Maximum flow rate of argon, recommended by technological instructions, is 5 l/to.min.

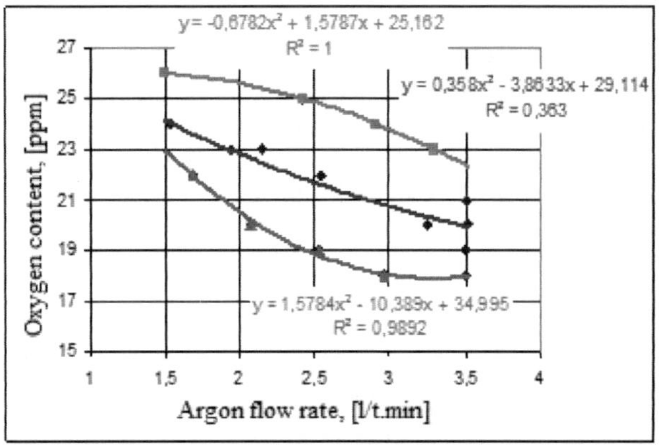

Figure 6. Correlation between oxygen content and the argon flow rate

The data has been processed in MATLAB computing program, in Figures 7-14 are presented the correlation surfaces, the plane representation of the level curves and the spatial representation of the level curves. Also, are presented the correlation equations between oxygen content - as a dependent parameter and, as independent parameters: vacuum process duration, vacuum process duration under high vacuum, steel temperature, temperature drop during the vacuum cycle, the argon flow rate.

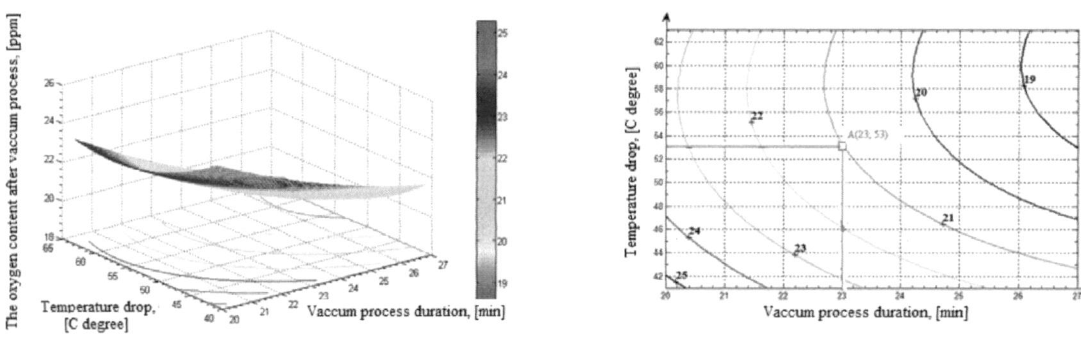

1) The correlation surface

2) Plane representation of the level curves

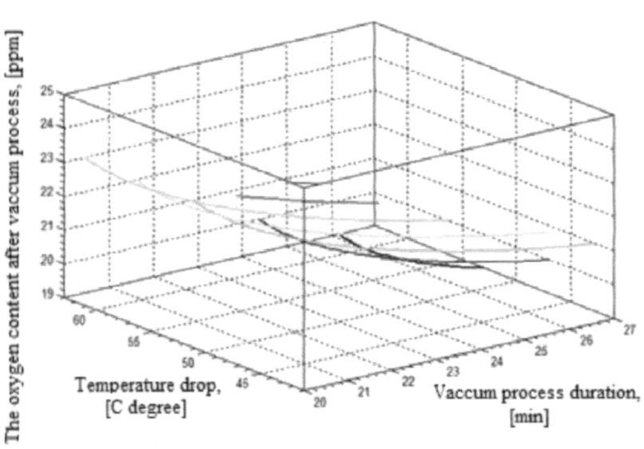

3) Spatial representation of the level curves

Figure 7. The oxygen content after vacuum process depending on the process duration and temperature drop

The correlation equation is:

$$Q_{O2}= 0.037001\cdot D_v^2+0.007779\cdot \Delta T^2-0.006621\cdot D_v\cdot \Delta T- 2.006263\cdot D_v-0.760839\cdot \Delta T+ 74.142629 \quad (6)$$

The correlation coefficient is:

$$R^2= 0.854582$$

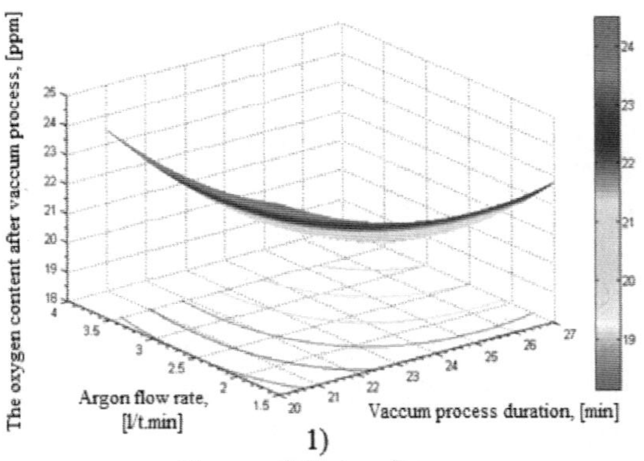

1)
The correlation surface

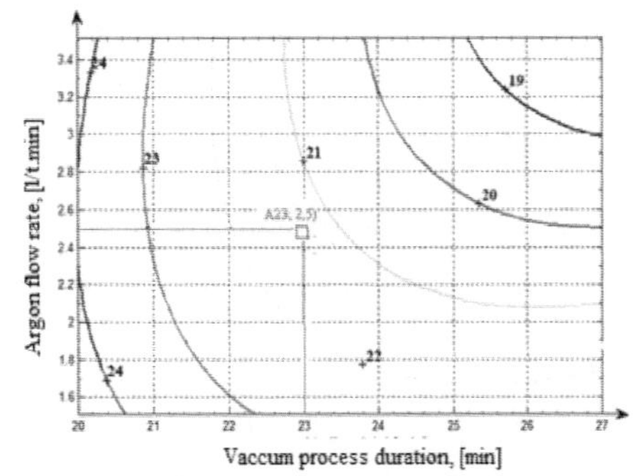

2)
Plane representation of the level curves

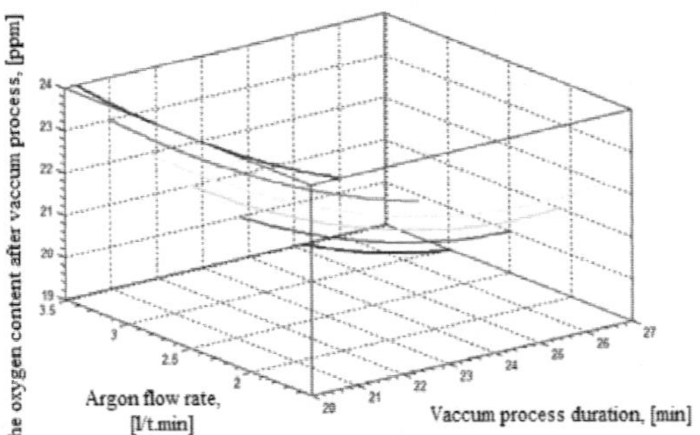

3)
Spatial representation of the level curves

Figure 8. The oxygen content after vacuum process depending on the vacuum process duration and argon flow rate

The correlation equation is:

$$Q_{O2}= 0.081470 \cdot D_v^2+0.454855 \cdot Q_{Ar}^2-0.319134 \cdot D_v \cdot Q_{Ar}-3.598831 \cdot D_v+ 4.055380 \cdot Q_{Ar}+66.342935 \quad (7)$$

The correlation coefficient is:

$$R^2= 0.848899$$

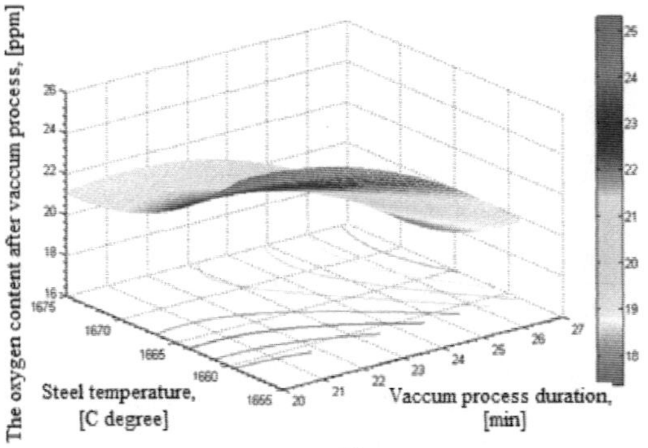

1)
The correlation surface

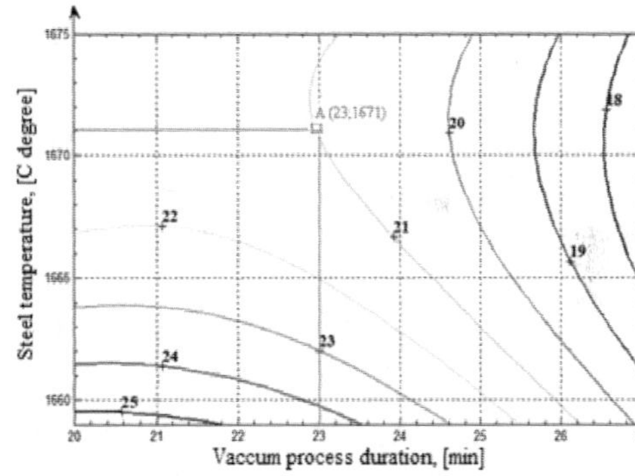

2)
Plane representation of the level curves

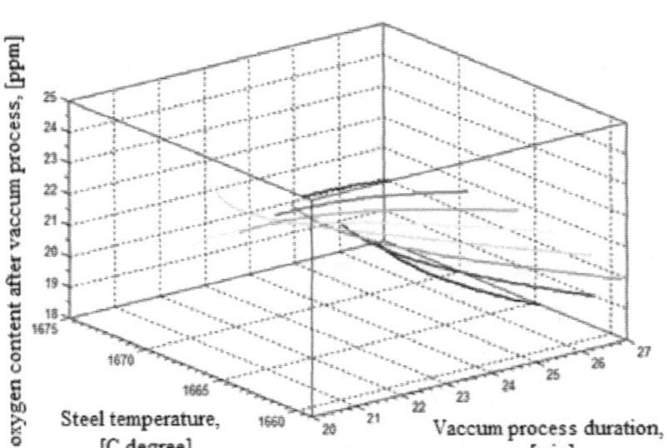

3)
Spatial representation of the level curves

Figure 9. The oxygen content after vacuum process depending on the steel temperature and the vacuum process duration

The correlation equation is:

$$Q_{O2}= -0.118565 \cdot D_v^2+0.019332 \cdot T^2+ 0.019183 \cdot D_v \cdot T-27.029464 \cdot D_v-65.099306 \cdot T + 54769.241855 \quad (8)$$

The correlation coefficient is:

$$R^2= 0.867153$$

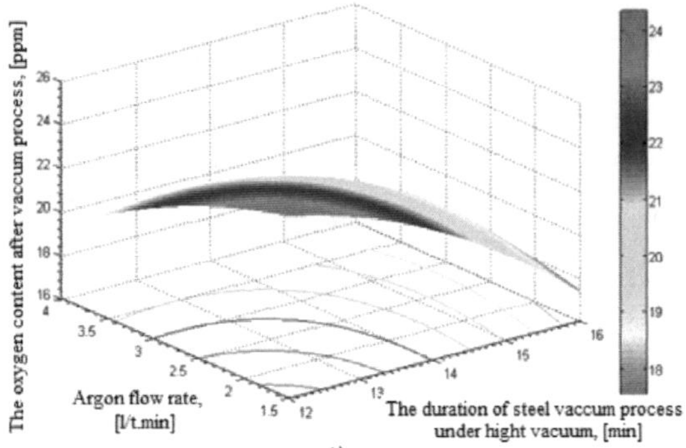

1)
The correlation surface

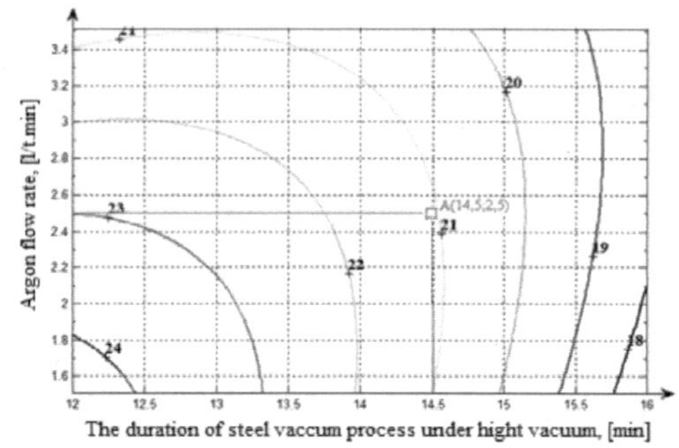

2)
Plane representation of the level curves

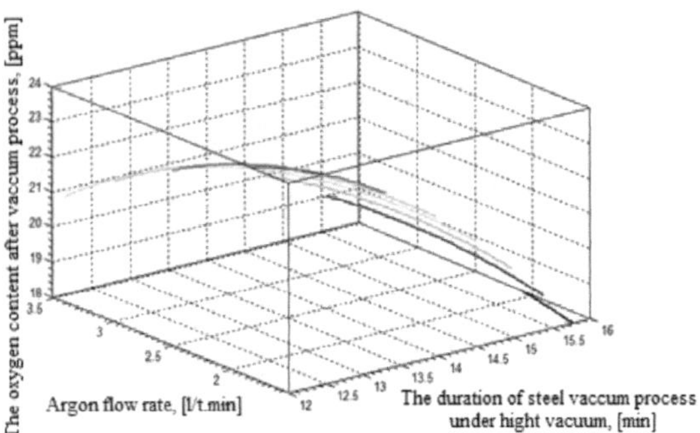

3)
Spatial representation of the level curves

Figure 10. The oxygen content after vacuum process depending on the argon flow rate and the duration of steel vacuum process under high vacuum

The correlation equation is:

$$Q_{O2}= -0.272637 \cdot D_{a.v}^{2} - 0.434313 \cdot Q_{Ar}^{2} + 0.5687389 \cdot D_{a.v} \cdot Q_{Ar} + 5.029522 \cdot D_{a.v} - 6.443282 \cdot Q_{Ar} + 3.659968 \quad (9)$$

The correlation coefficient is:

$$R^2 = 0.777176$$

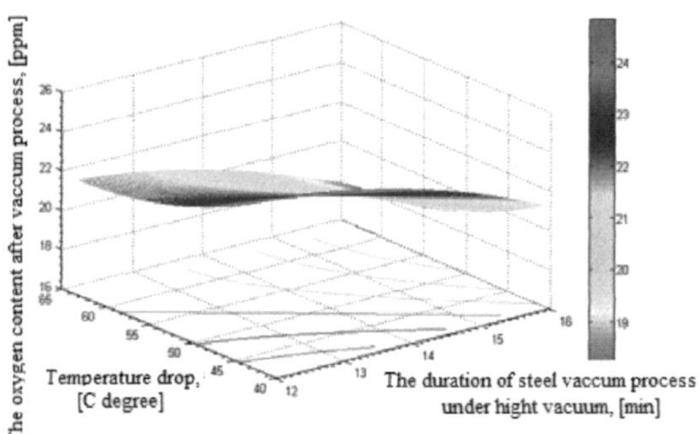

1)
The correlation surface

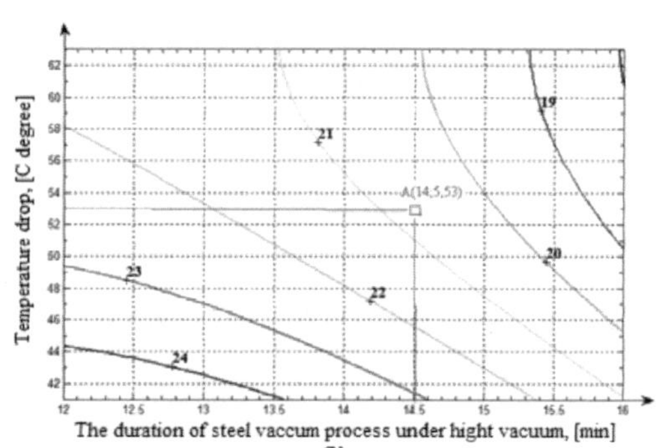

2)
Plane representation of the level curves

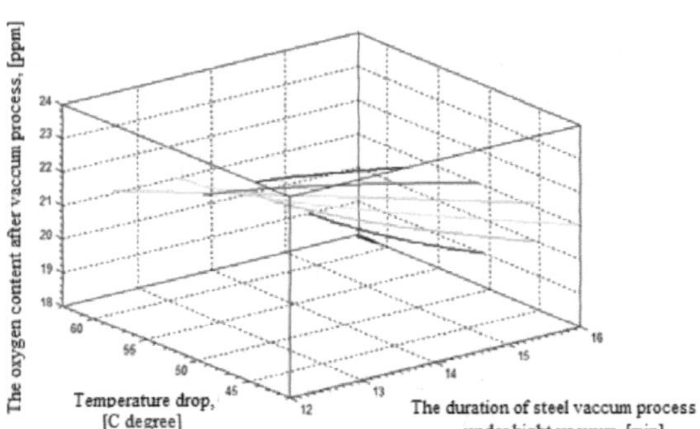

3)
Spatial representation of the level curves

Figure 11. The oxygen content after vacuum process depending on the temperature drop and the duration of steel vacuum process under high vacuum

The correlation equation is:

$$Q_{O2}= -0.171459 \cdot D_{a.v}^{2} + 0.005995 \cdot \Delta T^{2} - 0.001519 \cdot D_{a.v} \cdot \Delta T + 3.922981 \cdot D_{a.v} - 0.740859 \cdot \Delta T - 23.483942 \quad (10)$$

The correlation coefficient is:

$$R^2 = 0.809316$$

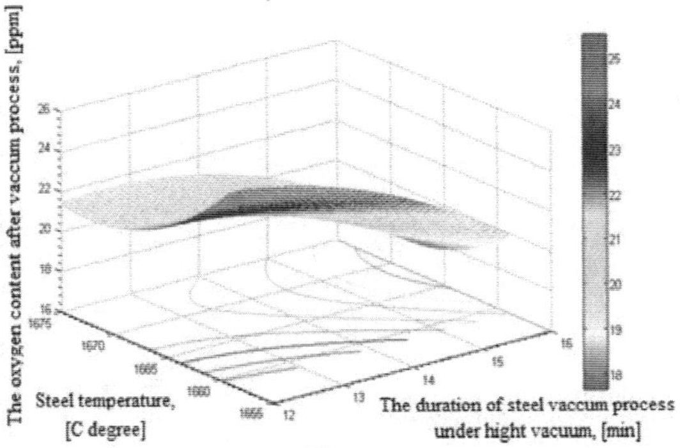

1)
The correlation surface

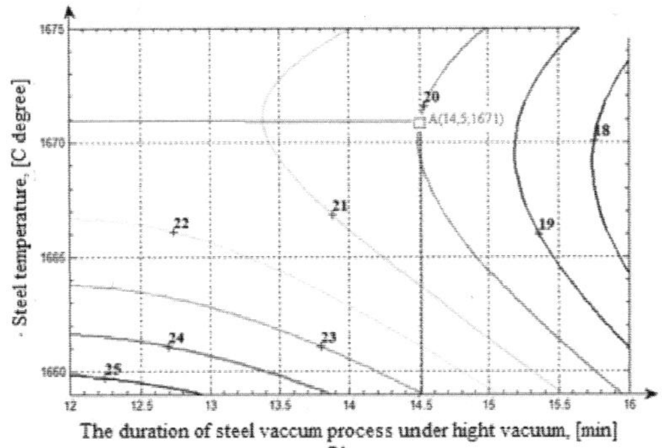

2)
Plane representation of the level curves

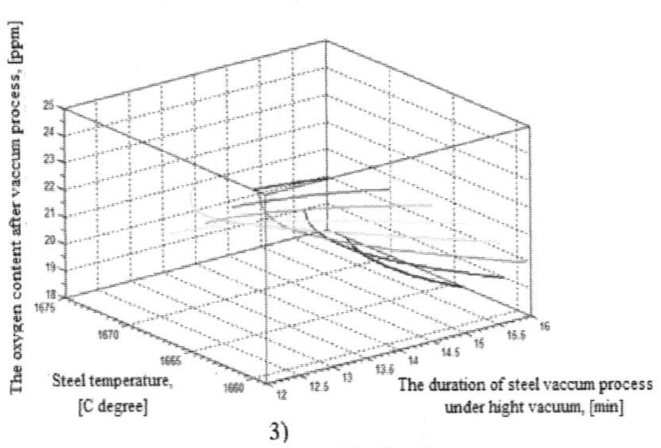

3)
Spatial representation of the level curves

Figure 12. The oxygen content after vacuum process depending on the steel temperature and the duration of steel vacuum process under high vacuum

The correlation equation is:

$$Q_{O2} = -0.276601 \cdot D_{a.v}^{2} + 0.024543 \cdot T^{2} + 0.041170 \cdot D_{a.v} \cdot T - 61.973022 \cdot D_{a.v} - 82.579114 \cdot T + 69438.211722 \quad (11)$$

The correlation coefficient is:

$$R^2 = 0.873801$$

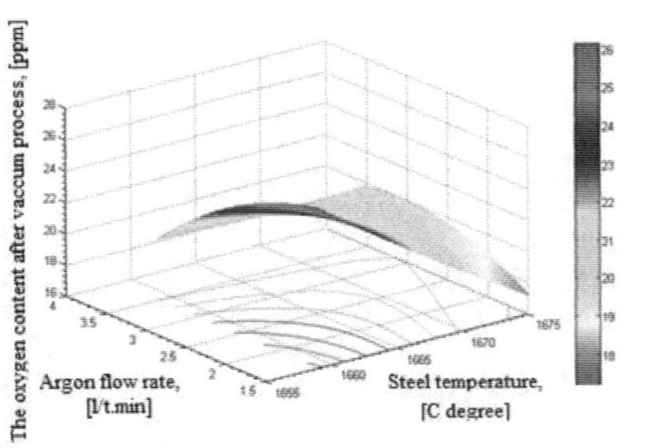

1)
The correlation surface

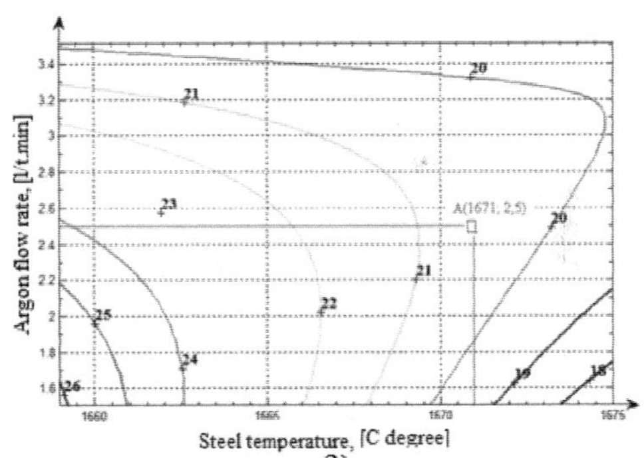

2)
Plane representation of the level curves

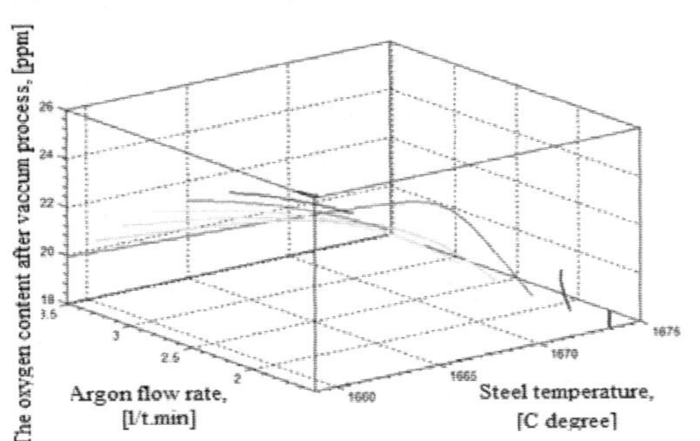

3)
Spatial representation of the level curves

Figure 13. The oxygen content after vacuum process depending on the steel temperature and argon flow rate

The correlation equation is:

$$Q_{O2} = 0.003447 \cdot T^{2} - 1.082846 \cdot Q_{Ar}^{2} + 0.275710 \cdot T \cdot Q_{Ar} - 12.464834 \cdot T - 455.100290 \cdot Q_{Ar} + 11217.090875 \quad (10)$$

The correlation coefficient is:

$$R^2 = 0.794606$$

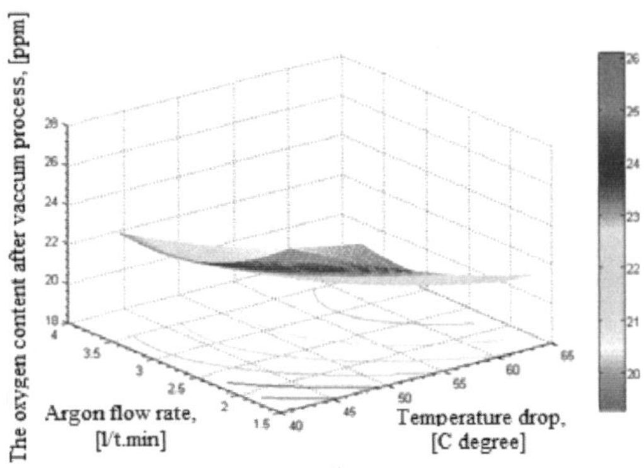

1)
The correlation surface

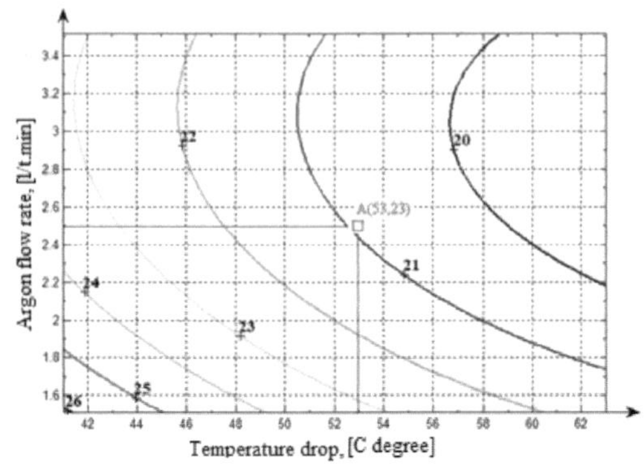

2)
Plane representation of the level curves

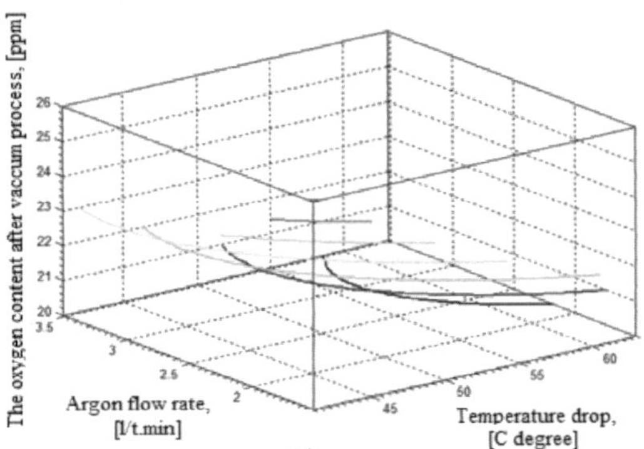

3)
Spatial representation of the level curves

Figure 14. The oxygen content after vacuum process depending on the temperature drop and argon flow rate

The correlation equation is:

$$Q_{O2} = 0.003921 \cdot \Delta T^2 + 1.090357 \cdot Q_{Ar}^2 + 0.016134 \cdot \Delta T \cdot Q_{Ar} - 0.632286 \cdot \Delta T - 7.550576 \cdot Q_{Ar} + 53.335814 \quad (13)$$

The correlation coefficient is:

$$R^2 = 0.755482$$

5. Conclusion

From the research data for study of the effect of deoxidation to the vacuum steel treatment, results the following conclusions:

- treatment of liquid steel in the LF and VD facilities, will allow an increase in purity and an advanced degassing of steel, reflected in the high quality of this;
- by processing the data in the Excel and MATLAB software, were obtained representative correlation equations, both in terms of the values of correlation coefficients and technology interpretation;
- obtained values for oxygen content, using multiple correlation equations, was under 21ppm (max 30ppm content), confirm, on the one hand, the validity of correlations obtained and, on the other hand, the possibility of using these in industrial practice;
- obtained equations are useful for industrial practice, they allowing choosing and correlation of the values for the independent parameters, so as to obtain the semi finished cast products (treated LF and VD facility) a lower oxygen content (obviously at the same time for H and N) [14], [15].

References

[1] Tripşa I and Pumnea C 1981 *Steel deoxidation*, Technical Publishing House, Bucureşti, Romania

[2] Nica Ghe, Socalici A, Ardelean E and Heput T 2003, *Technology to improve the steel quality*, Mirton Publishing House, Timişoara, Romania

[3] Birat J P 2016 Steel cleanliness and environmental metallurgy, *Mettalurgical Research & Technology* **113** 201

[4] Wang X X and Jiang Z H 2014 Deoxidation and Inclusion Control in Stainless Steel Refining, *Advanced Materials Research* **968** 146-150
[5] Chichkarev E A 2009 Improving steel deoxidation technology with the use of data on the activity of dissolved oxigen, *Metallurgist* **53**(9) 607-612
[6] Livshits D A, Isaev O B, Kostyrya I N, Kisletsa V V, Ganoshenko I V and Chichkarev E A 2011 Rational technology for the deozidation and out-of-furnace treatment of low-silicon steel, *Metally* **12** 1110-1112
[7] Heput T, Ardelean E, Socalici A, Osaci M and Ardelean M 2010 Steel deoxidation with synthetic slag, *Metalurgia International* **15**(5) 22-28
[8] Protasov A V, Sivak B A, Lukyanov A V, Nikitenko A S and Shchegolev N A 2010 Status and future of vacuum treatment of liquid steel development, *CIS Iron and Steel Review* **1** 15-18
[9] http://www.totalmateria.com/
[10] http://ispatguru.com/oxygen-and-steels/
[11] Vătăşescu M 2015 *Contributions on implementation of management systems, occupational health and safety in the steel industry*, Politehnic University Timişoara, Romania, Doctoral Thesis
[12] Malmberg K, Nzotta M, Karasev A and Jonsson P G 2013 Optimization of stirring conditions during vacuum degassing in order to lower inclusion content in tool steel, *Ironmaking & Steelmaking Process Products and Applications* **40**(3) 231-237
[13] Dilner D, Lu Q, Mao H, Xu W , Zwaag S and Selleby M 2014 Process-time Optimization of Vacuum Degassing Using a Genetic Alloy Design Approach, *Materials* **7**(12) 7997-8011
[14] Ardelean E, Ardelean M, Josan A and Pinca-Bretotean C 2014 Contribution on the influence of steel ladle processing (LF) upon the nitrogen removal rate, *IOP Conf. Ser.: Mater. Sci. Eng.* **57** 012001
[15] Dragoi F, Socalici A, Ardelean E, Popa E and Heput T 2011 Researches on the influence of the slags formed in the installations on the hydrogen removal efficiency, *Revista de Metalurgia* **47**(6) 477-484

IOP Conf. Series: Materials Science and Engineering **163** (2017) 012026 doi:10.1088/1757-899X/163/1/012026

GUI for studying the parameters influence of the electric arc model for a three-phase electric arc furnace

L Ghiormez[1], O Prostean[1], M Panoiu[2] and C Panoiu[2]

[1]Politehnica University of Timisoara, Department of Automation and Applied Informatics, Vasile Parvan Street, no. 2, 300223, Timisoara, Romania
[2]Politehnica University of Timisoara, Department of Electrical Engineering and Industrial Informatics, Revolution Street, no. 5, 331128, Hunedoara, Romania

E-mail: loredana.ghiormez@fih.upt.ro

Abstract. This paper presents an analysis regarding the modeling of the behavior for a three-phase electric arc furnace installation. Therefore, a block diagram is implemented in Simulink that represents the modeling of the entire electric arc furnace installation. This block diagram contains also the modeling of the electric arc which is the element that makes the electric arc furnace behaving as a nonlinear load. The values for the model parameters of the electric arc furnace installation are like the ones from the real installation taken into consideration. Other model parameters are the electric arc model ones. In order to study the influence of the parameters of the electric arc models, it is developed a Matlab program that contains the graphical user interfaces. These interfaces make connection with the models of the electric arc implemented in Simulink. The interfaces allow the user to modify parameters for each of the electric arc model. Current and voltage of the electric arc are the variables taken into account to study the influence of the parameters on the electric arc models. Waveforms for voltage and current of the electric arc are illustrated when a parameter of the model is modified in order to analyze the importance of this parameter on the electric arc model. Also, for each of the models is presented the voltage-current characteristic of the electric arc because this characteristic gives information about the behavior of the electric arc furnace installation.

1. Introduction

In our days the steel is used in many fields of industry, such as alimentary, constructions, automobile, chemically and marine. The liquid steel is obtained by the electric arc furnaces (EAFs) which can be of two categories: alternative current (AC) and continuous current (DC) furnaces [1]. In this paper is studied the behavior of a three-phase EAF [2] with direct action. In order to perform the control of this kind of plant it is necessary to study the electric arc and this is the purpose of this paper. In order to understand the model of the electric arc all the model parameters should be analyzed to notice the importance of each of them on the model. After knowing the model of the process can be developed control strategies. The models presented in this paper are developed by the authors of this paper in other researches [3-5].

2. Electric Arc Furnace Plant

In Figure 1 are illustrated the main constructive elements of a three-phase EAF [4] with direct action. This kind of EAF has three graphite electrodes which are supplied by a three-phase power transformer [6]. These three electrodes are used to melt the metals loaded in the furnace tank. When the current passes through the electrodes and there is a significant amount of electrical charges to the electrodes appears the phenomena of the electric arc discharging [7] formed between the electrodes and the metals. In order to study the behavior of the EAF it is necessary to study the electric arc because this is the nonlinear element of the circuit [8]. The EAF is supplied by a three-phase power transformer placed in a special chamber and flexible cables makes connection between this transformer and the electrodes of the EAF. Electrode arms support the three electrodes of the EAF and a hydraulic system is used to obtain different arc lengths by moving up or down the electrode arms.

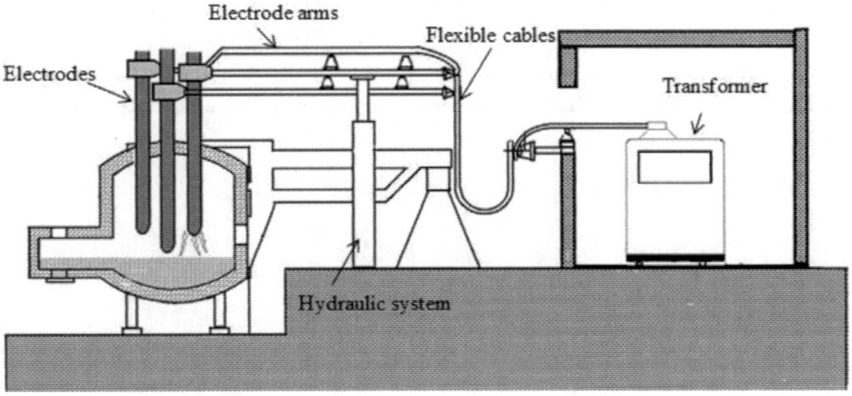

Figure 1. Main constructive elements of a three-phase EAF with direct action

3. Modeling of the Electric Arc

In Figure 2 is illustrated the power supply electric circuit of an EAF for a single phase. The EAF has a resistive character [9] this is why it can be considered as a variable resistance. In Figure 2 the electrical parameters are as follows: v is the alternative voltage from the secondary side of the furnace transformer for a single phase, I is the electric current for a single phase in the secondary side of the furnace transformer, r is the total electrical resistance for a single phase of the secondary side of the furnace transformer, L is the total electrical inductivity for a single phase of the secondary side of the furnace transformer, R_{arc} is the resistance of the electric arc, being reported at the voltage from the primary side of the furnace transformer. In Figure 3 are illustrated the waveforms for the supply voltage, v_S, the electric arc voltage, v_{arc} and the electric arc current, i_{arc}, as presented in the reference literature [10].

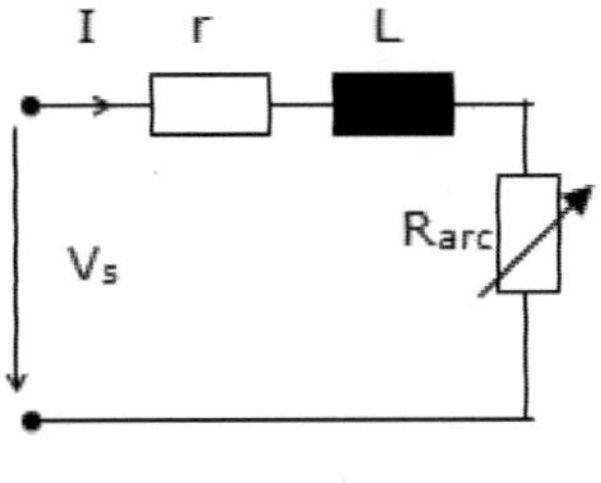

Figure 2. The power supply electric circuit of an EAF for a single phase

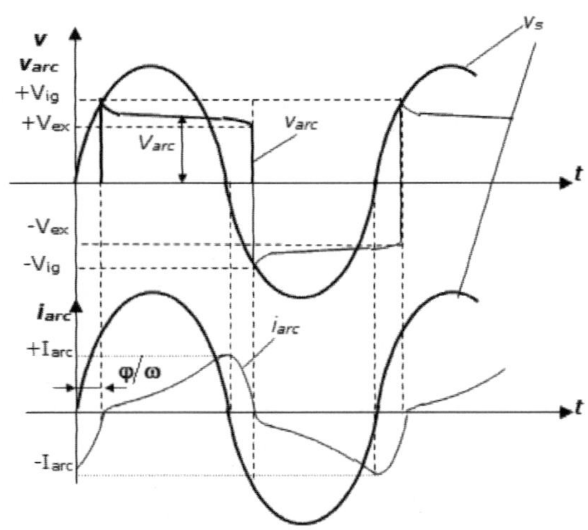

Figure 3. The waveforms for the supply voltage of the EAF, electric arc voltage and electric arc current

IOP Conf. Series: Materials Science and Engineering **163** (2017) 012026 doi:10.1088/1757-899X/163/1/012026

In this study were chosen the following values of the parameters: total resistance $r=0.47\ m\Omega$ and total reactance $X=5.5\ m\Omega$. Voltage from the secondary side of the furnace transformer was set as considered that the transformer has plot set on step *16* which correspond to the *894 V*. These values are taken from a real industrial plant.

2.1. Exponential Model of the Electric Arc

In (1) is presented the mathematical model of the electric arc and in (2-6) are illustrated the computation of some model parameters. In (1) v is the voltage of the electric arc, i is the current of the electric arc, v_{ex} is the extinction of the electric arc, v_{ig} is the ignition voltage of the electric arc, v_m is the average between the ignition and extinction voltage and R_1, R_2, R_3 are lines slopes of the voltage-current characteristic presented in Figure 4, this being the desired characteristic to be obtained with the model presented in (1).

$$v=\begin{cases} -v_{ex}+(i_3-i_4)\cdot R_3, & i<-i_4 \text{ and } di/dt\geq 0 \\ & \text{or } i<-i_4 \text{ and } di/dt<0 \\ -v_{ex}+(i+i_3)\cdot R_3, & i\in[-i_4,-i_3) \text{ and } di/dt\geq 0 \\ R_1\cdot i, & i\in[-i_3,i_1) \text{ and } di/dt\geq 0 \\ & \text{or } i\in[-i_1,i_3) \text{ and } di/dt<0 \\ v_{ex}+(v_{ig}-v_{ex})\cdot e^{(i_1-i)/i_2} & i\in[i_1,i_2) \text{ and } di/dt\geq 0 \\ v_m+(i-i_2)\cdot R_2 & i\in[i_2,i_4] \text{ and } di/dt\geq 0 \\ v_{ex}+(i_4-i_3)\cdot R_3 & i>i_4 \text{ and } di/dt\geq 0 \\ & \text{or } i>i_4 \text{ and } di/dt<0 \\ v_{ex}+(i-i_3)\cdot R_3 & i\in[i_3,i_4] \text{ and } di/dt<0 \\ -v_{ex}+(v_{ex}-v_{ig})\cdot e^{(i_1+i)/i_2} & i\in[-i_2,-i_1] \text{ and } di/dt<0 \\ -v_m+(i+i_2)\cdot R_2 & i\in[-i_4,-i_2] \text{ and } di/dt<0 \end{cases} \tag{1}$$

$$i_1=v_{ig}/R_1 \tag{2}$$

$$i_2=2\cdot i_1 \tag{3}$$

$$i_3=v_{ex}/R_1 \tag{4}$$

$$i_4=(v_m-v_{ex}+i_3\cdot R_3-i_2\cdot R_2)/(R_3-R_2) \tag{5}$$

$$v_m=(v_{ig}+v_{ex})/2 \tag{6}$$

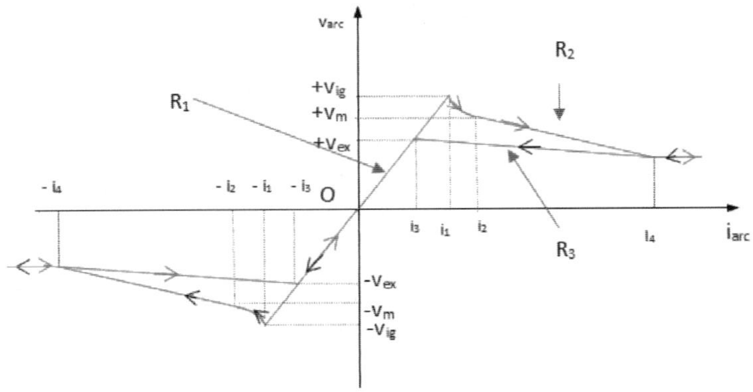

Figure 4. Desired voltage-current characteristic

In Figure 5 is presented the block diagram developed by the authors of this paper in Simulink and which was implemented by applying second Kirchhoff's theory in the electric circuit from Figure 2. *ARC* block contains the mathematical model presented in (1), *Source* block is the voltage from the secondary side of the furnace transformer, *R* is the total resistance of the circuit and *L* is the total inductivity of the circuit. All the model parameters values are modified from the graphical user interface (GUI) presented in Figure 6.

In Figure 6 is illustrated the GUI for the exponential model which allows user to modify the model parameters in order to notice their importance on the voltage and current of the electric arc. For each of the model parameter was chosen a range in which it can be varied. In order to study the influence of each of the model parameters on the voltage and current of the electric arc, each parameter was modified in two cases. First of all was set an initial value and then were chosen two different values as compared to the initial one, first after taking *800 samples* and the second after *1500 samples*. In one second was set to take *10000 samples*. When a parameter is modified the other parameters are set to the initial values.

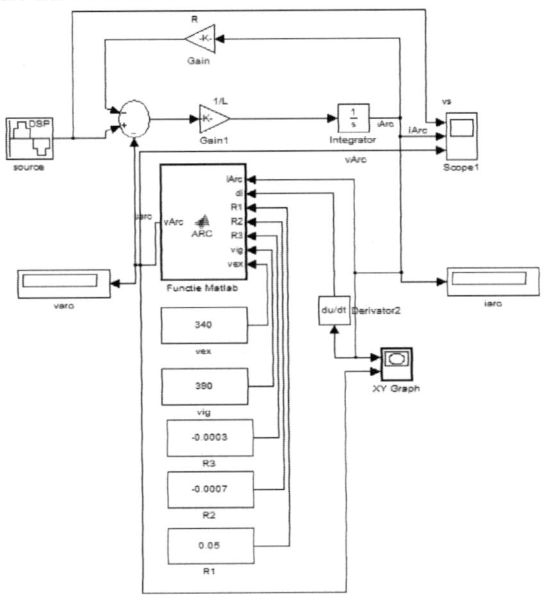

Figure 5. The block diagram for the exponential model of the electric arc

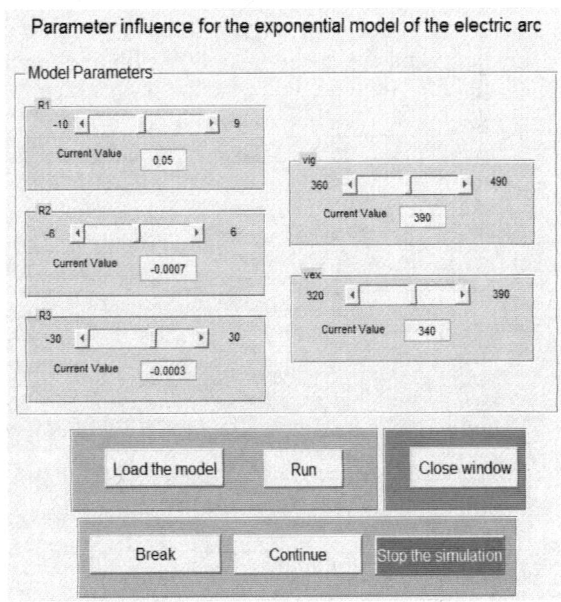

Figure 6. GUI for the exponential model of the electric arc

In Figure 7 is presented the voltage-current characteristic of the electric arc when the ignition voltage of the electric arc is modified and in Figure 8 is presented the voltage-current characteristic of the electric arc when the extinction voltage of the electric arc is modified. These two characteristics were obtained by plotting electric arc voltage against electric arc current. Because for the real plant cannot be measured the electric arc voltage, the voltage-current characteristic was represented also by plotting the voltage measured before the EAF against the current of the EAF.

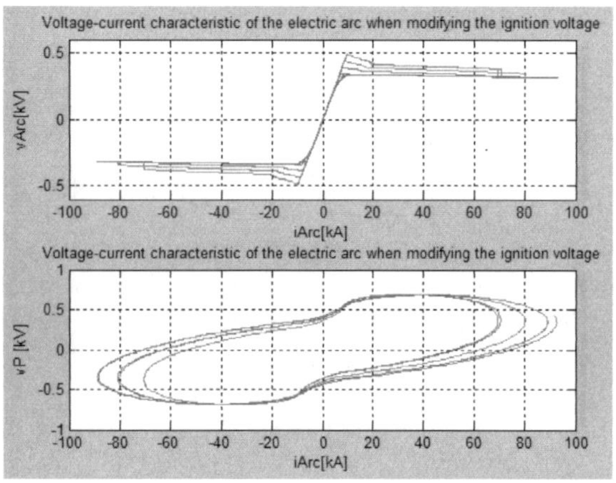

Figure 7. Voltage-current characteristic of the electric arc when modifying the ignition voltage

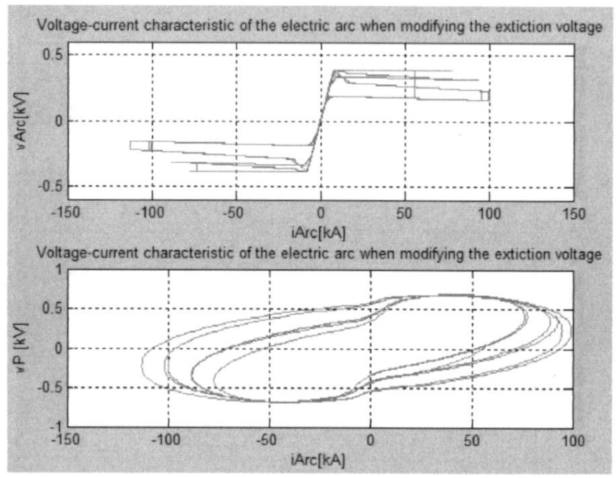

Figure 8. Voltage-current characteristic of the electric arc when modifying the extinction voltage

In Figure 9 are illustrated the waveforms for the electric arc current and for the electric arc voltage when modifying the ignition voltage. Initial value for the ignition voltage is *390 V*, after taking *800 samples* was modified to *490 V* and after *1500 samples* was set to *440 V*.

One can notice from Figure 9 that when the ignition voltage is increasing the electric arc current is decreasing and when the ignition voltage is decreasing the electric arc current is increasing. In both cases the electric arc voltage is distorted.

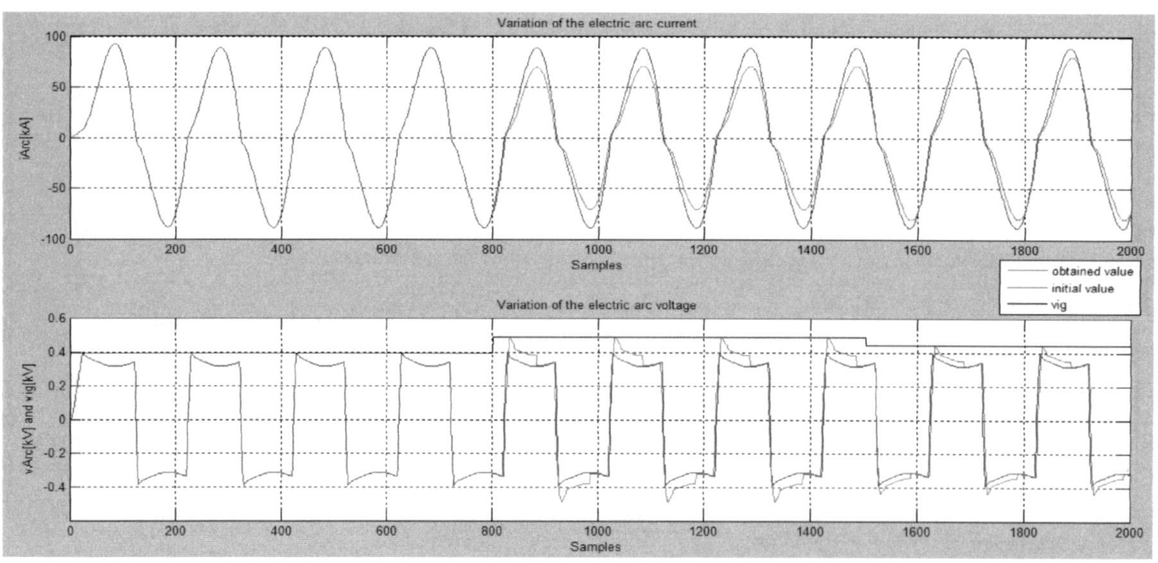

Figure 9. Variation of the electric arc current and voltage when is modified the ignition voltage

In Figure 10 are illustrated the waveforms for the electric arc current and for the electric arc voltage when modifying the extinction voltage. Initial value for the extinction voltage is *340 V*, after taking *800 samples* was modified to *390 V* and after *1500 samples* was set to *190 V*. One can notice that when the extinction voltage is increasing the electric arc current is decreasing and when the extinction voltage is decreasing the electric arc current is increasing. Regarding the electric arc voltage one can notice that if the ignition and the extinction voltages have the same value the electric arc voltage has a rectangular shape and when is a difference two large between the ignition and the extinction voltages the electric arc voltage is strongly distorted.

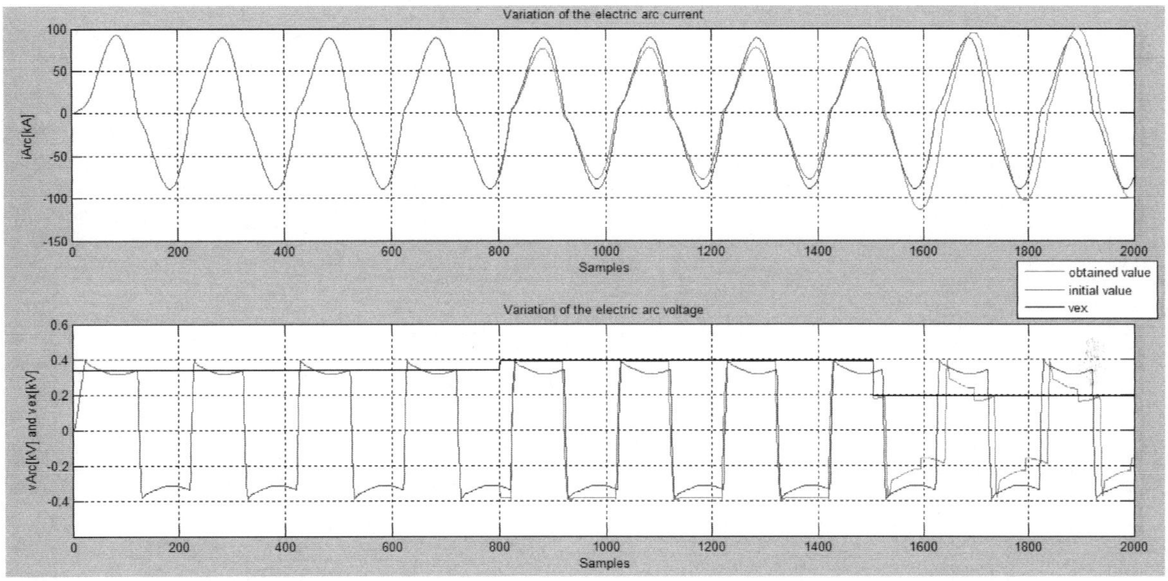

Figure 10. Variation of the electric arc current and voltage when is modified the extinction voltage

In Figure 11 is presented the voltage-current characteristic of the electric arc when the line slope R_1 is modified and in Figure 12 is presented the voltage-current characteristic of the electric arc when the line slope R_2 is modified. For both characteristics were used two representations as mentioned before.

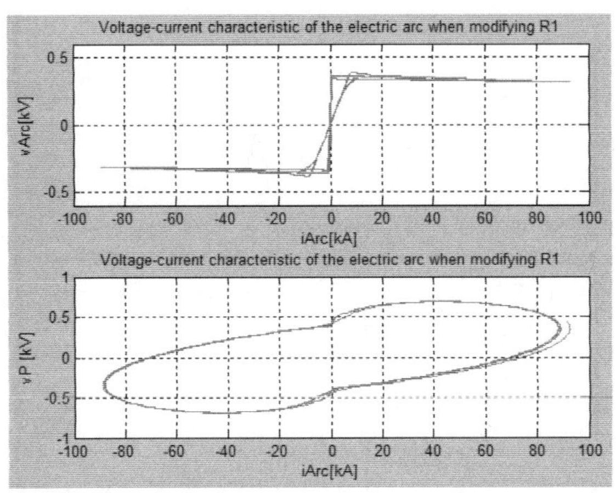

Figure 11. Voltage-current characteristic of the electric arc when modifying the line slope R_1

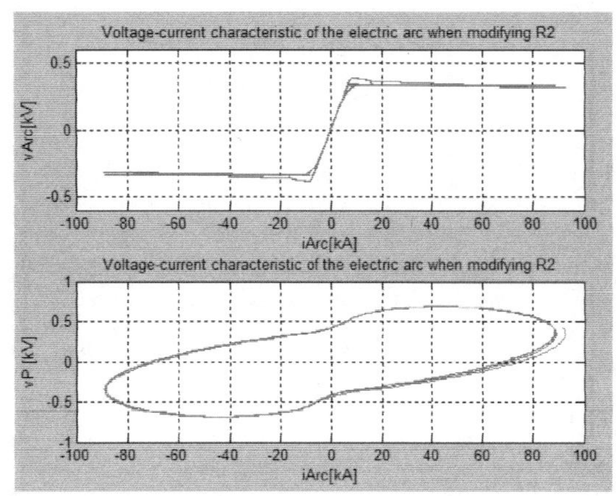

Figure 12. Voltage-current characteristic of the electric arc when modifying the line slope R_2

In Figure 13 are illustrated the waveforms for the electric arc current and for the electric arc voltage when modifying line slope R_1. Initial value for R_1 is *0.05*, after taking *800 samples* was increased with *2* and after *1500 samples* was decreased with *10*. One can notice that the amplitude of both voltage and current of the electric arc is the same, concluding that this parameter do not has a large influence on the amplitude on the electric arc voltage and current.

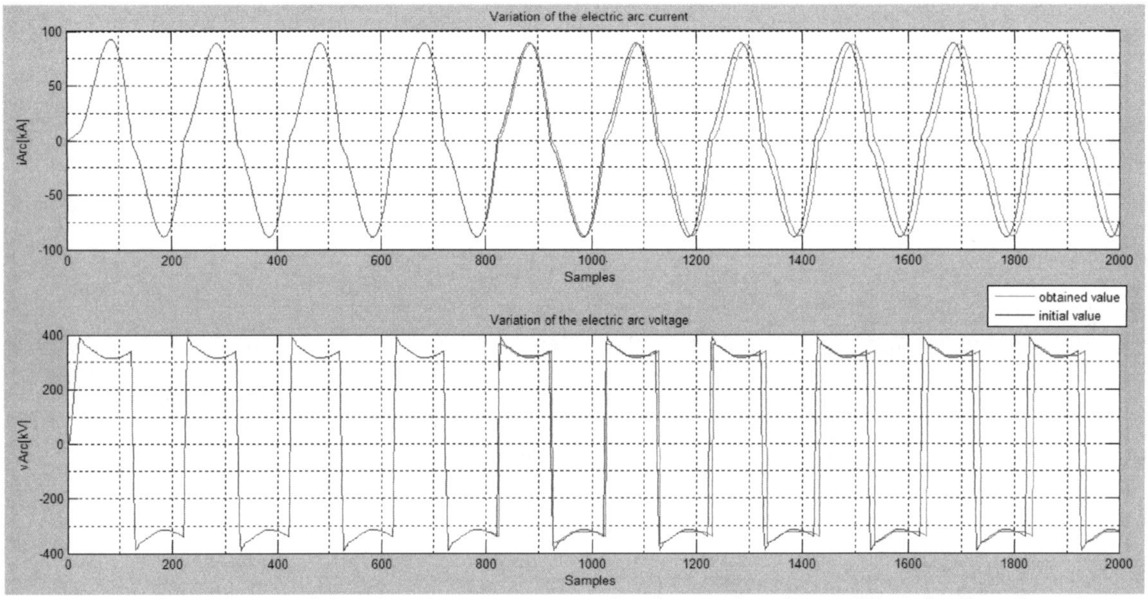

Figure 13. Variation of the electric arc current and voltage when is modified the line slope R_1

In Figure 14 are illustrated the waveforms for the electric arc current and for the electric arc voltage when modifying line slope R_2. Initial value for R_2 is *-0.007*, after taking *800 samples* it was increases with *5* and after *1500 samples* was decreased with *10*.

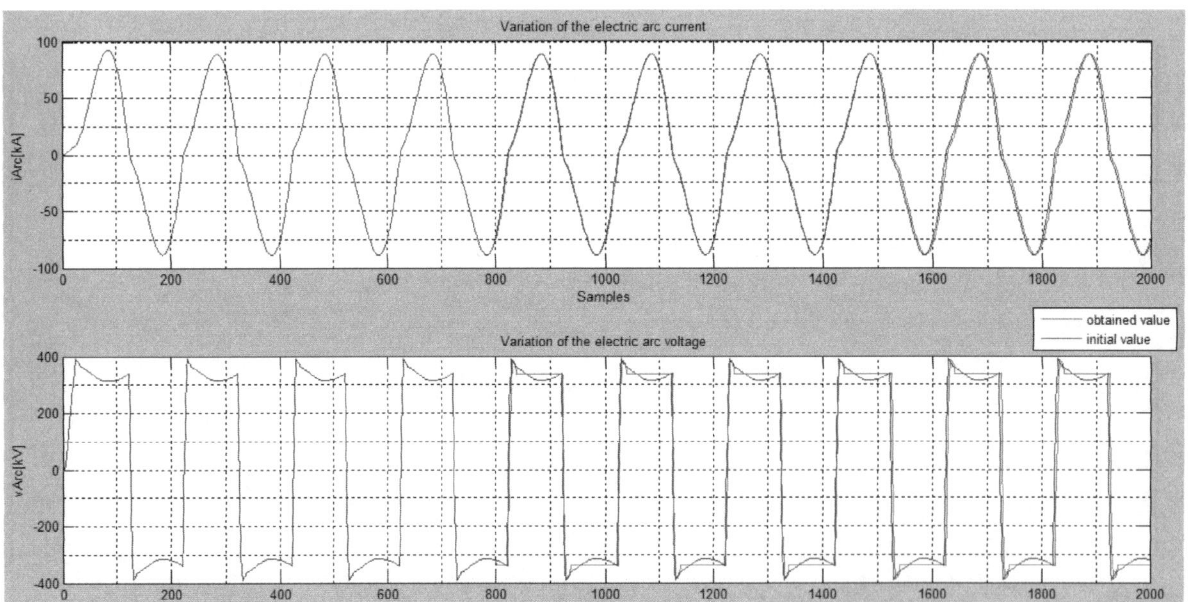

Figure 14. Variation of the electric arc current and voltage when is modified the line slope R_2

One can notice form Figure 14 that the amplitude of both voltage and current of the electric arc is the same, concluding that this parameter do not has a large influence on the amplitude on the electric arc voltage and current.

In Figure 15 is presented the voltage-current characteristic of the electric arc when the line slope R_3 is modified and in Figure 16 is presented the variation of this model parameter. For both characteristics were used two representations as mentioned before.

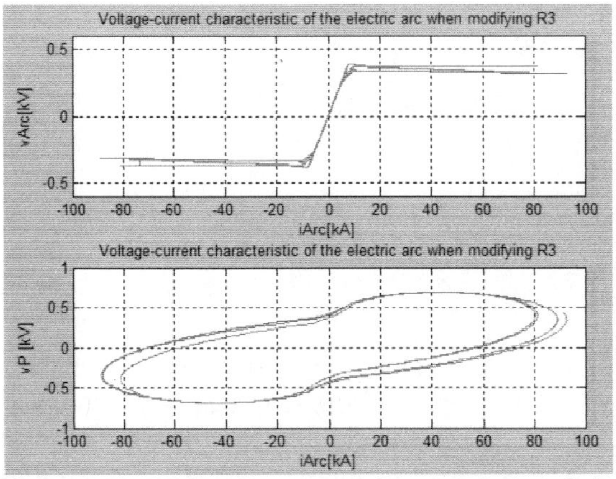

Figure 15. Voltage-current characteristic of the electric arc when modifying the line slope R_3

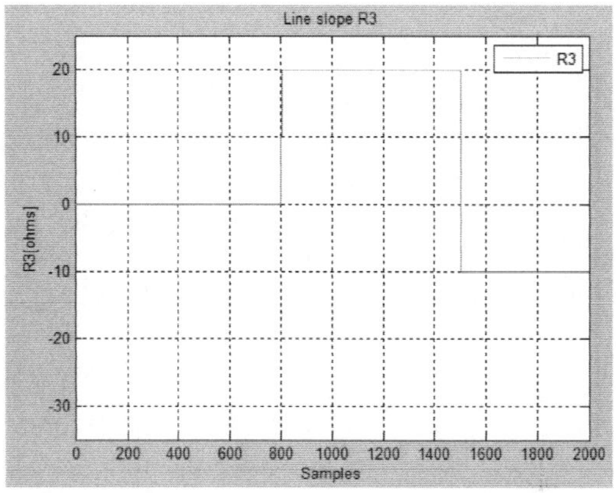

Figure 16. Variation of the line slope R_3

In Figure 17 are illustrated the waveforms for the electric arc current and for the electric arc voltage when modifying line slope R_3. Initial value for R_3 is *-0.003*, after taking *800 samples* it was increases with *20* and after *1500 samples* was decreased with *30.*

One can notice that the amplitude of the electric arc current is influenced in a small measure, but the electric arc voltage is more close to a rectangular shape.

In conclusion for the exponential model of the electric arc the ignition and the extinction voltage of the electric arc can influence in a large percentage the voltage and current of the electric arc.

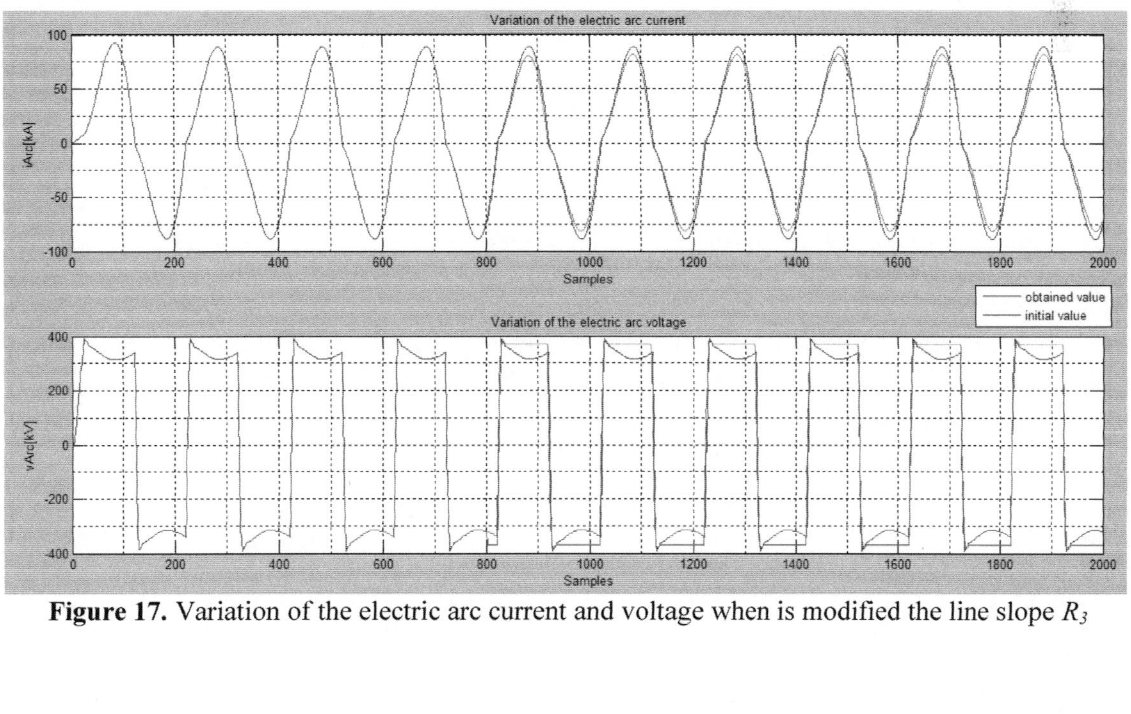

Figure 17. Variation of the electric arc current and voltage when is modified the line slope R_3

2.2. Hyperbolic Model of the Electric Arc

In (7) is presented the hyperbolic mathematical model of the electric arc. In this model v is the voltage of the electric arc, i is the current of the electric arc, v_{ex} is the extinction of the electric arc, C is a constant having value 190000, D is a constant having value *5000* and R_l is the line slope for the range $[-i_l, i_l]$ of the voltage-current characteristic presented in Figure 18, this being the desired characteristic to be obtained with the model presented in (7).

$$v = \begin{cases} i \cdot R_l,\ i < i_l \ \ and\ di \geq 0\ \ or\ i > -i_l\ \ and\ di < 0 \\ -v_{ex} + \dfrac{C}{D+|i|},\ i < -i_l\ \ and\ di < 0\ or\ di \geq 0 \\ v_{ex} + \dfrac{C}{D+|i|},\ i > i_l\ \ and\ di < 0\ or\ di \geq 0 \end{cases} \tag{7}$$

$$v_{ig} = v_{ex} + C/D \tag{8}$$

In Figure 19 is presented the block diagram developed by the author of this paper in Simulink and which was implemented by applying second Kirchhoff's theory in the electric circuit from Figure 2. *ARC* block contains the mathematical model presented in (1), *Source* block is voltage from the secondary side of the furnace transformer, R is the total resistance of the circuit and L is the total inductivity of the circuit. All the model parameters are modified from the graphical user interface (GUI) presented in Figure 20.

In Figure 20 is illustrated the GUI for the hyperbolic model which allows user to modify the model parameters in order to notice their importance on the electric arc voltage and current. For each of the model parameter was chosen a range in which it can be varied. In order to study the influence of each of the model parameters on the electric arc voltage and current, each parameter was modified in two cases. First of all was set an initial value and then were chosen two different values as compared to the initial one, first after taking *800 samples* and the second after *1500 samples*. In one second was set to take *10000 samples* similarly to the previous model. When a parameter is modified the other parameters are set to the initial values.

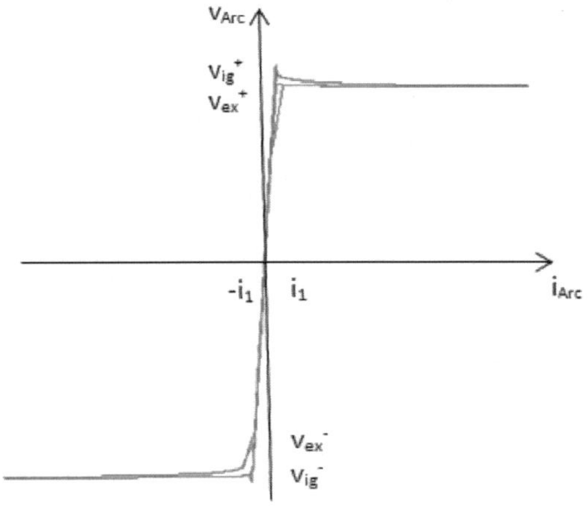

Figure 18. Desired voltage-current characteristic

IOP Conf. Series: Materials Science and Engineering **163** (2017) 012026 doi:10.1088/1757-899X/163/1/012026

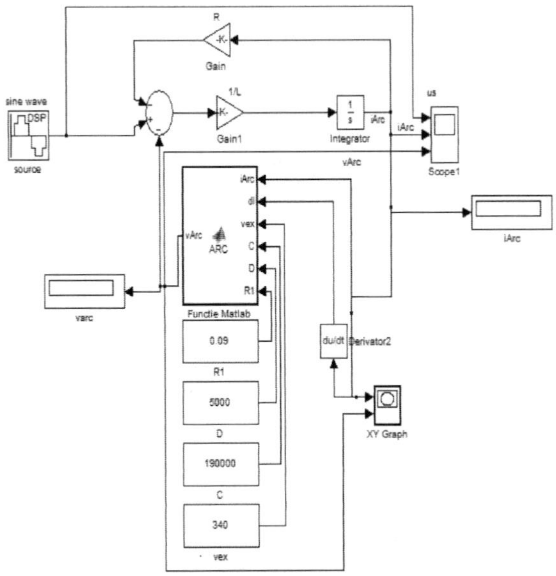

Figure 19. The block diagram for the hyperbolic model of the electric arc

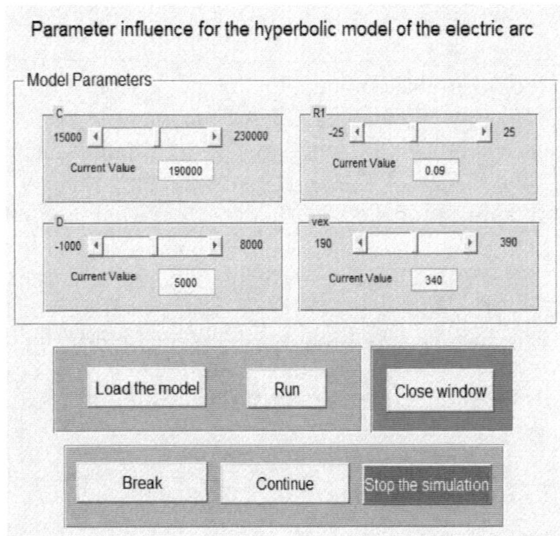

Figure 20. GUI for the hyperbolic model of the electric arc

In Figure 21 is presented the voltage-current characteristic of the electric arc when the extinction voltage of the electric arc is modified and in Figure 22 is presented the voltage-current characteristic of the electric arc when the line slope R_1 is modified. These two characteristics were obtained by plotting electric arc voltage against electric arc current. Because for real plant cannot be measured the voltage of the electric arc was represented also the voltage-current characteristic obtained by plotting the voltage measured before the EAF against the current of the EAF.

In Figure 23 are illustrated the waveforms for the electric arc current and for the electric arc voltage when modifying the extinction voltage. Initial value for the extinction voltage is *340 V*, after taking *800 samples* was modified to *390 V* and after *1500 samples* was set to *190 V*. One can notice that when the extinction voltage is increasing the electric arc current is decreasing and when the extinction voltage is decreasing the electric arc current is increasing.

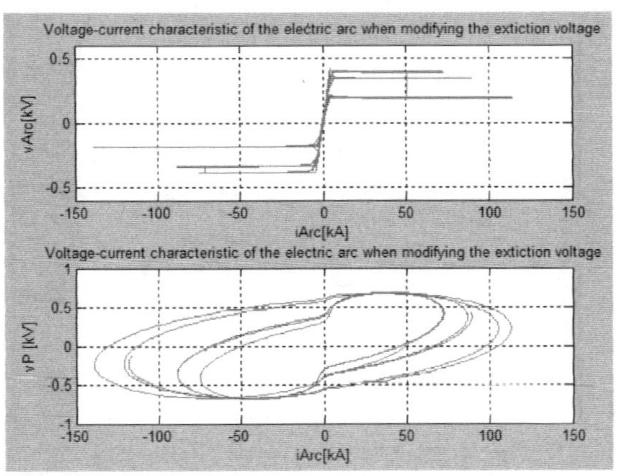

Figure 21. Voltage-current characteristic of the electric arc when modifying the ignition voltage

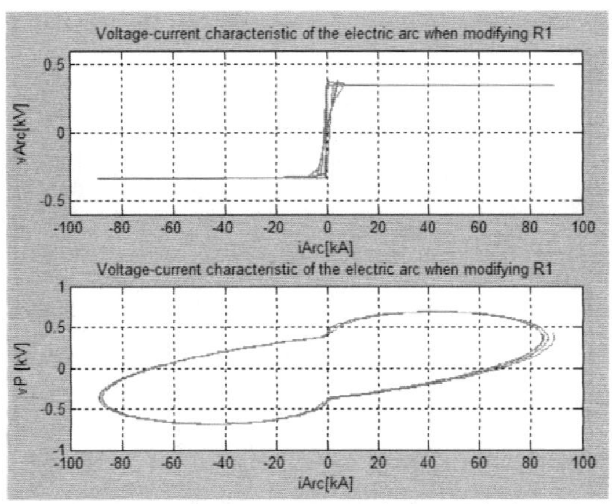

Figure 22. Voltage-current characteristic of the electric arc when modifying the line slope R_1

Regarding the electric arc voltage one can notice that if the extinction voltage is modified than the waveform of the electric arc voltage is maintaining the shape because the ignition voltage of the electric arc is computed taking into consideration the extinction voltage of the electric arc.

In Figure 24 are illustrated the waveforms for the electric arc current and for the electric arc voltage when modifying line slope R_l. Initial value for R_l is *0.09*, after taking *800 samples* was increased with *20* and after *1500 samples* was decreased with *40*. One can notice that the amplitude of both voltage and current of the electric arc is the same, concluding that this parameter do not has a large influence on the amplitude of the voltage and current of the electric arc.

In Figure 25 is presented the voltage-current characteristic of the electric arc when the constant C is modified and in Figure 26 is presented the voltage-current characteristic of the electric arc when the constant D is modified. For both characteristics were used two representations as mentioned before.

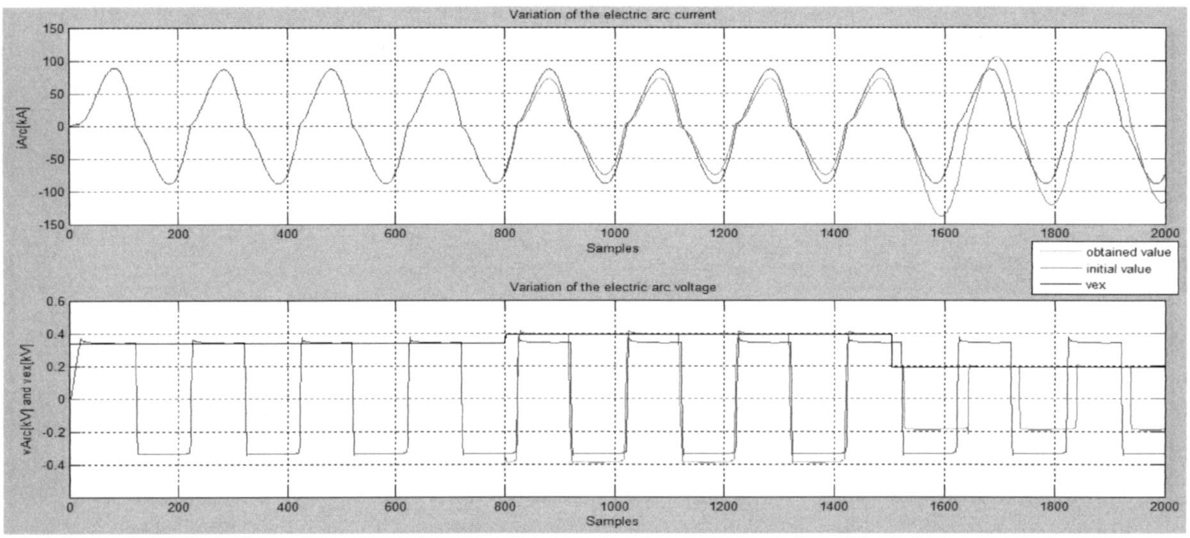

Figure 23. Variation of the electric arc current and voltage when is modified the extinction voltage

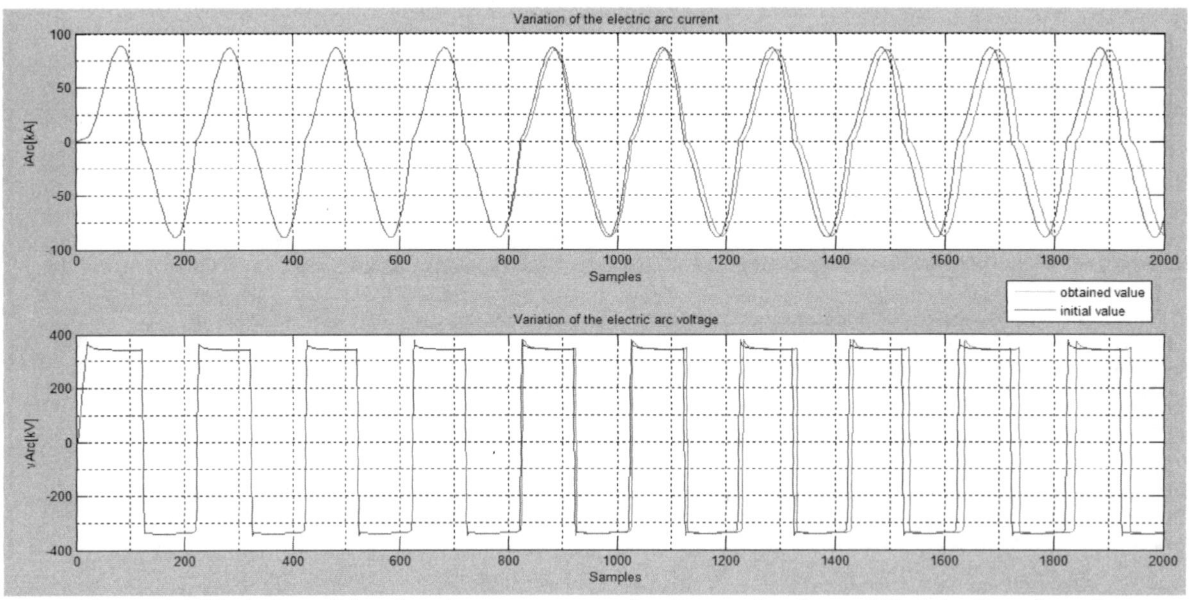

Figure 24. Variation of the electric arc current and voltage when is modified the line slope R_l

In Figure 27 are illustrated the waveforms for the electric arc current and for the electric arc voltage when modifying constant *C*. Initial value for *C* is *190000* when the derivative of the current is positive and *39000* when the derivative of the current is negative, after taking *800 samples* was set *C* to *230000* and after *1500 samples* was set to *40000* for the positive derivative of the current. One can notice that the amplitude of both voltage and current of the electric arc is the same, concluding that this parameter do not has a large influence on the amplitude of the voltage and current of the electric arc.

In Figure 28 are illustrated the waveforms for the electric arc current and for the electric arc voltage when modifying constant *D*. Initial value for *D* is *5000*, after taking *800 samples* was set *D* to *15000* and after *1500 samples* was set to *-1000*. One can notice that the amplitude of both voltage and current of the electric arc is the same, but when *D* has negative value the ignition voltage of the electric arc has a peak to large.

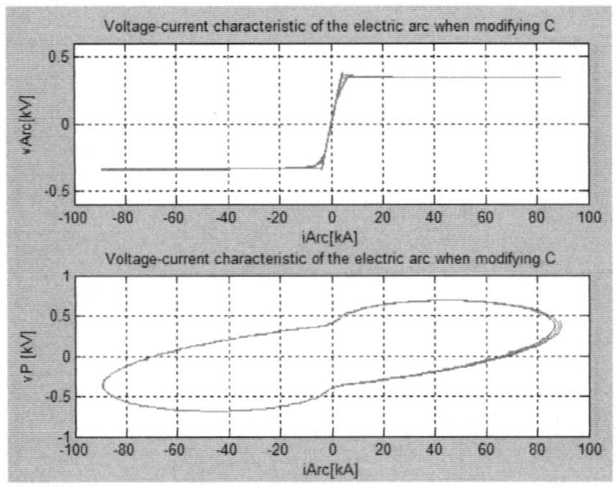

Figure 25. Voltage-current characteristic of the electric arc when modifying C

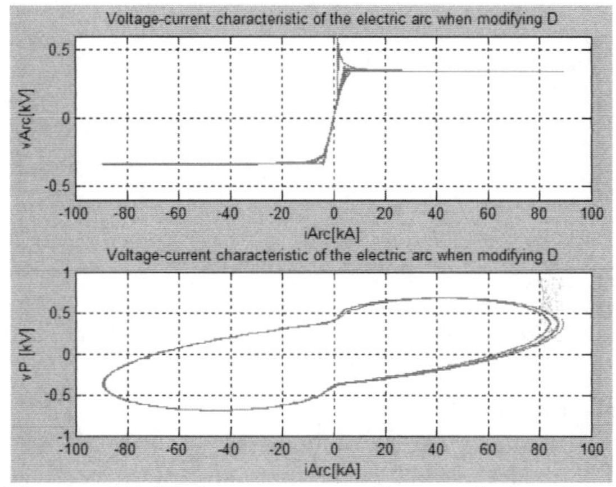

Figure 26. Voltage-current characteristic of the electric arc when modifying D

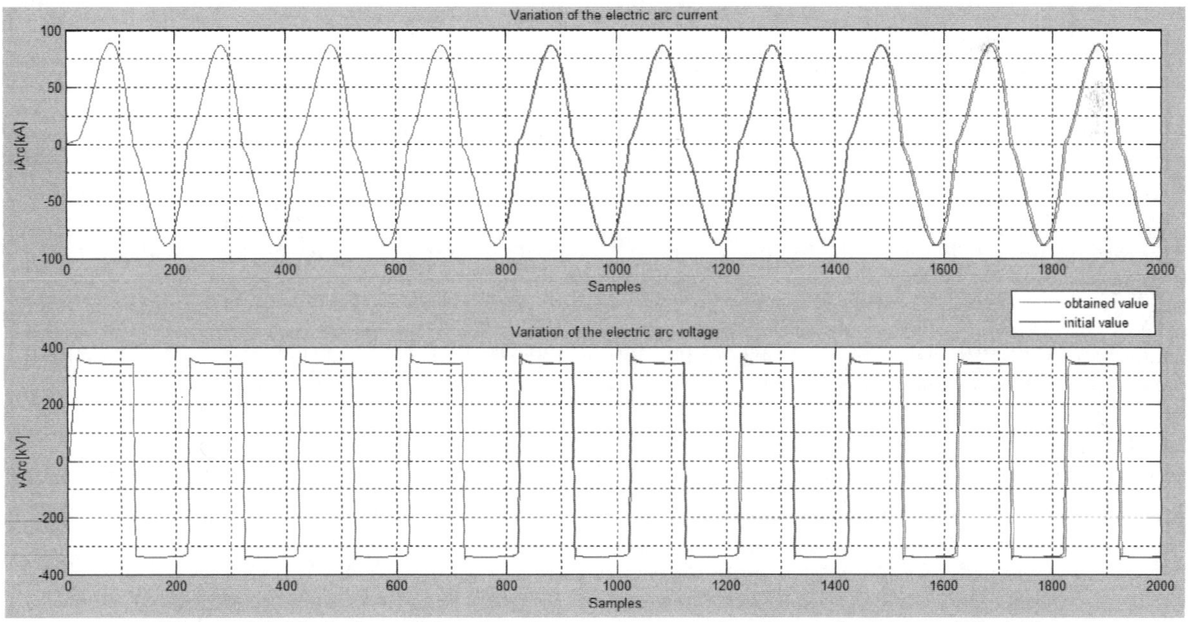

Figure 27. Variation of the electric arc current and voltage when is modified *C*

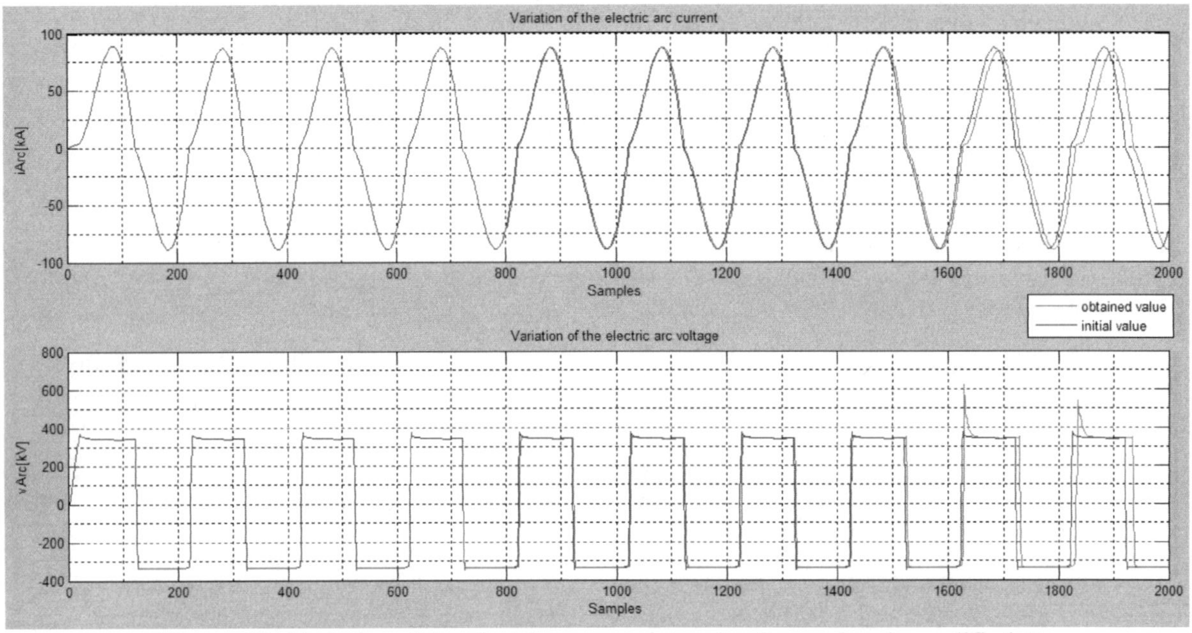

Figure 28. Variation of the electric arc current and voltage when is modified *D*

4. Conclusions

In this paper two models of the electric arc were developed in Matlab/Simulink in order to simulate the electric arc furnace behavior. For each of the model were implemented GUIs which allow modifying the parameters of each of the two models of the electric arc in order to study the importance of them on the voltage and current of the electric arc. Parameters of the electric arc furnace were set like the ones of the real installation taken into consideration.

By analyzing the results obtained from simulation one can conclude that in each model the significant parameters that influence the voltage and current of the electric arc are the ignition and the extinction voltage of the electric arc. Starting from the presented models can be developed other models of the electric arc.

Parameters influence in the electric arc modeling is important because it is required to know very well the model of the process in order to design control strategies for the entire installation, because for an EAF can be controlled the position of the electrodes in this way obtaining the desired arc power, increasing the productivity of liquid steel and reducing the damages caused by the breakage of the electrodes.

Acknowledgments. This work was developed through the Partnerships in Priority Areas - PN II, with the support of ANCS, UEFISCDI, Project No. 36/2012, Code: PN-II-PT-PCCA-2011-3.2-1519.

References

[1] Zhen W, Ninghui W, Tie L and Yong C 2012 3D numerical analysis of the arc plasma behavior in a submerged DC electric arc furnace for the production of fused MgO, *Plasma Sci. Technol.* **14** 321

[2] Yokoyama S, Arisawa R, Hisyamudin M N N, Murakami K, Maegawa A and Izaki M 2012 Applicability of Carbonated Electric Arc Furnace Slag to Mortar, *Journal of Physics: Conference Series* **352**(1) 012049

[3] Panoiu M, Panoiu C and Ghiormez L 2012 *Modeling of the Electric Arc Behaviour of the Electric Arc Furnace*, 5th International Workshop Soft Computing Applications SOFA 2012, Szeged,

Hungary, August 22-24, pp 261-271

[4] Ghiormez L and Prostean O 2013 *Parameters Influence in Electric Arc Modeling*, 8th IEEE International Symposium on Applied Computational Intelligence and Informatics SACI 2013, Timisoara, Romania, May 23-25, pp 209-213

[5] Ghiormez L, Panoiu M and Panoiu C 2012 *Harmonic Analysis of the 3 Phase Electric Arc Furnace using Models of the Electric Arc*, 3rd World Conference on Information Technology WCIT 2012, Barcelona, Spain, November 14-16, pp 424-430

[6] Balan R, Maties V, Hancu O, et al. 2007 *Modeling and control of an electric arc furnace*, in Control & Automation, Mediterranean Conference on MED'07, Athens, Greece, July 27 – 29, pp 1-6

[7] Lago F, Gonzalez J J, Freton P and Gleizes A 2004 A numerical modelling of an electric arc and its interaction with the anode: Part I. The two-dimensional model, *Journal of Physics D: Applied Physics* **37**(6)

[8] Panoiu M, Panoiu C, Osaci M, et al. 2008 Simulation result about harmonics filtering using measurement of dome electrical items in electrical installation on UHP EAF, *WSEAS Transactions on Circuits and Systems* **7**(1) 22-31

[9] Panoiu M, Panoiu C and Deaconu S 2008 *Study about the possibility of electrodes motion control in the EAF based on adaptive impedance control*, In Power Electronics and Motion Control Conference, EPE-PEMC 2008, Poznan, Poland, September 1-3, pp 1409-1415

[10] Panoiu M, Panoiu C, Iordan A and Ghiormez L 2013 Artificial neural networks in predicting current in electric arc furnaces, *IOP Conf. Ser.: Mater. Sci. Eng* **57** 012011

IOP Conf. Series: Materials Science and Engineering **163** (2017) 012027 doi:10.1088/1757-899X/163/1/012027

Study for the electric arc of alternative current at the single phase welding machine using the Matlab/Simulink environment

I Baciu[1], L Ghiormez[2] and C Vasar[2]

[1]Politehnica University of Timisoara, Department of Electrical Engineering and Industrial Informatics, Revolution Street, no. 5, 331128, Hunedoara, Romania
[2]Politehnica University of Timisoara, Department of Automation and Applied Informatics, Vasile Parvan Street, no. 2, 300223, Timisoara, Romania

E-mail: loredana.ghiormez@fih.upt.ro

Abstract. In this paper is presented a mathematical model of the electric arc for an alternative current welding machine of low power. The electric arc model is based on dividing the voltage-current characteristic of the electric arc in many functioning zones. For the model of the entire welding machine are used real parameters as the ones of the proper welding machine. The voltage and current harmonics spectrum that is obtained during the welding process is presented. Also, the waveforms for the current and voltage of the electric arc plotted against time and the voltage-current characteristic of the electric arc are illustrated. The electric arc is considered as being supplied by alternative voltage from the electrical power network using a single phase transformer which has the output voltage of 80 volts. The model of the welding machine is developed in Simulink and the variations of some parameters of the electric arc are obtained by modifying of them in a Matlab function. Also, in this paper is presented the total harmonic distortion for the voltage and current of the electric arc obtained during simulation of the welding machine.

1. Introduction

Welding is the technological process of making permanent joints of metal or nonmetallic components through the interaction of adjacent atoms. The welding process [1] is the introduction in a zone, by concentrating in time and space, of an amount of electrical energy in the welding area to bring out the atoms from their stable equilibrium state and approaching the adjacent atoms at equal or lower distances than the crystalline network parameter so they can recrystallize in a common network corresponding to a new stable state [2]. This energetic mechanism is illustrated in Figure 1.

Taking into consideration the energy sources which is performing local heating the welding processes are classified in welding with electrothermic, electrochemical and corpuscular energies [3].

In the case of welding with electrothermic energy the source of energy is the electric arc or the Joule effect. Welding with electric arc can be carried out by the following methods: with coated electrode, with shielding gas, under a stream or by other methods [4].

In this paper is studied the behavior of the electric arc in the welding process with coated electrode for a single phase welding machine of the alternative current. This study is performed by simulation in the Matlab/Simulink programming environment, by modifying the welding parameters. It is modified the supply voltage, the operating frequency of the power supply or the electric arc characteristics. The electric arc was modeled taking into consideration the voltage-current characteristic of the electric arc.

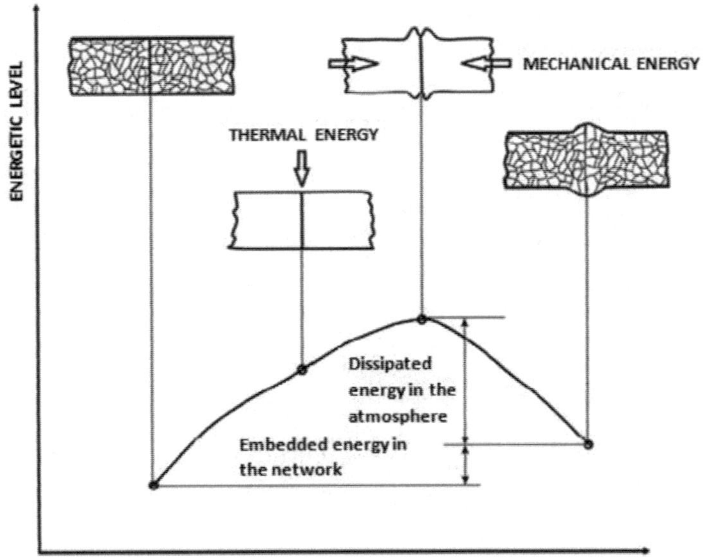

Figure 1. Welding phases

2. Presentation of the paper

2.1. Electric Arc of Alternative Current

Electric arc welding is a method of permanent joining of metal parts from an assembly, using the heat from the electric arc developed between the welding electrode and the material to be welded.

It is estimated that currently over 35% of steel production is assembled by electric arc welding processes and therefore can be considered that electric arc welding is one of the largest consumers of electricity.

Also, one can take into account that the welding machines are receivers which are causing important disturbances in the power network.

Opened electric arc is ignited when the electrode which is connected to one of the terminals of the welding machine's power source is brought into contact with the piece which is connected to the other terminal of the source. The supply source current (short circuit current) produces a strong heating of the irregularities on the contact surfaces, melting them.

By slow removal of the electrode, the molten weld metal stretches and shrinks its section. Current density increases and when is reached the temperature of boiling metal, the bridge breaks and appears the electric arc in metallic vapors slightly ionizable [2].

Electric arc extinction is achieved by removing the electrode from the piece (arc stretching), which results in an intense deionization of the discharged space.

The welding electric current that is developed between the electrode and the work piece, periodically changes its polarity, the role of cathode and anode alternating from the electrode to the work piece and backwards with a frequency f of the supply voltage. As a result, the current passes twice zero during a period of the supply voltage. Because of the deionization the electric arc is extinguish for a short time, when current passes by zero.

A particularity of the electric arc is the rectifying effect of the current, due to the different thermophysical properties between the electrode and the work piece. For this reason, the electric arc ignition conditions are different and the electric arc burns more in the half period in which the electrode is cathode. The presence of the continuous component reduces the welding quality because it decreases the depth of the welding penetration, so the magnetic circuit is saturated and appears the deforming power consumption.

2.2. Modeling of the electric arc characteristics

Considered as a circuit element, the electric arc has the properties of being a nonlinear resistor, characterized by a nonlinear dependence between the voltage and current intensity of the electric arc.

Voltage-current characteristics of the electric arc can be static or dynamic, as the intensity of the electric arc current varies and which can be very low (in particular null) or on the contrary can be very high.

Unlike the electric arc of continuous current, the electric arc of alternative current is only a quasi-stationary process which, at the length unit of the column [2], it is characterized by the powers equation presented in (1):

$$E_a i = p + \frac{dQ}{dt} \tag{1}$$

where Q is the energy of the arc column volume, E_a is the electric potential gradient, i is the current intensity, and p is the power transferred to the environment in the form of heat per unit of time. According to the Mayr advanced hypothesis, dependence of the conductance G for the arc column in relation to the content of energy Q can be expressed by the equation (2).

$$G = Ke^{\frac{Q}{Q_0}} \tag{2}$$

where K and Q_0 are constants. As for the column unit length, it can be written (3).

$$G = \frac{i}{E_a} \tag{3}$$

After applying logarithms and derivative in time and taking into account (1) and (2) is obtained equation (4).

$$\frac{1}{G}\frac{dG}{dt} = \frac{1}{T_a}\left(\frac{v_a i}{p\ell} - 1\right) \tag{4}$$

In (4) v_a is the voltage of the electric arc column of length l. Assuming a constant values, P_0 , for the power dissipated per unit length of the column and adopting the notation:

$$T_a = \frac{Q_0}{P_0} \tag{5}$$

where T_a is the time constant of the electric arc, differential equation (4) becomes:

$$\frac{1}{i}\frac{di}{dt} - \frac{1}{v_a}\frac{dv_a}{dt} = \frac{v_a i - P_0\ell}{P_0\ell T_a} \tag{6}$$

Relation (6) is also known as equation of electric arc in dynamic regime [2]. Considering that the current intensity through the electric arc is sinusoidal, like:

$$i(t) = \sqrt{2}I \sin \omega t \tag{7}$$

for the solution of differential equation (6) one can obtained the expression:

$$v_a(t) = \frac{\sqrt{2} P_0 \ell \omega t}{I\left[1 - \frac{\sin(2\omega t + \ )}{\sqrt{1 + (2\omega T_a)^2}}\right]} \tag{8}$$

where:

$$\varphi = arctg\left(\frac{1}{2\omega T_a}\right) \tag{9}$$

2.3. Characteristics of the electric arc supplied with alternative voltage

It can be considered that during the burning of the electric arc supplied with alternative voltage, the equivalent electric circuit of electric arc can be represented as illustrated in Figure 2.

In this figure parameters r and L represents the resistivity, respectively the inductivity of the equivalent electric circuit and R_A and v_A is the resistivity of the electric arc, respectively the voltage of the electric arc [5], [6].

Waveforms for the electrical parameters from the equivalent circuit are presented in Figure 3. One can obtain the following conclusions:

- The electric arc is a nonlinear receiver because during the burning of the electric arc the voltage of the electric arc is approximately the same and the current is varying;
- Electric arc has a resistive character because the electric arc voltage and electric arc current are in the same phase;
- The electric arc current passes two times zero for each period of the source voltage, so, the electric arc is extinguished and ignited with a double frequency as compared to the one of the applied voltage;
- The ignition voltage of the electric arc is greater as compared to the electric arc voltage;
- The electric arc current is periodical, but nonsinusoidal;
- The electric arc supplied with alternative voltage has a rectifying character [7].

In Figure 4 is presented the dynamic voltage-current characteristic of the electric arc supplied with alternative voltage which is illustrated in the reference literature [8], [9]. This characteristic is obtained taking into account the variation of the voltage and current of the electric arc presented in Figure 3. Electric arc current can be in one of the two regimes: interrupted or uninterrupted current. If the electric arc burning is in the interrupted current regime the functioning of it is unstable and the electric arc current waveform is strongly distorted. So, it is necessary that the electric arc burning to be without breaks.

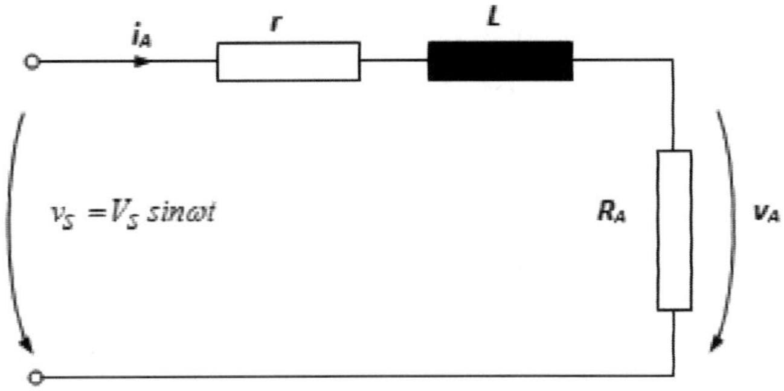

Figure 2. The equivalent electric circuit of electric arc supplied with alternative current

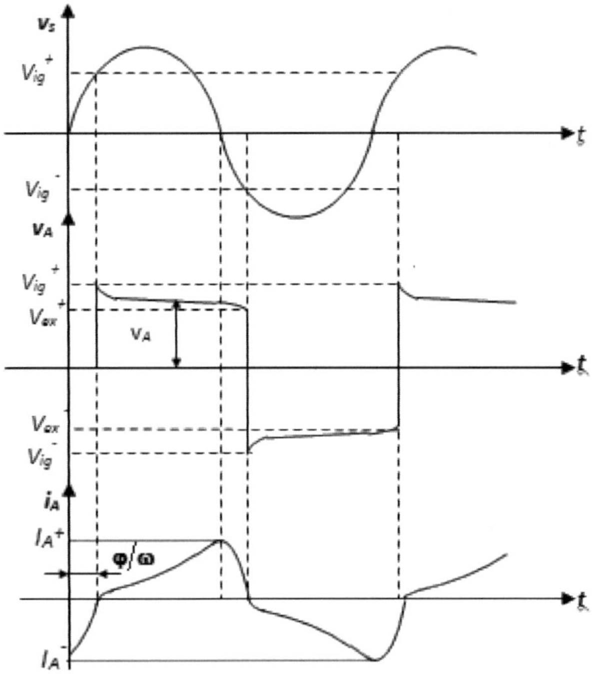

Figure 3. Waveforms for the electrical parameters in the electric arc circuit

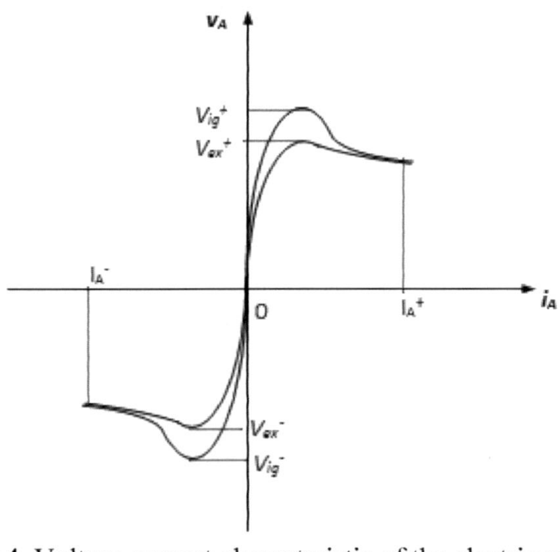

Figure 4. Voltage-current-characteristic of the electric supplied with alternative voltage

In order to burn the electric arc without breaks, in the assumption that $V_{ig}^{+} \cong \left|V_{ig}^{-}\right| \cong V_{ex}^{+} \cong \left|V_{ex}^{-}\right| \cong V_A$ and the supply voltage has a sinusoidal shape $v_s(t) = V_s \cdot \sin \omega t$ should be fulfilled the condition (10).

$$V_s \cdot \sin \quad \geq V_A \tag{10}$$

which leads to relation (11).

$$\frac{V_A}{V_s} \leq \sqrt{\frac{1}{1+\frac{\pi^2}{4}}} = 0{,}54\,, \tag{11}$$

Resulting that in order to have an uninterrupted current it is necessary that the welding machine has to operate having a power factor as illustrated in (12).

$$\cos\varphi \leq 0{,}85 \tag{12}$$

3. The implementation of the welding process in Matlab/Simulink

In Figure 5 is presented the Matlab/Simulink block diagram which corresponds for the equivalent electric circuit of the electric arc supplied with alternative current as illustrated in Figure 2. This block diagram was developed by using equation (13) which was obtained by applying second Kirchhoff's law for the electric circuit presented in Figure 2.

$$v_s = i(t)\cdot R + L\cdot\frac{di}{dt} + v_{arc} \tag{13}$$

In order to obtain the current of the electric circuit from equation (13) it is used relation (14).

$$i(t) = \frac{1}{L}\int V_L\,dt \tag{14}$$

Based on equations (13) and (14) is implemented the simulation of the welding process in Matlab/Simulink. In this simulation the electric arc is simulated using the relation (15), but implemented even for negative values of the parameters [10].

$$v_a = \begin{cases} i\cdot R_1 & 0 \leq i \leq i_1 \\ i\cdot R_2 + V_{ig}\cdot\left(1-\frac{R_2}{R_1}\right) & i_1 < i \leq i_2 \\ i_2\cdot R_2 + V_{ig}\cdot\left(1-\frac{R_2}{R_1}\right), & i > i_2 \end{cases} \tag{15}$$

Where

$$i_1 = \frac{V_{ig}}{R_1}, \tag{16a}$$

$$i_2 = \frac{V_{ex}}{R_2} - V_{ig}\cdot\left(\frac{1}{R_2}-\frac{1}{R_1}\right) \tag{16b}$$

For the block diagram illustrated in Figure 5 are used real values for the parameters of the welding machine like the ones of the equipment taken into consideration to study.

Also, the model parameters were chosen as the electric arc current shouldn't be in the interrupted regime and the voltage should follow the real waveforms of this parameter.

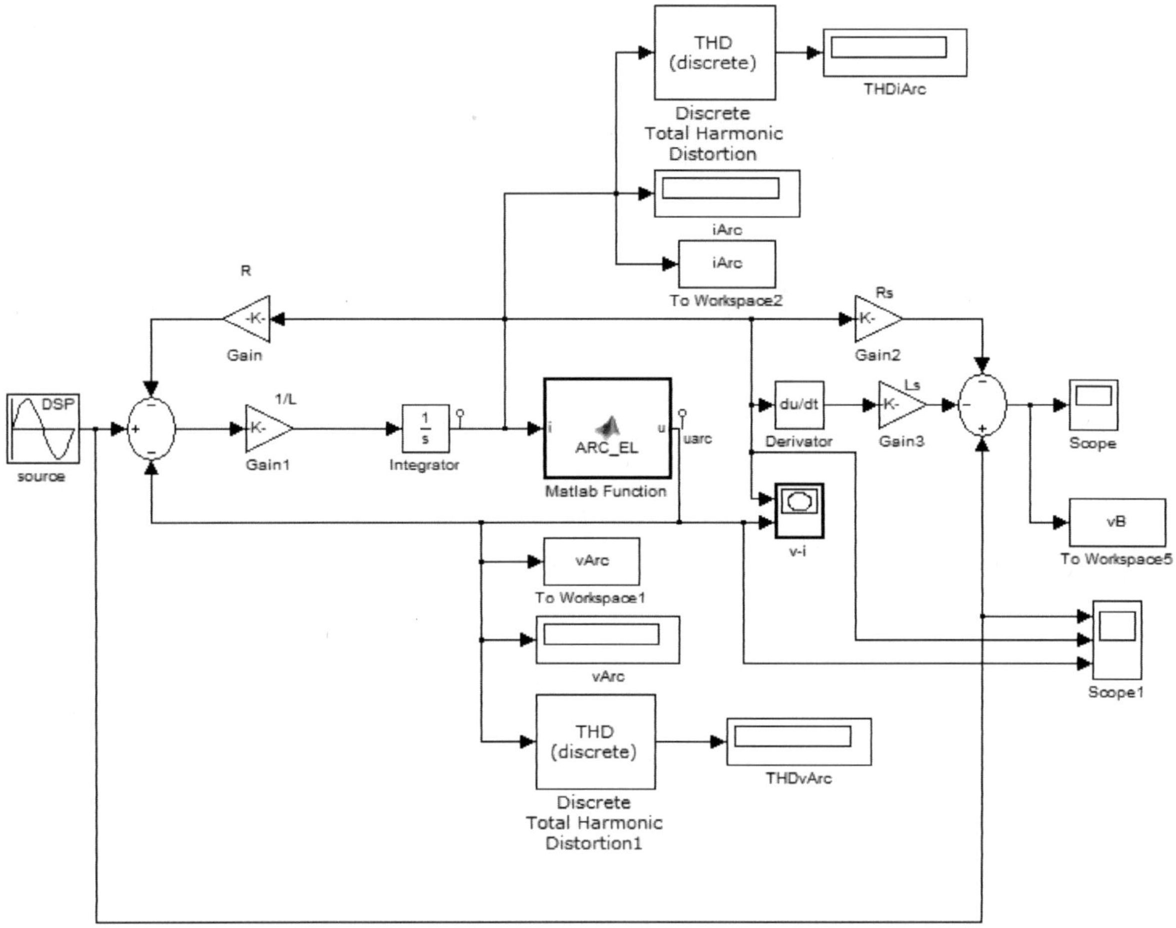

Figure 5. The implementation of the welding process in Matlab/Simulink

In Figure 6 are illustrated the waveforms for the electrical parameters of the electric arc: voltage of the power supply, the electric arc current, respectively the electric arc voltage. One can notice that the supply voltage of the welding machine has a sinusoidal shape and its amplitude is approximately *110 V*. The operating frequency of the power supply source is *50 Hz*. Also, one can notice the waveforms for the current of the electric arc which has approximately a sinusoidal shape and the operating regime is in the uninterrupted one. The voltage of the electric arc has approximately a rectangular shape. The waveforms for the current and voltage of the electric arc are corresponding for the ones illustrated in Figure 3.

Analyzing the voltage and current of the electric arc one can conclude that both electrical parameters are in the same phase with the supply voltage. The voltage is distorted and the current has a waveform close to a sinusoidal one, the electric arc having a resistive character. The current of the electric arc has amplitude of approximately *130 A*, the voltage of the electric arc is *60 V* and the extinction voltage is of *50 V*, these values corresponding to the prescribed ones in the electric arc model.

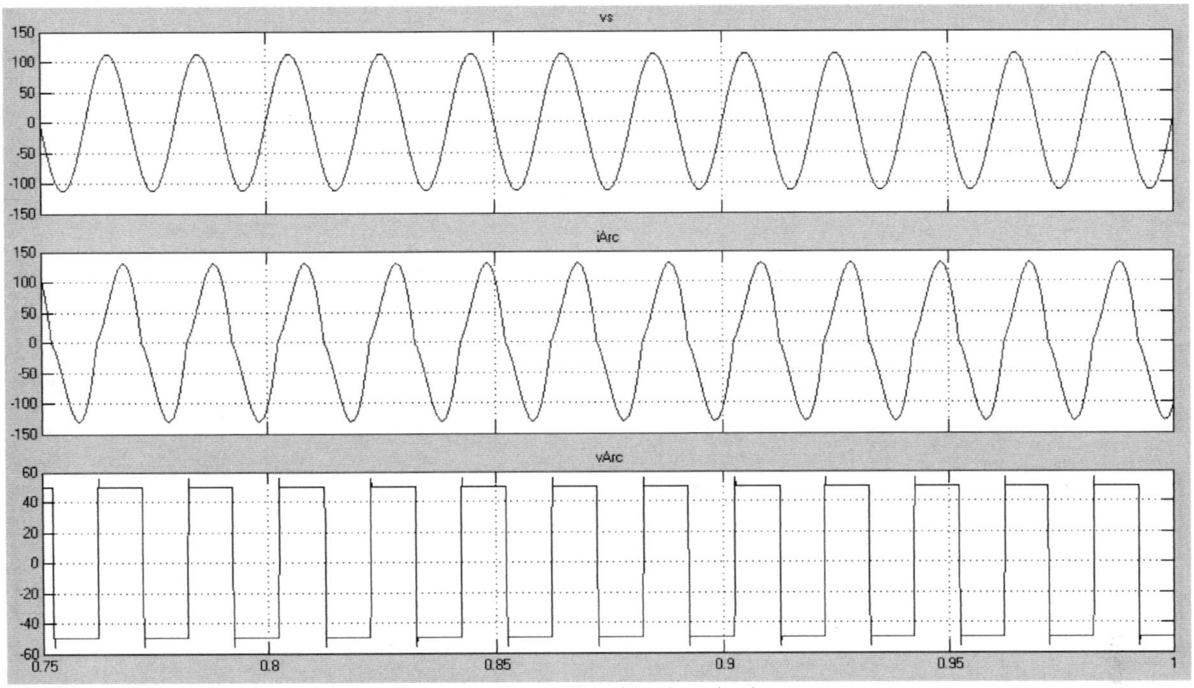

Figure 6. Waveforms for the electrical parameters

In Figure 7 is illustrated the voltage-current characteristic of the electric arc obtained for the values before presented. By analyzing of it one can conclude that the welding machine is a nonlinear circuit element, this could be linearized just on some operating zones as is evidenced by the voltage-current characteristic of the electric arc.

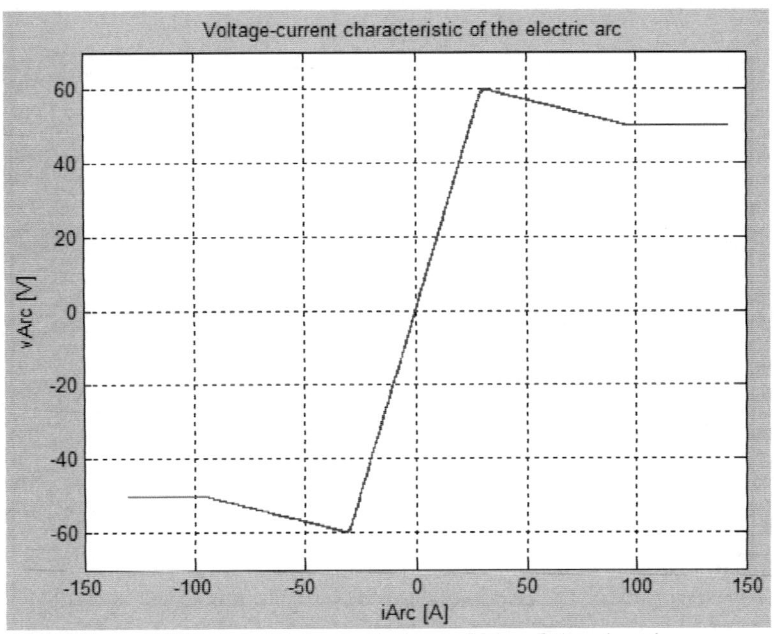

Figure 7. Voltage-current characteristic of the electric arc

In Figure 8 is illustrated the spectral characteristic of the electric arc current for the variations of the electric parameters illustrated in Figure 6. Current harmonics are presented on a semilogarithmic scale taking into consideration a range of frequency until *1000 Hz*.

In Figure 9 is presented the harmonics amplitude for the electric arc current using a different representation. One can notice that are presented only odd harmonics of current, the fundamental one having amplitude of approximately *110 A*.

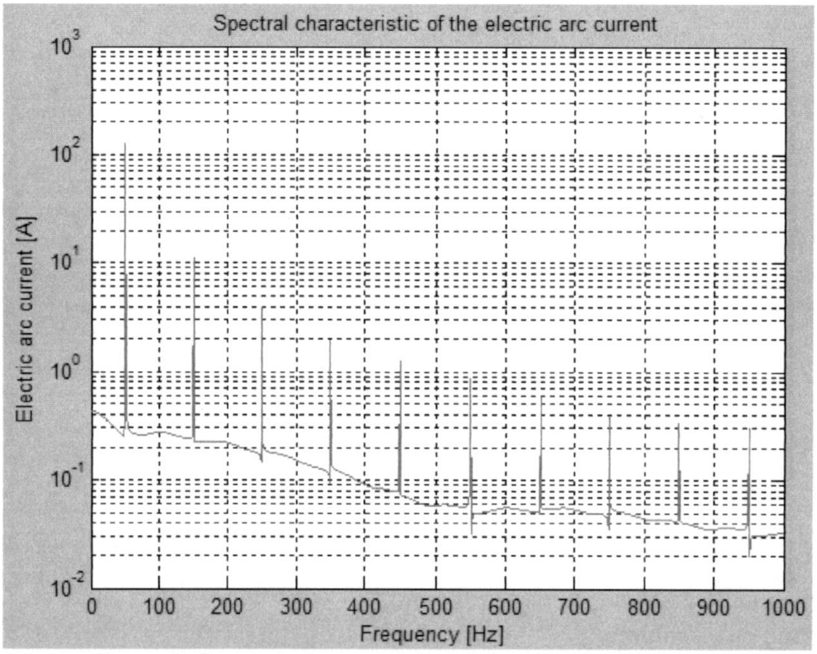

Figure 8. Spectral characteristic of the electric arc current

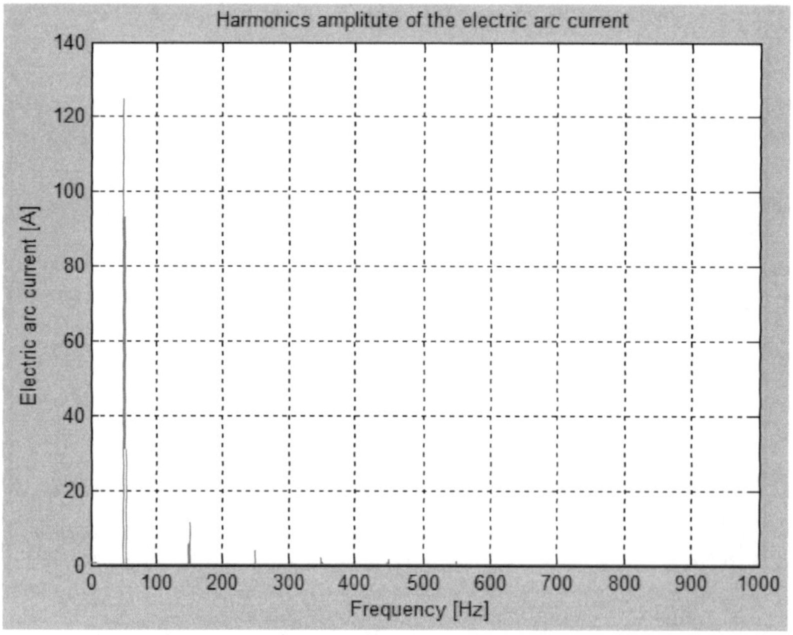

Figure 9. Harmonics amplitude of the electric arc current

In Figure 10 is illustrated the spectral characteristic of the electric arc voltage for the variations of the electric parameters illustrated in Figure 6. Current harmonics are presented on a semilogarithmic scale taking into consideration a range of frequency until *1000 Hz*.

In Figure 11 is presented the harmonics amplitude for the electric arc voltage using a different representation. One can notice that are presented only odd harmonics of voltage, the fundamental one having amplitude of approximately *64 V*.

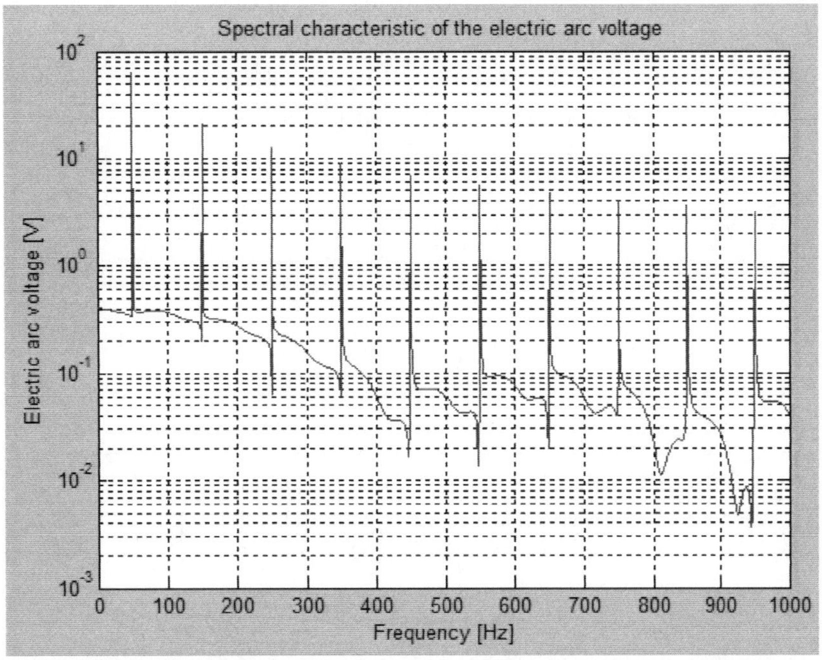

Figure 10. Spectral characteristic of the electric arc voltage

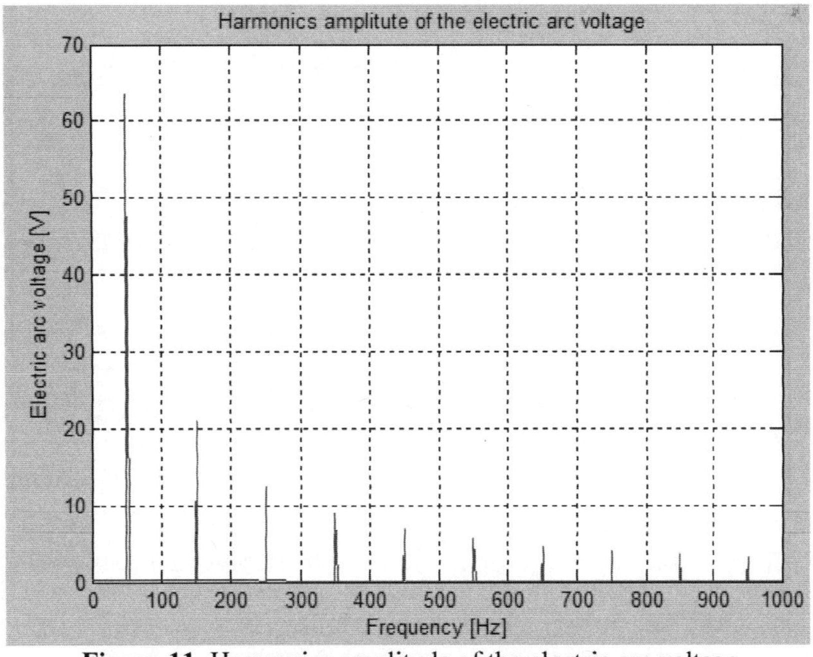

Figure 11. Harmonics amplitude of the electric arc voltage

IOP Conf. Series: Materials Science and Engineering **163** (2017) 012027 doi:10.1088/1757-899X/163/1/012027

Analyzing the Figures 8, 9, 10 and 11 one can conclude that are presented only the odd harmonics and the amplitude of them are decreasing while its order increase.

In order to obtain the influence of the supply voltage and of the power supply source operating frequency simulation were made. In Figure 12 is presented the case when the supply voltage was set to *50 V*, so the amplitude is approximately *70 V* and the operating frequency is *50 Hz*. In Figure 13 is presented the case when the operating frequency is set to *350 Hz* and the supply voltage is *50 V*. In Figure 14 is presented the case when the operating frequency is set to *50 Hz* and the supply voltage is *100 V*.

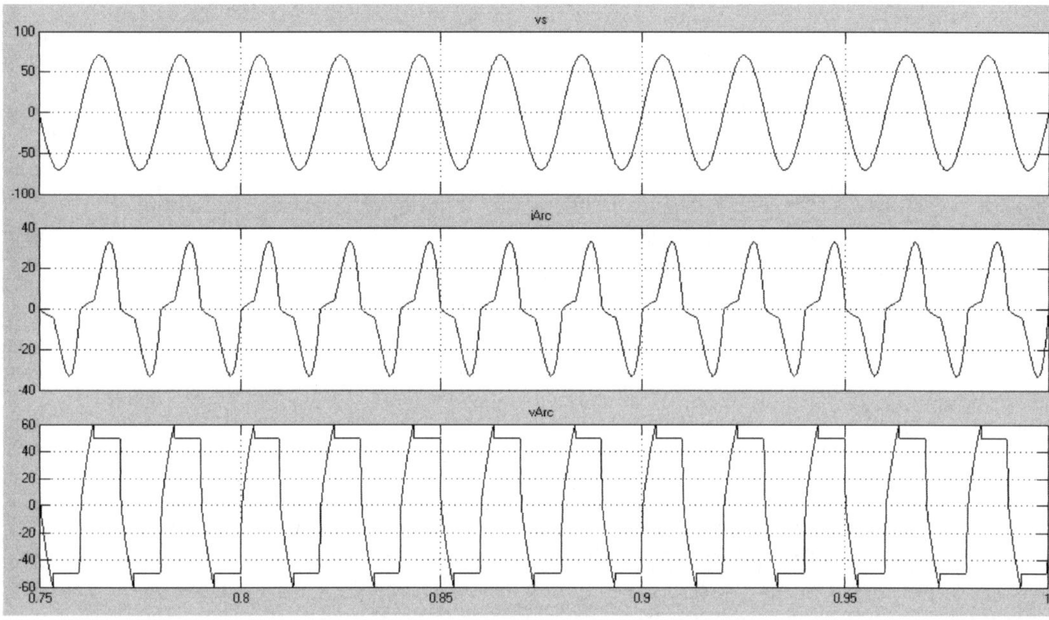

Figure 12. Waveforms for the electrical parameters for supply voltage of 50 V and the operating frequency is 50 Hz

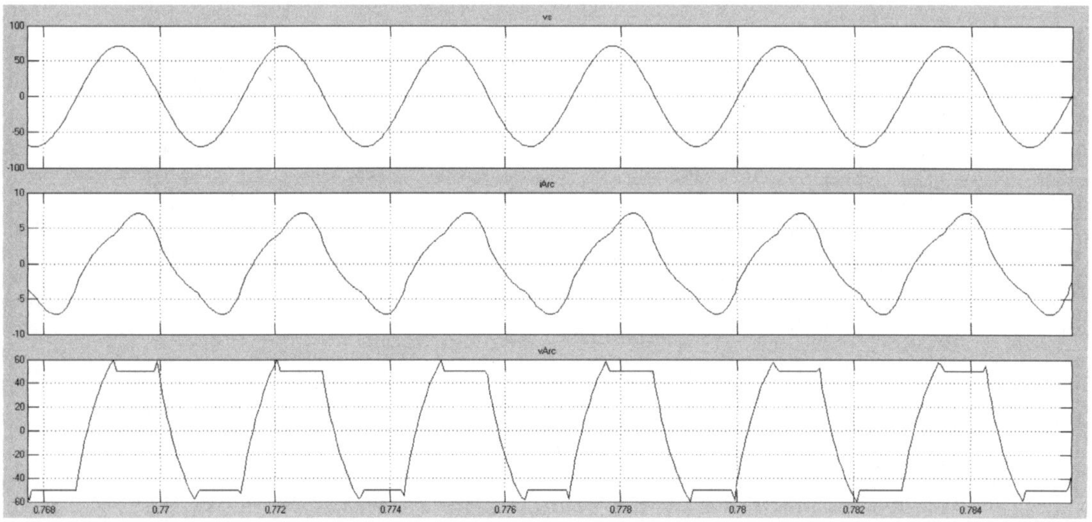

Figure 13. Waveforms for the electrical parameters for supply voltage of 50 V and the operating frequency is 350 Hz

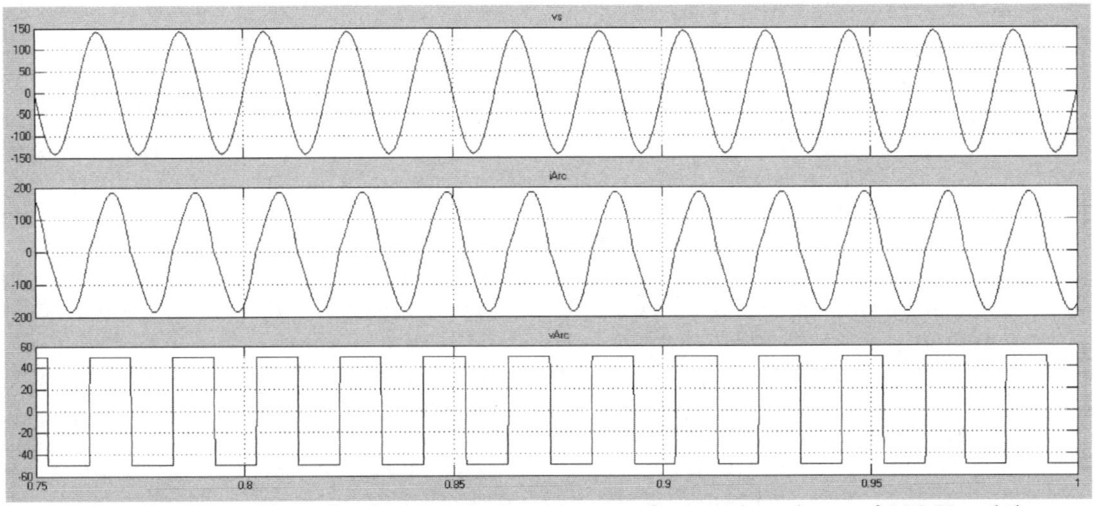

Figure 14. Waveforms for the electrical parameters for supply voltage of 100 V and the operating frequency is 50 Hz

Following the results obtained during simulation for different values of the power supply source one can conclude that appear a worsening of the electric arc characteristics while reducing the supply voltage value. In this situation it can be observed that the current of the electric arc is in the interrupted regime.

Also, if the operating frequency of the power supply source is above *350 Hz* it can be observed that the electric arc is in the stable area.

If the supply voltage is *100 V* and the operating frequency of the power supply source is *50 Hz* it can be noticed that the electric arc current waveform is approximately a sinusoidal one and the electric arc voltage waveform has a rectangular shape, so the arc is stable.

4. Conclusions

Following the experimental model simulation of the electric arc one can noticed the electric arc behavior for a welding machine of low power, this behavior illustrating the nonlinearity of the electric arc as presented in the voltage-current characteristic of the electric arc. By setting of both installation and model parameters with values as corresponding for the real welding machine is obtained an electric arc current having waveforms approximate to a sinusoidal and the electric arc voltage is strongly distorted. The THD factors for the voltage and current of the electric arc have considerable values so are necessary some actions in order to reduce them. So, the cost of such welding machine can be high and have to be taken into consideration the advantages for such of the investment.

One can conclude that are presented only the odd harmonics and its values are decreasing when the order of the harmonic increase.

The harmonics of the current and voltage of the electric arc were obtained by writing a Matlab program that was developed by the author of this paper. The values of the harmonics are obtained as concordant to the previsioned ones.

Following the simulations performed for different supply voltage values or different power supply operating frequency one can notice that in order to improve the characteristics of the electric arc it is necessary to increase the supply voltage or to increase the operating frequency of the power supply source. First solution is less advantageous remaining the solution with increasing the operating frequency. This aspect is fulfilled by developing of the frequency convertor welding machines.

Acknowledgments This work was developed through the Partnerships in Priority Areas - PN II, with the support of ANCS, UEFISCDI, Project No. 36/2012, Code: PN-II-PT-PCCA-2011-3.2-1519.

References

[1] Simončič S and Podržaj P 2012 Image-based electrode tip displacement in resistance spot welding, *Measurement Science and Technology* **23**(6) 065401

[2] Baraboi A and Adam M 2002 *Electrical equiments*, "GH. ASACHI" Publishing, Iasi, Romania

[3] Khafizov A A, Valiev R I, Shakirov Y I and Valiev R A 2014 Steel surface modification with plasma spraying electrothermal installation using a liquid electrode, *Journal of Physics: Conference Series* **567**(1) 012026

[4] Visan D 2008 *Welding technologies*, Galati, Romania

[5] Panoiu M, Panoiu C and Sora I 2006 Experimental research concerning the electromagnetic pollution generated by the 3-phase electric arc furnaces in the electric power supply networks, *Acta Electrotehnica* **47** 102

[6] Panoiu M, Panoiu C, Iordan A and Ghiormez L 2013 Artificial neural networks in predicting current in electric arc furnaces, *IOP Conf. Ser.: Mater. Sci. Eng* **57** 012011

[7] Panoiu M, Panoiu C, Sora I and Osaci M 2007 About the possibility of power controlling in the three-phase electric arc furnaces using PSCAD EMTDC simulation program, *Advances in Electrical and Computer Engineering* **7**(1) 38-43

[8] Panoiu M and Panoiu C 2008 *Modeling and simulation of the nonlinear processes in electrothermic*, Mirton Publishing, Timisoara, Romania

[9] Golovanov N and Sora I 1997 *Electrothermic and electrotechnologies*, vol I, Tehnica Publishing, Bucuresti, Romania

[10] Ghiormez L and Prostean O 2015 *Electric Arc Current Control for an Electric Arc Furnace Based on Fuzzy Logic*, 10th IEEE International Symposium on Applied Computational Intelligence and Informatics SACI 2015, Timisoara, Romania, May 21-23, pp 359 – 364

Small hydropower plants standardization, between myth and reality

A V Paraschivescu[1], K Ahmad-Rashid[2], F Popa[3] and B Popa[4]

[1]ConsultAbilities Ltd., Rwanda
[2]Electrical Engineering Department, Faculty of Engineering, University of Sulaimani, Kurdistan, Iraq
[3]ISPH Project Development, Bucharest, Romania
[4]University Politehnica of Bucharest, Department of Hydraulics, Hydraulic Machineries and Environmental Engineering, Bucharest, Romania

E-mail: bogdan.popa@upb.ro

Abstract. Many providers for small hydropower plants equipment have tried to standardize the components and even the entire equipment. So called "compact turbines" were launched on the market, ensuring the pre-designed solution of the modular components, but usually with lower efficiency then turbines specially designed for a certain site. For civil works it is possible to standardize some components, such as the powerhouse, the surge tank or the headrace, but not the intake and the weir.
Part of the hydropower plants can be standardized, but not the entire project, because there are a lot of variables that influence the design. Among these, the dimension, materials and design of the canal and the penstock are given by the hydrology, topography and the geology of the project's area.
This paper presents an attempt at standardization by using different heads and different installed flows. The case study is made on the Lukosi River from Tanzania, because there is a good hydrological database on power and energy calculation. For the powerhouse, pressure tower and intake dimensioning, the assumptions and materials considered cover all challenges that could appear in the geological and topographical structure of the project's area (worst case, most expensive). The study has highlighted African climatologic and hydrological conditions and the adapting of current technology to these conditions.

1. Introduction

An important question that remains viable in time is: "Can the small hydropower plants be standardized?"

Many equipment providers have tried to standardize the equipment or the components; many producers have launched on the market so called „compact turbines" that ensure the pre-design solution of the modular components. But will they reach the same efficiency as a site design turbine?

It is obvious that the water quality and the materials used imply certain conditions in order to use compact turbines. For civil works it is possible to standardize some of their components, such as the powerhouse, surge tank or headrace, but not the intake and the weir.

This article presents an attempt at standardization by using different heads and different installed flows. The case study is made on the Lukosi River from Tanzania, because the authors have a good hydrological database on power and energy calculation, as reference. The research study has been drawn up at the end of 2015.

1.1. Project location

For this research study, the location of the project was the Lukosi River, Tanzania, due to the particularities that the African river has, compared to the European rivers. Referring to standardization, this shall be applied everywhere in any topographical, geological and hydrological conditions.

Tanzania is a republic in East Africa, on the Indian Ocean (Figure 1). A diverse country in which almost 100 different languages are spoken, Tanzania was formed by the federation of the nations of Tanganyika and Zanzibar in 1964. The country's name is a combination of the first syllables of the component territories' names. Tanzania is bounded in the north by Kenya and Uganda; in the east by the Indian Ocean; in the south by Mozambique, Malawi, and Zambia; and in the west by the Democratic Republic of Congo (DRC), Burundi, and Rwanda. The country includes the islands of Zanzibar and Pemba, and other offshore islands in the Indian Ocean. Dar es Salaam is the executive capital and the largest city; the smaller city Dodoma is now the legislative centre of Tanzania and has been designated as the capital.

Geography. The landscape of mainland Tanzania is generally flat and low along the coast, but a plateau that has an average altitude of about 1,200 m (about 4,000 ft) constitutes the greater part of the country. Isolated mountain groups rise in the northeast and southwest. The volcanic Kilimanjaro (5,895 m/19,341 ft), the highest mountain in Africa, is located near the north-eastern border.

Three of the great lakes of Africa lie on the borders of the country and partially within it. Lake Tanganyika is located on the western border, Lake Victoria on the northwest, and Lake Malawi on the southwest. Lakes Malawi and Tanganyika lie in the Great Rift Valley, a tremendous geological fault system extending from the Middle East to Mozambique.

Iringa Region. Iringa region lies in the Southern Highlands of Mainland Tanzania. It stretches from the semi-arid central Tanzania in the north to the shores of Lake Nyasa in the South. The region is located between 7005' - 36032' South and 33047' – 36032' East. In the North, the Iringa region borders the Dodoma region, the Mbeya region to the West and the Morogoro region to the East. To the South, the Iringa region partly borders the Ruvuma region and partly Lake Nyasa. The region lies between latitudes 70 and 90 South of Equator, and between longitudes 32^0 and 35^0 East of the Greenwich Meridian. The Iringa region lies at an altitude of 475 masl (meters above sea level) with high peaks of 2,981 masl.

Climate. Tanzania is a big African country, overlooking the Indian Ocean, lying just south of the Equator. Most of the country is covered by a plateau, home to a large amount of world-famous touristic attractions, with several parks and nature reserves in the environment of the savannah. This plateau has a tropical climate, warm but tempered by the altitude. The thin flat coastal stretch is rather hot and humid throughout the year, especially from November to April. As regards the rain pattern, the country can be divided into four zones (Figure 2). In the north and east, excluding the region of Lake Victoria, there are two rainy seasons: one less intense, known as short rains season between October and December, and the other more intense, known as long rains season from March to May, with the peak in April.

There were two site visits in the study location: first between 16th September and 21st September 2015 and the second in October 2015. The rainy season was about to start, but everything was very wild and dry on these mountains. Lukosi River is a large river of Tanzania, with at least 10 m^3/s in the dry season.

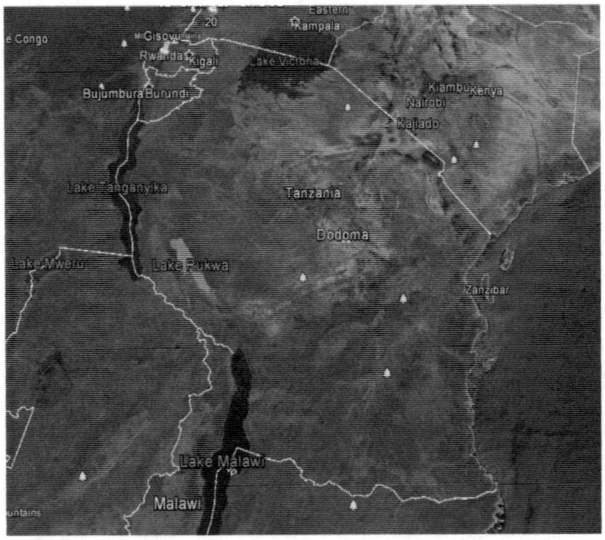

Figure 1. Tanzania

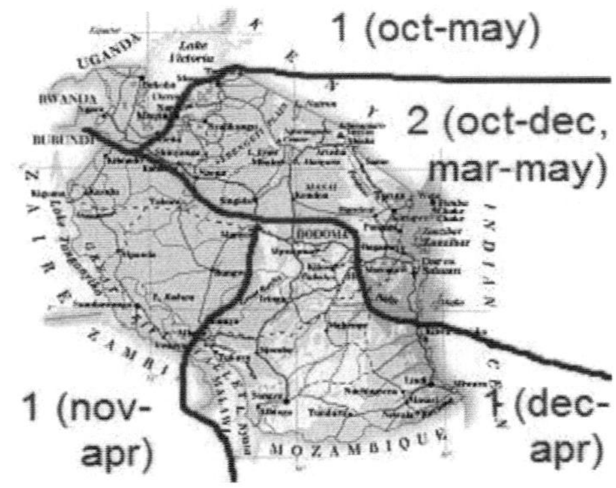

Figure 2. Tanzania seasons

2. Hydrological Data

2.1. Great Rwaha Basin

The Great Ruaha River is a river in south-central Tanzania that flows through the Usangu wetlands and the Ruaha National Park east into the Rufiji River. Its basin catchment area is 83,970 km². The population of the basin is mainly sustained by irrigation and water-related livelihoods such as fishing and livestock keeping. Great Ruaha is about 475 km long, its tributary basin has a catchment area of 68,000 km² and the mean flow is 140 m³/s.

The major rivers contributing to the Great Ruaha River are Lukosi (Figure 3), Yovi, Kitete, Sanje, Little Ruaha, Kisigo, Mbarali, Kimani and Chimala whereas the small ones include Umrobo, Mkoji, Lunwa, Mlomboji, Ipatagwa, Mambi and Mswiswi rivers.

Figure 3. Lukosi River

IOP Conf. Series: Materials Science and Engineering **163** (2017) 012028 doi:10.1088/1757-899X/163/1/012028

2.2. Historical Hydrological Data

The project area was visited twice, in September 2015 together with RPA representatives and in October 2015 together with CAL's experts. Two spot measurements were taken in the time of the site visits and presented in the Site Visit Report.

Available data and studies. The following information / studies were collected from the Developer:

- Preliminary Estimate of Water Resource Availability: Flow Duration Analysis of Stream Discharge Records recorded at Mtandika Gage Station from 1957 to 1987 On the Lukosi River, Tanzania, J.L. McAdoo, Hydrologist, February 2015;
- Mean annual rainfall map;
- Lower Lukosi River Elevation Profile from 90m DEM From Confluence with the Ruaha to Confluence with the Mlowa;
- Upper Lukosi River Elevation Profile from 90m DEM Confluence of the Lukosi with the Mlowa To Its Headwaters;
- Matindika - 1957 - 2010 data;
- Monitoring Stations_Metadata.

2.3. Measurements on River [2]

The finding of the mean flow was 12 m^3/s from visual estimation and measurement with the propeller (Figure 4). Lukosi River has a lot of tributaries, but in the dry season the tributaries are completely dry (Figure 5). Lukosi River's banks are almost non-accessible. The area is mostly rocky with few inhabitants and to build access roads will be a real challenge (involving blasting works). But this aspect will be analyzed in the geological survey that will be the next step of the hydropower development study.

Figure 4. Measurements of river using propeller

Figure 5. Lukosi River Tributaries in dry season

This measurement is somehow similar with the first two findings, presented above.

The river flow, in the dry season, is between 10 m^3/s and 12 m^3/s which is also the visual estimation.

Measurement with propeller (Figure 4).

Measurement with propeller was done one shot only because the river is loo large – about 20 m;

Depth: 1.24 m;

Average speed: 0.5 m/s;

River width: 20 m (approximation, impossible to measure);

Average area [m^2] = 24.8 m^2;

Return of flows: 12.4 m^3/s.

Measurement with bottle

Downstream of the first place measurements with an empty bottle were made;
Length of river sector: 15 m;
Width: 15 m;
Depth: 5 m;
Time for ¾ empty bottle: 61 s;
Area: 60 m^2; Speed: 150/61 = 0.205 m/s;
Flow: 12.3 m^3/s.

2.4. Mtandika Gauging Station. Previous hydrological study used in the project
The precise location of the gage station is a necessary piece of information for applying the flow duration analysis to potential hydropower sites in the Lukosi Basin other than at the gage. The location of the gage determines the total area of the rainfall catchment, based on elevation delineation.

Catchment area and catchment statistics of the watershed drained by the Lukosi River upstream from, and contributing flow to, the Mtandika Gauging Station were re-determined using the methodology and resources described below. The Mtandika location is originally described in the GRDC dataset as being located at: -7.56667 Decimal degrees (-South) Latitude (07° 34' 00.01" S); 36.433333 Decimal degrees East Longitude (36° 25' 59.93" E); 1,210 masl; 3,363 km^2 catchment area.

In February 2015 a study was performed: "Preliminary Estimate of Water Resource Availability: Flow Duration Analysis of Stream Discharge Records Recorded at Mtandika Gage Station from 1957 to 1987 on the Lukosi River, Tanzania", by J.L. McAdoo, hydrologist (Figure 6). According to this study, Lukosi River, at Matndika Gauging Station, is expected to have at least 21 m^3/s for 90% at its peak in April. In the rest of the time, 9 m^3/s is the assured average of flow produced by the river at least 90% of time.

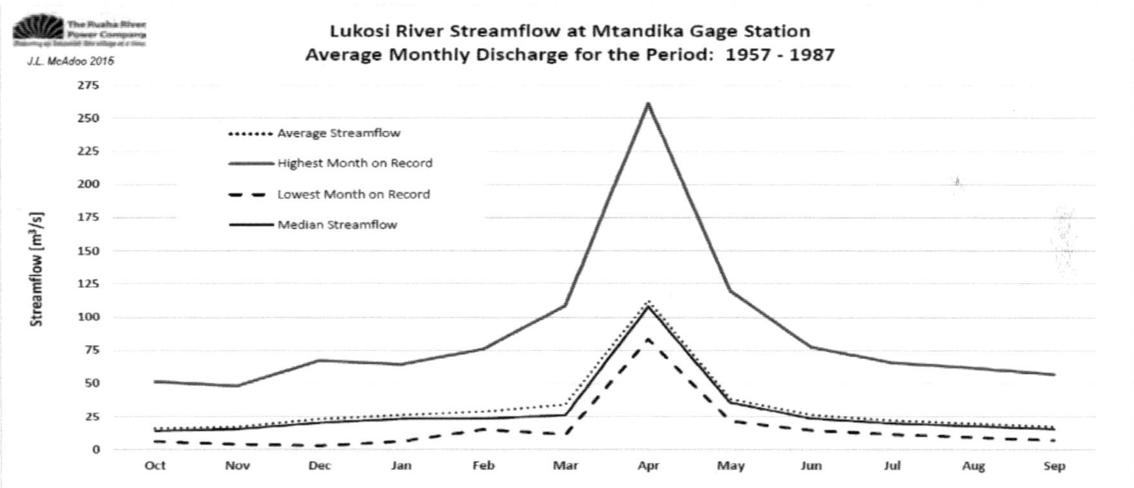

Figure 6. Lukosi River – Multiannual mean monthly flows for the period 1957 – 1987 (30 years)

<u>Lukosi River Flow Duration Analysis and Exceedance Probabilities</u>
Neither the average nor median flow rates represent the minimum amount of water flowing through the system 50% of the time. This is due to the fact that stream flow curves (hydrographs) do not follow a normal distribution. To find the actual duration of flows, it is therefore necessary to calculate the frequency statistics of each particular stream flow. To compute flow duration curves for each month, the raw daily discharge measurements at Mtandika Gauging Station were re-organized on a monthly basis, ranked, and assigned a frequency distribution.

The flow duration curve and the exceedance probability for each particular observation (o) are calculated according to the relation (1):

$$P_0 = 100 \times \left[M_0 \div (n+1)\right], \tag{1}$$

where:
P = probability that the flow will be greater or at least equal than a certain value (% of time),
M = the ranked position of the flow data point "o" (dimensionless),
n = the number of flow data points in the period of record.

The frequency of a particular flow rate occurring in the period of record is its expected probability of recurrence in the future, under the assumptions that (1) the existing 30 years of data capture the representative climate variability for the region and (2) that there are no statistically significant long-term trends affecting hydrologic controls, such as upstream water extraction, retention by dams, or any significant changes to watershed vegetation cover, land use, or climate. For viewing convenience, the stream flow values that are equal to or exceed 65% and 90% of the time, respectively, are summarized in Table 1 which is used for representing Figure 7.

Table 1. Monthly stream flows that correspond to 65% and 90% exceedance probabilities, Mtandika station, in m^3/s

Month	65%	90%
Jan	18.54	12.81
Feb	19.43	15.02
Mar	21.22	15.15
Apr	33.19	20.94
May	28.54	21.07
June	19.72	15.74
July	17.54	14.23
Aug	15.40	12.41
Sep	13.78	10.64
Oct	12.50	10.27
Nov	12.68	8.72
Dec	16.34	11.34

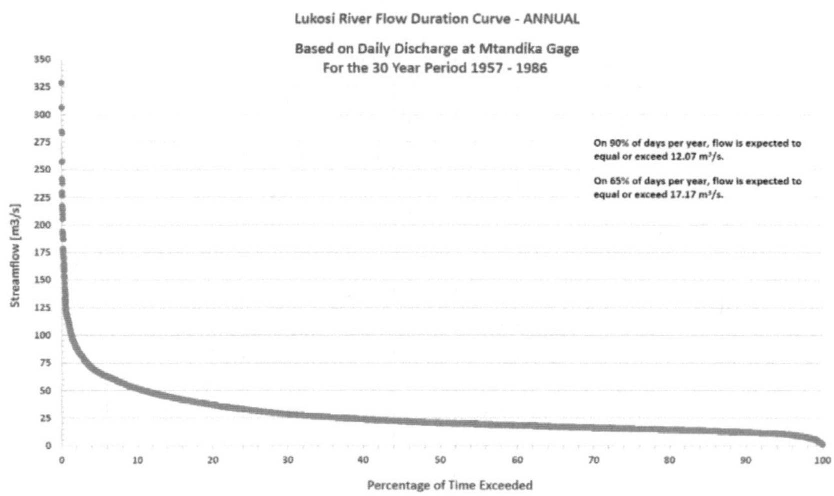

Figure 7. Lukosi River – Flow duration Curve for the period 1957 – 1987 (30 years)

3. Conceptual design criteria for standardization

Design criteria

1) fixed hydrostatic head scenarios of 50, 75, 100, 150 m,
2) annually variable water flow rate ranging between 3 to 10 m^3/s, and
3) installation of dual hydro turbines of identical model and capacity at each powerhouse.

Based on the design of each of the four powerhouses, the designer shall identify and specify appropriate types and configurations of hydro-turbine generating equipment and related electro-mechanical equipment.

Powerhouse Layout

There shall be provided four standardized powerhouse layout options. Layouts and drawings shall be provided for each one of the four standardized powerhouses. Each powerhouse layout shall include a control room, a room for auxiliary mechanical and electrical equipment, and an erection bay for the two hydro turbines. Each powerhouse design shall also include an appropriately sized, single transformer substation adjacent to each powerhouse to step up or down voltage as required for export to the grid or a mini-grid.

Steelworks and Penstocks

It has been developed a hydraulic grade-line spread sheet between data along selected reaches of the Lukosi River basin. Utilizing the grade-line, conceptual designs and fabrication considerations shall be provided for buried and above ground steel penstocks in three separate design scenarios to accommodate water flow rates of 3, 7, and 10 m^3/s.

Unit-of-length construction cost estimates shall be drawn up for the penstocks for each water flow rate scenario.

Diversion Weir and Intake Structure

Conceptual designs and construction details shall be provided for one or more alternatives appropriate to the Lukosi River basin for a concrete weir to divert river flow into a concrete RoR HEPP (run-of-river hydroelectric power plant) intake structure incorporating flow control gates, trash-racks, and sediment traps. These features shall be incorporated into digital drawings for use in the feasibility study. The vertical height of each diversion weir in these conceptual designs shall not exceed 5 m. Unit-of-area construction cost estimates shall be drawn up for the weir-face area together with an appropriate intake structure based on water flow rate scenarios of 3, 7, and 10 cubic meters per second.

Open Channel, Headraces or Canals

A conceptual design shall be developed for the construction of open channel headraces, or canals, which extend from the concrete diversion weir and intake structure to a head-works at the top of the penstocks. The conceptual design will be made with sufficient unit-of-length cost and performance estimations to compare the cost-benefit of open channel canals versus closed tubular conduits. The headrace canal designs shall include three separate design scenarios to accommodate water flow rates of 3, 7, and 10 cubic meters per second.

It was analyzed in a total of 12 variants, as in Table 2: 3 installed flow scenarios and for each of them, 4 head alternatives. It was chosen to have 3 scenarios due to similarities of civil works. For example, for the 3 m^3/s scenario, the intake, penstock or canal is the same. The difference lies in the thickness of penstock wall and the powerhouse equipment's area.

Table 2. Scenarios used in standardization project

Head [m]	Flow [m^3/s]	Scenario 1 3 m^3/s	Scenario 2 7 m^3/s	Scenario 3 10 m^3/s
50 m		Var 01	Var 02	Var 03
75 m		Var 04	Var 05	Var 06
100 m		Var 07	Var 08	Var 09
150 m		Var 10	Var 11	Var 12

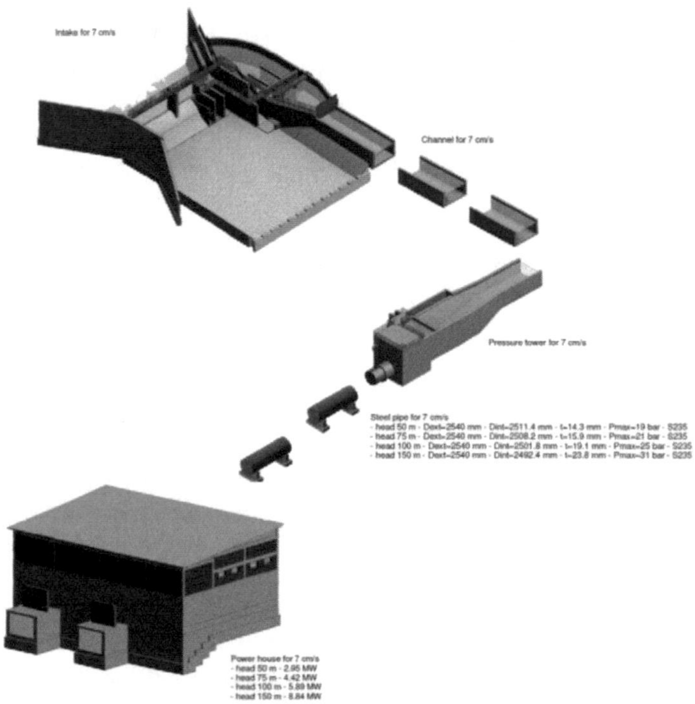

Figure 8. Scheme scenario for Lukosi River [1]

As it can be observed in the figure above, the scheme used for all scenarios is RoR (run-of-river) with weir/intake, desalting chambers, channel, fore bay, penstock and powerhouse [1].

4. Technical calculations [3], [4]

4.1. Weir/Intake and sand trap

Intake elements: Right bank recessed wall; Dam (spillway); Eco flow opening; Washing gate; Intake opening; Access channel; Sand trap; Overflow channel; Washing channel; Connection channel; Left bank recessed wall; Upstream concrete – stone slab; Energy dissipation basin; Gangway.

The calculation was done using the MathCAD software. An example is given in Figure 9.

The environmental flow was considered 10% of the mean flow. Since the space does not allow the presentation of the entire project's calculation, it being too large and complex, only a few examples shall be presented for intake. The results are shown in Table 3.

1. Intake opening dimensions

$Q_i = 3\ \frac{m^3}{s}$	installed flow
$h_o := 1.8 \cdot m$ $\quad b_o := 5.0 \cdot m$	chosen dimensions
$A := b_o \cdot h_o = 9\ m^2$	area of the opening
$d_{bars} := 2.5 \cdot cm$	rack bars diameter
$l_{bars} := h_o = 1.8\ m$	rack bars length
$o_{bars} := 5.5 \cdot cm$	distances betwen rack bar axis
$nr_{bars} := \frac{b_o}{o_{bars}} = 90.909$	numbers of bar
$A_{ef} := A - nr_{bars} \cdot d_{bars} \cdot l_{bars} = 4.909\ m^2$	active area of the opening
$V_{ef} := \frac{Q_i}{A_{ef}} = 0.611\ \frac{m}{s}$	incoming water speed
$V_{max} := 0.7 \cdot \frac{m}{s}$	maximum allowed speed
$A_{nec} := \frac{Q_i}{V_{max}} = 4.286\ m^2$	minimum area

Figure 9. Example of calculation for intake opening, scenario 1 [3], [4]

Table 3. Results of intakes calculations

	SCENARIO 1	SCENARIO 2	SCENARIO 3
Qmed [m³/s]	22.26	22.26	22.26
Qeco [m³/s]	2.22	2.22	2.22
Qmax [m³/s]	223.63	223.63	223.63
Rated (installed) flow [m³/s]	3.00	7.00	10.00
	Dam (spillway)		
Width [m]	14.25	14.25	14.25
Thickness [m]	0.40 – 4.70	0.40 – 4.70	0.40 – 4.70
	Eco flow opening		
Width [m]	1.00	1.00	1.00
Height [m]	0.72	0.72	0.72
	Washing gate		
Width [m]	1.80	1.80	1.80
Height [m]	2.80	3.30	3.30
	Intake opening		
Width [m]	5.00	8.00	11.50
Height [m]	1.80	2.30	2.30
	Water access channel into sand trap: Arc shape		
Length (interior) [m]	6.90	6.15	6.15

IOP Conf. Series: Materials Science and Engineering **163** (2017) 012028 doi:10.1088/1757-899X/163/1/012028

Length (exterior) [m]	12.55	14.40	18.8
Middle wall length [m]			5.5
Depth [m]	3.20	4.20	4.2
Depth (water) [m]	2.40	2.90	2.9
Width [m]	5.00	8.00	11.5
Flushing gate			
Width [m]	0.30	0.30	0.30
Length [m]	0.60	0.60	0.60
Flushing concrete step			
Height [m]	0.20	0.20	0.20
Width [m]	0.35 – 1.60	0.35 – 1.60	0.35 – 1.60
Sand trap basin:			
Number of rooms	2.00	3.00	4.00
Max retained particle	0.50	0.50	0.50
Length [m]	5.00	7.00	7.00
Height (total) [m]	3.60	4.35	4.35
Height (wet) [m]	2.80	3.35	3.35
Shape (top view) [m]	rectangular	Rectangular	rectangular
Room cross section [m]	trapezoidal	Trapezoidal	trapezoidal
Top width (one room) [m]	2.40	2.50	2.70
Bottom width (one room) [m]	0.80	1.10	1.30
Slope [%]	3%	3%	3%
Connection channel:			
Shape (top view)	trapezoidal	Trapezoidal	trapezoidal
Length [m]	2.75	5.50	9.80
Large base [m]	6.00	8.30	12.00
Small base [m]	4.00	4.40	5.00
Shape (cross section)	rectangular	Rectangular	rectangular
Height upstream (total) [m]	2.10	3.20	3.20
Height upstream (wet) [m]	1.60	2.20	2.20
Height downstream (total) [m]			3.50
Height downstream (wet) [m]			2.50
Inclination [%]			3%
Sand trap washing channel:			
Shape (top view)	Arc	Arc	Arc
Cross section	rectangular	Rectangular	rectangular
Width	2.50	2.50	2.50
Length	5.60	7.85	11.45
Height	1.00	0.95	0.95
Energy dissipation basin:			
Width	18.45	26.90	28.40
Length	20.50	27.00	27.00

IOP Conf. Series: Materials Science and Engineering **163** (2017) 012028 doi:10.1088/1757-899X/163/1/012028

	Gangway:		
Width	1.10	1.10	1.10
Length	20.70	20.70	20.70

3. Washing gate

$P_{gate} := 0.05\ m$ washing gate upstream step height

$b_{gate} := 1.8\ m$ washing gate width

$c_{gate} := 4.5\ m$ washing gate sidewalls length

$H_{apa} := h_o + 1\ m = 2.8\ m$ water height in gate

$g = 9.807\ \frac{m}{s^2}$

$m_q := \frac{0.3 + 0.55}{2} = 0.425$ usually flow coefficientflow coefficient (0.3 - 0.55)

$\sigma := 1$ drowning spillway coefficient

$\varepsilon := 1$ coefficient of contraction

$k := 1$ coefficient of skew

$$m_0 := \left(0.405 + \frac{0.0027}{H_{apa} \cdot \frac{1}{m}}\right) \cdot \left(1 + 0.55\left(\frac{H_{apa}}{H_{apa} + P_{gate}}\right)^2\right) = 0.621$$

$m_{gate} := m_0 \cdot \sigma \cdot \varepsilon \cdot k = 0.621$ flow coefficient calculated

$Q_{gate} := m_{gate} \cdot b_{gate} \cdot \sqrt{2 \cdot g} \cdot H_{apa}^{\frac{3}{2}} = 23.212\ \frac{m^3}{s}$ maximum flow for given H

Figure 10. Example of calculation for washing gate, scenario 1; [3], [4]

Since the river considered as reference is Lukosi, the intake dimensions are quite constant, but the openings and other items that are proportional with the rated flow are different.

4.2. Channel

The channel section that was considered is a rectangular section (to be easy calculated). An example for the calculation model is given in Figure 11 and the results are presented in Table 4.

13. Channel dimensions

$R := 0.8$ proposed hydraulic radius

$n := 0.014$ Manning's roughness coefficient

$y := 2.5 \cdot \sqrt{n} - 0.13 - 0.75 \cdot \sqrt{R} \cdot (\sqrt{n} - 0.10)$

$C := \frac{1}{n} \cdot R^{y} = 69.023$ Chezy coefficient

$i := \frac{0.1}{1000} = 0.0001$ slope of the channel

$V_d = 0.223 \frac{m}{s}$ velocity in sandtrap

$v_1 := C \cdot \sqrt{R \cdot i} = 0.617$ medium velocity in channel - Chezy

$v_2 := \frac{R^{\frac{2}{3}} \cdot i^{\frac{1}{2}}}{n} = 0.616$ medium velocity in channel - Bernoulli

$v_{med} := \frac{v_1 \cdot \frac{m}{s} + v_2 \frac{m}{s}}{2} = 0.616 \frac{m}{s}$

$Q_i = 3 \frac{m^3}{s}$ installed flow

$A_{ch\_min} := \frac{Q_i}{v_{med}} = 4.867\ m^2$ minimum channel area

$b_{ch} := 3.2\ m$ $h_{ch} := \frac{b_{ch}}{2} = 1.6\ m$ width of channel

$h_{ch} := 1.6\ m$ height of channel

$A_{ch} := b_{ch} \cdot h_{ch} = 5.12\ m^2$ channel area

$check_{channel\_area} := \text{if}\ (A_{ch\_min} < A_{ch}, \text{"Channel OK"}, \text{"Channel not OK"})$

$check_{channel\_area} = \text{"Channel OK"}$

$R_{ch} := \frac{A_{ch}}{b_{ch} + 2 \cdot h_{ch}} = 0.8\ m$ hydraulic radius

Figure 11. Channel calculation example [3], [4]

Table 4. Results of channels calculations

Channel		Option 1	Option 2	Option 3
Section	M/U	rectangular	Rectangular	rectangular
Width (outside)	m	4.00	5.20	5.80
Width (inside)	m	3.20	4.40	5.00
Depth (total)	m	2.10	2.70	3.00
Depth (wet)	m	1.60	2.20	2.50
Height	m	2.50	3.10	3.40
Inclination	‰	0.10	0.10	0.10
Walls thickness	cm	40.00	40.00	40.00
Base plate thickness	cm	40.00	40.00	40.00

4.3. Forebay

An example for the forebay calculation model is given in Figure 11 and the results are presented in Table 5.

14. Fore Bay calculation

$d := 0.5 \cdot mm$	minimum particle retained
$g := 9.80665 \cdot \frac{m}{s^2}$	gravitational acceleration
$\varphi_{sand} := 2000 \cdot \frac{kg}{m^3}$	density of particle (sand)
$\varphi_{water} := 998.23 \cdot \frac{kg}{m^3}$	density of water
$\mu_{water} := 0.001002 \cdot \frac{kg}{m \cdot s}$	viscosity of water
$V_t := g \cdot d^2 \cdot \frac{(\varphi_{sand} - \varphi_{water})}{18 \cdot \mu_{water}} = 0.136\ \frac{m}{s}$	settling speed
$h_{pt} := 3.5 \cdot m$	medium height of sandtrap between water level and slab
$b_{pt\_min} := \frac{h_{pt}}{1.25} = 2.8\ m$ $\quad b_{pt} := 4.2\ m$	width of sandtrap
$check_{bpt} := \mathbf{if}\ (b_{pt} > b_{ch}, \text{"OK"}, \text{"not OK"})$	$check_{bpt1} := \mathbf{if}\ (b_{pt} > b_{pt\_min}, \text{"OK"}, \text{"not OK"})$
$check_{bpt} = \text{"OK"}$	$check_{bpt1} = \text{"OK"}$
$A_{pt} := h_{pt} \cdot b_{pt} = 14.7\ m^2$	area of pt
$V_{pt} := \frac{Q_i}{A_{pt}} = 0.204\ \frac{m}{s}$	velocity in sandtrap of pt
$V_{maxpt} := 0.5\ \frac{m}{s}$	max velocity in sandtrap
$check_V := \mathbf{if}\ (V_{pt} < V_{maxd}, \text{"OK"}, \text{"not OK"})$	water speed in pressure tower has to be slower then the maximum allowed speed
$check_V = \text{"OK"}$	
$h_{sp} := 2.1\ m$	head of sand particle
$T_t := \frac{h_{sp}}{V_t} = 15.422\ s$	settling time

Figure 11. Forebay calculation example [3], [4]

Table 5. Results of forebays calculations

Fore Bay	M/U	Option 1	Option 2	Option 3
Concrete step				
Shape (top view)		arc with 4.7 radius	arc with 5.4 radius	arc with 7.50 radius
Height	m	1.40	1.40	1.40
Fine rack				
Inclination	deg	65	70	70
With	m	4.20	5.40	6.20
Height	m	2.00	2.50	2.50
Loading chamber				
Height	m	5.00	6.40	6.40
Height (wet)	m	4.50	5.90	5.90
Cross section		Trapezoidal	Trapezoidal	Trapezoidal
Width (top)	m	4.20	5.40	6.20
Width (base)	m	2.40	3.60	4.40
Inlet pipe				
DN (upstream)	m	2.00	2.80	2.80
DN (downstream)		1.60	2.50	2.50
Overpressure pipe		DN 300	DN 300	DN 300

4.4. Penstock

An example for the calculation model is given in Figures 12 - 13 and the results are presented in Table 6.

18. Pipe calculation - pressure tower to power house

$Q_i = 47550.967\ gpm$ — installed flow

$\rho_{water} = 62.317\ \frac{lb}{ft^3}$ — water density, lb/ft3

$v_{max} := \frac{48}{62.4^{0.33}} \frac{ft}{s} = 12.27\ \frac{ft}{s}$ — top speed $v_{max} = 3.74\ \frac{m}{s}$

$d_{min1} := \frac{0.73 \cdot (47550.967)^{0.5}}{(62.317)^{0.33}} \cdot in = 40.708\ in$ — minimum diameter v1

$d_{min2} := 0.321 \cdot \left(\frac{47550.967}{1^2}\right)^{0.434} \cdot in = 34.387\ in$ — minimum diameter v2

$d_{min} := \max(d_{min1}, d_{min2}) = 1.034\ m$ — minimum pipe diameter

$C_{HZ\_coef} := 150$ — Hazen Williams coefficient

$h_f := \left(42.7 \cdot \frac{47550.968}{150 \cdot 40.708^{2.63}}\right)^{1.852} \cdot ft = 0.647\ ft$ — frictition, ft H2O/100 ft

$h_f = 0.197\ m$ — frictition, m H2O/30.5 m

$D_{pipe} := 1.6\ m$ — pipe internal diameter - chosen

$check_D := \text{if}(D_{pipe} > d_{min}, \text{“OK”}, \text{“not OK”})$

$check_D = \text{“OK”}$

$H_{pipe} := \Delta H = 50\ m$ — aprox. gross head

$n := 0.012$ — Manning coeficient for steel pipe

Figure 12. Penstock calculation example (1/2)

$\rho_{water} = 998.23\ \frac{kg}{m^3}$ — water density

$\mu_{water} = 0.001002\ \frac{kg}{m \cdot s}$ — water dynamic viscosity

$v_{pipe} := \frac{Q_i}{\pi \cdot \left(\frac{D_{pipe}}{2}\right)^2} = 1.492\ \frac{m}{s}$ — velocity in pipe

$check_V := \text{if}(v_{pipe} > v_{max}, \text{“not OK”}, \text{“OK”})$

$check_D = \text{“OK”}$

$D_{pipe} = 62.992\ in$

$h_f := \left(42.7 \cdot \frac{47550.968}{150 \cdot 62.992^{2.63}}\right)^{1.852} \cdot ft = 0.077\ ft$ — frictition, ft H2O/100 ft

$h_f = 0.024\ m$ — frictition, m H2O/30.5 m

$R_{hp} := \frac{\pi \cdot \left(\frac{D_{pipe}}{2}\right)^2}{2 \cdot \pi \cdot \frac{D_{pipe}}{2}} = 0.4\ m$ — hydraulic radius

$R_e := \frac{\rho_{water} \cdot v_{pipe} \cdot R_{hp}}{\mu_{water}} = 594585.475$ — Reynold's number

$check_{laminar} := \text{if}(R_e < 2100, \text{“laminar flow”}, \text{“not laminar flow”})$

$check_{laminar} = \text{“not laminar flow”}$

$check_{transition} := \text{if}(2100 < R_e < 4000, \text{“transition flow”}, \text{“not transition flow”})$

$check_{transition} = \text{“not transition flow”}$

$check_{turbulent} := \text{if}(R_e > 4000, \text{“turbulent flow”}, \text{“not turbulent flow”})$

$check_{turbulent} = \text{“turbulent flow”}$

Figure 13. Penstock calculation example (2/2)

Table 6. Results of penstocks calculations

Penstock		Option 1	Option 2	Option 3
General:				
DN [mm]		DN 1600 mm	DN 2500 mm	DN 2500 mm
Distance between 2 saddles [m]		6.00 m	6.00 m	6.00 m
Wall thickness and material [mm]				
Head:	50 m	9.5 mm steel S235JR – max pressure PN 19	14.3 mm steel S235JR – max pressure PN 19	19.1 mm steel S235JR – max pressure PN 25
	75 m	11.9 mm steel S235JR – max pressure PN 24	15.9 mm steel S235JR – max pressure PN 21	22.2 mm steel S235JR – max pressure PN 29
	100 m	12.7 mm steel S235JR – max pressure PN 26	19.1 mm steel S235JR – max pressure PN 25	23.8 mm steel S235JR – max pressure PN 31
	150 m	15.9 mm steel S235JR – max pressure PN 32	23.8 mm steel S235JR – max pressure PN 31	23.8 mm steel S275JR – max pressure PN 36

Saddles			
Foundation			
Width	1.00 m	1.40 m	1.40 m
Length	3.00 m	3.60 m	3.60 m
Height	0.50 m	0.50 m	0.50 m
Pipe support			
Width	0.50 m	0.50 m	0.50 m
Length	2.00 m	2.90 m	2.90 m
Height	0.50 m – 0.90 m	0.50 m – 1.15 m	0.50 m – 1.15 m

4.5. Powerhouse

The powerhouse results are presented in Table 7.

Table 7. Results of powerhouses calculations

Option 1	Option 2	Option 3
Transformer room:	Transformer room:	Transformer room:
Width: 2.65 m	Width: 2.65 m	Width: 2.65 m
Length: 4.00 m	Length: 4.00 m	Length: 4.00 m
Height: 4.85 m	Height: 4.85 m	Height: 4.85 m
Channel (for oil spills)	Channel (for oil spills)	Channel (for oil spills)
Width: 0.50 m	Width: 0.50 m	Width: 0.50 m
Depth: 1.10 m	Depth: 1.10 m	Depth: 1.10 m
Length: 1.50 m	Length: 1.50 m	Length: 1.50 m
MV room:	MV room:	MV room:
Width: 2.00 m	Width: 2.00 m	Width: 2.00 m
Length: 4.00 m	Length: 4.00 m	Length: 4.00 m
Height: 3.75 m	Height: 3.75 m	Height: 3.75 m
Water exhaust channels:	Water exhaust channels:	Water exhaust channels:
Number: 2	Number: 2	Number: 2
Width: 3.40 m	Width: 4.55 m	Width: 4.55 m
Height (under base plate): 1.30 m	Height (under base plate): 3.10 m	Height (under base plate): 3.10 m
Inclination: 12‰	Inclination: 12‰	Inclination: 12‰
Concrete block:	Concrete block:	Concrete block:
Width: 3.00 m	Width: 4.00 m	Width: 4.00 m
Length: 11.95 m	Length: 3.00 m	Length: 3.00 m
Height: 4.00 m	Height: 5.50 m	Height: 5.50 m
Material: reinforced concrete	Material: reinforced concrete	Material: reinforced concrete

5. Conclusions [5]

It is difficult to give a response to the question: "Can the small hydropower plants be standardized?"

As it can be seen after this study, the hydropower plants are hard to be standardized, because the structures are variables of many external non-standardized factors; because the "fuel" used is variable and different from one case to the other and the locations' conditions are variable. It is not the fact that it can be said "here I will choose to install a hydropower plant". The hydropower plant needs to be installed in the best suitable position that will give the best returns for the investment made. Of course, here is not the case of hydropower plants with the installed capacity lower than 250 kW, for which there are some standardized options offered by the electromechanical providers (so called "package hydropower plants"). But even in this case the efficiencies and also the lifetime of these structures are not very clearly presented by the manufacturers.

Any investment in plants based on renewable energy sources, especially investments in hydropower plants which are much higher than in other sources, needs to lead to a power output that will assure the right returns. For this reason the best efficiency, both financial and technical, must be reached and this is not possible with full small hydropower development standardization.

References

[1] Stematiu D 2008 *Hydropower Developments*, Conspress Publishing House, Bucharest, Romania
[2] Giurma I 2004 *Special hydrology*, Politehnium Press, Iaşi, Romania
[3] Dumitrescu D and Pop R 1969 *Civil engineer's manual* (in Romanian: *Manualul Inginerului Hidrotehnician*), Vol. I and II, Technical Publishing House, Bucharest, Romania
[4] Popovici A and Suprovici P 1986 *Hydraulic structures. Applications*, Technical University of Civil Engineering of Bucharest, Romania
[5] Bolaert E 2004 *Standardization of civil engineering works of small hydropower plants*, Ecole Polytechnique Federale de Laussane, Switzerland

IOP Conf. Series: Materials Science and Engineering **163** (2017) 012029 doi:10.1088/1757-899X/163/1/012029

Advanced concepts and solutions for geothermal heating applied in Oradea, Romania

C Antal[1], F Popa[2], M Mos[3], D Tigan[3], B Popa[4] and V Muresan[5]

[1]University of Oradea, Department of Power Engineering, Oradea, Romania
[2]ISPH Project Development, Bucharest, Romania
[3]Oradea City Hall, Department of Internationally Financed Projects, Oradea, Romania
[4]University Politehnica of Bucharest, Department of Hydraulics, Hydraulic Machineries and Environmental Engineering, Bucharest, Romania
[5]SC PROARCOR SRL, Romania

E-mail: bogdan.popa@upb.ro

Abstract. Approximately 70% of the total population of Oradea benefits from centralized heating, about 55,000 apartments and 159,000 inhabitants are connected.
The heating system of Oradea consists of: sources of thermal energy production (Combined heat and power (CHP) I Oradea and geothermal water heating plants); a transport network of heat; heat distribution network for heating and domestic hot water; substations, most of them equipped with worn and obsolete equipment. Recently, only a few heat exchangers were rehabilitated and electric valves were installed to control the water flow.
After heat extraction, geothermal chilled waters from the Oradea area are: discharged into the sewer system of the city, paying a fee to the local water company which manages the city's sewers; discharged into the small river Peta; or re-injected into the reservoir. In order to ensure environmental protection and a sustainable energy development in Oradea, renewable sources of energy have been promoted in recent years. In this respect, the creation of a new well for geothermal water re-injection into the reservoir limits any accidental thermal pollution of the environment, while ensuring the conservation properties of the aquifer by recharging with geothermal chilled water.
The paper presents the achievements of such a project whose aim is to replace thermal energy obtained from coal with geothermal heating. The novelty consists in the fact that within the substation we will replace old heat exchangers, circulation pumps and valves with fully automated substations operating in parallel on both a geothermal system and on a primary heating system of a thermal plant.

1. Introduction

Among the renewable energies, a special place was and is still held by geothermal energy. Introducing the geothermal energy as a viable alternative to fossil fuels has been also supported, besides the technical and economic advantages they present, by the fact that 80 countries have significant geothermal resources. Geothermal energy is continuously available regardless of season and weather conditions.

Use of geothermal energy is divided into two categories: direct use and indirect use for conversion into electricity.

IOP Conf. Series: Materials Science and Engineering **163** (2017) 012029 doi:10.1088/1757-899X/163/1/012029

Direct use of geothermal energy means the use of the geothermal fluid thermal energy by heat transfer directly to a user or through another fluid. In general, areas of geothermal energy direct use are divided into 4 groups, namely: space heating and domestic hot water, agricultural uses (greenhouses, aquaculture, fisheries), balneology and industrial uses, [1].

Given the economic issues arising for space heating using classical fuel and the current energy crisis, there were developed many applications using geothermal water for space heating and domestic hot water. Such use of geothermal water is achieved in approximately 40 countries.

In 2007 the Romanian government adopted the "Romania's energy strategy for the period 2007-2020" which greatly emphasizes the development of renewable energy sources (including the geothermal one). The strategy is in line with the principles and EU directives (one of the objectives being the 20% renewable energy contribution to the EU 2020), [2].

Research of the hydro-geothermal deposits began in Romania in the years 1962-1965, by drilling the first well in the Western Plain. Up to now, within the geological research program, about 250 geothermal wells have been drilled and have given indications, wells performed by Transgex and Foradex companies.

The total installed capacity is of approximately 480 MWt, of which, today, 246 MWt is effectively used by exploiting 96 geothermal wells that are producing water with temperatures between 40 and 115 °C.

Main uses in Romania are: space heating and domestic hot water, balneology, greenhouse heating, wood drying, pasteurization of milk, flax and hemp smelting, intensive fish growth etc.

Overall energy savings annually achieved are of over 35000 toe. In balneal-therapeutic purposes there are used approximately 35 wells (with a total flow of 360 l/s, with water temperatures between 38-65 °C) providing the functioning of 16 spa resorts.

The repartition of the geothermal resource in Romania is presented in Figure 1, where in red are represented geothermal systems investigated through drilling and in mauve are represented perspective areas.

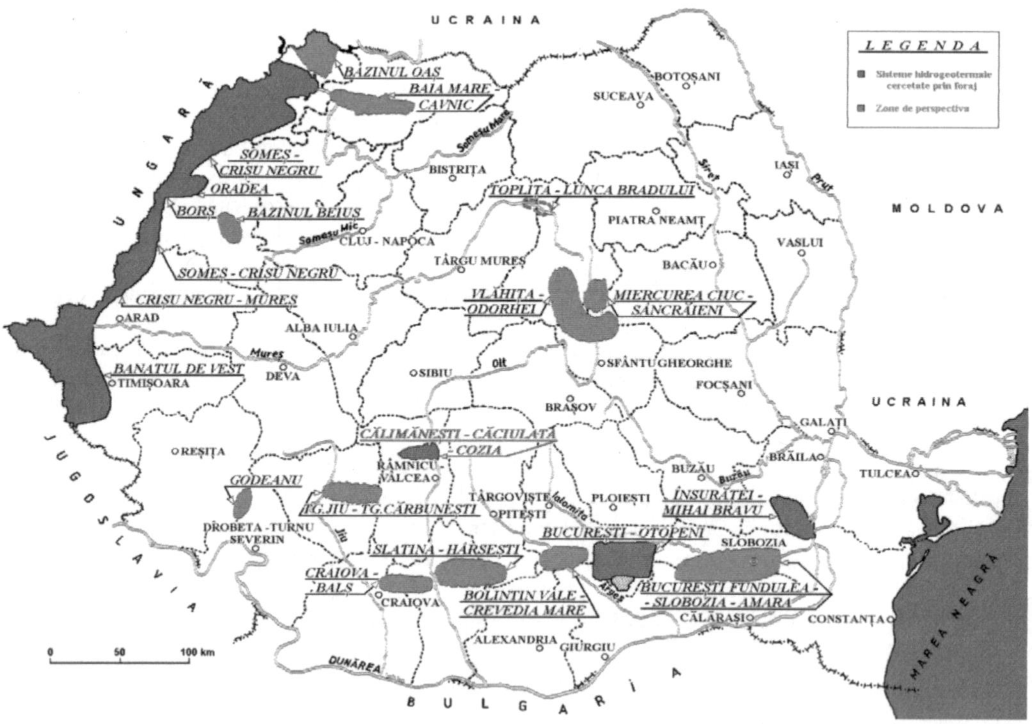

Figure 1. Repartition of geothermal resources in Romania

The largest geothermal reservoir in Romania is located in Bihor county, in the Oradea city. The installed capacity is currently about 30 MWt (13 production wells and an reinjection well), but it can be nearly tripled by generalizing production by pumping and reinjection in doublet systems (well production - heat exchangers - reinjection well). Geothermal potential in Oradea is approx. 220,000 Gcal/year, of which the average is used in the 100,000 Gcal/year.

Further development of geothermal water extraction from Oradea geothermal reservoir requires switching to pumping and water reinjection of geothermal waste heat. This can be achieved in the system with doublets, i.e. for each production well a reinjection well has to be installed (with the necessary equipment for reinjection), thus ensuring the maintenance of constant pressure and flow.

Usually, for increasing the production capacity, a deep well pump and relevant equipment have to be installed. By installing such a pump, the flow rate will increase. Geothermal water temperature will increase with higher flow rate.

Usually, the geothermal water is then conveyed to a tank for storing and degassing. Then, through a pump circulation, the geothermal water is transported to the heat station using insulated steel pipes. Here, through plate heat exchangers, the geothermal water is used to heat the water for space heating and to prepare the domestic hot water. After the heat is transferred, the waste geothermal water is discharged into the reinjection well, [3].

In order to automate an industrial process (geothermal complex within the project, in this case), the human operator must be replaced by an automatic control system that can turn, adjust and stop a process, capable of measuring variables in the system in order to obtain the needed output parameters. These objectives are achieved using a control system based on PLC (Programmable Logic Controller). Automation software, specific to each application, is implemented on the PLC using a PC [4].

To achieve fast and efficient links between the human operator and PLC, it is used a graphical interface that allows both permanent monitoring of system parameters and data storage and the possibility for operator to intervene fast and comfortable, if needed. The graphical interface allows the representation of the geothermal complex elements on the computer screen (temperature, pressure, flow, actuating valves, pumps, pipes, heat exchangers etc.). These charts are inactive but next to the transducers and actuators, there are small windows that give the momentary value of the parameters. These values are displayed in real time.

This system consists of transducers and actuators from the geothermal complex, programmable logic controller PLC, graphical interface and links between them, which is called SCADA system (Supervisory Control and Data Acquisition), is used for the geothermal complex in the project presented in this paper.

2. Advances in geothermal heating applied in Oradea

The subject of this paper was the object of a project financed from Icelandic funds called: *Best use of geothermal energy to produce heat for consumers connected to Substation PT 902 and reinjection of used geothermal water in thermal reservoir*, [5].

The justification of the need of this project starts from the fact that approximately 70% of the population from Oradea benefit from centralized heating, it means about 55,000 apartments and about 159,000 inhabitants connected.

Consumers connected to the centralized heating supply can be structured as follows:

- households: housing associations, apartments and individual houses;
- social-cultural institutions and businesses;
- hospitals, kindergartens, nurseries, schools;
- hotels, banks, shops.

The heating system of Oradea consists of:

- sources of thermal energy production: CHP I Oradea and geothermal water heating plants [2];
- transport network of heat (about 80 km);
- heat distribution network (about 369 km for heating and 177 km for domestic hot water DHW);
- substations: about 150 of which a large part are equipped with worn and obsolete equipment.

Recently, only a few heat exchangers were rehabilitated and electric valves were installed to control the water flow.

Due to the continuous increase in fuel prices, solutions must be found to best solve the problem of ensuring energy for space heating and for the preparation of domestic hot water to the inhabitants of Oradea in terms of cost and affordability. Oradea has a defined strategy for heating through a Master plan approved by City Council in 2010. It includes, besides the renovation of the combined heat and power plant using gas as fuel, renovation of the hot water transportation system and increase of geothermal energy intake.

In order to ensure environmental protection and a sustainable energy development in Oradea, renewable energy sources have been promoted in recent years. In this respect, the creation of a new well for geothermal water re-injection into the reservoir limits any accidental thermal pollution of the environment, while ensuring the conservation properties of the aquifer by recharging with geothermal chilled water, Figure 2.

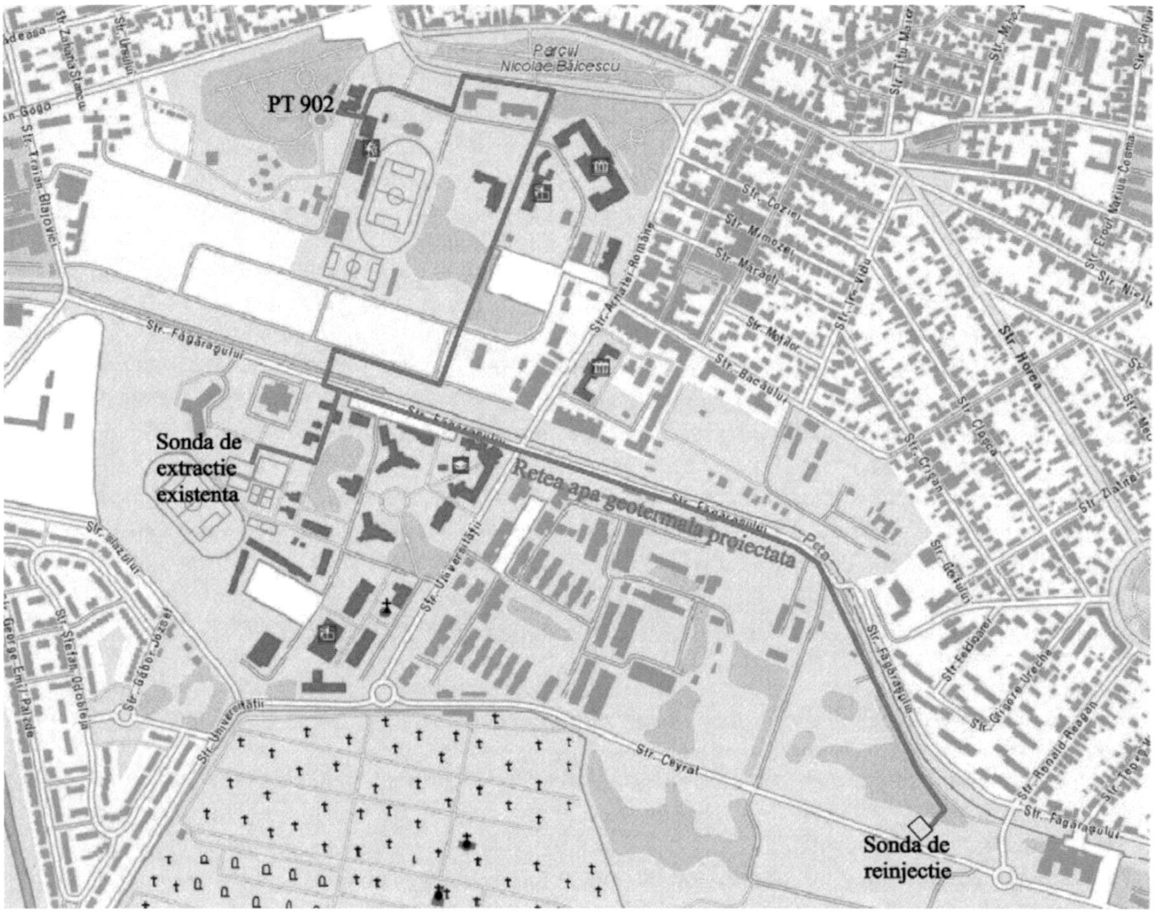

Figure 2. Designed network for geothermal energy use

After the heat extraction, the geothermal chilled water from Oradea area is:

- discharged into the sewer system of the city, for a fee paid to the local water company which manages the city's sewers;
- discharged into the small river Peta;
- or they are re-injected into the reservoir.

Peta is a thermal small river which springs from Baile 1 Mai and has a similar mineralization with the geothermal waters. For water that is discharged into the Peta small river, a fee is paid to "Romanian Waters" National Administration (ANAR). Geothermal chilled water from a single well only is re-injected into the reservoir. Nufărul geothermal plant is using geothermal water in the doublet system. Geothermal water is operated through well 4797 and after heat extraction is re-injected in well 4081. Although geothermal chilled water temperature comply with the standard temperatures at discharge, accidental situations may occur where the water temperature is higher.

The objectives of the project were:

- better use of geothermal energy sources available in the deposit of Oradea for the production of heat;
- improving environmental quality;
- reducing emissions of greenhouse gases;
- efficiency and rational use of primary energy resources.

The project's aim is to replace the thermal energy obtained from the coal with the geothermal heating obtained by the drilling well 4796 from the University of Oradea. The novelty is the fact that within the substation there will be replaced old heat exchangers, circulation pumps and valves with fully automated substations operating in parallel on both geothermal system and on primary heating system of CHP.

By implementing this project the following are expected to be achieved:

- the reduction of fossil fuel consumption by using geothermal energy;
- the reduction the cost of heat production for district heating;
- sustainable exploitation of geothermal water resources and providing the water pressure in the reservoir by injecting used geothermal water;
- increase of heating comfort for population and public institutions, consumers of the substation;
- reducing of CO_2 emissions by about 1,800 tons per year.

Regarding the project outputs, when completed, the project will materialize in rehabilitation and/or achievements of: technological equipment, networks, injection new well and re-injection station.

The technological equipment of PT902 substation will be provided with 2 fully automated substations operating in parallel on both geothermal system and primary heating system of CHP. The modules are provided in 2 stages (plate heat exchangers, heat circulation pumps in the secondary circuit, filling station - added). The installed power in substation is 2000 kW for heating and 800 kW for hot water.

Networks will be supplemented with geothermal water transport pipeline from 4796 well to PT 902 with diameter 150/250 and a total length of approximately 1,222 m and a transport pipeline for geothermal chilled water from PT 902 to the well re-injection, of polyethylene PE 160/200 with a total length of approximately 2,214 m.

The injection new well consists in the fact that all the extracted geothermal water will be re-injected, respectively 45 l/s, the depth of re-injection drilling being 2,900 m. Geothermal chilled water from substation PT 902 will be collected and transported via a PE 160 polyethylene pipe to the connecting pit of the geothermal chilled water to the discharge of waste water Oradea University substation, from where it will be transported to the re-injection probe through a PE 200 polyethylene pipe.

Works to the re-injection station implies: the purchase and installation of a 50 cubic meters tank; the purchase and installation of a pumping group with the installed flow of 45 l/s and the pumping height of 10 bars; fencing of re-injection station 15 x 15 linear meters.

It was estimated that the emissions of CO_2 will be reduced annually with 1,779 tons/year.

In terms of the thermal energy production, PT 902 geothermal substation will produce energy of 3,000 Gcal (3498 MW) per heating season (it means 172 days = 4128 hours) plus 150 Gcal (175 MW) per season, heating for hot water consumption.

The changes of the network scheme can be observed on Figures 3 to 6 and the legends are presented in Figures 7 and 8.

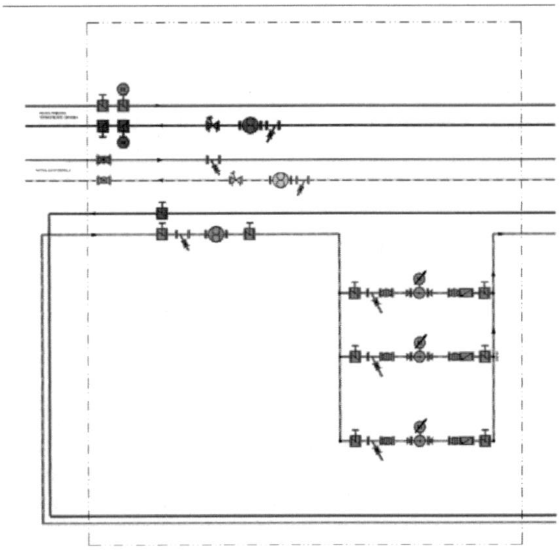

Figure 3. Pumping module

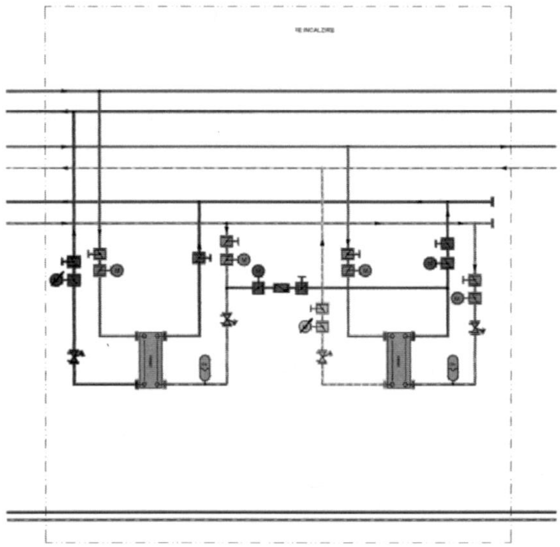

Figure 4. Heat exchanger module

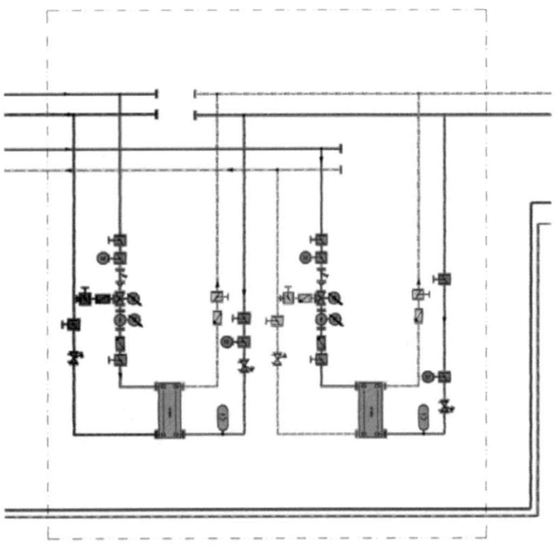

Figure 5. Hot water exchanger module

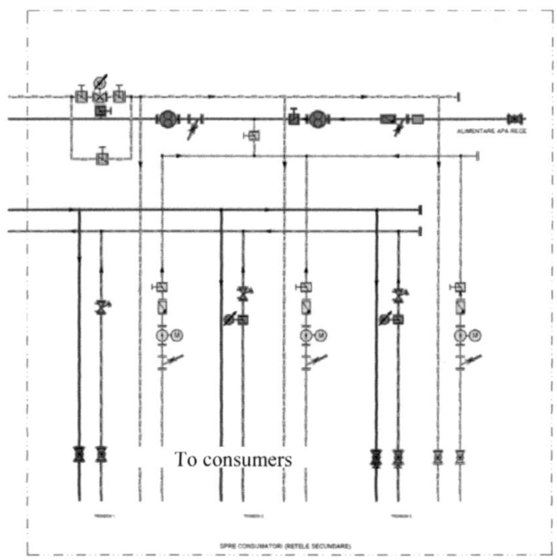

Figure 6. Distribution module

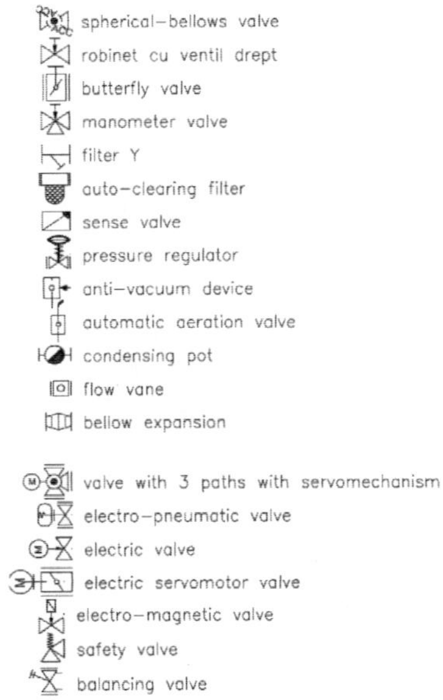

Figure 7. Symbols legend

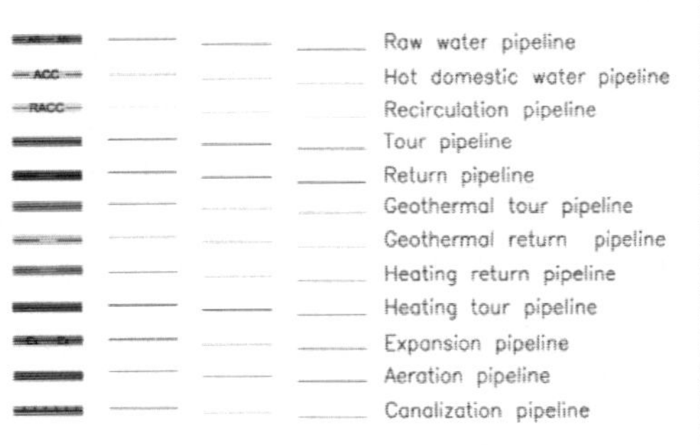

Figure 8. Pipelines legend

The direct beneficiaries of the project are:

- consumers of Substation PT 902 , respectively about 75 people living in the 52 buildings located on Matei Basarab, Suceava, Rimanoczy Kalman streets and buildings belonging to the Sports High School;
- the operator of the heating system as well as the costs of exploitation, production and transport of heat will be reduced by upgrading the equipment of PT 902. Indirect beneficiaries are residents of Oradea, as pollutant emissions (dust, CO_2, NOx) are reduced.

3. Conclusions

One of the advanced concepts presented in this paper could be considered the strategy of the use of primary agent, either the geothermal water or the hot water supplied by CHP I Oradea; also the use of heat exchangers separately for heating and hot domestic water, with the possibility to function at the same time in case of damage or interventions at heat exchangers.

Retrofitting of the PT 902 district heating substation permits a dual functioning which ensures thermal agent and hot water for consumers from two distinct sources: primary thermal agent produced by CHP I Oradea and geothermal hot water from 4797 geothermal well.

The new district heating substation consists of two heat exchangers units. The first unit, composed of two heat exchangers with an installed thermal power of 2 MW each, will produce secondary thermal agent for the consumers. The second unit is composed of two heat exchangers with an installed thermal power of 800 kW and will assure the production of domestic hot water.

Using two heat exchangers for hot water and secondary thermal agent production, provides greater security of supply of heat and domestic hot water. If one of the heat exchangers is in maintenance or a fault appears, the second heat exchanger can supply consumers with heat. The change between the two heat exchangers is made through a bypass.

Another advantage of the substation that increases security of supply is provided by the two primary agents used. If the geothermal well is in maintenance or a fault appears that disconnects the well, heat and domestic hot water will be prepared by using primary agent produced by CHP I Oradea. Also, if the supply of primary agent from CHP I Oradea is stopped, heat and hot water will be produced by using geothermal water from 4797 well.

Simultaneously, using pairs of heat exchanger for heat and domestic hot water production and two primary agents – geothermal water and primary agent from CHP I Oradea – ensure a greater flexibility in operation:

- each primary agent can be used for only one module. For example heat can be produced using primary agent form CHP I Oradea, and hot water can be produced using geothermal hot water;
- the installed capacity of the substation can be doubled by using all 4 heat exchangers.

Acknowledgements
The project "*Best use of geothermal energy to produce heat for consumers connected to Substation PT 902 and reinjection of used geothermal water in thermal reservoir*" was financed by Orkustofnun, OS – National Energy Authority in Iceland through Environment Fund Administration in the framework of the program RO06 Renewable Energy RONDINE.

References

[1] Dickson M H and Fanelli M 2003 *Geothermal energy: utilization and technology*, UNESCO Publishing, Renewable Energies series, Paris

[2] Bendea C, Antal C and Rosca M 2015 *Geothermal Energy in Romania: Country Update 2010-2014*, Proceedings World Geothermal Congress 2015, Melbourne, Australia, 19-25 April

[3] Maghiar T, Antal C and Cohut I 1998 Cascaded use of the geothermal water at the University of Oradea. The SCADA system, *Annals of University of Oradea, Power Engineering issue (in Romanian: Fascicola de Energetică)* **III** 133-140

[4] Zmaranda D 1995 *Automatic control and monitoring system for the district heating system at the University of Oradea*, United Nations University, Geothermal Training Programme, Iceland, report 16, 26 pp

[5] Presentation of the project "*Best use of geothermal energy to produce heat for consumers connected to Substation PT 902 and reinjection of used geothermal water in thermal reservoir*" financed by Icelandic Geothermal Engineering Ltd. (IGE), Háaleitisbraut 68, 103 Reykjavik, Iceland, Reg. no. 610995-2339, VAT no. 47 995, represented by Árni Gunnarsson, Managing Director

IOP Conf. Series: Materials Science and Engineering **163** (2017) 012030 doi:10.1088/1757-899X/163/1/012030

Vibration analysis of a hydro generator for different operating regimes

C Haţiegan[1], I Pădureanu[2], M Jurcu[2], M D Nedeloni[1], C O Hamat[2], C P Chioncel[1], S Trocaru[3], O Vasile[4], O Bădescu[5], D Micliuc[1], L (Filip) Nedeloni[2], A Băra[2] and L (Barboni) Haţiegan[2]

[1]Eftimie Murgu University of Resita, Department of Electric Engineering and Informatics, Traian Vuia str., no. 1-4, 320085 Resita, Romania
[2]Eftimie Murgu University of Resita, Department of Mechanical and Materials Engineering, Traian Vuia str., no. 1-4, 320085 Resita, Romania
[3]University of Bucharest, Faculty of Physics, Atomiştilor str., no. 405, 077125, Bucuresti, Romania
[4]Politehnica University of Bucharest, Department of Mechanics, Splaiul Independentei str., no. 313, 060042, Bucuresti, Romania
[5]Babeş-Bolyai University, Department of Educational Science, Sindicatelor str., no. 7, 400015, Cluj-Napoca, Romania

E-mail: c.hatiegan@uem.ro

Abstract. Based on experimental measurements, this paper presents the vibration analysis of a hydro generator that equips a Kaplan hydraulic turbine of a Hydropower plant in Romania. This analysis means vibrations measurement to different operating regimes of the hydro generator respectively before installing it and into operation, namely putting off load mode (unexcited and excited) respectively putting on load mode. By comparing, through the experimental results obtained before and after the operation of hydro aggregates are observed vibrations improvements.

1. Introduction

The hydro-generator i.e. hydraulic turbines are the main active components from a Hydro Power Plant. The most used hydraulic turbines of a Hydro Power Plant are Kaplan, Francis and Pelton turbines [1], [2]. For Francis and Kaplan turbines are recorded, results on the rotor-stator interaction [3] and vibration analysis for different operating modes of hydro aggregates [4], [5].

This paper presents the results of measurements on experimental vibration analysis of a hydro-generator operating a Hydro Power Plant in Romania. This Hydro Power Plant is equipped with vertical Kaplan turbines respectively synchronous generators. Due to the large number of hours of operation, its rehabilitation is needed both in terms of mechanical and electrical point of view [6], [7]. It is very important that before dismantling hydro aggregates for rehabilitation, to evaluate the actual performance of operation of hydro aggregates in terms of mechanical and electrical, and will settle all the necessary measures in terms of constructive and functional in order to bring parameters nominal hydro aggregates respectively.

From a mechanical point of view when checking vibrations, desired bringing set points of hydro aggregates and to improve its operation and in terms of vibration [8], which over time can lead to big problems in the functioning hydro-generators, have done research on experimental vibration rotor analysis vibration and temperature generator stator core [9] hydro aggregates vibration monitoring program using LabVIEW [10], namely research on vibration reduction [11]. Also, by measuring the vibration of hydro aggregates mechanical imbalance was followed, centering and focusing magnetic [12-14]. The signals were then analyzed using saved also LabVIEW software [15].

2. Used equipment and methods

For measuring the active power, tensions respectively currents across generator was used analyzer electrical parameters and process parameters that type VPA323 a laptop for data processing, as in Figure 1.

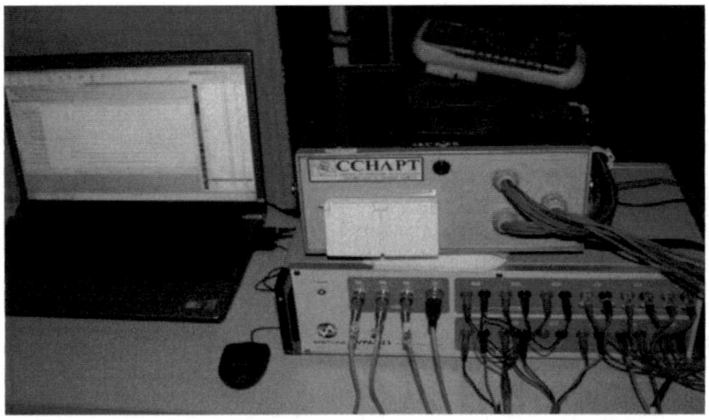

Figure 1. Parameter analyzer of the VPA323 type

The revolution speed was measured by a laser speed, the type QS30LDQ (Figure 2), and the vibration in the bearing was measured three vibration sensors mounted on the top bearing of the generator as in Figure 3, with the sensitivity 533.3 mV / g, 546.1 mV / g or 497.5 mV / g.

Figure 2. QS30LDQ laser speed

Figure 3. Vibrations sensor

The vibration analysis of the hydro aggregates through experimental checkups and measurements before the dismantling hydro aggregates supposed to repair and putting into service.

For the two types of measurements, hydro aggregates requested various operating modes namely: idle mode (unexcited and excited) respectively regime under load.

At these operating modes using LabVIEW software for all experimental measurements were recorded the following sizes: n rotor speed; higher radial thrust bearing vibration in the horizontal direction + x VLRAG_X; radial thrust bearing vibration higher radial + y V_{LRAG_Y}; lower radial bearing vibration direction + x V_{LAG_X}.

Subsequently, the signals were analyzed taken in situations considered using LabVIEW software. To view, it was performed a Fourier transform in order to highlight the range of variation of speed (amplitude) in that order harmonic frequency.

3. Measurement before dismantling the hydro-generator

The measurement results starting in transitional and steady unexcited before the dismantling of hydro-generator to repair (Figure 4) were analyzed using the LabVIEW program.

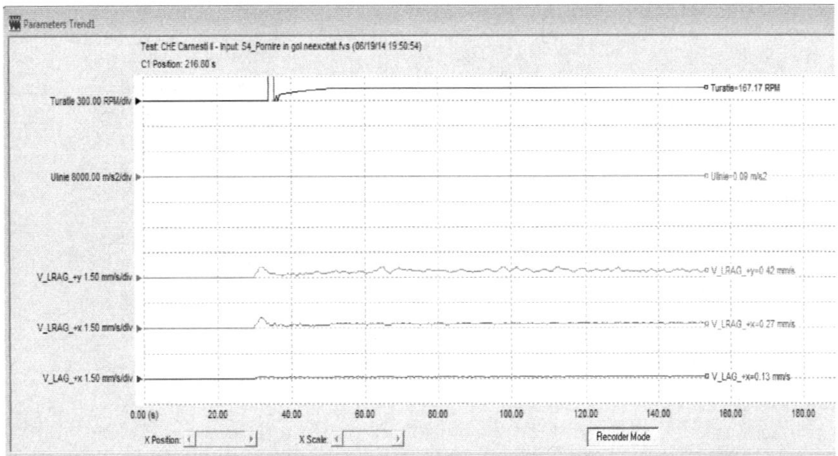

Figure 4. The variation of the measured parameters in transitory starting mode and in stable unexcited functioning

The values of the parameters measured in the starting phase before the dismantling the hydro-generator to repair his regime stabilized idle unexcited are presented in Table 1, and the values of the parameters measured at steady idle unexcited in Table 2.

Table 1. Parameters registered during the transitory starting mode

Time	Revolution	Vibrations		
t	N	$v_{LRAG\ x}$	$v_{LRAG\ y}$	$v_{LAG\ x}$
(s)	(RPM)	(mm/s)	(mm/s)	(mm/s)
0	0	0.33	0.35	0.11
1	71.67	0.22	0.15	0.08
2	86.73	0.17	0.19	0.09
3	97.09	0.29	0.18	0.08
4	110.8	0.25	0.26	0.13
5	110.8	0.25	0.26	0.13
6	128.28	0.26	0.22	0.11
7	131.71	0.31	0.21	0.11
8	141.81	0.3	0.17	0.08
9	150.24	0.28	0.21	0.1
10	158.23	0.34	0.23	0.09
11	162.47	0.36	0.25	0.13
12	167.18	0.4	0.31	0.1
13	167.23	0.3	0.28	0.13
Max. Value	167.23	0.4	0.35	0.13

Table 2. Measured parameters in a stable unexcited idle functioning mode

Time	Revolution	Vibrations		
t	N	$V_{LRAG\ x}$	$V_{LRAG\ y}$	$V_{LAG\ x}$
(s)	(RPM)	(mm/s)	(mm/s)	(mm/s)
1	167.2	0.65	0.28	0.14
2	167.2	0.68	0.29	0.13
3	167.2	0.68	0.29	0.13
4	167.2	0.67	0.31	0.15
5	167.19	0.65	0.32	0.15
6	167.19	0.61	0.31	0.13
7	167.19	0.57	0.3	0.14
8	167.2	0.53	0.29	0.13
9	167.21	0.49	0.27	0.12
10	167.21	0.44	0.25	0.12
11	167.22	0.38	0.26	0.12
12	167.22	0.38	0.26	0.12
13	167.22	0.42	0.26	0.12
14	167.23	0.46	0.3	0.12
V. Average	167.21	0.544	0.285	0.13

The vibration spectrum of steady unexcited measured parameters is shown in Figure 5, i.e. the spectrum of variation of speed (amplitude) in that order harmonic frequency.

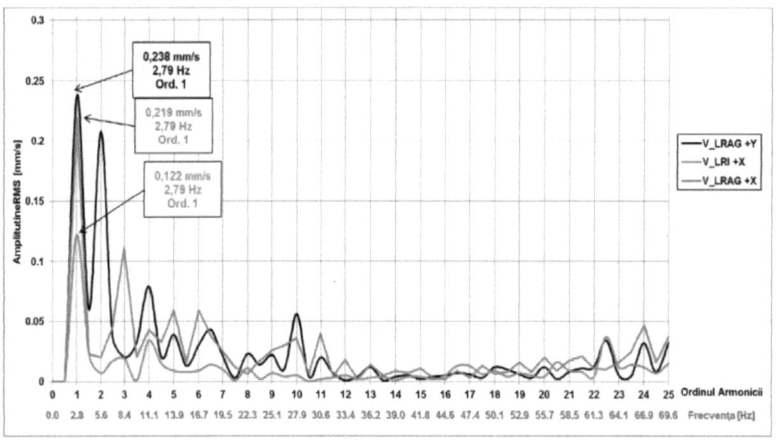

Figure 5. The vibrations spectrum of the measured parameters in stable unexcited mode

From Figure 5 is observed that maximum value is obtained for harmonic vibrations of order 1 axial- radial bearing higher along the horizontal direction + x, bearing higher axial-radial radial bearing lower respectively + y or + x direction.

It also notes that under unexcited idling vibration occurs due to sharp flow regimes and due to offsetting.

The regime idle excited, connecting the excitation voltage generator determines a nominal voltage stator, recorded the following parameters: rotor speed n, vibration thrust bearing radial upper horizontal direction + x V_{LRAG_X}, vibration thrust bearing radial higher radial + y V_{LRAG_Y}, lower radial bearing vibration direction + x V_{LAG_X}.

Tables 3 and 4 present the values of the parameters measured at steady idle excited that variation measured quantities during transitory exciting process.

Table 3. Measured parameters during the stable excited idle functioning mode

Time	Revolution	Vibrations		
t	n	$V_{LRAG\ x}$	$V_{LRAG\ y}$	$V_{LAG\ x}$
(s)	(RPM)	(mm/s)	(mm/s)	(mm/s)
1	167.2	0.82	0.8	0.31
2	167.2	0.82	0.8	0.31
3	167.2	0.82	0.8	0.31
4	167.2	0.81	0.8	0.33
5	167.19	0.81	0.8	0.34
6	167.19	0.81	0.8	0.34
7	167.19	0.81	0.8	0.34
8	167.2	0.81	0.81	0.35
9	167.21	0.81	0.81	0.35
V. Average	167.20	0.81	0.80	0.33

Table 4. The variation of the measured values during the transitory exciting process

Time	Revolution	Vibrations		
t	n	$V_{LRAG\ x}$	$V_{LRAG\ y}$	$V_{LAG\ x}$
(s)	(RPM)	(mm/s)	(mm/s)	(mm/s)
1	167.2	0.4	0.31	0.11
2	167.2	0.39	0.37	0.15
3	167.2	0.37	0.31	0.13
4	167.2	0.39	0.39	0.14
5	167.19	0.49	0.52	0.17
6	167.19	0.63	0.67	0.24
7	167.19	0.7	0.74	0.28
8	167.2	0.74	0.77	0.29
9	167.21	0.78	0.73	0.29
10	167.21	0.73	0.77	0.3

In Figures 6 and 7 there are plotted variation of measured parameters in regime starting and idling excited stabilized waveform regime that stabilized idle excited.

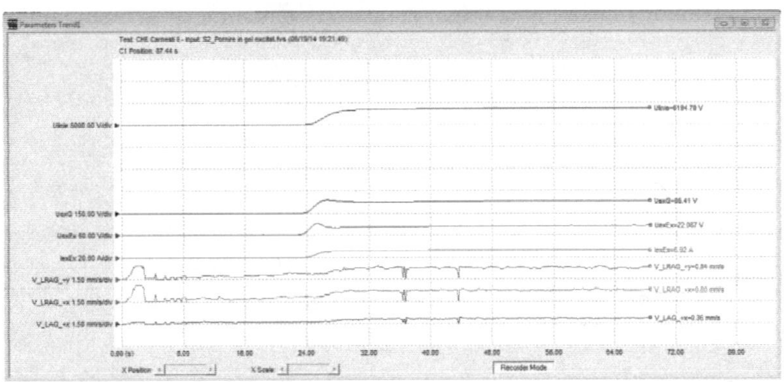

Figure 6. The variation of the measured parameters in starting mode and in idle excited functioning mode

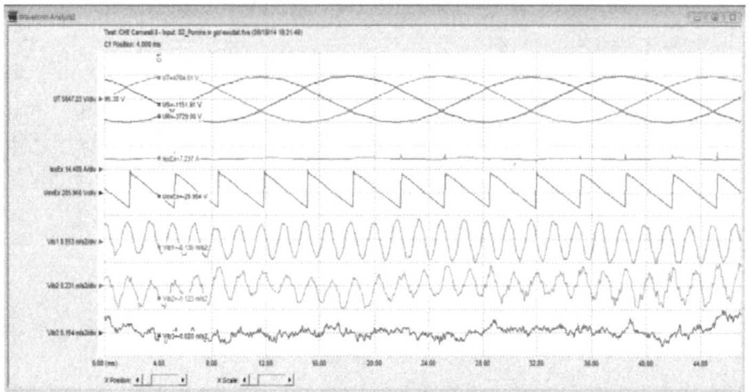

Figure 7. The wave form at stable idle excited functioning mode

Figure 8 represented the vibration spectrum of measured parameters i.e. unexcited steady speed variation spectrum (amplitude) in that order harmonic frequency. It is noted that the maximum value is obtained for harmonic vibrations of order 1 axial- radial bearing higher along the horizontal direction + x, bearing higher axial-radial radial bearing respective + y or + x lower direction.

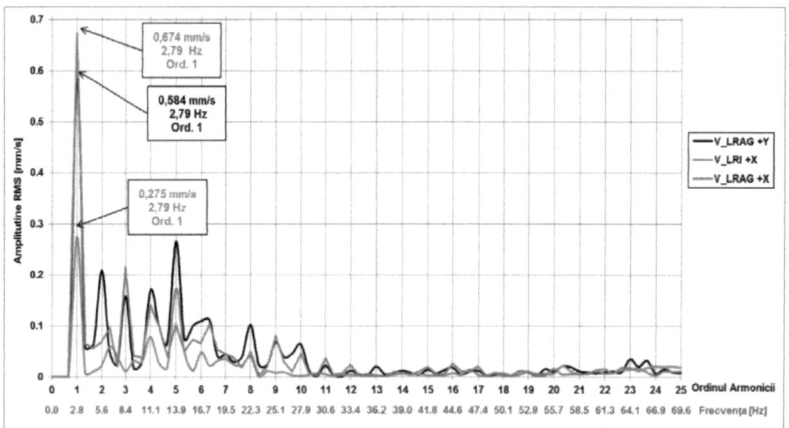

Figure 8. The vibration spectrum of the measured parameters in a stable excited mode

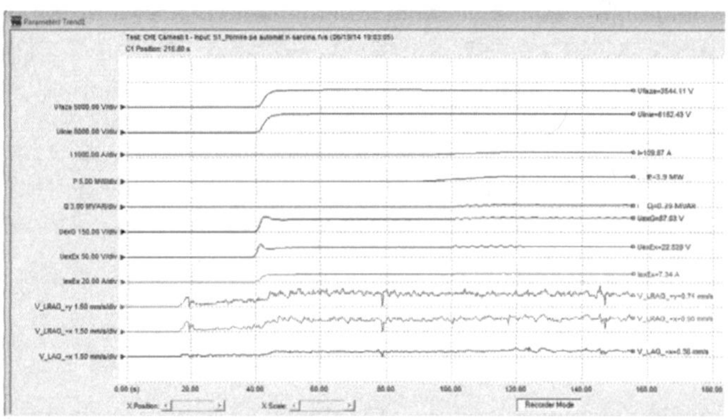

Figure 9. The variation of the parameters at start and in a stable mode, functioning in charge P=5.6 MW

In Figure 9 there are shown also graphically electrical and mechanical parameters measure to start hydro aggregates for the mode under load, i.e.: the rotor speed n, vibration thrust bearing radial upper horizontal direction + x V_{LRAG_X}, vibration thrust bearing radial superior radial + y V_{LRAG_Y}, lower radial bearing vibration direction + x V_{LAG_X}.

The values that are presented in Table 5 are the parameters for mechanical loading of 5.6 MW, and Table 6 presents the variation parameters for different values of active power: P = 1.6 ÷ 5.6 MW.

Table 5. The variation of the measured parameters at stable functioning in charge mode V, P=5.6 MW

Time	Revolution	Vibrations		
t	n	$V_{LRAG\ x}$	$V_{LRAG\ y}$	$V_{LAG\ x}$
(s)	(RPM)	(mm/s)	(mm/s)	(mm/s)
1	167.2	0.68	0.71	0.39
2	167.2	0.72	0.77	0.33
3	167.2	0.64	0.7	0.37
4	167.2	0.66	0.8	0.36
5	167.19	0.69	0.71	0.37
6	167.19	0.69	0.72	0.33
7	167.19	0.69	0.73	0.37
8	167.2	0.73	0.74	0.4
9	167.21	0.72	0.75	0.36
10	167.21	0.63	0.68	0.36
V. average	167.20	0.69	0.73	0.36

Table 6. The variation of the measured parameters in charged functioning at different power levels

Active power	Revolution	Vibrations		
P	n	$V_{LRAG\ x}$	$V_{LRAG\ y}$	$V_{LAG\ x}$
(MW)	(RPM)	(mm/s)	(mm/s)	(mm/s)
1.6 ÷ 5.6	167.2	0.68	0.81	0.38
	167.2	0.74	0.7	0.32
	167.19	0.57	0.71	0.34
	167.19	0.73	0.83	0.42
	167.19	0.66	0.7	0.43
	167.2	0.72	0.66	0.36

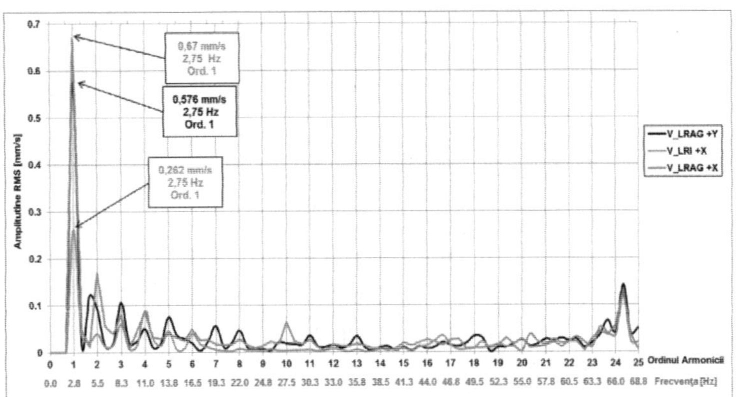

Figure 10. The vibration spectrum of the measured parameters stable charge

In Figure 10 is the vibration spectrum of parameters measured under load regime stabilized and the maximum value is obtained for harmonic vibrations of order 1.

4. Measurement upon starting the hydro-generator

The commissioning of hydro-generator were performed vibration measurements and checks for the following modes: idle (unexcited and excited) or under load.

By this test determines the performance of the hydro-generator mechanical assembly vibration measurements both in the transitional arrangements starting and steady idle. There have been following sizes: n rotor speed, vibration higher radial thrust bearing in the horizontal direction + x V_{LRAG_X}, higher radial thrust bearing vibration radial + y V_{LRAG_Y}, lowers radial bearing vibration direction + x V_{LAG_X}.

The measurement results in transitional and steady start after dismantling hydro-generator unexcited and operation are given in Figure 11. The commissioning of hydro-generator were performed vibration measurements and checks for the following modes: idle (unexcited and excited) or under load.

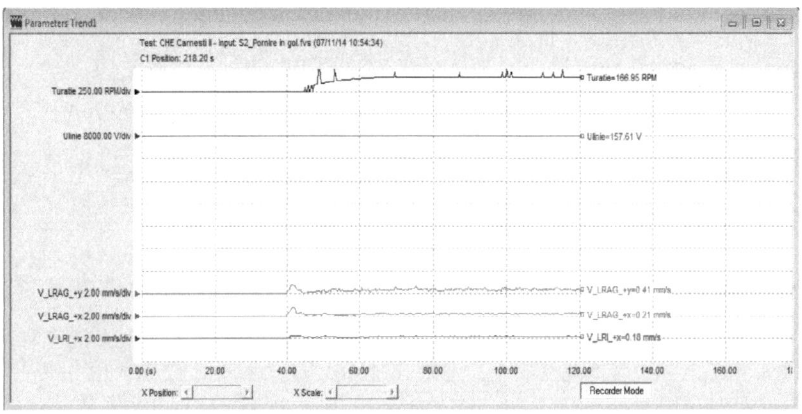

Figure 11. The variation of the parameters measured in transitory starting mode and idle unexcited functioning

Table 7. Parameters registered during the transitory starting mode

Time	Revolution	Vibrations		
t	N	V_{LRAG_x}	V_{LRAG_y}	V_{LAG_x}
(s)	(RPM)	(mm/s)	(mm/s)	(mm/s)
1	76.31	0.19	0.16	0.13
2	98.82	0.19	0.14	0.17
3	121.33	0.47	0.21	0.17
4	140.71	0.23	0.13	0.11
5	143.78	0.24	0.21	0.15
6	148.16	0.42	0.15	0.14
7	150.76	0.37	0.19	0.14
8	154.29	0.56	0.24	0.16
9	166.3	0.37	0.29	0.22
10	166.46	0.34	0.26	0.18
11	166.96	0.36	0.23	0.21
12	166.96	0.31	0.23	0.19
13	166.95	0.31	0.23	0.18
14	166.95	0.43	0.22	0.21

The values of the parameters measured in the starting phase after dismantling the hydro aggregate and commissioning of the regime idle unexcited stabilized are shown in Table 7, and the values of the parameters measured at steady idle unexcited in Table 8.

Table 8. Parameters measured in a stable mode of idle unexcited functioning

Time	Revolution	Vibrations		
t	n	$v_{LRAG\ x}$	$v_{LRAG\ y}$	$v_{LAG\ x}$
(s)	(RPM)	(mm/s)	(mm/s)	(mm/s)
1	166.7	0.59	0.21	0.18
2	166.95	0.34	0.26	0.17
3	166.79	0.32	0.24	0.2
4	166.95	0.37	0.25	0.19
5	166.95	0.44	0.23	0.2
6	189.91	0.36	0.25	0.19
7	166.95	0.48	0.27	0.17
8	166.95	0.43	0.22	0.21
9	166.95	0.39	0.23	0.2
10	166.95	0.42	0.25	0.24
11	166.7	0.41	0.27	0.19
12	166.95	0.37	0.28	0.22
13	166.78	0.49	0.23	0.2
14	166.78	0.49	0.23	0.2
V. average	168.52	0.421	0.244	0.197

The spectrum of vibration parameters measured at steady unexcited after dismantling of hydro aggregate and commissioning of are presented in Figure 12, and the maximum value is obtained for harmonic vibrations of order 1.

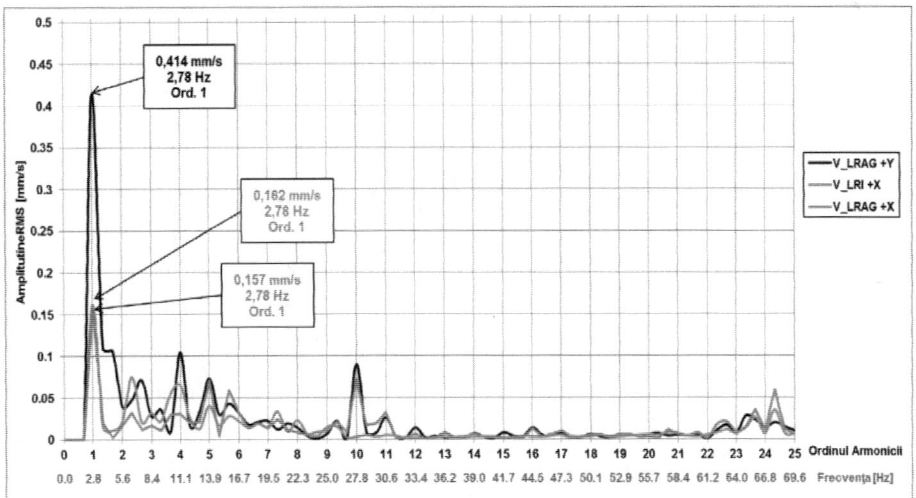

Figure 12. The vibration spectrum of the measured parameters in unexcited stable functioning mode

IOP Conf. Series: Materials Science and Engineering **163** (2017) 012030 doi:10.1088/1757-899X/163/1/012030

From Figure 12 it is observed that the maximum vibration along the axial-radial bearing higher radial + y respectively lower bearing radial direction + x is greater than before dismantling and axial - radial bearing higher along the horizontal direction is x + smaller.

In idle excited mode, connecting the generator excitation voltage determines a nominal voltage stator, rotor speed and registering vibrations in the chambers.

Figure 13 are plotted variation of measured parameters in regime starting and idling excited stabilized and is represented in Figure 7 wave-form stabilized and /or idle excited modes.

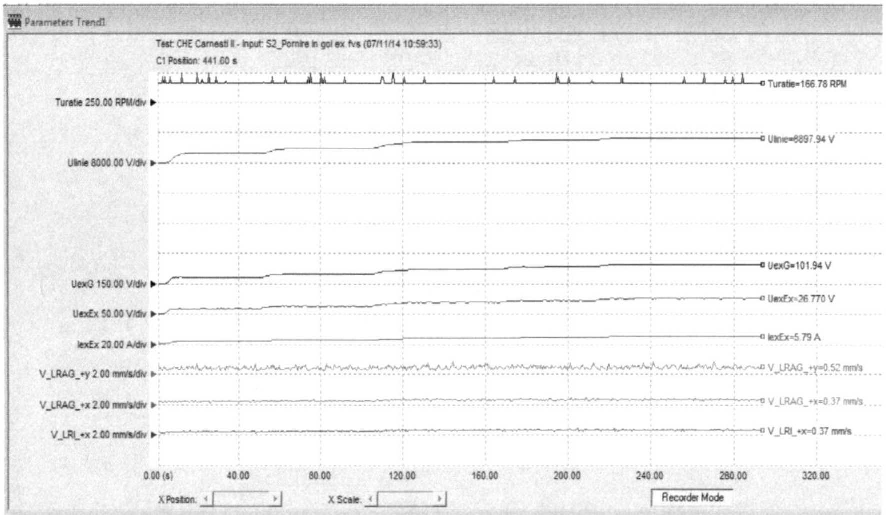

Figure 13. Parameter variation measured in the starting process and the idle stabilized excited mode

Parameters measured values steady idle excited are shown in Table 9 and the variation of measured values during transient excitement in Table 10.

Table 9. Parameters measured in stable idle functioning mode

Time	Revolution	Vibrations		
T	n	$v_{LRAG\ x}$	$v_{LRAG\ y}$	$v_{LAG\ x}$
(s)	(RPM)	(mm/s)	(mm/s)	(mm/s)
1	166.69	0.33	0.39	0.32
2	166.61	0.79	0.42	0.38
3	166.78	0.48	0.39	0.36
4	166.61	0.5	0.4	0.36
5	166.69	0.41	0.38	0.36
6	166.61	0.35	0.4	0.37
7	166.61	0.45	0.42	0.38
8	166.61	0.76	0.38	0.36
9	166.7	0.39	0.44	0.32
10	166.45	0.48	0.4	0.34
11	166.61	0.6	0.4	0.4
V. Average	166.63	0.50	0.40	0.36

Table 10. Variation of the measured values during the transitory exciting process

Time	Revolution	Vibrations		
T	n	$v_{LRAG\ x}$	$v_{LRAG\ y}$	$v_{LAG\ x}$
(s)	(RPM)	(mm/s)	(mm/s)	(mm/s)
1	0.4	0.26	0.21	0.4
2	0.41	0.23	0.24	0.41
3	0.4	0.25	0.22	0.4
4	0.47	0.27	0.22	0.47
5	0.45	0.28	0.25	0.45
6	0.55	0.31	0.27	0.55
7	0.55	0.4	0.3	0.55
8	0.48	0.31	0.35	0.48
9	0.43	0.4	0.37	0.43
10	0.61	0.39	0.38	0.61
11	0.43	0.41	0.33	0.43
12	0.43	0.33	0.32	0.43
13	0.48	0.39	0.36	0.48

Vibration spectrum parameters are measured at steady excited represented in Figure 14.

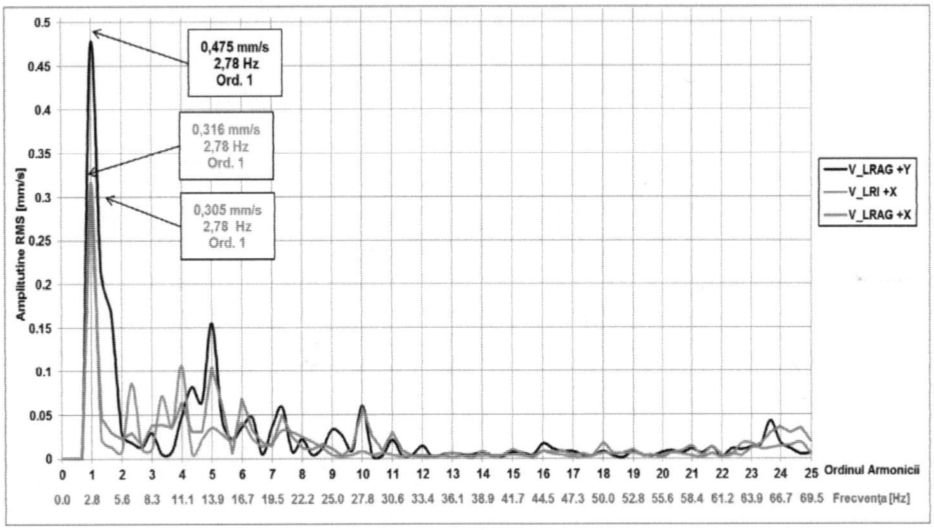

Figure 14. The vibration spectrum of the measured parameters an stable excited mode

From Figure 14 is observed that the maximum vibration along the axial-radial bearing higher radial + y respectively axial- radial bearing upper horizontal direction + x is less than before disassembly and lower radial bearing along the direction + x is bigger.

Attempts are considering establishing performance operation of the hydro vibration measurements regime under load. In Figure 15 are presented graphically electrical and mechanical parameters measured at the start hydro regime under load.

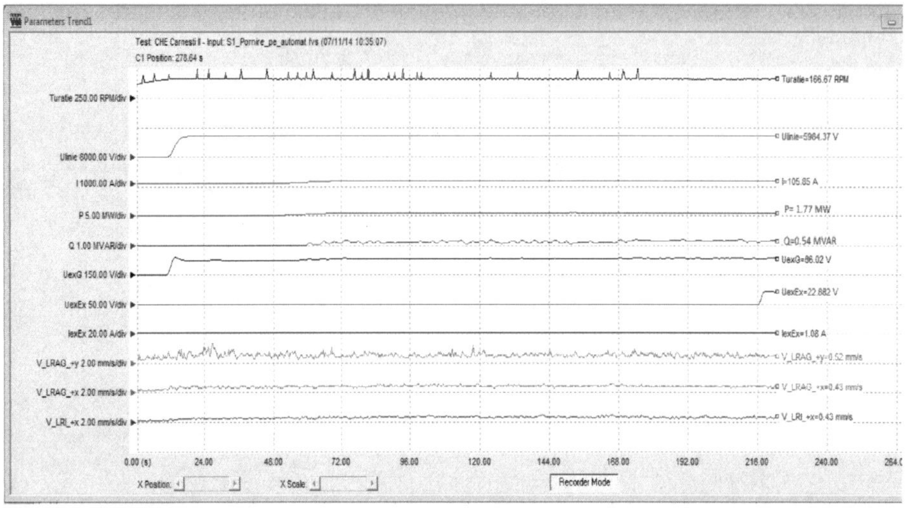

Figure 15. Variation of the parameters in excited and stable modes at charged functioning for P=1,77 MW

The values are presented in Table 11 parameters for electrical and mechanical loading of 5.6 MW, and Table 12 presents the variation parameters for different values of active power: P = 1.6 ÷ 5.6 MW.

Table 11. Variation of the measured parameters at charged stable functioning for P=5.6 MW

Time	Revolution	Vibrations		
t	N	$v_{LRAG\ x}$	$v_{LRAG\ y}$	$v_{LAG\ x}$
(s)	(RPM)	(mm/s)	(mm/s)	(mm/s)
1	166.63	0.48	0.46	0.4
2	166.71	0.52	0.49	0.36
3	166.71	0.39	0.46	0.37
4	166.63	0.43	0.37	0.37
5	166.63	0.46	0.4	0.35
6	166.63	0.44	0.39	0.4
7	166.47	0.6	0.4	0.43
8	166.46	0.36	0.47	0.36
9	166.63	0.4	0.36	0.39
10	166.79	0.68	0.43	0.39
11	166.46	0.46	0.38	0.36
12	166.46	0.46	0.44	0.43
13	166.63	0.45	0.38	0.36
14	166.79	0.6	0.42	0.34
15	166.63	0.51	0.42	0.37
16	166.46	0.75	0.41	0.36
17	196.21	0.61	0.43	0.38
18	166.63	0.41	0.44	0.39
19	166.17	0.37	0.4	0.37
20	166.71	0.49	0.41	0.42
V. Average	168.07	0.49	0.42	0.38

Table 12. Variation of the measured parameters at charged functioning at different power stages

Active power	Revolution	Vibrations		
P	N	$V_{LRAG\ x}$	$V_{LRAG\ y}$	$V_{LAG\ x}$
(MW)	(RPM)	(mm/s)	(mm/s)	(mm/s)
	167.8	0.54	0.48	0.41
1.6	166.7	0.39	0.33	0.31
÷	166.7	0.56	0.46	0.43
5.6	166.6	0.41	0.38	0.39
	166.7	0.55	0.42	0.34

Figure 16 represents the spectrum of vibration parameters measured under load regime stabilized and the maximum value is obtained for harmonic vibrations of order 1.

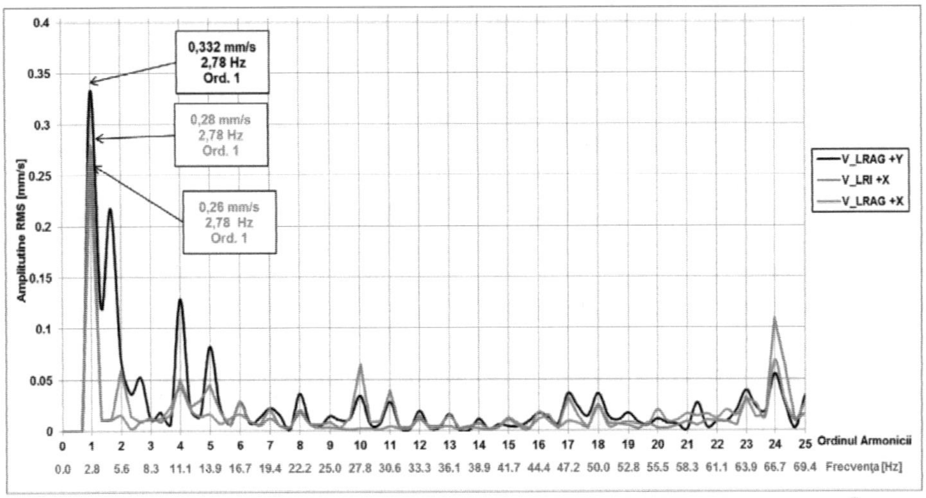

Figure 16. The vibration spectrum of the measured parameters in stable excited mode

Figure 16 represents the spectrum of vibration parameters measured under load regime stabilized and the maximum value is obtained for harmonic vibrations of order 1.

Figure 16 show that the maximum vibration along the axial-radial bearing higher radial + y respectively axial- radial bearing upper horizontal direction + x is less than before disassembly and lower radial bearing along the direction + x is bigger.

Based on experimental measurements made before and after repair the hydro aggregate is seen:

- At the level of the thrust bearing, vibration in the radial direction + y reaches the value of 0.8 mm/s to 0.385 mm/s in the empty unexcited;
- After repair maximum level of vibration in idle mode excited is 0.5 mm/s radial + x, to 0.421 mm/s value recorded under idle unexcited;
- The system under load is the tendency to standardize the level of vibration to the value of 0.49 mm/s;

5. Conclusions

In this paper based on experimental measurements performed, shows the following: vibrations in idle mode in the three camps are influenced largely by the state of magnetization of the rotor, the system under load is the tendency to standardize the level of vibration, under a throw of pregnancy there is a proper behavior in dynamic rotor over-speed during the transitional arrangements, observed that under unexcited idling vibration occurs due to sharp flow regimes and due to offsetting.

Acknowledgment

The work has been funded by the Sectoral Operational Programme Human Resources Development 2007-2013 of the Ministry of European Funds through the Financial Agreement POSDRU/159/ 1.5/S/132395.

References

[1] Bucur D M, Dunca G, Isbăşoiu E C, Călinoiu C and Roşioru O T 2012 Analysis of Operating Parameters During Normal and Transient Regimes af a High Head Hydro Power Plant, *U.P.B. Sci. Bull., Series D* **74**(1) 51-58

[2] Bucur D M, Dunca G and Călinoiu C 2012 *Experimental vibration level analysis of a Francis turbine*, Proceedings of 26th IAHR Symposium on Hydraulic Machinery and Systems, Beijing, China, August 19–23, pp 1-8

[3] Nicolet C, Ruchonnet N, Alligné S, Koutnik J and Avellan F 2010 *Hydroacoustic simulation of rotor-stator interaction in resonance conditions in Francis pump-turbine*, Proceedings of 25th IAHR Symposium on Hydraulic Machinery and Systems, Timisoara, Romania, September 20-24, pp 1-9

[4] Liu P, Chen X, Wang Q and Li D 2016 Analysis of rotor-stator interaction and vibration in high-head Francis turbines, *Journal of Hydroelectric Engineering* **35**(3) 91-98

[5] Veres M, Turcan R and Dumitrescu D 2011 Critical analysis of vibration sources of hydro aggregates in operating regime, *Journal of Sustainable Energy* **2**(4) 1-8

[6] Dunca G, Bucur D M, Isbăşoiu E C, Călinoiu C and Ghergu C M 2012 Vibrations level analysis during the operation of a high head hydropower plant, *U.P.B. Sci. Bull., Series D* **74**(1) 5966

[7] Dunca G, Isbăşoiu E C, Călinoiu C, Bucur D M and Ghergu C M 2008 *Vibrations level analysis during the operation of pumping station Gâlceag*, Proceedings of 5th National Conference of Romanian Hydropower Engineers, Bucharest, Romania, May 22-23, pp 181-190

[8] Půlpitel L, Veselý J and Mikulášek J 2009 *Comments to vibrations and pressure oscillations induced by the rotor stator interaction in a hydraulic turbine*, Proceedings of the 3rd IAHR International Meeting of the Workgroup on Cavitation and Dynamic Problems in Hydraulic Machinery and Systems, Brno, Czech Republic, October 14-16, pp 305-314

[9] Griscenko M and Vitols R 2015 *Stator core vibration and temperature analysis of hydropower generation unit at 100 Hz frequency*, Proceedings of 14th International Scientific Conference, Jelgava, Latvia, May 20-22, pp 383-388

[10] Griscenko M and Kulikovskis G 2014 *Vibration monitoring of power generation systems, using built-in Matlab functions for data analysis*, Proceedings of 13th International Scientific Conference, Jelgava, Latvia, May 29-30, pp 162-167

[11] Luo X, GUO P C, Liao W, Zheng X and Lu P 2006 *Vibration damping of hydraulic turbine unit —case study for Yantan Hydropower Plant*, Proceedings of the 4th WSEAS International Conference on Fluid Mechanics and Aerodynamics, Greece, August 21-23, pp 309-314

[12] Padureanu I, Jurcu M, Augustinov L, Hatiegan C, Răduca E and Pădeanu L 2015 Optimisation of the Start-up and Operation Regimes of Cooling Water Pumps of a High-Power Hydro Generator, *Analele Universităţii "Eftimie Murgu" Reşiţa* **22**(1) 345-358

[13] Padureanu I, Jurcu M, Augustinov L, Hatiegan C and Răduca E 2015 Implementation of an Automatic System for the Monitoring of Start-up and Operating Regimes of the Cooling Water Installations of a Hydro Generator, *Analele Universităţii "Eftimie Murgu" Reşiţa* **22**(1) 359-368

[14] Jurcu M R, Padureanu I, Padeanu L, Augustinov L and Hatiegan C 2015 Tests Regarding the Transitory Regimes of Putting off Load of the Hydroagregate, *Analele Universităţii "Eftimie Murgu" Reşiţa* **22**(2) 185-195

[15] Hatiegan C 2013 *Identification of Defects in Thin Elastic Plates by Means of Modal Analysis*, Eftimie Murgu" University of Resita, Romania, Doctoral Thesis

Determining the operating performance through electrical measurements of a hydro generator

C Haţiegan[1], C P Chioncel[1], E Răduca[1], C Popescu[2], I Pădureanu[3], M R Jurcu[3], D Bordeaşu[4], S Trocaru[5], F Dilertea[6], O Bădescu[7], I M Terfăloagă[1,8], A Băra[3] and L (Barboni) Haţiegan[3]

[1]Eftimie Murgu University of Resita, Department of Electric Engineering and Informatics, Traian Vuia str., no. 1-4, 320085 Resita, Romania
[2]Constantin Brâncuşi University of Târgu Jiu, Department of Electric Engineering, Eroilor str., no. 30, 210135 Targu Jiu, Romania
[3]Eftimie Murgu University of Resita, Department of Mechanical and Materials Engineering, Traian Vuia str., no. 1-4, 320085 Resita, Romania
[4]Aalborg University - Esbjerg, Department of Energy Technology, Niels Bohrs Vej str., no. 8, 6700 Esbjerg, Denmark
[5]University of Bucharest, Faculty of Physics, Atomiştilor str., no. 405, 077125, Bucuresti, Romania
[6]Politehnica University of Timisoara, Department of Electroenergetics, Vasile Parvan str., no. 2, 300223 Timisoara, Romania
[7]Babeş-Bolyai University, Department of Educational Science, Sindicatelor str., no. 7, 400015, Cluj-Napoca, Romania
[8]West University of Timisoara, Department of Computer Science, Vasile Parvan str., no. 4, 300223 Timisoara, Romania

E-mail: c.hatiegan@uem.ro

Abstract. This paper presents the results obtained by experimental measurements regarding the determining of the operating performance of a Hydro generator, through electrical measurements before the refurbishment of hydro aggregates. Experimental measurements were taken to determine the characteristic by putting off load and also by putting on load regimes for power between 1 and 5 MW of hydro generator. These electrical measurements are to establish appropriate measures in terms of constructive and functional in order to bring the nominal parameters of that hydro aggregate.

1. Introduction

In terms of operation efficiency, the technical and economic parameters for hydraulic machines operating in all industrial fields [1], a special attention should be given in the phase of design and technological development.

In the case of hydropower, which operates in a hydro power plant, a key factor is to maintain the temporal parameters that define the quality materials used in manufacturing [2] a defined by the number and duration of maintenance interventions respectively for their lifetime.

For example, the normal functioning of the hydraulic hydro turbines is influenced greatly in terms of vibration [3].

In this way, it can still be registered through various experimental measurements of hydropower through experimental laboratory research also through numerical simulations, the following:

- Testing and performance evaluation tests index, partial downloads and so on [4]. These measurements were made on a total of 18 small hydroelectric power plants;
- Monitoring the hydro aggregate parameters during operation is normal and transient [5]. Also, after dynamic balancing, certain parameters (voltage, rotation and vibration) were directly measured [6] or by indirect methods [7] and thermodynamic methods [8], where they recorded satisfactory measurements.

At smaller scales, the pumps, turbines, generators [9] and the phenomenon of partial discharge [10] or laboratory simulation respectively, were achieved with great interest [11-13];

- Research on cavitation respectively cavitation erosion, which is mainstay and acting destructively on materials used to manufacture hydraulic turbine rotors [14-16].

In conclusion, because of these features, the operation parameters inside of hydraulic turbines and hydro power plants respectively is influenced to a large extent and in time due to the number of hours of operation, imposed by overhauling the hydropower rehabilitation.

In this sense, this paper highlights several results obtained by measurement experiments on a high hydro power from a hydro power plant in Romania, in the feature idling and feature under load at different powers respectively partial discharge levels.

These results through tables and diagrams characteristic highlight the important parameters of hydro development. All measurements in this paper were made in accordance with applicable standards [17], [18].

2. Experimental measurement before dismantling the generator

To measure active power, tensions or currents respectively across generator electrical parameters analyzer was used and type VPA323 process parameters, a laptop for data processing, as in Figure 1. The hydro speed was measured with a laser speed sensor, such QS30LDQ.

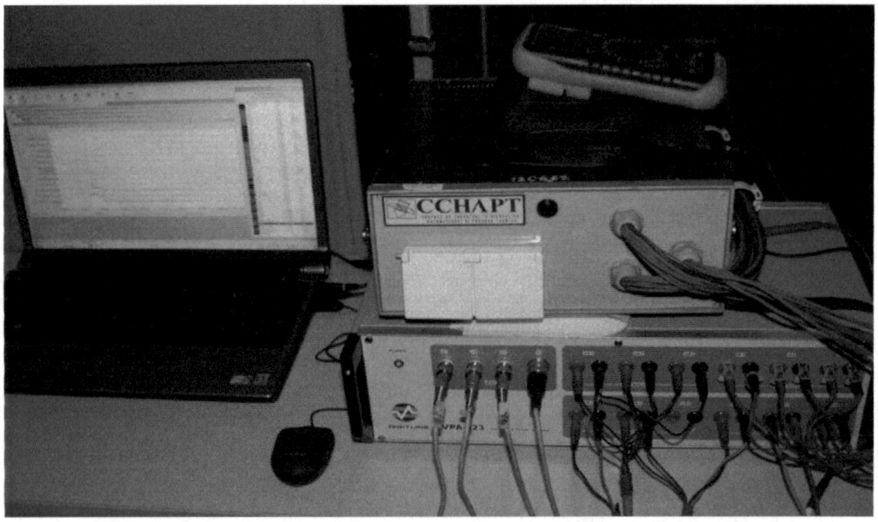

Figure 1. Analyzer of the parameters of the VPA323 type

2.1. Measurement electrical parameters steady idle excited and transient arousal

Before removing the rehabilitation of hydro, it is very important to assess the actual operating performance in electrical and mechanical terms. In this regard, certain electrical parameters were measured vibrations respectively (Figure 1).

The electrical parameter values are given in Table 1.

Table 1. Values of the electric parameters in a stable idle functioning environment as well as in a transitory excited mode

Time	Revolution	Values of the electric parameters							
t	n	Stable idle functioning mode				Transitory excited mode R			
[s]	[RPM]	U_{linie} [V]	U_{exG} [V]	U_{exEx} [A]	I_{exEx} [V]	U_{linie} [V]	U_{exG} [V]	U_{exEx} [A]	I_{exEx} [V]
1	167.2	6139.82	84.39	21.625	6.82	168.3	2.33	1.839	0.26
2	167.2	6141.66	84.39	21.67	6.83	256.08	9.62	8.66	0.68
3	167.2	6143.4	84.45	21.632	6.83	825.49	44.61	25.079	2.49
4	167.2	6144.93	84.49	21.7	6.83	2360.26	86.84	23.709	4.93
5	167.19	6146.21	84.45	21.712	6.83	3784.43	92.55	18.024	5.78
6	167.19	6147.61	84.5	21.646	6.84	4807.77	87.04	17.531	6.03
7	167.19	6149.24	84.56	21.753	6.84	5351.76	81.94	18.323	6.12
8	167.2	6150.82	84.58	21.665	6.84	5591.79	79.4	18.744	6.18
9	167.21	6152.07	84.59	21.63	6.85	5699.34	77.65	19.03	6.17
10	167.21	6153.21	84.62	21.682	6.85	5743.92	76.68	19.428	6.16
V. Average	167.20	6147.60	84.52	21.69	6.84	3458.91	63.86	17.036	4.48

Electrical parameters steady idling respective excited arousal phase are: line voltage generator U_{line}; phase voltage U_{phaz}; U_{exG} excitation voltage generator; U_{exEx} excitation voltage excitation and the excitation current exciter I_{exEx}.

Figures 2 and 3 show the change in regime starting measured parameters and idle respectively excited stabilized waveform regime stabilized idle excited.

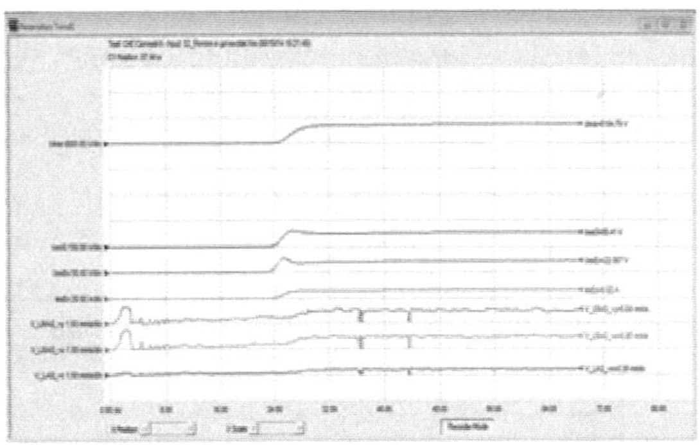

Figure 2. Measurement scheme for idle functioning

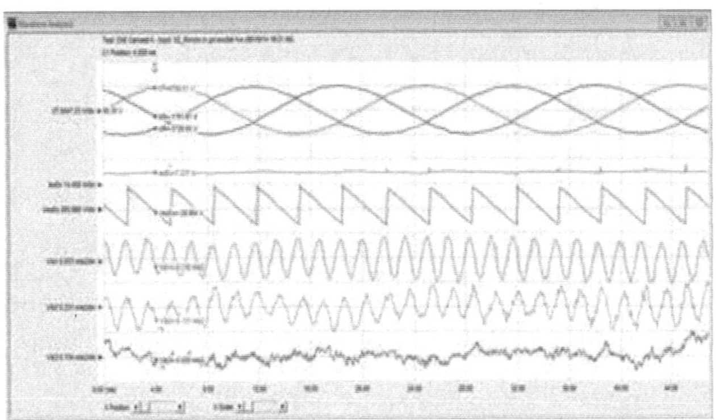

Figure 3. Measurement scheme for idle functioning mode

IOP Conf. Series: Materials Science and Engineering **163** (2017) 012031 doi:10.1088/1757-899X/163/1/012031

2.2. Characteristics of empty functioning

Feature idling is given by $U_0 = f(I_{exEx})$ for n = ct. (F = ct) and I = 0, where U_0 is the voltage across the generator under idling [19], [20]. Whereas under idle load current is zero (I = 0), the voltage across the generator will be equal to the electromotive voltage U_{eE}, flow-induced excitation current Φp determined I_{exEx} the winding rotor. The phasor diagram (Figure 4), the electromotive voltage U_{eE} is out of phase π / 2 after the flow of excitation respectively after excitation current, and Figure 5 is presented under the scheme as idle.

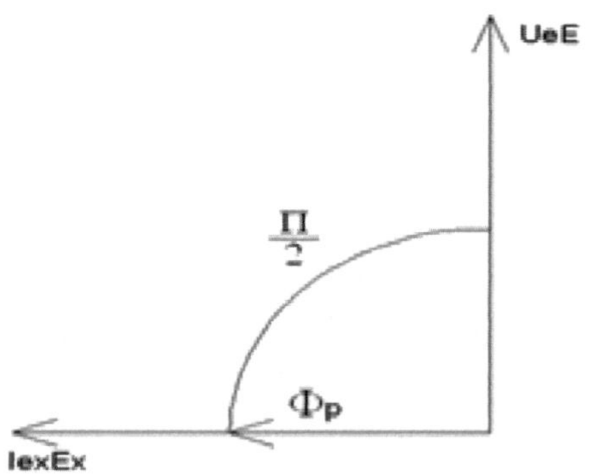

Figure 4. Phased diagram of the synchronous generator in idle functioning

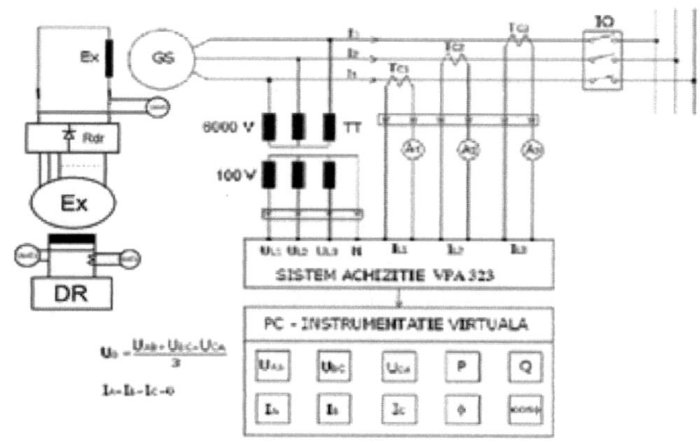

Figure 5. Measurement scheme in idle functioning mode

In conditions where the generator rotation is maintained constant (n = 167.2 rpm) which corresponds to the frequency f = 50 Hz, changes the excitation current of the excitation winding and I_{exEx} the corresponding values of the voltage read at the terminals U_0.

The graphical representation of the variation $U_0 = f(I_{exEx})$ is shown in Figure 6 and Figure 7 extrapolation.

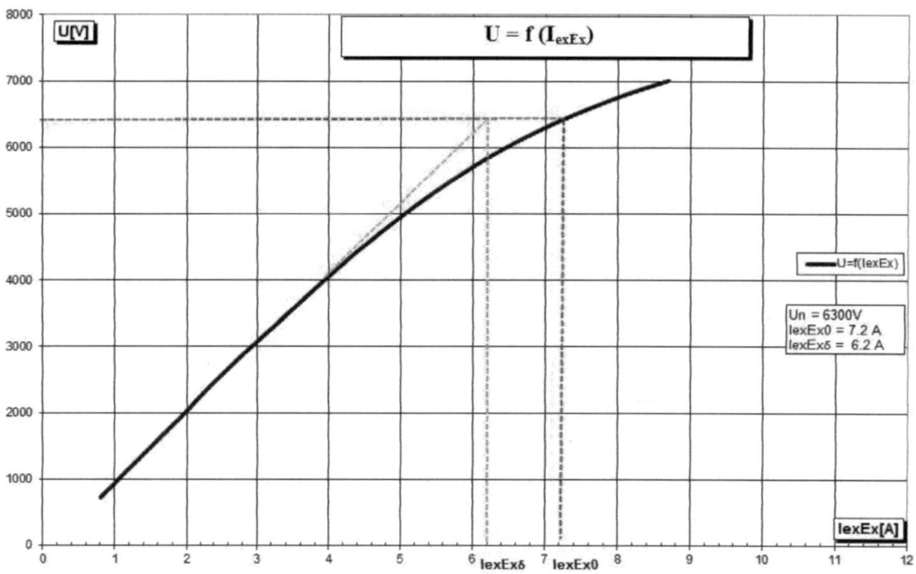

Figure 6. Idle functioning characteristic

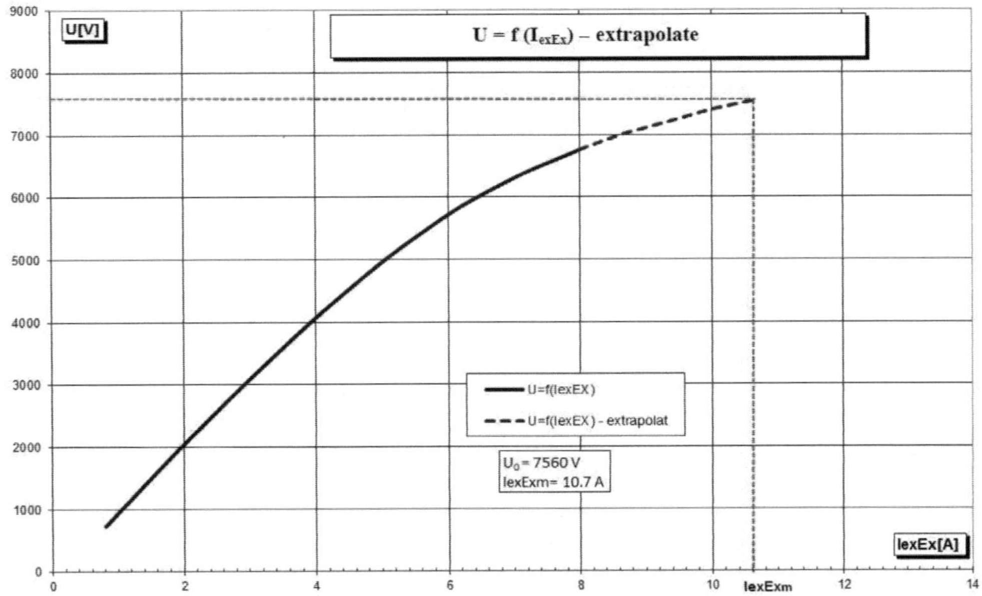

Figure 7. Idle functioning characteristic (extrapolated)

Of characteristics resulting idle following: excitation current gap: $I_{exEx\delta} = 6.2$ A; excitation current idle: $I_{exEx0} = 7.2$ A; maximum excitation current $U_0 = 1.2 \cdot U_n = 7560$ V, resulting in extended idling feature: $I_{exExm} = 10.7$ A.

The state of saturation of the magnetic circuit can be appreciated from the report:

$$\frac{I_{exExm}}{I_{exExo}} = \frac{10.7}{7.2} = 1.48 \quad (1)$$

where:

I_{exEx0} is the current excitation voltages corresponding to an idle $U_0 = A$,
I_{exExm} excitation current voltages corresponding to an idle $U_0 = 1.2 \cdot U_n$.

2.3. Characteristic of charged functioning (in V)

The characteristic V, $I = f\,(I_{exEx})$ for U = ct., f = ct., P = ct., corresponds to the phase diagram [19] in Figure 8. The condition of P = ct., requires the active component of current flowing I_A = ct. and place the tip of the variation of I_{exEx} (U_{eE}) is right (Δ). In the diagram, the machine is overexcited and debit inductive power network.

If you start to lower excitation current, the current drops to the minimum value when the generator is running toward the basket network cos φ ≈ 1; decreasing further excitation (under exciting the engine), the current rise again and cos φ is capacitive.

For a given power P, it is necessary that excitement does not fall below a certain minimum. In Figure 5, was represented by a dotted line and the minimum value in the synchronous operation U_{eEmin} is possible power not considered. The curve $I = f\,(I_{exEx})$ is obtained; it is in the form of a V.

For another power value P = ct. higher right (Δ) moves so as to increase the value of I_a. Excitation current value necessary to establish new, take is also higher toe and V-axis system (I_{exEx}, I) moving up and to the right.

In Figure 9 are the characteristics I = f (I_{exEx}) for different powers [21]. The curve represented by open, connecting the tips of the V curves separates from the viewpoint of reactive load, the operation of the generator inductive, capacitive generator of the (receiver inductive).

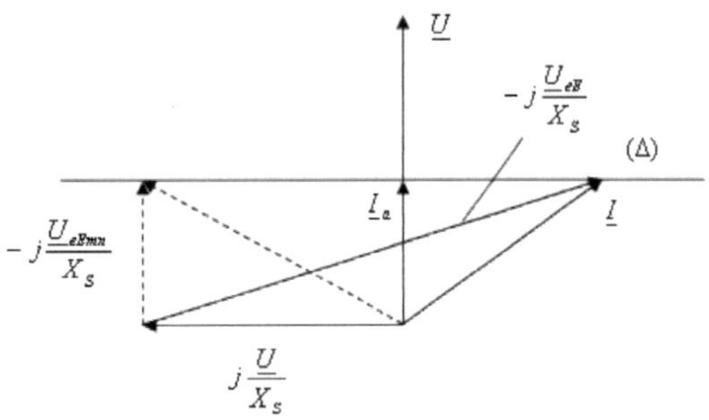

Figure 8. The phased diagram of a synchronous generator functioning at U=ct., f=ct., P=ct.

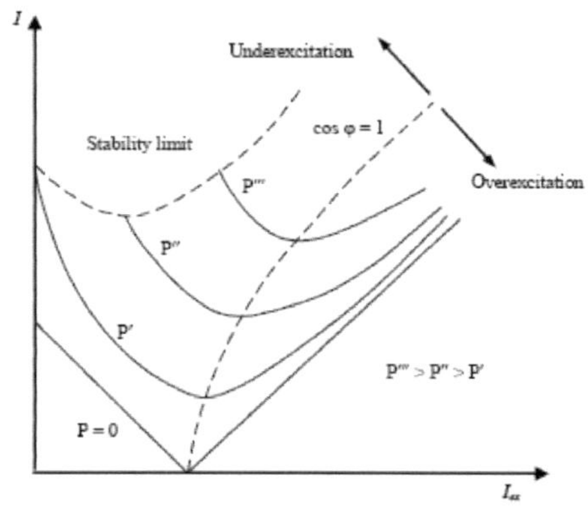

Figure 9. The V curves of a synchronous engine for different values P=ct.

The curve represented capacitive discontinued operation area indicates the limits to which may be deducted for various excitation power P = ct., such as synchronous operation is possible.

Measurement scheme load operating mode is shown in Figure 10.

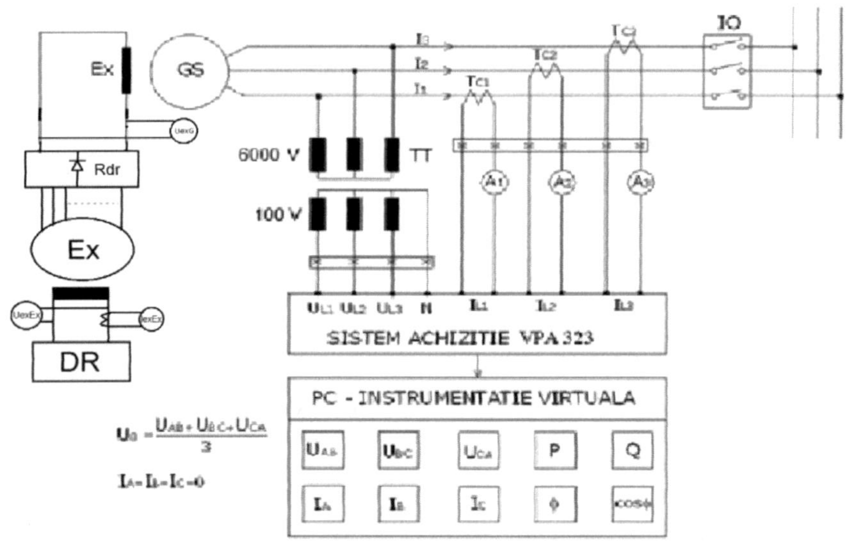

Figure 10. The measurement scheme for the charged functioning system

For different values of active power, load characteristics corresponding amounts that power. For different values of the excitation current, the corresponding values are read stator current, stator voltage, active and reactive powers.

These results are presented below in tables and graphs for power between 1 ÷ 5 MW. For these 5 attempts, the generator is connected on the network, measuring the corresponding values: line voltages U_{RS}, U_{ST}, U_{RT}; stator current I; I_{exEx} excitation current; U_{exG} excitation voltage generator; U_{exEx} exciter excitation voltage; active power; reactive power cos φ power factor respectively frequency f.

Tables 2 – 6 shows the values of the electrical parameters regime under load for power values from 1 through 5 MW cut-in feature that these powers (Figures 11 – 15).

Table 2. The values of the electric parameters in charged functioning mode for P=1 MW

Time	Values of the electric parameters								
t	f	P	Q	cos φ	U_{line}	I	U_{exG}	U_{exEx}	I_{exEx}
[s]	[Hz]	[MW]	[Mvar]	[-]	[V]	[A]	[V]	[A]	[V]
1	50	1.68	0.63	0.93	6164.26	111.15	89.13	24.096	7.5
2	50.03	1.59	1.83	0.66	6216.14	148.39	105.03	28.119	8.96
3	50.02	1.68	1.98	0.65	6221.32	153.2	107.58	28.864	9.02
4	50.02	1.65	2.4	0.57	6249.98	175.21	113.79	30.901	9.79
5	50.02	1.59	3.75	0.39	6314.61	244.51	133.84	37.207	11.97
6	50.01	1.59	3.9	0.38	6316.75	253.05	136.77	38.746	12.3
7	50	1.56	4.02	0.36	6323.87	259.94	138.38	38.966	12.51

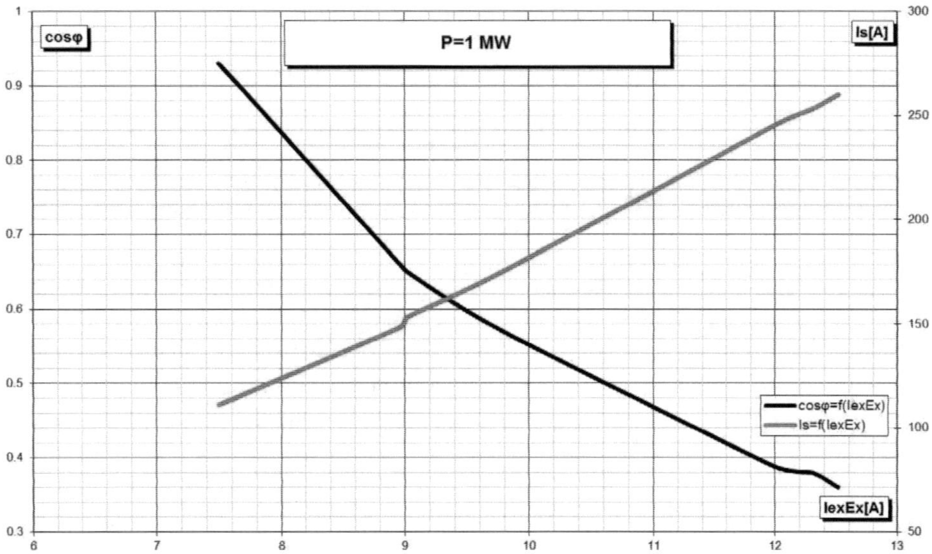

Figure 11. Charged functioning characteristic at P=1 MW

Table 3. Values of the electric parameters in charged functioning mode for P=2 MW

Time	Values of the electric parameters								
t	f	P	Q	cos φ	U_{line}	I	U_{exG}	U_{exEx}	I_{exEx}
[s]	[Hz]	[MW]	[Mvar]	[-]	[V]	[A]	[V]	[A]	[V]
1	50.01	1.98	3.69	0.47	6321.88	252.07	134.42	36.875	11.98
2	50.01	1.98	3.36	0.51	6306.9	235.2	129.41	35.165	11.37
3	50.01	1.98	3.3	0.52	6302.59	231.67	128.17	34.653	11.24
4	50.01	2.01	2.76	0.59	6278.49	205.93	120.07	31.524	10.34
5	50.01	2.01	2.64	0.6	6274.78	200.58	118.4	31.073	10.16
6	50.01	2.01	2.13	0.69	6246.18	177.64	110.86	28.953	9.39
7	50.01	2.01	2.04	0.7	6240.82	174.27	109.62	28.561	9.26

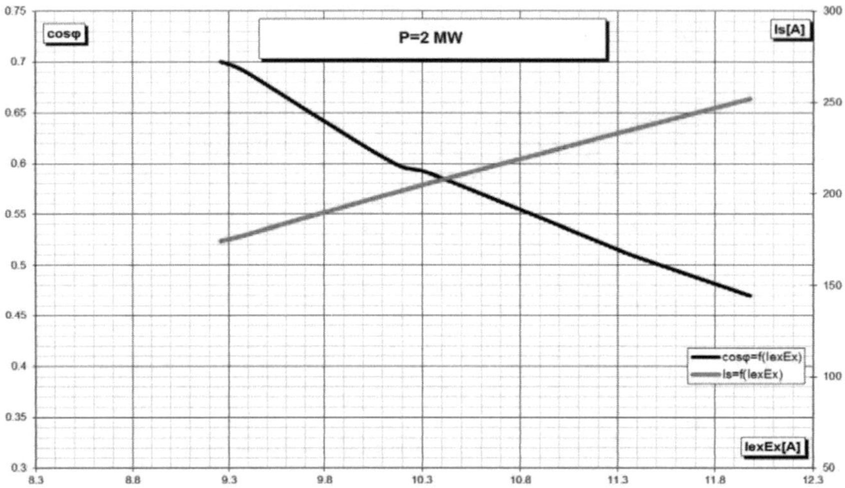

Figure 12. Charged functioning characteristic at P=2 MW

Table 4. Values of the electric parameters in charged functioning mode for P=3 MW

Time	Values of the electric parameters								
t	f	P	Q	cos φ	U_{line}	I	U_{exG}	U_{exEx}	I_{exEx}
[s]	[Hz]	[MW]	[Mvar]	[-]	[V]	[A]	[V]	[A]	[V]
1	50.01	3.21	0.48	0.99	6145.75	203.94	93.03	24.077	7.74
2	50.01	3.21	0.48	0.99	6151.16	203.3	93.56	24.359	7.78
3	50.01	3.18	1.77	0.87	6221.38	223.33	112.54	30.174	9.51
4	50.01	3.15	3.18	0.71	6289.11	271.67	132.52	36.753	11.68
5	50.01	3.12	3.66	0.65	6316.27	291.65	140.49	39.857	12.62
6	50.01	3.09	4.17	0.59	6344.93	313.1	147.41	42.152	13.55
7	50	3.09	4.2	0.59	6343.09	313.29	147.83	42.322	13.59

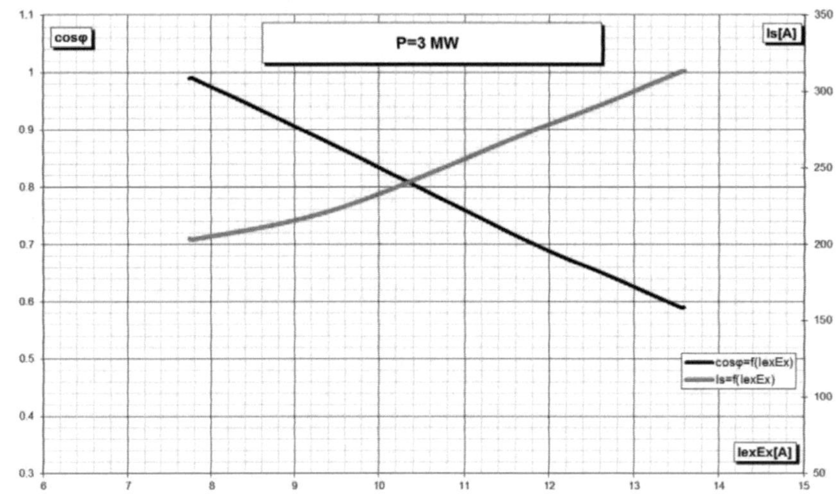

Figure 13. Charged functioning characteristic at P=3 MW

Table 4. Values of the electric parameters in charged functioning mode for P=4 MW

Time	Values of the electric parameters								
t	f	P	Q	cos φ	U_{line}	I	U_{exG}	U_{exEx}	I_{exEx}
[s]	[Hz]	[MW]	[Mvar]	[-]	[V]	[A]	[V]	[A]	[V]
1	49.99	4.35	3.96	0.74	6333.99	355.42	149.64	42.539	13.77
2	49.99	4.35	3.12	0.81	6287.11	326.23	136.48	37.236	12.08
3	49.99	4.38	2.79	0.84	6272.28	317.42	132.24	35.638	11.56
4	50	4.38	2.58	0.86	6257.94	311.34	128.69	34.278	11.15
5	49.99	4.38	1.56	0.94	6208.8	288.62	114.4	29.698	9.63
6	49.99	4.41	0.21	1	6142.25	277.7	96.34	24.37	7.96
7	49.99	4.38	0.12	1	6130.31	277.71	96.64	24.733	7.93

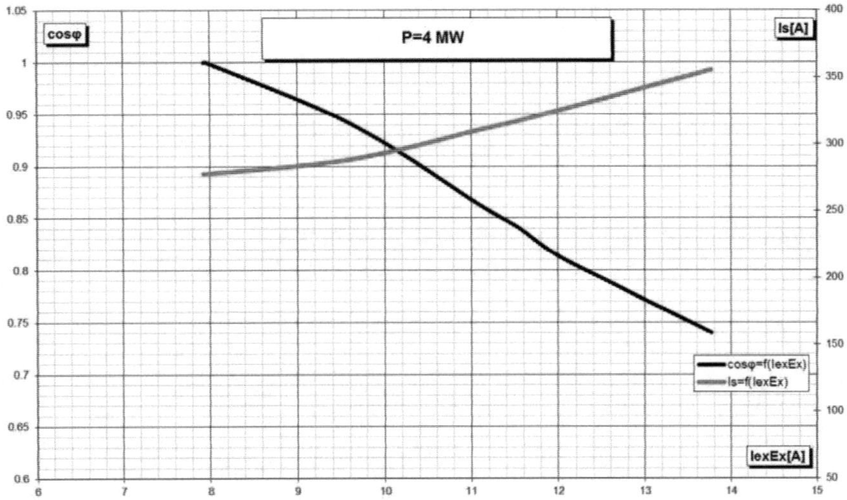

Figure 14. Charged functioning characteristic at P=4 MW

Table 5. Values of the electric parameters in charged functioning mode for P=5 MW

Time	Values of the electric parameters								
t	f	P	Q	cos φ	U_{line}	I	U_{exG}	U_{exEx}	I_{exEx}
[s]	[Hz]	[MW]	[Mvar]	[-]	[V]	[A]	[V]	[A]	[V]
1	49.99	5.22	0.78	0.99	6168.37	331.38	112.79	30.208	9.42
2	49.97	5.19	2.22	0.92	6241.46	348.72	132.06	36.292	11.48
3	49.98	5.19	2.91	0.87	6279.52	364.8	141.7	40.244	12.67
4	49.98	5.16	3.51	0.83	6311.05	380.61	150.56	43.408	13.81
5	49.98	5.13	3.81	0.81	6319.09	389.31	154.66	45.252	14.36
6	49.98	5.13	4.14	0.78	6338.88	400.51	159.6	47.387	15.07
7	49.98	5.16	4.23	0.77	6347.36	404.24	160.84	47.531	15.26

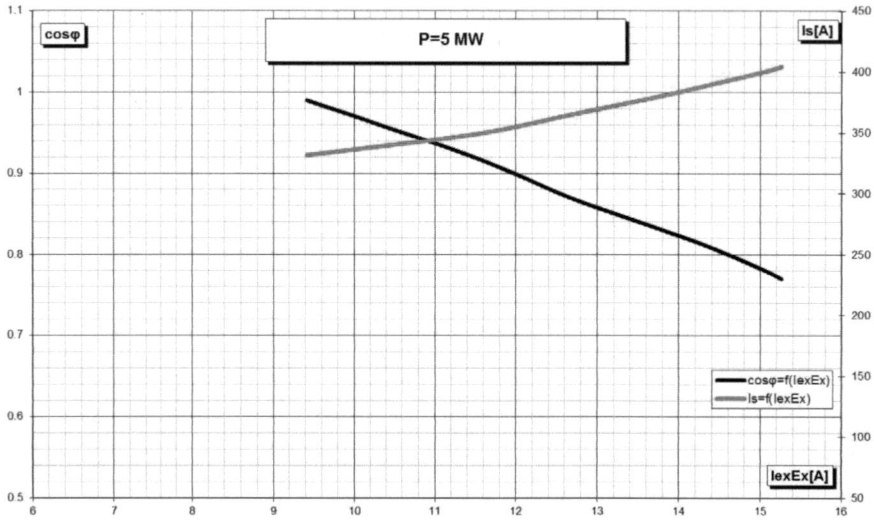

Figure 15. Charged functioning characteristic at P=5 MW

2.4 Measurement the electric parameters at charge surges for P=5.7 MW

In this case, the respective electrical parameters were measured vibration (Figure 16). The variation of electrical parameters for a total of 17 tests is given in Table 6 and in Figure 13.

Table 6. The electric parameters variation in charge surges at P=5.7 MW

Nr. crt.	I [A]	P [MW]	Q [Mvar]	U_{line} [V]	U_{exG} [V]	U_{exEx} [A]	I_{exEx} [V]
1	605.58	5.73	3.18	6297.12	150.55	42.794	13.71
2	605.37	5.73	3.18	6296.49	150.58	42.855	13.71
3	605.58	5.73	3.18	6295.51	150.6	42.685	13.71
4	605.87	5.73	3.18	6292.99	150.5	42.8	13.71
5	606.11	5.73	3.18	6292.58	150.57	42.798	13.71
6	607.06	5.73	3.18	6292.27	150.57	42.817	13.71
7	606.35	5.73	3.18	6294.29	150.6	42.873	13.71
8	569.75	5.07	2.82	6429.98	150.32	29.096	13.45
9	460.1	3.33	1.83	6772.53	109.53	-12.36	9.94
10	312.48	1.53	0.84	6913.14	63.47	-25.12	5.82
11	0.22	0	0	6839.79	17.27	-37.95	1.74
12	0.22	0	0	6279.36	-0.19	0.936	0.22
13	0.22	0	0	5669.69	-0.16	1.013	0.22
14	0.23	0	0	5037.49	-0.14	1.017	0.22
15	0.22	0	0	4409.26	-0.15	1.02	0.22
16	0.22	0	0	3807.45	-0.13	1.01	0.22
17	0.22	0	0	3250.15	-0.2	1.016	0.22

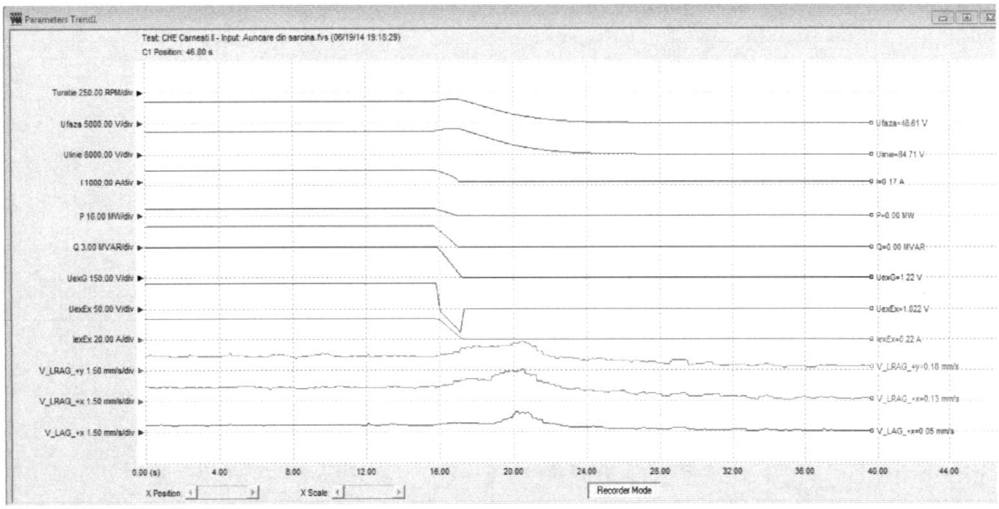

Figure 16. The variation of the measured parameters in charge surges at P=5.7 MW

3. Experimental measurement into operation

Into operation of the hydro aggregate, were made following tests and measurements: tests under idle mode respectively tests under load.

3.1. Tests under idle mode

For this type of tests, were certain operating conditions with the generator disconnected from the network. Besides vibrations analysis, following electrical parameters were measured depending on the rotor rotation, denoted by n: line voltage generator U_{line}; phase voltage U_{phaz}; U_{exG} excitation voltage generator; U_{exEx} excitation voltage excitation and the excitation current exciter I_{exEx}. The registration of these parameters is shown in Figure 14 respectively in Tables 7 and 8:

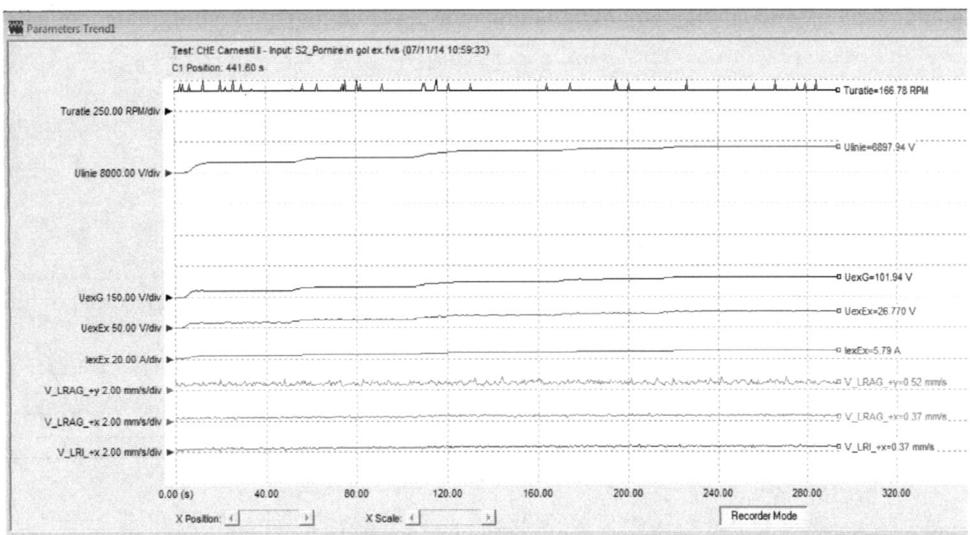

Figure 14. The variation of measured parameters in regime starting and idling excited stabilized

Table 7. Parameters measured at steady idle excited

Nr. crt.	n [RPM]	U_{line} [V]	U_{exG} [V]	U_{exEx} [A]	I_{exEx} [V]
1	166.69	6893.26	102.14	26.65	5.79
2	166.61	6897.3	101.99	26.606	5.79
3	166.78	6901.96	101.96	26.558	5.79
4	166.61	6901.37	102	26.604	5.79
5	166.69	6897.9	101.92	26.649	5.79
6	166.61	6897.84	102.06	26.733	5.79
7	166.61	6899.21	102.05	26.726	5.79
8	166.61	6898.2	101.96	26.587	5.79
9	166.7	6897.6	101.98	26.733	5.79
10	166.45	6897.4	102.25	26.858	5.8
11	166.61	6901.85	101.95	26.638	5.8
V. Medie	166.63	6898.54	102.02	26.67	5.79

Table 8. The variation of measured values during transient excitation

Nr. crt.	n [RPM]	U_{line} [V]	U_{exG} [V]	U_{exEx} [A]	I_{exEx} [V]
1	166.69	164.66	2.06	1.048	1.08
2	166.61	701.42	24.8	8.892	1.85
3	166.78	2074.39	37.6	8.651	2.49
4	166.61	2792.1	35.13	9.367	2.62
5	166.69	3170.12	45.34	12.26	2.97
6	166.61	4240.15	53.5	14.056	3.46
7	166.61	5285.17	71.06	19.134	4.22
8	166.61	6146.65	86.26	22.015	4.95
9	166.7	6357.64	88.25	22.917	5.1
10	166.45	6421.1	89.88	23.023	5.18
11	166.61	6612.04	96.5	25.613	5.47
V. Medie	166.63	3996.85	57.3	15.179	3.58

3.2. Tests under load

For this type of testing, were certain operating conditions with the generator disconnected from the network. Besides vibrations analysis, following electrical parameters were measured depending on the rotor rotation, denoted by n:

- line voltage generator U_{line};
- phase voltage U_{phaz};
- U_{exG} excitation voltage generator;
- U_{exEx} excitation voltage excitation;
- the excitation current exciter I_{exEx},
- active P and reactive Q power respectively the current stator I.

The registration of these parameters is shown in figure 15 respectively in Tables 9 and 10:

IOP Conf. Series: Materials Science and Engineering **163** (2017) 012031 doi:10.1088/1757-899X/163/1/012031

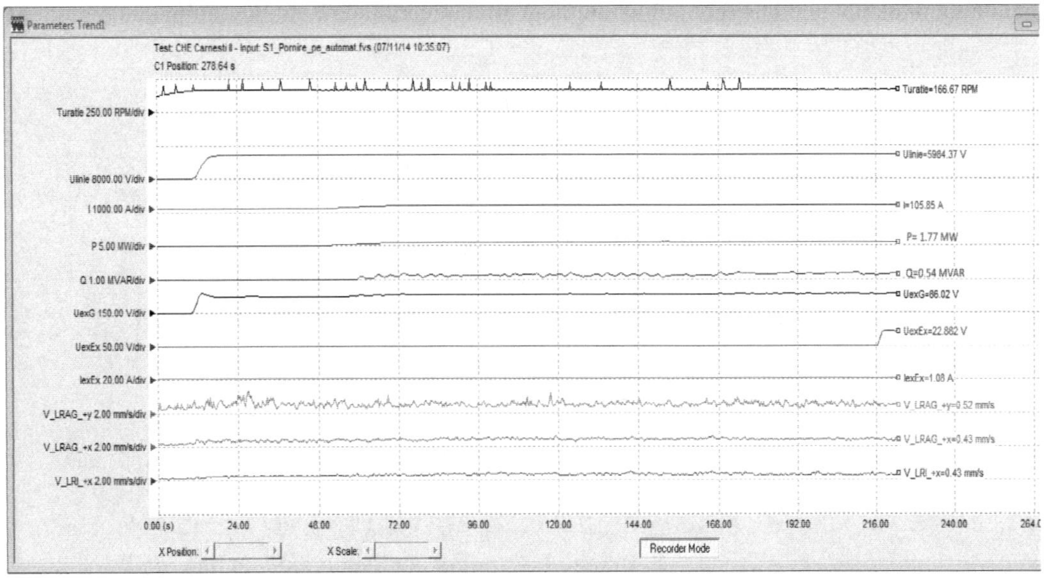

Figure 15. Variation of the parameters in excited and stable modes at charged functioning for P=1,77 MW

Table 9. Variation of the measured parameters at charged stable functioning for P=5.6 MW

Nr. crt.	n [RPM]	U_{line} [V]	I [A]	P [MW]	Q [Mvar]	U_{exG} [V]	U_{exEx} [A]	I_{exEx} [V]
1	166.63	6137.25	378.17	5.64	2.46	135.1	36.52	1.08
2	166.71	6129.71	376	5.64	2.34	133.54	36.291	1.08
3	166.71	6124.6	374.49	5.67	2.25	132.24	35.639	1.08
4	166.63	6119.6	372.79	5.67	2.16	131.24	35.561	1.08
5	166.63	6113.93	372.37	5.7	2.07	130.68	35.329	1.08
6	166.63	6111.16	371.25	5.7	2.01	129.35	34.415	1.08
7	166.47	6101.53	369.71	5.7	1.89	127.77	33.958	1.08
8	166.46	6100.4	367	5.7	1.77	125.96	33.285	1.08
9	166.63	6094.59	364.01	5.67	1.68	124.72	33.281	1.08
10	166.79	6089.96	363.81	5.67	1.59	124.04	32.962	1.08
11	166.46	6086.13	363.17	5.67	1.53	123.12	32.464	1.08
12	166.46	6084.27	361.36	5.67	1.41	121.45	31.866	1.08
13	166.63	6076.34	359.9	5.64	1.32	120.07	31.663	1.08
14	166.79	6071.79	358.41	5.64	1.23	119.27	31.43	1.08
15	166.63	6064.76	358.41	5.67	1.14	117.84	30.735	1.08
16	166.46	6063.82	357.92	5.67	1.08	117.38	30.995	1.08
17	196.21	6065.07	356.96	5.64	1.02	116.31	29.961	1.08
18	166.63	6065.97	355	5.64	0.93	114.97	29.835	1.08
19	166.17	6063.48	354.14	5.67	0.78	112.41	28.642	1.08
20	166.71	6050.12	353.93	5.67	0.63	110.9	29.167	1.08
V. Medie	168.07	6090.72	364.44	5.67	1.56	123.42	32.70	1.08

Table 10 presents the variation in parameters for different values of active power: P=1.6 ÷ 5.6 MW.

Table 10. The variation of measured values during transient excitation

Putere	n	U	I	P	Q	U_{exG}	U_{exEx}	I_{exEx}
[MW]	[RPM]	[V]	[A]	[MW}	[Mvar]	[V]	[A]	[V]
1.6	167.8	6111.33	231.77	1.68	3.36	127	34.89	1.08
2.6	166.7	5985.43	174.68	2.85	-0.24	81.79	21.57	1.08
3.6	166.7	6203.75	336.3	3.84	3.96	145.99	42.2	1.08
4.6	166.6	6004.45	308.66	5.01	0.06	98.07	25.35	1.08
5.6	166.7	6022.65	352.41	5.67	0.06	105.63	27.47	47.28

4. Conclusions

The voltage in no load mode, depending on the exciter excitation current increases approximately linearly.

For different power generator connected to the network, its observe the change of the excitation current in the rotor winding, determining the electrical parameters from the system under load.

Stator current regime under load depending on the exciter excitation current increases linearly and power factor depending on the exciter excitation current decreases linearly.

The regime under load there is a tendency to standardize the electrical parameters.

Under rid of pregnancy there is a rotor dynamics appropriate behavior during the transitional arrangements for over speed.

Acknowledgment

The work has been funded by the Sectoral Operational Programme Human Resources Development 2007-2013 of the Ministry of European Funds through the Financial Agreement POSDRU/159/ 1.5/S/132395.

References

[1] Wood A J and Wollenberg B F 1996 *Power Generation, Operation, and Control*, Second Edition, A Wiley-Interscience Publication, New York, USA

[2] ***2005 *Maşini electrice rotative, Partea 1: Valori nominale şi caracteristici de funcţionare* SR EN 60034-1:2005

[3] Nedelea V, Augustinov L and Liuba G 2011 Rotor-Stator Air Gap Unevenness Influence on Vibrations in Operation of High Power Hydrogenerators, *Analele Universităţii "Eftimie Murgu" Reşiţa* **18**(2) 193-200

[4] Verma H K and Kumar A 2007 *Performance Testing and Evaluation of Small Hydropower Plants* International Conference on Small Hydropower - Hydro, Sri Lanka, October 22-24, pp 1-9

[5] Bucur D M, Dunca G, Isbăşoiu E C, Călinoiu C and Roşioru O T 2012 Analysis of Operating Parameters During Normal and Transient Regimes af a High Head Hydro Power Plant, *U.P.B. Sci. Bull., Series D* **74**(1) 51-58

[6] Jurcu M R, Padureanu I, Padeanu L, Augustinov L and Hatiegan C 2015 *Tests Regarding the Transitory Regimes of Putting off Load of the Hydroagregate* Proceedings of Multi-Conference on System & Structures - SysStruc '15, Resita, Romania, September 24-26, pp 185-195

[7] Isbasoiu E C, Constantinescu M, Otomega D V, Safta C A, Neagoie C G and Ghergu C 2004 *Indirect Method to Calculate the Operating Flow on a Kaplan Turbine*, 6th International Conference on Hydraulic Machinery and Hydrodynamics, Timisoara, Romania, October 21-22, pp 113-118

[8] Patil S S, Verma H K and Kumar A 2010 *Efficiency Measurement of Hydro Machine by*

Thermodynamic Method, IGHEM-2010, IIT Roorkee, India, October 21-23, pp 181-189

[9] Caxaria G A, Sousa D M and Ramos H M 2011 *Small scale hydropower: generator analysis and optimization for water supply systems* World Renewable Energy Congress, Linkoping, Sweden, May 8-13, pp 1386-1393

[10] Yuan W, Chuangli G, Zhipeng W and Pengcheng D 2011 *Partial Discharge Online Monitoring for Hydro-generator* Computer Distributed Control and Intelligent Environmental Monitoring (CDCIEM), Changsha, China, pp 111 – 114

[11] Webster J G 2004 *Electrical Measurement, Signal Processing, and Displays*, CRC PRESS, New York, USA

[12] Stone G C and Warren V 2006 Objective Methods to Interpret Partial-Discharge Data on Rotating-Machine Stator Windings, *IEEE Transactions on Industry Applications* **42**(1) 195-200

[13] Haţiegan C, Răduca M., Frunzăverde D, Răduca E, Pop N and Gillich G R 2013 The modeling and simulation of the thermal analysis on the hydro generator stator winding insulation, *J Therm Anal Calorim* **113**(3) 1217-1221

[14] Manzana M E, Ghiban B, Bordeasu I, Ghiban N, Marin M and Miculescu F 2014 Structural Analysis of Steels by Cavitation Erosion, *Key Engineering Materials* **583** 28-31

[15] Bordeasu I, Popoviciu M O, Mitelea I, Ghiban B and Ghiban N 2013 Cavitation erosion resistance of two steels with the same percentage of Chromium and Nickel but different Carbon content, *IOP Conf. Ser.: Mater. Sci. Eng.* **57** 012007

[16] Nedelcu D, Nedeloni M D and Lupinca C I 2014 Cavitation Erosion Research on the X3CrNi13-4 Stainless Steel, *Materials Science Forum* **782** 263-268

[17] ***2001 *High-voltage test techniques-Partial discharge measurements,* The European Standard EN 60270:2001

[18] ***SR EN 60034-2-1 (IEC 34-2-1), SR EN 60034-2-2 (IEC 34-2-2) and SR HD 60364-4-41 (IEC 364-4-41)

[19] Biriescu M, Prostean G, Liuba G, Nedelea V M, Augustinov L, Madescu G and Mot M 2011 *Experimental model of a hydro generator with static excitation* International Conference on Computer as a Tool (EUROCON), Lisbon, Portugal, April 27-29, pp 1-4

[20] Novacescu F, Hatiegan C, Raduca M, Raduca E and Pop N 2012 A New Method for Testing the No-Load Work of an Electric One-Phase Transformer Using The Graphical Programming, *Scientific Bulletin of Politehnica University of Timisoara* **57**(2) 54-67

[21] Bucur D M, Dunca G, Călinoiu C and Isbăşoiu E C 2015 Experimental investigation over in site operational characteristics of a Hydro generator, *U.P.B. Sci. Bull., Series D* **77**(1) 205-212

Study regarding the density evolution of messages and the characteristic functions associated of a LDPC code

S Drăghici[1,2], O Proştean[1], E Răduca[2], C Haţiegan[2], I Hălălae[2], I Pădureanu[3], M Nedeloni[2] and L (Barboni) Haţiegan[3]

[1]Politehnica University of Timisoara, Department of Automation and Applied Informatics, Vasile Parvan str., no. 2, 300223 Timisoara, Romania
[2]Eftimie Murgu University of Resita, Department of Electric Engineering and Informatics, Traian Vuia str., no. 1-4, 320085 Resita, Romania
[3]Eftimie Murgu University of Resita, Department of Mechanical and Materials Engineering, Traian Vuia str., no. 1-4, 320085 Resita, Romania

E-mail: s.draghici@uem.ro

Abstract. In this paper a method with which a set of characteristic functions are associated to a LDPC code is shown and also functions that represent the evolution density of messages that go along the edges of a Tanner graph. Graphic representations of the density evolution are shown respectively the study and simulation of likelihood threshold that render asymptotic boundaries between which there are decodable codes were made using MathCad V14 software.

1. Introductive notions about LDPCs

LDPC (Low Density Parity Check) codes represent a class of block linear error correcting codes put forward by R.G. Gallager in 1963 [1] whose usefulness has been rediscovered in the last two decades.

The essential difference as opposed to the other classes of block codes consists in the fact that the control and parity matrix H is sparsely populated (sparse) with binary values of 1 logic, and a greater prevalence of values of 0 logic, which facilitates an easier decoding, because LDPCs are a class of codes that facilitates parallel calculus, they work well for high speeds of data transmission (WiMax radio channels, ATM technologies for data transfer, radio communications in outer space and so on). Even when the H matrix does not have values of 1 logic sparsely distributed in its structure, it can be brought to a sparsely populated shape/ pattern, by linear combinations of rows and columns. Furthermore the H matrix can be divided in sub-matrix blocks of H, which are sparsely populated blocks.

Through all this the main code is divided into sub-codes that are easier to decode, and thus the speed of data transmission remains very high even when multiple decoders are used. The LDPC codes are regular and irregular and are characterized by the set of parameters (n, d_v, d_c), where n is the lengh of the codeword. In regular LDPCs the H matrix is composed of exact d_v Hamming weights in colums and exact d_c Hamming weights in lines.

The LDPCs codes are represented by a bipartite Tanner graph (regular or irregular). In this kind of representation, in general, on the left hand side we have the variable nodes associated with bits in the code words, and on the right hand side, the parity control nodes, associated with the equations used to verify the parity. This two part graph is an ensemble characterised by two vectors [2-4]:

$$\lambda = (\lambda_1, \lambda_2, ..., \lambda_{d_v}) \text{ and } \rho = (\rho_1, \rho_2, ..., \rho_{d_c}) \tag{1}$$

Where λ_i – represents the fraction between the edges associated to the variable nodes with i degrees of freedom; ρ_j – represents the fraction of Tanner graph edges associated parity check nodes with j degrees of freedom; Taking into account the relationship (1) the rate r of a LDPC code is defined by [2], [3]:

$$r(\lambda, \rho) = 1 - \frac{\sum_{j=2}^{d_c} \frac{\rho_j}{j}}{\sum_{i=2}^{d_v} \frac{\lambda_i}{i}} = 1 - \frac{L}{N} \tag{2}$$

Where L is the total number of variable nodes, N is the total number of parity check nodes, d_v is the number of degrees of freedom of a variable node (which is the same for every variable node in the case of a regular LDPC), and d_c is the number of the degrees of freedom of a parity node.

A particular case is the regular Tanner - graph (associated to a regular LDPC), when [2], [3], [5]:

$$N \cdot d_v = L \cdot d_c \text{ and } r = 1 - \frac{d_v}{d_c} \tag{3}$$

In practice its usual for the variable nodes to be represented by circles, and those of parity by squares, in the graphic representation of a bipartite Tanner graph. In the LDPC code study relevant is also the number E of the bipartite Tanner graph edges. Thus, for a Tanner graph with L variable nodes, with each having i degrees of freedom, the total number of edges E_v, starting from these nodes is [2], [3]:

$$L \cdot \frac{\lambda_i / i}{\sum_{j \geq 2} \lambda_j / j} = L \cdot \frac{\lambda_i / i}{\int_0^1 \lambda(x) dx} \tag{4}$$

Where the x variable is a continuos or meshed variable specific to a probable continuos or meshed distribution, determined by the type of data transmission channel used.

At the same time, one calculates the number of E_c edges exiting the parity check nodes, so the rate of a LDPC code can be written thus:

$$r(\lambda, \rho) = 1 - \frac{\int_0^1 \rho(x) dx}{\int_0^1 \lambda(x) dx} \tag{5}$$

In the study of LDPCs binary symmetric channels are used (B.S.C.), binary erasure channels (B.E.C.), binary input aditive gaussian noise channels (B.I.A.W.G.N.C.), binary input Laplacean channels (B.I.L.C.), and others.

2. Obtaining the evolution density function

2.1. Presenting the characteristic function, the L.L.R. reports, and the cost functions

The decoding algorithms for LDPCs used to consist in passing the messages along the edges of a Tanner graph, from the variable nodes v towards the parity check nodes c, and back. For the graphic representation of this decoding procedure some cost function associated to the E_v and E_c edges of a Tanner graph are defined:

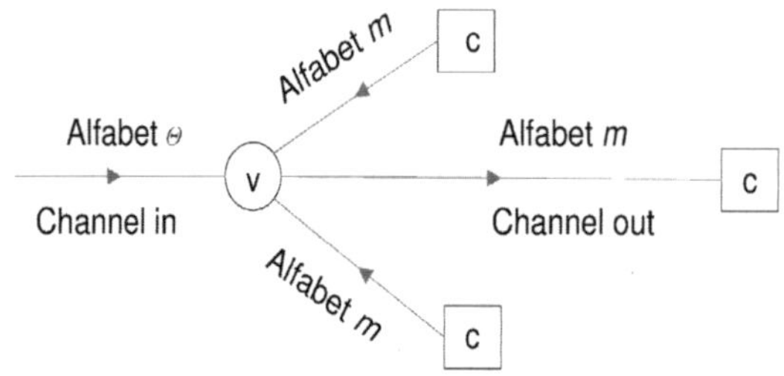

Figure 1. The cost functions m corresponding to the edges of a Tanner graph

In Figure1, θ is used to represent the function that corresponds to the initial message, v and c being variable nodes and parity check nodes. The cost functions depend on the iteration number, that is the message sent to a l – from the variable node v to the parity node c, is represented by the cost function $m_{vc}^{(l)}$. These are specific to the Gallager A and B decoding algorithms, which consist of evolution density transport and which supply the estimate number of incorrect messages that go through a Tanner graph at each iteration. This is how the notion of passing messages and entry and outpout alphabet appeared (see Figure 1). For mathematical reasons the cost functions of m parameter are defined by the log- likelihood ratios (L.L.R.) [2], [3], [6]:

$$m = \ln \frac{P(x=1/y)}{P(x=-1/y)} = \ln \frac{P(y/x=1)}{P(y/x=-1)} \tag{6}$$

In relation (6) Bayes rule is applied, and the random x and y variables can be, for example, the entrances and exits of a B.S.C. as the one represented in the next figure [2]:

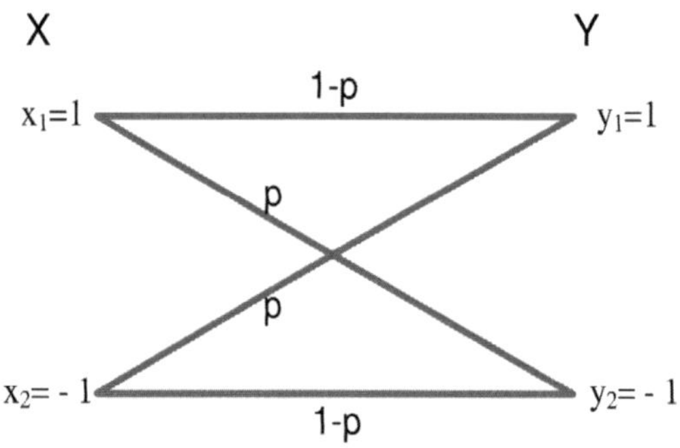

Figure 2. The structure of a B.S.C.

In Figure 2, p represents the likelihood of erroneous transmission of a symbol in a word code. The notation 1 was chosen in place of 0 logic because in the functions characterizing the evolution density of LDPCs the atypical function *sgn(x)* intervenes, which characterizes the error in decoding and defines the move from space $\{0, 1\}$ to space $\{-1, +1\}$ according to relation (7):

$$\operatorname{sgn}(x) \overset{def}{=} \begin{cases} 0, if\ x \geq 0; \\ 1, if\ x \leq 0; \end{cases} \tag{7}$$

It can be seen in $x = 0$ that *sgn(x)* is both zero and one with a probability of 1 / 2. From a mathematical standpoint it is practically an indetermination. Coming back to Relation (6), if we rewrite $m \overset{def}{=} L(x/y) = L(y/x)$ - according to Bayes rule, than:

$$\ln L(x/y_1, y_2, \dots, y_l) = \sum_{i=1}^{l} \ln L(x/y_i) \tag{8}$$

Where $y_1, y_2, \dots, y_l$ are independent random variable;
We note:

$$L \overset{not}{=} L(x_i / y_i) \text{ and } l \overset{not}{=} \ln L \tag{9}$$

Also, it is noted:

$$p \overset{not}{=} 2 \cdot P(x_1 = 0 / y_1) - 1,\ q \overset{not}{=} 2 \cdot P(x_2 = 0 / y_2) - 1 \tag{10}$$

The relation (10) determines the following relation:

$$2 \cdot P(x_1 \oplus x_2 = 0 / y_1, y_2) - 1 = p \cdot q \tag{11}$$

Where $x_1 \oplus x_2$ is the sum of module 2 over field *GF(2) - Galois Field* [7];
From the relation (9) and (10) we observe that:

$$2 \cdot P(x_i = 0 / y_i) - 1 = \frac{L-1}{L+1} = \tanh\left(\frac{l}{2}\right) \tag{12}$$

Taking into account relations (11) and (12) and extrapolating, we come to:

$$\ln L(x_1 \oplus \dots \oplus x_l / y_1, \dots, y_l) = \ln \frac{1 + \left(\prod_{i=1}^{l} \tanh\left(\frac{l_i}{2}\right)\right)}{1 - \left(\prod_{i=1}^{l} \tanh\left(\frac{l_i}{2}\right)\right)} \tag{13}$$

Where $l_i \overset{not}{=} \ln L(x_i / y_i)$;

As a consequence of relation (13) a function $\gamma(x)$ is defined, that characterizes the evolution density map [2]:

$$\gamma:\ (-\infty, +\infty) \to GF(2)\ \times\ [0, +\infty) \text{ and } \gamma(x) := (\gamma_1(x), \gamma_2(x)) := \left(\operatorname{sgn}(x), -\ln\tanh\left|\frac{x}{2}\right|\right) \tag{14}$$

The act of defining the function $\gamma(x)$ in relation (14) represents a functional transformed (Transform – T [9]) and makes the transition from the pairs space (x_i, y_j) – of code words, in the LLR space. In the LLR space the function *sgn(x)* defines the eroneous reception or not of a code word (determined by the presence or lack of modification of the reception signal), and the $\ln\tanh\left|\frac{x}{2}\right|$ function represents the amplitude of the received signal. Sugestive to those presented above are the following Figures [3], [8]:

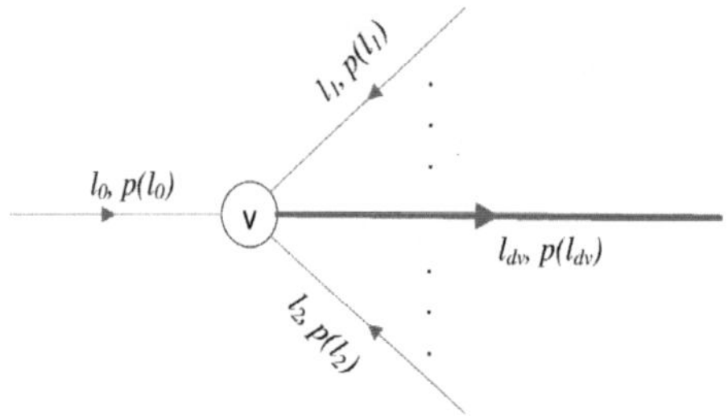

Figure 3. The evolution density corespondent to variable node *v*

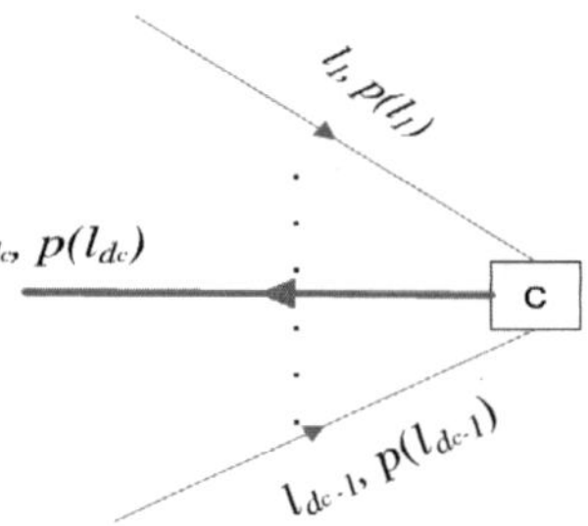

Figure 4. The evolution density corespondent to parity node *c*

It is apparent that:

$$\left(x_{d_c}, y_{d_c}\right)=\left(\sum_{i=1}^{d_c-1} x_i\,(\mathrm{mod}2), \sum_{i=1}^{d_c-1} y_i\,(\mathrm{mod}2)\right) \tag{15}$$

In consequence of transform – T defined in the relation (14), relation (15) can be written thus:

$$\tanh\left(\frac{l_{d_c}}{2}\right)=\prod_{i=1}^{d_c-1}\tanh\left(\frac{l_j}{2}\right) \tag{16}$$

Taking into account the relations (15) şi (16), the decoding process and the density map are represented suggestively in figure:

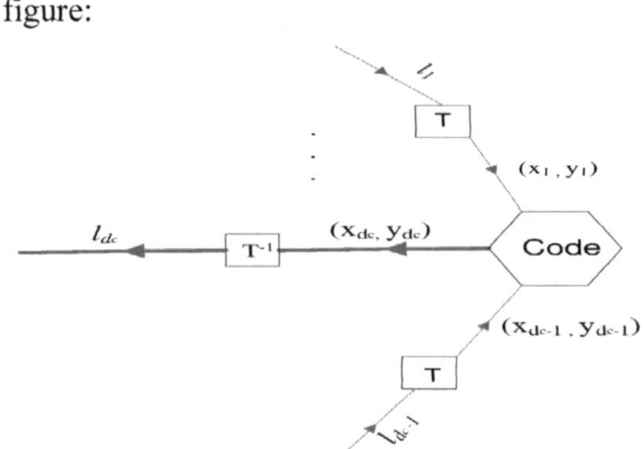

Figure 5. Transform-T; T^{-1} is transformed in reverse T

From Figure 5 it can be observed l_{d_v} is:

$$l_{d_v} = \sum_{i=1}^{d_v - 1} l_i \tag{17}$$

and l_i and l_{d_c} is the one given by the relations (13) and (16) respectively.

Returning to relation (14) which represents the evolution density map, it is demonstrated that [2] this function $\gamma(x)$ is the likely alottment density of a function i $F(x)$ defined by [2], [9]:

$$P\{z \in (-\infty, x]\} = F(x) \tag{18}$$

where z is a random variable with values between $(-\infty, +\infty)$, and $F(x)$ is continuous and nonnegative.

By defining a function $G(x)$ similar to $F(x)$, the convolution between the two functions in the real numbers domain R can be determined by [2], [9]:

$$(F \otimes G)(x) = \int_R F(x-y)dG(y) = \int_R G(x-y)dF(y) \tag{19}$$

One demonstrates [10], with the Lesbegue-Stieltjes relations used in cumulative distribution functions that the convolution of likelihood functions $F \otimes G$ corresponds to the convolution of their derivatives. Using the same Lesbegue-Stieljes relations we can define a function of G that associates the $F(x)$ to the evolution density map defined through $\gamma(x)$, thus [2], [11]:

$$G(s,x) := \chi_{\{s=0\}} \cdot G^0(x) + \chi_{\{s=1\}} \cdot G^1(x) \tag{20}$$

where $\chi_{\{s\}} = \chi_{\{\gamma_1(x) \in \{0,1\}\}}$ is the characteristic function associated to the set of elements $\{0, 1\}$ - namely to $\gamma_1(x)$ and $\chi_{\{s=0\}} = 1$ for $s = 0$ and is nil otherwise; also, $\chi_{\{s=1\}} = 1$ for $s = 1$ and is nil otherwise.

The function $G(s,x)$ shown in relation (20) is defined by the variation domain of the x variable with values in the $GF(2)\ x\,[0, +\infty)$ set and is also a characteristic function, by associating $F(x)$ to $\gamma(x)$. This is represented by means of an additional functional operator $\Gamma(s, F(x))$:

$$\Gamma(s, F(x)) = \chi_{\{s=0\}} \cdot \Gamma_0(F(x)) + \chi_{\{s=1\}} \cdot \Gamma_1(F(x)), \text{ where } \Gamma_0(F(x)) = 1 - F\left(-\ln\left|\tanh\left(\frac{x}{2}\right)\right|\right),$$
$$\text{and } \Gamma_1(F(x)) = F\left(\ln\left|\tanh\left(\frac{x}{2}\right)\right|\right). \tag{21}$$

As likelihood functions, the relations (21) can be written as:

$$\Gamma_0(F(x)) = P\{\gamma_1(z) = 0, \gamma_2(z) \le x\} = P\left\{z \ge -\ln\left|\tanh\left(\frac{x}{2}\right)\right|\right\} \text{ and}$$
$$\Gamma_1(F(x)) = P\{\gamma_1(z) = 1, \gamma_2(z) \le x\} = P\left\{z \le \ln\left|\tanh\left(\frac{x}{2}\right)\right|\right\} \tag{22}$$

This is important because if G_{z_1} and G_{z_2} are two characteristic functions of the Γ type, then a function marked by a random value $z = z_1 + z_2$ corresponds to a characteristic function determined by $G_{z_1} \otimes G_{z_2}$ [2].

The description of the evolution density for LDPC codes is achieved with an operator function of the $\Gamma(s, F(x))$ type, which is the characteristic function associated with the LDPC code. So, for a certain class of LDPC codes the group of characteristic functions specific to that class of codes is defined.

The function $\Gamma(s, F(x))$ has the quality of permitting its mirrored opposite.

Also, given that G and H are two characteristic functions of the Γ type, that satisfy the relations (20), then the convolution between the G and H distribution types is [2]:

$$G \otimes H = \chi_{\{s=0\}} \cdot \left[\left(G^0 \otimes H^0\right) + \left(G^1 \otimes H^1\right) \right] + \\ + \chi_{\{s=1\}} \cdot \left[\left(G^0 \otimes H^1\right) + \left(G^1 \otimes H^0\right) \right] \tag{23}$$

2.2. Describing the evolution density of the LDPC codes with the help of characteristic function

The $\int P_l$ and $\int Q_l$ functions, are associated to the probable densities given by the cost functions $m_{vc}^{(l)}$ and $m_{cv}^{(l)}$, where P_l and Q_l are probabilities. By observing Figure 1 the equations for the messages under belief propagation can be written [3]:

$$m_{vc}^{(l)} = \begin{cases} m_0, \ if \ l = 0 \\ m_0 + \sum_{c^* \in C_v \setminus \{c\}} m_{c^* v}, \ if \ l \geq 1 \end{cases}, m_{cv}^{(l)} = \gamma^{-1}\left(\sum_{v^* \in V_c \setminus \{v\}} \gamma\left(m_{v^* c}^{(l-1)}\right) \right) \tag{24}$$

where C_v is the parity node set that converge from the variable node v, V_c is the variable node set that converge from the c parity node, and m_0 is the initial likelihood density (that corresponds to an initial LLR for a $x = \pm 1$ bit).

By reanalyzing the (4) and (5), one can write:

$$\lambda(x) = \sum_{i \geq 2} \lambda_i \cdot x^{i-1} \text{ and } \rho(x) = \sum_{i \geq 2} \rho_i \cdot x^{i-1} \tag{25}$$

Taking into account the (14), (21), (22) and (25) relations, as well as the association between the characteristic function $F(x)$ and the function $\gamma(x)$, we obtain [2]:

$$Q_l = \Gamma^{-1}(\rho(\Gamma(P_{l-1}))) = \Gamma^{-1}\left(\sum_{i \geq 2} \rho_i (\Gamma(P_{l-1}))^{\otimes(i-1)} \right) \tag{26}$$

where Γ^{-1} is the reverse of the characteristic function Γ;

Relation (26) expresses the fact that the density of the message passed from check node c to variable node v at round l is equal to Q_l. Also we write that the density of the message passed from variable node v to check node c at round l is equal to P_l.

Taking into account the function $\gamma(x)$ definition and the fact that the convolution of the characteristic functions G_{z_1} and G_{z_2} has as correspondent the sum of the random variables z_1 and z_2, as well as the relation (24), it is determined that:

$$P_l = P_0 \otimes \lambda(Q_l) = P_0 \otimes \sum_{i \geq 2} \lambda_i (Q_l)^{\otimes(i-1)} \tag{27}$$

where P_0 is the density of the random variable describing the channel;

By combining the (26) and (27) relations, the evolution density for the LDPCs is obtained through a recursive mathematical formula with values of P_l and P_{l-1} described with the help of the Γ characteristic functions:

$$P_l = P_0 \otimes \lambda\left(\Gamma^{-1}(\rho(\Gamma(P_{l-1})))\right) \tag{28}$$

where l being the iteration value.

3. Representing and simulating the evolution density for a particular LDPC case

To being with we shall present graphically in MathCad 14 the likelihood thresholds (f0(x) and f1(x)):

$$f0(x) := -\ln\left(\left|\tanh\left(\frac{x}{2}\right)\right|\right) \tag{29}$$

$$f1(x) := \ln\left(\left|\tanh\left(\frac{x}{2}\right)\right|\right) \tag{29'}$$

Those functions determined by the relation (22) for different values of the random variable x, have the followings representation:

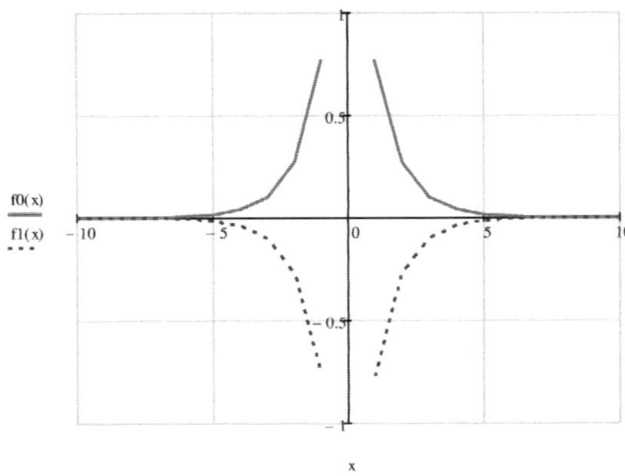

Figure 6. The simulation of the f0 and f1 functions for x with values between -10 ÷ 10

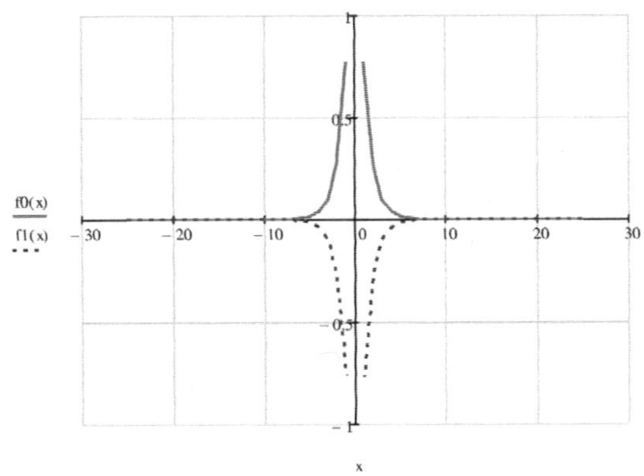

Figure 7. The simulation of the f0 and f1 functions for x with values between -25 ÷ 25

IOP Conf. Series: Materials Science and Engineering **163** (2017) 012032 doi:10.1088/1757-899X/163/1/012032

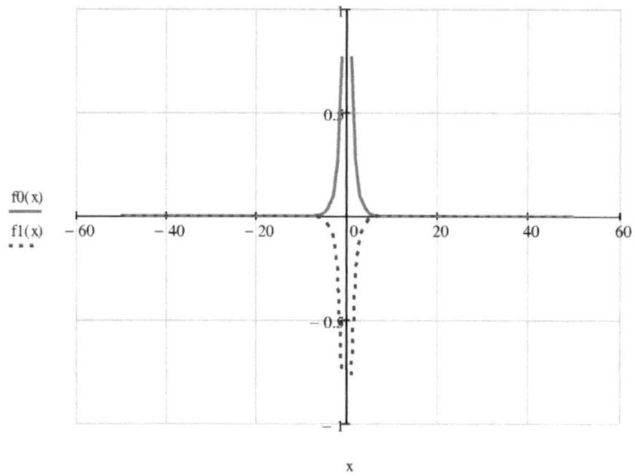

Figure 8. The simulation of the f0 and f1 functions for x with values between -50 ÷ 50

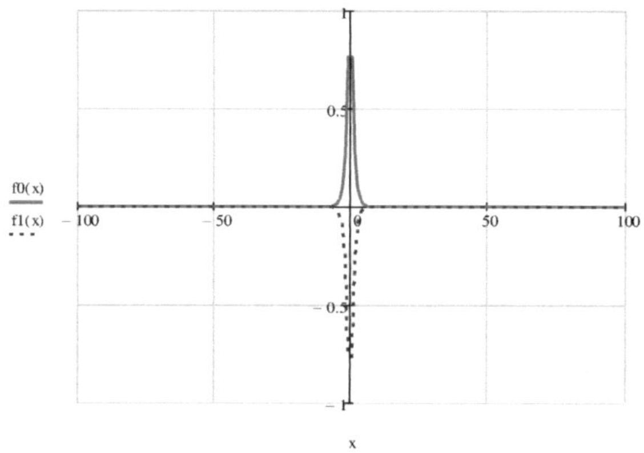

Figure 9. The simulation of the f0 and f1 functions for x with values between -100 ÷ 100

It can be observed that the larger the values of *x*, the more the likelihood thresholds go towards zero. Therefore, the bigger the code dimension, the better its decoding goes.

We simulate and represent the initial evolution density (that corresponds to the zero iteration) for an entrance binary channel which is affected by white Gaussian noise (BIAWGN), and that has the initial evolution density determined by the following relation [2]:

$$P_0(l)=\sqrt{\frac{\sigma^2}{8\cdot\pi}}\cdot e^{\frac{\left(l-\frac{2}{\sigma^2}\right)^2\cdot\sigma^2}{8}} \tag{30}$$

where σ represents the signal - noise ratio, and l the iteration value.

We have chosen to simulate the evolution density for the BIAWGN channel as well as for the case of a regular LDPC, because, in this context the evolution density of the messages has a continuous distribution, which makes it easier the represent in MathCad 14 [12], as opposed to a discrete distribution. In the simulation different values for σ will be given. Also, taking into account relation (28) we shall simulate P_0 for different iterative steps (P_{01}, P_{02} and P_{03}, P_{11}, P_{21} and P_{31}):

$\sigma_1 := 0.8,\ \sigma_2 := 0.5,\ \sigma_3 := 1$

e := 2.71, π := 3.1415, l := −10..50

$$P01(l) := \sqrt{\frac{(\sigma_1)^2}{8\cdot\pi}}\cdot e^{\frac{-\left[1-\frac{2}{(\sigma_1)^2}\right]\cdot(\sigma_1)^2}{8}} \qquad P02(l) := \sqrt{\frac{(\sigma_2)^2}{8\cdot\pi}}\cdot e^{\frac{-\left[1-\frac{2}{(\sigma_2)^2}\right]\cdot(\sigma_2)^2}{8}}$$

$$P03(l) := \sqrt{\frac{(\sigma_3)^2}{8\cdot\pi}}\cdot e^{\frac{-\left[1-\frac{2}{(\sigma_3)^2}\right]\cdot(\sigma_3)^2}{8}}$$

$$P11(l) := P01(l)^2,\ P21(l) := P01(l)^3,\ P31(l) := P01(l)^5$$

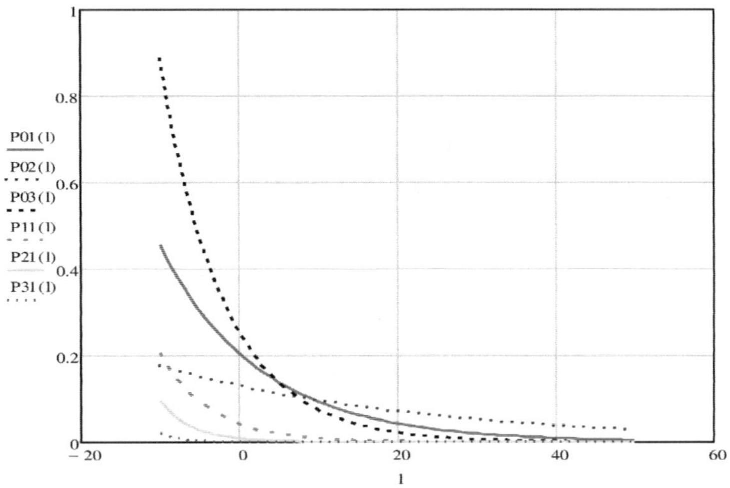

Figure 10. The simulation of the evolution density of messages

4. Results, conclusions, observations, future research directions

One notices that the more utterances there are in Figure 10, P_l decreases, therefore $\ln P_l$ increases, thus the evolution density of the messages that go through a Tanner graph also increases. The simulation was carried out for a regular LDPC $(d_v = 3, d_c = 6)$, with a ½ rate. A more simple LDPC was chosen, because the iteration computing according to the relation (28) is very complex. Aproximations have been carried out, taiking into accound that $F(x)$ is a function with a cumulative distribution [9]:

$$\lim_{x\to\infty} F(x) = 1 \tag{31}$$

Therefore we may consider $F(x)$ being constant on several portions.

For the same values $(d_v = 3, d_c = 6)$ we have considered in relation (28) $\lambda(x) = x^2$ and $\rho(x) = x^5$. Thus, we have simplified greatly the recursive relation of evolution density. As a consequence of this fact, the decoding algorithms are linear.

The evolution density represents the essence of the decoding process of LDPCs, the recursive function by which it is expressed being directly associated to a covering mode of a Tanner graph by the code words. It is an analysis method based on asymptotic edges.

Based on this principle all decoding algorithms (soft or hard) are implemented. The most known are the Gallager A and B [1] algorithms. In the last years there have been implemented algorithms taking into account the rapid Fourier and Laplace transformations of the characteristic functions, this being an actual and future study problem.

The study of the characteristic functions and of the evolution density is important because the implementing of all soft and hard decoding algorithms at LDPCs is carried out with their help.

Acknowledgment

The work has been funded by the Sectoral Operational Programme Human Resources Development 2007-2013 of the Ministry of European Funds through the Financial Agreement POSDRU/159/1.5/S/132395.

References

[1] Gallager R G 1963 *Low Density Parity-Check Codes*, M.I.T. Press, Massachusetts, USA

[2] Richardson T J, Shokrollahi A and Urbanke R L 2001 Design of Capacity-Approaching Irregular Low-Density Parity-Check Codes, *IEEE Transactions on Information Theory* **47**(2) 619-637

[3] Richardson T J and Urbanke R L 2001 The Capacity of Low Density Parity Check Codes under Message-Passing Decoding, *IEEE Transactions on Information Theory* **47**(2) 599-618

[4] Drăghici S Scientific report No. 1 – Ph.D. thesis, unpublished

[5] https://www.youtube.com/watch?v=_45M-dO99-M&list=PL85783E9E7B5F2C0D&index=31, existing at 27.01.2016

[6] Richardson T J and Urbanke R 2004 Fixed points and stability of density evolution, *Communications in Information and Systems* **4**(1) 103-116

[7] Draghici S, Proştean O, Răduca E, Anghel-Drugărin C, Chioncel C, Răduca M and Rudolf C 2015 The relations between the mathematic ideal, the generating polynome and the matrixes G, H used in encodings, *Procedia - Social and Behavioral Sciences* **197** 2032-2039

[8] Shokrollahi A 2003 *LDPC Codes: An Introduction* Technical Report Digital Fountain Inc. - Zürich

[9] https://en.wikipedia.org/wiki/Cumulative_distribution_function, existing at 27.01.2016

[10] https://en.wikipedia.org/wiki/Lebesgue%E2%80%93Stieltjesintegration

[11] https://www.youtube.com/watch?v=bGup-R--bEU&index=32&list=PL85783E9E7B5F2C0D, existing at 27.01.2016

[12] Macfarlan K H and McClain S T 2011 *The use of Mathcad functions in teaching an undergraduate aerospace propulsion course* ASME Turbo Expo Conference, Vancouver, Canada, June 6-10, pp 505-510

IOP Conf. Series: Materials Science and Engineering **163** (2017) 012033 doi:10.1088/1757-899X/163/1/012033

Analysis of synchronous and induction generators used at hydroelectric power plant

C M Diniş, G N Popa and A Iagăr

Politehnica University of Timisoara, Department of Electrical Engineering and Industrial Computer Science, Hunedoara, Romania

E-mail: corina.dinis@fih.upt.ro

Abstract. In this paper is presented an analysis of the operating electric generators (synchronous and induction) within a small capacity hydroelectric power plant. Such is treated the problem of monitoring and control hydropower plant using SCADA systems. Have been carried an experimental measurements in small hydropower plant for different levels of water in the lake and various settings of the operating parameters.

1. Introduction

In this age increasing demand of energy has forced us to look at the other options to make electrical energy, different from conventional means of exploiting energy as conventional sources of energy are exhaustible. The starting point energy crisis was the oil crisis [1], [2].

As the awareness came in early seventies also came the search for the alternatives. Also it was required that the alternatives should be such that, they should not be the major changes required in the system. Renewable energy emerged as the best option. The major renewable are hydro, solar, wind and biomass. The beauty of these renewable energy systems is that they are non-exhaustible [1-3].

Hydropower is one of the most promising available energy sources in the world. However, with limited potential and other problem like high construction cost, environmental concerns due to impounding of river water and long gestation period of large hydro have ceased to be offering any long-term solution for our increasing power demand problem. Small hydropower on the other hand, can be built in less time and are likely to create less environmental problems [2], [3].

Of all the non-conventional energy sources, small hydro represents the highest density resource and stands in the first place in generation of electricity from such sources throughout the world [3], [4].

Hydropower plants are complex installations where hydraulic energy of natural or artificial water falls is converted into mechanical energy through hydraulic turbines and then into electricity in power generators.

The potential to produce electricity depends on the drop and water flow. Their operating principle is to transform the potential energy of water abstraction in lakes into mechanical energy. From the lake, through the penstock, water falls on a hydraulic turbine blades rotating shaft. This drives the electric generator, which converts mechanical energy into electricity [3].

After the power produced, hydro are of three kinds: hydroelectric plants with an installed capacity exceeding 100 kW, small hydropower plants with installed capacity of 5 to 100 kW and micro plants built on brooks, do not have the Barrage and installed capacity under 5kWh [3].

A hydraulic turbine is composed of three main bodies:

- a distributor that prints the fluid speed and direction convenient size for the attack rotor in desired optimal conditions with minimal load losses;
- a rotor fitted with blades or buckets that aims to transform the hydraulic energy into mechanical energy;
- a vacuum device, which recovers the form of pressure, the kinetic energy of the water that has to exit from the rotor and draining downstream. This piece is missing in action turbines.

Table 1. Classification of types of turbines

Turbine type	Fall of, m		
	Big (150... 2000 m)	Medium (50...150 m)	Small (3...50 m)
Action	Pelton Turgo	Banki Turgo	Banki
Reaction	-	Francis	Propeller Kaplan

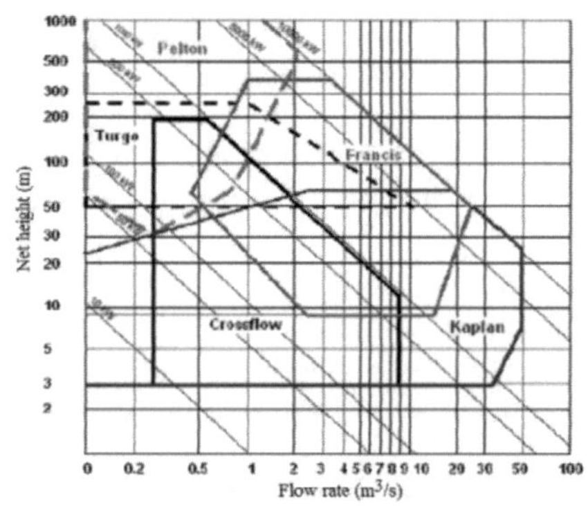

Figure 1. Pressure-flow characteristics of different types of turbines

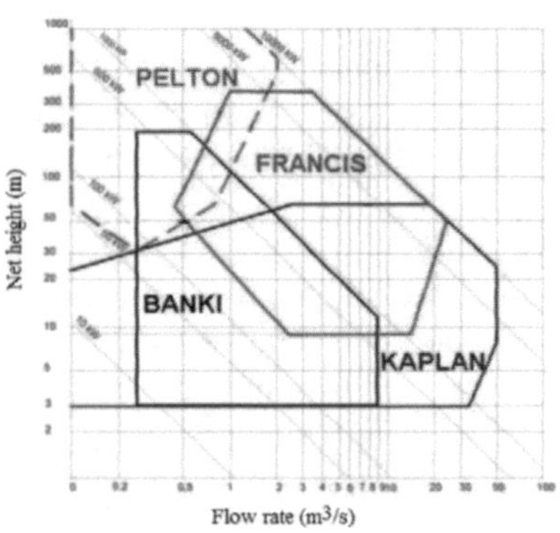

Figure 2. Nomogram of selecting the turbines for small hydropower plant

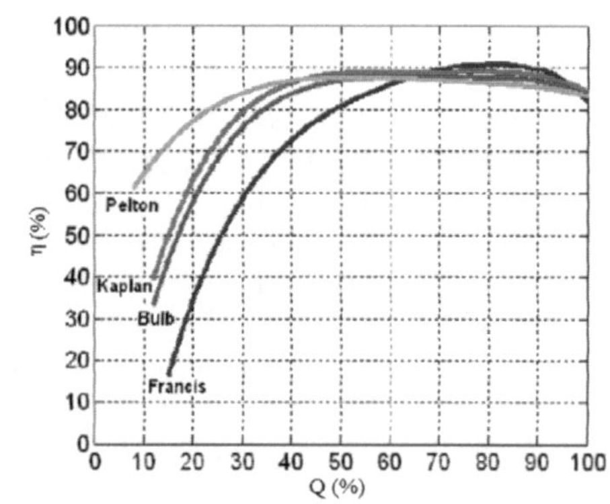

Figure 3. Efficiency-flow rate characteristics of different types of turbines

Type selection, turbine geometry and dimensions depend mainly on falling, the fluent flow and rotor speed (Figure 1). Turbine used for small or medium failures are most often the reaction and include turbines Francis and Kaplan turbines with blades fixed or variable. The turbines used for high dropping arrangements are acting. These include Pelton turbines, Turgo and Banki (cross flow) (Figures 2, 3) [3].

Before the existence of automation, qualified personnel operated the equipment manually called as the manual system where the automatic system reads the information on the equipment status operation and then activates commands or controls to optimize the output production. Therefore the need for automation for a small hydroelectric power plant is to improve the efficiency, productivity and the operating management of the system which solves the problem of production needs [1], [2], [5].

2. The main equipment used in hydropower plant

At hydropower plant to which were made measurements were use two generators: induction made by Marelli (Figure 4) having characteristics P_n = 232 kW, cosφ = 0.9, n = 1500 rpm and synchronous made by AVK (Fig.5) with the characteristics P_n = 950 kW, cosφ = 0.9 and n = 600 rpm [6-8].

Figure 4. Induction generator

Figure 5. Synchronous generator - side view

Figure 6. Vertical Francis turbine FV 335/330

Figure 7. Horizontal Francis turbine FO 285/720 – spiral room and director device

Also in the same hydropower plants uses a vertical Francis turbine FV335/330 (Figure 6) and a horizontal Francis turbine FO 285/720 (Figure 7) powered by penstock from a barrage situated within the distance of about 350 m and a net height of 30 m. Vertical Francis turbine coupled to the induction generator and executes 1500 rpm and horizontal Francis turbine coupled to the synchronous generator and runs at 600 rpm [6-8].

IOP Conf. Series: Materials Science and Engineering **163** (2017) 012033 doi:10.1088/1757-899X/163/1/012033

3. Monitoring and control of a hydropower plant using SCADA system

The automation system is realized modular PLCs Schneider: TWD LCDE 40DRF – module CPU (24xDIN + 16Xdout), TM2 DDI 32DK – extension module (32xDIN), TM2 AMI 4LT – extension module (4xAIN), TM2 DDO 16TK – extension module (16xDOUT), TM2 AMI 8HT extension module (8xAIN), TM2 ARI 8LT – extension module (8xAIN), TM2 AVO 2HT – extension module (2xAOUT) and software [1], [4], [9].

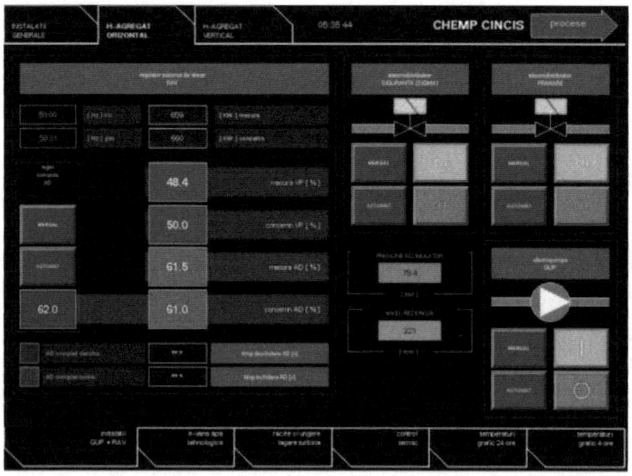

Figure 8. SCADA system interface

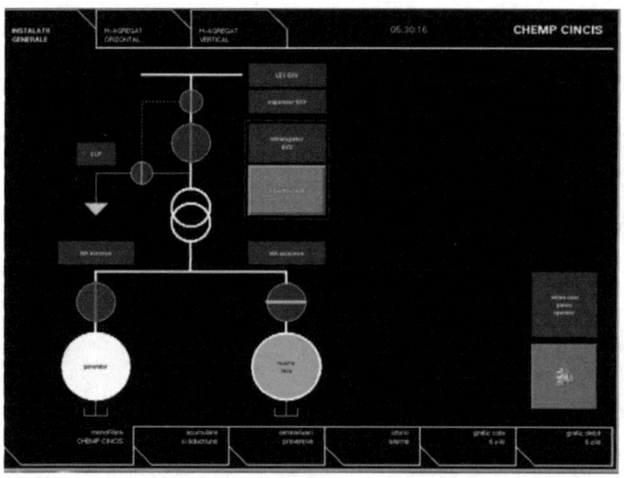

Figure 9. The interface for single wire electric diagram of hydropower station

In Figure 8 is shown SCADA system interface from hydro plant for horizontal turbine is connected to the synchronous generator AVK type. From here it can run the command to the device director. Generally, director device works in automatic mode and helping it move, depending on the water level and pressure in penstock. There is the risk not to achieve production set when the device director remains locked. In this case, it will switch to manual mode and adjust the extent to unlock. At the top of the interface seeks instruction (power provided by the operator) and measure (what power achieved by generator) [1], [4], [5], [10].

IOP Conf. Series: Materials Science and Engineering **163** (2017) 012033 doi:10.1088/1757-899X/163/1/012033

At the time shown in Figure 9 HA circuit breaker is coupled to the network, this is indicated by the red vertical line in the figure, and induction generator is disconnected from the network, this is indicated in the figure by the horizontal green line. "Cold reserve" indicates that the generator does not qualify the conditions to startup. When all conditions are met it will show "hot reserve". At the top the diagram continues with oil transformer (Trafo 1250 kVA; 20/6/0,4 kV) this raising voltage of 4 kV to 6 kV. The grounding separator (CLP) can be closed only if the disconnector (bar) is open, and the disconnector (bar) can be closed only if the grounding separator (CLP) is open. The diagram ends with 6 kV underground lines. From this interface is the option to disconnect the transformer from the red button [9–12].

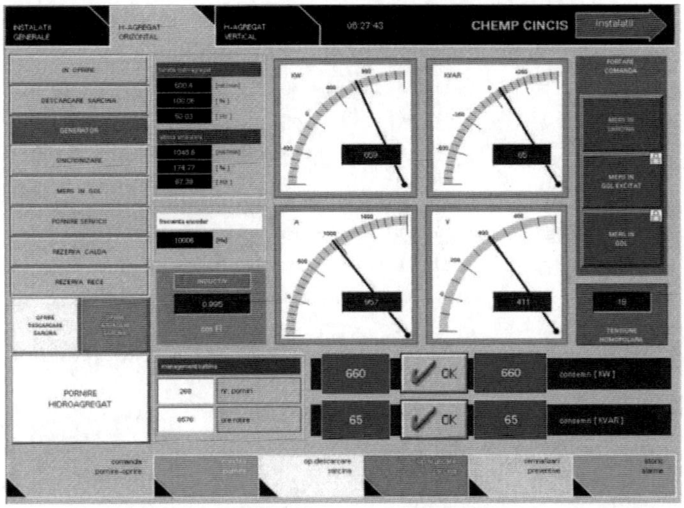

Figure 10. The interface for viewing the main electrical parameters (active power, reactive power, voltage, current, power factor) to horizontal hydro-aggregate and turbine FO management

In interface shown in Figure 10 the operator has the ability to start and stop the horizontal hydro-aggregate. Also, the operator can adjust the active power and reactive power of synchronous generator. The reactive power of synchronous generator must not exceed 10% of its active power.

In the middle of interface can be viewed using virtual instruments placed on interface, the active power and reactive power of synchronous generator, and the operating current and voltage generator.

Also, for this interface it can view the speed at which the hydro-aggregate, the speed at which the final product packaging of hydro-aggregate, power factor of the synchronous generator, and the number of starts and hours of rotation of the synchronous generator.

In interface shown in Figure 11, the operator has an opportunity to start and stop vertical hydro-aggregate unit and can adjust the active power of induction generator. In the middle interface can be viewed using virtual instruments placed on interface the active power induction generator and reactive power as well as voltage and current on which they operate the generator. Also, this interface can view the speed on which they operate hydro-power vertical unit, the speed at which the final product packaging hydro-power and the number of starts and hours of rotation of the induction generator [9-12].

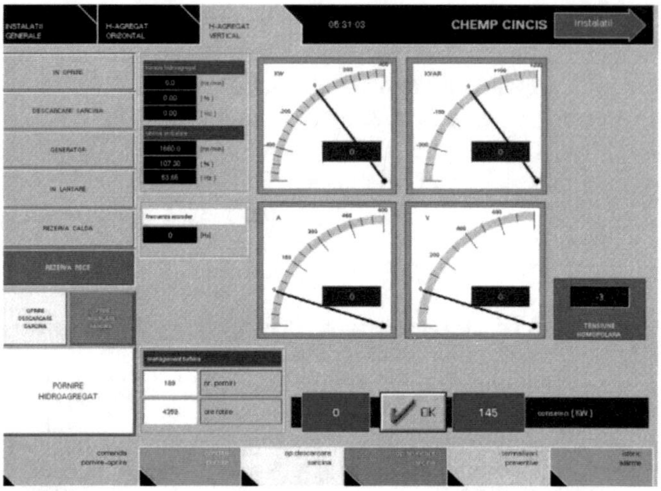

Figure 11. The interface for viewing the main electrical parameters (active power, reactive power, voltage, current, power factor) to vertical hydro-aggregate and turbine FV management

Figure 12. Command and control equipment enclosures for SCADA system from hydro plant

Figure 13. Contactors cabinets for both generators (synchronous and induction)

4. Experimental results

In the hydropower plant analyzed operation of AVK synchronous generator and operation of induction generator Marelli, at different levels of water in the lake and various settings of the operating parameters. Hydro- aggregates from hydro plant can operate in two modes: automatic mode - when the unit director of the turbine is automatically adjusted according to pressure inside penstock, and manual mode - when the unit director of the turbine is adjusted by the operator (manual) [6-8].

For operation of AVK synchronous generator at the lake by 292.40 mdM (meters over sea) (N1) and 290.50 mdM (N2), in manual mode, adjusting of the opening director device from 25% to 100%, the ambient temperature being 15 °C and in the hydropower plant hall being 22 °C have been obtained measured values from Table 2.

Table 2. Parameters measured when operating synchronous generator of hydropower plant

AD (%)	P_1 (kW)	P_2 (kW)	Q_{t1} (m^3/s)	Q_{t2} (m^3/s)	p_1 (bar)	p_2 (bar)
20	169	147	1.10	1.03	3.00	2.85
25	323	236	1.38	1.41	2.99	2.80
30	374	302	1.65	1.69	2.97	2.77
35	400	352	1.91	1.92	2.94	2.74
40	436	392	2.16	2.11	2.90	2.71
45	489	428	2.41	2.30	2.85	2.67
50	550	468	2.67	2.51	2.80	2.62
55	606	512	2.93	2.75	2.75	2.55
60	648	546	3.17	2.93	2.70	2.50
65	671	552	3.35	2.96	2.67	2.49
70	680	555	3.50	2.93	2.64	2.50
75	684	553	3.63	2.92	2.60	2.50
80	696	554	3.75	2.93	2.55	2.50
85	650	553	3.87	2.94	2.51	2.53
90	645	555	4.00	2.93	2.50	2.54
95	640	546	4.05	2.97	2.43	2.54
100	640	542	4.09	3.00	2.39	2.53

Due to some calculation errors in the design of penstock and the FO turbine is observed (in Figure 14) from the power curve that after opening the unit director at bigger value than 80%, the achieved power across generator drops, water consumption increases and the pressure inside penstock decreases.

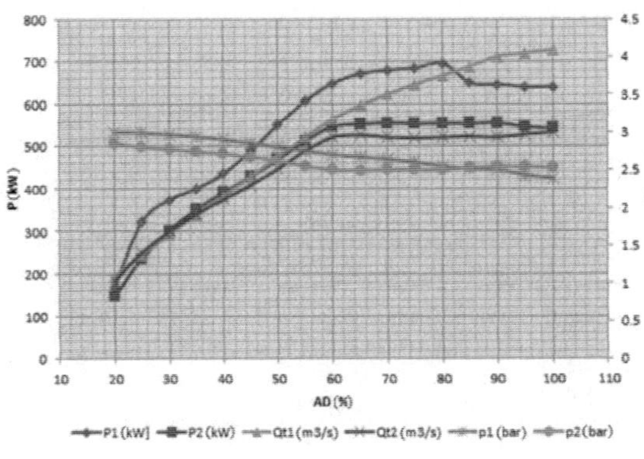

Figure 14. Diagrams for P, Q_t, p for hydro-aggregate-FO, depending on the level and AD opening (N_1=292.40mdM and N_2=290.50 mdM)

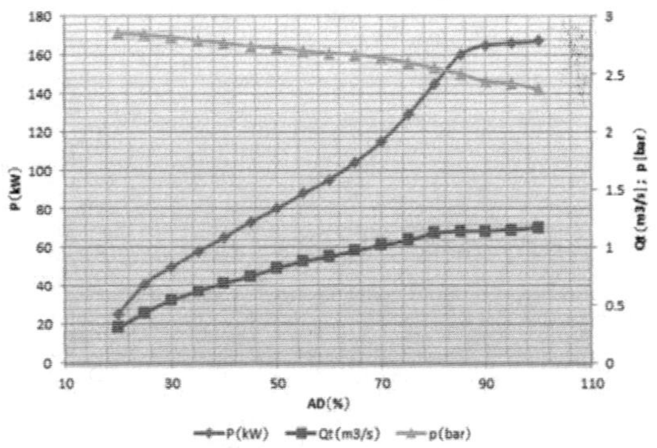

Figure 15. Diagrams for P, Qt, p for hydro-aggregate-FV, depending on the level and AD opening (N_3=289.80 mdM)

For operation Marelli induction generator at lake level of 289.80 mdM (meters over sea) (N3) in manual mode, adjusting the opening unit director from 20% to 100%, the ambient temperature being 10 °C, and in small hydropower plant hall being 15° C (Figure 15) were obtained measured values from Table 3.

For the operation of synchronous generator AVK type in parallel (simultaneously) with induction generator Marelli at lake level of 293 mdM in manual mode, adjusting the opening unit director from 20% to 100%, ambient temperature being 20 °C, and in small hydropower plant hall temperature being 27 °C (Fig. 16) were obtained measured values shown in Table 4.

Table 3. Parameters measured when operating induction generator of hydropower plant

AD (%)	P (kW)	Q_t (m^3/s)	p (bar)
20	25	0.30	2.85
25	41	0.43	2.84
30	50	0.54	2.82
35	58	0.62	2.79
40	65	0.69	2.77
45	73	0.75	2.74
50	80	0.82	2.73
55	88	0.88	2.70
60	95	0.92	2.68
65	104	0.97	2.67
70	115	1.02	2.64
75	129	1.06	2.60
80	145	1.12	2.55
85	160	1.14	2.50
90	165	1.14	2.44

Table 4. Measured parameters for parallel operation of the generators from hydro power plant

AD_{FO} (%)	AD_{FV} (%)	$P_{HA\text{-}FO}$ (kW)	$P_{HA\text{-}FV}$ (kW)	P_{total} (kW)	Q_t (m^3/s)	p (bar)
20	20	133	36	169	1.10	2.85
25	25	193	52	245	1.38	2.80
30	30	247	66	313	1.65	2.73
35	35	297	80	377	1.92	2.67
40	40	344	92	436	2.19	2.66
45	45	389	105	494	2.53	2.62
50	50	433	117	550	2.97	2.56
55	55	477	128	605	3.30	2.50
60	60	511	137	648	3.60	2.50
65	65	528	142	670	3.71	2.45
70	70	536	144	680	3.86	2.45
75	75	541	146	687	3.95	2.44
80	80	545	147	692	4.04	2.43
85	85	549	148	697	4.15	2.43
90	90	547	149	696	4.22	2.42
95	95	545	148	693	4.35	2.39
100	100	544	146	690	4.44	2.33

For the operation of synchronous generator AVK type at lake level of 293.05 mdM (N4) in automatic mode, within 24 hours, with two different power set on the operator panel, the ambient temperature being 15 °C, and in the hydropower plant hall temperature 22 °C were get measured parameters shown in Table 5.

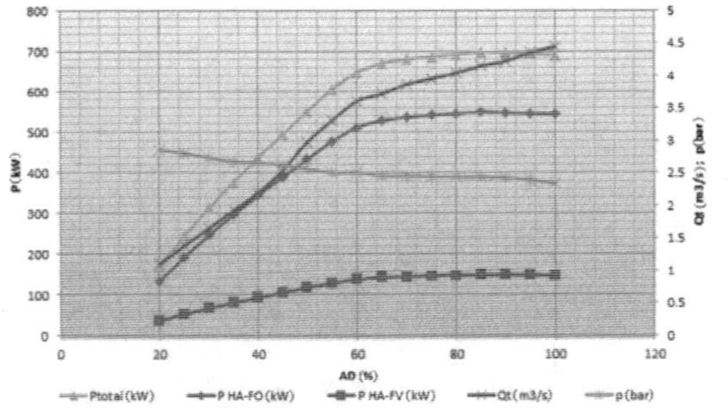

Figure 16. Diagrams for P, Qt, p for hydro-aggregate-FV and hydro-aggregate -FO (operating in parallel) depending on water level and opening AD (N_4=293 mdM)

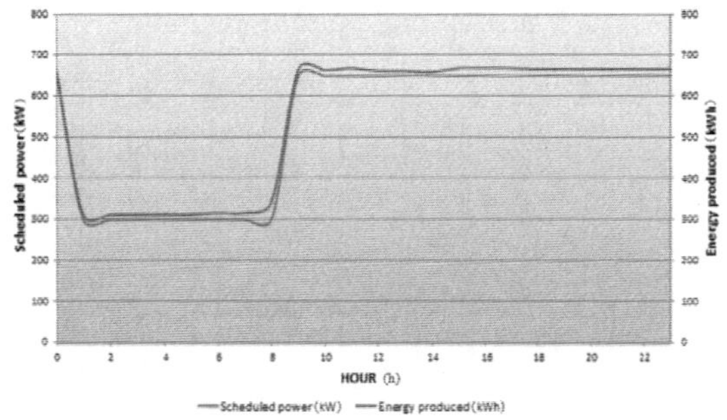

Figure 17. Diagrams for power programmed and produced energy for operation hydro-aggregate-FO in automatic mode at the level N_4=293.05 mdM, within 24 hours

Table 5. Measured parameters for the functioning hydro-aggregate FO in automatic mode, within 24 hours

Hour	Scheduled power [kW]	P_{HA-FO} [kW]	Energy produced [kWh]	Q_t [m^3/s]	p [bar]
0	650	315	664	1.80	3.10
1	300	315	312	1.80	3.10
2	300	315	312	1.80	3.10
3	300	315	312	1.80	3.10
4	300	315	312	1.80	3.10
5	300	315	312	1.80	3.10
6	300	315	316	1.80	3.10
7	300	315	316	1.80	3.10
8	300	665	343	1.80	3.10
9	650	665	667	3.30	2.84
10	650	665	664	3.30	2.84
11	650	665	669	3.30	2.84
12	650	665	662	3.30	2.84
13	650	665	662	3.30	2.84
14	650	665	660	3.30	2.84
15	650	665	669	3.30	2.84
16	650	665	669	3.30	2.84
17	650	665	669	3.30	2.84
18	650	665	667	3.30	2.84
19	650	665	667	3.30	2.84
20	650	665	667	3.30	2.84
21	650	665	667	3.30	2.84
22	650	665	667	3.30	2.84
23	650	665	667	3.30	2.84

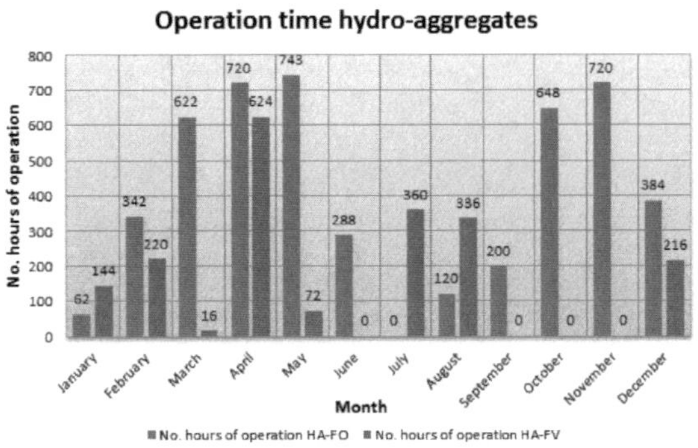

Figure 18. Monthly chart of number of operating hours for both hydro-aggregates from the hydropower plant

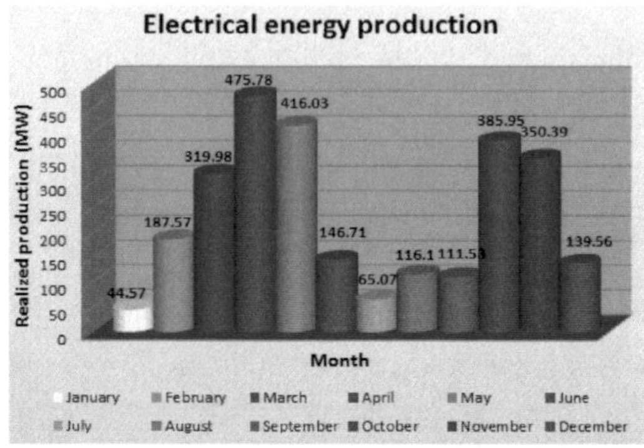

Figure 19. Monthly chart production electric energy achieved with both hydro-aggregates from the hydropower plant

5. Conclusions

Hydraulic energy conversion into electrical energy is not polluting, requires relatively low maintenance costs, no problems with fuel and is a long-term solution [3].

Hydropower plants have the lowest operating costs and highest lifetime compared to other types of power plants. There are more than a century of experience in the development and operation of hydropower plants, which makes them reach levels high technical performance [3].

Advantages of hydropower plants consist of low cost electricity, requires no fuel, requires low maintenance and operating personnel, do not pollute, they have high operational safety and high efficiency.

By using hydropower plants can be solved free public lighting and supply of electricity for public buildings. Small hydropower plants can benefit from green certificates, so basically they can sell at a price of 3-4 times greater energy produced if connecting to the national system electricity power supply. Also, small hydropower plants does not pollute nature, does not emit carbon dioxide, and the maintenance and operating costs are very low [3].

SCADA systems provide major added reliability and productivity at the level factories in every type of industry, and implicitly to hydropower plants, offering both operators and process engineers or managers, a full overview of all production equipment. Viewing the production values of both real-time and in the past, it helps increase the speed and quality in decision-making, offering operators a reliable tool in performing the daily tasks [1], [2], [4], [5], [9-12].

By implementing algorithms for instant alarm for exceeding the normal parameters of the production is increase the reliability and efficiency of the hydropower plant equipment, operators can avoid timely damage which could lead to production stops. Also, the presented solution enables the operator to replace the quickly the elements that could be damaged (sensors, transducers, actuators, etc.).

It is observed that the measurements made in the hydropower plant functions both hydro-aggregates only when the water level exceeds 293 mdM (meters over sea). When the lake level is very low (less than 290 mdM quota) will operate only vertical hydro-aggregate, and when the lake level is medium (between 290 mdM and 293 mdM quota) will only operate horizontal hydro-aggregate unit.

References

[1] Rajeswari V, Rajeshwari Y and Padma Suresh L 2012 Real- Time Implementation of Hydroelectric Power Plant Using PLC and SCADA, *International Journal of Engineering Research and Applications* **2**(3) 899-902

[2] Gupta R, Singh S N and Singal S K 2007 *Automation of Small Hydropower Station,* International Conference on Small Hydropower – Hydro, Sri Lanka, October 22-24, pp 1-9

[3] Popa B and Paraschivescu A V 2007 *Introducere în utilizarea energiei apelor*, Editura Politehnica Press, București, Romania

[4] Guozhen H, Tao C, Changsong C and Shanxu D 2009 *Solutions for SCADA System Communication Reliability in Photovoltaic Power Plants*, IEEE 6th International Conference Power Electronics and Motion Control IPEMC'09, Wuhan, China, May 17-20, pp 2482-2485

[5] Kaur S, Kathpal N and Munjal N 2015 Role of SCADA in Hydro Power Plant Automation, *International Journal of Advanced Research in Electrical, Electronics and Instrumentation Engineering* **4**(10) 8085-8090

[6] Deaconu S I, Deaconu R and Latinovic T 2013 High Power Static Converters in Industry Applications, *Annals of Faculty Engineering Hunedoara – International Journal Of Engineering* **XI**(4) 301-394

[7] Tutelea L N, Deaconu S I and Popa G N 2015 Dual stator winding variable speed asynchronous generator: optimal design and experiments, *IOP Conf. Ser.: Mater. Sci. Eng.* **85** 012010

[8] Tutelea L N, Muntean N, Deaconu S I and Cunţan C D 2016 Dual stator winding variable speed asynchronous generator: magnetic equivalent circuit with saturation, FEM analysis and experiments, *IOP Conf. Ser.: Mater. Sci. Eng.* **106** 012029

[9] Diniş C M and Popa G N 2014 Measurements in SCADA System Used at a Wastewater Treatment Plant, *Annals of Faculty Engineering Hunedoara – International Journal Of Engineering* **XII**(4) 207-215

[10] Glattfelder A H, Husser L, Dörfler P and Steinbach J 2009 *Control Systems, Robotics and Automation Vol XVIII,* Encyclopedia of Life Support Systems (EOLSS), Automatic Control for Hydroelectric Power Plants, pp 283-314

[11] Mathew J, Shankar S S et al. 2014 *Implementation of high availability SCADA system for superheater steam temperature control in a 210 MW thermal power plant*, IEEE International Conference on Electronics, Computing and Communication Technologies, January 6–7, Bangalore, India, pp 1-6

[12] Lakhoua M N and Jbira M K 2012 Project Management Phases of a SCADA System for Automation of Electrical Distribution Networks, *IJCSI International Journal of Computer Science* **9**(2) 157-162

IOP Conf. Series: Materials Science and Engineering **163** (2017) 012034 doi:10.1088/1757-899X/163/1/012034

An experimental comparative analyse between three low capacity PLCs

G N Popa, C M Diniş, A Iagăr and I Popa

Politehnica University of Timisoara, Department of Electrical Engineering and Industrial Informatics, 5 Revolution Street, Hunedoara, 331128, Romania

E-mail: gabriel.popa@fih.upt.ro

Abstract. PLCs are used increasingly more often in industrial applications and beyond. PLCs from different companies involved different programming languages. Some of the PLCs can be programmed in all languages, while others do not. For three types of PLCs was made an application for a traffic light with priority. PLCs usually have a general and a special set of instructions that only some of these are found from a PLC to another. Programs were made with specific instruction set of respectively PLC. The programs were carried out in three languages. For a case study, it makes a comparative analysis between three programming languages.

1. Introduction

Control process aims to increase the competitiveness of a product which can be done directly, through the cost and quality, and or indirectly by improving working conditions [1-3].

Programmable logic controllers (PLCs) are the main devices of today's industrial automation. They are more and more often included in different automation. PLCs used industrial microcontroller systems (in more recent times it meet microprocessors instead of microcontrollers), where hardware and software are designed to industrial environment. Basically, PLCs are simple microcomputers are used to solve problems of sequential logic. The goal of PLCs is replacing classical electrical diagrams conducted in wired logic with logic circuits and relays. PLCs can be used successfully in most industrial applications of small or medium complexity [4-8].

The first application of PLCs was done in the automotive. The goal was to replace the traditional control electrical diagrams made with logic circuits and relays, with cheaper equipment, more reliable and more flexible. In this way, the number of electrical connections between electrical equipment was significantly reduced [1], [5].

The first language was used for programming PLCs was Ladder Diagram (LAD), because it has the advantage that it can be used by people who have no training in informatics. Over time, there were hundreds of companies producing PLCs, each has developed its own variant of LAD, different to some extent from each other. International Electrotechnical Commission (IEC) conducted a standard IEC 1131 (later IEC 61131) for standard of instructions and blocks used at PLCs languages [9-12].

They were performed and other programming languages besides LAD, graphical languages for beginner users, and textual languages for the advanced users. So, have appeared PLCs that have general functions, that are found at the most basic instructions used in all PLCs, and special functions, which enabled the introduction of special instructions that may be different from one producer to another [2], [3]. However, working with a PLC must begin with knowledge of instructions and blocks used in programming [13], [14].

2. PLCs operation and software

Programs for PLCs are made cyclic (Figure 1) [1]. When the program starts, it checks the status of the PLC inputs. PLC's inputs are connected to sensors (for measurement of electrical and non-electrical quantities), which can be digital (on a bit), numerically (on the more bits, eg. 8) or analogue (eg. continuous voltage). Sensors in automation, depending on the type of construction, are connected to the inputs of the PLC. Status of inputs is written in image input table (which may be register or a file).

Thereafter, the program itself is performed to scan and solve the program lines depending on the state of inputs. The results are written in the output image table (register or file). After implementing the program, results from output image table are transferred to the output where actuators are connected. Actuators can be ordered digital, numerically or analogue. Then check the internal memory, speed of execution and operation. Also in this phase is made communication with other devices (eg. an operator panel). After that, the process is repeated. Depending on the type of PLC's, some stages of the operating cycle, may be missing or may be repeated (eg. in the same cycle of realization of the program can scan multiple times the inputs).

The scan time of PLC depending on the speed of the microcontroller (or microprocessor), the length of the program, the type of instructions executed, conditions inside the program. The PLC computes the scan time each time from the first instruction to the END instruction. Typical scan time data include the maximum scan time and the last scan time.

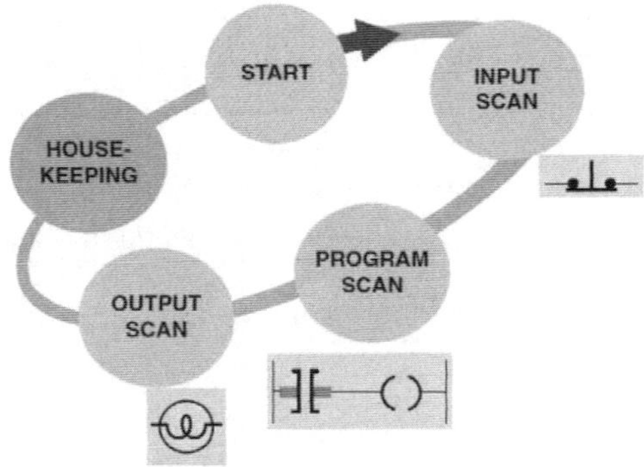

Figure 1. Typical program scan cycle for a PLC

PLCs can be programmed with specific programming languages (Figure 2). There are textual and graphical languages languages. Basically, graphical languages are easier than textual languages. Instruction List language (IL) has similar language with assembler language using specific mnemonics of PLC. Structured Text language (ST) is high level language, and is similar with C or Pascal languages.

Graphic languages are easier to use, especially for non-specialists in informatics. Ladder Diagram (LAD or LD) is similar with control circuit diagrams of various installations, but are rotated with 90^0. Programming in this language lends to those who worked in electrical installations (electricians). LAD used the so-called networks. Function Block Diagram (FBD) closely resembles the layout of the control electronic circuits and them interconnection. It is easy to use by those who have worked with electronic devices (electronics).

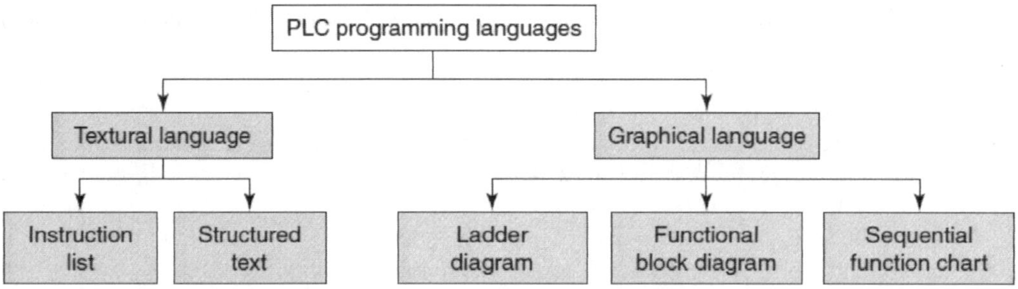

Figure 2. Types of PLC programming languages

Perhaps, most simple graphic language used in PLCs is Sequential Function Chart (SFC) which is similar to a flow chart that can be written by anyone. An SFC is composed with steps-actions and transitions-receptivity (Figure 3) [1]. Not all PLCs can be programmed in all languages; some may accept only one programming language (eg. IL by PS3 PLC from Klöckner Moeller).

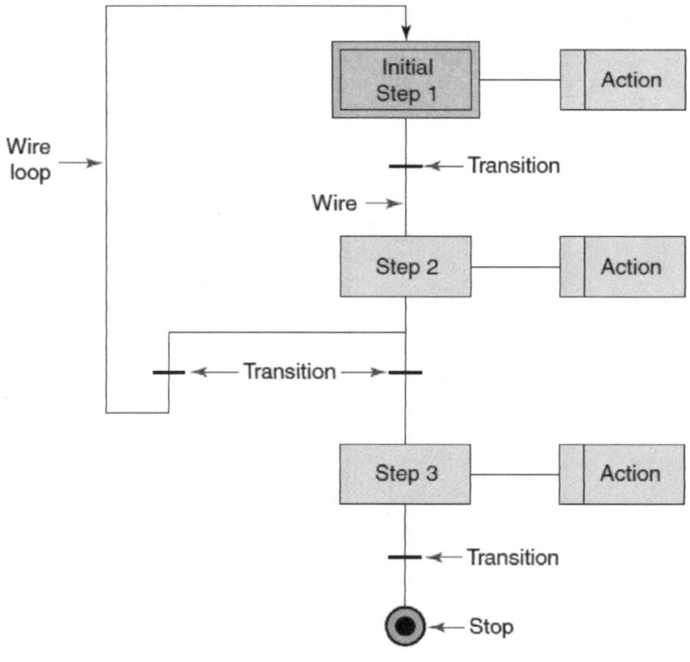

Figure 3. Example of SFC language

3. Comparative analyses between three PLCs. A case study

At comparative study were used three small capacity PLCs.

The first PLC from Klöckner-Moeller is PS3. It is an old PLC (constructed in three variants) that allows programming in IL on programmable console (PRG 3S). There use specific mnemonics for programming. It has 16 digital inputs, 8 digital outputs - relays (other variant has 16 digital outputs - transistors), 4 analogue inputs and 1 analogue outputs. At programming can be used 32 on-delay timers, 32 counters, clocks, comparators, etc. It does not support extension modules, but they can be connected into network with multiple PLCs.

The second PLC is S7 214 Siemens (there are designed four variants). There are two families: CPU210 (oldest) and CPU 220 (newer). S7 214 can program the PLC in three different languages (IL, LAD or FBD, with the possibility of automatic conversion from one to another) using a RS232 to RS

485 converter. The PLC supports general functions and special functions (on-delay timers, off-delay timers, with retention timers, comparators, up-down counter, comparators, clocks, etc.). It has 10 digital inputs and 8 digital outputs. It does not have analogue inputs/outputs (which can be a major drawback in some applications). It supports extension modules (with digital or analogue inputs and outputs, with specialized modules to connect the temperature sensor module, for controlling stepper motors, communications, etc.). S7 214 can connect to industrial networks through specialized modules.

The third type of PLC is LOGO! Siemens (designed in several variants). General supports functions and special functions (on-delay timers, off-delay timers, pulses, comparators, trigger, RS flip-flop, counters, etc.). It has 12 digital inputs and 8 digital outputs. It does not have analogue inputs /outputs, but can connect with various extension modules. Programming can be done on PC or with built console (in FBD).

To make comparative analysis is considered an application with traffic lights, with the primary road and secondary road, with priority for pedestrians crossing the main path.

An intersection has two paths, a primary road and a secondary road (Figure 4). Traffic lights at the corners of the intersection have three lights: red, yellow, and green. Traffic lights are synchronized, so that if the primary road is the green light, there is red light on secondary road and vice versa. The transition from green to red is both ways via the yellow warning light. The transition from red to green is without passing through intermediate yellow light. Since the primary road passing many cars, light green on the primary road stays on for longer time than the green light on the secondary road. To eliminate the disadvantage that pedestrians wishing to cross the main road staying too long, was installed at every traffic lights a switch (normally open contact) that allows those who wish to cross the main road, to request, after some time, passing red light on the main road. Pressing the switch (S_p) has the effect of shortening the waiting time to a value that is equal to the secondary road. The four switches generate a single signal, connected to one input of PLC, which is 1 if the switch is pressed and 0 otherwise.

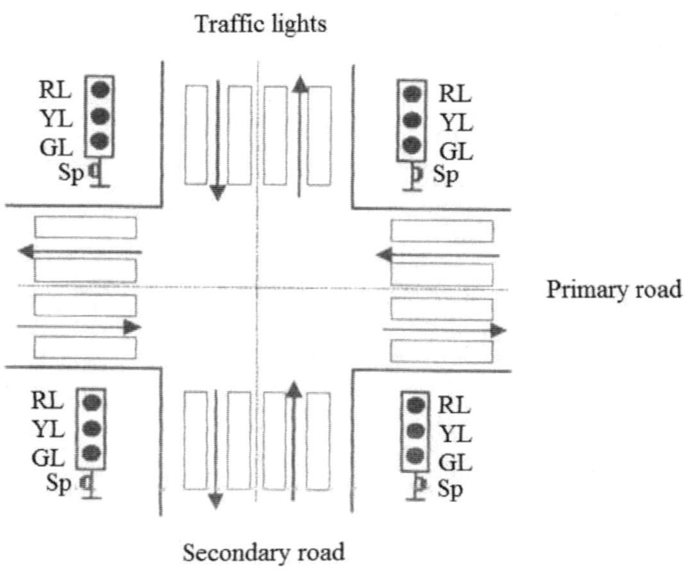

Figure 4. Driving traffic lights at an intersection with priority

Table 1. The acronyms for traffic lights on primary and secondary roads

RLP	Red Light Primary	RLS	Red Light Secondary
YLP	Yellow Light Primary	YLS	Yellow Light Secondary
GLP	Green Light Primary	GLS	Green Light Secondary

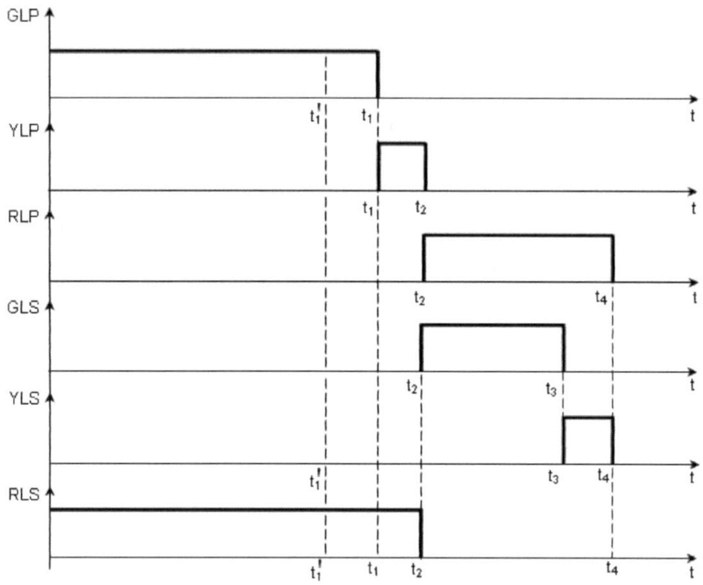

Figure 5. Timing diagrams for traffic lights

In Figure 5 are timing diagrams for this application. The significance of the output is shown in Table 1. It was considered: $0\text{-}t_1=3$ min.; $t_1'\text{-}t_1=10$ s; $t_1\text{-}t_2=10$ s; $t_2\text{-}t_3=1$ min.; $t_3\text{-}t_4=10$ s. The application starts at 0. t_1' is the time when push the S_p switch on traffic lights by pedestrians; $t_1' \in [0\text{-}t_1]$.

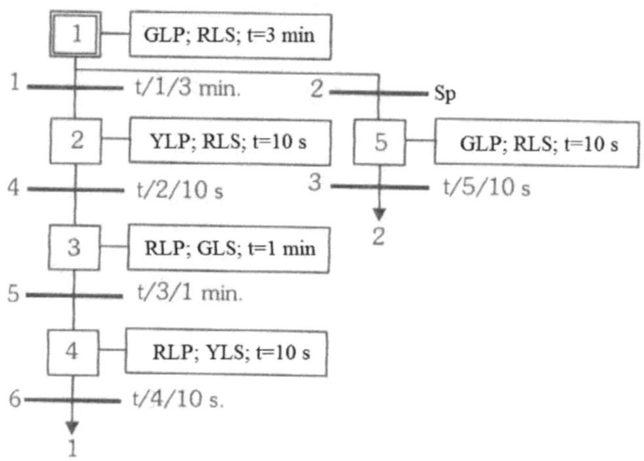

Figure 6. Graph automation for application

The graph of Figure 6 is implemented in the control unit. The left branch is typically traveled for traffic lights. If a pedestrian push the switch from one of the traffic lights, when green light is on the primary road, the program no longer made on the left branch and the right branch will go. The application has two inputs (one input for turning traffic lights with normally open contact with mechanical locking; at another input of PLC is connected four normally open contact switches, parallel connections) and six outputs (corresponding red, yellow, green lights for the primary and secondary roads). Each PLC's output has a normally open contact switch that is connected in series with the lamps of the traffic lights.

3.1. Using of PLC PS3 from Klöckner-Moeller

The input I0.0 is used to turning on the traffic lights. I0.1 connects the four switches on the four traffic lights. Table 2 shows output variables used for PS3 PLC.

Table 2. The output variables for PS3 PLC

LP	Q0.10	RLS	Q0.15
YLP	Q0.9	YLS	Q0.14
GLP	Q0.8	GLS	Q0.13

```
000  L      NMO.0
001  A      I0.0
002  =      M1.0
003  L      I0.1
004  S      M1.2
005  L      M1.2
006  JC     8
007  TR1
       S    :     M1.0
       STP  :
       I  W :     KW1800
       EQ   :     M1.1

008  TR2
       S    :     M1.2
       STP  :
       I  W :     KW100
       EQ   :     M1.3

009  L      M1.1
010  O      M1.3
011  =      M1.4
012  TR3
       S    :     M1.4
       STP  :
       I  W :     KW100
       EQ   :     M1.5

013  TR4
       S    :     M1.5
       STP  :
       I  W :     KW600
       EQ   :     M1.6

014  TR5
       S    :     M1.6
       STP  :
       I  W :     KW100
       EQ   :     M1.7

015  L      M1.0
016  S      Q0.13
017  S      Q0.10
018  R      Q0.14
019  R      Q0.15
020  R      Q0.8
021  R      Q0.9
022  L      M1.4
023  S      Q0.14
024  S      Q0.10
025  R      Q0.13
026  R      Q0.15
027  R      Q0.8
028  R      Q0.9
029  L      M1.5
030  S      Q0.15
031  S      Q0.8
032  R      Q0.13
033  R      Q0.14
034  R      Q0.9
035  R      Q0.10
036  L      M1.6
037  S      Q0.15
038  S      Q0.9
039  R      Q0.13
040  R      Q0.14
041  R      Q0.8
042  R      Q0.10
043  R      M1.2
044  L      NI0.0
045  R      Q0.13
046  R      Q0.10
047  L      M1.7
048  =      M0.0
```

Figure 7. IL language for PS3 PLC

Program listing is given in Figure 7. Implementing programs in IL requires dexterity and experience. PLC programming is done the hardest in this language. Programming was done by a programmable console (PRG3S), which made it difficult to troubleshoot program. Significant limitations on program PLC in IL, arise of using one type of timer (on-delay timer, with time base of 100 ms). It used intermediate memory M1.0, M1.4, M1.5, M1.6, and M1.7 for setting two groups of lamps for traffic lights (via primary and secondary roads) according to the timing diagrams from Figure 5.

For timers were used: TR1 for $0\text{-}t_1=3$ min.; TR2 for $t_1'\text{-}t_1=10$ s; TR3 for $t_1\text{-}t_2=10$ s; TR4 for $t_2\text{-}t_3=1$ min.; TR5 for $t_3\text{-}t_4=10$ s. Timers are set by a bit, and the end of time is set other bit.

3.2. Using PLC S7 214 Siemens

The inputs I0.0 and I0.1 have the same meaning as the previous application. The output variables are presented in Table 3. The program was carried out in the LAD. Program listing is given in Figure 8. Programming was done more easily than when using PLC PS3. PLC programming on PC's, and then transfer the program through cable greatly relieved debugging the program. Network programming is intuitive, allowing the sharing of more complex issue in simple problems to be solved. Although it can use several types of timers (on-delay timer, on-delay timer with retention, off-delay timer) were embedded in the application only on-delay timer with time base of 100 ms (there are others timers with resolution of 10 ms and 1 ms).

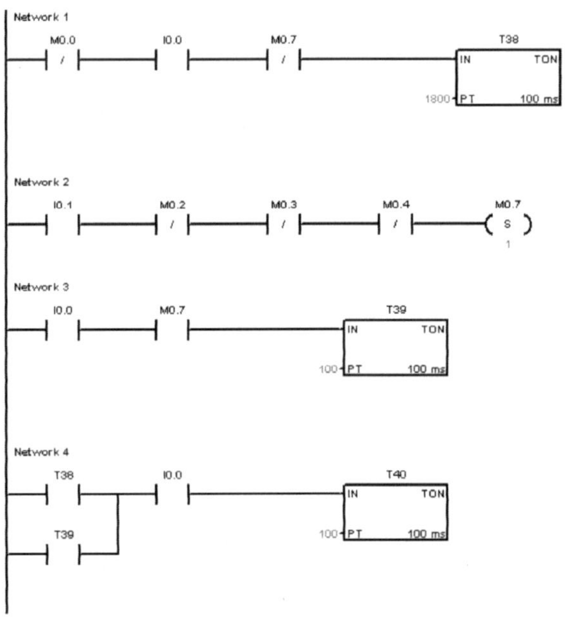

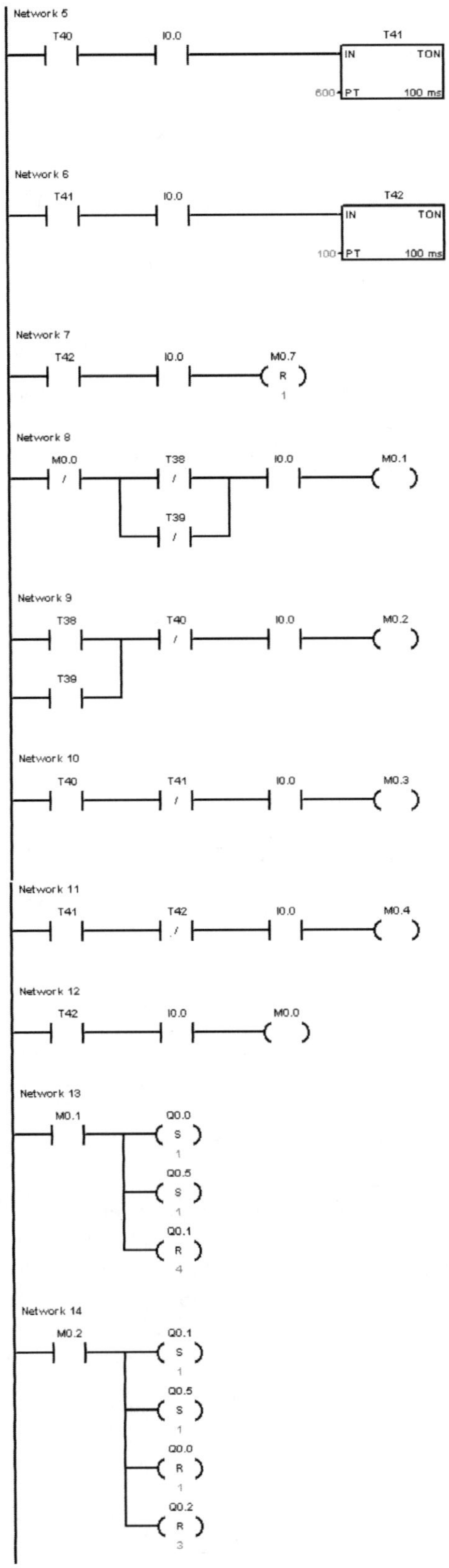
Network 5
T40
I0.0
T41
IN
TON
600
PT
100 ms
Network 6
T41
I0.0
T42
IN
TON
100
PT
100 ms
Network 7
T42
I0.0
M0.7
R
1
Network 8
M0.0
T38
I0.0
M0.1
T39
Network 9
T38
T40
I0.0
M0.2
T39
Network 10
T40
T41
I0.0
M0.3
Network 11
T41
T42
I0.0
M0.4
Network 12
T42
I0.0
M0.0
Network 13
M0.1
Q0.0
S
1
Q0.5
S
1
Q0.1
R
4
Network 14
M0.2
Q0.1
S
1
Q0.5
S
1
Q0.0
R
1
Q0.2
R
3

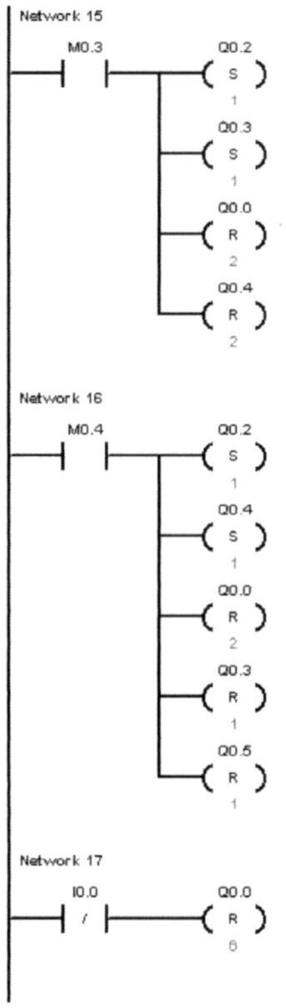

Figure 8. LAD language for S7 214 Siemens

The program was divided into program areas, each with a well-established operation (Figure 5). It has used the technique of activation, deactivation of steps (Figure 6). For timers were used: T38 for 0-t_1=3 min.; T39 for t_1'-t_1=10 s; T40 for t_1-t_2=10 s; T41 for t_2-t_3=1 min.; T42 for t_3-t_4=10 s.

Table 3. The output variables for S7 214 PLC

RLP	Q0.0	RLS	1. Q0.3
YLP	Q0.1	YLS	2. Q0.4
GLP	Q0.2	GLS	3. Q0.5

3.3. Using PLC LOGO! Siemens

The input I1 has significance of I0.0, I2 has significance of I0.1 in the previous application. Output variables are shown in Table 4. The program was carried out in the FBD. Program listing is given in Figure 9. The easiest programming (and most intuitive) is performed in FBD. The plurality of blocks, especially special ones, enables easy program. It used AND logic blocks with detection on rising and falling edge for inputs and memories, even when moving from one steps to another. At the same time, it was used memory (M1-M5) for setting and resetting the steps. Programming was done relatively easily and is intuitive in view of the timing diagram from Figure 5. It uses timers that perform off-delay time. For timers were used: B01 for 0-t_1=3 min.; B05 for t_1'-t_1=10 s; B07 for t_1-t_2=10 s; B13 for t_2-t_3=1 min.; B15 for t_3-t_4=10 s.

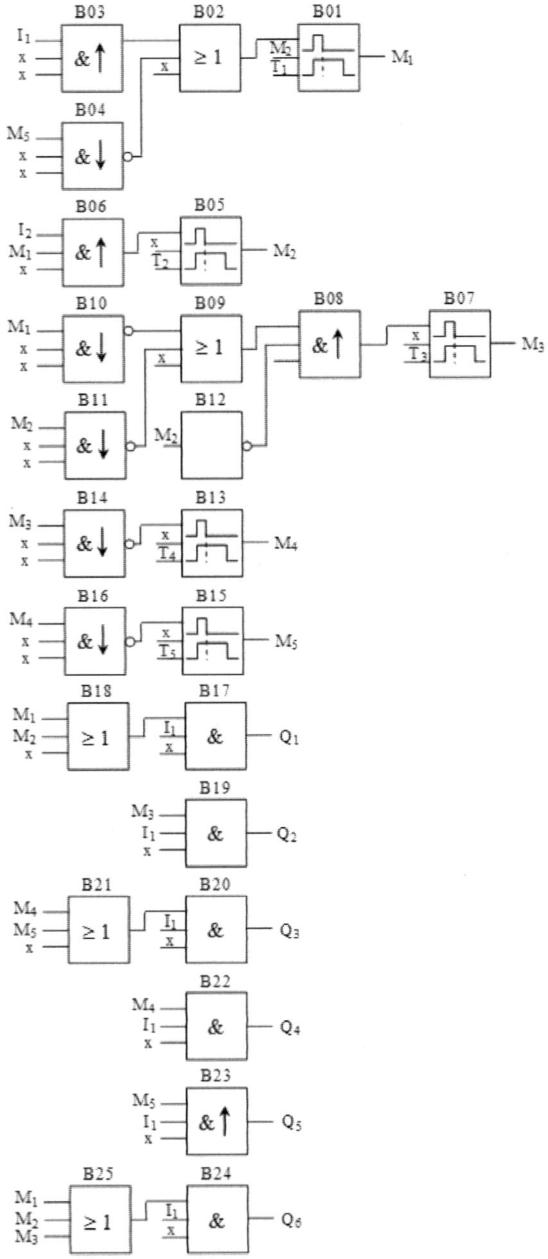

Figure 9. FBD language for LOGO! Siemens

Table 4. The output variables for LOGO! Siemens

RLP	Q3	RLS	Q6
YLP	Q2	YLS	Q5
GLP	Q1	GLS	Q4

4. Conclusion

From comparative analysis of the three types of PLCs (PS3 from Klöckner-Moeller, S7 214 Siemens and LOGO! Siemens) results:

- PLCs are flexible in development of applications;
- It is easier to program in the language that was used more;
- Programming is made easier in graphic languages (LAD/FBD);
- Are useful PLCs that can be programmed in any language (eg. S7-200);
- PC programming (eg. for S7-200 and LOGO!) is more elegant than the one on programmable console (eg. PS 3).

References

[1] ***2011, *Basic of PLC Programming*, The McGraw-Hill Companies, Inc.
[2] Collins K 2010 *PLC Programming for Industrial Automation*, CRC Press
[3] Matik N 2008 *Introduction to PLC Controllers*, MikroElectronika
[4] Costano M 1997 *Programmable Logic Controllers*, Edward Arnold
[5] Mărgineanu I 2005 *PLCs*, Albastră Publishing House, Cluj-Napoca, Romania
[6] Cunţan C D, Baciu I and Osaci M 2015 Operational Study of a Frequency Converter with a Control Sequence, Utilizing Xilinx Software, *Acta Polytehnica Hungarica* **12**(6) 201-212
[7] Baciu I and Cunţan C D 2012 *Operation Analysis of a Frequency Converter with Control Realized in LabView*, Proceedings of IEEE ICIT 2012, Athens, Greece, March 19-21, pp 432-437
[8] Schütz C et al. 2013 Development of PLC-Based Software for Increasing the Dependability of Production Automation Systems, *IEEE Transactions on Industrial Informatics* **9**(4) 2397-2406
[9] Bolton W 2010 *Ladder and Functional Block Programming*, www.newnespress.com 453-481
[10] Heuser B V et al. 2013 Evaluation of a UML-Based Versus an IEC 61131-3-Based Software Engineering Approach for Teaching PLC Programming, *IEEE Transactions on Education* **56**(3) 329-335
[11] Adiego B F et al. 2015 Applying Model Checking to Industrial-Sized PLC Programs, *IEEE Transactions on Industrial Informatics* **11**(6) 1400-1410
[12] Isik M F et al. 2014 *Monitoring and Control of PLC Based Motion, Control Systems via Device-Net*, 16th International Power Electronics and Motion Control Conference and Exposition, Antalya, Turkey, September 21-24, pp 963-966
[13] Frey G and Litz L 2000 *Formal Methods in PLC Programming*, Proceedings of the IEEE Conference on Systems Man and Cybernetics SMC 2000, Nashville, USA, October 8-11, pp 2431-2436
[14] Berger H 2011 *Automatisieremn mit STEP 7 in AWL und SCL*, Siemens

IOP Conf. Series: Materials Science and Engineering **163** (2017) 012035 doi:10.1088/1757-899X/163/1/012035

Experimental analysis of direct current corona discharge

A Iagăr, C M Diniş and G N Popa

Politehnica University of Timisoara, Department of Electrical Engineering and Industrial Informatics, 5 Revolution Street, Hunedoara, 331128, Romania

E-mail: angela.iagar@fih.upt.ro

Abstract. This paper presents the experimental analysis of current-voltage characteristic of DC corona discharge. In experiments were used an electrostatic discharge system, consists of an active brass electrode, and a passive copper electrode (plane). A high-voltage source has been used to produce the positive or negative DC discharge. To analyse the influence of active electrode surface were used a conical and a spherical electrode, with small radius of curvature. The gap between the electrodes was varied by moving the passive electrode. Current-voltage characteristics of DC corona discharge were obtained at a constant distance between the discharge electrodes, by changing the amplitude of supply voltage. Corona current and voltage were acquired using a digital oscilloscope. Experiments allowed the study of the effects due to the active electrode and the air gap between the electrodes, on the onset voltage and the current of corona discharge.

1. Introduction

Direct current (DC) corona discharge has numerous applications, such as in electrostatic precipitation (ESP), electrostatic separation, photocopying, laser print, generation of ozone, sanitization of pool water, decomposition of toxic gases, etc. [1-9].

Industrial ESPs reduce the emissions of fumes and dust, playing an important role to maintain a clean environment. ESPs are used in electric power plants, cement production, chemical processing and domestic air cleaning [2-4], [8].

The process of corona generation in the air at atmospheric conditions requires a non-uniform electrical field, which can be obtained by using a small diameter electrode (active or emission electrode), energized from a high-voltage supply, and a metallic plate (used as collecting electrode) connected to the ground [5], [6].

Corona discharge presents the following major advantages: can be used easily in forming non-thermal plasmas, with high rates for gases flowing between electrodes, is stable at atmospheric pressure and operates at low currents and power consumption [5], [9].

The disadvantage is that the active volume of the ionization region, surrounding the high electric field electrode, is very much smaller than the total discharge volume, between the electrodes [1], [6], [7], [9].

The wires-to-planes is the most used configuration in ESPs. It consists of high-field parallel active wires, located midway between the collecting electrodes. When the gas flows through the discharge electrodes, the ions produced by the corona discharge (near the active electrodes) charge the dust particles, which are driven toward the collecting electrodes, where their charges is neutralized, dust particles being thus collected [1].

IOP Conf. Series: Materials Science and Engineering **163** (2017) 012035 doi:10.1088/1757-899X/163/1/012035

The onset voltage of corona discharge is a prior condition for calculating the corona current at a given voltage applied to the discharge electrodes [6].

This paper aims at experimental analyzing the current-voltage characteristics of positive and negative DC corona discharges.

2. Experimental setup

In experiments were used an electrostatic discharge system (ESD, Figure 1), consists of an active brass electrode, and a passive copper electrode (plane).

To analyse the influence of active electrode surface were used a conical electrode and a spherical electrode, with small radius of curvature. The gap between the electrodes was varied between 10...30 mm, by moving the passive electrode.

In order to change widely the voltage on discharge system, was used an autotransformer (AT, Figure 1). Variable low voltage was applied to the primary of an elevator transformer (T, Figure 1) whose secondary was connected to a rectifier bridge (P, Figure 1). This high-voltage source has been used to produce the positive or negative DC discharge. The active electrode was connected to the DC high-voltage installation which has a voltage between 0 and ±12 kV.

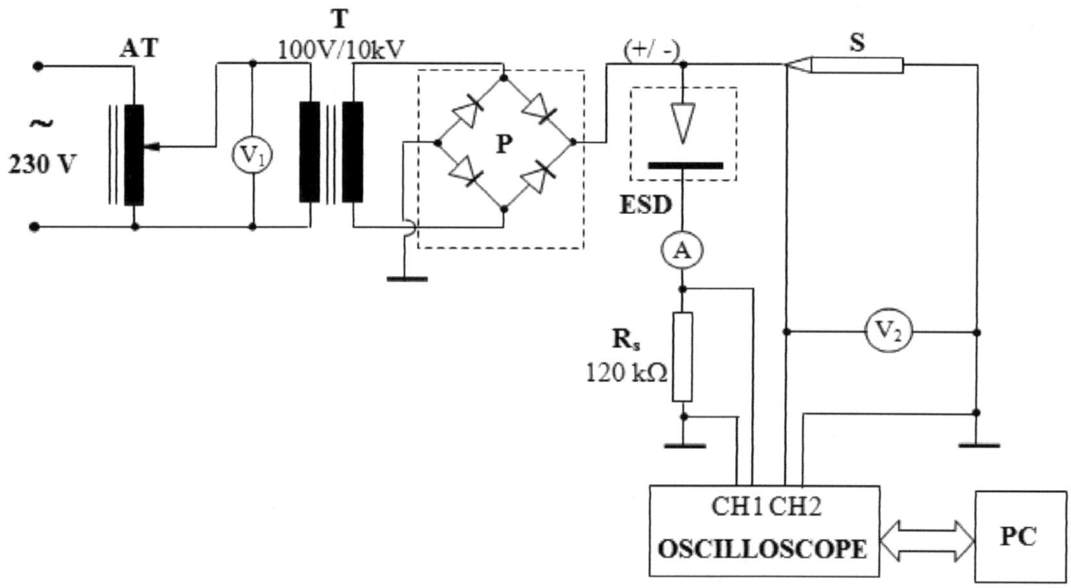

Figure 1. Electrical diagram of experimental setup

Figure 2. High-voltage probe (resistive divider) used to measure the DC voltage

The corona discharges operate at low currents (μA) and high voltages (kV). Measuring of DC voltage was performed using a high-voltage probe (resistive divider, S, Figure 2). The value of the discharge current was measured with a micro-ammeter.

3. Waveforms of voltage and current at negative and positive corona discharges

Corona current and voltage were acquired using a digital oscilloscope (25 MHz). The data is then transferred to a personal computer (PC). For the measurement of the corona current was used a resistance load of R_s=120 kΩ.

IOP Conf. Series: Materials Science and Engineering **163** (2017) 012035 doi:10.1088/1757-899X/163/1/012035

a b

Figure 3. Conical electrode (a) and spherical electrode (b), with small radius of curvature, used to analyse the influence of active electrode surface

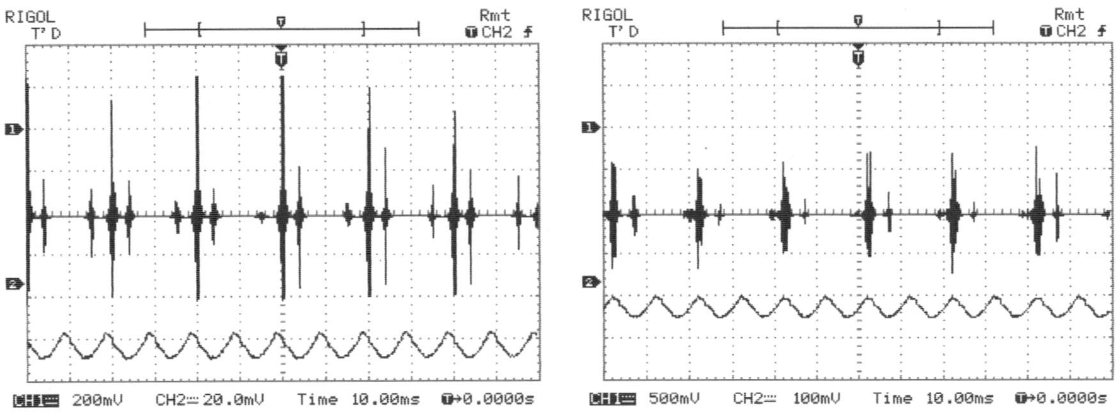

a b

Figure 4. Waveforms of voltage and current at negative corona discharge in the case of conical (peak)-plane electrodes at d=25 mm: (a). U^{-}=5 kV, (b). U^{-}=10 kV

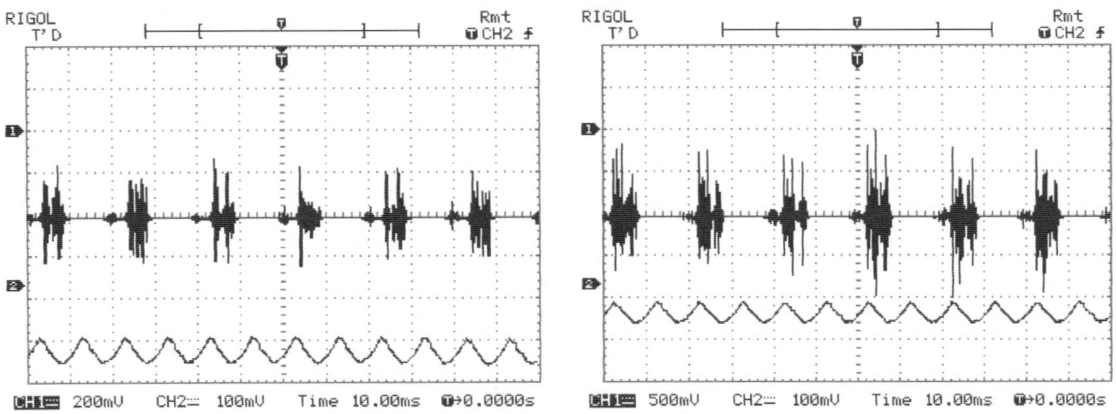

a b

Figure 5. Waveforms of voltage and current at negative corona discharge in the case of conical (peak)-plane electrodes at d=30 mm: (a). U^{-}=5 kV, (b). U^{-}=10 kV

IOP Conf. Series: Materials Science and Engineering **163** (2017) 012035 doi:10.1088/1757-899X/163/1/012035

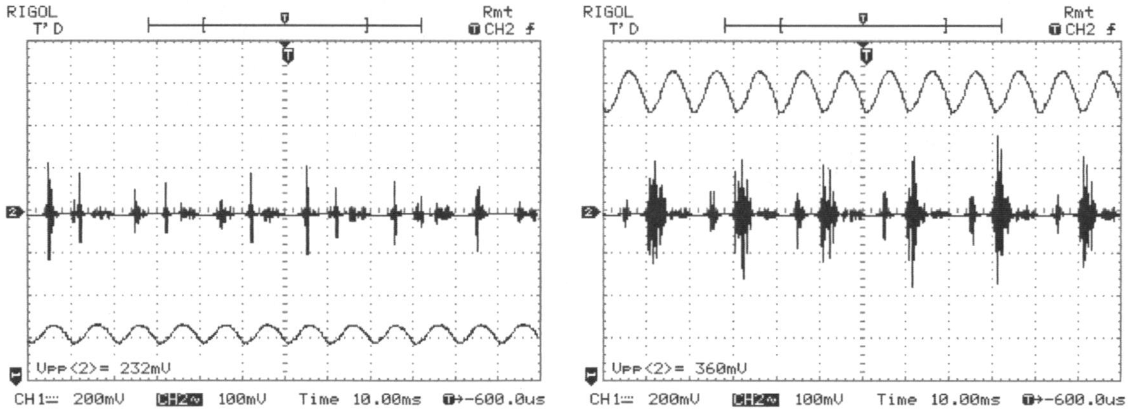

a b

Figure 6. Waveforms of voltage and current at positive corona discharge in the case of conical (peak)-plane electrodes at d=25 mm: (a). U^+ =5 kV, (b). U^+ =10 kV

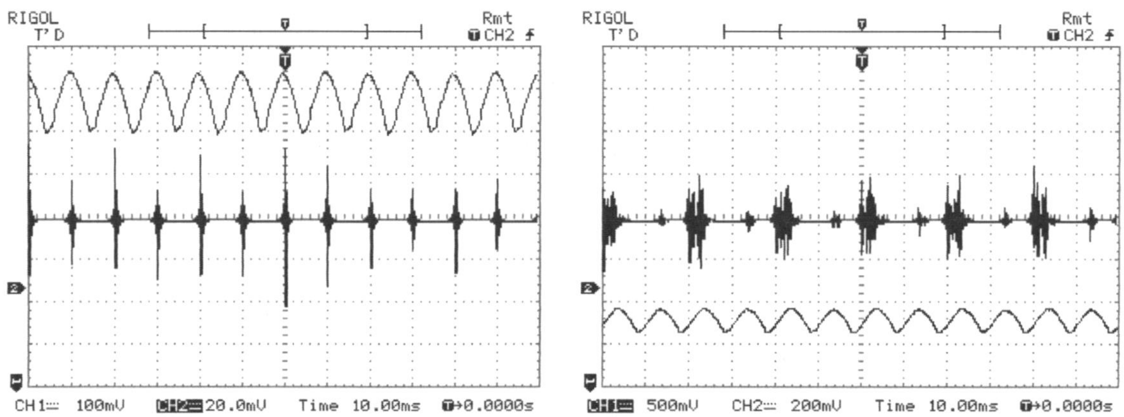

a b

Figure 7. Waveforms of voltage and current at positive corona discharge in the case of conical (peak)-plane electrodes at d=30 mm: (a). U^+ =5 kV, (b). U^+ =10 kV

The waveforms of voltage and current at negative and positive corona discharges are shown in the Figures 4-7. Measurements were carried out for different voltage values (5 kV and 10 kV) applied to the discharge electrodes (conical-plane), at distances of 25 mm and 30 mm between them.

Notations have the following meanings: d - distance between the electrodes; U^+ - applied voltage to the discharge electrodes, when active electrode was connected to the positive polarity of DC high-voltage installation; U^- - applied voltage to the discharge electrodes, when active electrode was connected to the negative polarity of DC high-voltage installation.

Partial electrical discharges are shown clearly in the corona current waveforms from Figures 4-7.

4. Experimental analysis of DC corona discharge current-voltage characteristics

Further are present the current-voltage (I-V) characteristics of positive and negative Corona discharges, for different distances between the electrodes, and different configurations of them.

These characteristics were obtained at a constant distance between the discharge electrodes, by changing the amplitude of supply voltage.

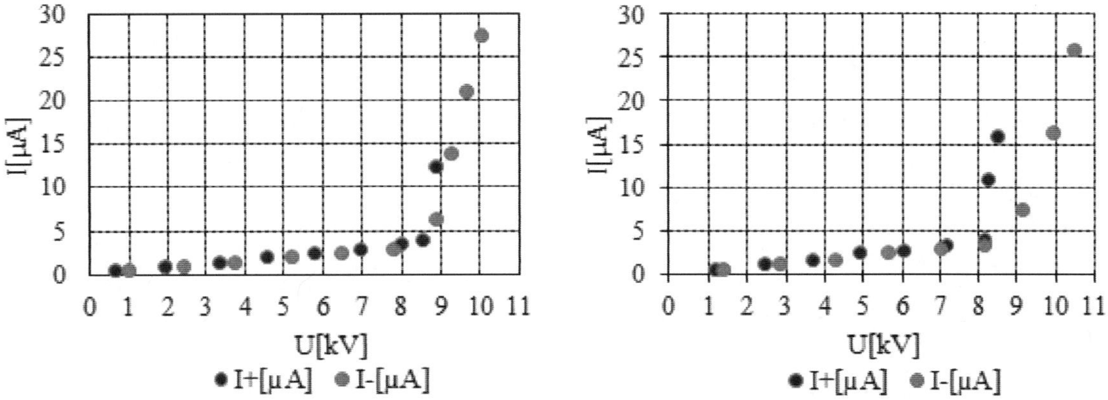

a b

Figure 8. DC corona discharge I-V characteristics for conical-plane electrodes: (a). d=11 mm, (b). d=12 mm

If the distance between the discharge electrodes (conical-plane) was d=11 mm (Figure 8.a), it is found that current intensity as a function of applied voltage is in agreement with the Ohm's law, for voltages in the range of 0-8 kV (current increases linearly with the applied voltage), for both polarities (positive and negative). The positive corona discharge occurs at a voltage of 8.3 kV, and the negative corona discharge occurs at a voltage of 8.84 kV.

If the distance between the discharge electrodes (conical-plane) was d=12 mm (Figure 8.b), positive corona discharge occurs at a voltage of 8.5 kV, and negative corona discharge occurs at a voltage of 9.12 kV.

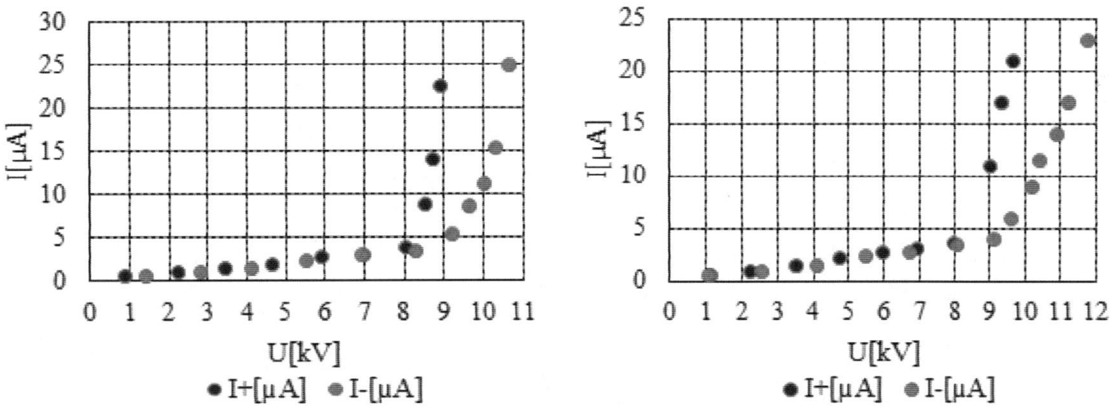

a b

Figure 9. DC corona discharge I-V characteristics for conical-plane electrodes: (a). d=14 mm, (b). d=15 mm

At d=14 mm (Figure 9.a), positive corona discharge occurs at a voltage of 8.52 kV, and negative corona discharge occurs at a voltage of 9.2 kV. At d=15 mm (Figure 9.b), positive corona discharge occurs at a voltage of 9 kV, and negative corona discharge occurs at a voltage of 9.62 kV.

As seen from Figures 8-12, the onset voltage and corona discharge occur at increasing tensions, if the inter-electrode gap increases, at both polarities.

Corona current decreases as the gap increases, for the same measurement conditions.

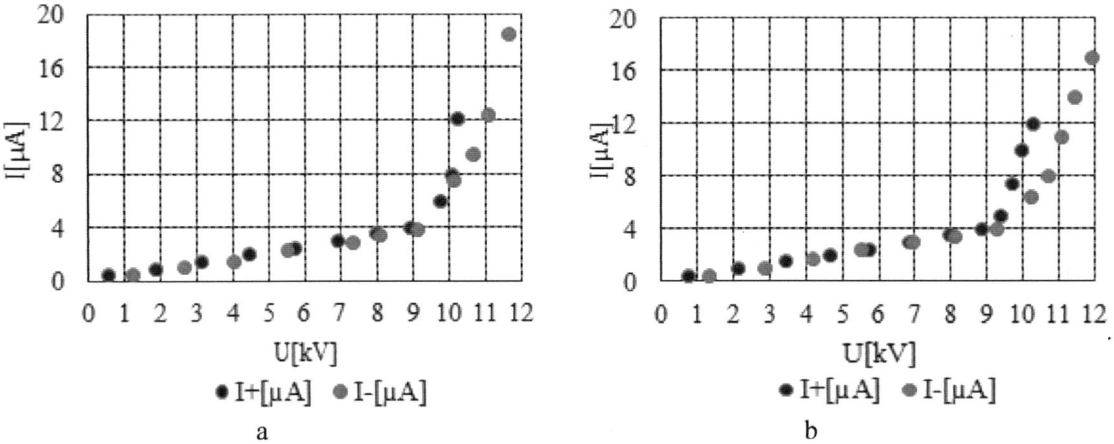

Figure 10. DC corona discharge I-V characteristics for conical-plane electrodes: (a). d=16 mm, (b). d=17 mm

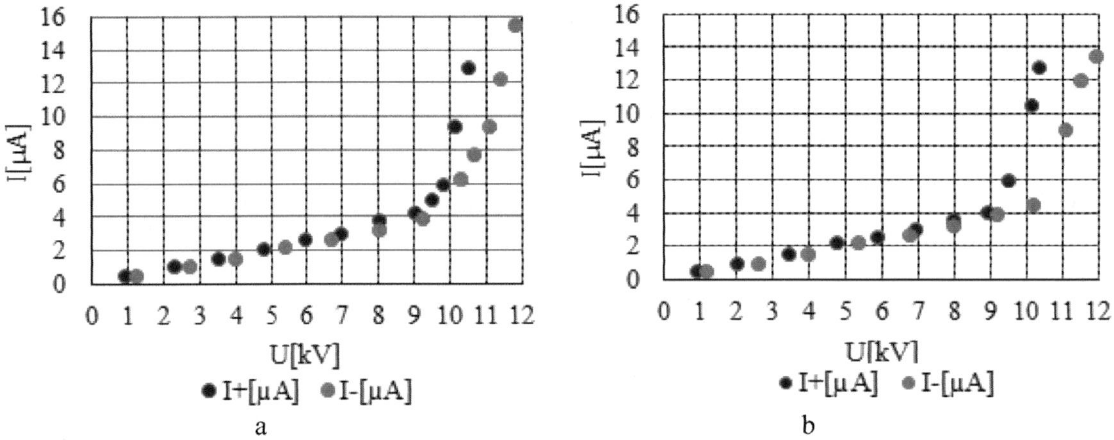

Figure 11. DC corona discharge I-V characteristics for conical-plane electrodes: (a). d=18 mm, (b). d=19 mm

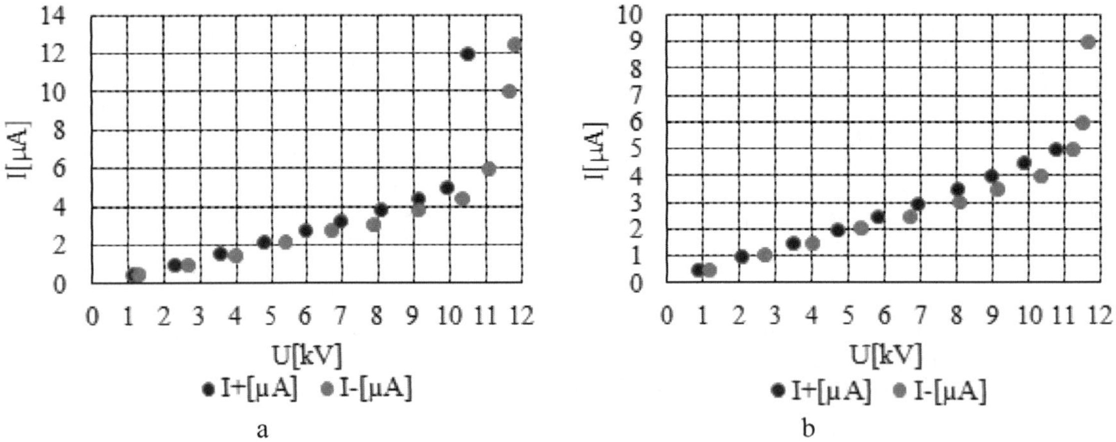

Figure 12. DC corona discharge I-V characteristics for conical-plane electrodes: (a). d=20 mm, (b). d=21 mm

IOP Conf. Series: Materials Science and Engineering **163** (2017) 012035 doi:10.1088/1757-899X/163/1/012035

Significant differences between the positive (+) and the negative (-) corona discharges are observed at small values of the inter-electrodes gap (d=11...21 mm), in the case of conical-plane electrodes (Figures 8-12).

Corona onset voltage and electric discharge occur at higher voltages in the case of negative (-) polarity, compared to the positive (+) polarity.

Negative corona discharge is more intense than positive (negative current values are higher than positive current values at the same distance between the electrodes). Also, it is found that negative corona discharge is more stable than positive corona discharge.

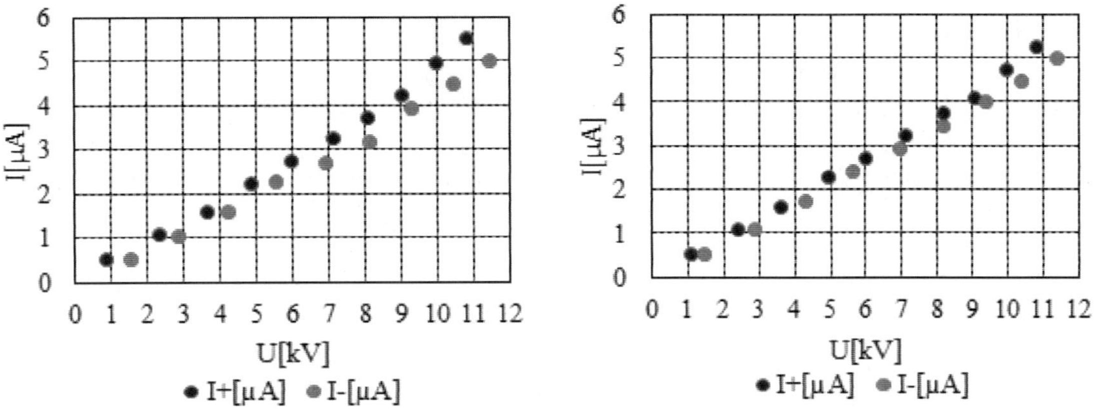

a b

Figure 13. DC corona discharge I-V characteristics for conical-plane electrodes: (a). d=22 mm, (b). d=25 mm

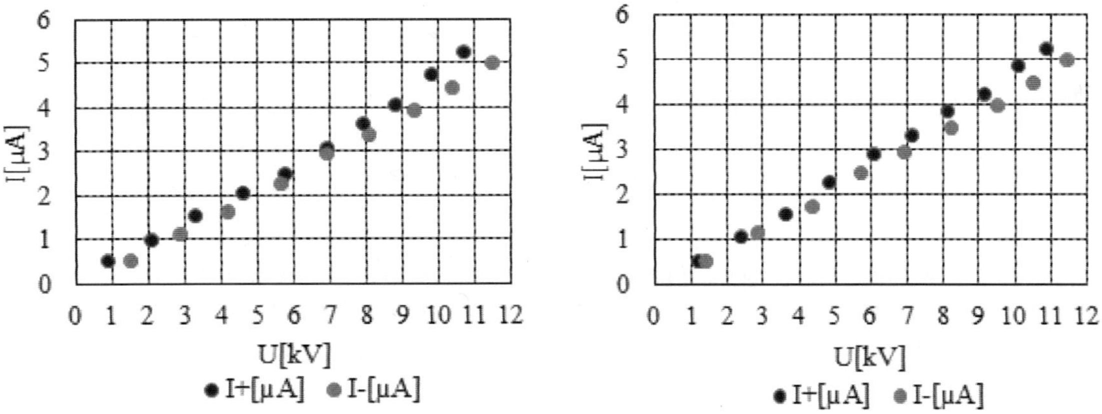

a b

Figure 14. DC corona discharge I-V characteristics for conical-plane electrodes: (a). d=28 mm, (b). d=30 mm

Variation of the currents is linear at large distances between the conical-plane electrodes (d=22...30 mm); because the discharge is not autonomous, differences between the currents occur at positive and negative polarity in these situations, are not significant (Figures 13 and 14).

Because conical-spherical electrodes were not collinear in experiments, negative corona currents were smaller than the currents occurring between conical-plane electrodes, at the same distances (Figure 15).

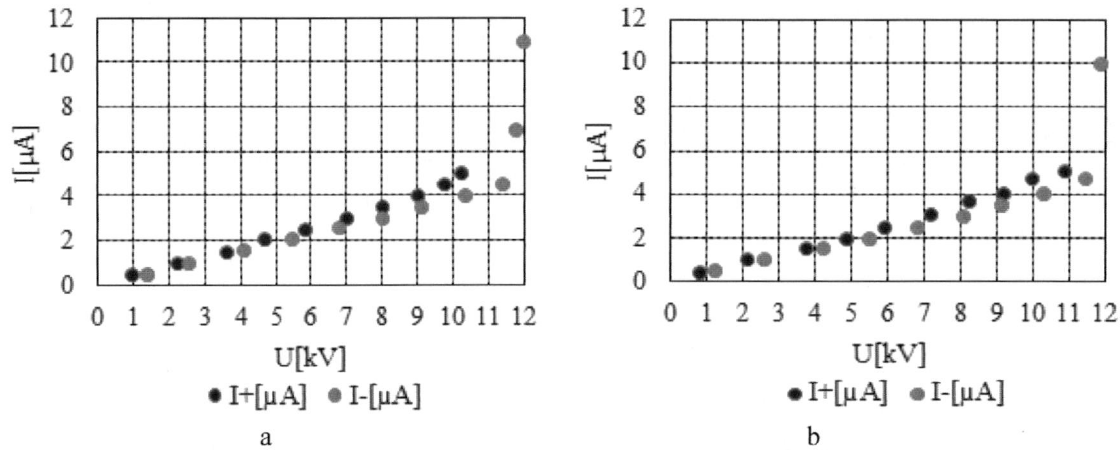

Figure 15. DC corona discharge I-V characteristics for conical-spherical electrodes: (a). d=11 mm, (b). d=12 mm

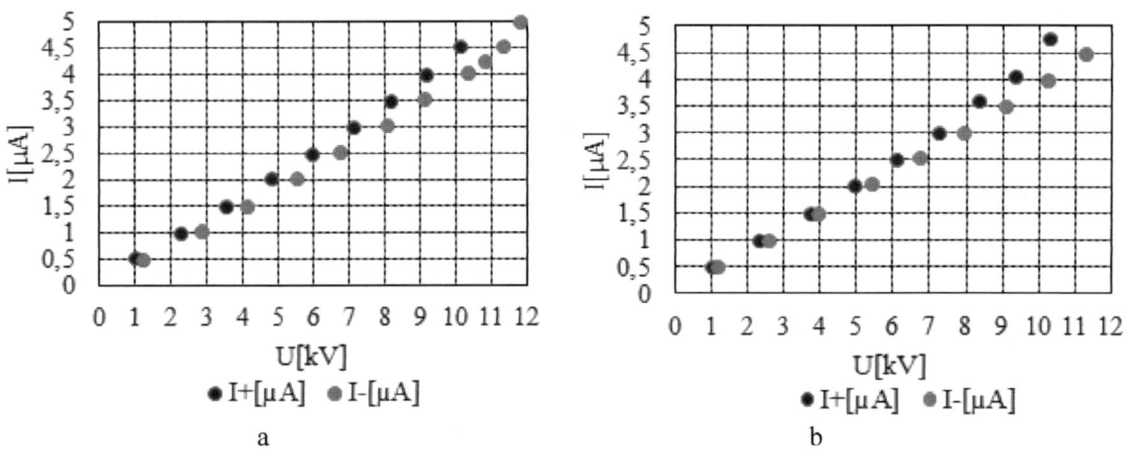

Figure 16. DC corona discharge I-V characteristics for spherical-plane electrodes: (a). d=10 mm, (b). d=20 mm

Figure 16 shows that corona current is much lower for spherical-plane electrodes than for conical-plane electrodes, in the same measuring conditions, because corona discharge is even more intense as the electrode surface emission is lower.

5. Conclusions

Experiments allowed the study of the effects due to the active electrode and the air gap between the electrodes, on the onset voltage and the current of DC corona discharge.

The onset voltage of corona discharge is influenced by the active electrode surface and the inter-electrode gap, for the same applied voltage.

At negative polarity, the onset voltage of corona discharge is greater than that of the positive polarity, for a given geometry.

Experimental results show that corona current is affected by the voltage between the electrodes, their shape and the inter-electrode gap. Significant differences between the positive and the negative corona discharge are observed at small values of the inter-electrodes gap (d=11...21 mm), DC corona currents being nonlinear functions of the applied voltages.

Currents through the inter-electrode gap increase linear with the applied voltage at large distances between the electrodes (d=22...30 mm); differences between the currents occur at positive and negative polarity in these situations are not significant, because the discharge is not autonomous.

Onset voltage and corona electric discharge occur at increasing tensions, if the inter-electrode gap increases, in the case of conical-plane electrodes (at both polarities), but corona current decreases as the inter-electrodes gap increases.

Surface of active electrode decisively influences the appearance of corona discharge. Corona discharge is even more intense as the surface of active (emission) electrode is less.

Negative corona discharge is more intense and more stable than positive. This is due to the existence of a difference between the positive and negative onset voltage; also the mobility of negative ions is higher than that of positive ions. Therefore, most industrial applications, including ESPs, use negative corona discharge.

Experimental study of DC corona discharge current-voltage characteristics is very useful to understand the phenomena that occur in wires-to-planes configuration of ESPs.

References

[1] Chang J S, Kelly A J and Crowley J M 1995 *Handbook of Electrostatics Processes*, Marcel Dekker, NewYork, USA

[2] Popa G N, Diniş C M, Deaconu S I and Iagăr A 2012 *A Prospective on Power Analyse of Three Sections Plate-Type Electrostatic Precipitator Supplies*, Proceedings of the 11th WSEAS International Conference on Instrumentation, Measurement, Circuits and Systems, Rovaniemi, Finland, April 18-20, pp 49-54

[3] Popa G N, Vaida V, Iagăr A and Diniş C M 2008 *Simulations and Measurements of Some Parameters from Plate-Type Electrostatic Precipitators with Three Sections*, Proceedings of 2nd WSEAS/IASME International Conference on Energy Planning, Energy Saving, Environmental Education, Corfu, Greece, October 26-28, pp 106-112

[4] Popa G N, Vaida V, Iagăr A and Diniş C M 2008 Improvement of Technological and Electric Performances for Plate-Type Electrostatic Precipitators with Three-Sections, *WSEAS Transactions on Power Systems* **8**(8) 557-566

[5] Chang J S, Lawless P A and Yamamoto T 1992 Corona discharge processes, *IEEE Transactions on Plasma Science* 19(6) 1152 - 1166

[6] Ziedan H, Tlustý J, Mizuno A, Sayed A and Ahmed A 2010 Corona current-voltage characteristics in wire-duct electrostatic precipitators, Theory versus Experiment, *International Journal of Plasma Environmental Science and Technology* **4** 154-162

[7] Ziedan H, Syed A, Mizuno A and Ahmed A 2010 Onset voltage of corona discharge in wire-duct electrostatic precipitators, *International Journal of Plasma Environmental Science and Technology* **4** 36-44

[8] Nouri H, Zouzou N, Moreau E, Dascalescu L and Zebboudj Y 2012 Effect of relative humidity on current-voltage characteristics of an electrostatic precipitator, *Journal of Electrostatics* **70**(1) 20-24

[9] Hakim Ait Said, Massinissa Aissou, Hamou Nouri and Youcef Zebboudj 2014 Effect of Wires Number on Corona Discharge of an Electrostatic Precipitators, *Journal of Electrical Systems* **10** 392-405

IOP Conf. Series: Materials Science and Engineering **163** (2017) 012036 doi:10.1088/1757-899X/163/1/012036

Single-phase frequency converter

I Baciu and C D Cunțan

Politehnica University of Timisoara, Department of Electrical Engineering and Industrial Informatics, 5 Revolution Street, Hunedoara, 331128, Romania

E-mail: corina.cuntan@upt.ro

Abstract. The paper presents a continuous voltage inverter - AC (12V / 230V) made with IGBT and two-stage voltage transformer. The sequence control transistors is achieved using a ring counter whose clock signal is obtained with a monostable circuit LM 555. The frequency of the clock signal can be adjustment with a potentiometer that modifies the charging current of the capacitor which causes constant monostable circuit time. Command sequence consists of 8 intervals of which 6 are assigned to command four transistors and two for the period break at the beginning and end of the sequence control. To obtain an alternation consisting of two different voltage level, two transistors will be comanded, connected to different windings of the transformer and the one connected to the winding providing lower voltage must be comanded twice. The output of the numerator goes through an inverter type MOS and a current amplifier with bipolar transistor.To achieve galvanic separation, an optocoupler will be used for each IGBT transistor, while protection is achieved with resistance and diode circuit. At the end there is connected an LC filter for smoothing voltage variations.

1. Introduction

The static power converters, designation used to define the power electronic devices, constitute a complex technique that includes the power switching, control, regulation and conversion by using power semiconductors, diodes, thyristors, transistors, etc., as well as auxiliary circuits for voltage measurement and protection.

Most of the current technological processes, applied in various industries, require increasingly electrical drives as cheap and robust as possible.

The frequency converters transform the alternating current of a certain frequency into alternating current of another frequency.

They have a wide range of technical applicability, since there are many receivers requiring a power source with a different frequency from the network frequency (fluorescent lamps, induction furnaces, etc.) or with variable frequency (AC electric motor drive systems).

Depending on the power circuit structure, the frequency converters can be indirect frequency converters or direct frequency converters [1].

The indirect frequency converters change the frequency in two steps, through the process of recovery and inversion. These converters can be found in the literature with the alternative designation of frequency converters with DC voltage intermediate circuit.

In contrast, the direct frequency converters enable changing the frequency in a single step, without the prior conversion of AC into DC. Both types of converters can be designed as converters driven from the supply network (autonomous) or independent (autonomous).

Based on the switching mode, the direct frequency converters can be classified in direct frequency converters with natural switching and direct frequency converters with forced switching.

The power part of the converter is realised with controlled semiconductive power devices (thyristors, transistors) or, uncontrolled (diode). This devices working in comutation conditions, having the role of a swich so, resulting in a steady state formed by a peryodic succesion of transistor conditions. Successive opening and closing of these switches is done by logic imposed by the principle of operation of the converter. This logic is insured by the electronic comand circuit. So, all the converters contain a part of power (strenght) and a commanding part [2].

The converters provide the conversion of large quantities of energy.

This requires the main sizing criteria to be the yield, which leads to differences between the power electronics and the signal electronics, where the main aim is to obtain an accurate output signal.

2. Problem formulation

2.1. Interface circuits and galvanic isolation circuits

The electronics power converters or, in general, the power electronics systems, include a part of force and a part of command and control.

The force part that ensures the circulation of the energy mainstream operates at high voltages and electric potentials, being subjected to intense electromagnetic disturbances. The command and control part operates with logic or analogue low amplitude signals. The drivers of the power semiconductor devices are placed on the border between the two parts. The communication is done in both directions by means of logic signals. Thus, a control circuit will contain, in its turn, a logical block capable to handle and process these logic signals, and an interface block, which can perform one or more of the following functions [3]:

- Reception of the logic control signals from the hierarchically superior control block, their reformation where they are affected by disturbances, and adaptation of the logic level to the one required by the control circuit of the power semiconductor;
- Sending logic signals of appropriate level, used for the reverse communication in various situations or states seized by the control circuit;
- Galvanic isolation between the control circuit and the numerical control system.

The galvanic isolation elements, optocouplers or signal transformers, can be placed either at the entry of the control circuit of the power devices or before the final stage of the transistor control circuit. Figure 1 shows an alternative embodiment, in which the galvanic isolation elements (optocouplers) are arranged at the driver entrance, the point where the driver interfaces with the numerical control system (microprocessor - µP).

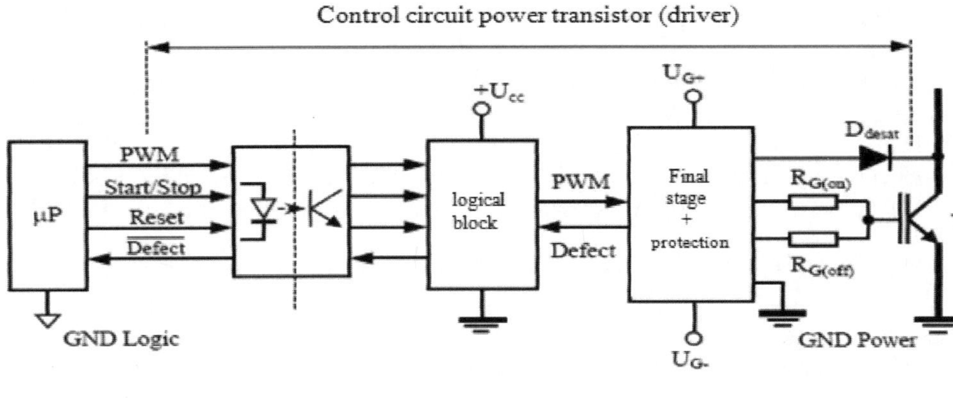

Figure 1. Implementation of the galvanic isolation function at the entrance of the power transistor control circuit

2.2. The power supply structures of the control circuits

Another important issue to be taken into account when the control circuits are designed is the selection of power supply options and sources.

In the example shown in Figure 2, the driver can operate if it is supplied with four different voltages: U_{cc1}, U_{cc2}, U_{G+} and U_{G-}.[4] A solution to simplify the power supply problem is the use of the same positive voltages, either in the logic part or in the final stage of signal processing ($U_{cc1} = U_{G+}$). For this, we must abandon the galvanic isolation and optimise the voltage amplitudes for a proper operation of both parts. Not always we can find solutions in this regard, because a complex control scheme requires certain standard voltage values for the circuits used in the logic part, and other values for the grid voltages.

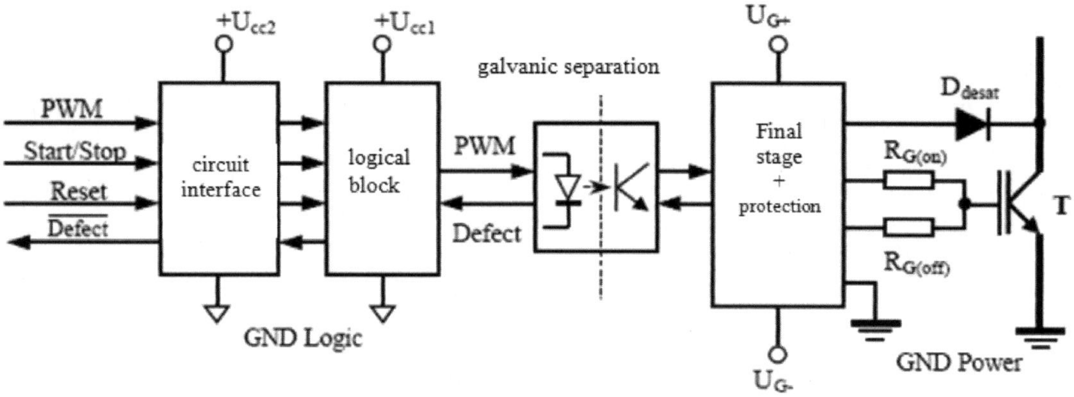

Figure 2. Power supply option for the complex control circuit blocks

For example, to obtain an effective amplitude of + 15V on the MOS grid of the power device, the voltage U_{G+} should be slightly higher to compensate the voltage on the T_{on} transistor, when this transistor is under conduction.

If the galvanic isolation is made before the final stage, the grid voltages (U_{G+}, U_{G-}) must come from a source isolated from the source that supplies the voltages to the other blocks of the control circuit.

Similarly, in the logic part of the control circuit, it might be necessary multiple power supply voltages. It is possible that some blocks, which include the specialised IC, to require a voltage U_{cc1} = +15V, and the interface circuit with the external control system to work with TTL - level logic signals, U_{cc2} = +5V.

From the foregoing, it can be said that, to solve the power supply problem faced by a control circuit, there are necessary two galvanic isolated sources, of which at least one must be able to provide multiple voltages. The problem is further complicated if the force structure contains several power transistors. At some of these transistors, the control terminal relates to a floating ground. For solving the problem of the power supply sources required by the control circuits which are part of a complex electronic power system, the designer has two choices [5]:

- Using a stabilised multi-source with a sufficient number of galvanically isolated outputs. The source may be linear, with network transformer having several isolated secondary windings. This is a classic solution that is poor in terms of gauge, mass and energy efficiency. The modern option is to use stabilised switching sources with multiple isolated outputs, such sources eliminating the disadvantages of the linear ones;
- Using certain integrated circuits or specialised control modules, specifically designed to reduce the number of sources and supply voltages.

2.3. Surge protection circuits for transistors

For circuit analysis blocking protection, the parasitic inductances in the converter have been neglected, so the surges generated when blocking the current flow through the transistor have been neglected as well. The surge protection can be obtained using RCD-type circuits, as the one shown in Figure 3(a), where it has been considered that the parasitic inductances can be replaced by an equivalent inductance.

At baseline, the transistor is open and the voltage VC_{0V} across the capacitor in the surge protection circuit is equal to V_i (the capacitor is precharged).

When blocking, assuming that the time interval required for cancelling the current flow through the transistor is very small, the current flowing through the inductor L_α is maintained at the value of I_0, the load current flow circuit being closed through the leakage diode D_f. For this state, the equivalent circuit is shown in Figure 3(b), where the diode D_f is shorted, and the transistor is blocked.

The energy stored in the parasitic inductances is now transferred through the diode D_{0V} into the capacitor C_{0V}, and charges it up. The capacitor voltage increases beyond the value of V_i.

In this state, the overvoltage ΔV_{Cov} occurs also across the transistor:

$$\Delta V_{Cov} = \Delta V_{CE} \tag{1}$$

and can be determined by replacing the precharged capacitor with an equivalent circuit, as shown in Figure 3(c).

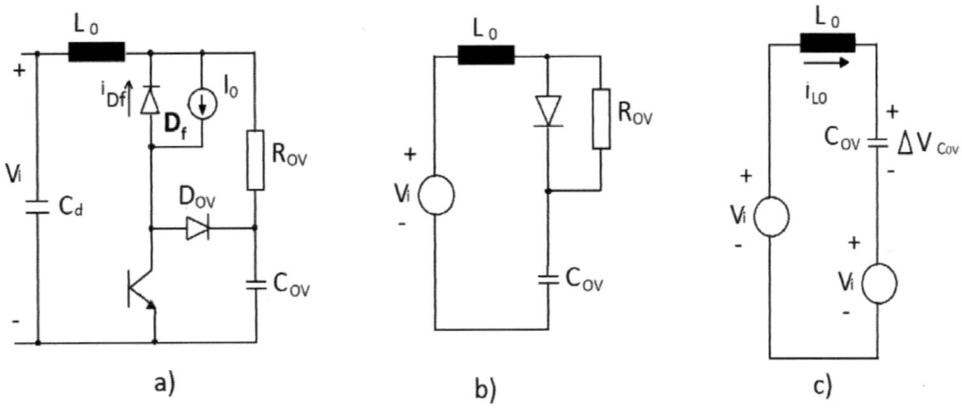

Figure 3. (a). Surge protection circuit; (b), (c). Equivalent circuits when blocking the transistor

The equation (1) shows that a high value of the capacitor C_{0v} would minimise the overvoltage ΔV_{CEmax}. Once the inductor current reached zero, the capacitor is discharged through the resistor R_{0V}, with the time constant $R_{0v}C_{0v}$, which must be small enough to enable the voltage on C_{0v} to fall to about V_i, thus making the protection circuit operational before the transistor is blocked again [6].

2.4. The control circuit

The commanding part is realised with a monostable circuit as a tact signal for a ring counter that generates the commanding sequence for the commutation elements from the invertors scheme.

The monostable circuit is realised with a LM555 circuit which provides the adjustable frequency command pulse. The frequency can be modified from a potentiometer. The output pulses are tact signals for a counter, realised with a CD4017 which provides sequences of 8 intervals necessary for commanding the invertors transistors.

IOP Conf. Series: Materials Science and Engineering **163** (2017) 012036 doi:10.1088/1757-899X/163/1/012036

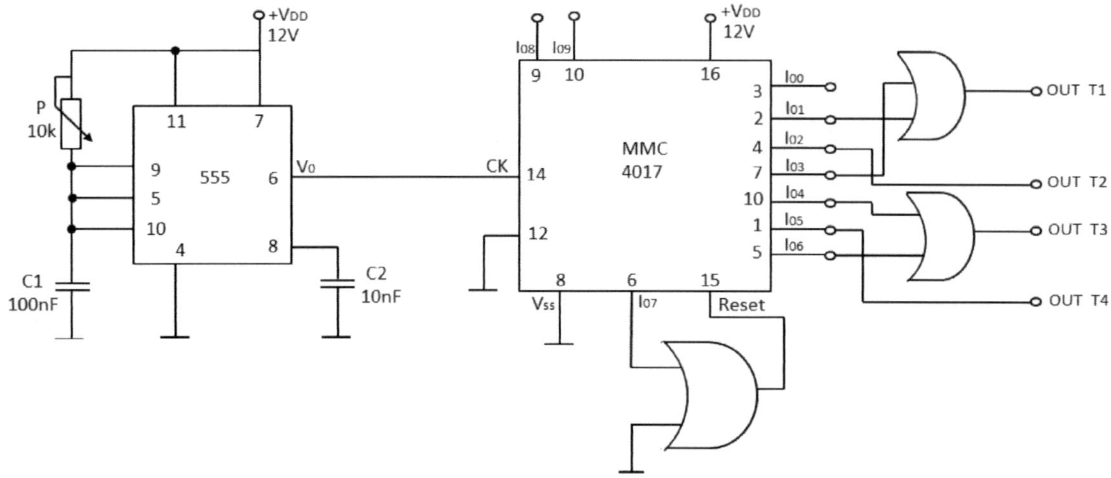

Figure 4. The control circuit of the voltage inverter

Four IGBT transistors are used, which command a four primary winding transformer which provide two steps of voltage at the output [7].

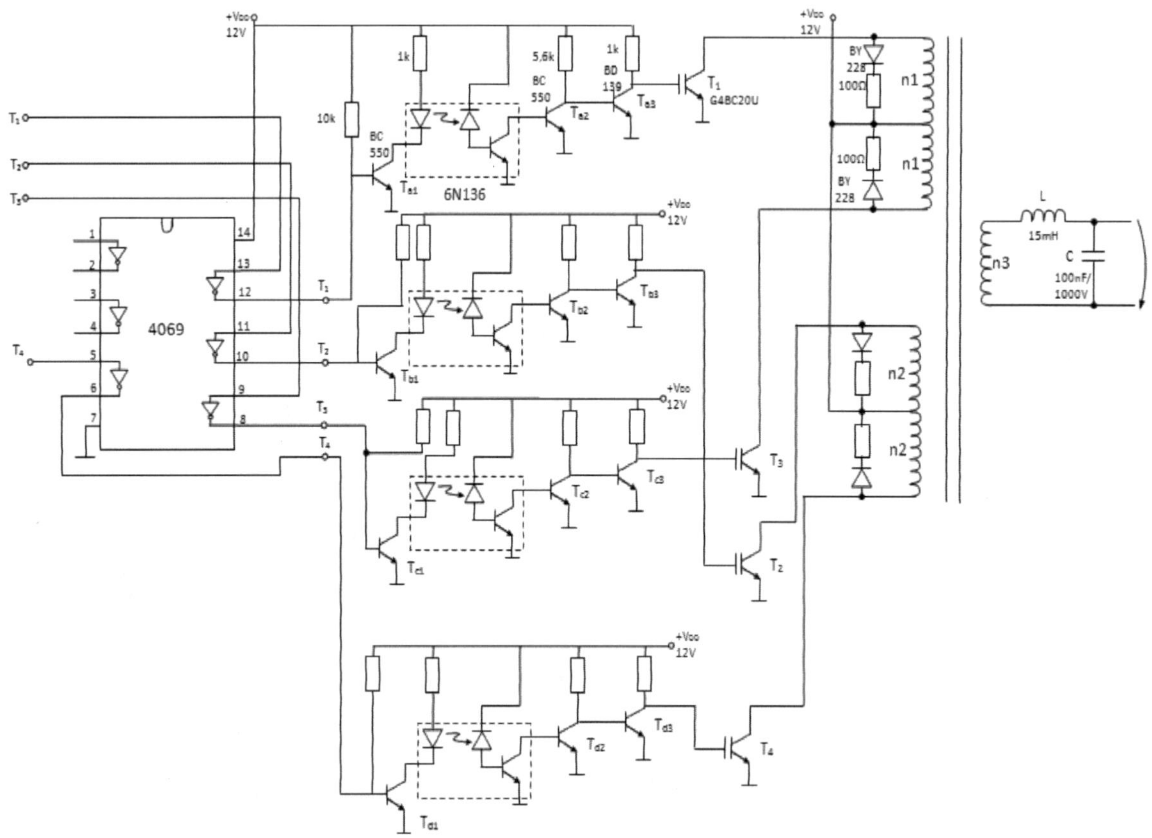

Figure 5. Single-phase inverter wiring diagram

The commanding sequence from the counter is brought to the IGBT grid through a MOS(4069) inversor. Each output is connected to a current amplifier with a bipolar transistor(BC550) to provide enough current for the optocupler used for the galvanic separation of the comand circuit on the power circuit. The signal obtained from the optocuplers output is amplified into curent with same bipolar transistor (BC 550) following to be inversed again with a BD 139 stage transistor. This second reversal is necessary to amplify both current command and therefore a faster switching transistors and IGBT for controlling the phase of the output sequence from numbers that will produce output voltage of the synthetic form. For a form as close to one using sinusoidal output filter LC [8].

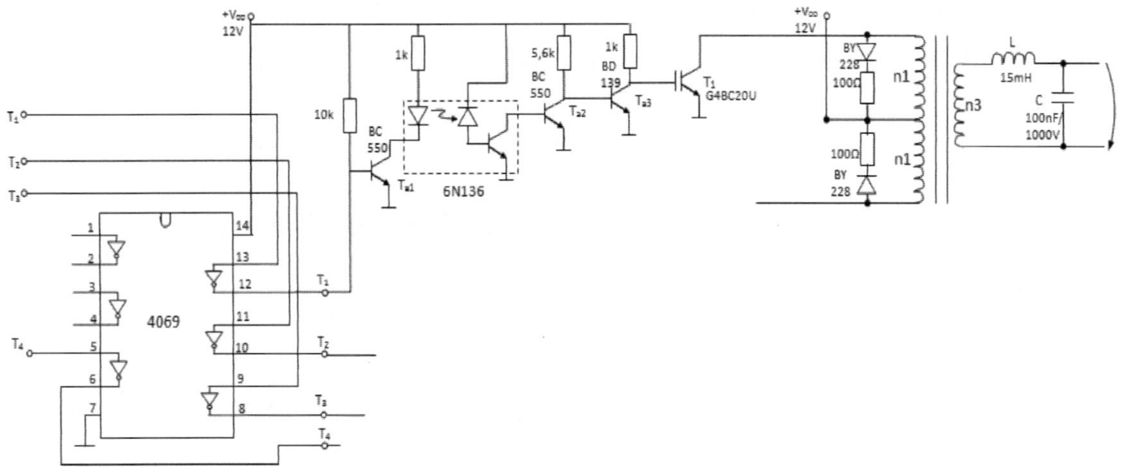

Figure 6. Wiring diagram for control of a IGBT transistor

3. Problem solutions

The proper functioning of the circuit is highlighted by presenting the wave forms, in different points of the scheme, as it follows.

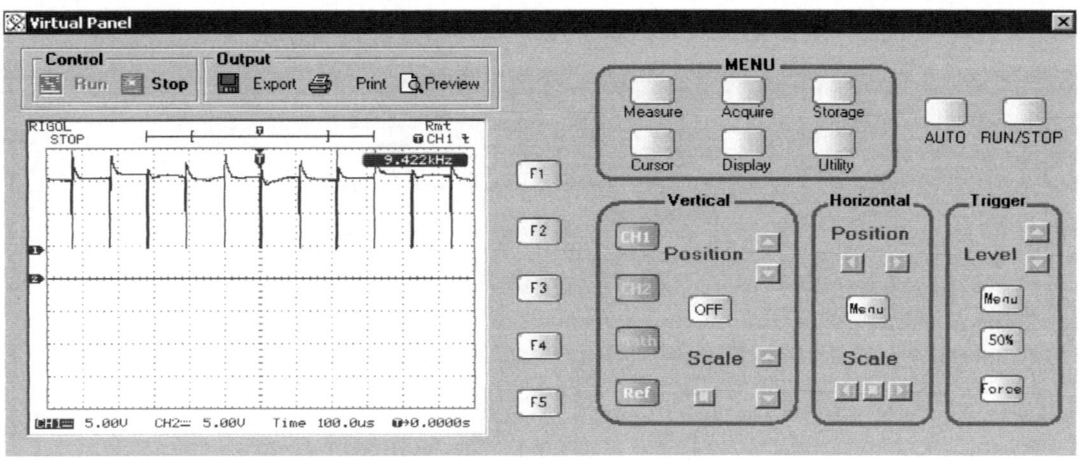

Figure 7. Tact pulses for the counter

IOP Conf. Series: Materials Science and Engineering **163** (2017) 012036 doi:10.1088/1757-899X/163/1/012036

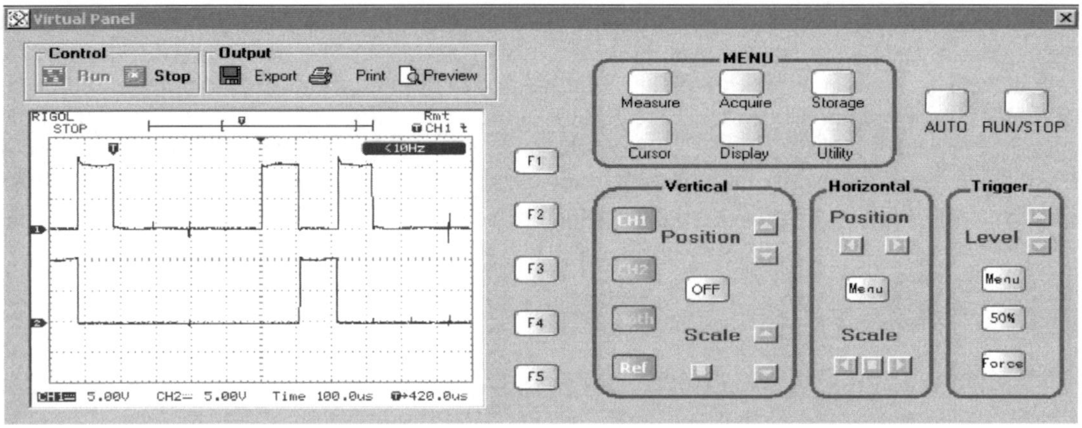

Figure 8. Comand impulses for optocupleurs one and two

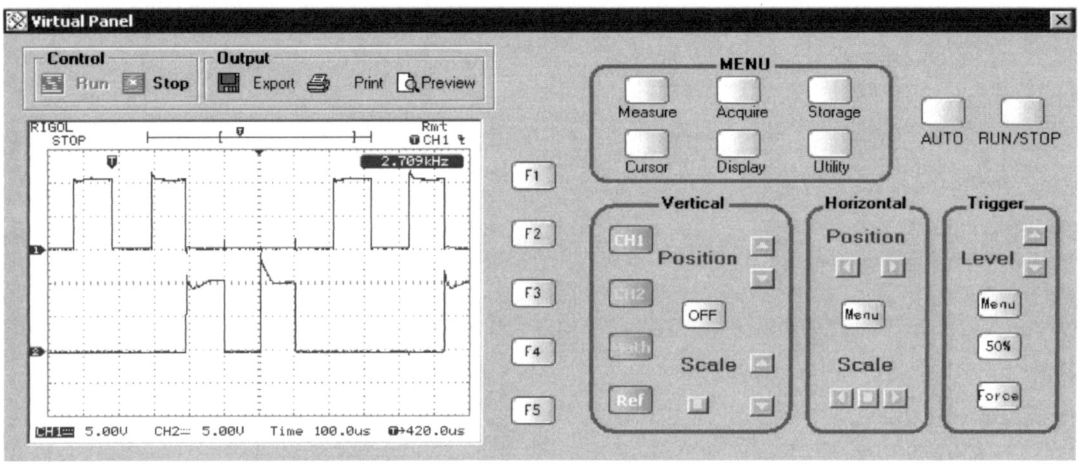

Figure 9. Comand impulses for optocupleurs one and three

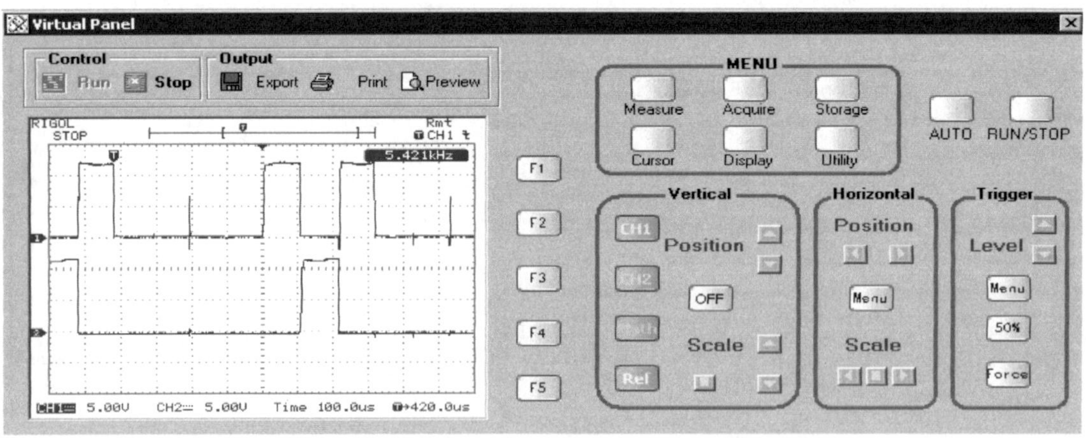

Figure 10. Command impulses for transistors T1 and T2

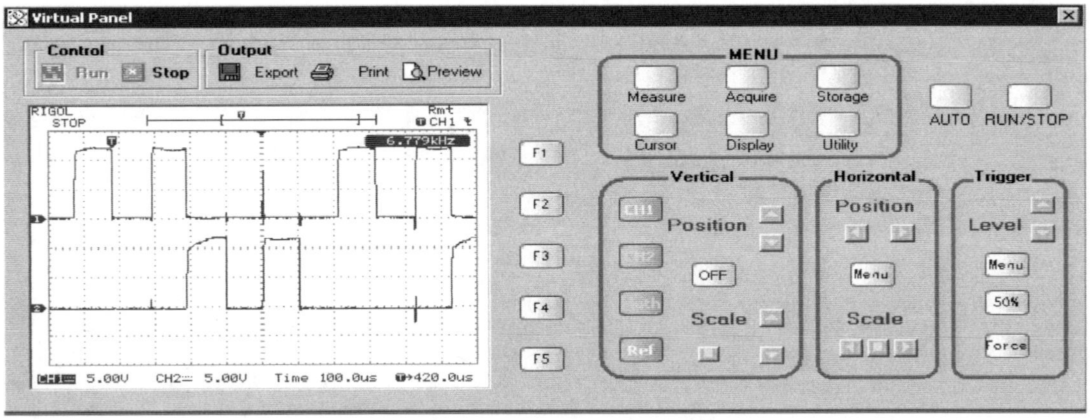

Figure 11. Command impulses for transistors T1 and T3

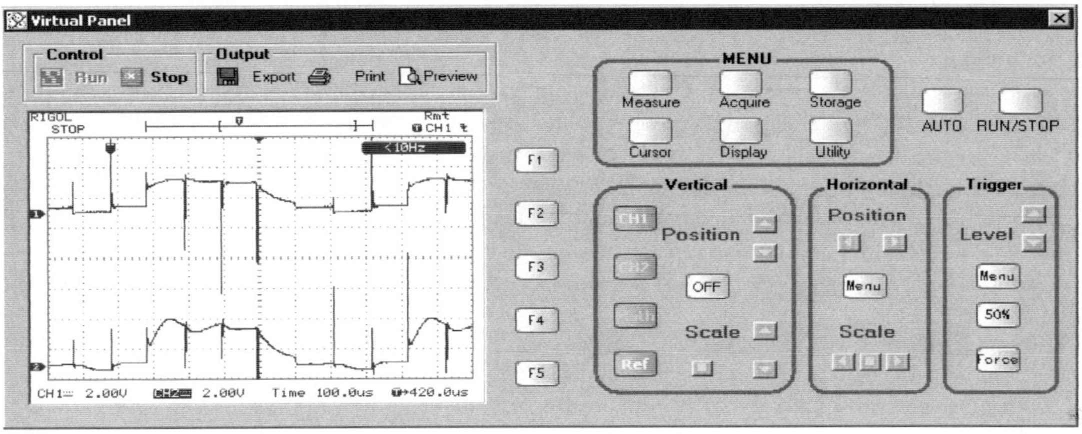

Figure 12. Collector - emitter voltage for T1 and T2

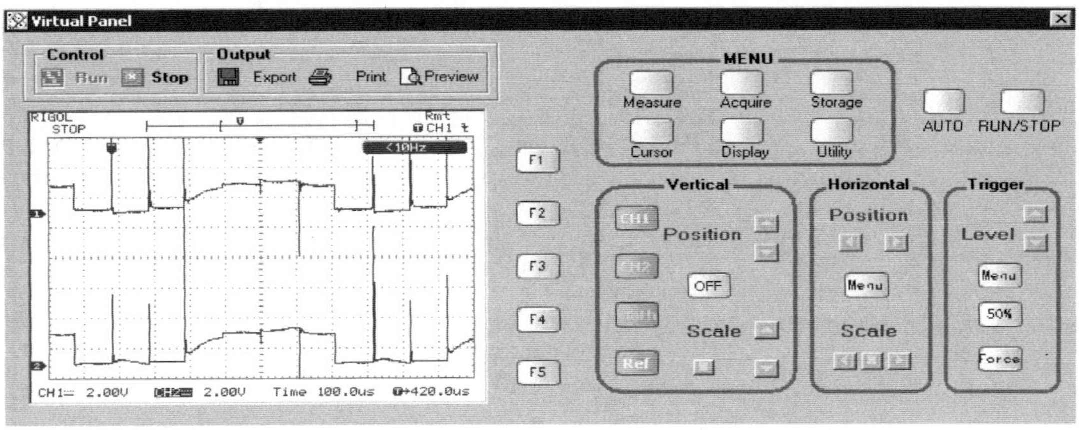

Figure 13. Collector - emitter voltage for T3 and T4

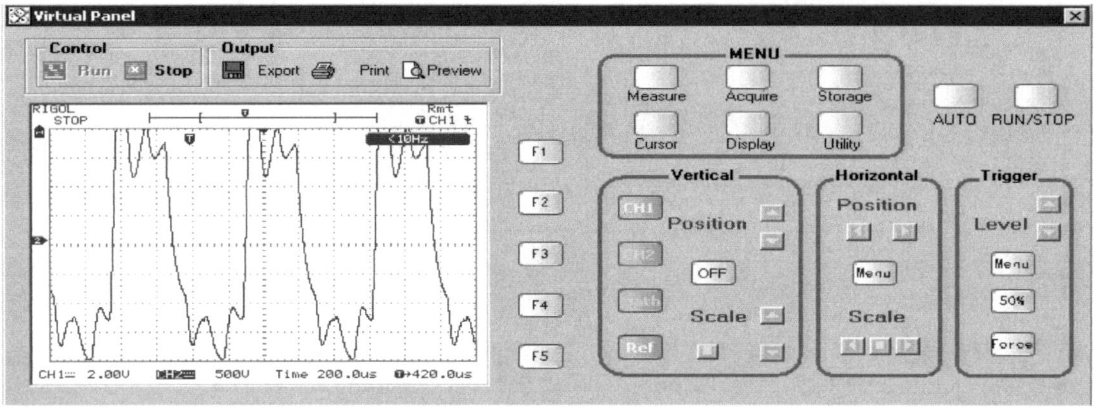

Figure 14. Output voltage

4. Conclusions

It appears an output voltage which no sudden jumps and shape is similar to sinusoid. Command sequence is under full waveform, which requires the use of large radiating surface radiators.

This was necessary because it was used a transformer laminations ferro silicon.

During the conduction of a transistor output voltage is variable as shown in figure.

The solution to this problem stands in realizing an impulse commanding sequence instead the one with full wave and replacing the transformer with a ferrite core one.

References

[1] Tamer Khatib T N, Mohamed A, Amin N and Sopian K 2010 An efficient maximumpower point tracking controller for photovoltaic systems using new boost converter design and improved control algorithm, *WSEAS Transactions on Power Systems* **5**(2) 53- 63

[2] Alexa D and Hrubaru O 1989 *Applications of the static power converters*, Technical Publishing, Bucharest, Romania

[3] Măgureanu R and Micu D 1994 *Static frequency converters in drives with induction motors*, Technical Publishing, Bucharest, Romania

[4] Lascu D 2004 *Techniques and circuits for active correction of the power factor*, Western Publishing, Timişoara, Romania

[5] Ionescu F, Florincău D, Niţu S, Six J P, Delarue Ph and Boguş C 1998 *Power Electronics. Static Converters*, Technical Publishing, Bucharest, Romania

[6] Deaconu S I, Topor M, Popa G N and Bistrian D 2010 Experimental Study and Comparative Analysis of Transients of Induction Motor with Soft Starter Startup, *Advances in Electrical and Computer Engineering* **10**(3) 27-33

[7] Popa G N, Popa I, Diniş C M, Deaconu S I and Abrudean C 2016 *Control of Single-Phase Inverter with PIC Microcontroller,* The XVI National Conference "Prof. Dorin Pavel - The Founder of Romanian Hydropower", Since and Engineering, Sebes, Romania, June 10-11, pp 447-457

[8] Iagăr A, Popa G N and Diniş C M 2015 Analysis the electrical parameters of a high-frequency coreless induction furnace, *IOP Conf. Ser.: Mater. Sci. Eng.* **85** 012013

International Conference on Applied Sciences (ICAS2016) IOP Publishing
IOP Conf. Series: Materials Science and Engineering **163** (2017) 012037 doi:10.1088/1757-899X/163/1/012037

Experimental research on mathematical modelling and unconventional control of clinker kiln in cement plants

S Rusu-Anghel

Politehnica University of Timisoara, Department of Electrical Engineering and Industrial Informatics, 5 Revolution Street, Hunedoara, 331128, Romania

E-mail: stela.anghel@fih.upt.ro

Abstract. Analytical modeling of the flow of manufacturing process of the cement is difficult because of their complexity and has not resulted in sufficiently precise mathematical models. In this paper, based on a statistical model of the process and using the knowledge of human experts, was designed a fuzzy system for automatic control of clinkering process.

1. Introduction

The analytical identification of the process during the fabrication of the cement is difficult because this process is highly complex and they have helped us obtain some accurate mathematical methods who allow an effective conventional management.

According to the current evolution of the calculation techniques and of the process' management principles, the experimental identification based on statistic data is an alternative we should consider.

The most usual methods are the system, the linear, and the discrete models who enclose the effect of the random disturbance, such as in Figure 1 [1].

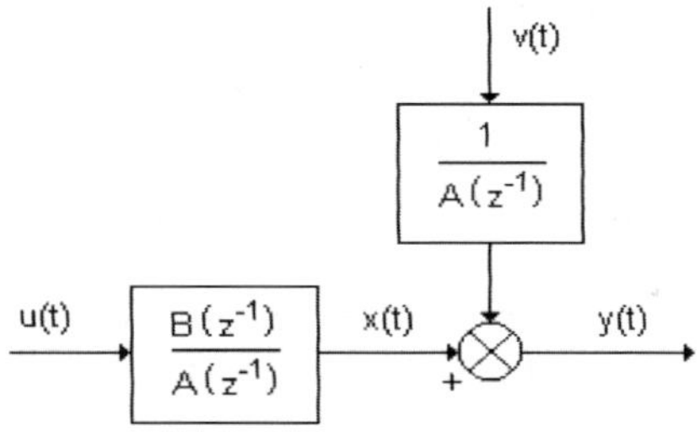

Figure 1. Discrete models [1]

Synthetically speaking, z(t) is uptaken to this process (the parameter is the discreet time t). Because y(t) contains some noise (in addition), it is considered as such (it encloses a determination element x(t)). The statistic features of the output we have measured - y(t) - are determined by the statistic features of the noise z(t), or v(t), according to Figure 1. Due to the fact that the identification experiment does not use – for economic and engineering reasons – all the accomplishment of the noise, and/or either of the y(t) output, it does not acknowledge the identification methods of the output, because they need some statistic properties (hypothesis) which help them identify the PFD pattern. This is not possible after only one significant experiment (for off-line identification methods) or after continuous process, but they have the same properties (for on-line identification methods).

In [2] was presented experimental statistical model drafted for a clinkering kiln. This model will be used in this paper to analyze the work regimes.

It considered the following variables:

- for input: - fuel flow - with significant effect in heat transfer and, indirectly, on features of chemical reactions in the kiln;
  - material flow;
- for output: - temperature of the combustion gas in the first heater;
  - actuating motor power of the kiln - is correlated with combustion zone temperature and quality of the clinker.

Although there are other possible variables, the variables mentioned are the most important and the identification process has been reduced to them.

Sizes measured in the process are represented in Figure 2.

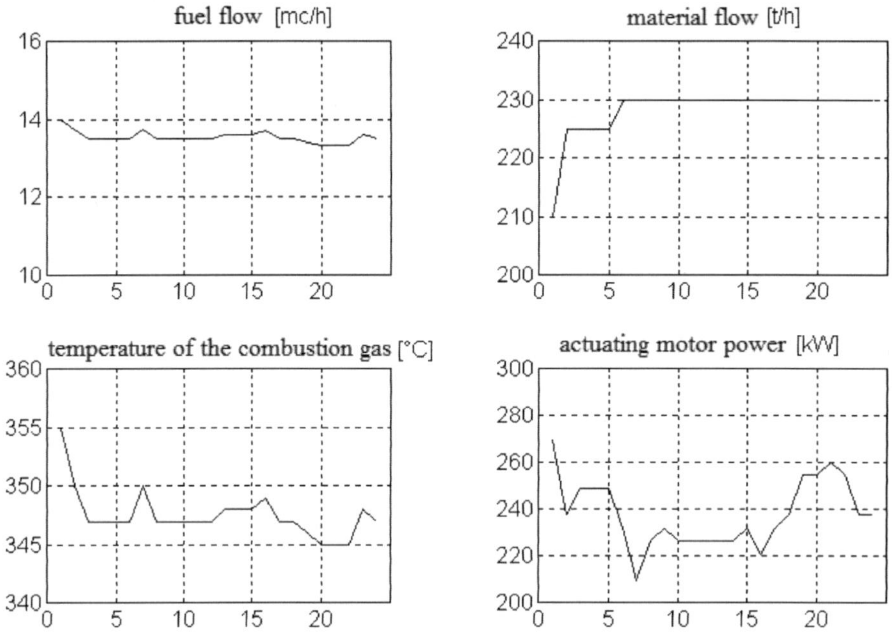

Figure 2. Representation of process inputs and outputs [2]

The proper shaping-up has been performed according to the CMP criterion – using two retrogressive methods – linear and non-linear II-type [2-4].

a) *Linear retrogressive*

If we use a linear subordination of the output size, according to the input sizes, we have:

$$Y_1 = A_{11}X_1 + A_{12}X_2 + B_1 , \quad (1)$$

$$Y_2 = A_{21}X_1 + A_{22}X_2 + B_2 . \quad (2)$$

By using the method of the smallest squares, the constant values are:

- for output size $Y_1(t)$: $A_{11} = 11.04$; $A_{12} = -0.1055$; $B_1 = 222.2$

The precision indicator of the regress pattern is: $R^2 = 0.9866$

- for input size $Y_2(t)$: $A_{21} = -85.92$; $A_{22} = -3.784$; $B_2 = 2265$

The precision indicator of the regress pattern is: $R^2 = 0.8266$

Thus, the pattern is:

$$\begin{bmatrix} Y_1 \\ Y_2 \end{bmatrix} = \begin{bmatrix} 11.04 & -0.1055 \\ -85.92 & -3.784 \end{bmatrix} \cdot \begin{bmatrix} X_1 \\ X_2 \end{bmatrix} + \begin{bmatrix} 222.2 \\ 2265 \end{bmatrix} \quad (3)$$

b) *II- degree polynomial regress*

If we use a II-type non-linear subordination of the output size, according to the input sizes, we have:

$$Y_1 = C_{11}X_1 + C_{12}X_2 + C_{13}X_{12} + C_{14}X_{22} + C_{15}X_1X_2 + C_{10} \quad (4)$$

$$Y_2 = C_{21}X_1 + C_{22}X_2 + C_{23}X_{12} + C_{24}X_2 + C_{25}X_1X_2 + C_{20} \quad (5)$$

By using the method of the smallest squares the constant values are:
- for output size $Y_1(t)$:
C_{11} = -27.36; C_{12} = 1.628; C_{13} = 2.8; C_{14} = 0.001288; C_{15} = -0.1642; C_{10} = 273.6
The precision indicator of the regress is: R^2 = 0,9988
- for output size $Y_2(t)$:
C_{21} = -298; C_{22} = 170.9; C_{23} = 62.97; C_{24} = -0.1891; C_{25} = -6.506; C_{20} = -16317
The precision indicator of the regress is: R^2 = 0.8576
Thus, the pattern is:

$$\begin{bmatrix} Y_1 \\ Y_2 \end{bmatrix} = \begin{bmatrix} -27.36 & 1.628 & 2.78 & 0.001288 & -0.1642 \\ -298 & 170.9 & 62.97 & -0.1891 & -6.506 \end{bmatrix} \cdot \begin{bmatrix} X_1 \\ X_2 \\ X_1^2 \\ X_2^2 \\ X_1X_2 \end{bmatrix} + \begin{bmatrix} 273.6 \\ -16317 \end{bmatrix} \quad (6)$$

The number of processed data was not too high, resulting one degree of accuracy of about 0.9 for model determined, sufficient for its intended purpose.

Along with the experience of human experts, this model will allow the determination sets of rules for the synthesis of a fuzzy control system.

2. Control of clinkering process

The objectives are: bringing and maintaining the operating parameters of the kiln to allow optimal values for increasing production and reduction of specific consumptions (energy and materials).

2.1. The structure of the control system [2], [5], [6]

The structure of the proposed system for automatically controlling the clinkering process is shown in Figure 3.

The process is managed by a PLC (AP) equipped with modules I/O and network-connected to the higher level. Is proposed the use of smart sensors will communicate with the PLC via a local network. AP comprises a fuzzy controller whose the algorithm is described in the following paragraph. The other equipments used in the scheme are typified.

2.2. Control algorithm steps

a) Read the necessary parameters of process management: torque (CM); the temperature in the clinkering zone (TZ); CO content output cyclones tower (COG); the kiln speed (TC); fuel flow (DJDC); feed rate flour branches A and B (AFA, AFB); blinds position VRA, VRB (DJVRA, DJVRB); oxygen to stage IV (OX). Required: coupling to the system of calculation of ten analog inputs.
b) Processing parameters sampled from the process as required calculation of input quantities.

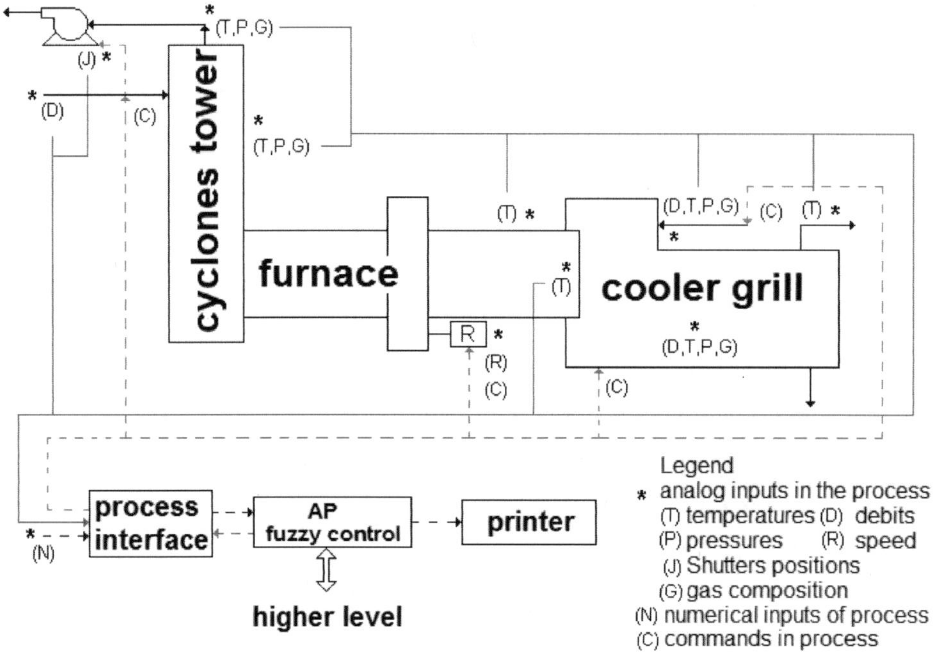

Figure 3. The structure of the proposed system for automatically controlling the clinkering process [2]

c) Establish process condition at regular intervals (about 10 minutes) from the launch management strategy. System checks framing all input quantities in normal (N). If all input quantities are found in normal state, system becomes operational and launches first leading decision or adjust calculated output quantities and their new values. If one or more input quantities are not in normal situation, the system senses situation and launch an error message. The operator takes over control the process for to bring it in optimal condition.

d) During automatic control the system updates his status and validates a special situations (eg separation of shell).

e) Execution of orders in process is done at the times set by the system. For the implementation of controls, is necessary connect to the computering system of outputs for: flow supply flour arm A and arm B (CFAA CAFB), kiln speed (CTC), position blinds VRA and VRB (CDJVRA, CDJVRB), position valve for fuel flow adjustment (CDJDC).

3. Fuzzy algorithm design

For the synthesis of fuzzy algorithm was use a fuzzy controller developed starting from to on PIC 16C74 microcontroller and dedicated software for introducing rules, configuration settings, the shape of the membership functions, simulation. The program is functional only if Fuzzy controller is connected to the PC via the serial interface [2], [7], [8].

The program's main menu is presented with the default image activated (Figure 4). At the top of is entered the program name and the name loaded system. Bar following contains the main menu. Additionally, the left side of the bar is a visual function, with which subroutines at main program can be called.

3.1. Information input data

a) TZ - temperature in the clinkering area, °C

3-status: S – low 1100 ÷ 1300 °C; N – normal 1300 ÷ 1400 °C; R – high 1400 ÷ 1500 °C

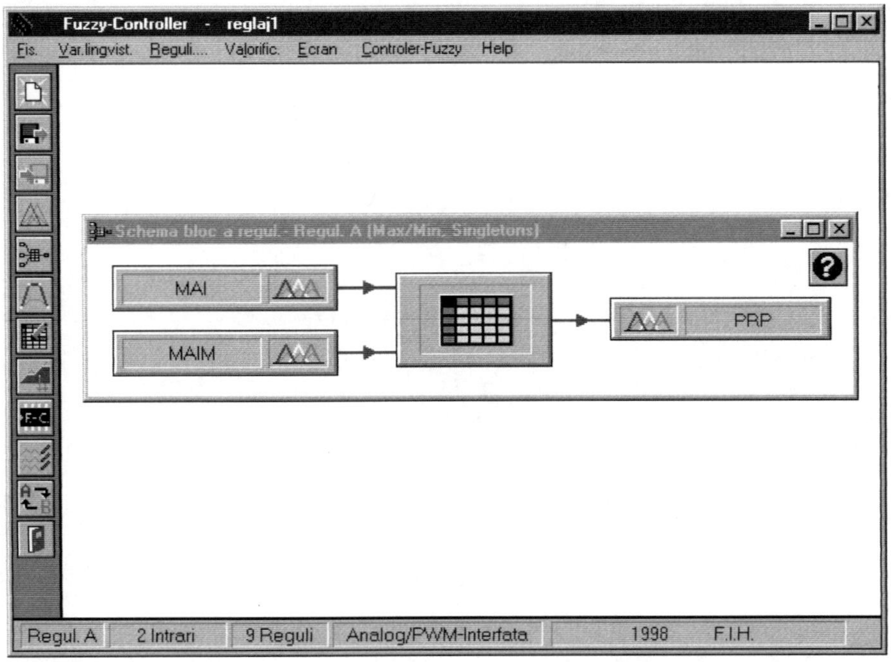

Figure 4. The program's main menu with the default image activated

b) PTZ - average slope on temperature in the area clinkering, degrees

3-status: status -1 (-50 ÷ -15); status 0 (-15 ÷ 15); status 1 (15 ÷ 50)

PTZ characterize temperature trend in the area combustion.

c) PC - average slope on torque, degrees

5-status: status 0 (normal operation: -10÷10); status 1 (heating: 10÷20);
status 2 (separation of shell: 20÷50); status -1 (cooling: -10÷ -20);
status -2 (strong cooling: -20÷ -50)

Torque defines the thermal load of the kiln. The delimitation status ranges is done by tracking into running, because every kiln is defined from this point of view specifically.

d) AC - torque deviation from the reference value, %

2-status: status – (-25 ÷ 0); status + (0 ÷ 25)

AC defines the location of torque compared to the reference value, which the system it establishes and it updates in running.

e) TC - kiln speed, rot/min

2-status: status S (low: 1.3 ÷ 1.7); status N (normal: 1.7 ÷ 1.8)

Speed is correlated with flour supply. In general, the supply increase/decrease at the same time with increase/decrease speed.

f) NC - correlation level of speed with flour supplying, %

2-status: status N (normal: 0 ÷ 90); status R (high: 90÷100)

The correlation speed-supply with flour is specific to each kiln and is represented by a family of speed-flow curves. R - means that it has reached the maximum level on supply (without increasing speed).

IOP Conf. Series: Materials Science and Engineering **163** (2017) 012037 doi:10.1088/1757-899X/163/1/012037

By moving from one family with lower index at another with higher index can achieve a supply increase from stage to stage without changes in speed.

g) DC - fuel flow - fuel oil, l/h

2-status: status N (normal: 10000 ÷ 15000); status R (high: 15000 ÷ 16000)

The "DC" data was introduced to provide information on the possibility of increasing the flow of fuel, if conditions of the kiln permit. For example, where TZ is low (S) and a maximum "DC" (R), may not bring the kiln to a normal state by increasing the flow rate of the fuel supply, it will be necessary reduce the feed rate of the flour.

h) COG - CO content in the flue gas to output of cyclones tower, %

2-status: status N (normal: 0 ÷ 0.45); status R (high: 0.45 ÷ 0.6)

Maximum 0.6% CO is the value that triggers the electrostatic precipitator. CO value is a measure of the combustion quality and condition for electrostatic precipitator protection.

Membership functions for inputs are shown in Figure 5.

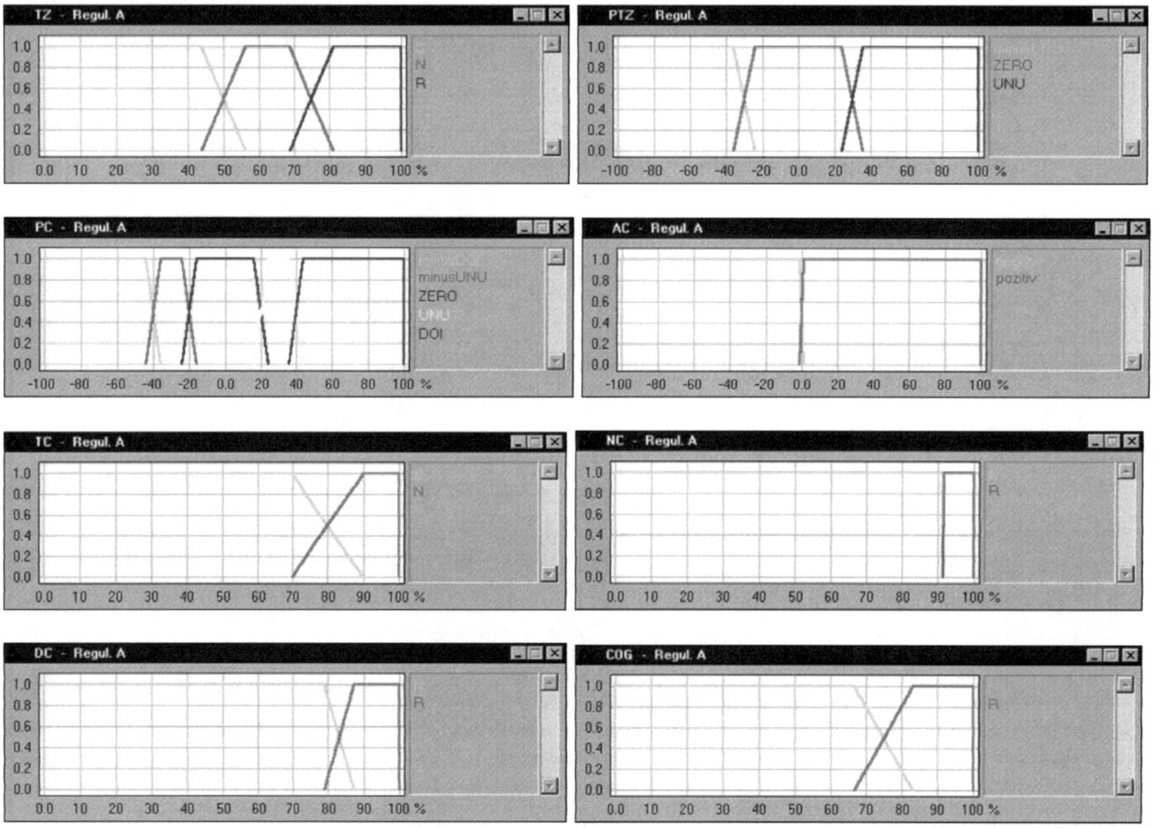

Figure 5. Membership functions for inputs

3.2. Information output data

a) dCDC - fuel flow adjustment (by adjusting valve), l/h

5 values: CM - increase much (400); CP - increase little (200); M – maintain (0);
SP - decrease little (-200); SM - decrease much (-400)

IOP Conf. Series: Materials Science and Engineering **163** (2017) 012037 doi:10.1088/1757-899X/163/1/012037

b) dCNC - adjusting the level of correlation on flour supply and kiln speed
3 values: C – increase (1); M – maintain (0); S – decrease (-1)

c) dCTC - adjustment the kiln speed, rot/min
5 values: CM - increase much (0.2); CP - increase little (0.1); M – maintain (0); SP - decrease little (-0.1); SM - decrease much (-0.2)

d) dAFA, dAFB - flour feed rate, the branches A and B, t/h
These flows are calculated based on the fuzzy value obtained for speed according *3.1.* f

e) TIMP - relaunch time of control, minutes
3 values: A (12); B (16); C (18)

f) DJVRA, DJVRB - opening a blinds blowers: VRA and VRB, %. It is provided with a separate control loop, depending on the content of O2 in the fourth step.

Membership functions for outputs are shown in Figure 6.

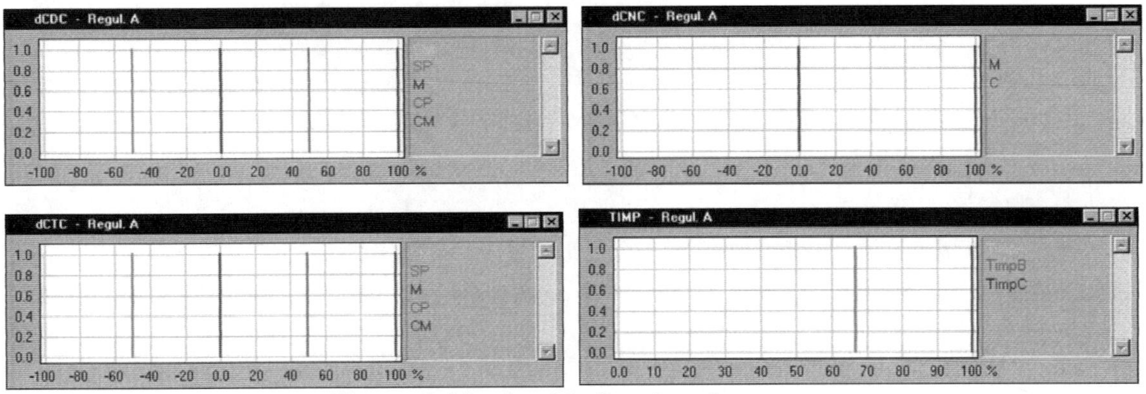

Figure 6. Membership functions for outputs

3.3. Control rules

Rule number = 5 * 2 * 3 * 3 * 2 * 2 * 2 * 2 = 1440 rules

An example of the control rules is given in Figure 7 and in Figure 8 shows the simulation of fuzzy control for the dCDC size of command.

Editor baza reg.

Tabela	If								then	
	TZ	PTZ	PC	AC	TC	NC	DC	COG	dCDC	Pondere
1.Regula	u. R	u. ZERO	u. ZERO	u. pozitiv	u. N	u. R	u. N	u. R	SP	1.0
2.Regula	u. S	u. ZERO	u. ZERO	u. pozitiv	u. N	u. R	u. N	u. R	CP	1.0
3.Regula	u. N	u. ZERO	u. ZERO	u. pozitiv	u. N	u. R	u. N	u. R	M	1.0
4.Regula	u. R	u. UNU	u. ZERO	u. pozitiv	u. N	u. R	u. N	u. R	SM	1.0
5.Regula	u. S	u. UNU	u. ZERO	u. pozitiv	u. N	u. R	u. N	u. R	M	1.0
6.Regula	u. N	u. UNU	u. ZERO	u. pozitiv	u. N	u. R	u. N	u. R	SP	1.0
7.Regula	u. R	u. minusUNU	u. ZERO	u. pozitiv	u. N	u. R	u. N	u. R	M	1.0
8.Regula	u. S	u. minusUNU	u. ZERO	u. pozitiv	u. N	u. R	u. N	u. R	CM	1.0
9.Regula	u. N	u. minusUNU	u. ZERO	u. pozitiv	u. N	u. R	u. N	u. R	CP	1.0

Regul. A | 8 Intrari | 9 Reguli | SI / SAU => Conexiune Si / Sau

Figure 7. Example of the control rules

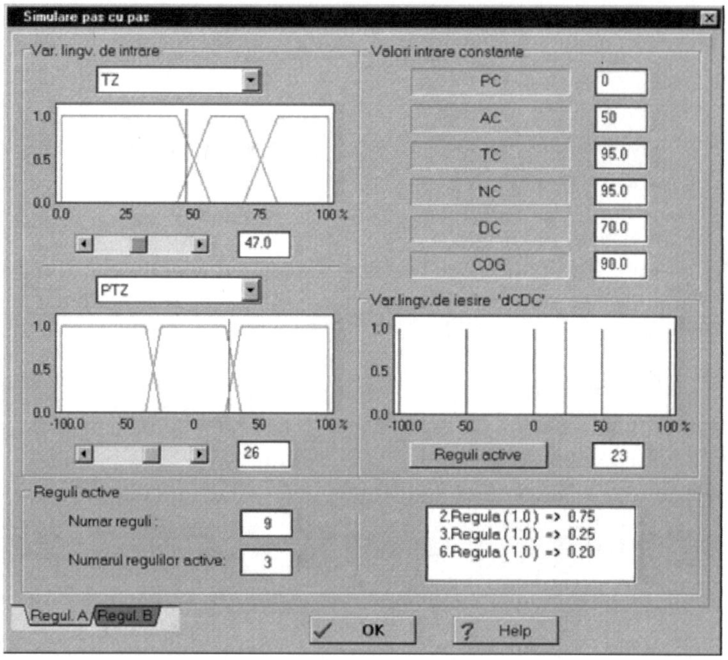

Figure 8. Simulation of fuzzy control for the dCDC size of command

3.4. Static characteristics of designed fuzzy controller

Input data: TZ (temperature in the area clinkering), PTZ (average slope on temperature in the area clinkering), PC (average slope on torque), AC (torque deviation from the reference value), TC (kiln speed), NC (correlation level of speed with flour supplying), DC (fuel flow), COG (CO content in the flue gas output cyclones tower).

The output: dCDC (fuel flow adjustment), dCNC (adjusting the level of correlation on flour supply and kiln speed), dCTC (adjustment the kiln speed), AFA, AFB (flour feed rate, the branches A and B), dJVRA, dJVRB (opening blinds Blowers VRA and VRB).

Was chosen for representation only the function dCDC = f (TZ, PTZ, PC, AC, TC, NC, DC, COG) specified in the table inference given in Figure 6. Control surface obtained in this case is shown in Figure 9, considering variables TZ, PTZ (charts reasons, other inputs were considered constant).

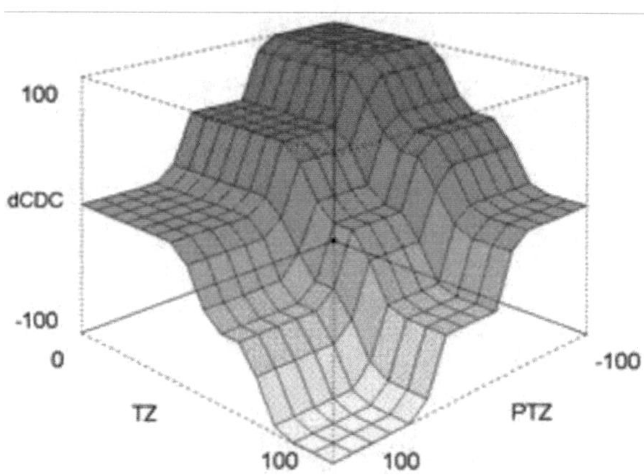

Figure 9. Control surface for function dCDC = f (TZ, PTZ)

In Figure 10 is illustrated plan static characteristic representation in coordinates (TZ, PTZ) through curves dCDC = const. This corresponds to dCDC inference table.

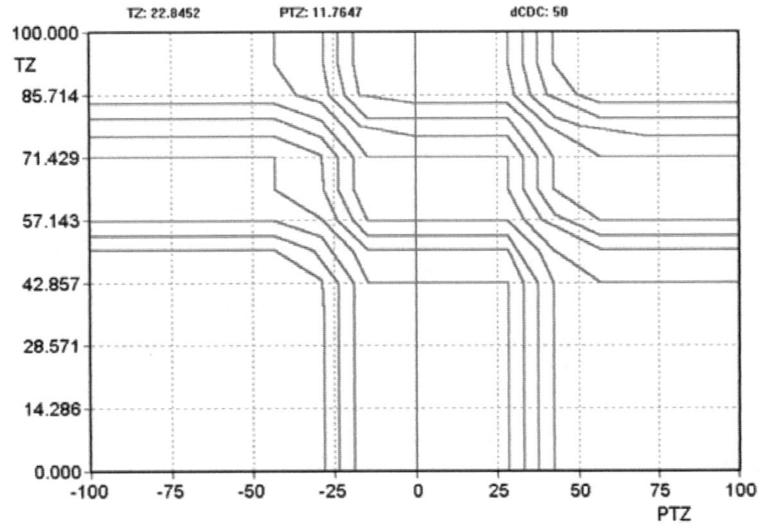

Figure 10. Static characteristic

The control system has been designed so that database, including the rules can always be improved, with the possibility of online testing. This flexibility allows continuous improvement of automat control, based on experience in the current running. After loading the rules base, the controller can work independently in the plant.

4. Conclusion

Usually, the lack of a firm mathematical model of the process is not an impediment for synthesis of a fuzzy automatic control. In relation to classical control, fuzzy control can be based and focused heavily on experience of human operator, experience which the Fuzzy controller can "shape" more accurately than a conventional controller.

To complex processes, it is recommended to have a mathematical model to can assess correlations between significant characteristic parameters of the process.

This is why was performed the approximate model of the clinkering kiln. Since deterministic models may not take into account, in this case, all real phenomena from the kiln, it was done by a statistical identification. Using advanced methods of calculation, although the number of data processed was not too high, resulted a degree of accuracy of about 0.9 for kiln model, sufficient for its intended purpose.

A special problem is to establish the control rules and the decisions to be taken automatically, thanks to their direct technological implications and the multitude of practical situations that may arise. These issues directly influences the quality of expert system.

Consequently, there have been set very carefully control rules, based on the technological implications, the consultation of a large bibliography, discussions with specialists technologists.

In addition to the control algorithms it has been proposed a structure for the automation schema, using PLC which subsequently will be integrated into the overall system a production tracking. Were established thus sizes that must be measured in the process, the number of inputs / outputs, of the sampling interval and the necessary equipment. Through these specifications are emphasized practical aspects of theoretical research that enable their application in practice.

One of the great advantages of the proposed system is that the software has been designed so that the database, including the rules can always be improved, with the possibility of online testing. This flexibility allows continuous improvement of automat control, based on experience in the current running.

References

[1] Calin S and Dumitrache I 1984 *Reglarea numerica a proceselor tehnologice*, Editura Tehnica, Bucuresti, Romania

[2] Anghel S 1998 *Cercetari privind modelarea si conducerea proceselor din fabricile de ciment*, Politehnica University of Timisoara, Romania, Doctoral Thesis

[3] Alic C, Miklos C C and Miklos I Z 2009 *Upon the actual tendencies in modelling and simulating the behavior of the pantograph-catenary pairing, Machine design*, Anniversary volume of the Faculty of Technical Sciences NoviSad, pp 85-90

[4] Miklos I Z, Miklos I, Alic C I and Miklos C C 2008 *CAM mechanisms-modelling and analysis with Autodesk Inventor*, KOD 2008 Proceedings of The 5th International Symposium About Shaping, Industrial And Product Design, Novi Sad, Serbia, April, pp 385-388

[5] Colina E, Falconi M, Morocho V, Medina J and Mora A 2015 *Design of a Supervisory Control System for a Clinker Kiln Operation*, Quito, Ecuador, July 14-16, pp 387-391

[6] Shahriari K and Tarasiewicz S P E 2009 *A model-based approach for clinker rotary kiln monitoring and instrumentation*, IEEE Cement Industry Technical Conference Record, Palm Spring, California, USA, May 29-June 5, pp 1-11

[7] Voloşencu C 1997 *Reglare Fuzzy şi neuronală*, Editura Eurobit, Timişoara, Romania

[8] Costea C R, Silaghi H M, Zmaranda D and Silaghi M A 2015 Control System Architecture for a Cement Mill Based on Fuzzy Logic, *International Journal of Computers Communications & Control* **10**(2) 165-173

IOP Conf. Series: Materials Science and Engineering **163** (2017) 012038 doi:10.1088/1757-899X/163/1/012038

Experimental determination of the filling coefficient for an aspirated spark-ignition engine

S Rațiu, V Alexa, I Kiss and V Cioată

University Politehnica Timisoara, Faculty of Engineering Hunedoara, Department of Engineering and Management, 5 Revolution Street, 331128 Hunedoara, Romania

E–mail: sorin.ratiu@fih.upt.ro

Abstract. This study aims at determining, by experiment, the filling coefficient of a spark-ignition, normal aspirated engine, with carburettor. For this purpose, a pilot plant was designed for measuring the pressure at various points on the route, simulating a stationary air flow regime by means of a vacuum pump. Measurements were made for various lifting heights of the intake valve and various opening positions of the throttle body, thus highlighting how their influence on the pressure loss and on the filling coefficient.

1. Introduction

The components of fresh load entering the engine cylinders during the intake stroke are the fresh air and – in systems with external mixture formation – the fuel (gasoline), in the form of vapours, found in suspension in it. Most of the amount of fresh air enters the cylinders through the channel controlled by the throttle. Additionally, the fresh load can be sucked into the engine through the vapour recirculation system, if it exists. The air mass found in cylinders after closing the intake valve is the decisive factor in relation to the mechanical work produced by cyclically burning of the fresh load, with direct influence on the engine torque. Consequently, the measures to increase the maximum torque and the maximum power almost always involve creating the conditions for obtaining a maximum possible filling ratio [1].

When using carburettors as equipment for achieving the optimal dosage of the fuel mixture, the intake system was characterized by manifolds able to feed one or more cylinders. As a result of its geometry, this system has significant losses of fresh load, leading to lower maximum power and momentum.

On the intake route, two kinds of losses are recorded:

- *Pressure losses* or *gasodynamic losses* caused by the existence of hydraulic resistances on the suction line, which can be quantified using the well-known relation:

$$\Delta p = \xi \cdot \rho \cdot \frac{w^2}{2} \tag{1}$$

where ξ is the pressure loss coefficient characteristic to each component of the intake system, w is the flow rate of the fresh fluid, and ρ is its density.

- *Heat losses* caused by fluid heating from the suction line walls, which determine the final temperature to be higher than the engine inlet temperature, the temperature increase resulting in density decrease and, hence, affecting the filling.

To establish a single criterion for quantifying the filling performance, we must compare the amount of load that actually enters the cylinder with the one that might enter if there were no losses.
The *filling degree* or *filling ratio*:

$$\eta_v = \frac{C}{C_0} \tag{2}$$

is the ratio between the load *actually retained* in the cylinder (C), after closing the last device for stopping the access of the engine fluid to and from the cylinder, and the amount of load *that might be retained* in the cylinder (C_0), in the condition recorded when entering into the engine, per cylinder capacity unit [2].

This study focuses only on the loss of pressure caused by the existence of hydraulic resistances, for finding them by experimental measurements and highlighting their effects on the process of filling the engine cylinders with fresh fluid.

2. Experimental research. Equipment and procedures

This study focuses on theoretical analysis, design and practical realisation of a gasodynamic test stand for providing information about how the fresh gases flow through the manifolds, during the intake process [3]. The purpose of using this device is to find the characteristic parameters that emphasize the perfection of filling.

2.1. How to choose of the intake system model

To materialise the gasodynamic studies, we decided to analyse the intake system of a Dacia engine, type 810.99, equipped with CARFIL 32 IRM carburettor, WEBER licence, centrally positioned on the intake manifold. The cylinder head, cast from aluminium alloy, has two valves per cylinder (one inlet and one outlet). The combustion chamber is wedge-type, with the volume of 35.6 cm^3. The diameter of the intake valve disk is 33.5 mm [4]. The reason for choosing this intake system is that it has the highest complexity in terms of the gasodynamic resistances found on the route between the air filter and the interior of cylinder.

2.2. Research methodology

To study the intake process in gasodynamic terms, it is necessary to create the same mechanism governing the gas flow through an internal combustion engine. This mechanism is given by the pressure difference between the interior of cylinder and the outer environment.

In this situation, we considered, as methodology for determining the gas flow parameters, the measurement of the pressure differences created between the outer environment and various points, specially chosen, located on the intake route, including the interior of cylinder, for various lifting heights of the intake valve and various positions of the throttle. With the aid of the measured pressure difference values, we can find the gas flows passing through the intake system and cylinders. With the aid of the measured flows and the theoretically determined flows, on the basis of the geometry of the components crossed by gases, we can find a number of characteristic values of the intake process.

Subsequently, we can determine the following parameters: filling coefficient, flow coefficient of the intake valve, momentary flow coefficient of the hole provided by the intake valve port, resistance coefficient of the intake route, efficiency of flow in the vicinity of the valve, etc. In this article we are focused on filling coefficient.

2.3. Presentation of the experimental stand

For studying the gas flow, it is necessary to create a depression along the intake route. In accordance with this requirement, the gasodynamic research device must be capable of producing negative pressure.

The stand should be able to measure the pressure difference between the external environment and various points located along the intake route, including the interior of cylinder. Considering these aspects, the embodiment includes the following elements:

- equipment for obtaining the vacuum (vacuum pump);
- device for measuring the pressures and pressure differences;
- carburettor, centrally mounted on the intake manifold;
- cylinder head with holder;
- transparent cylinder;
- device for lifting the intake valve;
- device for measuring the lifting height of the intake valve;
- device used to operate the throttle body and to measure its angular position.

To ensure a good accuracy of the experimental measurements, a depression of 150 - 600 mmH_2O is required. The depression change should be continuous, for providing constant values for any geometry of the components to be studied. When studying the air flow through the intake valve, depending on its lifting height from the seat, it is required to change the test pressure for reading the small flows circulated by gases. For this reason, the vacuum pump is equipped with devices able to change the circulated flows and the values of the created depressions.

The measurement of testing pressure values is carried out by using an assembly of manometers with vertical tube filled with liquid. We choose this solution because it satisfies the required measurement accuracy and, in addition, it provides a good sensitivity when reading the values representing small pressure changes. The liquid used in the measuring tube is water.

Figure 1. Experimental stand

The route of the pressure ports is shown in Figure 2. They were positioned so as to measure the pressure drop on each representative section of the intake route.

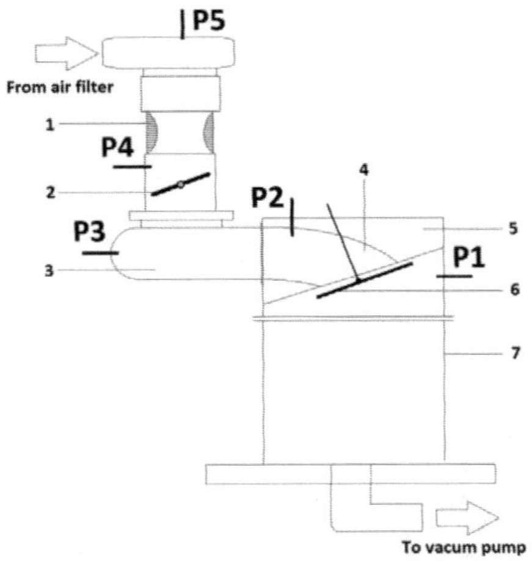

Figure 2. Intake route with the positioning of the pressure ports

The significance of the parts presented in Figure 2 is as follows: 1 – carburetor diffuser, 2 – throttle body, 3 – intake manifolds, 4 – intake channel of the cylinder head, 5 – cylinder head, 6 – intake valve, 7 – cylinder, P1…P5 – pressure ports.

The pressure ports are made in the form of holes in cylindrical wall (Figure 3), over which the air stream flows. A pressure port has the diameter (ϕ) of 1.5 mm, no burrs, its axis is normal to the cylindrical or plain wall, smooth, placed sideways or sideways-upwards (to avoid possible collection of gaseous inclusions). The flexible pipe is made of transparent plastic material [5].

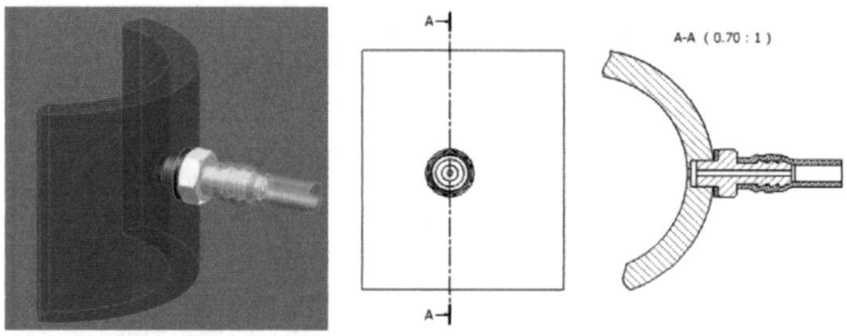

Figure 3. Pressure port in cylindrical wall

We considered 4 positions of the throttle body: 10% open, 25% open, 50% open and 100% open. For each position of the throttle body, we made measurements for 4 lifting heights of the throttle body, i.e. 0.1 mm, 2 mm, 4 mm, 9 mm (maximum lifting height). All this time, the depression produced by the vacuum pump was kept constant. For each flow regime, as defined by all the above combinations, 3 sets of measurement have been made, the final values representing their arithmetic mean.

3. Results and conclusions

Hereinafter, we are going to present the results and conclusions of the experimental measurements, specifying that the measured pressure values are relative to the atmospheric pressure.

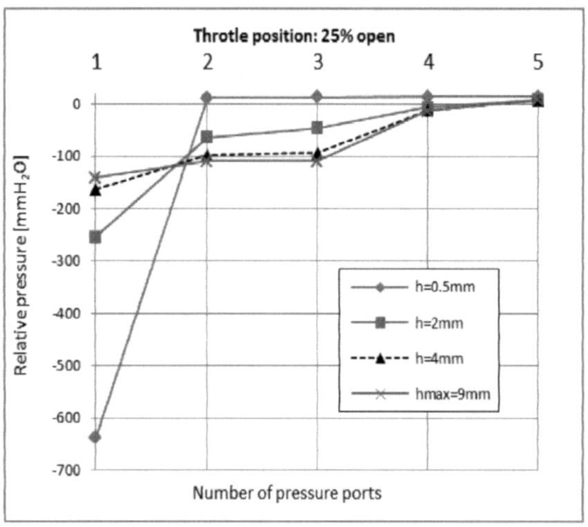

Figure 4. Relative pressure at the 5 ports, for the throttle body position "25% open"

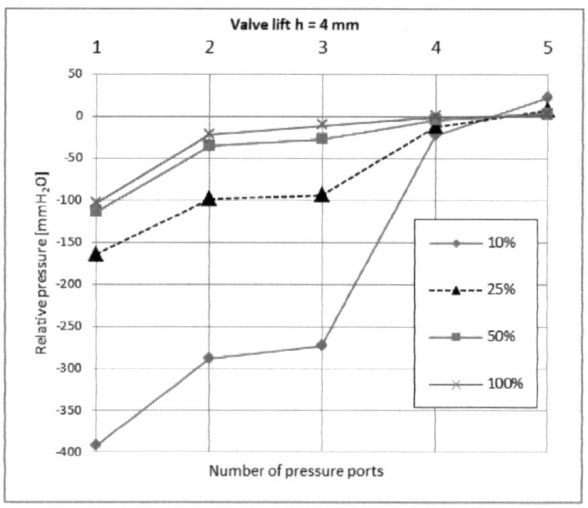

Figure 5. Relative pressure at the 5 ports, for the valve lifting height h = 4 mm

In Figure 4, we can see that the depression values in the measuring points on the intake route, at the same throttle position (25% open), increase with increasing lifting height of the valve (h). On the contrary, the pressure differences between two consecutive measuring points (pressure drop or pressure loss) decrease with increasing lifting height of the valve. This is because, with increasing lifting height of the valve, the flow area in the vicinity of the valve increases and, consequently, the flow rate of the fluid decreases. A low flow speed means lower pressure losses.

Figure 5 shows that the depression values recorded in the measuring points, when the lifting height of the valve is maintained constant (h = 4 mm), are decreasing as the throttle opens. We can also see that the loss of pressure when passing in the vicinity of the throttle (pressure difference between the ports no. 3 and 4) is lower the more open the throttle is.

If we focus on the pressure port P1, located in the combustion chamber (Figure 2), we can say that, at small lifting heights of the intake valve, due to the extremely high pressure losses, at maximum throttle openings, the depression in the combustion chamber is maximal (Figure 6). Based on these values of depression, we can calculate the filling coefficient whose variation, depending on the valve lifting height and throttle body position, is shown in Figure 7. We notice that the filling coefficient increases with increasing the valve lifting height, regardless of throttle position. Also, the filling coefficient increases as the throttle opens. This means that the engine cylinders filling with fresh fluid are better at height loads [6], [7].

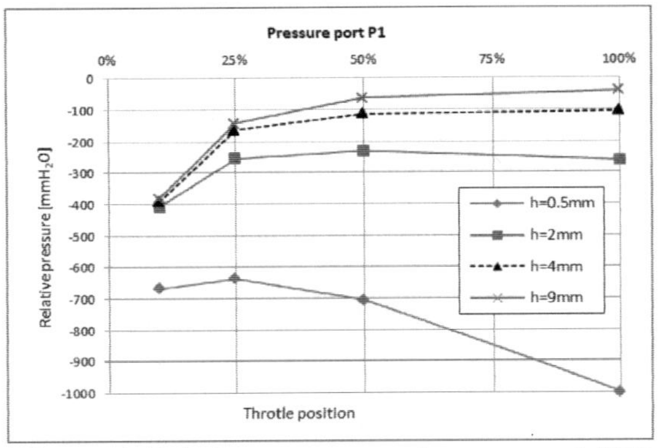

Figure 6. Relative pressure at the port no. 1 (P1) versus the valve lifting height and throttle body position

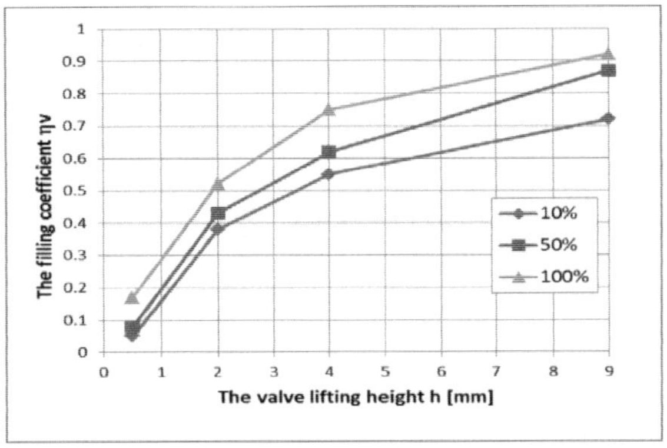

Figure 7. The filling coefficient η_V versus the valve lifting height and throttle body position

References

[1] ***2007 *Automotive Handbook*, 7-th Edition, Robert Bosch GmbH

[2] Raţiu S and Mihon L 2008 *Motoare cu ardere internă pentru autovehicule rutiere – Procese şi Caracteristici* [Internal combustion engines for road vehicles - Processes and Characteristics], Mirton Publishing House, Timişoara, Romania

[3] Leonte V L 2014 *Contribuţii privind optimizarea procesului de umplere la motoarele cu ardere internă* [Contributions to optimise the filling process at the internal combustion engines], Gheorghe Asachi Technical University of Iasi, Romania, Doctoral Thesis

[4] Mondiru C 1990 *Autoturisme Dacia - Technical book*, Technical Publishing House, Bucharest, Romania

[5] Raţiu S 2009 *Motoare cu ardere internă pentru autovehicule rutiere - Procese şi Caracteristici, Experimente de laborator* [Internal combustion engines for road vehicles - Processes and Characteristics. Laboratory experiments], Mirton Publishing House, Timişoara, Romania

[6] Abdul R I and Rosli A B Semin 2008 An Investigation of Valve Lift Effect on Air Flow and Coefficient of Discharge of Four Stroke Engines Based on Experiment, *American Journal of Applied Sciences* **5**(8) 963-971

[7] Ramesh Kumar C and Nagarajan G 2012 Investigation of Flow During Intake Stroke of a Single Cylinder Internal Combustion Engine, *ARPN Journal of Engineering and Applied Sciences* **7**(2) 180-186

International Conference on Applied Sciences (ICAS2016) IOP Publishing
IOP Conf. Series: Materials Science and Engineering **163** (2017) 012039 doi:10.1088/1757-899X/163/1/012039

Influence of intake air temperature on internal combustion engine operation

C Birtok-Băneasă, S Rațiu and T Hepuț

University Politehnica Timisoara, Faculty of Engineering Hunedoara, Department of Engineering and Management, 5 Revolution Street, 331128 Hunedoara, Romania

E–mail: sorin.ratiu@fih.upt.ro

Abstract. This paper presents three methods for reduce thermal losses in the intake system with improvement of airflow and thermal protection. In the experiment are involved two patented devices conceived by the author and one PhD theme device: 1- Dynamic device for air transfer, 2-Integrated thermal deflector, and, 3-Advanced thermal protection. The tests were carried on different vehicle running in real traffic and in the Internal Combustion Engines Laboratory, within the specialization "Road vehicle" belonging to the Faculty of Engineering Hunedoara, component of Politehnica University of Timişoara. The results have been processed and compared whit the ones obtained without these devices.

1. Air temperature influence on fuel consumption and exhaust emissions

Recent studies show significant differences in fuel consumption and exhaust emissions depending on the temperature of the intake air. It is well accepted that the combustion and emissions characteristics are highly rely on the oxygen availability and fuel properties. The excess of oxygen resulted in very lean mixture that lead to the instable combustion and misfire. In fact, lean mixture tends to promote a longer ignition delay and a slower burning rate resulted in longer combustion duration. This condition creates longer time for heat transfer from combustion to the end gas that lead to the knocking phenomenon. However, the engine operates with deficiency of oxygen resulted in rich mixture that leads to the higher unburned gaseous fuel due to shorter ignition delay and faster burning rate.

Therefore, it is vital to make sure the charged air entered into the combustion chamber was sufficient of oxygen in order to promote complete combustion. The previous researches also showed that the air intake temperature has a significant role to increase the combustion efficiency, stability and reduced exhaust emissions [1-3].

Based on previous study, fuel economy can be improved through several strategies such as introduce electronic devices into the gasoline engine. In general, lower fuel consumption and less emission can be achieve by having complete combustion in the combustion chamber. The increased oxygen availability in the combustion chamber can promotes complete combustion and oxidation of both CO and HC emissions. Controlled air intake temperature is one of the possible methods to promote higher combustion efficiency and improved combustion stability. Air intake temperature is important to ensure higher amount of fuel involves in the combustion process. Besides that, air intake temperature will also control the exhaust emissions and combustion process [1].

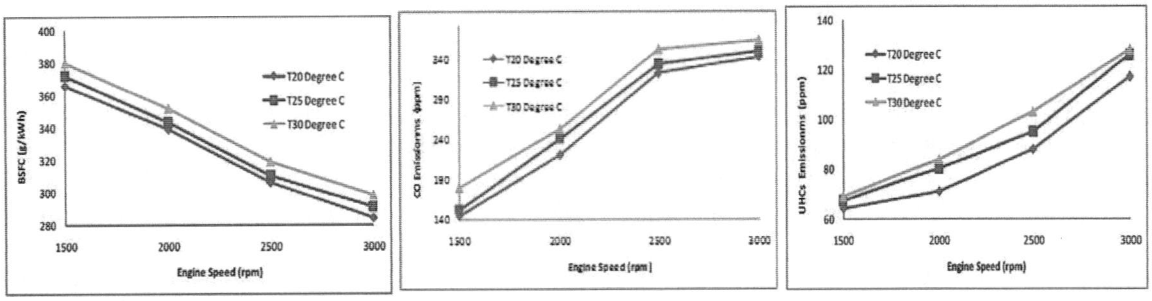

(a) (b) (c)

Figure 1. a - BSFC with variation air intake temperature at various of engine speeds and constant engine load [1], b - CO Emission with variation air intake temperature at various of engine speeds with constant engine load [1], c - UHCs Emission with variation air intake temperature at various of engine speeds with constant engine load [1]

Brake Specific Fuel Consumption (BSFC):
Figure 1.a shows the result of BSFC when the engine operating at 1500 rpm until 3000 rpm with a variation of air intake temperatures (20°C, 25°C and 30°C). In all engine test conditions the BSFC decreases as the engine speed increases due to air turbulent effects resulted in an improved fuel mixture and complete combustion. At higher engine speed, a larger portion of gaseous fuel to be involved in the oxidation process that leads to higher reaction activity resulted in lower BSFC. The results also show that the BSFC decreases as the air intake temperature decreases [1].

The ignition delay period is reduced with lower air intake temperature due to higher oxygen availability. This allows combustion occurs at the end of compression stroke and early of expansion stroke which convert larger fraction of the fuel energy to the useful work. The results show that the effect of air intake temperature is dominant at higher engine speed compared with lower engine speed. This is due to higher injected fuel at higher engine speed that requires more oxygen to complete the combustion.

The highest value of brake specific fuel consumption was 380 g/kW.hr at higher air intake temperature 30°C, which was 4 % higher than the lowest air intake temperature 20°C at the same engine speed of 1500 rpm. But, when the engine speed increase up to 3000 rpm, the value of BSFC reduce about 22 % from 380 g/kW.hr to 298 g/kW.hr. The higher air intake temperature resulted in lower oxygen concentration will lead to a small negative effect on the combustion rate and BSFC [1].

Carbon Monoxide (CO) Emission:

Figure 1.b illustrates the carbon monoxide emissions of the engine operating at different engine speed and constant engine load. In fact, CO emissions are greatly influenced by fuel–air mixing and the air-fuel ratio. It is observed that the concentrations of carbon monoxide keep increase directly proportional due to increase of engine speed. The engine speed was set-up from 1500 rpm to 3000 rpm. The result shows that 341 ppm of CO emissions was produced when the engine operates with 20°C of air intake temperature at 3000 rpm of engine speed. The value was 6 % lower than the engine operates at 30°C of air intake temperature [1]. The behaviour of CO emission is consistent with the quality of the combustion process and poor oxidation of mixture. The higher air intake temperature tends to reduce oxygen availability resulted in an unstable combustion where partial burn and misfire may take place.

It also was found that, the emissions of CO decreased with the decrease of air intake temperature regardless of engine speed. The possible reason is due to higher oxygen availability at lower temperature resulted in complete fuel mixing process and complete combustion [1].

Unburned Hydrocarbons (UHCs) Emission:

Figure 1.c shows the UHCs emissions trend for a gasoline engine operating with different engine speeds and constant engine load at variation of air intake temperatures. Ideally, the fuel should be evaporated and mix with air before initiates the combustion process. The mixing process is highly dependent on the fuel droplet size, heat and oxygen availability. The trend of UHCs emissions was similar to that observed in previous results of CO emissions. The UHCs emission level was increased with the increase of the engine speed is due to high amount of fuel involved during the combustion process. However, the UHCs emissions decreased with the decrease of air intake temperature. This is due to high combustion temperature caused by pre-mixed burning and high rate of oxidation of UHCs at lower air intake temperature. The lower air intake temperature is also possible to ignite combustion at rich region mixture [4].The higher UHCs emissions are produced by the engine operating at 30°C of air intake temperature is due to poor mixing process and incomplete combustion. The behaviour of UHCs emission is consistent with the amount of unburned gaseous fuel in the exhaust gas. This value is 128 ppm which is 10 % higher than the engine operates with 20°C of air intake temperature at the same engine test condition. The UHCs emissions at high engine speed 3000 rpm was higher about 85 % than at lower engine speed 1500 rpm for all variation of air intake temperature because of influence by the reducing of oxygen concentration that supplied to combustion chamber and high amount of injected fuel.

UHCs emissions are generally produced due to incomplete combustion of a carbon containing fuel. Lower excess oxygen concentration results in rich air– fuel mixtures at different locations inside the combustion chamber [5]. This heterogeneous mixture does not combust completely and results in higher UHCs emissions. In addition, the longer ignition delay due to the oxygen deficiency caused by increased air intake temperature will lead to poor UHCs oxidation process.

In conclusion we can say that the fuel consumption is improved with the decrease of air intake temperature regardless of engine speed. Also, carbon monoxide and unburned hydrocarbons were decreased with the decrease of air intake temperature, again regardless of engine speed. So, based on the studies presented before, the lower air intake temperatures resulted in lowest fuel consumption and reduced exhaust emissions regardless of engine speed. This conclusion leads us to the solution presented in next session of the article.

2. Presentation of the methods and devices

A cold air intake (CAI) is an assembly of parts used to bring relatively cool air into a car's internal-combustion engine. Most vehicles manufactured from the mid-1970s until the mid-1990s have thermostatic air intake systems that regulate the temperature of the air entering the engine's intake tract, providing warm air when the engine is cold and cold air when the engine is warm to maximize performance, efficiency, and fuel economy. With the advent of advanced emission controls and more advanced fuel injection methods modern vehicles do not have a thermostatic air intake system and the factory installed air intake draws unregulated cold air. Aftermarket cold air intake systems are marketed with claims of increased engine efficiency and performance. The putative principle behind a cold air intake is that cooler air has a higher density, thus containing more oxygen per volume unit than warmer air [6].

2.1. Dynamic Device for Air Transfer (DDAT) pat. No. RO2009 00028

Dynamic device for air transfer (Figure 2) refers to a device designed to collect and transfer air from outside the engine compartment on the air filter, to make a laminar, concentrated flow of air and lowering its temperature for increase the volumetric efficiency of internal combustion engine. The Dynamic Air Transfer Device consists of: external collection diffusers (one or several), connection joints and external axial collector (one or two passages). We can say that the advantages of this device are: the air transfer to the filter has a concentrated laminar flow; the low air temperature allows a more efficient internal cooling of the ICE; a slight effect of overfeeding is created, which is proportional to the speed of the car; an increased volumetric efficiency; a higher burning yield of the combustion mix lower levels of polluting emissions; dynamic admission; makes possible to shorten the distance between the filter and admission gallery.

(a) (b)

Figure 2. Dynamic device for air transfer
a - operation principle, b – virtual model, made in Catia V5

Pressure variation measurements (Figure 3) have been carried out at different speeds both in the area where the air is transferred from the DDAT to the filter and the area where the air is absorbed by the filter (left and right side) without the DDAT. The measurements were performed with the digital manometer TESTO 510 (0-100hPa). A significantly higher air collection and transfer effect is observed with DDAT than in the case of the simple absorption performed by the super-absorbing filter [7], [8].

2.2. Integrated thermal deflector pat. no. RO201000026

The integrated thermal deflector (Figure 4) it's designed to protect the air filter area from the thermal radiations generated by the cooling radiator and combustion engine. The integrated deflector has the form of a separating wall between air filter area and rest of engine compartment.

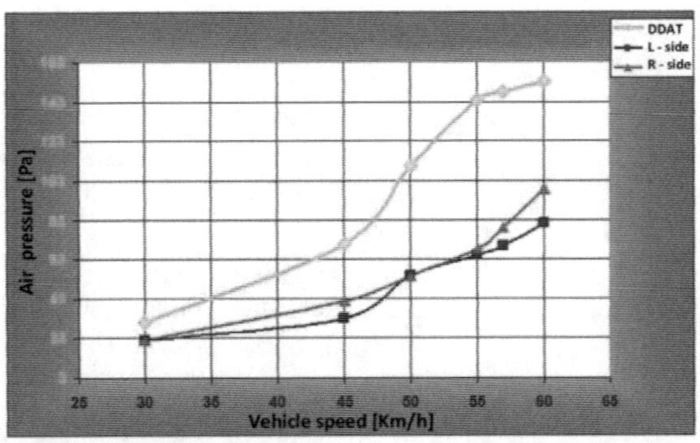

Figure 3. Intake pressure values

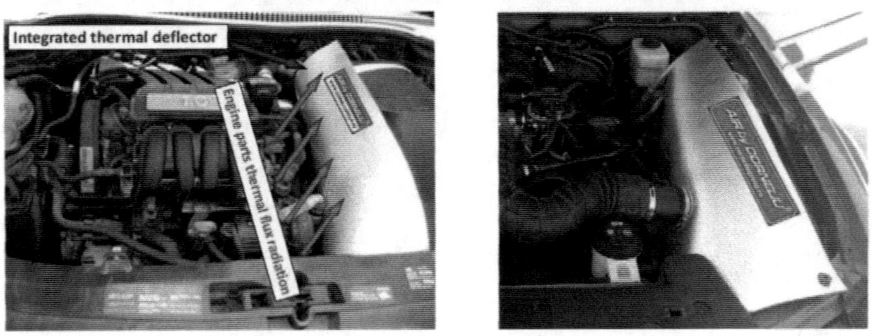

(a) (b)

Figure 4. Integrated thermal deflector
a – operation principle, b - real model

The usage of the integrated thermal deflector has the following advantages: the air flow generated by cooling radiator (i.e. thermal radiations) is deflected outside the air filter area; the temperature of the air filter is maintained at an optimum level (i.e. their over-heating is avoided).
Temperature variation measurements (Figure 5) have been carried out at different speeds both in the area of the air filter and outside the area of the air filter whit the integrated thermal deflector mounted on the site [7].

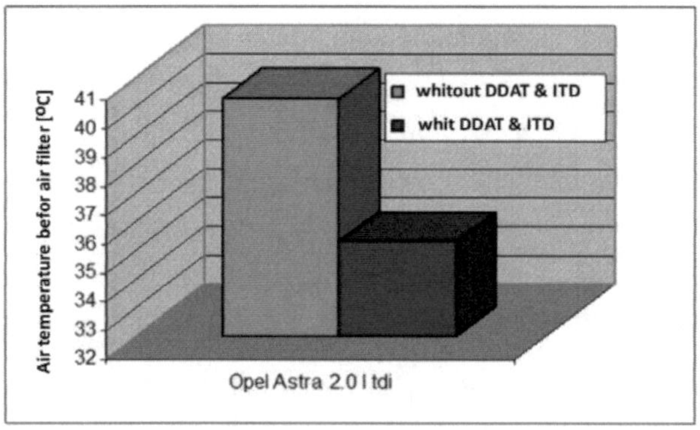

Figure 5. Intake air temperature values

2.3. Advanced thermal protection
Because of the position in the engine compartment the intake (Figure 6) is highly expose to thermal radiations from the cooling radiator, exhaust pipe and engine itself which is a disadvantage, the intake air it heated and the result is a lower density, thus containing less oxygen per volume unit than cold air.

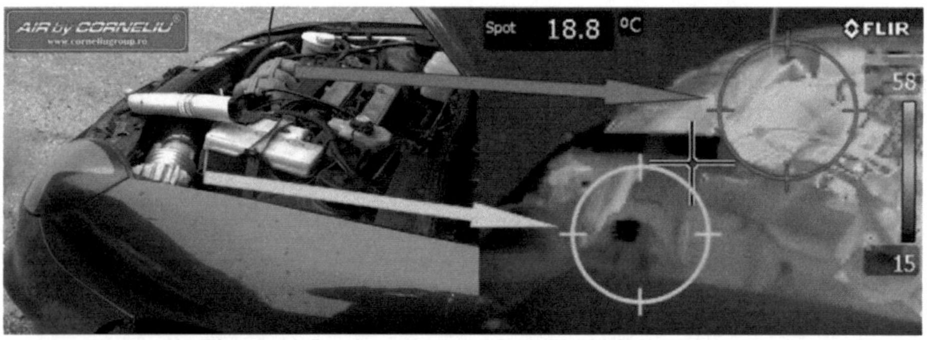

Figure 6. Intake temperature - Thermal imager

One method of reducing thermal losses is to insulate the intake whit a new kind of material. The principal scope is to resource and development a new insulates material from composite, natural and organic materials based from recycling materials.

Early comparative tests were carried out on sections (Figure 7) of intake with or without layers of insulation materials. The tests were performed with various layers (Figure 8) of insulation materials such as silicone, carbon fibre, cork and various mixtures.

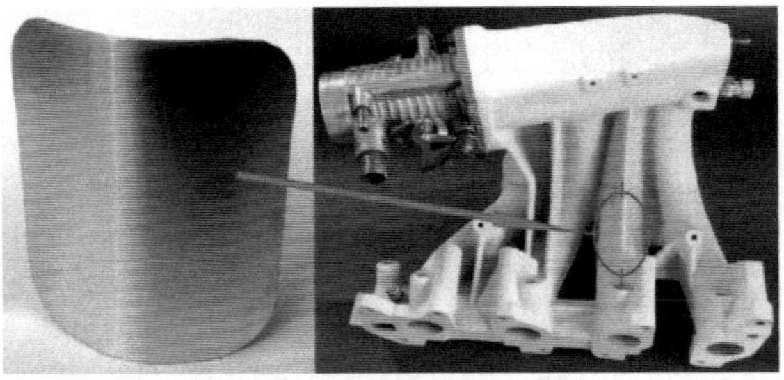

Figure 7. Intake sections probe

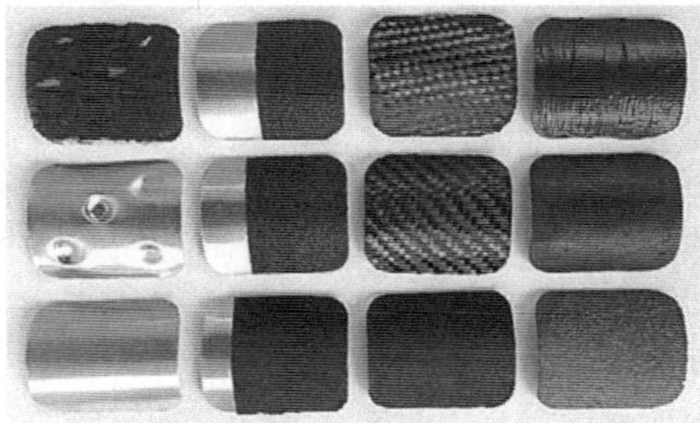

Figure 8. Intake sections coated

Comparative tests (Figure 9) on sections from intake coated with an insulating layer revealed a significant improvement regarding thermal losses.

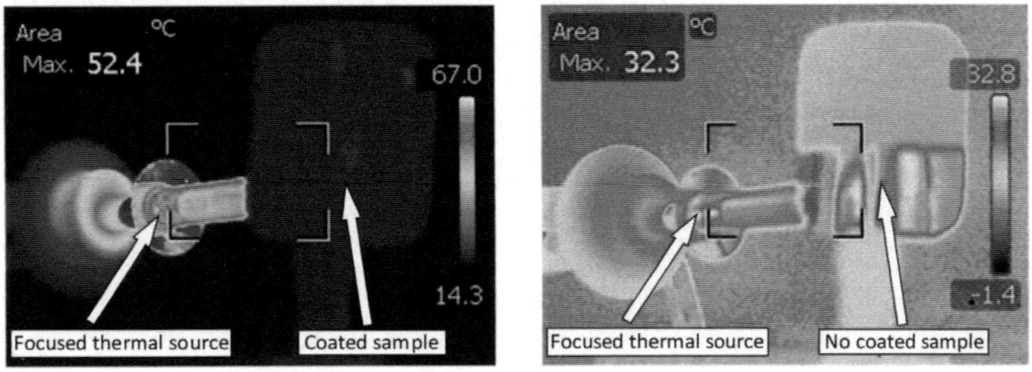

(a) (b)

Figure 9. Comparative thermal imager tests;
a – coated intake section probe, b – no coated intake section probe

3. Conclusions

In conclusions, we can say that these methods can reduce thermal losses in the intake system by providing a higher pressure air flow while maintaining a low temperature along intake system, contributing to increasing the filling efficiency η_V of the engine cylinders. Also improving the filling efficiency η_V means a lower specific fuel consumption and lower exhausts pollutants.

References

[1] Nik R A, Hazimi I, Zeno M, Asiah A R and Hazim S 2015 Effects of air intake temperature on the fuel consumption and exhaust emissions of natural aspirated gasoline engine, *Jurnal Teknologi* **76**(9) 25-29

[2] Mamat R, Abdullah N R, Xu H M, Wyszynski M L and Tsolakis A 2009 *Effect of Fuel Temperature on Combustion and Emissions of a Common Rail Diesel Engine Operating with Rapeseed Methyl Ester*, SAE Technical Paper 2009-01-1896

[3] Rizalman M, Ahmad F Y, Abdul A A, Amir A and Nik R A 2013 Effect of EGR and Air Temperature on Exhaust Emission of a Diesel Engine Operating with Biodiesel, *Journal of Biobased Materials and Bioenergy* **7** 461-463

[4] Rakesh K M and Avinash K A 2011 Experimental Investigation on the Effect of Intake Air Temperature and Air–fuel Ratio on Cycle-to-Cycle Variations of HCCI Combustion and Performance Parameters, *Journal of Applied Energy* **88** 1153-1163

[5] Ramzi H A, Ibraheem Al-B, R and Talabany Z J 2013 Effect of Ambient Air Temperature on Specific Fuel Consumption of Naturally Aspirated Diesel Engine, *Journal of Science and Engineering* **1** 1-7

IOP Conf. Series: Materials Science and Engineering **163** (2017) 012039 doi:10.1088/1757-899X/163/1/012039

[6] Knowling M 2007 Shielded Sucker: A cheap and easy air intake improvement - with no CAI pipe in sight, AutoSpeed. Retrieved 2013-04-15

[7] Birtok–Băneasă C and Raţiu S 2011 *Admisia aerului în motoare cu ardere internă – Filtre supraaspirante – Sisteme dinamice de transfer* [Air intake in internal combustion engines – Super absorbing filters – Dynamical Air Transfer Systems], Publisher Politehnica, Timişoara, Romania

[8] Raţiu S 2009 *Motoare cu ardere internă pentru autovehicule rutiere – Procese şi caracteristici, Experimente de laborator* [*Internal combustion engines for road vehicles – Processes and Features – Laboratory experiments*], Publisher Mirton, Timişoara, Romania

IOP Conf. Series: Materials Science and Engineering **163** (2017) 012040 doi:10.1088/1757-899X/163/1/012040

Advanced study of thermal behaviour of CSZ comparing with the classic YSZ coating

A Dragomirescu[1], N Constantin[2], A Ştefan[1], V Manoliu[1] and R Truşcă[3]

[1]National Institute of Aerospace Research and Development 'Elie Carafoli', Department of Materials, Bucharest, Romania
[2]University Politehnica of Bucharest, Department of Material Science, Bucharest, Romania
[3]University Politehnica of Bucharest, Department of Science and Engineering of Oxide Materials and Nanomaterials, Bucharest, Romania

E-mail: dragomirescu.alina@yahoo.com

Abstract. Thermal barrier coatings (TBC) are advanced materials typically applied to metal surfaces subjected to extreme temperatures to protect them and increase their lifetime. Ceria stabilized zirconia ceramic layer (CSZ) is increasingly used as an alternative improved as replace for classical TBC system - yttria stabilized zirconia - thanks to superior properties, including mechanical and high resistance to thermal corrosion. The paper describes the thermal shock testing of two types of thermal barrier coatings used to protect a nickel super alloy. For the experimental procedure, it was used plate samples from nickel super alloy with a bond coat and a ceramic top coat. The top coat was different: on some samples, it was used YSZ and on others CSZ. Ni based super alloys have good corrosion resistance in reducing environments action, but poor in oxidizing conditions. Extreme environments can lead to loss of material by oxidation / corrosion, along with decreased mechanical properties of the substrate due to damaging elements which diffuses into the substrate at high temperatures. Using laboratory equipment, the TBC systems were exposed repeatedly to extreme high temperatures for a short time and then cooled. After the thermal shock tests, the samples were morph-structured characterized using electronic microscopy to analyze the changes. The experimental results were compared to rank the TBC systems in order of performance.

1. Introduction

The last decades' challenge for industrial gas turbines and aircraft engines is to increase the maximum operating temperature without overheating the metal parts during operation. In this way, the life of the materials used in turbine must be raised by increasing the resistance to corrosion at high temperatures, such improvement reducing the cost of the turbine. "Hot parts" of co-generative systems, and the turbo aircraft are subject to tribological demands from erosion, at speeds over Mach 3, thermal shock, abrasive wear, etc. The efficiency of gas turbines, for either industrial production of energy or marine propulsion for aircraft applications, has improved constantly for years. These advances have occurred, largely because they have been found means to operate the gas generator section of the engine at higher temperatures. The need for performance increasingly higher advanced turbine engines will continue asking the operating efficiency even higher, higher operating life duration and reduced emission levels [1].

Of all the wear factors acting simultaneously, thermal shock acts the most disruptive. In the case of hot parts of turbo engines, its temperature varies depending on the operating conditions at takeoff, landing, cruising speed, landing failure, in-flight engine shutdown, etc. Successive transition from one operating mode to another, lead to changes between 5-20%, in terms of thermal parameters. Variation of temperature produce high tensions in parts material and it can eventually lead to the plastic deformation of them, to the delamination of protective layers and to the damaging of the components and subassemblies.

One of the solutions consists in coating of combustion chamber and turbine blades with a thermal barrier coating (TBC). It protects and isolates the metal components of the gas turbine engine hot gas flow allows for higher combustion temperatures and improving engine efficiency. In the case of thermal shock, for testing materials were innovated and built on international level specific installations that simulate actual working conditions keeping in this way the similarity for working conditions and the processes involved.

2. Experimental procedure

2.1. TBC systems

A classic thermal barrier coatings (TBC) is consists in two ceramic layers (the bond coat and the top coat) deposited over metallic parts in order to provide their thermal insulation from the combustion gases. In this way, thermal barrier coatings allow the substrate metallic parts to achieve higher efficiency of combustion engines (internal combustion engines and gas turbine engines) due to an increase of their operating temperature [2].

General structure of TBC

Usually, thermal barrier coatings have four layers (Figure 1):

a. the external ceramic layer (Top Coat) - ensures low thermal conductivity. Classic candidate for this layer is oxide ZrO_2 stabilized with Y_2O_3 (YSZ - yttria stabilized zirconia) material as this material solution shows one of the lowest values of thermal conductivity at high temperatures: 2.3 $Wm^{-1}K^{-1}$ at 1000° C and a thermal expansion coefficient value of 10.7×10^{-6} K^{-1}, which reduces thermal stress [3]. Usually, the thickness of a ceramic external layer is between 250 and 375μm [4].

Intensive research [5], [6], have demonstrated over the time that, of the terms of thermal expansion coefficient, a better alternative to the YSZ ceramic layers are the ones of zirconia doped with ceria and yttria (CSZ). Expansion coefficient of YSZ is 10.7×10^{-6} K^{-1} recorded values at room temperature for up to 1000°C, and the same coefficient for typical super alloys used as substrate and for the classic Bond Coat (NiCrAlY) registered values of 16×10^{-6} K^{-1}, respectively 17.5×10^{-6} K^{-1} for the same temperature range [5]. This notable difference coefficient values recorded at the three components of the TBC can lead to high thermal tensions that have the ability to degrade over time the protective system. Thus, these tensions could influence the life of TBC systems. An alternative is adding CeO_2 as a stabilizing factor, which would also result in obtaining a coefficient of thermal expansion closer than the recorded value of to the substrate and Bond Coat. Sodeoka et al. [6] observed that by adding CeO_2 is obtained a coefficient of thermal expansion amounting to 12.5×10^{-6} K^{-1} in the temperature range of 25-800°C which helps to reduce the tensions.

b. bonding layer (Bond Coat) is typically Ni(Co)CrAlY. The thickness of this layer is between 75 and 125μm. Contact layer has a high content of aluminium and chromium which helps improve resistance to oxidation and corrosion hot. The content of yttria contributes to reducing thermal conductivity value.

c. oxide layer barrier TGO (thermally grown oxide layer) - this develops as a result of the temperature rise generated on the surface. The oxide grows at the interface between the ceramic layer and bond coat, as a result of oxidation of the bond coat during the oxidation and thermal shock. Its role is to inhibit the oxidation process of bond coat.

d. nickel-based alloys and cobalt-based alloys are typically used as substrate materials. Other alloys that are suitable to be used as substrate are structures made of steel, titanium alloy or Ti-Al alloys of systems, structures made of niobium alloys etc.

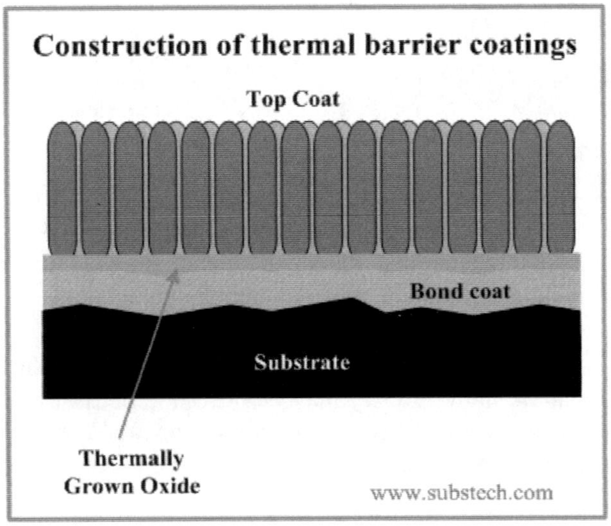

Figure 1. The structure of a TBC sistem

Generally, most TBC are produced by a method of thermal spraying air atmosphere APS (Air Plasma Spray) or using HVOF, HVAF or EB-PVD methods. Thermal barrier layers deposited on the elements of turbine combustion chambers are made by thermal spraying methods with the use of plasma guns and the rotating blades are covered through technology EB-PVD. A global characterization of the thermal barrier coatings and their degradation mechanisms require a detailed description of the properties, microstructure, and the role of individual components contained in a composition of the layers of TBC, especially the bonding layer and the ceramic layer.

2.3. Deposition methods

2.3.1. APS (Air Plasma Spray). A characteristic of the thermal spraying processes is highly concentrated source of energy, wherein the coating material is supplied in the form of powder, fiber or rod. The coating material is melted and accelerated towards the substrate, forming coverage. The coating consists of several stacked flattened drops that are strengthening consecutively square and lashed each other. Due to high kinetic energy of the droplets, they spread on the substrate, forming a blanket incrementally. Plasma spray gun consists of two electrodes. Figure 2 presents a plasma spray gun with a tungsten cathode located in the interior of the anode copper which is water cooled. A gas mixture of argon and hydrogen is injected into the annular space between the two electrodes. To begin the process, a DC electric arc between two electrodes is stabilized. Arcing occurs gas ionization, atoms lose electrons become positive ions. The electrons moving at high speed at the anode, while the ions moving to the cathode. On their way, the electrons and neutral atoms collide with atoms and molecules of gas. Therefore, the electric arc continuously transforms the gas into plasma (a mixture of high-energy ions and electrons). Plasma is electrically neutral and is generally characterized by a very high temperature [7]. The kinetic energy of plasma (mostly given by free electrons) is converted into heat energy during collisions between ions, electrons and atoms. Thus, the plasma is capable of producing temperatures up to about 104 ° C [8]. The hot gas exits the gun nozzle at high speed. The powder is fed into the plasma layer. The powder particles are melted and propelled by hot gas on the substrate surface [9]. When molten particles strike the substrate surface, they form dots flattened by scattering, cooling and solidification. These dots gradually build coating.

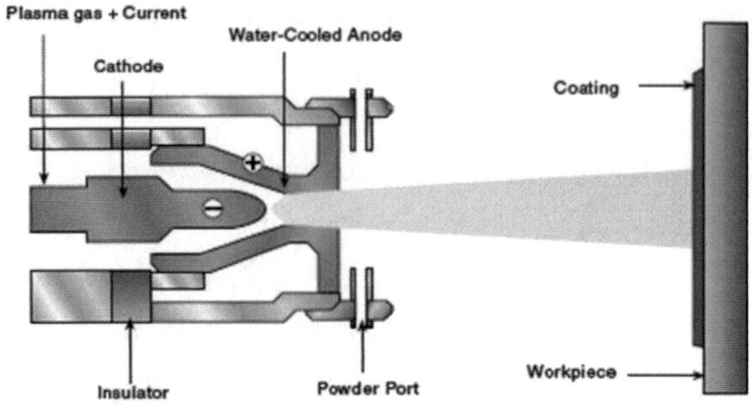

Figure 2. The scheme of plasma spray gun [12]

The level of porosity of the plasma sprayed ceramic coatings is generally in the range of 3-20% [10]. For thermal barrier coating, high porosity may be desirable because the thermal conductivity of the layer decreases as the presence of porosity [11]. Certain applications require low porosity; this may be achieved by spraying conditions are controlled to speed up the droplets.

2.3.2. HVAF. During this process, the combustion gas is mixed with ambient air in a combustion chamber, the combustion causing a pressure above 10 bars. Due to this pressure, the flame velocity exceeding 600 m / s. The powder may be fed axially into the HVOF combustion chamber under high pressure or by the side of the Laval nozzle where the pressure is lower. Fuel gas (propane, propylene or hydrogen) and air are supplied at high pressure, combustion occurs outside the nozzle but in the layer of the supplied compressed air. The compressed air tightens and accelerates the flame and acts as a cooling agent for the spray gun. The material to be deposited can take the form of a powder or rod.

HVAF deposition procedure using material in powder form is using a gaseous fuel and air for projecting powder at supersonic speeds. The particles reach a speed of 800 m / sec, while remaining at a temperature below the melting point. Since the melting limit is not reached, the carbides are not damaged, nor does it undergo dissolution in the metallic binder.

In the technique HVAF of electric arc and the material in the form of rods, the wire is melted by the operation of an electric arc and a feed supersonic combustion air (propane, propylene, LPG) for atomization of the molten metal and to increase the speed to give the fines. Cape gun contains a toroidal combustion chamber with a ceramic area insertion for controlling catalytic combustion process where propylene is burned. The flue gases without oxygen forms a supersonic flow (which will not oxidize the molten metal), which is placed in the direction of the arc (Figure 3).

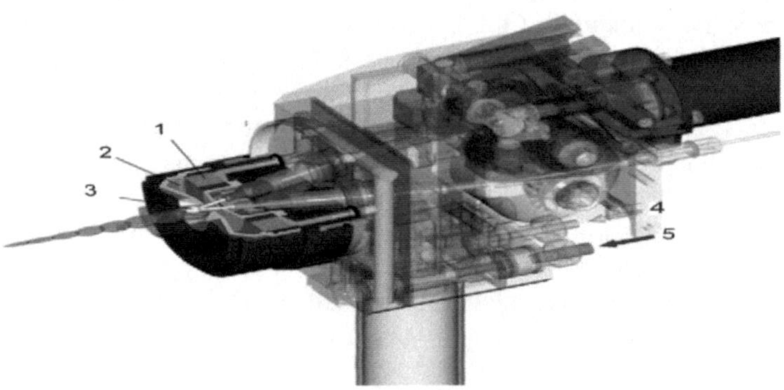

Figure 3. HVAF gun: 1-ceramic insertion; 2-combustion chamber; 3- electric arc; 4- fuel gas; 5-compressed air [13]

2.4. Thermal shock testing

Testing procedure

In the experimental campaign were carried out tests of thermal shock on dedicated QTS2 facility (Figure 4) the property of the Materials Compartment laboratory of INCAS Bucharest (National Institute for Aerospace Research and Development "Elie Carafoli") on various samples with protective ceramic layers. Thermal shock test consists essentially of a fast heating (less than 2 minute) from room temperature to the test temperature (1200° C), maintaining the temperature of the specimen at this level for 3-4 minutes after which it is rapidly air cooled to ambient temperature, with compressed air (~ 9 bar) for 1 minute. The date acquisition in the time of the tests is performed by Lab VIEW program (Figure 5).

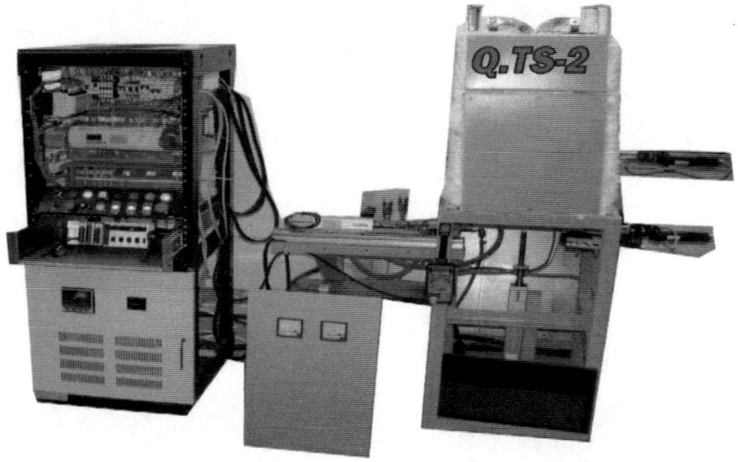

Figure 4. QTS2 thermal shock facility

Materials

There were tested two types of ceramic layer: no. 7 YSZ specimen – APS deposition ($ZrO_2 8\% \ Y_2O_3$) and no. 34 CSZ specimen – HVAF deposition technique ($ZrO_2 24\% \ CeO_2 2.5\% \ Y_2O_3$).

For the deposited ceramic layers, it was used a bond coat of NiCoCrAlY (deposited by APS technique) on a metallic support of Nimonic 90. Number of cycling has been counted up to exfoliation of about 20-30% of the deposited area, above which is considered the ceramic surface can no longer perform thermal protective function.

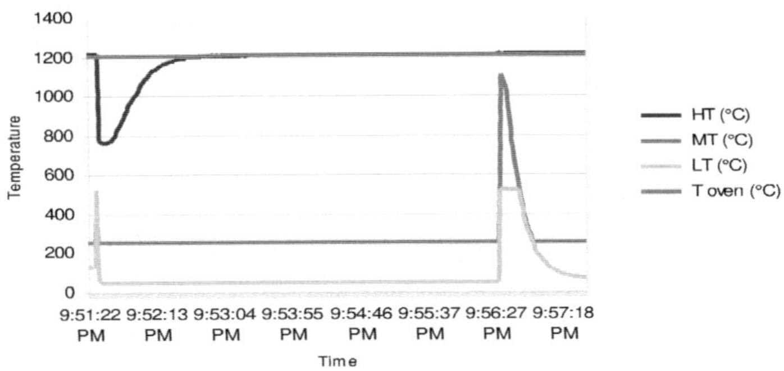

Figure 5. The temperature Lab VIEW representation during a thermal shock test (heating – dwell - cooling)

Following are the few specimens tested in the conditions above.

Thermal shock tests were observed better behavior of the specimens with the CSZ (Figure 7) compared with YSZ specimens (Figure 6).

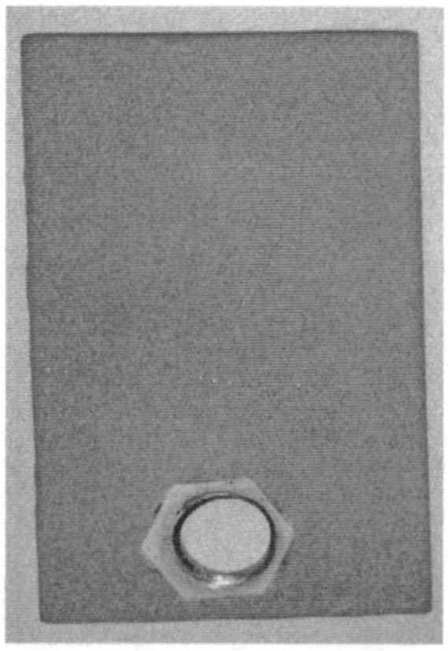

Figure 6.a) The no. 7 specimen initially appearance (before thermal testing)

Figure 6.b) The no. 7 specimen appearance after the testing program (75 thermal cycler)

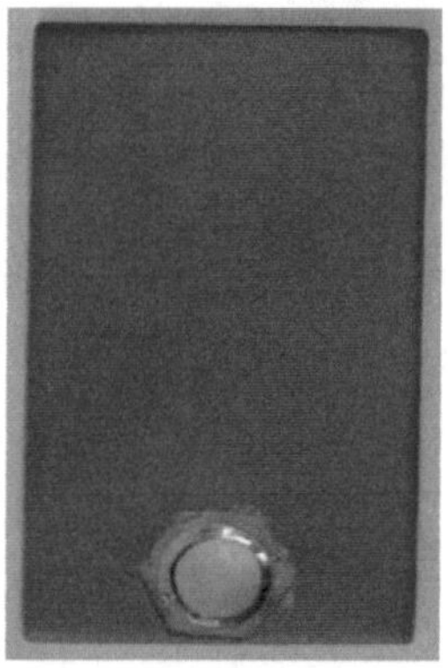

Figure 7.a) The no. 34 specimens initially appearance (before thermal testing)

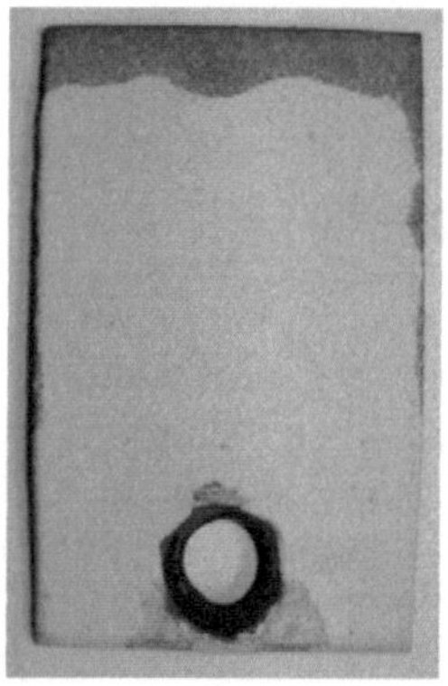

Figure 7.b) The no. 34 specimen appearance after the testing program (100 thermal cycler)

2.5. Microstructure analisys

Microscopic analysis was performed in the case of thermal shock test specimens in areas where no macroscopic delamination was observed.

For the microscopic characterize were choose sections presenting all protection systems components, after that the samples were specifically embedded and processed for analysis.

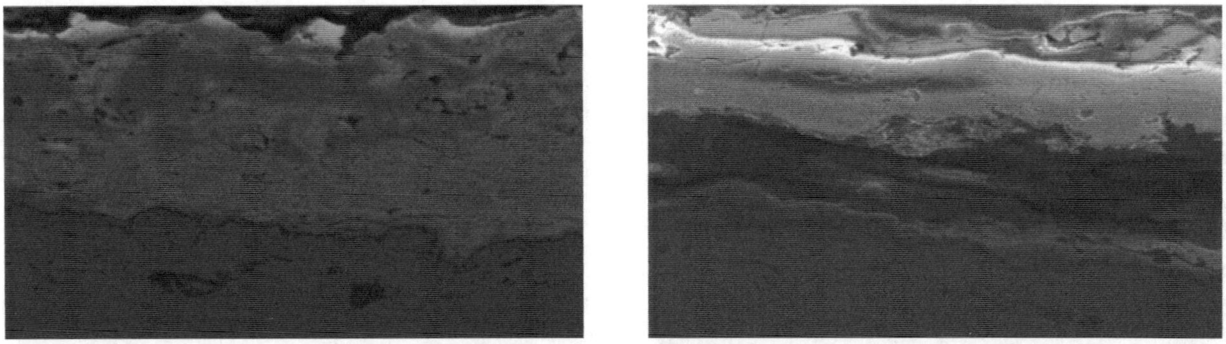

Figure 8. Microstructure of specimen no. 7 YSZ before and after testing (mag. 2000x)

Figure 9.a) Microstructure of specimen no. 34 CSZ before testing (mag. 1000x)

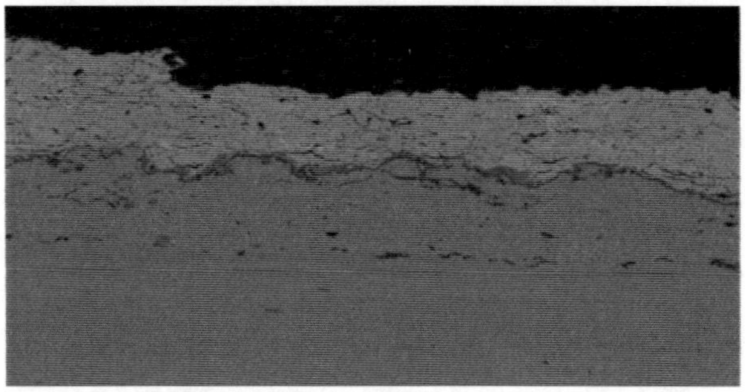

Figure 9.b) Microstructure of specimen no. 34 CSZ after testing (mag. 1000x)

Can be notice the formation of cracks due to temperature differences the specimens are repeatedly subjected during thermal tests. Also, seen in Figures 8, 9.a and 9.b, is an increase of porosity of ceramic layers.

Qualitative and quantitative analysis of protection systems highlights the formation of the fourth layer component present in TBC structures: TGO - thermally growth oxides, which form at the top coat/bond coat interface having a protective role. Further qualitative analyses are given for two types of protection systems in the two cases: before and after exposure to temperature specimens to observe changes that occur in the material structure due to thermal shock tests.

It is noticed that the aluminium migrates from bond coat layer and focuses on interface between the bond coat and ceramic layer after the sample have been exposure to temperature (Figures 10.a and 10.b). So, the main element that forms the growing thermal oxide layer (TGO) is aluminium. It reacts with oxygen to form Al_2O_3. Comparing sizes of TGO layer in case of the two test pieces it is notice an increase of the TGO layer size proportional to the number of exposures to temperature. The oxide layer form develops once with prolonged exposure to high temperatures.

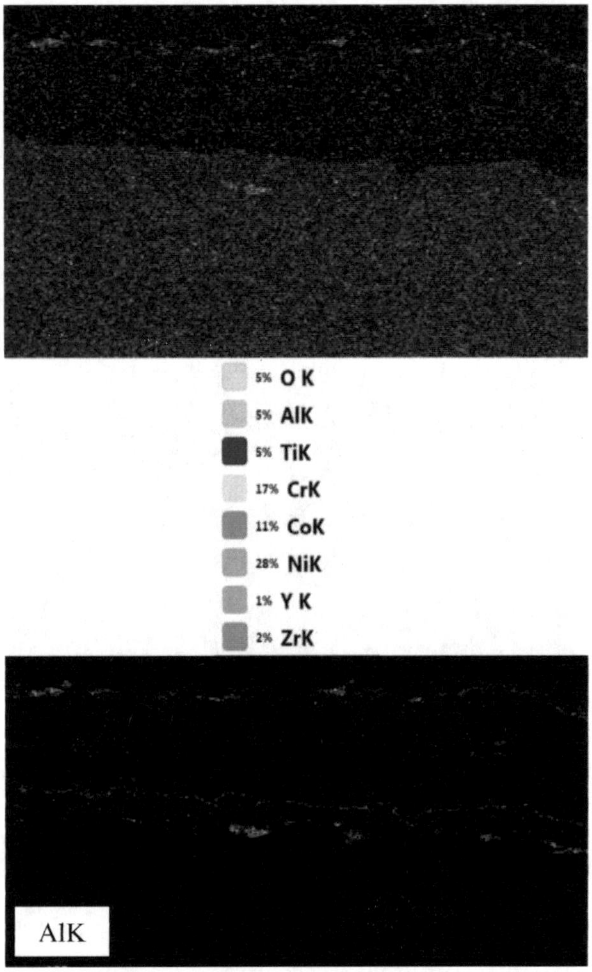

Figure 10.a) Element mapping in section of specimen no.7 YSZ before testing (mag. 2000x)

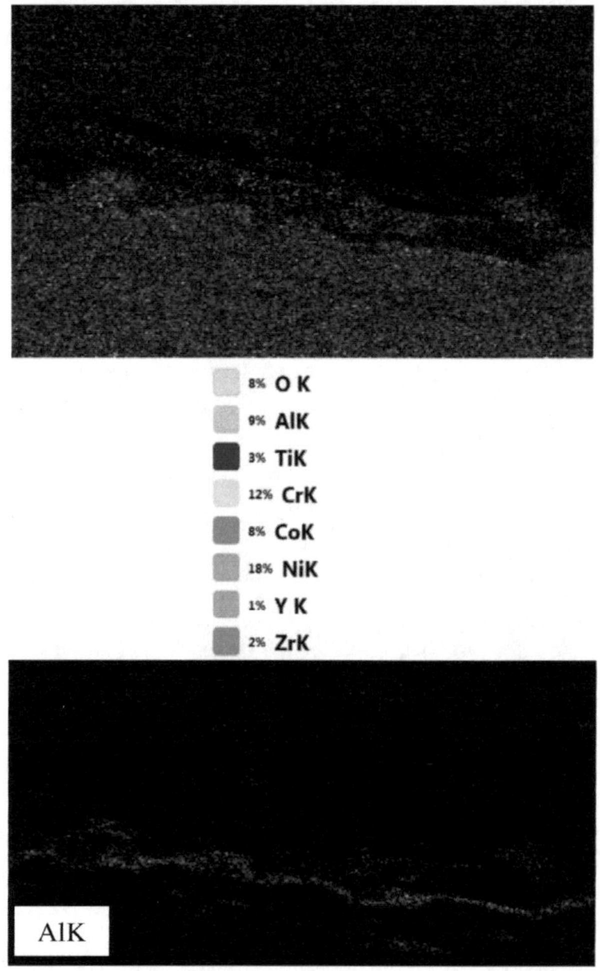

Figure 10.b) Element mapping in section of specimen no.7 YSZ after testing (mag. 2000x)

3. Conclusions

Tests carried out on the installation QTS2 are reproducible, safe and helps assess the TBC layers for high-tech applications (aeronautics, space, energy, etc.). In the course of our experiments we used two types of ceramic layers with the use of the same substrate - Nimonic 90 and the same Bond Coat NiCrAlY. For ceramic layers were used two types of powders based on zirconia (ZrO2): yttria partially stabilized zirconia YSZ and ceria stabilized zirconia (CSZ).

Comparing test results for those two types of TBC structures observed a higher resistance of CSZ (YSZ structures have shown the protective structure degradation after 75 thermal shock tests and the structures suffered delamination after about 100 tests).

After analysing the microstructure is observed in both cases forming an oxide layer (TGO) following exposure to the temperature, this being proportional to the number of tests. From qualitative analysis, it appears that this oxide is largely in the form of aluminium oxide. Another observation after microscopic analysis is good adhesion between layers.

References

[1] ***1996 *Coatings for High-Temperature Structural Materials - Trends and Opportunities,* Committee on Coatings for High-Temperature Structural Materials, Commission on

Engineering and Technical Systems, National Research Council, Washington DC
[2] http://www.substech.com
[3] Padture N, Gell M and Jordan E H 2002 Thermal Barrier Coatings for Gas-Turbine Engine Applications, *Science* **296** 280-284
[4] Jones R L 1996 *Thermal barrier coatings, in Metallurgical and Ceramic Coatings*, Chapman & Hall, London, UK
[5] Va β en R, Kerkhof G and Stöver D 2001 *Mater. Sci. Eng., A* **303**(1-2) 100
[6] Sodeoka S, Suzuki M, Ueno K, Sakuramoto H, Shibata T and Ando M1997 Thermal and mechanical properties of ZrO_2-CeO_2 plasma-sprayed coatings, *J. Therm. Spray Technol.* **6**(3) 361-367
[7] Fauchais P and Vardelle M 1994 Plasma Spraying: Present and Future, *Pure & Applied Chemistry* **66**(6) 1247-1258
[8] Apelian D, Paliwal M, Smith R W and Schilling W F 1983 Melting and Solidification in Plasma Deposition - Phenomenological Review, *International Metals Reviews* **28**(5) 271-294
[9] Heimann R B 1996 *Plasma-Spray Coating Principles and applications*, VCH Verlagsgesellschaft, Weinhein, Germany
[10] Matejka D and Benko B 1989 *Plasma Spraying of Metallic and Ceramic Materials*, John Wiley & Sons
[11] Schlichting K W, Padture N P and Klemens P G 2001 Thermal Conductivity of Dense and porous yttria-stabilized zirconia, *Journal of Materials Science* **36**(12) 3003-3010
[12] http://www.amphardchrome.co.uk/plasma-spraying.php
[13] Petre D, Iovanaş R and Pascu A 2009 Studies on the applicability of the high velocity thermal spray used in the automotive industry, *Bulletin of the Transilvania University of Braşov* **2**(51) 187-192

IOP Conf. Series: Materials Science and Engineering **163** (2017) 012041 doi:10.1088/1757-899X/163/1/012041

The influence of stiffening ribs on the natural frequencies of butterfly valve disks

C Ursoniu[1], D Pepa[1], M Tufoi[2] and G R Gillich[2]

[1]UCM Resita, Department of Mechanical Engineering, Golului str., no.1, 320053, Resita, Romania
[2]"Eftimie Murgu" University of Resita, Department of Engineering and Management, Piata Traian Vuia str., no.1-4, 320085, Resita, Romania

E-mail: cursoniu@gmail.com

Abstract. In this paper a study regarding the influence of the ribs shape on the dynamic behavior of butterfly valves, in terms of natural frequency variation, is presented. This behavior is important because the valve disk vibrates due to fluid flow when it is fully or partially open. If the disk is "locked in", which means that frequency of oscillation is equal to the frequency of vortex shedding, the negative effect of resonance occurs, and harming of the structure is expected. The phenomenon is undesired and can be avoided by designing the disk in order to have the natural frequencies higher as the shedding frequencies. The study is performed via the finite element method (FEM) and first concerns in finding the proper disk thickness for the valve's geometrical input parameters by static analysis. Afterward, modal analysis on disks with stiffness ribs of various shapes and positions is made. As a result, guidelines for designing the disk's stiffening elements are provided.

1. Introduction

Valves regulate either the pressure or the flow of the fluid in a circuit. This duty involves starting and stopping flow, controlling the flow rate or the fluid pressure, preventing back-flow, or diverting flow [1]. Various aspects, as the resistance and flow coefficients, corrosion and cavitation, noise and vibration produced by the flow through valves, are research topics of numerous authors [2-6].

Figure 1. Butterfly valve with ribbed disk

Butterfly valves have the closing mechanism element in the form of a disk. These valves are generally preferred because they have a lower price as other valve designs. Also, being lighter in weight, less support is required. The disc is positioned in the center of the pipe. A rod, passing through the disc is connected to an actuator on the outside of the valve, as depicted in Figure 1. Rotating the actuator turns the disc either parallel or perpendicular to the flow, the valve being in this way fully open or closed.

When the valve is fully open, the disc allows an almost unrestricted passage of the fluid. The valve may also be opened incrementally to throttle flow. In these cases, the disk is made to vibrate by the fluid flow; if one of its natural frequencies approach the shedding frequency resonance occurs.

The aim the research presented herein is to find the optimal disk shape topology, i.e. the disk thickness and the ribs geometry and position. The essential role of the valve is to totally restrict passage of the fluid. This desire is fulfilled if the maximum static deformation of the disk does not exceed a critical value. In addition, the lowest natural frequency should exceed the shedding frequency, in order to avoid significant vibration amplitudes, with negative consequences on the valve supports.

In prior researches we analyzed the influence of the mass and stiffness changes on the natural frequencies of beams and plates [7-10]. In this paper we present a study regarding the effect of stiffeners on the natural frequencies of circular plates fixed in two points, and conclusions regarding the optimum shape and position of these elements are traced.

2. The finite element analysis

The analysis is performed on a butterfly valve disk, as that presented in Figure 2. The disk, with a constant cross-section, has attached two ribs in the central position and two hubs located near the rim, placed in opposite position on the diameter.

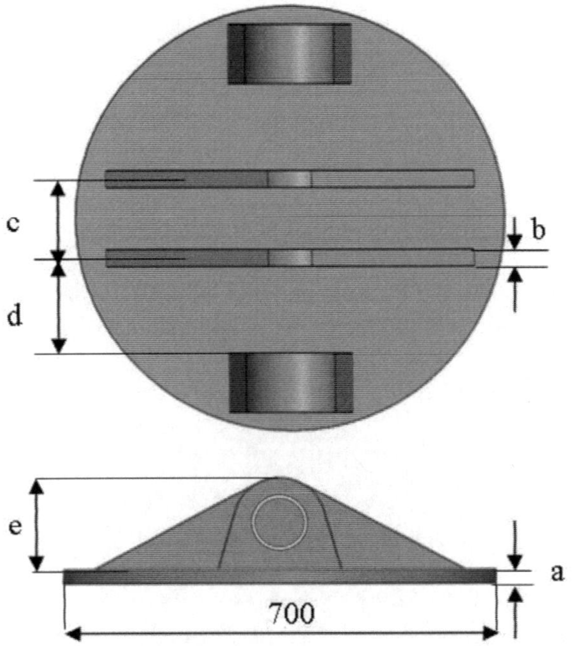

Figure 2. The butterfly valve disk with highlighted essential parameters

The finite element analysis is made using the *ANSYS* simulation software, in particular the static analysis respectively the modal analysis module. A circular plate whit ribs, fixed on the two hubs, as shown in Figure 3, is used for analysis. The plate has the radius $D = 700$ mm and a constant thickness a, with values set between 10 and 30 mm. Involved material is AISI 1045 structural steel, with the physical parameters presented in Table 1.

Table 1. The mechanical properties for AISI 1045 steel

Yield strength [N/mm²]	Elastic modulus [N/mm²]	Poisson's ratio [-]	Mass density [kg/m³]
530	205	0.3	7,850

From the scheme presented in Figure 2, one can observe that there are four independent parameters proposed to be change in order to find the role of each of them on the total deflection and natural frequencies. This results due to the dependency between parameters c and d, which provides the constant sum $c + 2d = 410\,\text{mm}$. Herein, the parameter a is altered during the static analysis and the optimal value is chosen. Then, for the identified thickness a, in the modal analysis the parameters b, c and e are changed. Table 2 shows the considered values of the geometrical parameters, with bold being marked the reference values.

Table 2. Values for the five geometrical parameters taken into consideration in the analysis

a [mm]	B [mm]	c [mm]	d [mm]	e [mm]
10	10	**100**	**155**	100
12	15	110	150	110
15	20	120	145	120
20	25	130	140	130
25	**30**	140	135	140
30	35	150	130	150
	40	160	125	**160**

In Figure 3 the boundary conditions and applied loads for the static analysis are presented. With green arrows the fixing condition in the hubs holes is marked, while the red arrows represent the applied pressure of 1 MPa. Under pressure the disk deforms; this deformation should be limited, in order to stop the water flow properly.

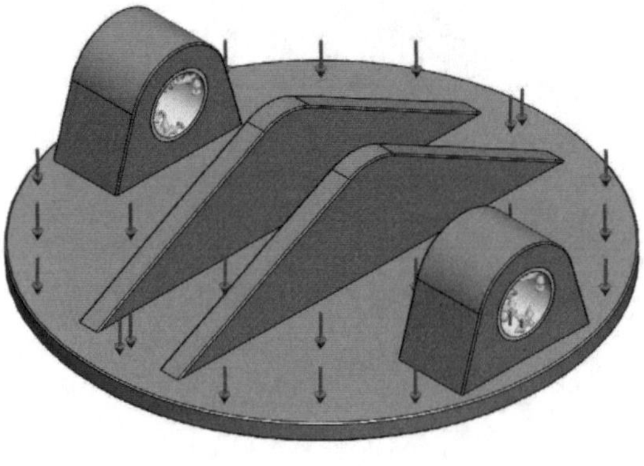

Figure 3. Fixing conditions and applied loads for the static analysis

Figure 4. Applied mesh for the butterfly disk assembly

To obtain reliable results *Standard Mechanical* elements with twenty characteristic points are used, the maximum edge dimension of one element being 3 mm. This mesh is applied both for static as well as for modal analysis. A number of around 4300000 nodes and 2300000 elements have resulted for all disk geometries; the fine element dimension ensures accurate results, especially necessary in the case of the modal analysis. Figure 4 present the applied mesh for the butterfly disk with ribs.

3. Results and discussion

Parametric studies were made for numerous disk configurations, defined by combinations of parameters *a-e*, which influence the rib geometry and position. The main findings are presented in next subsections.

3.1. Static analysis

First, a static analysis was performed in order to find the optimal disk thickness, which ensures a restricted deformation ($\delta < 1.5$ mm). For this reason the parameter representing the disk thickness a is iteratively increased from 10 mm to 30 mm, resulting a number of six analysis cases. For all for these cases the other parameters are indicated in the first row of Table 2. The simulation results are presented in Table 3.

Table 3. Deflection vs. disk thickness

A [mm]	δ [mm]
10	10.011
12	3.489
15	1.739
20	1.675
25	1.050
30	0.711

Having a look in Table 3 one can observe that a reasonable disk deflection is attained for the thickness of 25 mm. The deflection occurs at the disk central position, as depicted in Figure 5. The stress distribution for this case is presented in Figure 6. One can observe that the regions subjected to highest stress are around the hub. This happens because, for this analysis, a sudden cross-section change is present.

Figure 5. Disk displacement field for the applied pressure of 1 MPa

Figure 6. Stress distribution for the applied pressure of 1 MPa

The graphical representation of the deflection dependency on the disk thickness is represented in Figure 7. Note that all other parameters are these marked with bold in Table 2.

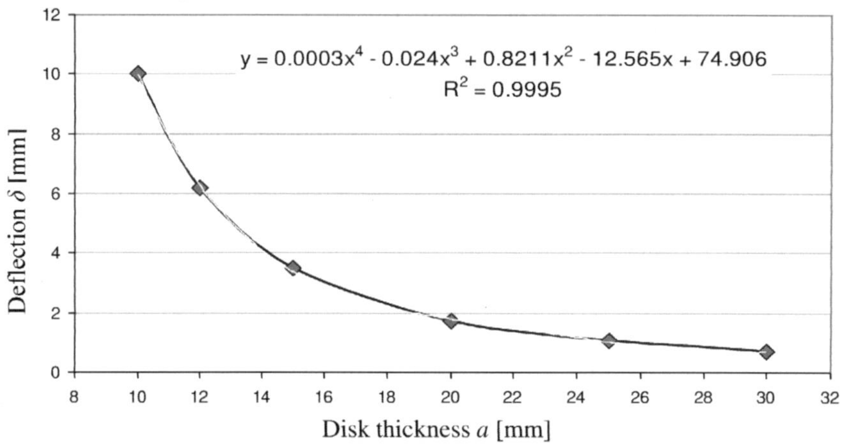

Figure 7. Graphical representation of deflection dependency of the disk thickness

Summarizing, the thickness of 25 mm will be used in the modal analysis, even if changing the rib's position and geometry the maximum achieved deflection of around 1 mm is easily altered.

3.2. Effect of rib thickness on the natural frequencies

To understand and quantify the effect of the rib thickness, parametric modal analysis was performed on the valve disk for several values of *c*, indicated in Table 2. The results for the first three vibration modes are presented in Table 4 and the frequency evolution is depicted in Figure 8.

Table 4. Frequency evolution in respect to rib thickness

Thickness b	[mm]	10	15	20	25	30	35	40
Frequency f_1	[Hz]	264.26	262.06	259.89	257.96	256.22	254.59	252.95
Frequency f_2	[Hz]	360.31	357.82	350.1	341.85	333.96	326.74	320.32
Frequency f_3	[Hz]	439.71	606.6	658.05	664.66	675.2	680.19	671.07

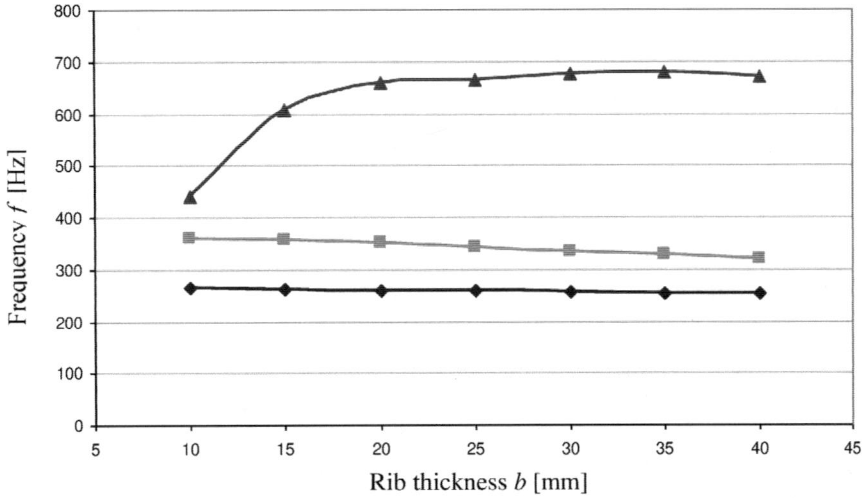

Figure 8. Graphical representation of deflection dependency of the rib thickness

One observes that the first two natural frequencies slowly decrease by rib thickness increase, while the superior modes increase dramatically for low values of *b*, and maintain approximately constant for higher rib thickness values.

3.2. Effect of inter-rib distance on the natural frequencies

To parametric modal analysis of the rib thickness influence on the valve disk frequencies was performed for the values of *c*, indicated in Table 2. The results for the first three vibration modes are presented in Table 5 and the frequency evolution is depicted in Figure 9.

Table 5. Frequency evolution in respect to inter-rib distance

Distance c	[mm]	100	110	120	130	140	150	160
Frequency f_1	[Hz]	258.14	260.2	262.41	264.8	267.33	270.02	272.93
Frequency f_2	[Hz]	336.3	338.8	341.47	344.31	347.27	350.35	353.58
Frequency f_3	[Hz]	672.47	669.65	666.85	664.19	661.64	659.29	657.32

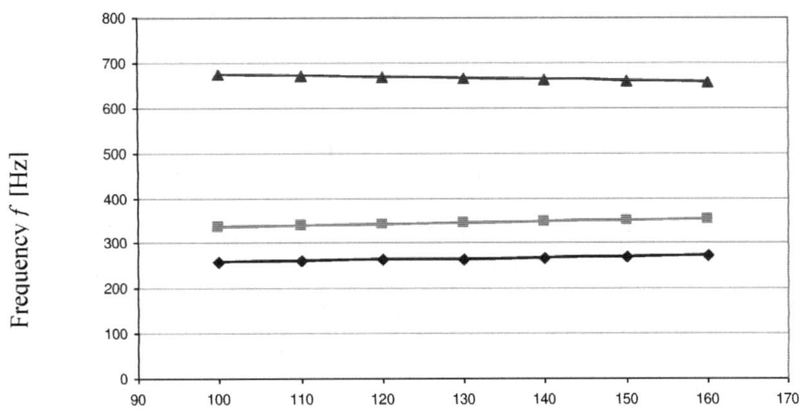

Figure 9. Graphical representation of deflection dependency of the inter-rib distance

One observes that the first two natural frequencies slowly increase by inter-rib distance increase, while the superior modes slowly decrease. This happens because the first two modes are weakly dependent of the transverse rigidity provided by the ribs, while the superior modes do not depend on parameter *c*. In this last case the mass increase defines the frequency change.

3.3. Effect of rib height on the natural frequencies

The parameter *e*, i.e. the rib height, was also altered and modal analysis performed. The values, indicated in last column in Table 2, vary between 100 and 160 mm. Again, the other parameters are maintained constant and are indicated with bold in the same table. Achieved results are indicated in Table 6 and illustrated in graphical form in Figure 10.

Table 6. Frequency evolution in respect to rib height

Height *e*	[mm]	100	110	120	130	140	150	160
Frequency f_1	[Hz]	264.66	263.75	262.89	262.09	261.39	260.82	260.42
Frequency f_2	[Hz]	352.81	350.24	347.69	345.34	343.36	341.81	340.82
Frequency f_3	[Hz]	680.57	680.2	679.72	679.21	678.73	678.32	678.01

In this case all the natural frequencies slowly decrease by decreasing the rib height. This happens due to low contribution to the rigidity of the upper rib sector (in fact low stress values are observed), combined with the effect of frequency decrease due to reduced mass which is associated to rib height decrease.

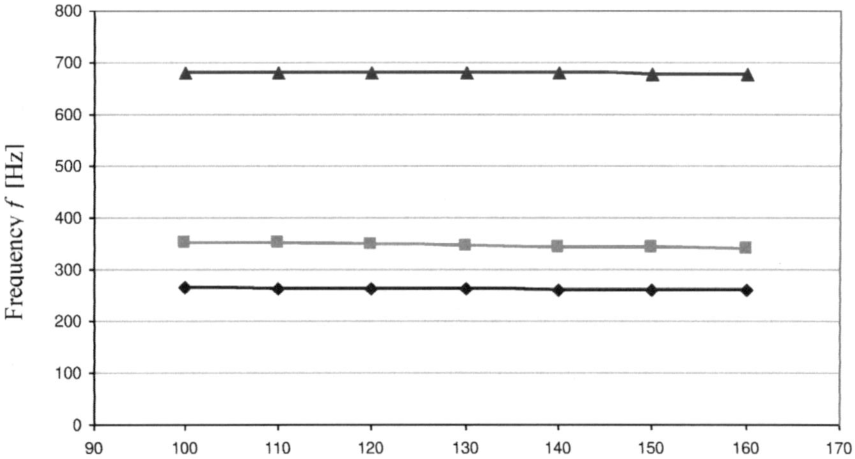

Figure 10. Graphical representation of deflection dependency of the rib height

To have a comprehensive image on the phenomenon, and clearly understand the effect of the geometrical parameters *b-e* upon the changes induced to the natural frequencies, the first six mode shapes are presented in Figure 11. Relevant are the displacements in each mode shape; lowest value being depicted in blue color, highest value in red color.

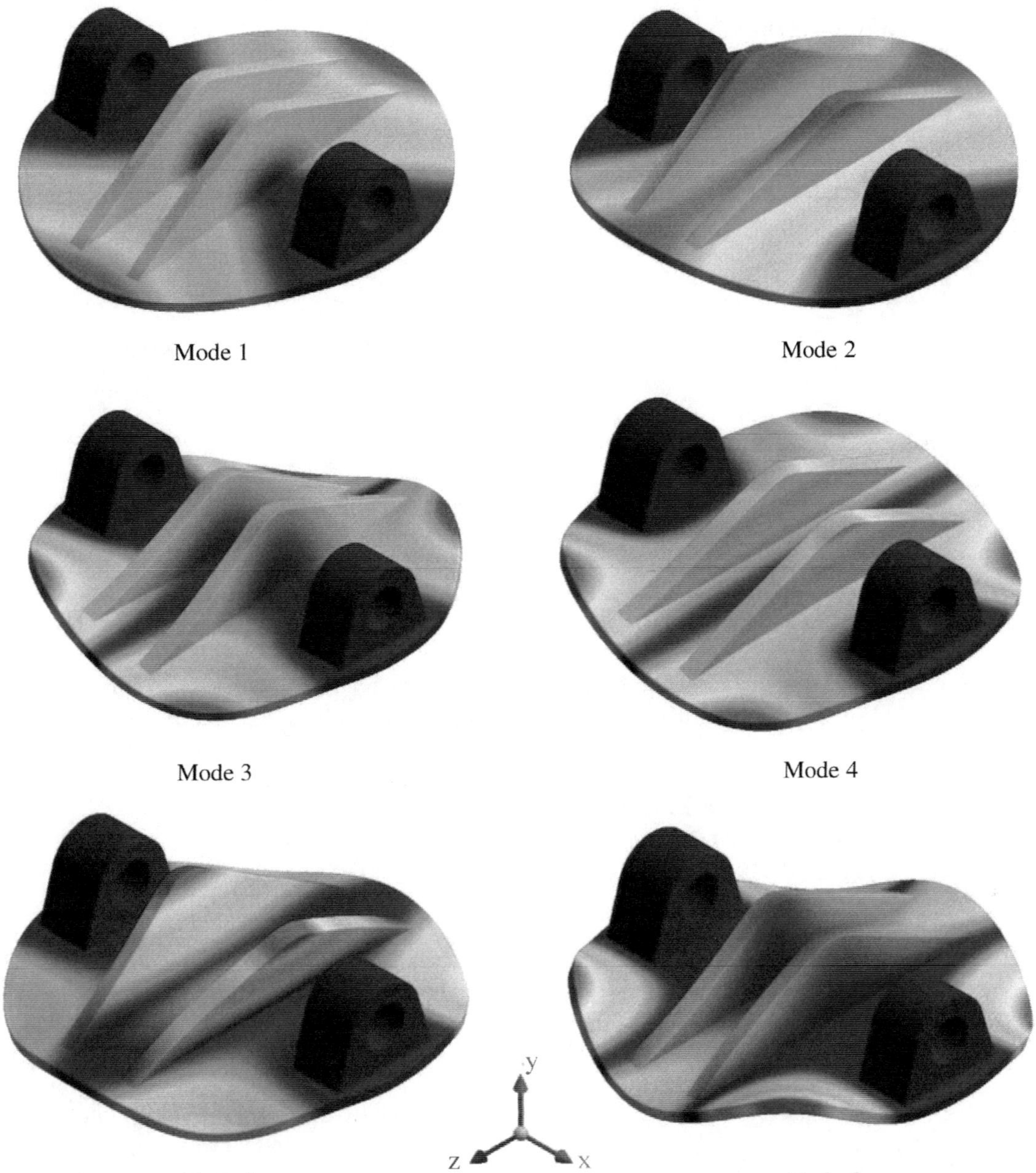

Figure 11. Graphical representation of deflection dependency of the rib height

4. Conclusion

In this paper the effect of rib geometry on the natural frequencies of a butterfly valve disk is analyzed. It was found that the natural frequencies can be controlled by the ribs shape, dimension and position. First, it was deduced that a higher mass ensures a frequency decrease, which is an obvious conclusion. It certifies the correctness of the study. Regarding the rib height, it controls the rigidity in the z direction, thus influencing the corresponding modes (e.g. vibration modes 1 and 2). Because shorter ribs are associated to lower mass, the effect of frequency increase due to stiffening is lightly reduced by the additional rib mass.

IOP Conf. Series: Materials Science and Engineering **163** (2017) 012041 doi:10.1088/1757-899X/163/1/012041

Regarding the rib position, it influences the disk rigidity in the *x* direction, which is orthogonal to the ribs, consequently controlling frequency of some particular modes as vibration mode 4 and 5.
As a general conclusion, the natural frequencies of the valve disk can be moved to a desired domain by properly setting the rib geometrical parameters. This helps avoiding the shedding frequencies by increasing the natural frequency, so that the negative effect of resonance does not occur.

Acknowledgement
The work has been funded by the Sectorial Operational Program Human Resources Development 2007-2013 of the Ministry of European Funds through the Financial Agreement POSDRU/159/1.5/S/132395

References
[1] Smith P O and Zappe R W 2004 *Valve Selection Handbook: Engineering fundamentals for selecting the right valve design for every industrial flow application*, Fifth Edition, Elsevier, Amsterdam, The Nedelands
[2] Dietrich H and Schwarz V 1970 *The Optimal Application of Butterfly Valves*, VAG Armaturen GmbH, translation from J Schif und Hafen, No 22, Seehafen-Verlog, Erik Blumenfeld, Hamburg, August, pp 740-741
[3] Powell W *Study of resonant Phenomena in Pilot Operated Safety Relief Valves*, Report no 2-0175-51, Anderson, Greenwood
[4] Hill R, Lee E H and Tupper S J 1951 Method of Numerical Analysis of Plastic Flow in Plain Strain and Its Application to the Compression of a Ductile Material between Rough Plates, *J. Appl. Mech.* June 49
[5] Tufoi M, Hatiegan C, Vasile O and Gillich G R 2013 Dynamic Analysis of Thin Plates with Defects by Experimental and FEM Methods, *Romanian Journal of Acoustics and Vibration* **10**(2) 83-88
[6] Gillich G R, Praisach Z I, Iancu V, Furdui H and Negru I 2015 Natural Frequency Changes due to Severe Corrosion in Metallic Structures, *Strojniski Vestnik - Journal of Mechanical Engineering* **61**(12) 721-730
[7] Gillich G R, Praisach Z I, Abdel Wahab M and Vasile O 2014 Localization of transversal cracks in sandwich beams and evaluation of their severity, *Shock and Vibration* 607125
[8] Gillich G R and Praisach Z I 2013 Detection and Quantitative Assessment of Damages in Beam Structures Using Frequency and Stiffness Changes, *Key Engineering Materials* **569** 1013-1020
[9] Tufoi M, Gillich G R, Praisach Z I, Ntakpe J L and Hatiegan C 2014 An analysis of the dynamic behavior of circular plates from a damage detection perspective, *Romanian Journal of Acoustics and Vibration* **11**(1) 41-46
[10] Gillich G R, Praisach Z I, Onchis-Moaca D and Gillich N 2011 *How to correlate vibration measurements with FEM results to locate damages in beams*, Proceedings of the 4th WSEAS International Conference on Finite Differences - Finite Elements - Finite Volumes - Boundary Elements, Paris, France, April 28-30, pp 76-81

IOP Conf. Series: Materials Science and Engineering **163** (2017) 012042 doi:10.1088/1757-899X/163/1/012042

Single-phase PM synchronous motor simulation with Matlab/Simulink

O Chiver[1], L Neamt[1], E Pop[1], M Horgos[1], Z Erdei[1], A Pop-Vadean[2] and P P Pop[2]

[1]Technical University of Cluj-Napoca, Electrical, Electronic and Computer Engineering Department,V. Babes str., no. 62A Baia Mare, Romania
[2]Technical University of Cluj-Napoca, Department of Mechatronics and Machine Dynamics, Cluj-Napoca, Romania

E-mail: olivian.chiver@cunbm.utcluj.ro

Abstract. The permanent magnet (PM) synchronous motors have a major advantage compared to the asynchronous motors, namely that there are no losses in the rotor. As regarding the single-phase motors must be considered the possibility to obtain the starting torque. A simple and inexpensive constructive solution is the motor with tapered air-gap, which can develop a starting torque (although specific starting torque is small) in a well-defined direction, simply by powering the stator winding from the network of industrial frequency. Starting torque occurs due to the fact that the axis of the rotor field created by PM in the rest position differs from the axis of the stator because of the tapered air gap. Therefore this motor is more often used in low power household appliances. The development and the implementation in Matlab / Simulink of the motor model will be presented in this paper. The model is based on a mathematical model with lumped parameters, parameters that are determined through calculations and finite elements analysis. The motor behavior will be studied, both in stationary and dynamic operation. The accuracy of the model is verified by comparing simulation results with results obtained by measurements.

1. Introduction

It is known that the cheapest energy and with the smallest negative effect on the environment is the saved energy. For this reasons the new standards have imposed the continuous increase of energy conversion efficiency. In order to increase the efficiency of conversion of electrical energy into mechanical energy through the motors only few measures are possible: the using of new materials with improved qualities; the using of new constructive solutions concretized in new motor types; in some cases, the reducing of electrical and magnetic loads by increasing the geometrical dimensions.

As regarding the first solution, limitations arise because of relatively low technical progress in the field. In the last case, the new costs which will be higher should be considered. Restrictions regarding enclosing the motors in a certain class of sizes also will arise in this case. The conception of a new type of motors is always of great interest, but they must have a proven advantage (for certain applications), and usually, they must be simple and cheap.

Until recently, usually in small power electrical drives, asynchronous motors were used due to their robustness and simplicity. In recent years there is a trend of replacing these motors with different types of synchronous or D.C. brushless motors. The permanent magnets (PMs) motor and switched reluctance motor are among the most commonly used synchronous motors. In order to supply the switched reluctance motors, special sources are required, and for this reason, the drives with these motors are usually more expensive. In case of PMs synchronous motors, there are solutions that can be powered directly from industrial frequency network. Generally they are with three-phase and the starting process is performed in asynchronous operation.

In case of small power drives, when the motor is supplied from the single-phase network, a very simple solution is the PM synchronous motor with tapered air-gap, Figure 1.

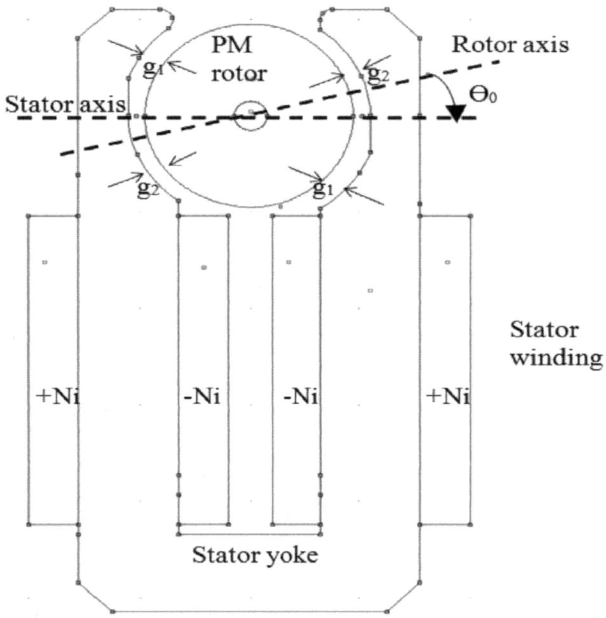

Figure 1. Cross section of the single-phase PM synchronous motor with tapered air gap

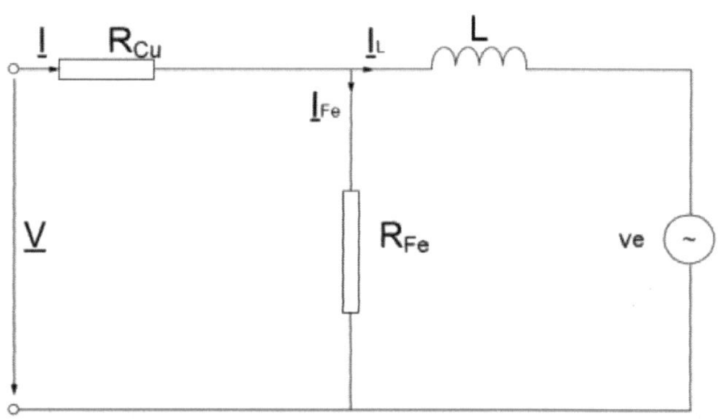

Figure 2. Electrical equivalent circuit

This motor has the great advantage that it is able to develop starting torque based on the initial phase shift existing between the stator field axis and the axis of the rotor flux produced by PM. This initial phase shift is due to the uneven air gap, consisting of two main areas: one with large air gap, g1, and another with small air gap, g2. If the load torque is smaller than the initial torque, the rotor begins to move and it synchronizes after a few half-cycles.

Electrical equivalent circuit of single-phase synchronous motor with PMs and tapered air-gap is presented in Figure 2 [1]. The material of PMs is Ferrite type, with parallel magnetization.

Based on electrical equivalent circuit, the motor equations can be written [1-3],

$$V = iR_{Cu} + i_{Fe}R_{Fe} \tag{1}$$

$$i_{Fe}R_{Fe} = L\frac{di_L}{dt} - v_e \tag{2}$$

$$i_L + i_{Fe} = i \tag{3}$$

The induced voltage is given by relation,

$$v_e = -\frac{d\Psi}{dt} = -\frac{d\Psi}{d\theta}\frac{d\theta}{dt} = 2N\phi_{rp}sin(\theta - \theta_0)\frac{d\theta}{dt} \tag{4}$$

The motion equation is,

$$J\frac{d\Omega}{dt} = 2Ni\phi_{rp}sin\theta - T_{rp}sin[2(\theta - \theta_0)] - T_l \tag{5}$$

It has been noted: V – rated voltage; R_{Cu} – winding resistance; RFe – resistance corresponding to the iron losses; i – stator current; i_{Fe} – current corresponding to the iron losses; i_L – current corresponding to the magnetic flux produced by stator winding; v_e – voltage induced by PM; ψ – total flux of PM linked by stator winding; N – number turns of one stator coil; ϕ_{rp} – peak rotor flux; θ – angle between stator and rotor axis; θ_0 – initial angle (rest pozition of the rotor); J – total moment of inertia of the rotor; Ω – rotor speed; T_{rp} – peak reluctance torque; T_l – load torque.

In the previous equations it was assumed that the reluctance torque and the flux created by PM have a sinusoidal variation depending on the rotor position (θ). This assumption is not far from reality as it can be seen in Figure 5 and Figure 6.

2. Finite elements analysis

In order to obtain some of the motor parameters required for implementation of the motor model in Matlab/Simulink, the numerical model of the motor has been created in FEMM and magnetostatic analysis has been performed. The PMs properties have been chosen so that the PM flux to be the same with the magnetic flux determined based on the induced voltage, experimentally recorded in no load operation, the machine being driven by another motor, Figure 3 [4].

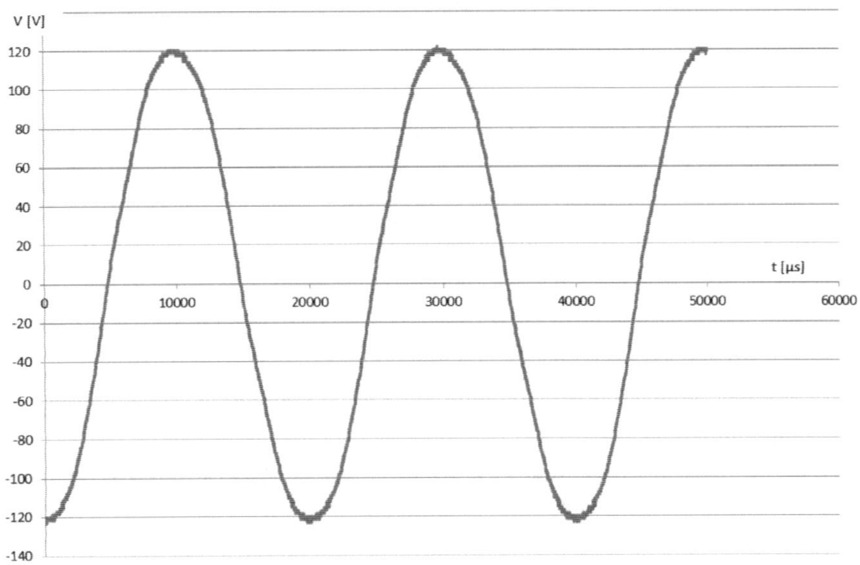

Figure 3. Recorded voltage induced by PMs

IOP Conf. Series: Materials Science and Engineering **163** (2017) 012042 doi:10.1088/1757-899X/163/1/012042

Figure 4 shows the flux lines and magnetic flux density map for two rotor positions, without stator winding current: the rotor axis at 900 to the axis of the stator (a), and rotor axis aligned with stator axis (b).

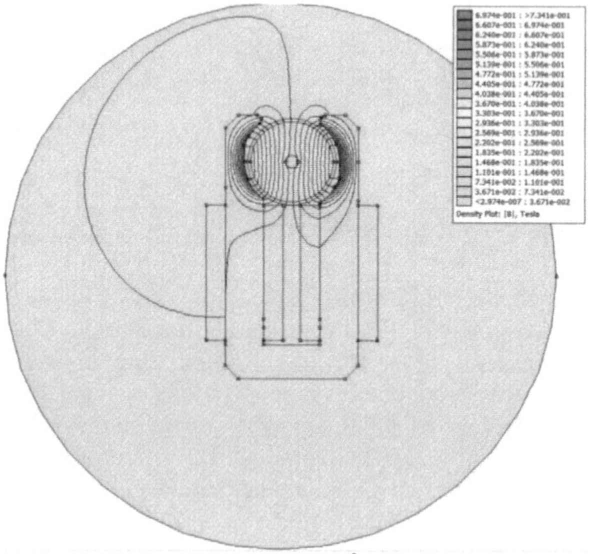

(a) 90^0

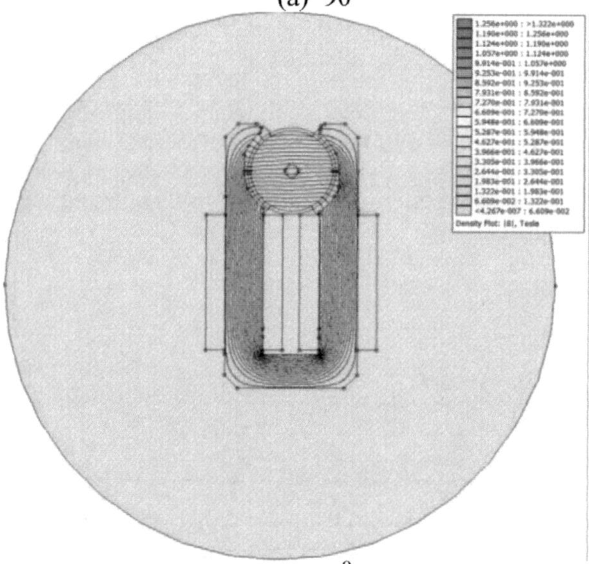

(b) 0^0

Figure 4. Flux lines and flux density map

With FEMM, the torque was obtained by Maxwell Stress Tensor Method [5], for different rotor positions. The dependence Tr=f(θ) is showed in Figure 5. From Tr=f(θ), the rest position of the rotor is also obtained. This corresponds to the angle for which the reluctance torque is zero. The dependence of the total flux of PM linked with stator winding by rotor position is presented in Figure 6.

The stator winding resistance is measured without problems, but during operation, its values changes as the winding heats.

A method for analytically determination of the stator winding inductance, depending on the stator geometry and materials, is presented in [1], [6], [7].

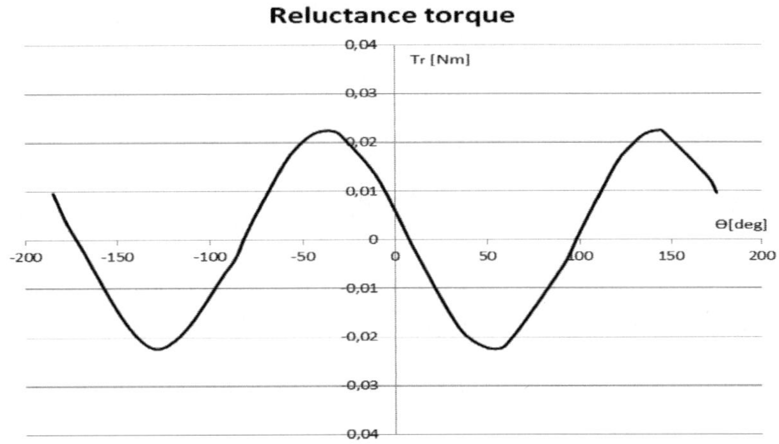

Figure 5. $T_r=f(\theta)$

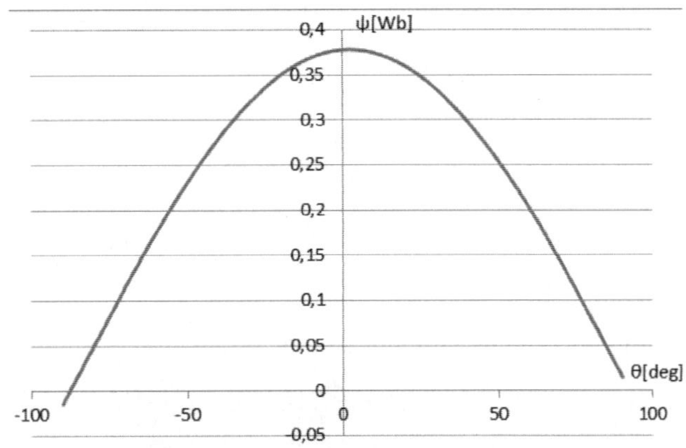

Figure 6. $T_r=f(\theta)$

The inductance can also be measured ([1], [8]) or determined by finite elements analysis. It should be noted that it depends on the rotor position (θ), Figure 7.

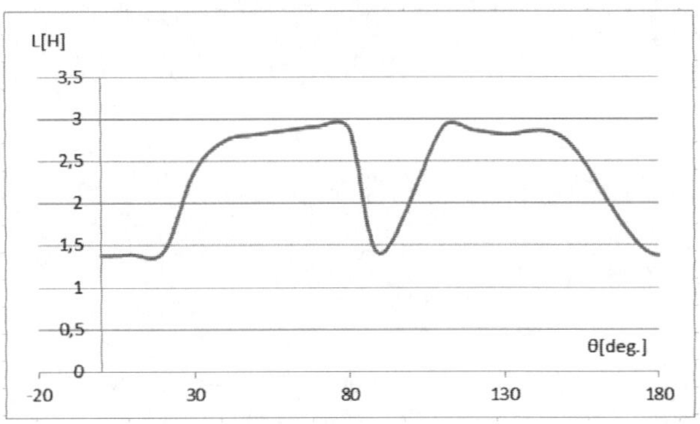

Figure 7. $L=f(\theta)$

3. Motor modeling in Simulink

Once the rotor parameters have been determined, the model can be implemented in Matlab/Simulink. Based on motor equations a model in Matlab/Simulink has been made, it being presented in Figure 8.

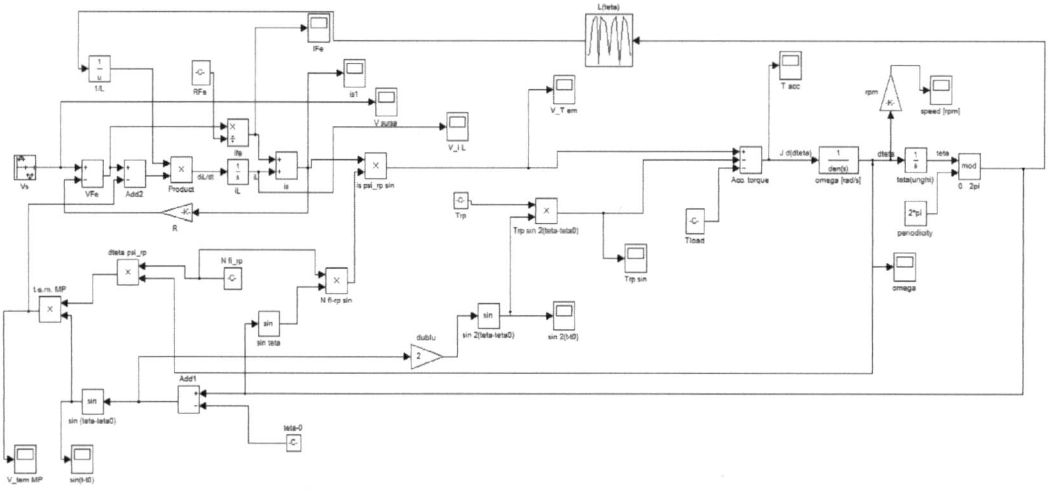

Figure 8. Single-phase PMs synchronous motor with tapered air-gap model

A voltage source has been used for generation of the sinusoidal voltage signal, and a lookup table for variable stator inductance [9]. The other blocks are common for Simulink. For visualization of different parameters, many oscilloscopes have been used.

4. Results and conclusion

The simulations have been performed for five seconds. In the following the obtained results regarding some of the motor characteristics are presented.

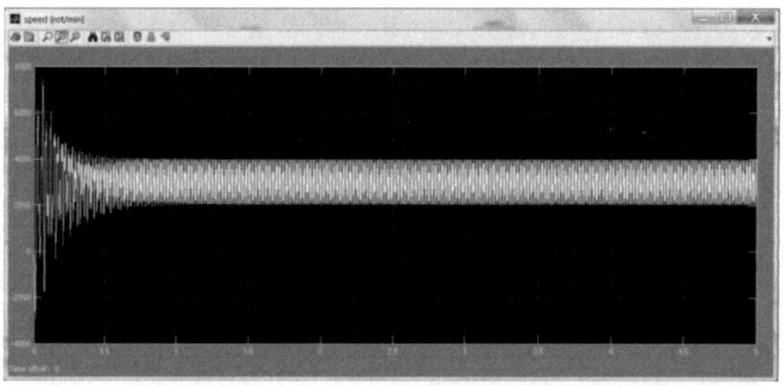

(a)

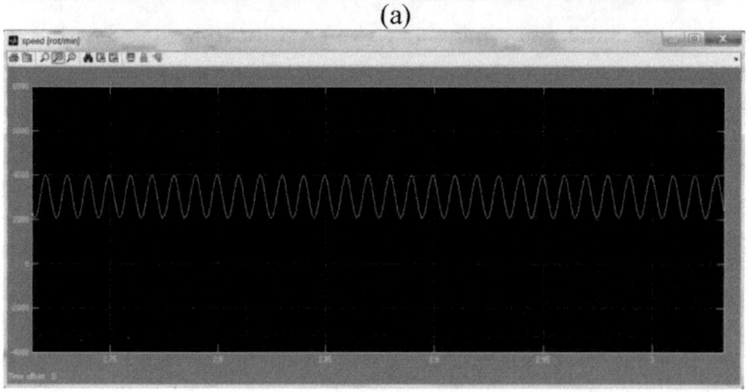

(b)

Figure 9. Speed [rot/min] variation in time

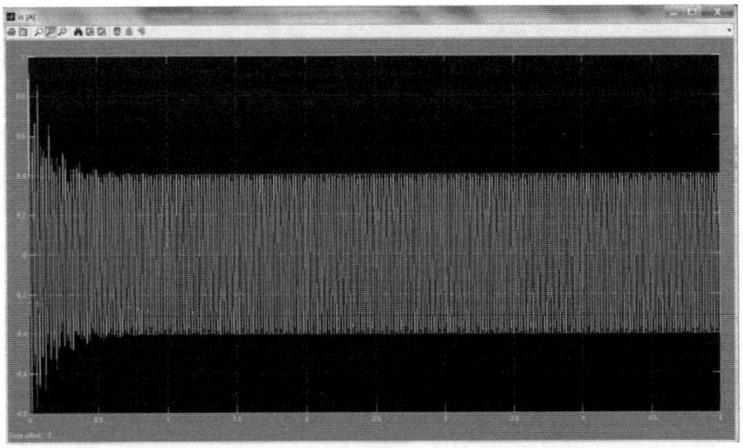

(a)

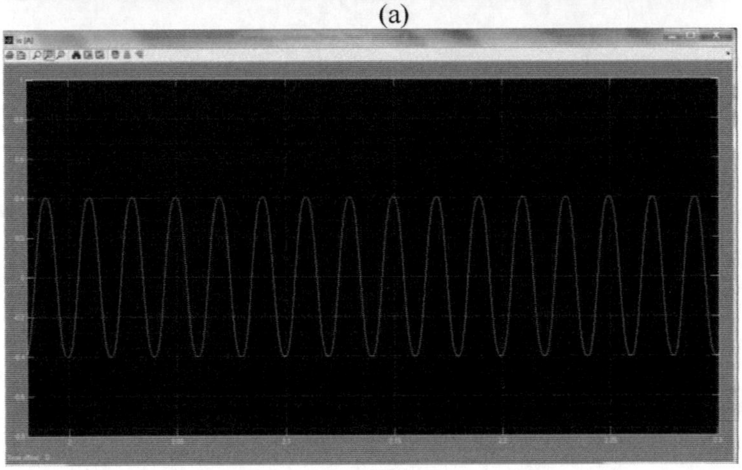

(b)

Figure 10. Stator current [A] variation in time

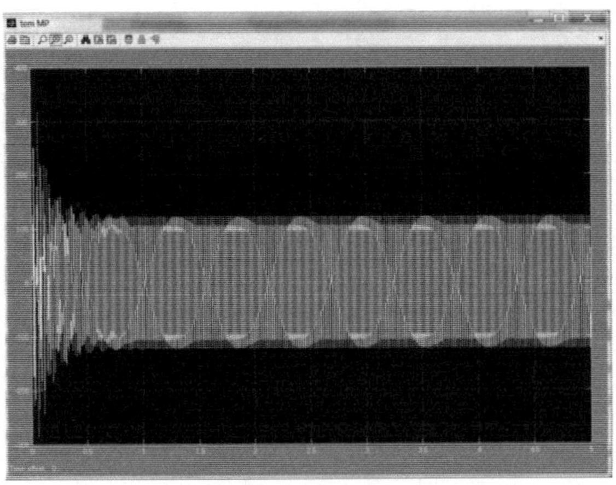

(a)

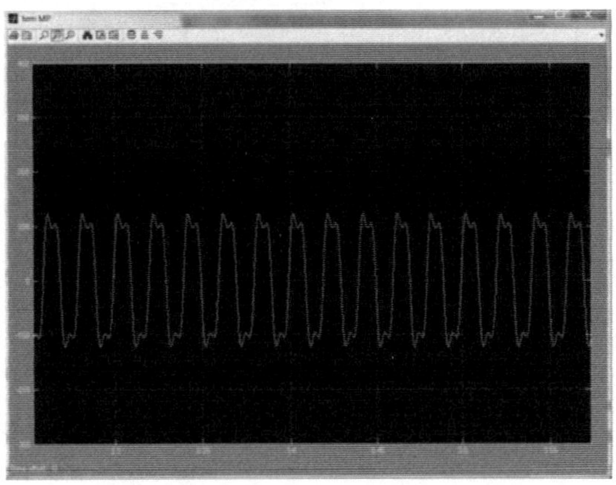

(b)

Figure 11. Induced voltage [V] variation in time

Some conclusions concerning the presented results must be done. The rotor speed is not constant, but the motor runs with an average speed equal to that of synchronism (3000 rot/min).

The induced voltage is not the same with de recorded voltage. The one recorded has a sinusoidal shape, and that obtained by simulation is deformed. Also, the stator current in no load operation, obtained by simulation, has constant amplitude, while the recorded no load current varies around of an average value, Figure 12.

At this time the accuracy of the model is not satisfactory. Although the average values seem to be alright, the model will be studied to achieve results like those recorded. This will be done in a next paper.

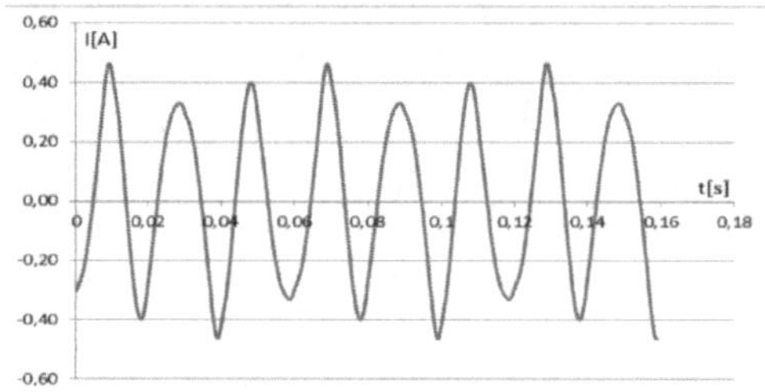

Figure 12. Recorded no load current

References

[1] Ertan H B, Bülent D and Capolino G A 2005 Calculation of parameters of single-phase PM motor for design optimization, *IEEE Transactions on Energy Conversiono* **20**(3) 538-548

[2] Min F, Yang Y, Chen Y and Jin H 2013 *Influence of step-air gap on line-start performance of u-shape single-phase permanent magnet motors,* 2nd International Conference on Measurement, Information and Control, Harbin, China, August 16-18, pp 1046-1050

[3] Andriollo M et al 2008 *Performance assessment of a single-phase PM synchronous motor for small fan appliances*, International Conference on Electrical Machines, Vilamoura, Portugal, September 6-9, pp 1-6

[4] Chiver O, Neamt L, Pop E, Barz C and Pop Vadean A 2016 *Identification of parameters of single-phase PM synchronous motor*, 16th International Conference on Environment and Electrical Engineering (EEEIC), Florence, Italy, June 7-10

[5] *** FEMM User`s Manual

[6] Pereira L A, Locatelli E R, Zolet G and Canalli V M 2001 *Single phase permanent-magnet motors - part I: Parameter determination and mathematical model*, IEEE International Electric Machine and Drives Conference, Cambridge, Massachusetts, June 17-20

[7] Pereira L A, Locatelli E R, Zolet G and Canalli V M 2001 *Single phase permanent-magnet motors - part II: Performance and test results*, IEEE International Electric Machine and Drives Conference, Cambridge, Massachusetts, June 17-20

[8] Ertan H B and Şahin I 2012 *Evaluation of Inductance Measurement Methods for PM Machines*, XXth International Conference on Electrical Machines (ICEM), Marseille, France, September 2-5, pp 1672-1678

[9] Iepure I L, Tutelea L and Boldea I 2008 *FEM analysis and control of a tapered airgap single phase PMSM*, 11th International Conference on Optimization of Electrical and Electronic Equipment OPTIM 2008, Braşov, Romania, May 22-14

IOP Conf. Series: Materials Science and Engineering **163** (2017) 012043 doi:10.1088/1757-899X/163/1/012043

Mechanical and thermal analysis of the internal combustion engine piston using Ansys

V G Cioată, I Kiss, V Alexa and S A Rațiu

Politehnica University of Timisoara, Engineering and Management Department, 5 Revolution Street, Hunedoara, 331128, Romania

E-mail: cioata.vasile@upt.ro

Abstract. The piston is one of the most important components of the internal combustion engine. Piston fail mainly due to mechanical stresses and thermal stresses. In this paper is determined by using the finite element method, stress and displacement distribution due the flue gas pressure and temperature, separately and combined. The FEA is performed by CAD and CAE software. The results are compared with those obtained by the analytical method and conclusions have been drawn.

1. Introduction

The piston is an important component of the internal combustion engines, due to the role it fulfils, but also due to the complex loads which act upon it. Upon the piston act both thermal and mechanical loads. The piston's resistance to the thermal and mechanical loads limits the power of the internal combustion engine.

For this reason, during the design stage of the piston, there is a great emphasis on determining the stresses and deformations due to the mechanical and thermal loads, an aspect which is revealed in numerous scientific papers dealing with this issues. Studying these papers [1 – 6], show that the authors have used CAD software for 3D modeling of the piston, and CAE software to determine the stresses and deformations in the piston, which allow finite element analysis and the study of the product behavior in terms of mechanical-thermal loads.

By using the finite element method, this paper determines the stresses and deformations of the piston subjected to mechanical loads (flue gas pressure) and thermal loads (flue gas temperature), considering that they act independently and simultaneously (mechanical – thermal load).

The results are compared with those obtained by the analytical method and conclusions are drawn.

Autodesk Inventor Professional is used to obtain the 3D model of the piston and Ansys Workbench V.15 is used for the finite element analysis.

2. Initial data

This study considers a flat head piston, with two compression segments and a lubrication segment. In order to prevent the heat flow orientation from the piston head to the first segment, the canal of the fire segment is placed below the bottom of the piston. To the same end, the port-segment area and the bottom of the piston are broadly joined up. The length of the piston's mantle must be enough in order to provide a good guidance, low side pressure and to limit the tilting.

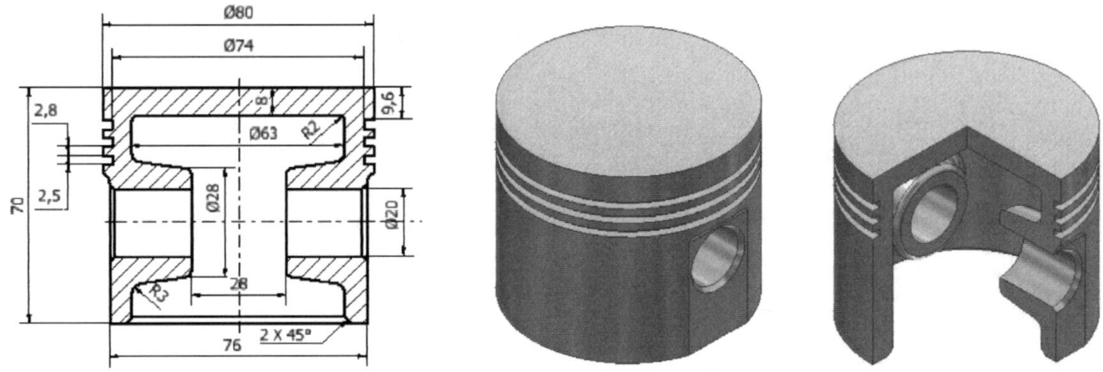

(a) (b)

Figure 1. Dimensions of the piston (a) and 3D model of the piston (b)

Based on the initial data and recommendations from literature [7], [8], we established the main dimensions of the piston, after which we built the 3D model (Figure 1, a and b).

The material used for the piston is a silicon based aluminum alloy. Eutectic alloy type ATC12CuMgNi, having the properties on table 1, will be used.

Table 1. Properties of the material

Properties of the material	Value
Thermal conductivity	Figure 2
Thermal Expansion	$2{,}3 \cdot 10^{-5}$ $1/^{0}C$
Specific Heat	875 $J/kg^{0}C$
Density	2770 kg/m^3
Young's Modulus	$7{,}1 \cdot 10^4$ MPa
Poisson's Ratio	0,34
Tensile Yield Strength	280 MPa
Tensile Ultimate Strength	310a

3. Analytical calculation of the piston

The piston head is the most used area, in thermal and mechanical terms, as it comes into contact with the flue gases and takes over pressure forces at the same time. For this reason, the analytical calculation of the stresses from the piston head, which arise due to this loads, is used.

3.1. Analytical calculation of the mechanical stresses in the piston head

Mechanical stresses (σ_{mec}) were calculated analytically, using the formulas and terms described in [7], [8]. The piston head was considered as a circular plate embedded to the contour given by the inner diameter of the piston head, having a constant thickness and a uniformly distributed load, given by the maximum pressure of the gas.

In such plate, radial stresses (σ_r^m) and tangential stresses (σ_t^m) appear, whose extreme values are obtained in the center of the plate (c) and in the embedment (m), on the bottom surfaces (b) and the top surfaces (t).

IOP Conf. Series: Materials Science and Engineering **163** (2017) 012043 doi:10.1088/1757-899X/163/1/012043

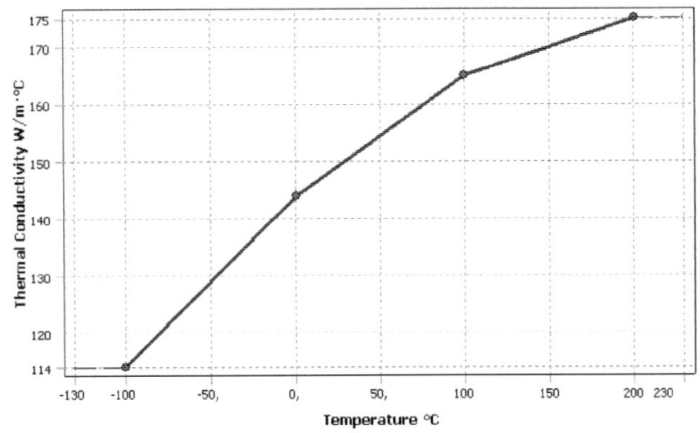

Figure 2. Thermal conductivity

3.2. Analytical calculation of the thermal stresses in the piston head

Thermal stresses (σ_{ter}) are generated by the uneven distribution of temperatures in the piston head and by the way it expands free or restricted. In the case of a flat piston, the maximum temperature is reached in the center of the piston and the temperature decreases towards the extremities.

The thermal stress of the head piston was established as follows: the stresses (σ_{r1}^{t}) and (σ_{t1}^{t}) were determined, considering the piston head as a circular plate, free on the contour, with constant thickness, subjected to an axially symmetric temperature field. It was taken into account that the side wall of the piston prevents the free expansion of the head, creating a pressure on the surface of the piston, which causes radial stresses (σ_{r2}^{t}) and tangential stresses (σ_{t2}^{t}). Considering a linear temperature variation on the thickness of the piston head, there are bending stresses on the radial (σ_{r3}^{t}) and tangential (σ_{t3}^{t}) direction, equal between them. Tangential and radial thermal stresses are determined by superposition of effects, obtaining σ_r^t şi σ_t^t.

3.3. Analytical calculation of mechanical-thermal stresses in the piston head

Mechanical-thermal stresses ($\sigma_{mec-ter}$) that appear in the piston head σ_r şi σ_t are also determined by applying the principle of superposition effects.

Table 2. Results of the analytically calculus, in MPa

Type of Stress			Position (m)	Position (c)
Mechanical Stress σ_{mec}	Radial σ_r^m	(t)	63,95	-42,84
		(b)	-63,95	42,84
	Tangential σ_t^m	(t)	21,74	-42,84
		(b)	-21,74	42,84
Thermal Stress σ_{ter}	Radial σ_r^t	(t)	-15,63	-25,21
		(b)	-7,11	-16,69
	Tangential σ_t^t	(t)	17,11	-25,21
		(b)	26,93	-16,69
Mechanical – Thermal Stress $\sigma_{mec-ter}$	Radial σ_r	(t)	48,32	-68,05
		(b)	-71,06	26,15
	Tangential σ_t	(t)	38,85	68,05
		(b)	5,25	33,94
Equivalent Stress	$\sigma_{echiv\ MSS}$ (Tresca)	(t)	48,32	68,05
		(b)	76,31	33,94
	$\sigma_{echiv\ Mises}$ (Mises)	(t)	44,34	68,05
		(b)	73,82	30,79

In the design calculation, it is recommended to use equivalent stress, determined using a theory of resistance. Due to the fact that in the center of the plate, as well as in the marginal section, there is a state of two dimensional stress space, the maximum shear stress theory can be used. If tensions σ_r și σ_t have the same sign, the maximum stress is considered in an absolute value and it is compared to the allowable stress. If the two stresses have different signs, the sum of the absolute values will be calculated and compared to the allowable stress. At the same time, the equivalent stress calculated using von Mises theory customized for plane stress state, can be used.

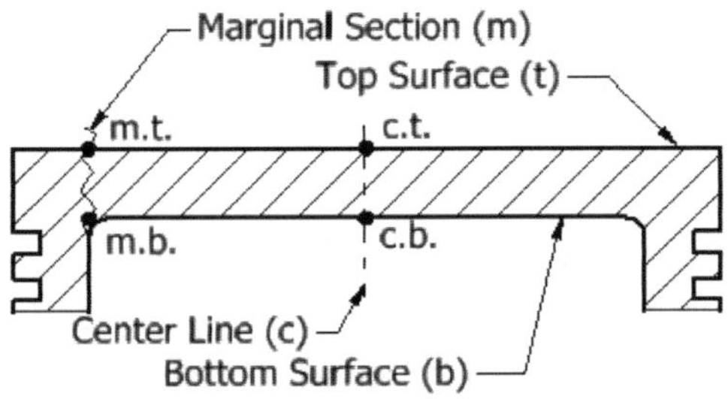

Figure 3. Notations used in paper

The results obtained through the calculation are shown in Table 2, using the following notations: (t) – top surface of the piston head, (b) – bottom surface of the piston head, (m) – marginal, (c) – center of the piston head (Figure 3), $\sigma_{echiv\ MSS}$ - the maximum shear stress and $\sigma_{echiv\ Mises}$ - von Mises stress.

4. Finite element analysis of the piston

4.1. Finite element analysis of the piston subjected to mechanical loads

To perform the finite element analysis of the piston when upon it acts the pressure of the gasses, a structural analysis using Ansys Workbench V.15.1 takes place.

At this stage, the analysis of the piston is a linear static one, when small changes in rigidity happen, there are no changes in the direction of the loading, the materials remain within linear elastic area and small deformations and stresses are produced.

The model of the piston is made in Autodesk Inventor and saved in this program as *.sat, and then imported in Ansys Workbench.

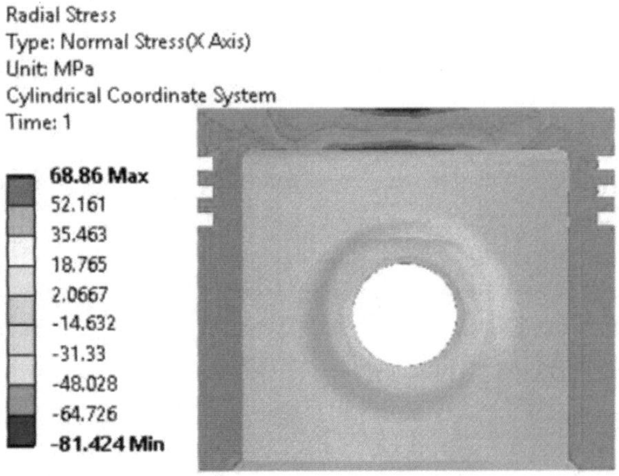

Figure 4. Radial stress due to mechanical loads – in section

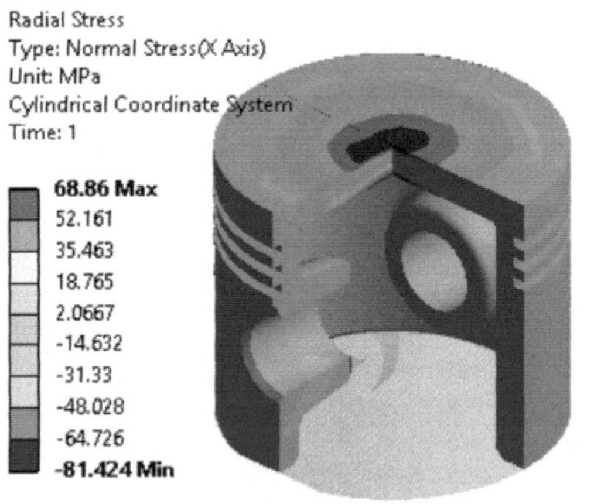

Figure 5. Radial stress due to mechanical loads – in 3D model

The pressure of the gasses is the main mechanical stress which the piston is being subjected to, which in this paper is considered to be equal to 6.5 MPa, normally applied on the top surface of the piston head. Also, all the degrees of freedom of the bores for the bolt are bound.

The results are obtained as stresses fields and deformations fields. For the present situation, due to the fact that the radial and tangential normal stresses in the piston head were determined analytically, the value fields of these dimensions were displayed in the areas of interest.

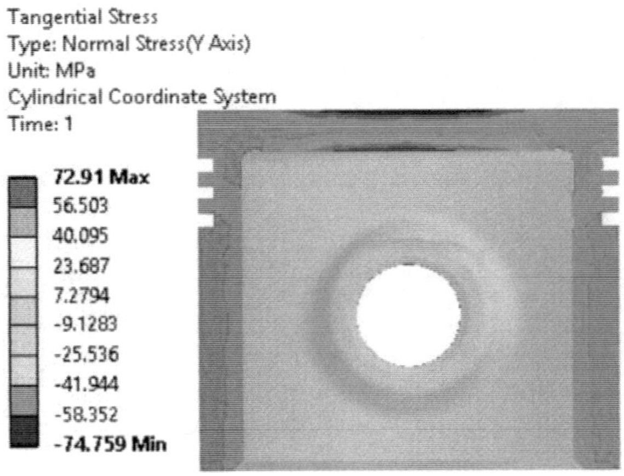

Figure 6. Tangential stress due to mechanical loads – in section

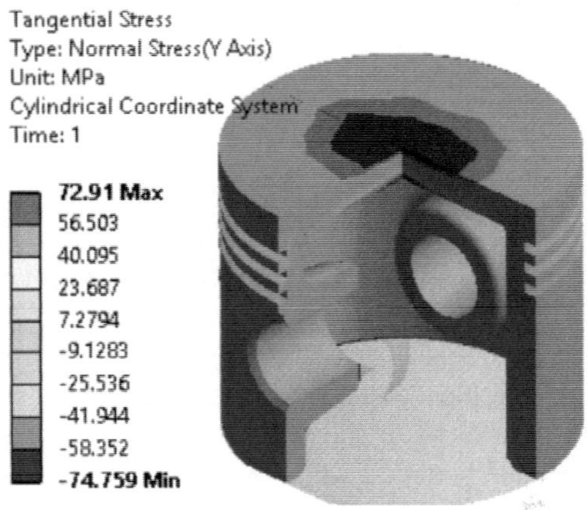

Figure 7. Tangential stress due to mechanical loads – in 3D model

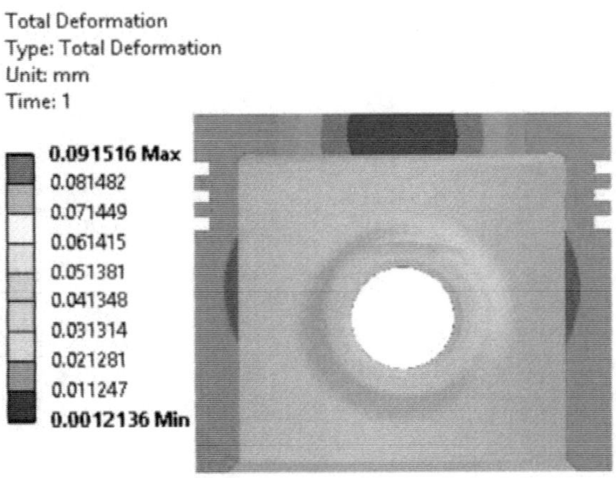

Figure 8. Total deformation due to mechanical loads – in section

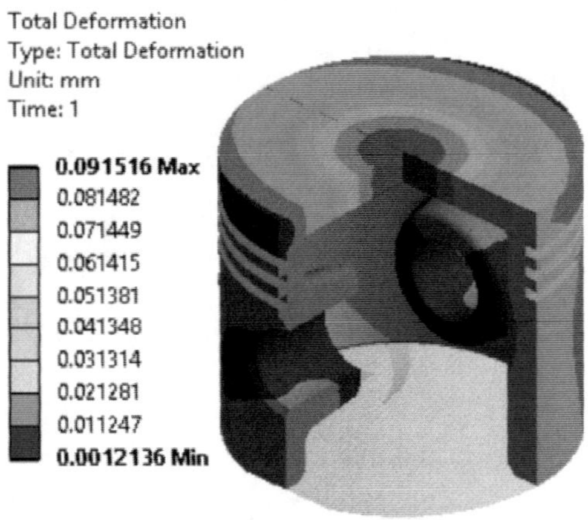

Figure 9. Total deformation due to mechanical loads – in 3D model

The field of the normal radial stress is shown in Figures 4 and 5, as follows: in section through the axis of the piston, with a perpendicular plane on the axis of the bore for the bolt (Figure 4) and in ¾ section (Figure 5). The normal tangential stresses are show in the same manner in Figures 6 and 7. The total directional deformations are shown in section in Figure 8 and in ¾ section in Figure 9.

4.2. Finite element analysis of the piston subjected to thermal loads

The designed piston operates in high temperatures conditions, amounted to 200…250^{0}C. Hereinafter, we present the results obtained from the thermal analysis of the piston, in order to determine the piston temperature field in the piston body and thermal stresses due to different temperatures in the piston body.

IOP Conf. Series: Materials Science and Engineering **163** (2017) 012043 doi:10.1088/1757-899X/163/1/012043

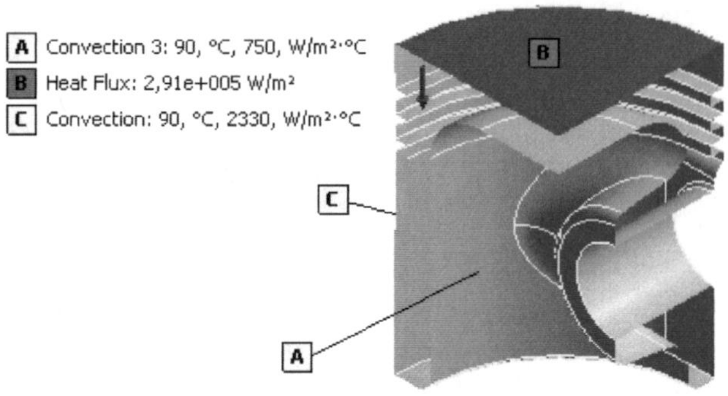

Figure 10. Loads and constraints for thermal analysis

The analysis is a steady – state thermal one.

Upon the piston acts a heat flux ($291 \cdot 10^3$ W/m^2), which is applied to the upper surface of the piston head. The heat transfer between the piston and the oil film between the piston and the cylinder is made by convection, case which requires a convection heat transfer coefficient equal to 2330 W/m$^2\cdot$C, at a temperature of 90^{0}C, and between the inner surfaces of the piston and oil, a coefficient equal to 750 W/m$^2\cdot$C, having the same temperature as the oil – Figure 10.

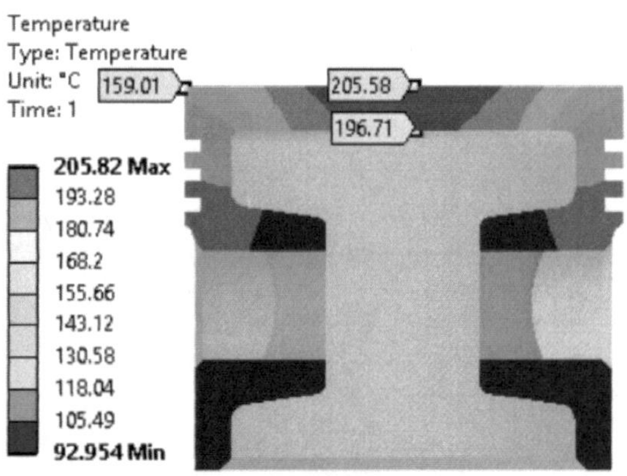

Figure 11. Temperature field in piston due to the heat flux

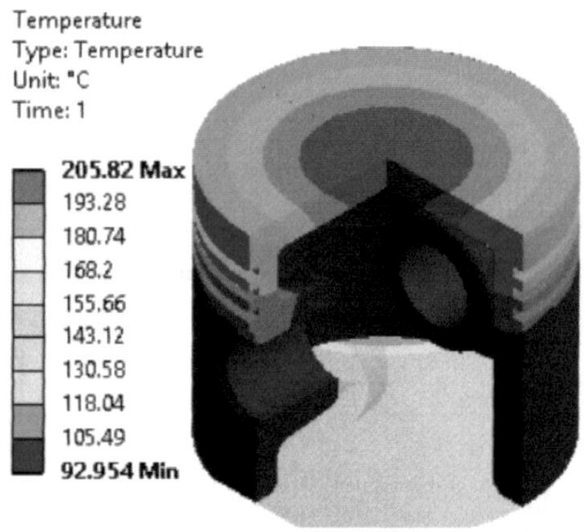

Figure 12. Temperature field in piston due to the heat flux

The results are shown in the following figures. Thus, the temperature field in the piston body is shown in Figures 11 and 12, noticing that the thermal flow determines a maximum temperature on the top surface of the piston head, which decreases towards its edge.

Thermal load, as well as mechanical load, determine thermal stresses and deformations inside the piston body. These can be determined by performing a static analysis using the thermal conditions of the performed steady analysis.

As a constraint, the binding of the degrees of freedom of the bores for bolt of the piston is required, and the results are obtained as thermal stresses and deformations.

The results are shown in the following figure, as follows: the ones related to normal radial stresses are shown in Figures 13 and 14, the ones related to tangential normal stresses are shown in Figures 15 and 16, and the ones related to total deformations in the set of Figures 17 and 18.

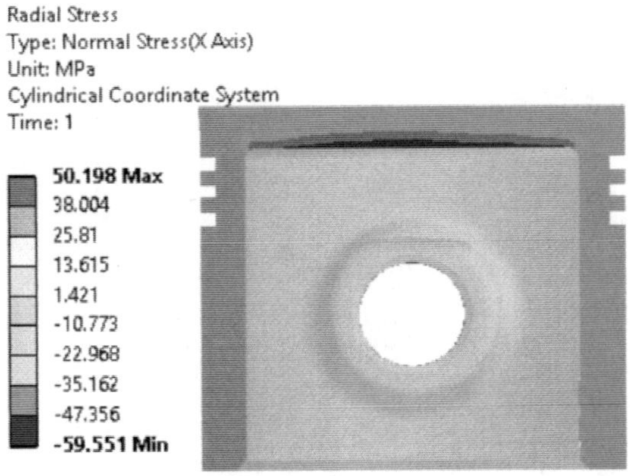

Figure 13. Radial stress due to thermal loads – in section

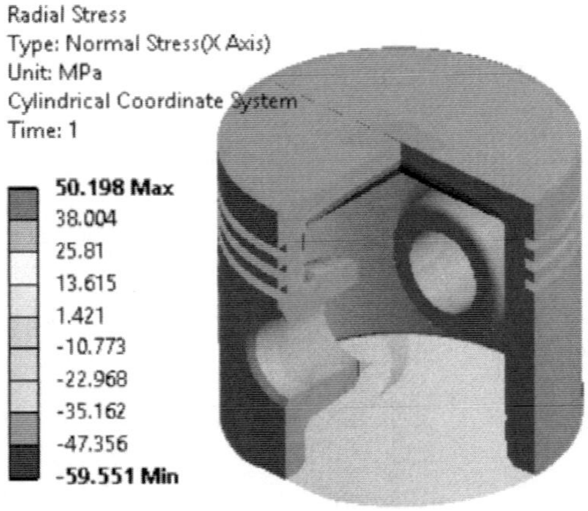

Figure 14. Radial stress due to thermal loads – in 3D model

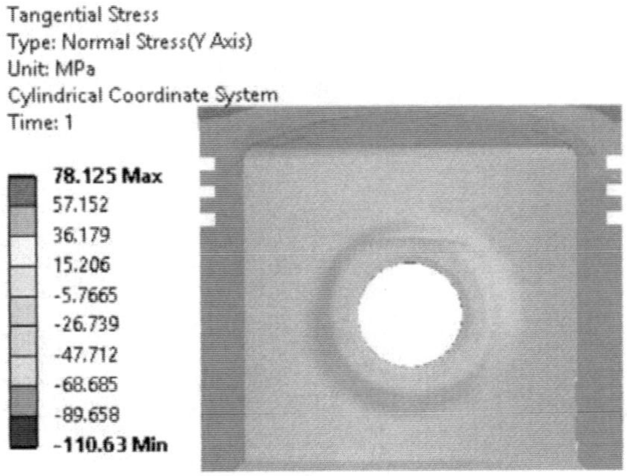

Figure 15. Tangential stress due to thermal loads – in section

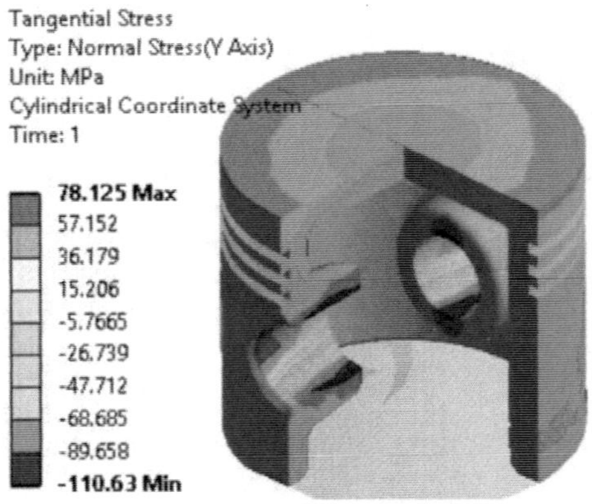

Figure 16. Tangential stress due to thermal loads – in 3D model

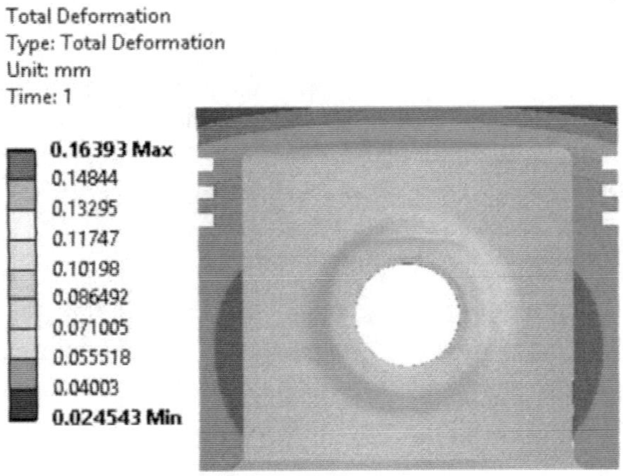

Figure 17. Total deformation due to thermal loads – in section

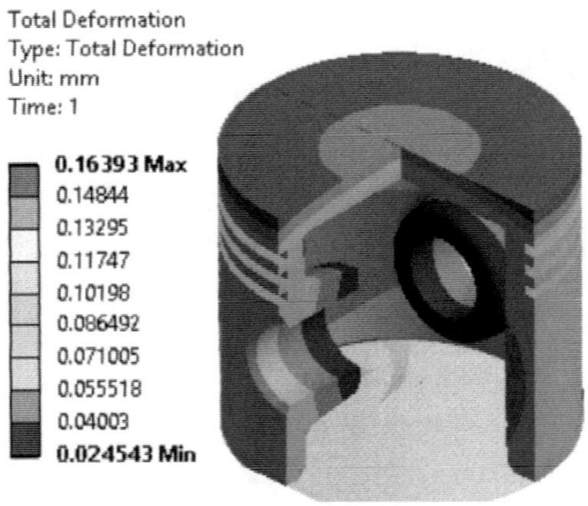

Figure 18. Total deformation due to thermal loads – in 3D model

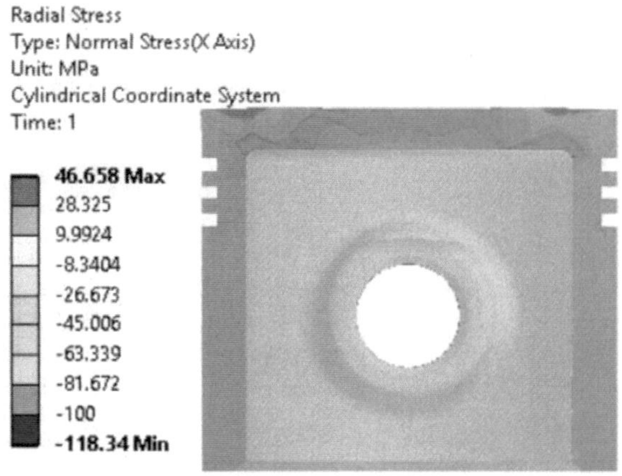

Figure 19. Radial stress due to mechanical - thermal loads – in section

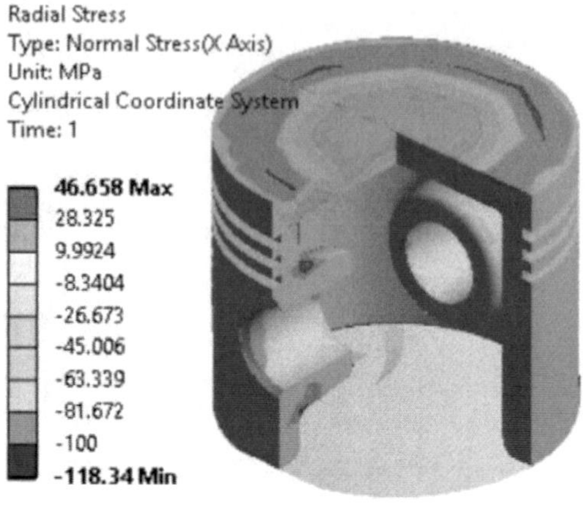

Figure 20. Radial stress due to mechanical - thermal loads – in 3D model

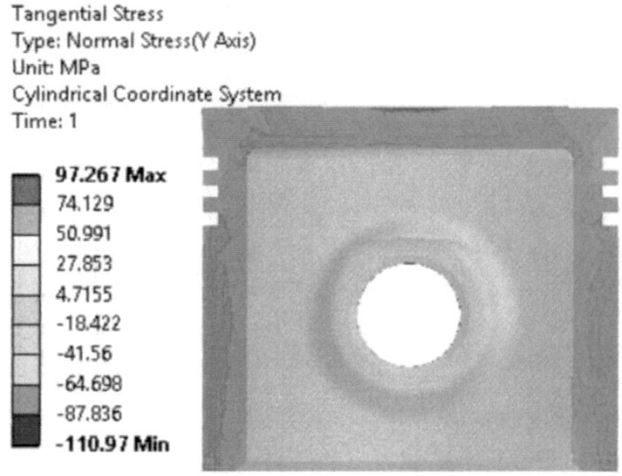

Figure 21. Tangential stress due to mechanical - thermal loads – in section

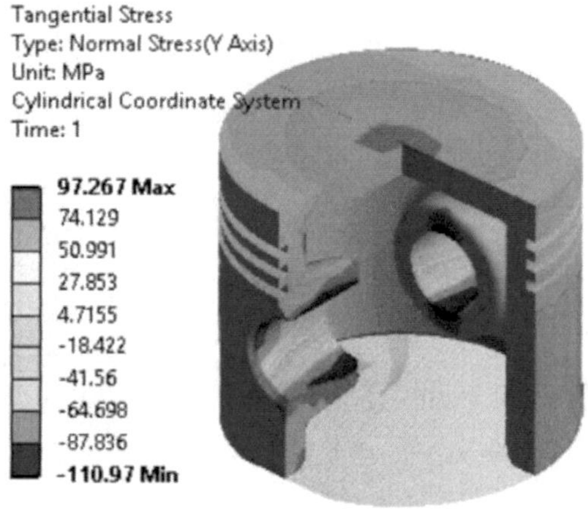

Figure 22. Tangential stress due to mechanical - thermal loads – in 3D model

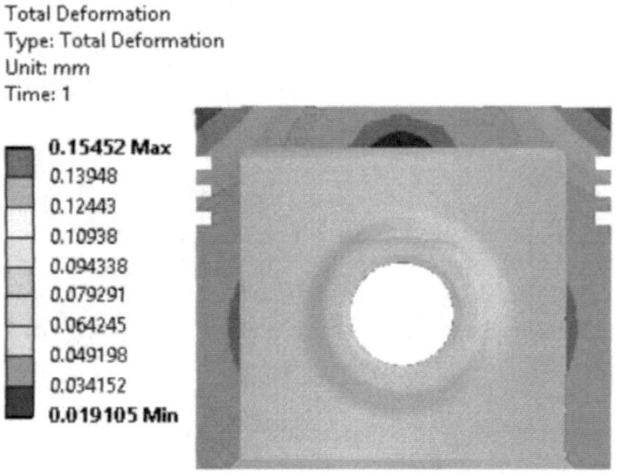

Figure 23. Total deformation due to mechanical - thermal loads – in section

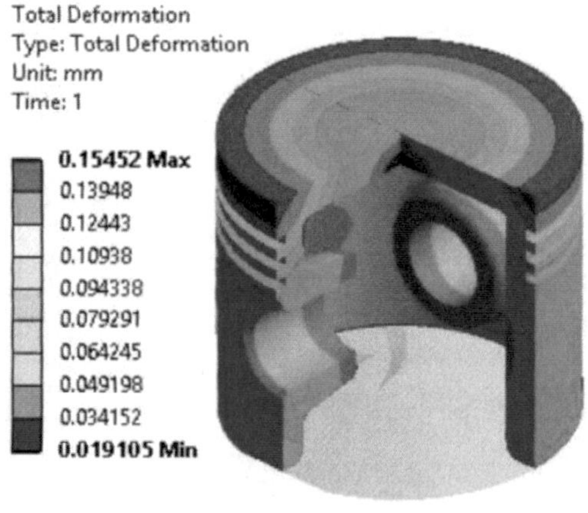

Figure 24. Total deformation due to mechanical - thermal loads - in 3D model

IOP Conf. Series: Materials Science and Engineering **163** (2017) 012043 doi:10.1088/1757-899X/163/1/012043

4.3. Finite element analysis of the piston subjected to combined loads: mechanical and thermal

For this analysis, in addition to the thermal analysis conditions, it is considered that a pressure equal to the maximum gas pressure (6.5 MPa) acts on the top surface of the piston head acts.

The results obtained emphasize the influence of both types of loads on the stresses and deformations which appear in the considered piston body. The radial stresses are shown in Figures 19 and 20, tangential stresses are shown in Figures 21 and 22, and Figures 23 and 24 display the total deformations of the analyzed piston.

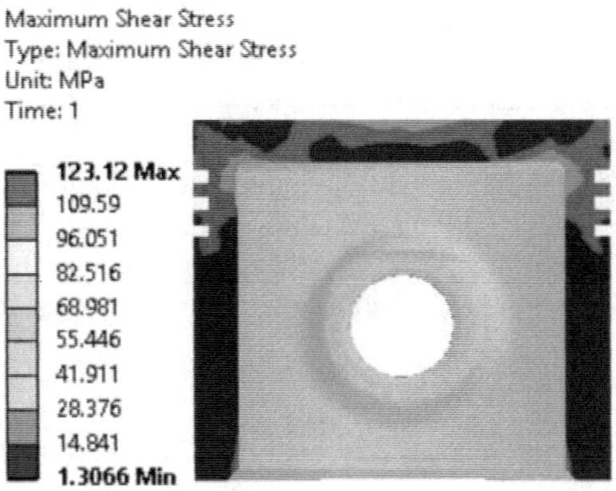

Figure 25. Maximum Shear Stress - in section

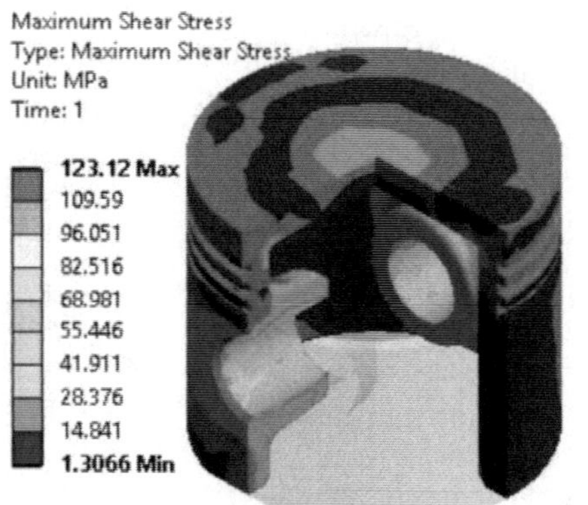

Figure 26. Maximum Shear Stress - in 3D model

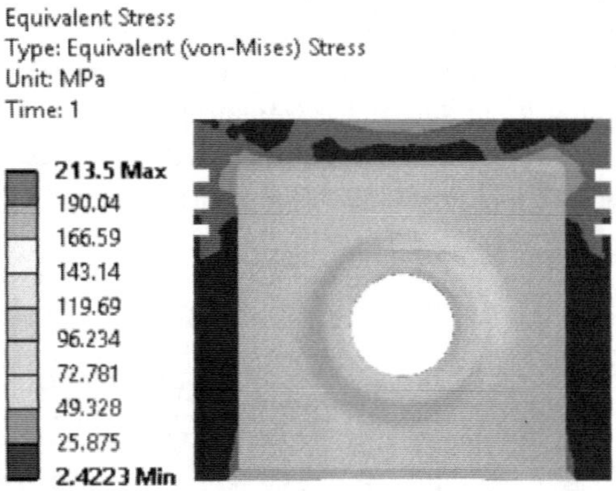

Figure 27. Von Mises Stress - in section

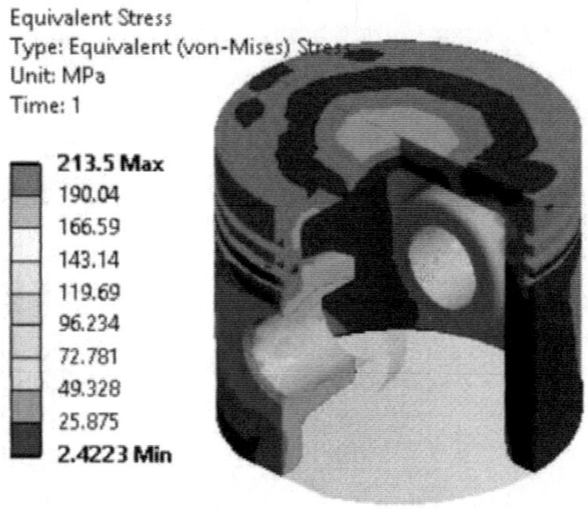

Figure 28. Von Mises Stress - in 3D model

Equivalent stresses are also determined (Figures 25, 26), calculated using the thrid theory of resistance (Maximum Shear Stress theory), recommended theory in the case of a plane state of stress, respectively using the 4th theory (von Mises), customized for the plane state of stress (Figures 27, 28).

5. Analysis of the results

Analyzing Figures 4…7, we notice that radial and tangential stresses caused by mechanical loads in the interest areas (at the edge of the piston head and in the center of the piston head, on the top and bottom surfaces), obtained through finite element analysis, have similar values as the ones obtained through analytical calculation method. Simultaneously, we can note that the obtained stresses are similar in terms of signs.

Figures 8 and 9 show that the maximum total deformation of the piston head caused by mechanical load is 0,09 mm and it is positioned in the center of the piston head.

IOP Conf. Series: Materials Science and Engineering **163** (2017) 012043 doi:10.1088/1757-899X/163/1/012043

The field of temperature in the piston is shown in Figure 11. It is noticed that the maximum value of the temperature is in the center of the piston head and it decreases towards the edge of the piston head, the difference between the center and the edge of the piston head being of 45...50^{0}C, and the difference between the top surface of the piston and the bottom surface being around 8...10^{0}C, which confirms both the analytical calculation and data from the literature [7], [8].

In the Figures 13...16, it is noticed that both radial and tangential stresses have, on the top and bottom surfaces, values and signs corresponding to the results obtained through the analytical calculation. The total deformations caused by thermal loads (Figures 17 and 18) are maximum on the top surface of the piston head (0.164 mm), on the edge, and have the opposite deformations sense due to mechanical load.

Regarding the mechanical-thermal load, the radial and tangential stresses (Figures 19 – 22) correspond in terms of values and signs with those determined analytically.

In the case of this complex load, the equivalent stresses are of interest, because in the design stage these stresses are compared to the allowable stress of the material the piston is made of.

It is noted that the values of the equivalent stresses calculated with the theory of the maximum shear stresses (Figures 25 and 26), in the piston head, do not exceed 55 MPa, a value much lower than the allowable stress of the material (80...120 MPa).

For the same purpose it is important to determine the von Mises equivalent stresses (Figures 27 and 28). Also, it is noticed that in the piston head, the von Mises equivalent stresses do not exceed the allowable stress of the material.

Analyzing Figures 25...26 and table no. 3, we can see that the values of the equivalent stresses calculated using the finite element method are very similar to the ones calculated using the analytical method. Also, it can be stated that both types of equivalent stresses can be used for the designing of the piston, their values being sensibly equal, and the distribution is identical.

The maximum total deformation of the piston due to mechanical-thermal load is lower than the maximum total deformation due to thermal load, it is found at the edge of the top surface of the piston head and has the same sense as the deformation due to thermal load (Figure 29). It can be concluded that the deformation due to mechanical load cancels in part the deformation due to thermal load.

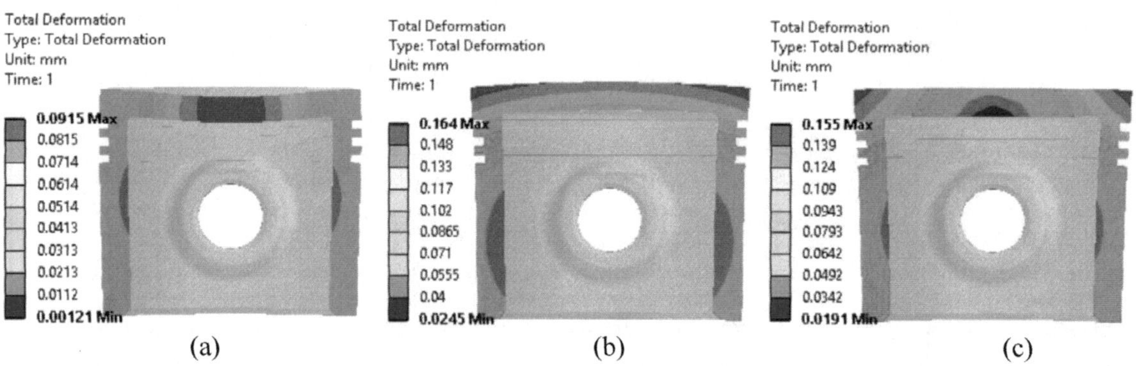

Figure 29. Total deformation: (a) – due to mechanical loads, (b) – due to thermal loads, (c) – due to mechanical - thermal loads

6. Conclusions

Due to the constructive complexity and complex mechanical and thermal loads the piston is subjected to, the accurate determination of the unitary stresses and deformations is required. The classical methods of calculation of the resistance of the piston (determination by calculation of the unitary stresses) are based on simplified representations of the geometric shape and loads.

Using the finite element analysis for the calculation of the unitary stresses and deformations, enables the detailed analysis of every area of the analyzed structure, under the conditions of obtaining an adequate precision of the results, results which can be successfully used in the stage of the optimization of the piston's shape.

By using CAD and CAE software in designing parts/assemblies of parts, the designer has the opportunity to analyze various constructive variants in a short time, thereby optimizing the designing work.

Regarding the results obtained using the two designing methods, it is noted that the obtained values are similar in the case of stresses, which indicates the fact that the three dimensional model, the loads and the constraints have been properly defined, and the finite element analysis is recommended for use.

By using the finite element method, stresses and deformations are obtained relatively quickly and important and complex information can be achieved in terms of the behavior of the piston during operation.

References

[1] Srinadh M and Rajasekhara Babu K 2015 Static and Thermal Analysis of Piston and Piston Rings, *International Journal of Engineering Technology, Management and Applied Sciences* **3**(8) 51-58

[2] Bhagat A R and Jibhakate Y M 2012 Thermal Analysis and Optimization of IC Engine Piston using finite element method, *International Journal of Modern Engineering Research (IJMER)* **2**(4) 2919-2921

[3] Wang Y X, Liu Y Q and Shi H Y 2010 Finite Element Static and Dynamic Analysis for a Piston, *Advanced Materials Research* **97-101** 3323-3326

[4] Singh P and Pramanik D 2015 Structural and thermal analysis of a C. I. engine piston of different materials using FEM technique, *MR International Journal of Engineering and Technology* **7**(1) 41-48

[5] Sharma S K, Saini P K and Samria N K 2013 Modelling and analysis of radial thermal stresses and temperature field in diesel engine valves with and without air cavity, *International Journal of Engineering, Science and Technology* **5**(3) 93-105

[6] Bhattacharya S, Basu A, Chowdhury S and Upadhyaya Y S 2014 Analysis of piston of two stroke engine, *International Journal of Research in Engineering and Technology* **3**(6) 642-648

[7] Raţiu S 2010 *Motoare cu ardere internă pentru autovehicule rutiere. Mecanismul motor. Construcţie şi calcul*, Mirton Publishing House, Timişoara, Romania

[8] Gaiginschi R and Zatreanu G 1995 *Motoare cu ardere internă, constructie şi calcul*, "Gh. Asachi" Publishing House, Iaşi, Romania

International Conference on Applied Sciences (ICAS2016) IOP Publishing
IOP Conf. Series: Materials Science and Engineering **163** (2017) 012044 doi:10.1088/1757-899X/163/1/012044

Transdisciplinary integration and interfacing software in mechatronic system for carbon sequestration and harvesting energy in the agricultural soils for rewarding farmers through green certificates

P P Pop[1], A Pop-Vadean[1], C Barz[2] and T Latinovic[3]

[1]Technical University of Cluj Napoca, Department of Mechatronics and Machine Dynamics, Muncii str., no. 103-105, 400641 Cluj Napoca, Romania
[2]Technical University of Cluj Napoca, Center University of North from Baia-Mare, Department of Electrical Engineering Electronics and Computers, dr. Victor Babes str., no.62A, 430083 Baia Mare, Romania
[3]University of Banja Luka, Faculty of Mechanical Engineering, Republic of Srpska, Bosnia and Herzegovina

E-mail: poppaulpetrica@yahoo.com

Abstract. In this article we will present a transdisciplinary approach to carbon sequestration in agricultural soils. The software provides a method proposed to measure the amount of carbon that can be captured from different soil types and different crop. The application has integrated an intuitive interface , is portable and calculate the number of green certificates as a reward for farmers financial support for environmental protection. We plan to initiate a scientific approach to environmental protection through financial incentives for agriculture fits in EU rules by taxing big polluters and rewarding those who maintain a suitable environment for the development of ecological and competitive agriculture.

1. Introduction

Chemical element that has the greatest importance in the world is C (carbon). Is found in all organic substances.He play a very important role in developing and sustaining life on Earth and by adjusting the amount of the atmosphere may improve environmental conditions. Through photosynthesis, plants can transform CO_2 and H_2O in carbohydrates and O_2 (oxygen). This process is bio-energy and nuclear plant uses solar energy as the main fuel. That is, plants produce carbon-containing organic substances from water and inorganic carbon in the presence of light. Our idea is to integrate the technological process in agriculture as a source of carbon sequestration in plants and soil and to financially reward farmers by measuring the amount of carbon sequestered annually. Such amount may be converted to measured and these green certificates can then be redeemed. CO_2 in air, water and soil is an mineral nutrient needs of plants. An important consequence of this feature is to release O_2 molecule.

Figure 1. Photosynthesis

Formula photosynthesis: $6CO_2 + 6H_2O + light \rightarrow C_6H_{12}O_6 + 6O_2$

Photosynthesis takes place at night and the plants just breathe. During 24 hours resulting in breathing CO_2 production is less than that of O_2 resulted from photosynthesis during the day. Plants produce O_2, yet are large sources of CO_2 (as converted). During photosynthesis, plant materials possess a number of mechanisms typical to decarbonate dioxide fixation (Figure 1). In plants, such as photosynthesis is determined by the number of carbon atoms in an organic molecule formed during the first CO_2 fixation. Humification is the degradation of organic matter of roots, crop residues, leaves falling and the results produce a certain amount of C and humus, which gives the degree of soil fertility. Humus is formed as a result of biochemical processes of transformation of dead organic substances. In fact, the process is completed by the action of soil bacteria. Soil organic carbon is the largest reservoir in interaction with the atmosphere and is estimated to between 1500 and 2000 Pg C to a meter deep (2450 about two meters deep). Inorganic carbon has a somewhat lower value Pg 750, but is captured in more stable forms, such as carbonates. Vegetation (Pg 650) and atmosphere (750 Pg) sequestres considerable amounts lower than the ground. The flow of terrestrial carbon or organic carbon in soil and atmosphere is important and can be positive (sequestration) or negative (CO_2 emission). We can say that the plants and the soil is humus huge reserves of carbon dioxide and from this point of view plays an important role in the heating processes of the soil but it has great importance for soil fertility preservation at a satisfactory level.

Figure 2. Roots and MOS for carbon storage

2. Storage of CO_2 [3]

Terrestrial ecosystems and oceans take much of the CO_2 (pools) but about 3.4 Gt C accumulates in the atmosphere each year. CO_2 is dissolved in water and thus reaches rivers, lakes, seas and oceans, which is in this way huge storage tanks of CO_2. The majority (2/3) of terrestrial carbon is found in soil and roots. 1500 Gt C estimated organic world, some more protected against decomposition. Reserve C below the ground in forests, permanent grassland and other permanent ecosystem (mountain areas) remains almost intact ecosystem as long as the changes do not suffer (Figure 2). There is no guarantee that aquatic and terrestrial mainframe systems mentioned above remain active in absorbing C when limits are reached. There are big differences between arable soils and grassland soils, in terms of their storage capacity for carbon.

The first 10 cm of the basement levels are important to C. When performing work on grassland soils in arable soils for processing in the first few years after conversion lost almost half of reserve C. This phenomenon is well highlighted in the experiment conducted at Rothamsted-Johnston Highfield [1]. The transformation of arable land into grassland, requires several years (more than 50) to return to the same level the percentage of C of a permanent grassland.Garwood et al [2] demosntrat clearly have the first 10 cm in the basement of a grassland containing a percentage of C nearly double compared with arable land.

3. Calculation deposits C

Samples must be taken from different depths. For a soil arable, they must be at 20, 40, 60 and 100 cm and soils grass: 10, 20, 50 and 100 cm (because most concentration of C was found in the top 15 centimeters basement) [4], [5], [6], [7].

C in organic substance (MOC) of soil is distributed by an exponential curve:

$$C_z = C_b + (C_0 - C_b).e^{-k.z} \tag{1}$$

C_z = density C (g MOC.cm^{-3} ground) at depth z.
C_0= density C to a depth 0
C_b = density C at a depth b
k: constant.

The density to be calculated: % C (g MOC.g^{-1}sol) x density of soil ρ_d.

From this expression (1) to calculate the constant k, which expresses the C exponential decrease with depth. A big factor C corresponds to a decrease rapidly during the first 10-20 cm soil depth. The study dealt factor k Mestdagh I. [8] in his doctoral thesis, and the influence of agricultural practices on soil density and storage MOC. From this study it can be concluded that the Belgian soils that are heavy texture (clay) contain 50% more MOC and 25% more compared to sandy soils and limestone.

By integrating expression (1) between 0 and 100 cm^{-1} C ha get values up to 1 m depth (Figure 3):

deposit

$$\int_0^{100}(C_b + (C_0 - C_b).e^{-k.z})dz = 100C_b - k^{-1}(C_0 - C_b)e^{-k.100} - (-k^{-1}(C_0 - C_b)e^{-k.0})$$
$$= 100C_b + k^{-1}(C_0 - C_b)(1 - e^{-k.100})$$

deposit

$$C_{0-100cm} = (1 - e^{-k.100})(C_0 - C_b).k^{-1} + 100C_b$$

In Article 3.4 of the Kyoto Protocol additional anthropogenic activities (human induced) give opportunity to be calculated as CO_2 storage pools:

ΔGES (2008-2012) -5 * ΔGES (1990) (2)

(2008-2012 = periodic 5 ans)

(GES: greenhouse gas)

For arable and grassland soils is accepted additional human activities (agriculture) such as re-cultivation of vegetables, reseeding grassland management systems etc. It is estimated that deposits C meadows between 3% and 26% of global deposits. C rainforest reserve is 19%.

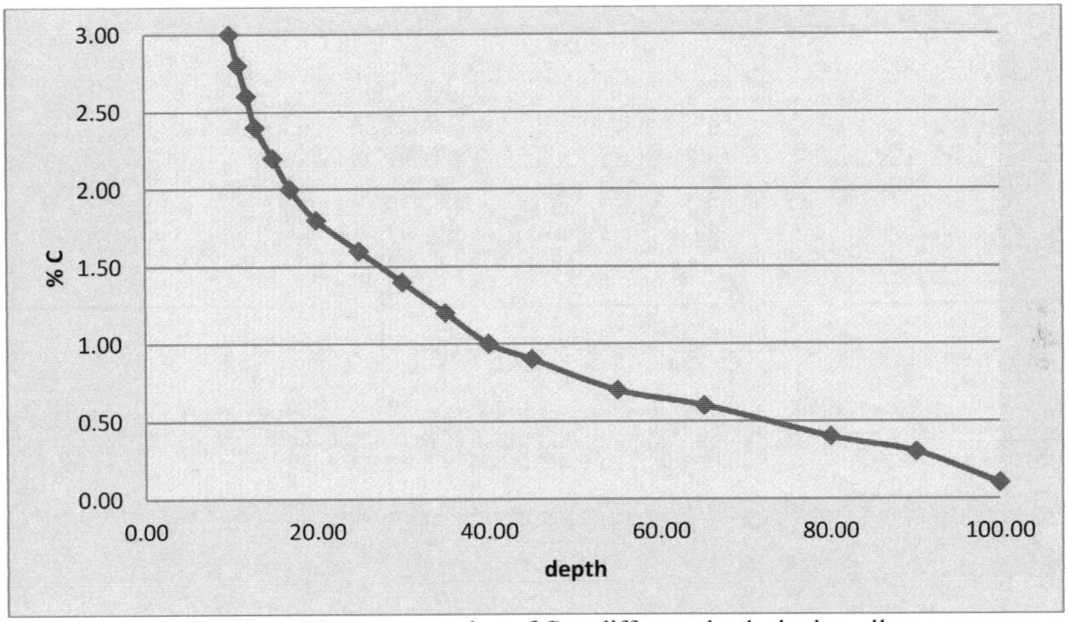

Figure 3. The concentration of C at different depths in the soil

4. Experimental measurements of carbon deposits

Soil samples were taken from Veţel town located 30 km from Deva, Hunedoara County, Romania. They covered all soil types and crops where samples were taken. Laboratory analysis show following data [6] (Table 1, Table 2).

Table 1. Laboratory analysis for soil MOC (town Veţel)

Sample No.	Sample	Deph	MOC
1	A0	10	2.15
	C1	11	2.03
	C2	12	1.85
2	A0	15	1.75
	C	16	1.65
3	A0	18	1.53
	C	19	1.46
4	A0	20	1.42
	C	21	1.42

5	A0	22	1.39
	C	23	1.39
6	A0	24	1.37
	C	28	1.36
7	A0	29	1.35
	C	30	1.35
8	A0	32	1.32
	BV	33	1.31
9	A0	46	1.29
	BV	51	1.29
	C	53	1.28
10	A0	55	1.22
	BVW	62	1.22
	CW	65	1.17
11	A0	69	1.17
	BTW	70	1.15
12	A0	71	1.13
	BTW	86	1.07
	C	90	1.05
	C1	92	1.02

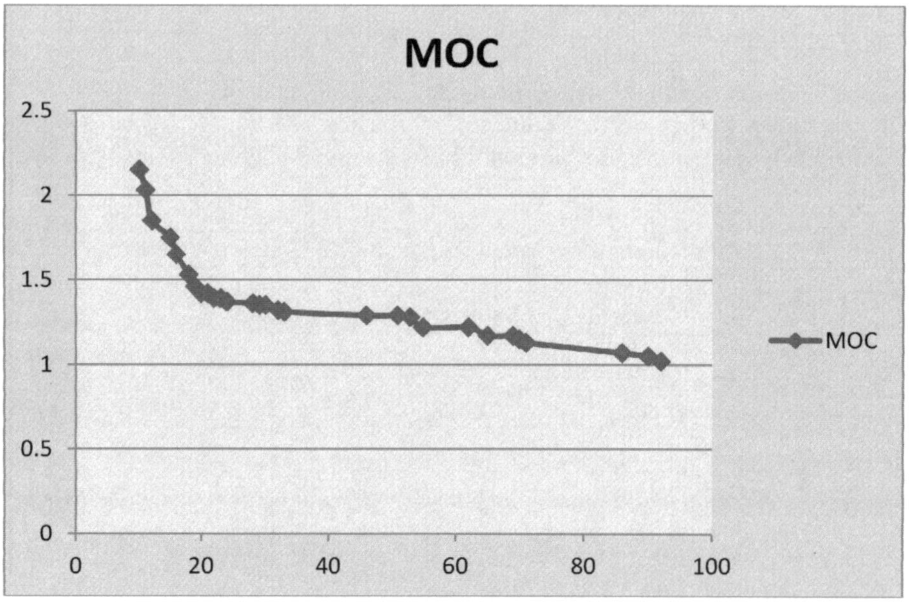

Figure 4. Values MOS at various depths in the soils of the area Veţel village (Table 1)

Table 2. Laboratory analysis for soil carbon deposits (town Veţel)

Sample No.	Sample	Deph	Carbon%
1	A0	12	6.40
	C1	14	6.10
	C2	16	5.48
2	A0	18	5.22
	C	22	4.85
3	A0	24	4.70
	C	26	4.24
4	A0	28	3.80
	C	30	3.45
5	A0	34	3.20
	C	42	2.90
6	A0	52	2.60
	C	62	2.30
7	A0	64	2.20
	C	68	2.05
8	A0	70	2.00
	BV	72	1.90
9	A0	74	1.70
	BV	76	1.50
	C	78	1.40
10	A0	80	1.25
	BVW	84	1.10
	CW	92	0.80
11	A0	94	0.70
	BTW	97	0.60
12	HAVE	104	0.50
	BTW	105	0.40
	C	115	0.20
	C1	122	0.10

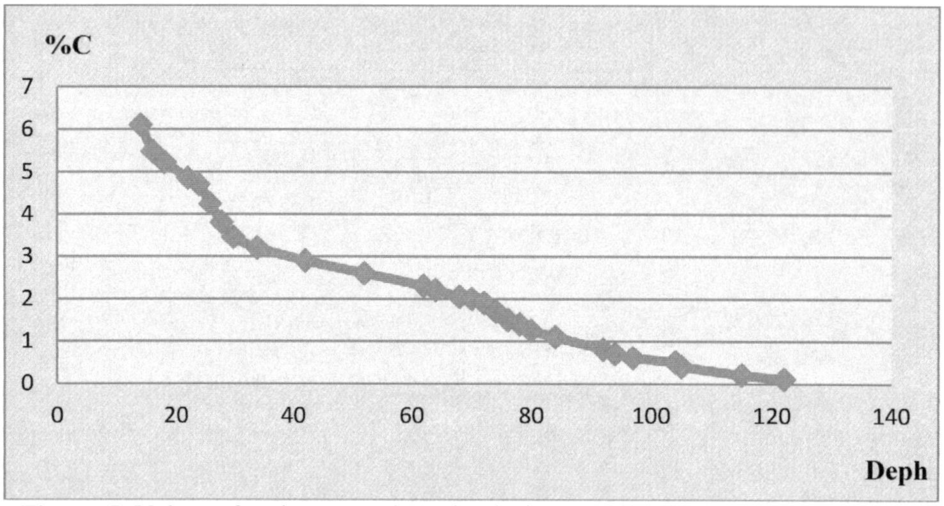

Figure 5. Values of carbon at various depths in the soils of the area Veţel village (Table 2)

We note that carbon values recorded in soils that were sampled are close to those of a general nature were recorded in the European Union. Laboratory tests and their modeling shows that Romanian soils are part of European stimulate the agriculture desire to become a factor of stability in world climate. These results encourages us improve the software that transform farmers efforts to sequester carbon in cash. The development money will boost agriculture responsible and respectful with the environment (Table 1, Figure 4, Table 2, Figure 5).

5. Calculating carbon flux in soil

Calculating the carbon content in different situations once the program follows a formula CESAR (Carbon Emission and Sequestration by Agricultural Land Use) [7], [8].

Flow of year t-1 to t in soils in arable and grassland is:

$$F_C = c*h_c*(Y_{t-1}/H - Y_{t-1}) - r_a*C_{t-1}$$

Meadows for grazing (recycling C through animal droppings)

$$FC = c * [hc * (Y_{t-1} / H - Y_{t-1}) + h_f * f * Y_{t-1}] - r_a * C_{t-1}$$

Of organic material in soil: MOS (humus) of organic material > 1 year
FC = flow to the MOS carbon (humus) $t_C.ha^{-1}\ a^{-1}$ c = fraction C in MOS $t_C\ t_{m.s}{}^{-1}$
h_c = coefficient of humification crop residues (dimensionless) (Table 3).
Y_{t-1} = biomass harvested in t a^{-1} tm.s ha^{-1} (Table 5).
H biomass harvested as part of annual net production of culture (dimensionless) (Table 4).
r_a = degree of decomposition annual MOS a^{-1}
h_f = coefficient of humification (dimensionless)
f = dry matter of feces as part of the dry matter harvested.

Parameters can be found in the literature:
$c = 0.58\ t_C\ t_{m.s}{}^{-1}$
$r = 0.029\ a^{-1}$
hf = 0.44
f = 0.25

Table 3. Value for hc parameter

Nr.Crt	Crop	Value
1	herbaceous crops	0.33
2	Grain	0.31
3	Potatoes	0.22
4	Sugar beet	0.21
5	hemp, sunflower, etc.	0.33

Table 4. Value for H parameter

Nr.Crt	Crop	Value
1	herbaceous crops	0.45
2	cereals harvested	0.67
3	free cereal straw	0.46
4	potatoes and sugar beet	0.69
5	for hemp	0.42
6	flower sun	0.92

Table 5. Value for Y_{t-1} parameter

Nr.Crt	Crop	Value	Country
1	herbaceous plants	11 a^{-1} tm.s ha^{-1}	Belgium, Netherland, Germany and Romania
2	grain + straw	9.7 a^{-1} tm.s ha^{-1}	France
		12 a^{-1} tm.s ha^{-1}	Belgium, Netherland
		10 a^{-1} tm.s ha^{-1}	Germany, France
		9 a^{-1} tm.s ha^{-1}	Romania

Table 6. Net carbon flux results, calculated with specific formulas

Nr.Crt	Crop	Value
1	Grassland	0.52 t C $ha^{-1}y^{-1}$
2	arable land	- 0.84 t C $ha^{-1}y^{-1}$
3	conversion of arable land to grassland	1.44 t C $ha^{-1}y^{-1}$
4	application of sludge (10 tons ha-1)	1.50 t C $ha^{-1}y^{-1}$
5	works fewer	0.25 t C $ha^{-1}y^{-1}$
6	straws leaving the soil after harvest crops	0.15 t C $ha^{-1}y^{-1}$
7	average temperature ± 1 °	- 0.05 t C $ha^{-1}y^{-1}$
8	CO2 + 0.2%	0.01 C $ha^{-1}y^{-1}$

6. Software to monitor soil carbon sequestration

The software has a connection interface, allowing the user or administrator with specific credentials and each having its operating rights program (Figure 6).

The application has a database that allows to add all information on existing carbon deposits in the ground and to monitor the flow of carbon annually (Figure 7). They are placed on the data where they can be identify. Then you can upload data on each field he owns farmer and type of crop (cereals, potatoes, fodder plants, etc.) on the field that has a certain type of soil (loam, sandy, loamy, etc.) (Table 6).

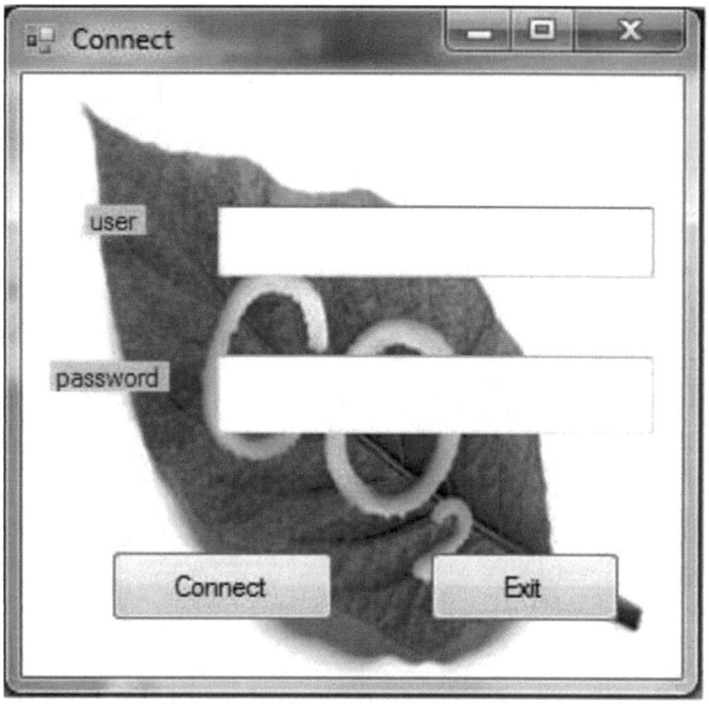

Figure 6. Module connection

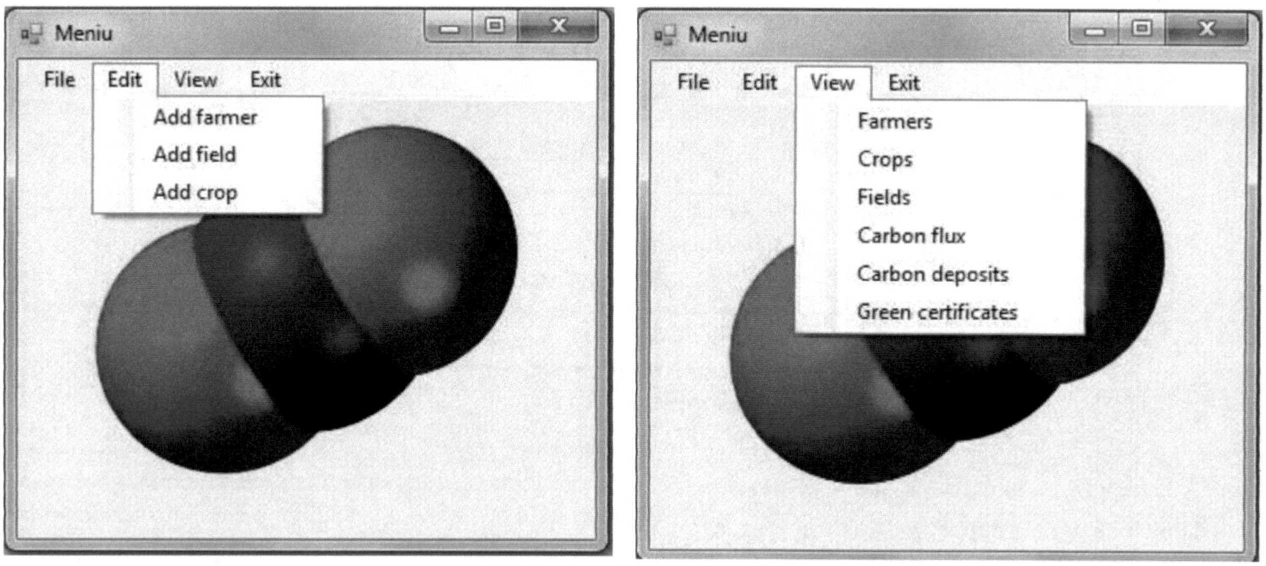

Figure 7. Application Menu

Data for the storage of carbon deposit on existing information comes from laboratory tests on the samples taken from the ground [6] (Figure 8).

IOP Conf. Series: Materials Science and Engineering **163** (2017) 012044 doi:10.1088/1757-899X/163/1/012044

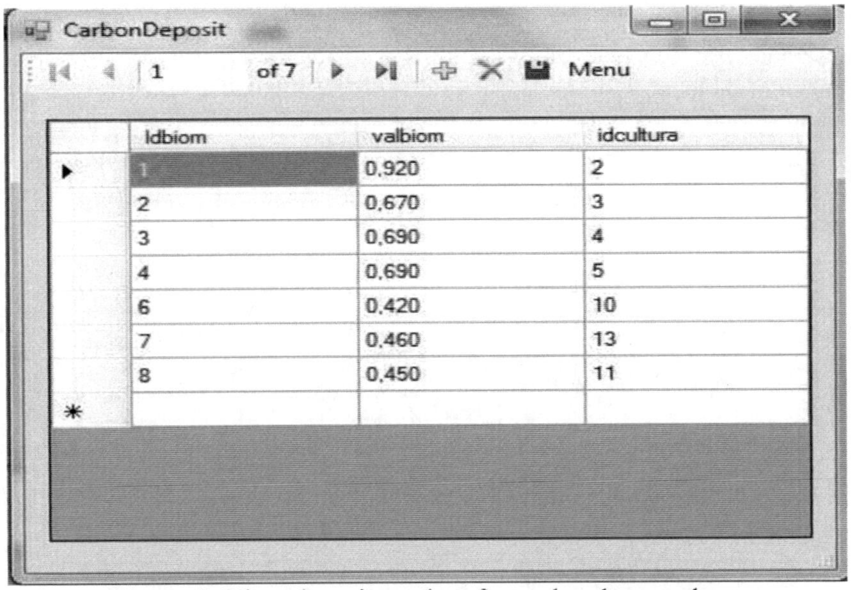

	Idbiom	valbiom	idcultura
▸	1	0,920	2
	2	0,670	3
	3	0,690	4
	4	0,690	5
	6	0,420	10
	7	0,460	13
	8	0,450	11
*			

Figure 8. View deposite carbon for each culture, values for ha

Regarding the annual flow of carbon from the soil data are obtained as follows: after applying the above formula, based on field verification of declarations provided by the farmer and the supporting documents (Figure 9). Information provided by the farmer shall be supported by certificates of origin for seed culture existing in the plot but also for agricultural work performed or any other intervention in field.

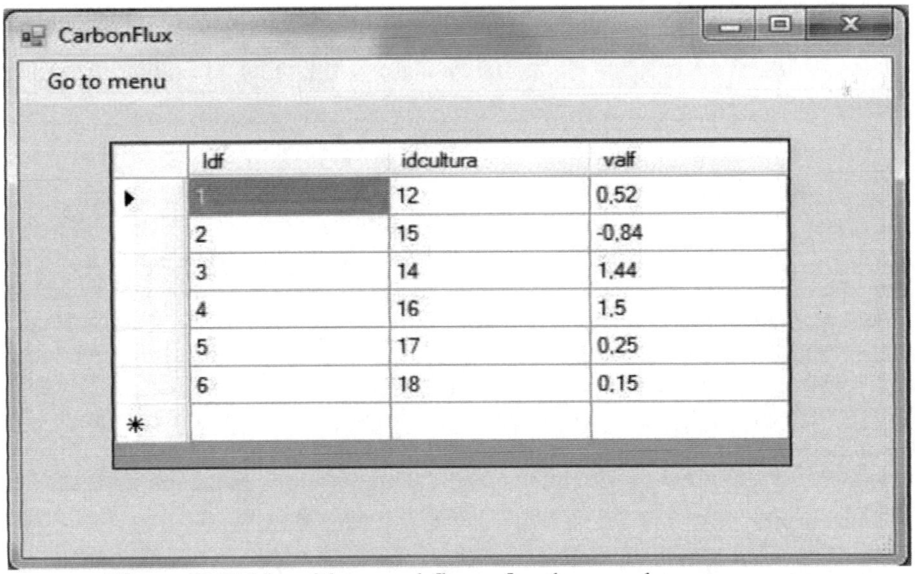

	Idf	idcultura	valf
▸	1	12	0,52
	2	15	-0,84
	3	14	1,44
	4	16	1,5
	5	17	0,25
	6	18	0,15
*			

Figure 9. Table with annual flow of carbon per ha per crop per year

After obtaining these data the program calculates the number of green certificates (Figure10) incumbent on every farmer in the agricultural year in question. This calculation complies with environmental law in force and can be quickly adapted to any change in the law. By law the application can convert these green certificates in cash. Money can automatically turn farmers accounts. The software is designed so as to be consistent with the Agency for Payments and Intervention in Agriculture at the European level.

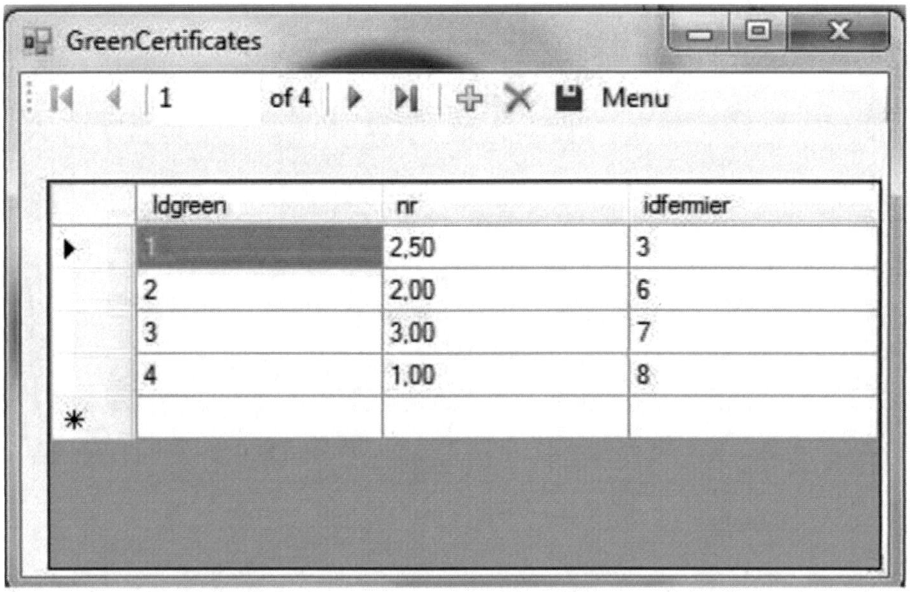

Figure 10. Database number of allowances issued each farmer

He can share data between these databases to optimize human and material costs and resources involved in this. The technology used is Microsoft Visual Studio programming environment with programming language C# [9], [10], [11] and Microsoft SQL Server database [12]. The program can run on any PC or smart device equipped with Windows operating system (Windows XP, Vista, 7, 8, 10). We are considering extending it on Apple and Android platforms.

7. Conclusion

We approached transdisciplinary global environmental problem that can be solved through sustainable development agriculture. This agriculture which uses innovative ideas can become sustainable in the long term in terms of environmental protection. Agriculture and informatics especially mechatronics can provide sustainable solutions through knowledge and can transform a field emitter and environmentally reckless in one profitable and friend of nature. Biological nutrition can be so healthy. By integrating new technologies we can protect people from the economic or social crises and even self-destruction.

We will refine these ideas by integrating new technologies in agriculture on a large scale. The next stage involves extending the software used on most computer platforms, intuitive menu improvement, integration of domain specific automation equipment and not least harvesting energy potential. We want to achieve and how to plan for the long-term carbon sequestration to support farmers. We will thus achieve an even better their outreach through planning and coordinating. We hope to offer through our approach to optimize predictability and rational use of natural resources as a source of food and as a carbon sequestration rezervoir.

IOP Conf. Series: Materials Science and Engineering **163** (2017) 012044 doi:10.1088/1757-899X/163/1/012044

References

[1] Johnston A E 1973 *The effects of ley and arable cropping systems on the amounts of soil organic matter in Rothamsted and Woburn Ley arable experiments*, Rothamsted Experimental Station Annual Report for 1972

[2] Garwood E A, Clement C R and Williams S E 1977 Leys and soil organic matter, *J. Agric. Sci. Camb.* **98** 342-353

[3] Carlier L, Rotar I and Vidican R 2009 Fixarea dioxidului de carbon: favorabilă pentru producţia culturilor agricole şi pentru mediu, *ProEnvironment* **2** 73-83

[4] Carlier L, Rotar I, Vlahova M, Vidican R and Atanassov A 2007 Good agricultural and environmental condition of agricultural land: cross compliance of the new EU policy, *Bulletin USAMV-CN* **63** 13-23

[5] Rotar I and Vidican R 2005 The storage of carbon in systems of meadows, *Bulletin USAMV-CN* **61** 29-33

[6] Benea L 2016 *Capture and storage of carbon dioxide: a metod for countering climatic change,* submitted

[7] Vleeshouwers L M and Verhagen A 2002 Carbon emission and sequestration by Agricultural land use: a model study for Europe, *Global Change Biology* **8** 519-530

[8] Mestdagh I 2005 *Carbon sequestration in different Flemish grassland ecosystems,* University of Ghent, Belgium, Doctoral Thesis

[9] Donis M 2006 *Programming Microsoft Visual C# 2005: The Language,* Microsoft Press, Seatle, USA

[10] Jhon S 2010 *Microsoft Visual C# 2010. Step by Step,* Microsoft Press, Redmond, Washington, USA

[11] Paul K 2002 *Advanced C# Programming,* McGraw Hill, Osborne, Australia

[12] ***https://msdn.microsoft.com/enus/library/dd831853 (v=vs.100).aspx

IOP Conf. Series: Materials Science and Engineering **163** (2017) 012045 doi:10.1088/1757-899X/163/1/012045

Capture and storage of Carbon dioxid: a method for countering climatic changes

L M Benea

Politehnica University of Timisoara, Department of Engineering and Management, 5 Revolution Street, Hunedoara, 331128, Romania

E-mail: laura.benea@fih.upt.ro

Abstract. One of the options aimed at preventing climatic changes is the capture and storage of carbon dioxide, a method with a great potential for reducing greenhouse gases.
Capturing and storing carbon dioxide in the soil involves new benefits for the communities in the respective areas. Those benefits also follow from the fact that the organic compound has an essential factor in the soil, determining its properties.
The paper presents several results concerning the determination of the quantity of carbon dioxide in different types of soil and it is intended to be the beginning of the process of data collection and the analysis of the reserves and the flow of carbon.

1. Introduction

One of the options that aims to prevent climate change is the capture and storage of carbon dioxide, a method with high potential for reducing greenhouse gas emissions.

The main greenhouse gases are water vapor (H_2O), carbon dioxide (CO_2), methane (CH_4), nitrous oxide (N_2O) and ozone (O_3).

Water vapor is at the origin of 55% of the greenhouse effect. Additional carbon dioxide released by human activity is responsible for 55% of the increase in the greenhouse effect.

Terrestrial ecosystems and oceans absorb much of the CO_2 emissions. The majority (2/3) terrestrial carbon is found within the soil surface (soil and roots). The flow of terrestrial carbon or organic carbon in soil and atmosphere is important and can be positive (sequestration) or negative (CO_2 emission).

The plants, through photosynthesis, transform CO_2 and H_2O in carbohydrates and oxygen:

$$6\ CO_2 + 6\ H_2O + \text{light} \rightarrow C_6H_{12}O_6 + O_2$$

An important consequence of photosynthesis is the release of O_2 molecule.

Photosynthesis stops at night, but the plant breathes both day and night. The amount of CO_2 resulting in breathing is less important than that of O_2 resulting from photosynthesis during the day.

Soil humification process occurs (the process of degradation of organic matter, roots, crop residues, leaves falling) which affords a certain amount of carbon and humus. The quality of the humus obtained depends on the physico-chemical conditions of the soil in question (pH, oxygen content, the content of the base parts, or calcium cation).

There are big differences between arable soils and grassland soils, in terms of their storage capacity for carbon.

Arable crops are recognized by their specificity:

- erosion caused by water and wind
- annual crop rotation and the disturbance a live of soil
- soil compaction by heavy machines (spreading of fertilisers, seeds and harvests)
- Ground work, etc.

For the pasture there are:

- more organic substance (roots)
- period of growth and photosynthesis longer
- permanent coverage
- higher retention water
- presence of ruminants (more emissions of different greenhouse gases), etc.

This paper presents experimental results on the analysis of pedological soil Veţel village, Hunedoara county.

2. Experimental part

2.1. The geographical framework of the village of Veţel (Hunedoara County)

Veţel village is located in the central part of Hunedoara county, consisting of the following localities: Veţel - communal residence, Boia Bârzii, Bretelin, Căoi, Herepeia, Leşnic, Mintia, Muncelu Mic, Muncelu Mare, Runcu Mic. The surface of the village is 11389 ha. Veţel village is considered one of the most developed village in the county of Hunedoara [1].

Geographic position Veţel the village is limited by values 45° 54' 00" degrees north latitude and 22°48' 00"east longitude.

Shared access is on Route 7, which runs through the village. The main watercourse that runs through the territory of the village is the Mureş river. The administrative territory of the village Veţel id bordered to the North by the Mureş river and Ilia, Brănişca and Şoimuş, to the west is bordered by the communes of Ilia, Dobra, and to the south with the municipalities of Peştiş, Cerbăl and to the east of the village Cârjiţi and the town Deva.

The relief is varied and fall in Poiana Rusca massif consisting of crystalline schists with a complex geological structure and fragmentation pronounced. In terms of topography Veţel village is uneven, showing different levels of elevation 179.80 MN (Mures meadow in the locality Leşnic) to share the locality 780,00 MN Muncelu Mare. The vegetation is characterized predominantly hilly forests of beech, hornbeam and oak and grassland. Soils fall within the brown forest, sylvan podzolic and brown-yellowish.

Hydrology

The hydrographic network of the Mures river gathers all the waters flowing in the area. The valleys are mostly narrow and deep with active slope processes. At the bottom of the basin lands appear moderate and strong land erosion.

Geotechnical characteristics

From the survey shall be recorded in the area following stratification:

a) Shares between the land surface and considered to ± 0.00 - 0.50 topsoil
b) Among shares - 0.50 - 2.50 with alternate gray brown clay with calcareous concretions viscous plastic - land shrinkable
c) Under elbow - 2.50 appears the rock, appearing isolated layers of gravel and sand
d) Groundwater appears steady to share - 3.0 m
e) The minimum depth of the foundation is - 1.0 m
f) Allowable pressure of 200 kPa or 2.0 kgf / cm²

Land use

Veţel village covers an area of 11389 hectares, representing about 2% of the county of Hunedoara. In this area included:

a) 4085 ha - agricultural land accounting for 35.86% of the commune, of which: 1264 hectares of arable land 1520 ha pastures, hayfields 1298 ha and 3 ha of orchards

b) 6647 ha - forestry fund, respectively 58.36% of the area of the commune
c) 657 ha - non-agricultural lands occupied by courtyards, buildings, roads, unproductive lands, waters and reed.

2.2. Soil characterization

To characterize the soil of Veţel commune, they were determined following indicators:

- *pH* (Potentiometric method). It is determined in aqueous solution or brine, in various ratios of soil extract to water. The current method for determining the potentiometric pH is in aqueous solution with a ratio of soil: water of 1: 2.5. [3]
- *Determination of carbonate anion* from the aqueous extract, [3]
- *Hydrolytic acidity, Ah* (Kappen method). [3]
- *Determination of base cation exchange, SB* (Kappen method). [3]
- *The degree of saturation in bases VAH.* It is calculates by formula:

$$Vah\ [\%] = SB / (SB + Ah) \times 100 \tag{1}$$

where,
B - amount of exchangeable cations [me / 100 g soil]
Ah - hydrolytic acidity [me / 100 g soil]

- *Cation Exchange Capacity T.* It is given by formula:

$$T\ [me / 100\ g\ soil] = SB + Ah \tag{2}$$

where,
SB - The amount of exchangeable cations [me / 100 g soil]
Ah - hydrolytic acidity [me / 100 g soil]

- *Index nitrogen,* IN. It is given by:

$$IN = H \times VAH / 100 \tag{3}$$

where,
VAH - base saturation level
H% humus

- *Determination of soil organic carbon and humus content estimation* (Wet oxidation method and dosage titration Walkley-Black).[3]
- *Determination of phosphorus in cell extract sodium bicarbonate* (Olsen method). [4]

Samples must be taken from different depths. For arable soil, they must saddle be at 20, 40, 60 and 100 cm, and grass soils: 10, 20, 50 and 100 cm. The highest concentration of C is found in the first 15 centimeters of soil.

It is possible to calculate C from organic matter (MOC) of the soil on the basis of an exponential curve [2], [5]:

$$C_z = C_b + (C_0 - C_b)\cdot e^{-kz}\ (1) \tag{4}$$

where: Cz= Density C (g MOC·cm-3soil) at depth z
C_0 C = density at a depth 0
C_b C = density of the depth b
k = constant expressing exponential decrease of C with depth

The density should be calculated: % C (g MOC·g^{-1}soil) x density of soil □$_d$(G soil·cm^{-3}).

In the context of the fight against climate change and the Kyoto protocol, an important issue is represented by the creation of reserves of carbon important and quantifiable in agricultural soils from around the world.

According to data from the literature [5-8], [10], the calculation of the reserves of carbon stocks, by origin and the type of soil, to a depth of 1 meter is realized in the following way:

- General average: 30 - 120 t C ha^{-1};
- Arable crops in a year: 43 t C ha^{-1} (varying between 35 and 55 t C ha^{-1});
- Grapevine: 32 t C ha^{-1} (25-42 t C ha^{-1}) ;
- Soils abandoned: 45 t C ha^{-1} (32-58 t C ha^{-1}) ;
- Permanent grassland and forests: 70 t C ha^{-1} (54-88 t C ha^{-1});

IOP Conf. Series: Materials Science and Engineering **163** (2017) 012045 doi:10.1088/1757-899X/163/1/012045

- Meadows of altitude and swamps: 100 t C ha^{-1} (87 -115 t C ha^{-1}) (low temperature - lack of oxygen);
- Sandy soils: 30 t C ha^{-1};
- Clay soils: 120 t C ha^{-1};

3. Results and discussions

To characterize the soil Veţel, they took samples of more than 500 points, different profile and at different depths. Some of the results are shown in Table 1.

Agricultural development over time has led to consumption of land carbon stocks. As bulk of cultivated soils, there is a reduction in productivity. This can be changed by changing the type of agriculture practiced.

Guo and Gifford [7] conducted a study in 2002 calculated the effect of different substances used in soil on the concentration of carbon in the soil. From this study it follows that it loses much C through the conversion of grassland or forests into arable land, and on the other hand, it wins much C by the transformation of the surface soil in forests or meadows.

Arable soils are sources of CO_2.

The meadows are reserves of CO_2.

Kyoto Protocol and facilities for carbon sequestration in the soil are good opportunities for achieving this goal. Soils may retain carbon emissions, providing benefits to soil quality, crop and environment, in order to prevent erosion and to add value to biodiversity.

Farmland and pastures have the potential for soil carbon sequestration, provided that the crop be able to increase carbon accumulation and productivity.

Carbon sequestration involves the possibility of a double-win solutions [5], and new benefits for local communities. Attention should be drawn to the governmental authorities on the potential benefits and the need of initiating a process of data collection and analysis of carbon flux and reserves, carefully chosen from different sites at a given scale.

These benefits result by the fact that organic substance is a key factor in soil, which determines a multitude of properties or functions property related soil to buffer the effect of sustainability.

In Veţel, agricultural land represents 35.86% of the area (11.1% is arable and 24.7% pasture and hayfield). The humus content of the soil Veţel village is small and decreases with increasing soil depth. At depths greater than 50 cm is not found in soil humus.

Not all samples are ground carbonates. In samples containing carbonates can be seen that the content increases with increasing depth of sampling.

The phosphorus content in the soil could be found at shallow depths; 40-50 cm phosphorus is not found.

Nitrogen is present in small quantities and on small areas.

Cation exchange capacity is between 16.47 and 28.11.

4. Conclusion

The most important source of raw material is the soil organic plant material submitted annually, the material is only partially decomposed fallen recently.

The organic substance plays an essential role in soil, causing its properties have a buffer effect on sustainability. Basically, that contribute organic matter to soil fertility is humus (organic matter stable) is a form of carbon that recalcitrant to biodegradation, with stability in soil ranging between 50 and 2000 years.

Long-term evolution of the stock of humus in the soil is subject to two fundamental categories of processes: processes of humidifying, mineralization and erosion processes. They act conjugate at the top of the soil, resulting in a progressive, regressive or reserve humus complex.

Table 1. Indicators of soil samples from the village Veţel

Profile	Depth [Cm]	pH	H	P	SB	Ah	N	T	V	carbonates
4	8-36	5.50	1.46	29	11.48	9.39	1.56	20.87	55	-
	50-110	7.26	1.25	-	19.50	4.87	0.02	24.37	80	-
24	8-23	5.30	1.35	29	10.90	8.44	1.78	19.00	56	-
	24-48	5.66	1.19	22	11.48	9.39	1.20	20.87	55	-
	60-100	5.13	-	-	11.64	8.08	1.52	19.7	59	-
32	7-20	4.97	1.88	29	10.40	7.70	1.55	28:10	37	-
	40-110	5.46	1.59	-	10.90	6.43	1.83	16.47	60	-
38	7-34	6.10	1.45	26	14.80	6.96	-	21.76	68	-
	50-100	5.97	1.36	-	14.28	7.78	-	22.06	65	-
62	8-34	6.30	1.57	6	13.56	6.23	-	19.79	68	-
	50-100	6.08	1.29	-	13.96	6.56	-	20.52	68	-
116	8-32	5.60	1.58	10	14.28	7.78	-	22.06	65	-
	50-64	6.15	1.39	-	13.96	6.50	-	20.52	68	-
	64-84	6.22	-	-	18.64	8.37	-	27.01	69	-
149	7-26	8.46	1.36	60	-	-	-	-	90	2.40
	40-100	9.26	-	-	-	-	-	-	-	4.85
239	7-45	6.61	1.42	6	13.56	5.81	-	19.37	70	-
	60-90	6.26	-	-	19.00	8.15	-	27.15	69	-
	100-120	6.10	-	-	16.40	8.83	-	25.23	65	-
248	7-26	5.17	1.13	8	19.36	8.75	-	28.11	68	-
	37-50	5.23	1.08	16	13.08	10.00	-	23.08	56	-
283	8-20	5.70	1.83	7	19.36	8.75	-	28.11	68	-
	40-67	5.74	1.6	-	14.02	6.29	-	20.31	69	-
	67-80	5.60	-	-	16.44	10.52	-	26.96	61	-
336	7-40	8.43	1.48	24	-	-	-	-	83	1.25
	40-72	8.60	1.15	-	-	-	-	-	-	1.30
	82-130	8.75	1.28	-	-	-	-	-	-	1.40
350	7-24	8.20	1.15	34	25.2	3.76	-	28.96	87	2.47
	40-100	8.63	-	-	28.1	3.47	-	31.57	89	2.55
	110-120	8.20	-	-	26.8	3.65	-	30,45	88	2.40
396	7-34	9.65	1.42	54	-	-	-	-	-	4.87
	45-110	9.22	-	-	-	-	-	-	-	4.20
414	0-22	7.55	1.63	124	19.2	4.67	-	23.87	80	-
	22-38	7.40	1.22	120	17.56	4.40	-	21.96	79	-
	50-80	7.45	-	-	19.60	3.82	-	23:42	83	-
431	7-24	7.80	1.53	26	16.80	3.86	-	20,66	81	0.7
	34-105	7.68	1.17	-	19,20	4.67	-	23.87	80	0.55
	115-130	8.36	-	-	19.60	3.82	-	23:42	83	1.04
455	7-42	5.53	1.20	18	19.02	8.79	1.34	27.81	68	-
	52-110	5.70	-	-	12.20	10.63	1.46	22.83	54	-
503	7-38	8.32	1.46	34	-	-	-	-	85	0.10
	48-98	8.64	-	-	-	-	-	-	-	0.10
524	0-22	9.76	1.38	44	-	-	-	-	93	4.30
	22-44	9.64	1.25	56	-	-	-	-	-	4.52
	55-100	9.86	-	-	-	-	-	-	-	5.13

Degraded lands are lands that, due to the action of human activities (farming and forestry inappropriate, potentially toxic elements pollution) or natural (erosion under the influence of climatic factors) lost organic matter.

The accumulation of soil organic matter is beneficial both for crop production and for the fixation of carbon dioxide and reducing the impact of his action as a result of greenhouse gas. In this regard it has been developed a number of mechanisms by which governments, regional institutions and international organizations (such as the World Bank) provides financial support for measures to net carbon accumulation in the soil.

Carbon sequestration involves the possibility of a double-win solutions and new benefits for local communities. Therefore it is necessary to initiate a process of data collection and analysis of reserves and carbon flux at the national level.

Lack of knowledge and data on national practices regarding quantitative analysis related to the measurement and interpretation of field data, generate problems in terms of assessing the carbon flux.

Total deposit of carbon present in the grasslands ecosystem is far lower than in some forest ecosystems located in the basement but the higher meadows. Generally the carbon content of the soil of pastures is much higher compared to other crops.

Veţel village has 24.7% of the area occupied by pastures and hayfields and 58.36 forestry fund.
Cropland, grassland and forests in the village Veţel have potential for soil carbon sequestration, which generates reducing the amount of greenhouse gases and increase productivity.

References

[1] Erdoiu N, Pintilie D, Pintilie M and Orga V 2002 *Veţel-village complex monographic study*, Eurodidactica Publishing House, Cluj Napoca, Romania

[2] Pop P P, Pop-Vadean A, Barz C and Latinovic T 2016 *Transdiciplinary integration and interfacingmechatronic system software for Carbon Sequestration in the agricultural and energy harvesting soils for rewarding farmers through green certificates*, submitted

[3] ***2011*Chemical and microbiological analysis methods MARD*, National Institute of Research and Development for Soil Science and Agricultural Chemistry Environmental Protection, Sitech Publishing House, Craiova, Romania

[4] ***1964 *Research methods of soil*, RPR Academy Press, Bucuresti, Romania

[5] Carlier L et al. 2009 Fixing carbon dioxide: favorable for crop production Agriculture and Environment, *ProEnvironment* **2** 73-83

[6] Batjes N 1996 Total carbon and nitrogen in the soils of the world, *European Journal of SoilScience* **47** 151-163

[7] Guo L B and Gifford R M 2002 Soil carbon stocks and land use change: a meta analysis, *Global Change Biology* **8** 345-360

[8] Johnston A E 1973 *The effects of ley and arable cropping systems on the trace of soil Rothamsted and Woburn organic matter in arable experiments Ley*, Rothamsted Experimental Station Annual indexes for 1972, Part 2: 131-159

[9] ***2004 Règlement (EC) N ° 796/2004 of the Commission, *Journal officiel de l'Union européenne* **L 141** 18-58

[10] Schulze E D Valentini R and Sanz M J 2002 The long way from Kyoto to Marrakesh: Implications of the Kyoto Protocol Negotiations for global ecology, *Global Change Biology* **8** 505-518

Temperature controlling system using embedded equipment

R Rob, G O Tirian and C Panoiu

Politehnica University of Timisoara, Department of Electrical Engineering and Industrial Informatics, 5 Revolution Street, Hunedoara, 331128, Romania

E-mail: raluca.rob@fih.upt.ro

Abstract. Present paper describes the functionality of a temperature controlling system using PIC 18F45K22 microcontroller. The ambient temperature is acquired with LM35 analogue sensor. The microcontroller program is realized with MikroC compiler and it is able to control the speed of a cooling fan with dc motor. The speed can be increased in functioning with the increasing of the ambient temperature.

1. Introduction

Temperature measurement in industry today includes a wide variety of needs and applications. To meet this range of needs of the process control, the industry has developed a large number of sensors and devices in order to accomplish this demand.

Temperature measurement is one of the most common situations.

This paper presents the achievement of a data acquisition system for monitoring and controlling temperature inside a system developed around a core microcontroller type.

2. Theoretical considerations

As it is shown in Figure 1, the microcontroller reads the output code of the analogue to digital converter (ADC) where the temperature sensor is connected. The input signal is controlled by a voltage divider.

In order to control accurately the process temperature without external operator involvement, a temperature control system is based on a controller that accepts a temperature sensor such as a thermocouple or RTD at its input. Its purpose is to compare the current temperature with the set point and then it provides an output to a control element. The controller is part of the whole control system and the entire system should be considered in selecting the appropriate controller.

The following items should be considered:

-Type of sensor input (thermocouple, RTD) and temperature range
-Type of output required (electromechanical relay, SSR, analogue output)
-Algorithm which is necessary to control the application using the microcontroller
-The number and type of results (heat, cold, alarm limit) [1], [2].

3. Description of the temperature controlling system

The system is composed by PIC 18F45K22 microcontroller which is working from EasyPICv7 development board, a LM35 temperature sensor and a cooling fan with dc motor.

3.1. PIC 18F45K22 microcontroller

PIC18F45K22 microcontroller is a microcomputer integrated into a single chip available in 40-pin capsule with the following features:

-Full operation 5.5V (PIC18F2XK22 / 4XK22), frequency 1.8V low-voltage operation of 3.6V available (PIC18LF2XK22 / 4XK22), software auto-programmable power-on reset (POR), power-up timer (PWRT) and Oscillator Start-up timer (OST) Watchdog timer type (WDT) oscillator and on-chip software enables protection of code programmed in-circuit Serial Programming.

Analog features which must be taken into consideration are the followings: Analog to digital conversion mode (ADC) Resolution 10 bit, 17-channel analogue input (PIC18F / LF2XK22), 28-channel analogue input (PIC18F / LF4XK22) capacity self-acquisition, conversion is available during Sleep mode. Type sensing voltage high / low, the unit of measure load time (CTMU) for support *mTouch*: up to 28 channel button, sensor or the input cursor, the comparator analogue: two analog comparators, comparator pins for input. They are accessible and configurable from the outside. Peripheral Features:

Comes 24/35 I / O pins and only one input pin: The intensity of electric current 25 mA / 25 mA, the individual Interrupt pin, three external interrupt pins, up to seven timer modules, up to four timers 16-bit / counters with precursor. Up to three pairs 8-bit timers / counters, low-power oscillator Timer1 Module (CCP) one, two or four PWM outputs, selectable polarity, auto-shutdown and Auto-Restart [3], [4].

3.2. EasyPICv7 development board

EasyPIC V7 (Figure 2) is a complete development system for development and testing of applications for PIC microcontroller. This tool provides a complete development platform that supports high quality PIC devices (P10, P12, P16 and P18).

Featuring electronic integrated circuits such as USB2.0 programmer with mikroICD support, Touch Panel Controller, Keyboard, Port Expander, Serial COG display, support DS1820 and more. EasyPICV7 board offers many practical features for users to develop various embedded applications using the most popular component in electronics.

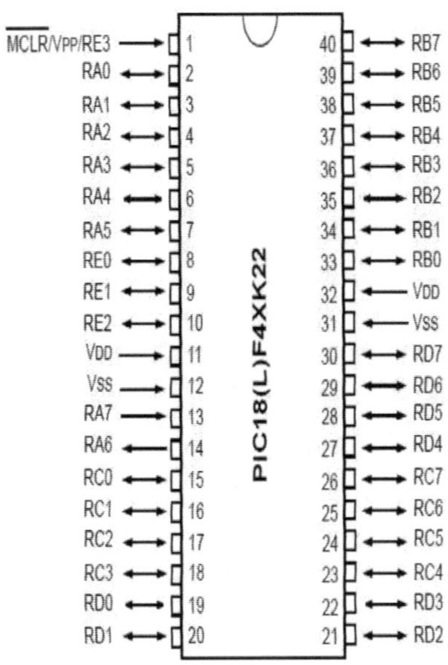

Figure 1. PIC18F45K22 microcontroller. Pin configuration

Figure 2. EasyPICv7 development board

3.3. LM35 temperature sensor

LM35 presented in Figure 3 has the following features:
-Calibration directly in degrees Celsius (°C).
-Guaranteed accuracy of 0.5°C to +25°C.
-Measurement range from -55° to 150°C.
-Suitable for remote applications.
-Supply voltage from 4V to 30V.
-Current consumption 60 saddle.
-Self-heating during operation 0.08 ° C (airborne capsule).
-Low output impedance of 0.1 Ohm 1 mA load.

The series circuits are LM35 precision temperature sensors, whose output voltage is directly proportional to Celsius (° C) temperature.

LM35 thus has an advantage over linear temperature sensors calibrated in degrees Kelvin (K), because the user is forced to use those to get a scaling constant in degrees Celsius convenient calibration requires no external. LM35 to provide accurate ± 0.25 ° C at 25 ° C and ± 0.75 °C over the range of temperature.

The output impedance low linear output and internal calibration make LM35 circuit easily interfaced with other circuits reading or checking It can be powered from sources Half-wave (+ V, 0V), or sources double-alternation (+ V, 0V ,-V). Very low power consumption of LM35 is 60 saddle triggers a self-heating of less than 0.1 ° C in the air.

3.4. DC motor controlling

The cooling fan will be controlled by the microcontroller using PWM technique.

Varying motor speed is accomplished by increasing or decreasing the period in which the control signal is 1 logical. Therefore, at a 100% duty cycle, the motor has a maximum speed. At a duty cycle of 50%, the dc motor speed is half of the nominal speed (Figure 4).

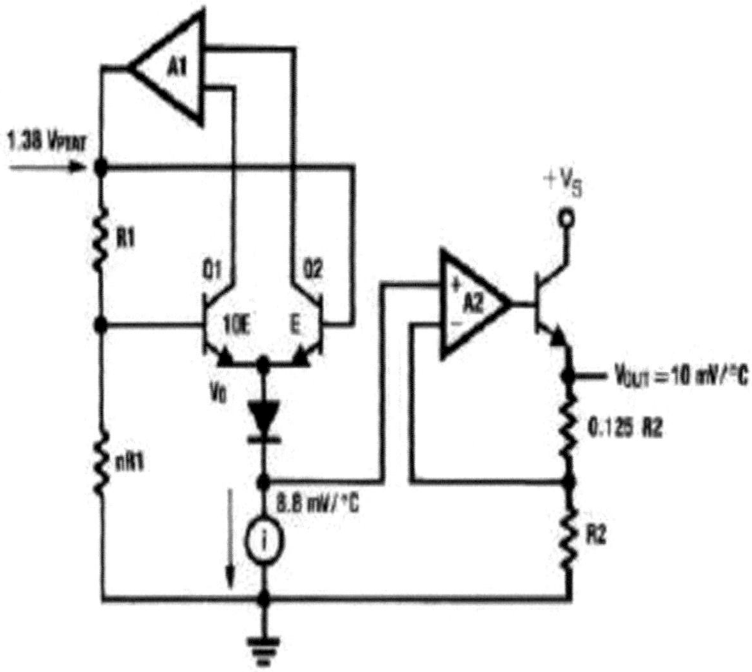

Figure 3. LM35 temperature sensor

$$f = \frac{t_{ON}}{t_{ON} + t_{OFF}} \cdot 100 \quad [\%]$$

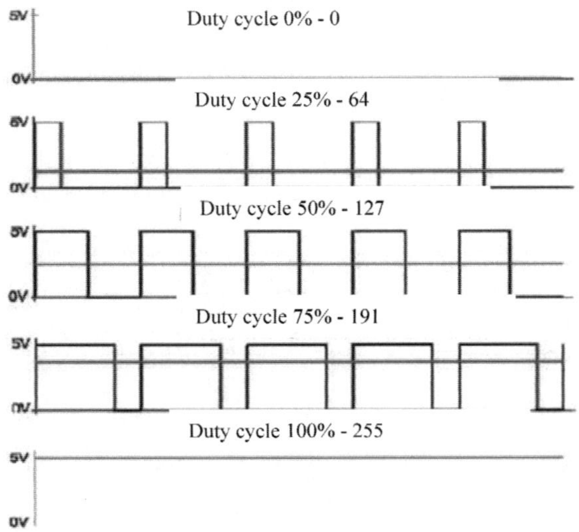

Figure 4. Motor speed

In order to supply the whole electric, a dc motor driver L293d is used (Figure 5).

L293D integrated circuit can control dc motors powered by 35V and 600 mA max. Each of the two bridges H has two input terminals (input) and a terminal activation (Enable). When EN terminal is connected to 5V, H bridge is active. If input IN1 is logical 1 (5V) and IN2 input is logic 0 (0V), the motor rotates; if the states of the two inputs are reversed, the motor will rotate in the opposite direction. When both inputs are logical 0, the dc motor stops and if both are logical 1, the motor shaft is braked.

The advantages for PWM technique are the following:

- Power dissipation per power electronic devices working in switching is much smaller than the power dissipated on regulatory elements series;

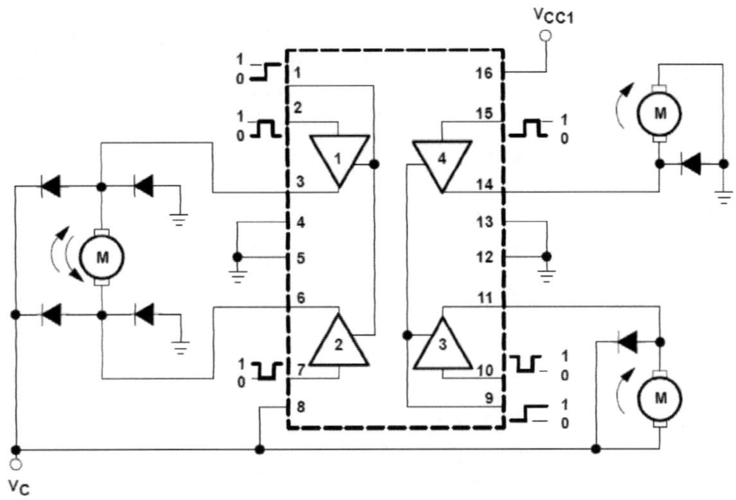

Figure 5. L293d motor driver

- Power consumption is even lower as the working electronic device power switching performance approaching an ideal electrical switch;
- As a consequence of the above, we have high energy efficiency, low volume, possibility of incorporating light into automation schemes, which is an important requirement of power electronic circuits;
- Through duty cycle control, PWM control circuitry can provide output voltages from a wide range;
- Stabilized output voltages are isolated from the input voltage.

The algorithm implemented for the temperature controlling is presented in Figure 6 [5], [6]. Figure 7 presents the simulation bloc scheme using Proteus which is a simulation environment. The Proteus scheme for this application is presented in Figure 8. In the followings, the source code of the application is presented.

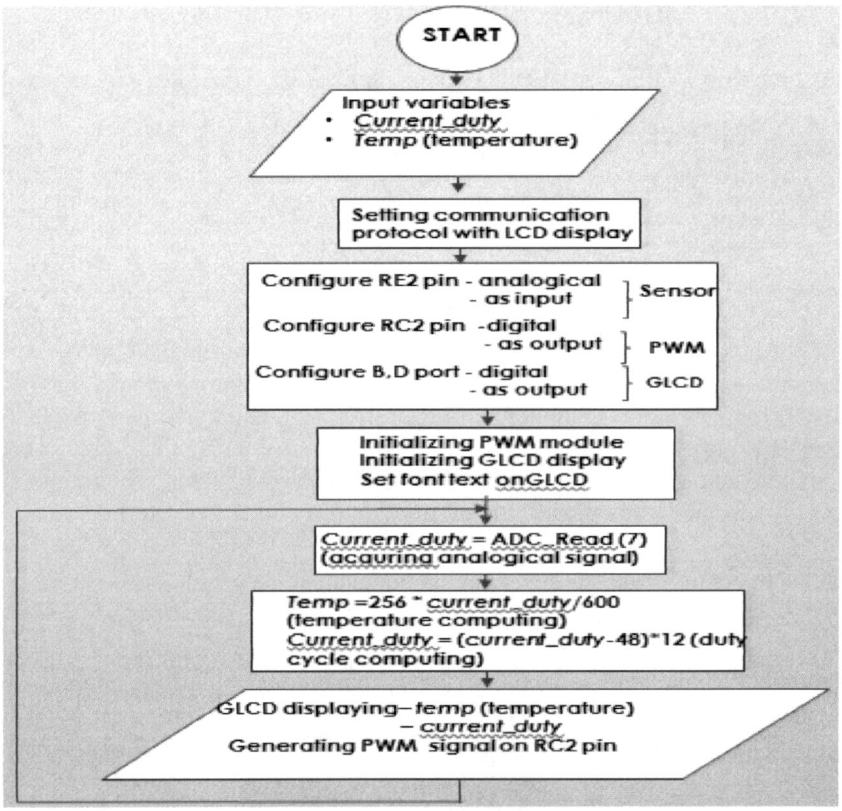

Figure 6. The application algorithm

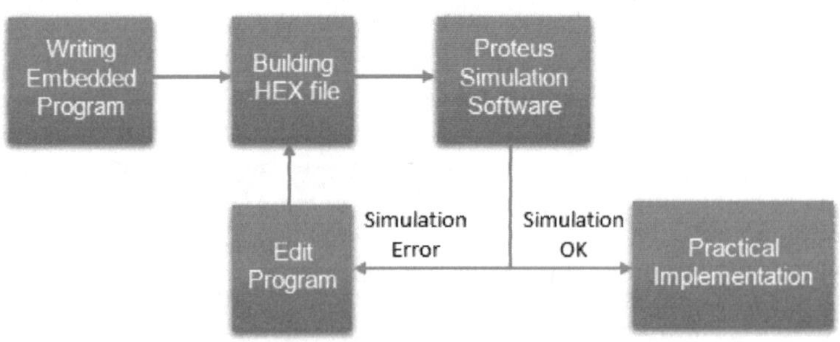

Figure 7. Simulation block scheme

Source code of the application:

```
char GLCD_DataPort at PORTD;
sbit GLCD_CS1 at LATB0_bit;
sbit GLCD_CS2 at LATB1_bit;
sbit GLCD_RS  at LATB2_bit;
sbit GLCD_RW  at LATB3_bit;
sbit GLCD_EN  at LATB4_bit;
sbit GLCD_RST at LATB5_bit;
sbit GLCD_CS1_Direction at TRISB0_bit;
sbit GLCD_CS2_Direction at TRISB1_bit;
sbit GLCD_RS_Direction  at TRISB2_bit;
sbit GLCD_RW_Direction  at TRISB3_bit;
sbit GLCD_EN_Direction  at TRISB4_bit;
sbit GLCD_RST_Direction at TRISB5_bit;
unsigned long current_duty;
unsigned long temp;
unsigned long frecv;
unsigned short factor[4]  ;
void main() {
  ANSELA = 0;        // Configure AN pins as digital
  ANSELB = 0;
  ANSELC = 0;
  ANSELD = 0;
  ANSELE = 0x04;                    //  RE2 as analog
  TRISE = 0x04;                     // RE2 as input
  LATA = 0xA0;
  TRISA = 0x0F;                     // configure PORTA pins as input
  LATB  = 0;                        // set PORTB to 0
  TRISB = 0;                        // designate PORTB pins as output
  LATC  = 0;                        // set PORTC to 0
  TRISC = 0;                        // PORTC as output
  PWM1_Init(500);      // Initialize PWM1 module at 500kHz
  current_duty = 0;  // initial value for current_duty
  frecv = 5000;
   Glcd_Init();
   Glcd_Fill(0x00);
   Glcd_Set_Font(Font_Glcd_System5x7,5,7,32);
   Glcd_Write_Text("Temperatura", 0, 2, 1);
   Glcd_Write_Text("Factor de umplere", 0, 3, 1);
   PWM1_Start();                    // start PWM1
   PWM1_Set_Duty(current_duty);       // Set current duty for PWM1
   do
  {
    current_duty = ADC_Read(7);         // Acquire analogical signal using ADC
   // current_duty=current_duty*5000/1023; // Conversion analogical signal in mV
   // current_duty=current_duty*255/5;      // Conversion analogical signal in mV
   temp = 256*current_duty/600;
   ByteToStr(temp, factor);
   Glcd_Write_Text(factor, 105, 2, 1);
    current_duty = (current_duty-48)*12;
   PWM1_Set_Duty(current_duty);       // Set current duty for PWM1
```

```
    ByteToStr(current_duty, factor);
    Glcd_Write_Text(factor, 105, 3, 1);
  } while(1);
  }
```

LM35 sensor used in this application is powered by a 5V voltage source. It output pin is connected to an analogue input pin (RE2) of the microcontroller. This is to take the analogue signal coming from the temperature sensor and to generate the PWM control signal of the cooling fan motor.

When the temperature sensed by the sensor reaches a threshold value which can be set by the user through the program, it actuates a cooling fan. The system works as shown below in Figure 8: PWM signal generated by the microcontroller will have 5V amplitude and duty cycle will be controlled by the acquired temperature. Thus, on reaching the required temperature, parameter Duty cycle will increase from 0% threshold at 100% to a value above the threshold, also chosen by the user. Increasing the duty cycle of the PWM signal, the system will increase the speed of the cooling fan.

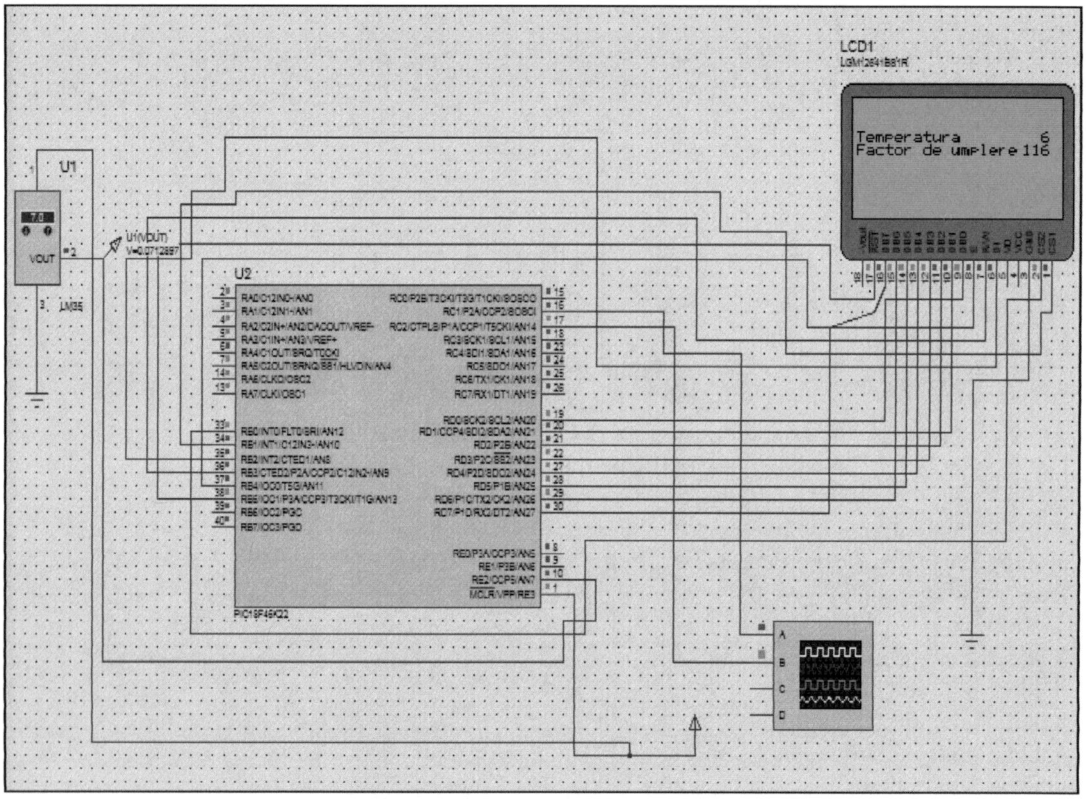

Figure 8. Simulation scheme using Proteus

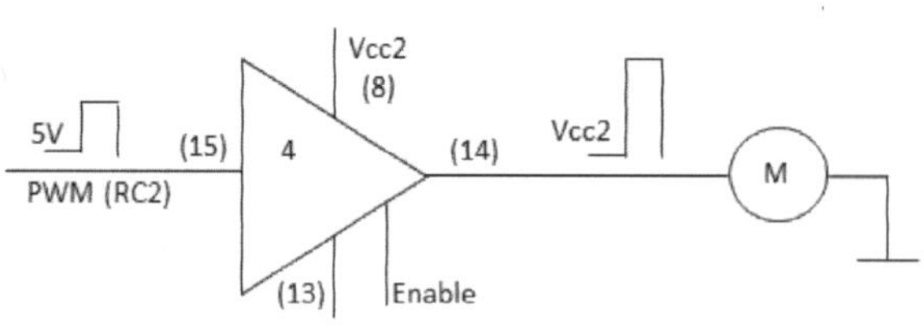

Figure 9. L293d pin configuration

Power driver L293D is supplied at terminal 16 (Vcc1) with voltage of 5V and at pin 8 with voltage of 5-36 V. The pins 12 and 13 are connected to ground. The enable pin is 8 (at '0' logic, the cooling fan does not work). The PWM command is connected to pin 15 and the dc motor supply is achieved between terminal 14 and ground.

4. Conclusions

This paper proposes a very simple, accessible and reliable method for temperature controlling. PIC microcontrollers permit a very wide range of applications and together with an appropriate development board is able to reduce the hardware circuits. Using development boards in applications, the improvement of the software application can be very accessible.

Acknowledgement
This work was supported by a grant of the Romanian National Authority for Scientific Research and Innovation, CNCS – UEFISCDI, project number PN-II-RU-TE-2014-4-1788.

References

[1] Mitescu M and Susnea I 2005 *Microcontrollers in practice,* Springer, Berlin Heidelberg New York
[2] Popa M 2008 *Sisteme cu microcontrolere orientate pe aplicaţii*, Orizonturi Universitare, Timişoara, Romania
[3] Calcutt D, Cowan F and Parchizadeh H 2004 *8051 Microcontrollers: An Applications Based Introduction*, Newnes Elsevier
[4] Rusu-Anghel S, Lihaciu I C and Rusu-Anghel N 2015 State feedback control of a robotic arm, *Annals of the Faculty of Engineering Hunedoara* **13**(1) 75-82
[5] Iordan A E and Panoiu M 2014 A new approach in the design of an interactive environment for teaching Hamiltonian digraphs, *IOP Conf. Ser.: Mater. Sci. Eng.* **57** 012012
[6] Muscalagiu I, Vidal J, Cretu V, Popa H E and Panoiu M 2007 The Efects of Agent Synchronization in Asynchronous Search Algorithms, *Lecture Notes in Artificial Intelligence* **4496** 53-62

IOP Conf. Series: Materials Science and Engineering **163** (2017) 012047 doi:10.1088/1757-899X/163/1/012047

Research on the rolling moment in the symmetrical and asymmetrical rolling process

V Alexa, S A Rațiu, I Kiss and C G Cioată

Politehnica University of Timisoara, Department Engineering and Management, 5 Revolution Street, Hunedoara, 331128, Romania

E-mail: vasile.alexa@fih.upt.ro

Abstract. Research distribution the rolling moments symmetrical and asymmetrical report presents great importance both in theory and to introduce clarifications to the calculation of rolling resistance line assemblies. Clarifying individuals of metallic material deformation between the rolls single cylinder diameters act of any difference of work and analysis of advance and delay phenomena. Torque drive value for each of the rolling cylinders was done by reducing the thickness of the laminate samples, an experimental facility located in the laboratory of plastic deformation of the Faculty of Engineering Hunedoara. The analysis of research results show that in terms of power consumption for deformation and safety equipment in operation is rational for mills which require such a difference between the work rolls to execute about one cylinder operated.

1. Introduction

The methods for measuring the rolling forces are classified according to the type of equipment and the possibility to determine them for a particular type of rolling mill, as follows [1]:

- direct measurement of the forces occurring during the rolling process, using resistive transducers placed in the point in which these forces are acting:
- measurement of deformations or strains that develop in a certain part of the working stand, with subsequent recalculation of these deformations based on the values of the rolling forces.

In addition to the forces, the rolling moments represent one of the most important parameters, whose knowledge is necessary for the design and operation of the rolling lines. The electric motor torque, developed during the metal material deformation, determines the loading level of the machine, as well as the resistance of some parts and assemblies.

The assessment of motor workload by taking into account the power consumption for deformation is not complete, because this parameter does not characterise the electrical and mechanical capacity of the equipment.

To solve the usability and to increase the level of exploitation and driving of the equipment, the next three parameters must be analysed [1]:

1. *the rolling forces* – which determine the strains in the major subassemblies of the working stand (rolls, frame, bearings);
2. *the moment developed during the rolling process* - which determines the motor strain and the resistance of its parts, so that the maximum torque is limited by the motor capacity and the resistance of its parts;

3. *the quadratic mean of the motor current* – which determines the level of motor warming.

2. Experimental research. Equipment and procedures

The load level of the motor, according to the moment value, is determined by comparing the value of the reduced moment acting at the motor shaft ($M_{mot.}$) with the value of the rated moment $M_{nom.}$.

The maximum moment for reciprocating motors equals to (2.5 - 3.0) M_{nom}, and for the non-reciprocating motors, (1.5 - 2.0) M_{nom}.

The ultimate goal being the determination of the moments that are developing during the deformation process, we used two methods for analysis, i.e. [1], [2]:

1. *Determination of the electric motor torque;*
2. *Measurement of the deformations of universal bars that convey the moment to the rolls.*

2.1. Determination of the electric motor torque

Knowing the value of the motor moment, we can calculate the rolling moment, because:

$$M_{mot.} = M_{lam.} + M_{mg.} + M_{fr.} \pm M_{din.} \quad [kN \cdot m] \tag{1}$$

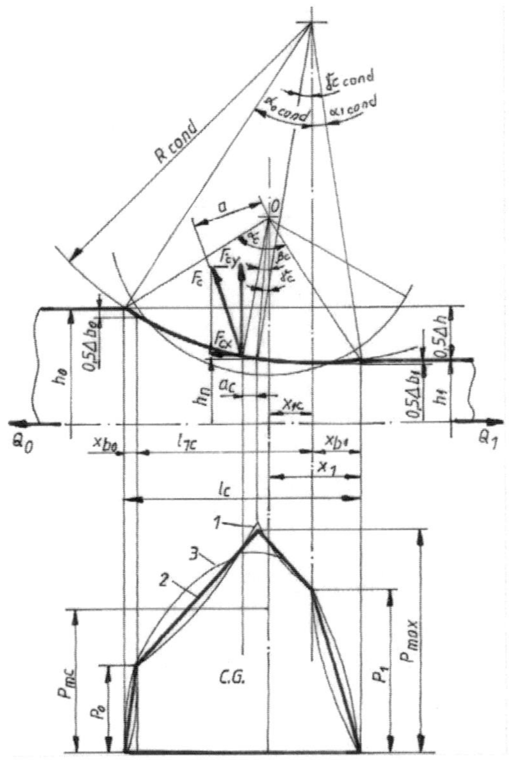

Figure 1. Diagram used to determine the neutral angle, material movement and rolling moment

To calculate the rolling moment, we used the balance equation, where the total moment at the roll barrel M_c, required for carrying out the plastic deformation at symmetric rolling (Figure 1), is determined by the tensile forces Q_0 and Q_1:

$$M_c = 2F_c \cdot a = 2F_{cy} + 2F_{cx}R \tag{2}$$

or if:

$$F_{cy} \cong F_c \tag{3}$$

$$M_c = 2F_c \cdot \psi_c \cdot l_c + R(Q_0 + Q_1) \tag{4}$$

Where:

- F_c, F_{cx}, F_{cy}, - represent the rolling force, its horizontal and vertical component;
- l_c - the contact arc length, calculated by considering the elongation and elastic deformations of the roll barrels;
- R – radius of the non-deformed rolls;
- Q_0, Q_1 – total anterior and posterior elongation of the strap;
- a, a_c – arms of the forces F_c, F_{cy};
- Ψ_c – coefficient that characterises the position of the resultant force application point on the contact arc, which is a complex function that depends on:
  - level and variation character of the normal and tangential strains;
  - reduction and hardening of the strap material;
  - coefficient of friction;
  - neutral angle and material elongation;
  - dimensions and elastic properties of the roll barrel material.

This coefficient is theoretically determined by considering the elastic deformation of the rolls and the strap, the uneven distribution of the contact normal strains and elongations, using the equation:

$$\Psi_c = \frac{M_{stn}}{p_{mc} \cdot l_c} \tag{5}$$

where:

- M_{stn} – static moment of the contact normal stress diagram in relation to the rotation centre of the rolls;
- p_{mc} – the contact average normal stress, calculated by considering the elastic deformations of the rolls (i.e. roll barrel mounted on the rolls);

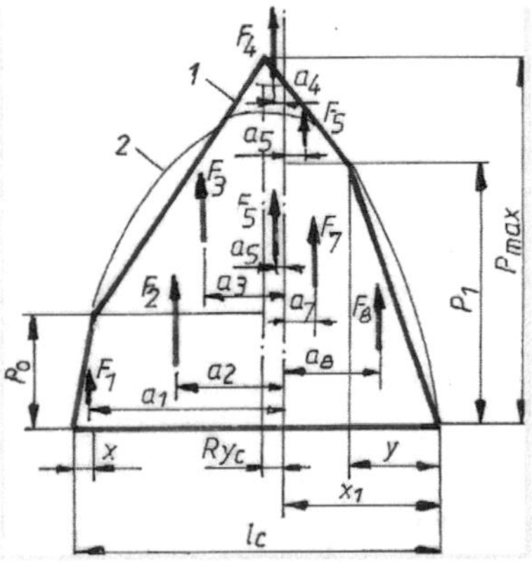

Figure 2. Diagram used to determine the Ψ_c coefficient

We determine this coefficient by two types of curves [3]:

a. *The normal contact stress diagram* is described by a continuous function consisting of linear segments (*curve 1,* Figure 2); assuming that the diagram area is divided into 8 simple geometric shapes, we use the Varignon's theorem:

$$M_{stn} = \sum_{i=1}^{8} M_i \tag{6}$$

The result of jointly solving the equation is:

$$\Psi_c = \frac{1}{6}\left[\frac{p_0}{p_{mc}}\left(1 - \frac{\gamma_c}{\alpha_c} - \frac{x_1}{l_c}\right)\left(1 - \frac{x_{1b}}{l_c}\right) - \frac{p_1}{p_{mc}}\left(\frac{\gamma_c}{\alpha_c} + \frac{x_1}{l_c}\right)\left(1 - \frac{x_{0b}}{l_c}\right) + 2\left(1 + \frac{\gamma_c}{\alpha_c} + \frac{x_{1b}}{l_c} - 2\frac{x_1}{l_c} - \frac{x_{0b}}{l_c}\right)\right] \tag{7}$$

b. *If the normal stress diagram is described by the curve 2* (Figure 2), then the static moment is determined by:

$$M_{stn} = \int_{R\gamma_c}^{l_c - x_1} p_{max}\left[1 - \left(\frac{x - R\gamma_c}{l_c - x_1 - R\gamma_c}\right)^2\right] x dx + \int_{l_c}^{R\gamma_c} p_{max}\left[1 - \left(\frac{x - R\gamma_c}{x_1 + R\gamma_c}\right)^2\right] x dx \tag{8}$$

Considering that $p_{max} = 1.5 p_{m.c}$, by jointly solving the equation we obtain:

$$\Psi_c = 0{,}125\left(3 + 2\frac{\gamma_c}{\alpha_c} - 6\frac{x_1}{l_c}\right) \tag{9}$$

These relationships are mostly used to theoretically determine the Ψ_c coefficient at rolling. The research regarding the accuracy of determination showed that, at cold rolling with great reductions ($\varepsilon > 20\%$), the values obtained for Ψ_c are close, but these values have large errors in case of small reductions. Therefore, the relationship x is the preferred one, because it provides a much higher accuracy thanks to the use of continuous linear segments function.

2.1. Measurement of the deformations of universal bars that convey the moment to the rolls

The second method based on the measurement of universal bars deformation, bars that convey the moment of the rolls, is based on the use of electrotensometric resistive transducers for the purpose of measuring the main strains [4].

The bearing holders of the inferior roller were modified for recording the lateral efforts so that the respective captors could be installed incorporated perpendicularity on the bearing's axis.

On the surface of captors were stuck tensometric stamps bound in deck, stamps that modify their dimension under the action of the effort to be measured.

These dimensional modifications of tensiometer stamps are generating variations of their electric resistance, that are proportional to the deformation efforts and the measuring of the forces is limited to the measuring of these resistance variations. The transducers are resistive, of grid type, have an electrical resistance of 203.2 Ω and the base width of 16 mm.

They were sticked using BF2-type glue, and their connection was made in bridge to enable us to apply the parallel resistance method, Figure 3, [1].

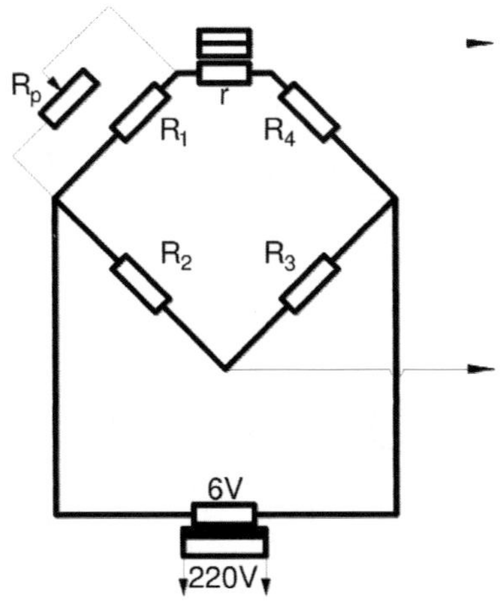

Figure 3. Tensometric bridge for measurements

Through the additional resistance *r*, connected in series with one of the inactive arms, the following condition is met:

$$(R_1 + r) \cdot R_3 > R_2 \cdot R_4 \quad (10)$$

so that the bridge is heavily unbalanced.

For balancing the tensometric bridge, we connect in parallel with R_1+r the resistance R=100KΩ, which in turn it is connected in series with the potentiometer R_p = 100 kΩ:

$$R_{ech1} = \frac{R_1(R + R_p)}{R + R_p + R_1} \quad (11)$$

This value is higher than R_1, so the bridge is unbalanced in this direction. For the extreme situation (potentiometer shorted):

$$R_{ech2} = \frac{R_1 \cdot R}{R + R_1} < R_1$$

This value is higher than R_1, so the bridge is unbalanced in this direction. For the extreme situation (potentiometer shorted):

$$R_{ech2} = \frac{R_1 \cdot R}{R + R_1} < R_1 \quad (12)$$

the bridge being unbalanced in the opposite direction.

In this situation, the equilibrium is found between R and R+R_p. Change in resistance in these extreme situations:

$$\Delta R = R_{ech1} - R_{ech2} = \frac{R_1(R + R_p)}{R + R_p + R_1} - \frac{R_1 \cdot R}{R + R_1} \quad (13)$$

Assuming that the rolling forces are F=1000 kN, in the active arms occur variations in resistance transducers ΔR=2.44 kΩ. Imposing the equilibrium condition to the bridge, the value of the resistance R+R_p connected in parallel, can be determined at equilibrium, resulting the value of 114 kΩ, which was confirmed during the balancing process.

The research for this theme purpose have been made on a 170 mm reversing two-high rolling mill, created and installed in the no conventional technologies and plastic deformation laboratory of the Engineering Faculty from Hunedoara [3].

An experimental installation formed of: special construction rollers, bearings, punctiform captors for lamination pressure, lamination forces captors and lateral pressure captors it was created for research in condition of technological similitude symmetrical and asymmetrical process.

In Figure 4 it is presented in overview the mentioned installation, with the way force captors are assembled in order to determine the lateral efforts in the longitudinal asymmetrical lamination but also to show the author's contribution regarding method of experimentation [1].

The symmetric and asymmetric rolling process was carried out by equipping the work rolls with segments adjusted with various radii, which have led to the following ratios between the diameters of the upper D_s and lower D_i rolls: $\frac{D_s}{D_i} = \frac{170}{170}; \frac{160}{180}; \frac{150}{190}; \frac{140}{200}$ mm

Figure 4. Montages punctifom captors for recording the pressure of contact

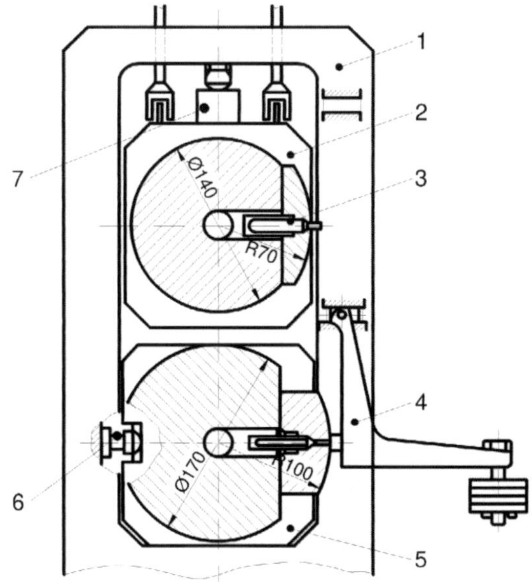

Figure 5. Diagram for measuring the deformations

1- the rolling mill stand; 2 - higher bearing; 3 - punctiform sensor; 4 - calibration device; 5 - lower bearing; 6 - sensor for the measurement of lateral forces; 7- sensor for the measurement of the rolling forces.

Usually, the tensiometer stamps of a forces captor are bound in deck. The deck has on a diagonal it is measured a electrical signal – proportional of the applied effort – and for recording of the measured values this signal is recorded by an oscillograph. The oscillograph is a the type N-700, having 14 channels, the impulses on recorded are a scale of 120 mm width, heaving 4 cm/s moving speed of the paper band.

3. Results and conclusions

When rolling thick samples (h_0 = 6; 12 mm), with the increasing degree of reduction ε, the resultant point of application is moving away from the plane through the centres of the rolls.

When rolling thin samples (h_0 = 1; 2 mm), with the increasing degree of reduction ε, the resultant point of application is approaching that plane.

These phenomena can be explained as follows:

- when rolling thick samples with low degree of deformation, the pressure along the deformation zone is distributed almost uniformly;
- when rolling thin samples, as a result of increasing degree of deformation, the influence of the stress and strain hardening state is intensified, leading to the movement of the resultant force point of application towards the exit plane of the metal material from the rolls.

For the symmetrical process, when rolling aluminium, this dependence is given as total moment M = f(ϵ), because in the ideal case of symmetry of all deformation conditions, the moments developed by the upper and lower rolls should have the same value. In practical cases, there is a difference between the moments of the upper and lower rolls, which is explained by a complicated dependence of the moment distribution on many factors of the deformation process.

Thus, experiments were carried out with the upper roll not driven, measuring the moment at the lower roll which characterises in this case the total moment required for rolling. A comparative assessment of the results obtained at symmetric rolling with both rolls driven and the case when the upper roll is not driven is shown in Figure 6.

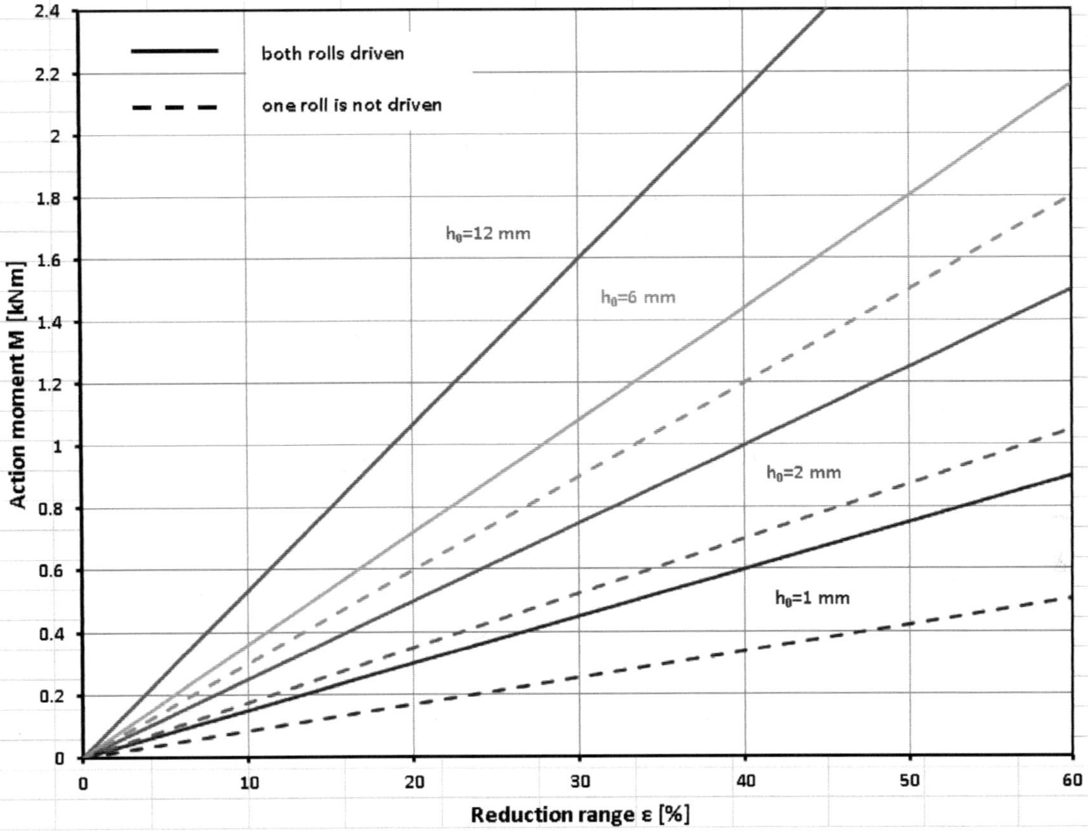

Figure 6. Moments versus reduction at the symmetric rolling ($\frac{D_s}{D_i} = \frac{170}{170}$ mm) of Al samples, with both rolls driven and when one roll is not driven

From the comparative analysis of the last two dependences, the same behaviour is resulting in terms of the distribution of moments, without any influence of the metal material nature, but in both cases when the upper roll is not driven, the moment required for deformation is much lower than when both rolls are driven.

For example, for a reduction ϵ = 40% and h_0 = 1 mm, the difference is 23%, and for the same reduction and h_0 = 6 mm, the difference is 14%.

This difference indicated above is taken as a percentage in relation to the moment consumed for the deformation of the same volume of metal material when both rolls are driven, with the result that the deformation of the metal material by symmetric rolling when one of the rolls is not driven is much more economical, because the moment required for deformation is lower.

For the asymmetric process, the moments are presented as separate functions for the upper and lower rolls, depending on reduction: $M_s = f(\epsilon)$ and $M_i = f(\epsilon)$. Figure 7 shows the dependences obtained by processing the data from asymmetric rolling of the aluminium samples, when the rolls ratio is $\frac{D_s}{D_i} = \frac{160}{180}$ mm.

This analysis shows that, within the reduction range 0 – 35%, the smaller diameter roll acts as a brake, because the developed moment has a negative value.

For samples of different thicknesses, at different reductions, the moments between rolls become equal and, with decreasing the thickness of samples, the intersection point between the moments of upper roll (whose diameter is smaller) and lower roll (whose diameter is greater) moves to the right, where the reductions are larger.

By increasing the difference between the working diameters of the rolls$\left(\frac{D_s}{D_i} = \frac{140}{200}\ \text{mm}\right)$, we note that the moment at the smaller-diameter roll is almost always zero at small reductions, and if we remove the friction moment value, it becomes negative.

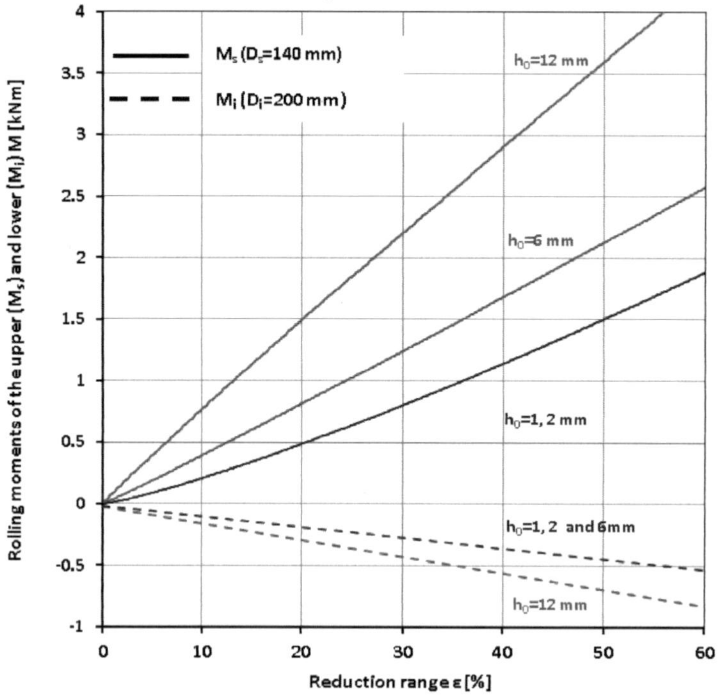

Figure 7. Rolling moments of the upper (M_s) and lower (M_i) driven rolls versus the reduction, at asymmetric rolling $\left(\frac{D_s}{D_i} = \frac{160}{180}\ \text{mm}\right)$

When both rolls used for rolling are driven, for $\frac{D_s}{D_i} = \frac{140}{200}$ mm we noted the followings:

- the curves of the moments corresponding to the two rolls do not intersect;
- by increasing the reduction grade within this range, the roll with a smaller diameter opposes a growing resistance against the movement of the larger-diameter roll;
- the moment developed by the larger-diameter roll increases considerably because, in addition to the moment required for the metallic material deformation, it must overcome the braking action of the smaller-diameter roll – with the result that the practical use of two driven rolls, with greater difference between the working diameters, is not rational;

Figure 8 shows the comparison of moments for $\frac{D_s}{D_i} = \frac{140}{200}$ mm, in two cases:
(1) both rolls driven, and (2) upper roll not driven, obtaining the following results:

- the required moment for deformation when the smaller-diameter roll is not driven is much more reduced, as follows: if $\epsilon = 40\%$, the difference is 29.4% for $h_0 = 1$ mm and 25% for $h_0 = 6$ mm. This is related to the moment required for the deformation of the same volume of metallic material when both rolls are driven;

- either from the point of view of energy consumption for deformation or operational safety, it is rational to use rolling equipment with such difference between the working diameters of the rolls only when a single roll is driven.

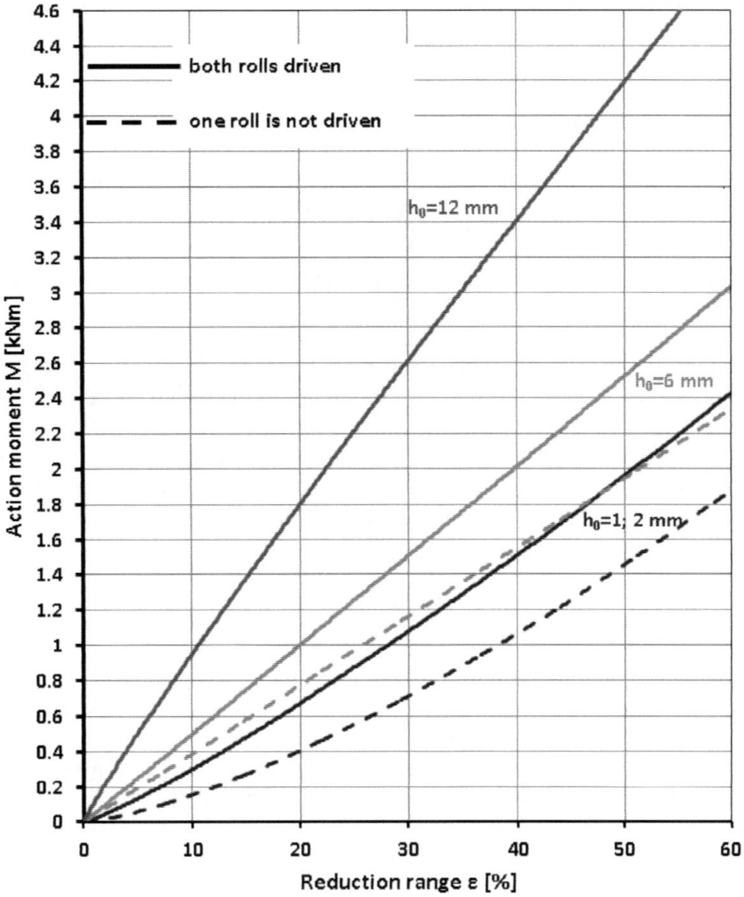

Figure 8. Driven moments M_s and M_i versus reduction, at asymmetric rolling $\left(\frac{D_s}{D_i} = \frac{140}{200} \text{ mm}\right)$

References

[1] Alexa V 2002 *Contributions to research asymmetric rolling process*, Politehnica University of Timişoara, Romania, Doctoral Thesis

[2] Alexa V, Ratiu S and Kiss I 2005 Theoretical and experimental analysis of the rolling strain in the deformation area, *Metalurgia International* **5** 20-28

[3] Alexa V, Ratiu S and Kiss I 2016 Metal rolling – Asymmetrical rolling process, *IOP Conf. Ser.: Mater. Sci. Eng.* **106** 012019

[4] Alexa V, Ratiu S and Popa G 2010 *Method and program for automatic calculation forces and momentum to asymmetric longitudinal rolling,* 12th WSEAS International Conference on Mathematical and Computational Method in Science and Engineering MACMESE 10, Algarve, Faro, Portugal, November 3-5, pp 303-306

IOP Conf. Series: Materials Science and Engineering **163** (2017) 012048 doi:10.1088/1757-899X/163/1/012048

The catalytic microwave synthesis of biodegradable polyester polyols based on castor oil and l-lactide

D Kojić[1], T Erceg[2], N Vukić[2], V Teofilović[2], I Ristić[2], J Budinski-Simendić[2] and V Aleksić[3]

[1]University of Business Engineering and Management, Faculty of Engineering, Despota Stefana Lazarevića bb, 78000 Banja Luka, Republic of Srpska, Bosnia and Herzegovina
[2]University of Novi Sad, Faculty of Technology, Bulevar cara Lazara 1, 21000 Novi Sad, Serbia
[3]University of East Sarajevo, Faculty of Technology, Karakaj bb, 75400 Zvornik, Republic of Srpska, Bosnia and Herzegovina

E-mail: kojic.d@hotmail.com

Abstract. Various strategies for achieving a functional poly(lactic acid) (PLA) have been developed such as ring-opening copolymerization with a functional monomer, the use of functional initiator and various post polymerization modifications. It is possible to obtain the star shaped polymer using natural oil with at least three OH groups as an initiator. It was estimated that despite of low-molecular mass of star-shaped PLA, the hydrophobic castor oil central core influenced the slow degradation rate in the case of injectable biomedical application. The star-shaped polymers with low-molecular-mass have a lower melt viscosity correlated with linear counterparts. In soft tissue reparation the polymer viscosity increases with fluid body contact and the solid implant can be formed. To ensure liquid state at injection temperature the low molar mass polymer is favorable. There is a particular size for each macromolecular chains at which chain entanglement occurs. In this work the influence of the l-lactide (LA) and the castor oil (CO) contents on the size of biodegradable branched polyester polyols was studied. The average molecular masses of synthesized polymers were estimated by GPC procedure. In sample formulations the [LA]/[CO] ratios were from to 113 to 533. Mn values for obtained polymers were from 5000 to 20000 Da. The molecular mass distribution for the resulting polymers was between 1.09 and 1.37.

1. Introduction

The polymer materials derived from non-petrochemical resources have commercial and academic importance. Over the last decade, there has been explosive growth in the renewable polymer fabrication, culminating with the manufacturing of polymers such as glucose-derived poly (lactic acid) (PLA) and natural oil-based polyurethanes. Now, with the ecological focus on sustainability, there is a direction to build the polymeric materials from agricultural resources [1].

PLA is thermoplastic aliphatic polyester that can be derived from inexpensive corn and sugar beets. This polymer is fully biodegradable and has good mechanical properties, including a high tensile strength. However, the industrial applications of PLA are limited at present due to the undesirable properties such as low heat resistance, low impact strength, low injection –crystallization rate, high brittleness, and a low barrier to gases. The low impact strength and high brittleness are the result of large crystal presence.

Different procedures have been made to enhance its flexibility by mixing with low-molecular-mass plasticizers or other polymers. The use of branched polymer will reduce the greenhouse gas emissions compared with PLA products containing commercially available petroleum-based plastificators. Materials based on natural resources have influenced the significant progress in medicine [2]. Nonlinear copolymers exhibit different characteristics in comparison with their linear counterparts. Biobased polyesters have a hydrophobic backbone with hydrolytically labile anhydride and/or ester bonds that may hydrolyze to monomers when placed in a physiological medium [3].

The main feature of star-shaped polymers is the compact structure and the multiple functionality convenient for some applications. PLA is an aliphatic polyester mainly used for packaging, pharmacy, medical devices, and tissue engineering applications. The injectable biodegradable polymers (liquid or pastes) are used for soft tissue reparation procedures. In a preferred embodiment, the polymer composition contains liquid or pasty hydroxy fatty acid-based copolyesters, polyester-anhydrides, or some combinations. The viscosity of the polymers increases upon contact with body fluid to form a solid implant suitable for soft tissue repair. To ensure that the polymer is a liquid at temperatures for injection application, it must have a low molar mass.

There exists a particular size for each polymer at which chain entanglement occurs (the critical entanglement length). Below this value the melt viscosity is typically Newtonian. Thus it is important to ensure that the size is below critical value to ensure a low viscosity during flow through a needle. Molecular mass also influences the melting point of polymer (melting point is decreasing as macromolecular size increases). For PLA, branching is determined to influence rheological properties such as complex viscosity and elastic modulus and its melt processing [4]. If the polymer is semi-crystalline, macromolecular size can also be designed to ensure that it is amorphous at operating temperature (37° C). To ensure that star-shaped polymer can pass through a needle, the material have a low glass transition temperature [5]. Various strategies for achieving a functional PLA have been developed, such as ring-opening copolymerization with a functional monomer, the use of a functional initiator, and various post polymerization modifications.

One interesting aspect of the ring-opening reaction procedure is the flexibility in different topology designing using multifunctional initiators. It is possible to obtain the star shaped polymer using natural oil with at least three OH groups as an initiator. It was estimated that despite low-molecular mass of star-shaped poly(lactic acid), the hydrophobic castor oil central core influenced the slow degradation rate in injectable materials application. The viscosity of polymer can be regulated by the initiator type. Star-branched polyester polyol can be synthesized using the oil with at least three OH groups in molecule (castor oil is convenient). Hydroxyl groups at castor oil can initiate ring opening polymerization of lactide. The microwave polymerization provides a new procedure for enhancing economic advantages through energy saving. Chemically homologous chains with linear and branched structures may reveal different properties. The increase of castor oil content (hydrophobicity) decreased the polymer degradation rates. It was estimated that despite low-molecular mass of star-shaped poly(lactic acid), the hydrophobic castor oil (triglyceride of ricinoleic acid) central core influenced the slow degradation rate in injectable materials application. Branched polyester polyol bearing a castor oil core is also using for plastification of poly (lactic acid) [6],[7].

The fabrication of multiphase polymers from renewable resources faces many challenges. Self-assembling polymers such as copolymers are produced by controlled polymerizations, such as anionic, cationic or living radical polymerizations. It can be quite difficult to apply these techniques to many of the available renewable resource monomers due to the lack of appropriate functionality for polymerization, the presence of functionality that is incompatible with these techniques, or the heterogeneity of the feedstock The majority of all-renewable block and graft copolymers found contain the aliphatic polyester polylactide as one component, largely due to the relative ease of the controlled syntheses of such thermoplastic materials. Macroinitiators containing hydroxyl groups used for thering-opening polymerization of lactide, copolymerization of lactide with other cyclic monomers (or alternatively of lactic acid with other monomers through condensation polymerizations), and coupling procedures to fabricate polymers of various architecture are studied in ref [8]. The microwave-irradiated reactions are influenced the higher polymer purity. To develop a technique of microwave-assisted polymerization of L-lactide and castor oil, to efficiently and easily prepare star-shaped polyester polyol the ring opening polymerization of L-lactide using microwave irradiation was investigated. The lactic acid oligomers are polar molecules, thus they can absorb microwave energy.

In this study the goal was the synthesis of a branched poly(lactic acid) in the presence of stannous octoate as catalyst in a bulk.

2. Experimental part

2.1. Materials

Monomer (3S)-cis-3,6-dimethyl-1,4-dioxane-2,5-dione (L-Lactide) was used in the ring-opening polymerization (purchased from Sigma-Aldrich). Castor oil with the hydroxyl number 170 mg KOH g^{-1} and acid value 1.27 mg KOH g^{-1} was purchased from Merck Chemical Co and used as received. Tin(II) 2-ethylhexanoate (95%) from Aldrich was used as polymerization catalyst. All solvents were obtained from Merck Chemical Co. Castor oil content for samples with desired molar mass for star-like architecture was calculated according [9].

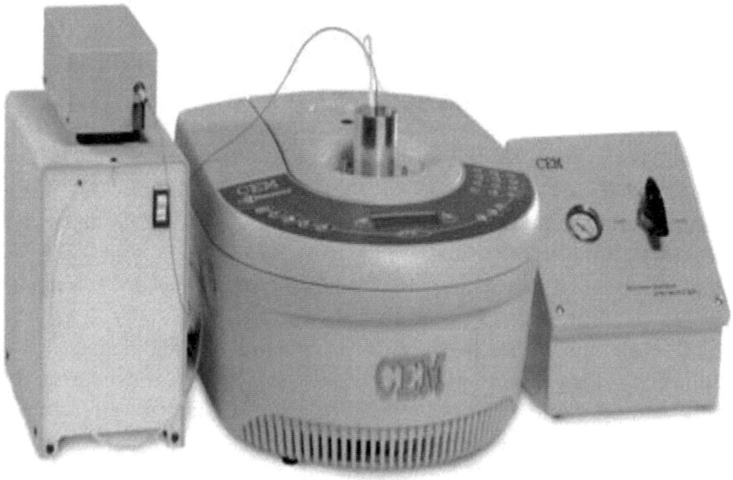

Figure 1. The picture of used "Single-mode" microwave reactor

2.2. Polymerization method

The polyester polyol was prepared by "core-first" method which involves the L-lactide polymerization using multifunctional initiator natural castor oil. Polymerization was carried out in microwave reactor. The molar feed ratio [LA]/[CO] was from to 113 to 533. To dry L-lactide (pre-crystallized from methanol) the catalyst tin(II) 2-ethylhexanoate mixed with toluene was added. The mixture was homogenized. The evaporation of toluene was managed during 12 h at 60 °C in vacuum. After that appropriate amounts of castor oil were added in reaction mixture. The reactants mixture was placed to glass ampoule. The synthesis procedure was performed in a conventional microwave oven (Figure 1). The frequency and the power applied were 2.45 GHz and 150 W, respectively. Final product was the polymer with OH groups ended poly(L-lactide) chains. The prepared polymer could be transformed into other end-groups chains (e.g. the more stable ones) or used for different polymer architectures.

2.3. Characterization of obtained polymers

The molecular masses of obtained polymers were determined by GPC method using Aglient 1100 Series system. The column ZORBAX PSM 300 was used. Tetrahydrofuran was used as eluent. Poly(styrene) standards were used for instrument calibration. Fourier transform infrared spectra, FTIR, were obtained using Bomem Hartmann & Braun MB-series. Samples were milled with KBr and tablets were prepared using vacuum press. Thermal properties of samples was investigated by differential scanning calorimetry (DSC, TA Instruments Q20). Hermetically sealed aluminium pans containing 3-5 mg of sample were used for measurements. The non-isothermal melt-crystallization was scanned from 20 to 220 °C with heating rate 5 °C min^{-1}.

3. Results and discussion

Co-polyester based on lactid acid and castor oil was synthesized. The conversion of 90% under the applied synthesis procedure was attained. The structure of prepared material is given in Figure 2. Molecular masses and molecular mass distribution are known as the main characteristics of polymers, since they have significant effects on different properties. For small sized polymers the mass of initiator comprises a significant portion content of the overall mass. The polymer molar masses and molar mass distributions Q assessed by GPC measurements are given in the Table 1 for samples prepared during different reaction times. For prepared samples the feed ratio [LA]/[CO] was varied from to 113 to 533 influenced different polymer size (Mn values from 5000 g mol^{-1} to 20000 g mol^{-1}). It was determined that the size of obtained star-shaped polyesters increased with increase of feed ratio. The molecular mass distribution of the resulting polymer was between 1.09 and 1.37. The chemical structure of obtained samples was estimated by FTIR spectroscopy. In Figure 3 are the FTIR spectrum of synthesized materials. The vibrations at 1189 cm^{-1} (asymmetrical valence) and symmetrical valence vibrations (C-O-C) of the aliphatic chain at 1093 cm^{-1} were shifted, compared with L-lactide bonds at 1276 and 1099 cm^{-1}. The peak at 1759 cm^{-1} was due to the aliphatic ester group (valence vibration of C=O), while the peak at 1454 and 1382 cm^{-1} originated from asymmetric and symmetric banding vibration of C-H from CH_3. Bond at 1269 cm^{-1} showed the overlapping C-H bending vibration and C-O-C stretching vibration (C-O-C). Existence of that peak confirmed L/lactide polymerisation.

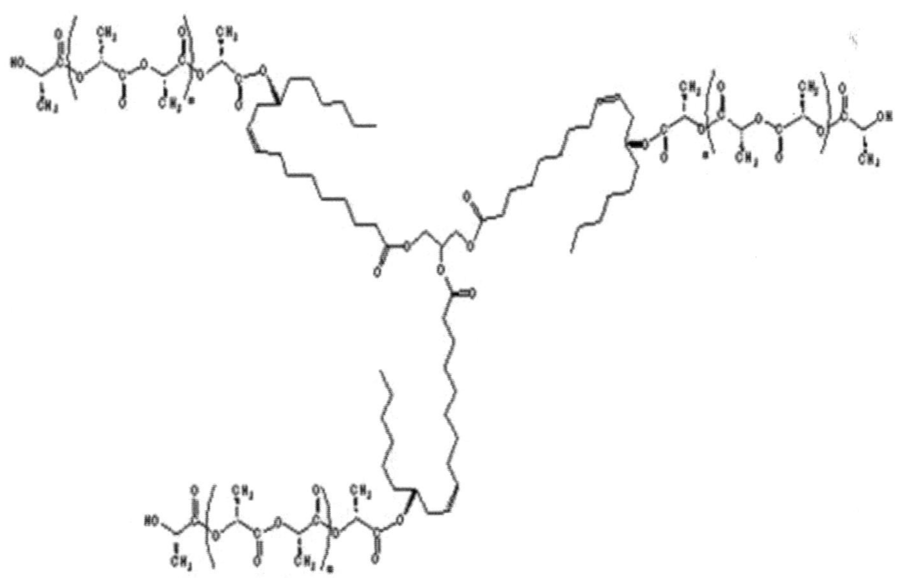

Figure 2. The structure of star-shaped polyester polyol prepared from castor oil as initiator and monomer L-lactide via ring-opening polymerization

The valence vibration bond (C=O) of aliphatic ester are splitted into two bonds (at 1759 cm^{-1} in PLA chains, and at 1748 cm^{-1} in the initiator castor oil). Bonds from asymmetric (2943 cm^{-1}) and symmetric (2856 cm^{-1}) νCH_2 vibrations with additional weak shoulder around 2993 cm^{-1} from νCH_3, are consequences of terminal methyl group of castor oil fatty acid. Strong apsorption at 3480 cm^{-1} originated from terminated hydroxyl groups, νOH, at poly(L-lactide) chains, which confirmed our presumption about polymerization mechanism. By regulating star-shaped polyester synthesis time and feed ratio of reactants it was possible to obtain the appropriate molecular mass of obtained polymer (very important for its degradation kinetics in the human body). However, before such product can reach the market, pharmaceutical-grade material quality and the fabrication costs must be estimated. In Figure 4 are given DSC curve of samples with molar feed ratio [LA]/[CO]s 533,1 synthesized in microwave reactor for different reaction time. In Table 2 are given all DSC data assessed for samples of polyester based on different feed ratio of monomer L-lactide and initiator castor oil.

Table 1. The number average molar mass (Mn) and polydispersity values (Q) for samples based on different feed ratio of monomer L-lactide and initiator castor oil

Sample name	Mass feed ratio	Molar feed ratio [LA]/[CO]	Reaction time (min)	Q	Mn (Da)
PLA-CO-I10	4,53	113,8	10	1.24	4567
PLA-CO-I15	4,53	113,8	15	1.37	5012
PLA-CO-I20	4,53	113,8	20	1.31	5588
PLA-CO-II15	6,85	168,8	15	1.02	6981
PLA-CO-II20	6,85	168,8	20	1.07	7358
PLA-CO-II25	6,85	168,8	25	1.09	7068
PLA-CO-III25	15,62	406,5	25	1.10	13598
PLA-CO-III30	15,62	406,5	30	1.09	14638
PLA-CO-III40	15,62	406,5	40	1.17	14687
PLA-CO-IV30	21,25	533,1	30	1.14	19876
PLA-CO-IV40	21,25	533,1	40	1.12	20933
PLA-CO-IV50	21,25	533,1	50	1.18	21692

IOP Conf. Series: Materials Science and Engineering **163** (2017) 012048 doi:10.1088/1757-899X/163/1/012048

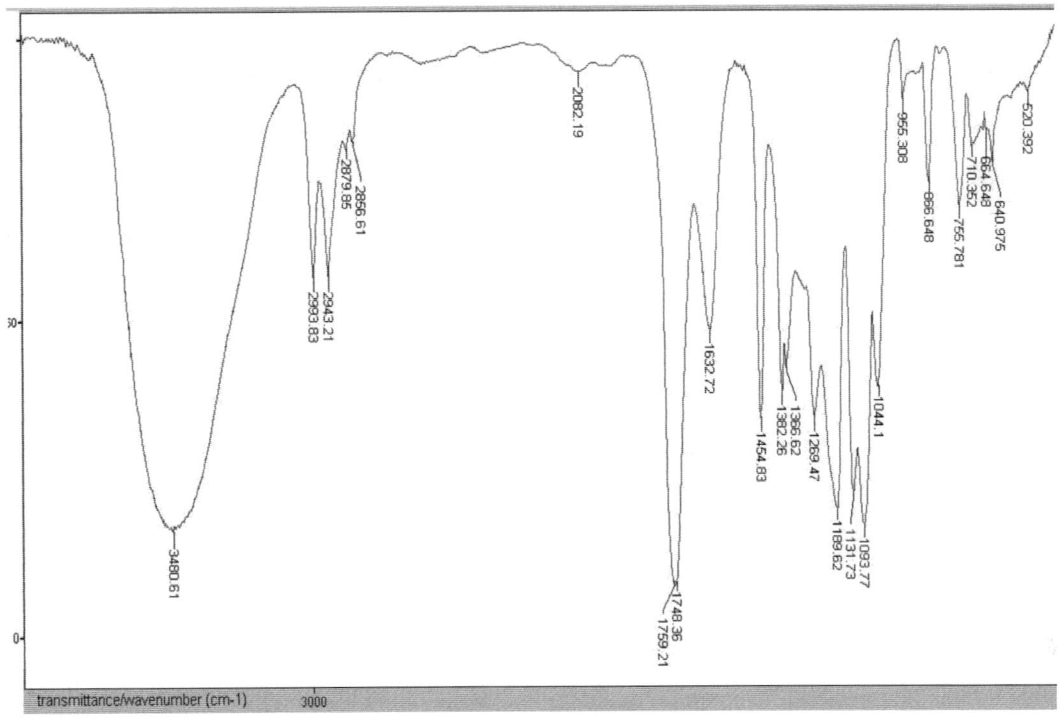

Figure 3. FTIR spectrum of synthesized samples based on lactid acid and castor oil

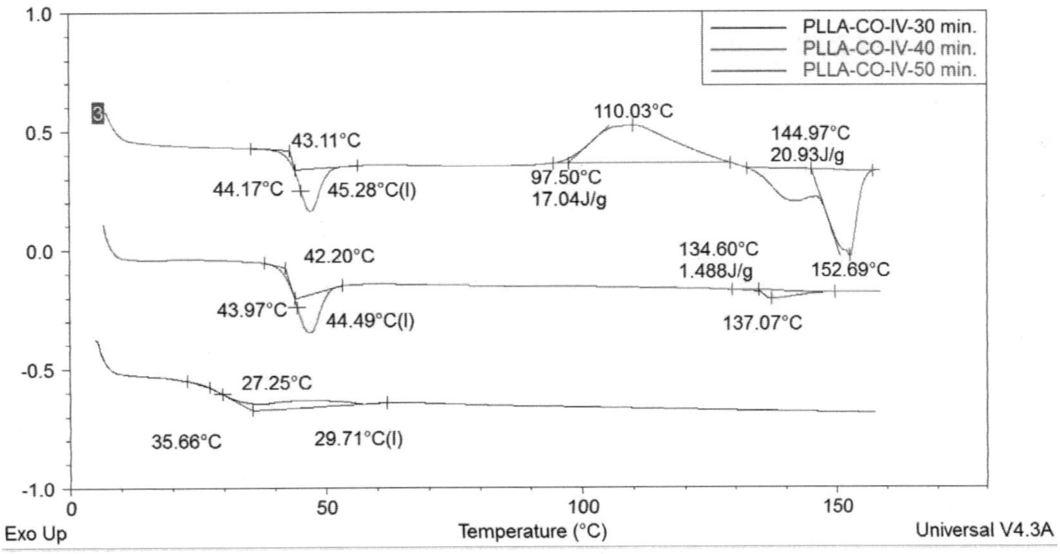

Figure 4. DSC curve of samples with molar feed ratio [LA]/[CO]s 533,1 synthesized in microwave reactor for different reaction time

Table 2. The DSC data assessed for polyester based on different feed ratio of monomer L-lactide and initiator castor oil

Sample name	Reaction time (min)	Molar feed ratio [LA]/[CO]	Tg [°C]	Tc [°C]	Tm [°C]
PLA-CO-I10	10	113,8	40.64	/	/
PLA-CO-I15	15	113,8	38.37	74.99	130.69
PLA-CO-I20	20	113,8	40.81	74.95	138.12
PLA-CO-II15	15	168,8	15	39.63	133.9
PLA-CO-II20	20	168,8	20	38.95	/
PLA-CO-II25	25	168,8	25	40.47	/
PLA-CO-III25	25	406,5	45.35	-	147.34
PLA-CO-III30	30	406,5	46.30	-	147.28
PLA-CO-III40	40	406,5	44.13	-	147.01
PLA-CO-IV30	30	533,1	35.66	/	
PLA-CO-IV40	40	533,1	43.97	137.07	
PLA-CO-IV50	50	533,1	44.17	152.69	

4. Conclusion

Using renewable feedstock for the production of new polymers can have both environmental and economic benefits. To design biodegradable branched polyester polyols based on L-lactide monomer and the castor oil initiator we developed a bulk synthesys microwave procedure using tin(II)2-ethylhexanoate as a catalyst. The star shaped structures of prepared materials was analysed using FTIR spectroscopy. Using the gel permeation chromatography method it was assessed that the size of prepared polyesters was affected by the reactant feed ratio. Obtained materials can be used for injection application if the molecular mass is below critical value to ensure a low viscosity for flowing through a needle. The feed ratio of the monomer and the initiator [LA]/[CO] from to 113 to 533 influenced number average molal mass from 5000 Da to 20000 Da. The polydispersity values were between 1.09 and 1.37. Obtained star-shaped polymers with smaller molar mass can be used for injection application if the molecular mass is below critical value to ensure a low viscosity for the flow through a needle.

Acknowledgement

Authors wish to express the gratitude to the Ministry of Education, Science and Technological Development Republic of Serbia, Projects No. III 45022 for financial support.

References

[1] Williams C K and Hillmyer M A 2008 Polymers from renewable resources: a perspective for a special issue of polymer reviews, *Polymer reviews* **48**(1) 1-10

[2] Peppas N A and Langer R 1994 New challenges in biomaterials, *Science* **263**(5154) 1715–1720
[3] McKee M G, Unal S, Wilkes G L and Long T E 2005 Branched polyesters: Recent advances in synthesis and performance, *Progress in Polymer Science* **30**(5) 507-539
[4] Amsden B G 2010 Liquid, injectable, hydrophobic and biodegradable polymers as drug delivery vehicles, *Macromolecular Bioscience* **10**(8) 825–835
[5] Sokolsky-Papkov M and Domb A J 2008 Stereoisomeric effect on in vitro drug release from injectable poly(lactic acid-co-castor oil), *Polymer for advanced technologies* **19**(6) 671–679
[6] Nouri S, Dubois C and Lafleur P G 2015 Synthesis and characterization of polylactide with different branched architectures, *Journal of Polymer Science, Part B: Polymer Physics* **53**(7) 522–531
[7] Slivniak R and Domb A J 2005 Lactic acid and ricinoleic acid based copolyesters, *Macromolecules* **38**(13) 5545-5553
[8] Robertson M L, Hillmyer M A, Mortamet A C and Ryan A J 2010 Biorenewable multiphase polymers, *MRS Bulletin* **35**(03) 194-200
[9] Tsujimoto T, Haza Y, Yin Y and Uyama H 2011 Synthesis of branched poly(lactic acid) bearing a castor oil core and its plasticization effect on poly(lactic acid), *Polymer Journal* **43**(4) 425-430

IOP Conf. Series: Materials Science and Engineering **163** (2017) 012049 doi:10.1088/1757-899X/163/1/012049

Cooperation between humans and robots in fine assembly

C K Jalba, P Konold, I Rapp, C Mann and A Muminovic

Ulm University of Applied Sciences, Production Engineering and Production Economics, Ulm, Germany

E-mail: klausjalba@t-online.de

Abstract. The development of ever smaller components in manufacturing processes require handling, assembling and testing of miniature similar components. The human eye meets its optical limits with ongoing miniaturization of parts, due to the fact that it is not able to detect particles with a size smaller than 0.11 mm or register distances below 0.07 mm - like separating gaps. After several hours of labour, workers cannot accurately differentiate colour nuances as well as constant quality of work cannot be guaranteed. Assembly is usually done with tools, such as microscopes, magnifiers or digital measuring devices. Due to the enormous mental concentration, quickly a fatigue process sets in. This requires breaks or change of task and reduces productivity. Dealing with handling devices such as grippers, guide units and actuators for component assembling, requires a time consuming training process. Often productivity increase is first achieved after years of daily training. Miniaturizations are ubiquitously needed, for instance in the surgery. Very small add-on instruments must be provided. In measurement, e.g. it is a technological must and a competitive advantage, to determine required data with a small-as-possible, highest-possible-resolution sensor. Solution: The realization of a flexible universal workstation, using standard robotic systems and image processing devices in cooperation with humans, where workers are largely freed up from highly strenuous physical and fine motoric work, so that they can do productive work monitoring and adjusting the machine assisted production process.

1. Handling small components

The development of ever smaller components during the manufacturing process requires the handling, assembling and testing of similar miniature components. The human eye meets its limits with actual miniaturization of optical sizes and as well as handling becomes difficult limits, because the human eye is not able to detect particles with a size of less than 0.11 mm or register distances below 0.07 mm.

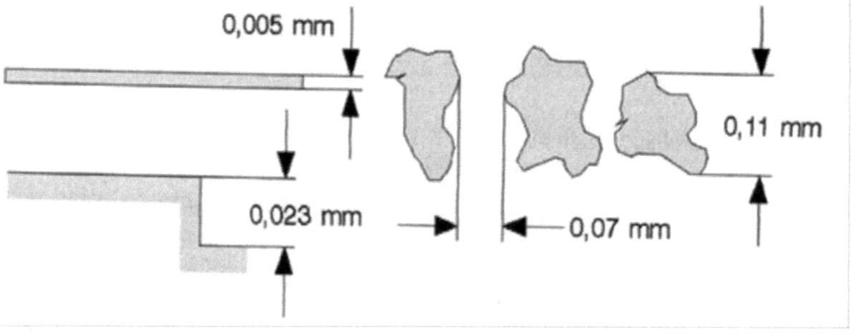

Figure 1. Detection ability limit of the human eye without aids

Employees can not differentiate after several hours color differences accurately and also ensure consistent quality.

Figure 2. Precision assembly at Endress + Hauser

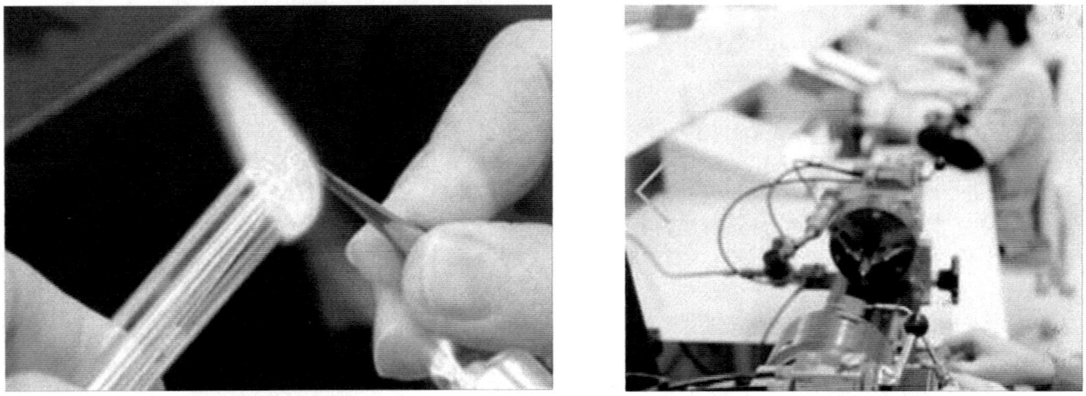

Figure 3. High precision resistor assembly at Endress + Hauser [1]

Assembly is usually made with tools, such as microscopes, magnifiers or digital measuring devices.

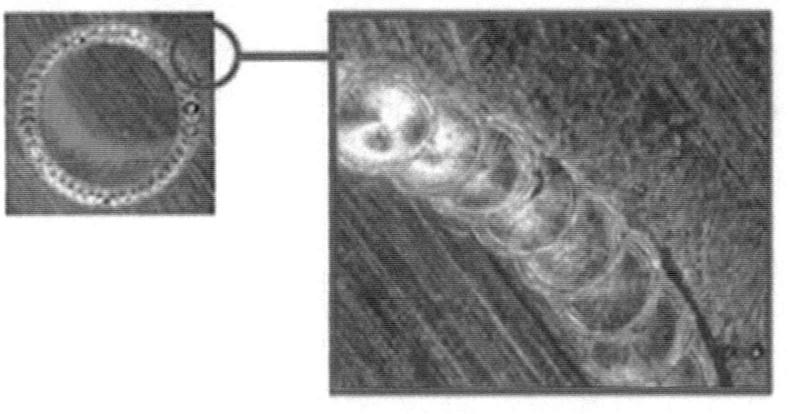

Figure 4. Enlarged micro welds [2]

Due to the enormous concentration during work, employees quickly get tired. This requires breaks or change of job and reduces productivity. Dealing with handling devices such as grippers, guide units and actuators for assembling the components requires a time consuming training process. Often succeed not until years of working with the miniature components to increase productivity.

Miniaturizations are needed for example in the surgery, because it must be often provided with additional very small instruments.

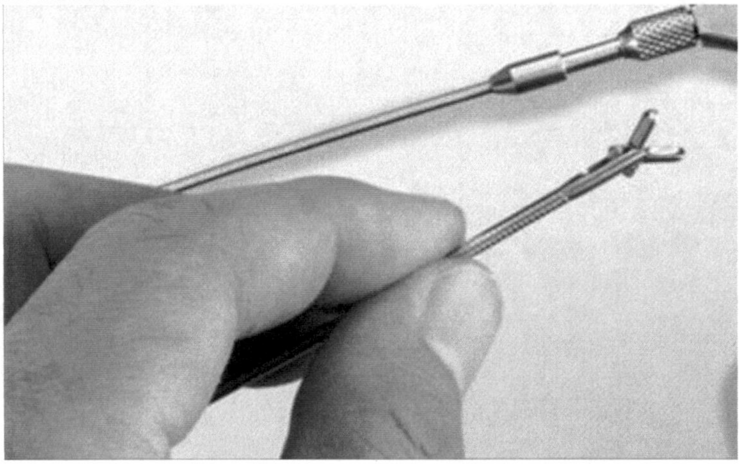

Figure 5. Small gripper used in the surgery produced by Bacher, Germany [3]

In measurement it is imperative and a competitive advantage, if the required data can be determined with a sensor as small as possible, with highest possible resolution.

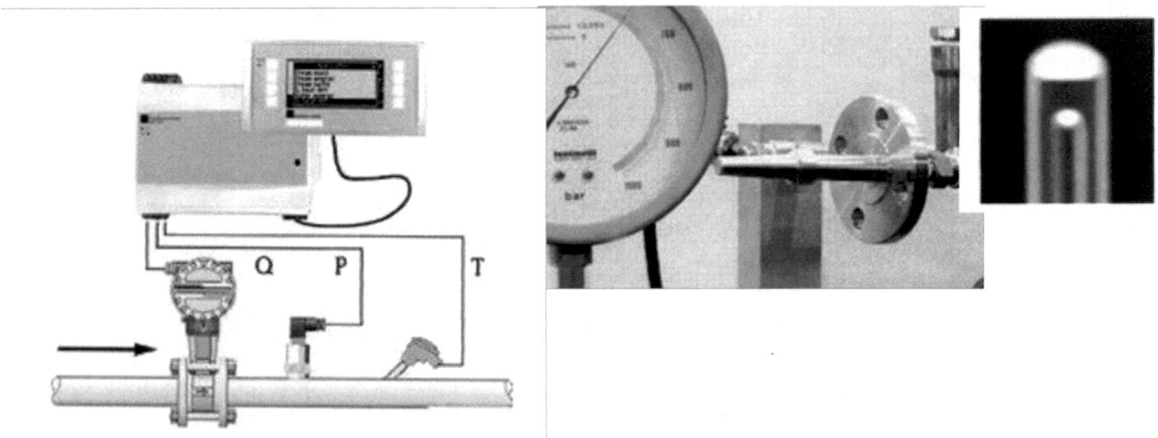

Figure 6. Measurement, Sensor in the flow technology, pressure, temperature [1]

2. Solution

To solve these types of problems we thought the realization of a flexible universal workstation through the use of standard robotics systems and image processing devices in cooperation with humans, where workers are largely free from the highly strenuous physical and fine motoric work, so that they can monitor and adjust the production process.

For the assembly of such small components we made contact with companies in southern Germany, namely: Endress + Hauser AG (sensor technology) [1], Bacher Medizintechnik GmbH [3] and Johann Deufel Chirurgische Instrumente GmbH [2] (both manufacturer and supplier of medical technology).

All these companies had the same problems: assembly of very small components, precision processes, long training of employees and cost pressure from low-wage countries.

They had, together with our University, the goal to realize a flexible human-robot workplace for assembly of small components, but with low cost standard devices.

The first objective was to establish a universal experimental workstation, expandable for series production. In this workstation the operator can respond quickly to changes, the robot can work quick and accurately, the image processing system can deliver high resolution pictures for object detection, exact positioning and process documentation.

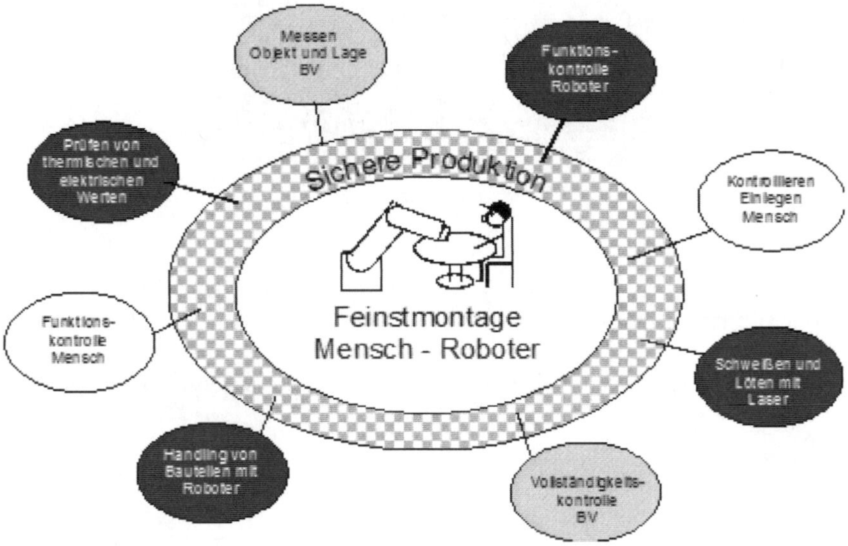

Figure 7. Workstation: Cooperation humans and robots

The next step was to find out which kind of technology was available in our technical university and then to decide what components we needed additionally to our devices.

We knew that robots have to work behind safety fences and other protective devices.

Figure 8. Robot cells at the University of applied sciences, Ulm

Figure 9. Robot cell in an exhibition [4]

3. Safety Eye

But because of the cells humans cannot work in direct cooperation with industrial robots. When for example a cell door is opened the robot will stop immediately.

A solution for this inconvenience is to implement a virtual safety system. For this aim we used the Safety Eye from the German company Pilz GmbH [5], developed in cooperation with Daimler AG.

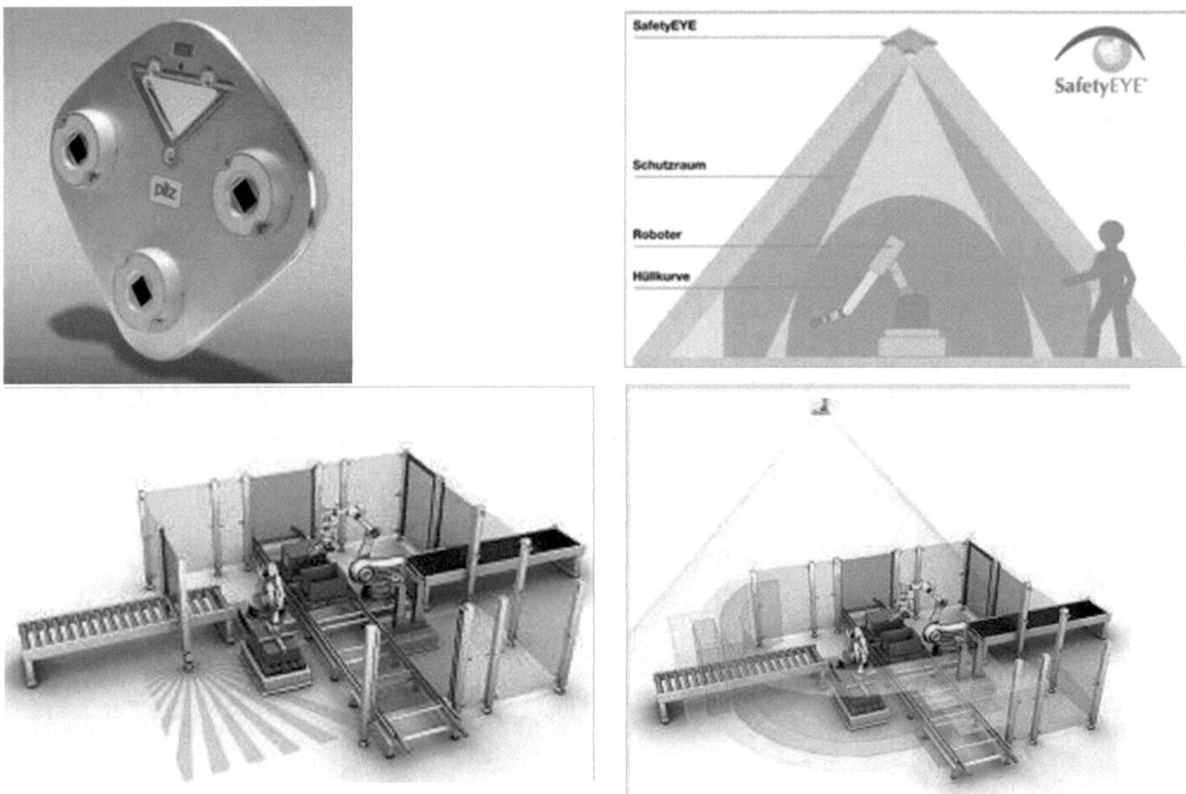

Figure 10. Safety Eye System from Pilz, Germany [5]

In this case operators, can walk up to the vicinity of the robots. The Safety Eye can detect every movement in the neighborhood of the robots. In front of a distance where robots can be dangerous for humans the robots reduce their working speed and an attention signal is switched on. If the operator ignores the signal and moves further towards the robot, the system stops immediately and a red light and an acoustical signal are switched on.

The safety system is designed as a redundant system, in the case if one of the operating systems has a dysfunction.

4. NeuroCheck

Other components of the work station are industrial camera systems. With them we were able to visualize objects, so that robots were able to search and detect the objects they needed for the assembly process.

Figure 11. Robots operating with cameras, University of applied sciences, Ulm [6]

The most used camera system in our experiments was NeuroCheck from Neurocheck GmbH, in Stuttgart, Germany. NeuroCheck [7] is used in many industrial applications in companies like: Daimler, BMW, Robert BOSCH, ZF, Philips, Federal Money Printing House, etc.

With NeuroCheck it is possible to identify barcodes, check object positions, to compare and recognize patterns, measure components, inspect surfaces, process colours, or to control and guide robots.

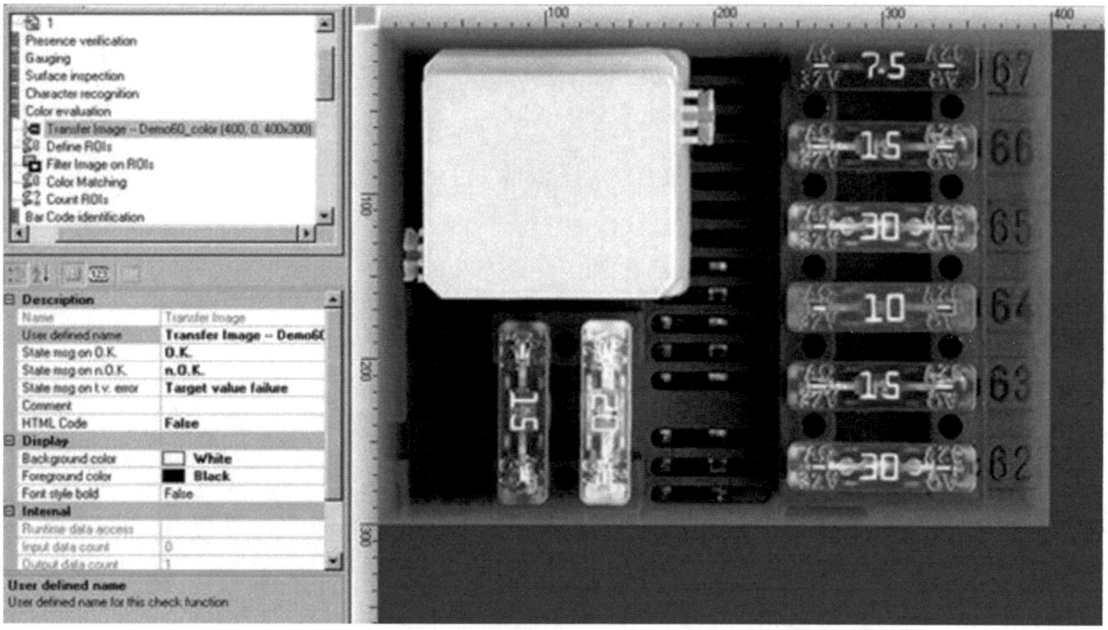

Figure 12. Color evaluation with NeuroCheck [7]

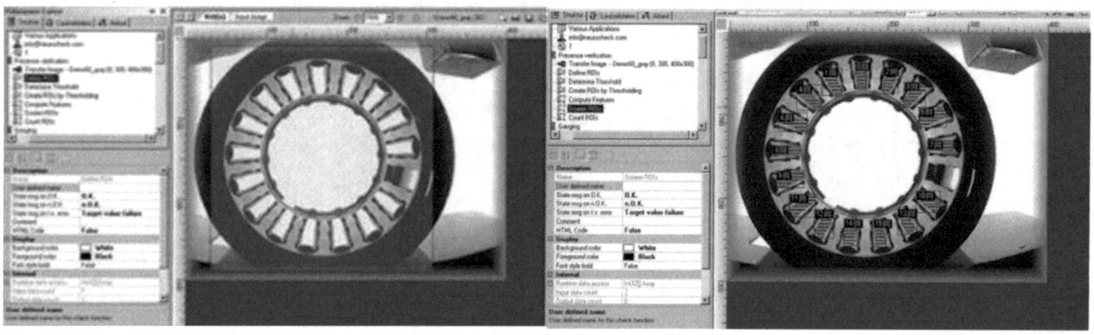

Figure 13. Presence verification with NeuroCheck [7]

Other advantage of NeuroCheck is the ability to use it for quality inspection, because it can inspect and record the whole automated production with a customized documentation.

5. Robots

In the workstations we used as well industrial robots from KUKA [8] as robots from Stäubli [9] for fine positioning.

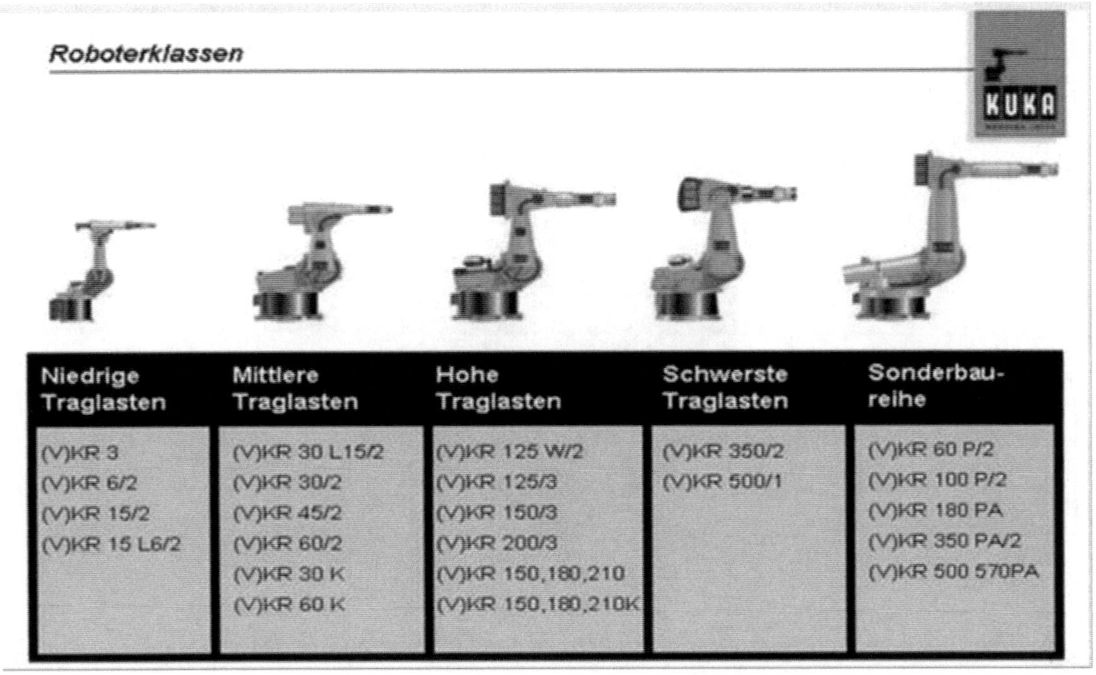

Figure 14. Industrial Robots from KUKA, Germany [8]

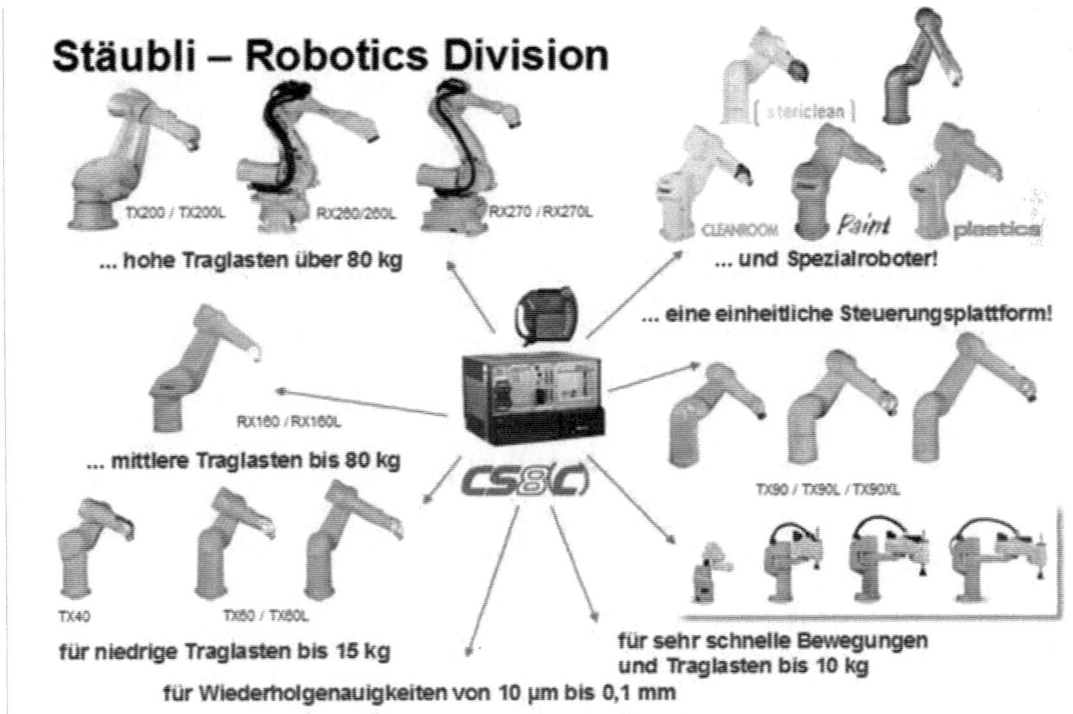

Figure 15. Robots from Stäubli, Germany/Switzerland/France [9]

IOP Conf. Series: Materials Science and Engineering **163** (2017) 012049 doi:10.1088/1757-899X/163/1/012049

6. x-y Fine Positioning

To reach precisions under 0,1 mm we used x-y-tables as they are used for micromachining, besides cameras and precise robots to position in the x - y direction.

Figure 16. Positioning device in X-Y direction

7. Conclusion

With this kind of device it was possible to get precisions under 10 μm. The precision was enough to place the objects in the desired position.

The project showed that companies can use common robots and technical devices for high precision applications and can realise workstations where robots can work together with operators.

References

[1] *Endress + Hauser, Messtechnik GmbH + CoKG*, Colmarer Straße 6, 79576 Weil am Rhein,Germany http://www.de.endress.com/de

[2] *Johann Deufel Chirurgische Instrumente GmbH*, Schreiberstraße 3, 78600 Kolbingen, Germany http://www.deufel.eu/

[3] *Bacher Ewald Medizintechnik GmbH*, Hinteres Öschle 6, 78604 Rietheim-Weilheim,Germany, http://www.ewald-bacher.de

[4] *MOTEK*, Internationale Fachmesse für Produktions- und Montageautomatisierung, 70629 Stuttgart-Flughafen, Messe Stuttgart, Germany, http://www.motek-messe.de/motek

[5] *Pilz GmbH & Co.* KG, Nellingen, Felix-Wankel-Straße 2, 73760 Ostfildern, Germany, https://www.pilz.com/de-DE

[6] *Hochschule Ulm,* Prittwitzstraße 10, 89075 Ulm, Germany, https://studium.hs-ulm.de/de/Seiten/Homepage.aspx

[7] *NeuroCheck GmbH*, Aldingen, Neckarstraße 76/1, 71686 Remseck am Neckar, http://www.neurocheck.de

[8] *KUKA AG*, Zugspitzstraße 140, 86165 Augsburg, Germany. http://www.kuka-robotics.com/germany/de/company

[9] *Stäubli Bayreuth GmbH*, Theodor-Schmidt-Straße 19, 95448 Bayreuth, Germany, http://www.staubli.com/index.php?id=22&L=2

IOP Conf. Series: Materials Science and Engineering **163** (2017) 012050 doi:10.1088/1757-899X/163/1/012050

Intelligent clutch control with incremental encoder to improve wear issues of an intercept pendulum in real time

C K Jalba[1], R Diekmann[1] and S Epple[2]

[1]Ulm University of Applied Sciences, Production Engineering and Production Economics, Ulm, Germany
[2]Kempten University of Applied Sciences, Faculty of Electrical Engineering, Kempten, Germany

E-mail: klausjalba@t-online.de

Abstract. A pendulum impact tester is a technical device which is used to perform plasticity characterizations of metallic materials. Results are calculated based on fracture behavior under pendulum impact loadings according to DIN 50115, DIN 51222/EN 10045. The material is held at the two ends and gets struck in the middle. A mechanical Problem occurs when testing materials with a very high impact toughness. These specimen often do not break when hit by the pendulum. To return the pendulum to its initial position, the operator presses a service button. After a delay of approximately 2 seconds a clutch is activated which connects the arm of the pendulum with an electric motor to return it back upright in start position. At the moment of clutch activation, the pendulum can still swing or bounce with any speed in any direction at any different position. Due to the lack of synchronization between pendulum speed and constant engine speed, the clutch suffers heavy wear of friction. This disadvantage results in considerable service and repair costs for the customer. As a solution to this problem this article presents a customized technical device to significantly increase the lifetime of the clutch. It was accomplished by a precisely controlled activation of the clutch at a point of time when pendulum and motor are at synchronized speed and direction using incremental encoders.

1. Pendulum impact tester

A pendulum impact tester is a technical device which is used to perform plasticity characterizations of metallic materials. [1]

Figure 1. Pendulum impact tester by Zwick/Roell [2]

Results are calculated based on fracture behavior under pendulum impact loadings according to DIN 50115, DIN 51222 / EN 10045 [1], [3], [4]. The specimen testing temperature range is between -190°C and +250°C.

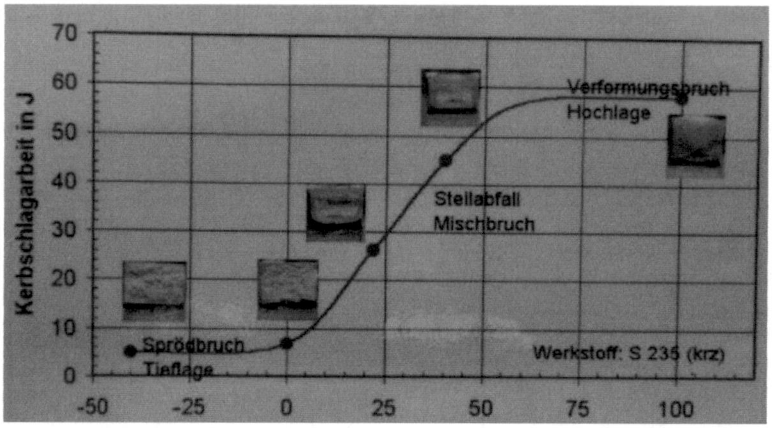

Figure 2. Impact strength as a function of temperature

By Charpy tests the metal is held at the two ends and gets struck in the middle.

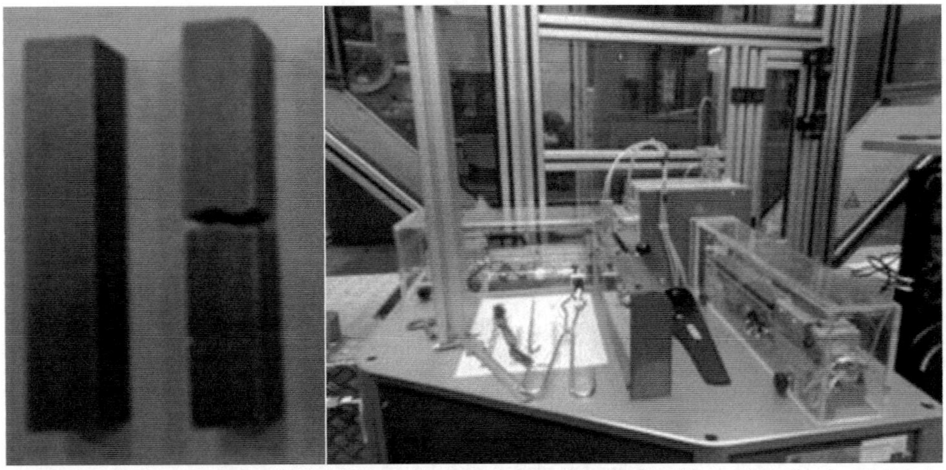

Figure 3. Pattern sample before and after the break (left), Clamping device (right)

2. Problems
To return the pendulum to its initial position, the operator presses a service button.

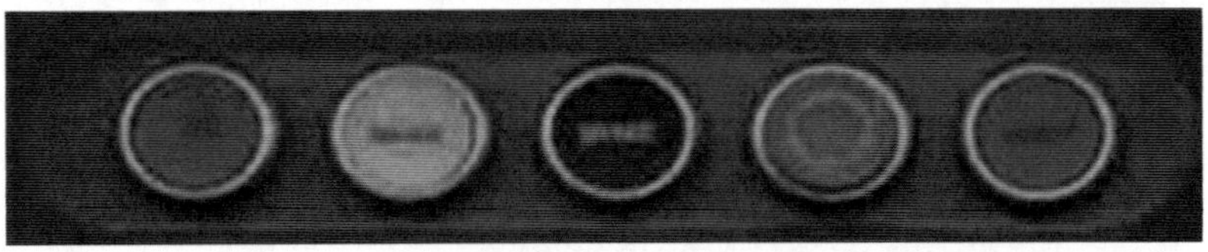

Figure 4. Control panel with start and service button

After a delay of approximately 2 seconds a clutch is activated which connects the arm of the pendulum with an electric motor to return it back upright in start position.

Figure 5. Electric motor and gear box, Laboratory Hochschule Ulm, Germany

At the moment of clutch activation, the pendulum can still swing or bounce with any speed in any direction at any different position. Due to the lack of synchronization between pendulum speed and constant engine speed, the clutch suffers heavy wear of friction. This disadvantage results in considerable service and repair costs for the customer.

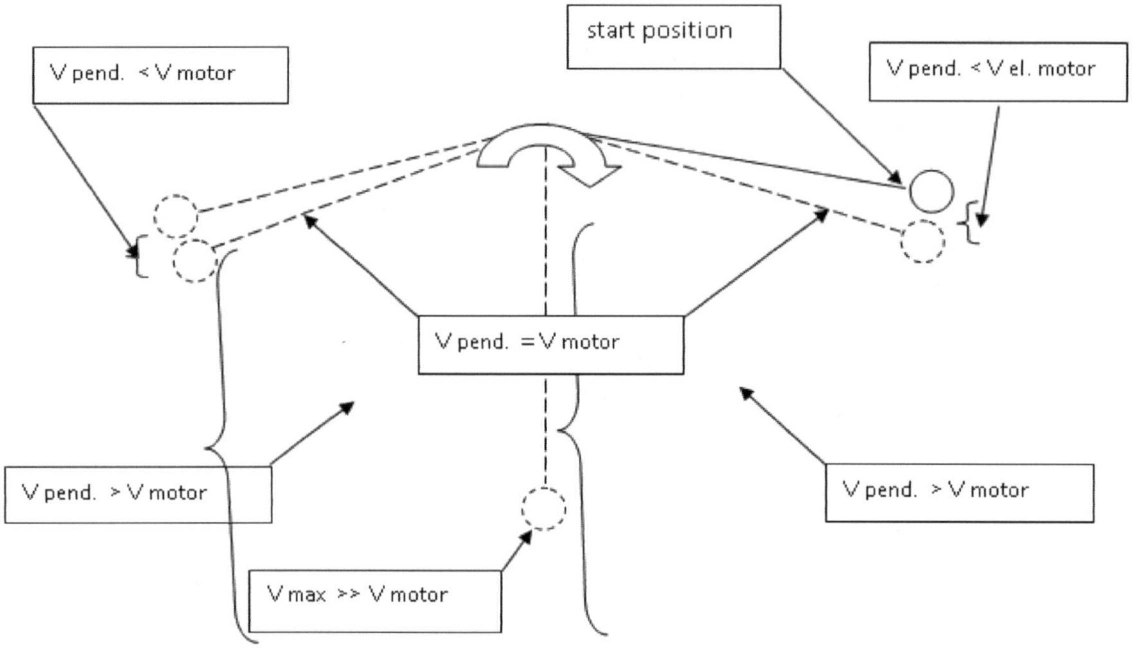

Figure 6. Pendulum in motion

A mechanical problem occurs when testing materials with a very high impact toughness are not broken like in the figure below. In this case the pendulum bounces backwards and the clutch suffers excessive wear due to the opposite turning directions between Pendulum and motor. (The clutch is then deteriorated by coming back of the pendulum to the start position.)

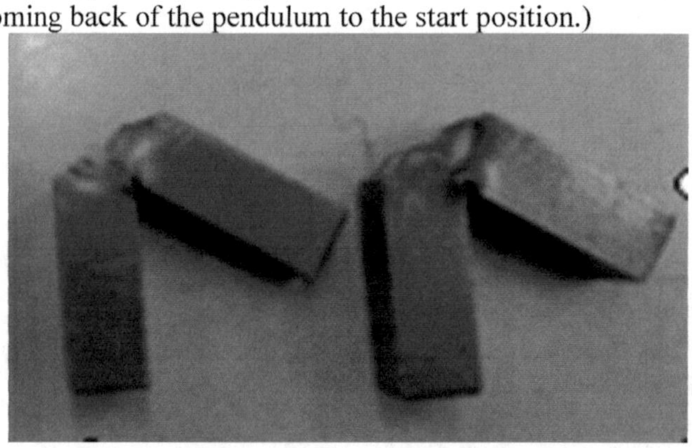

Figure 7. Patterns of tough metal

3. Solution

As a solution to this problem is a customized technical device to significantly increase the lifetime of the clutch. It was accomplished by a precisely controlled activation of the clutch at a point of time when pendulum and motor are at synchronized speed and direction using incremental encoders.

4. Basic requirements

At the time of design of the technical device the following basic requirements had to be considered:

- the device must be designed as an economic retrofit kit,
- must be easy to install,
- it must replace a delay relay,
- it has to fit in the hitherto existing housing,

Figure 8. Housing with delay relay

Figure 9. Housing with the new logic unit

IOP Conf. Series: Materials Science and Engineering **163** (2017) 012050 doi:10.1088/1757-899X/163/1/012050

- scheduled for 50/60 Hz,
- left or right pendulum locking,
- switching logic for different swing types,
- rugged design

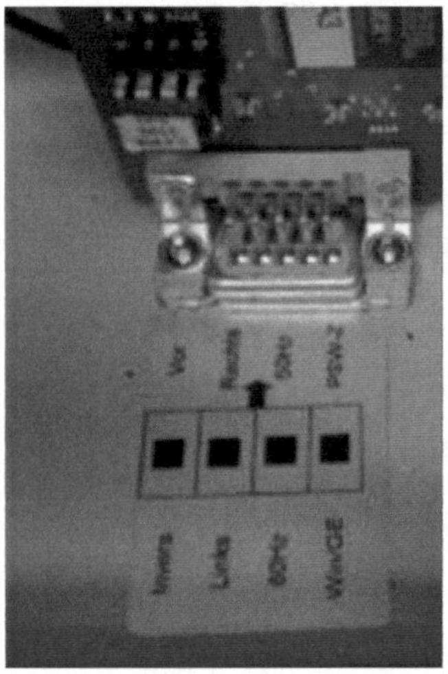

Figure 10. Switching pins

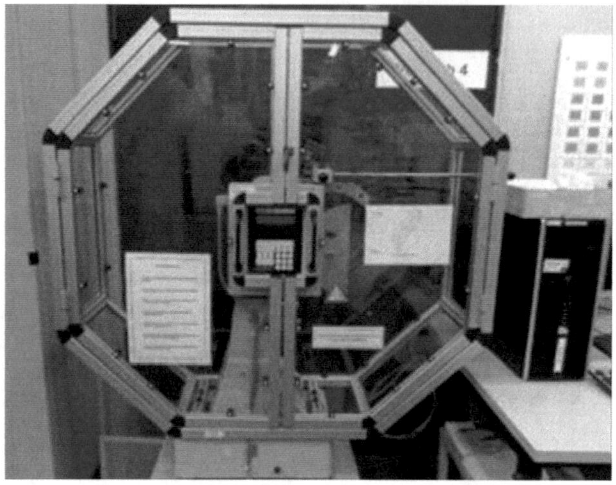

Figure 11. A smaller type of an intercept pendulum

5. Preliminary design of the digital circuit

At the design stage of the circuit following problems had to be solved:

- synchronization of the pendulum speed with engine speed when the clutch engages
- selection of hard- and software
- bidirectional evaluation of the pendulum motion
- measurements in real time
- freely programmable circuit

For the control could be selected between the following options: Analog circuits, microprocessors, PLCs, IC – circuits, and EPLD – Blocks.

After trading off the pros and cons, the decision was made for EPLD – Blocks.

To implement the task the development program MAX + plus II was chosen [6]. This software is suitable for both electronics beginners as well as professional developers.

A user friendly interface and the sophisticated digital database allow rapid and sophisticated realization of reliable digital solutions. The programmed digital circuits can be checked for their operability by corresponding simulation surfaces and then transferred to programmable hardware components.

The directional evaluation of the Pendulum motion was solved by using an incremental encoder, placed on the axle of the pendulum. With the incremental encoder, the speed of the pendulum could also be measured.

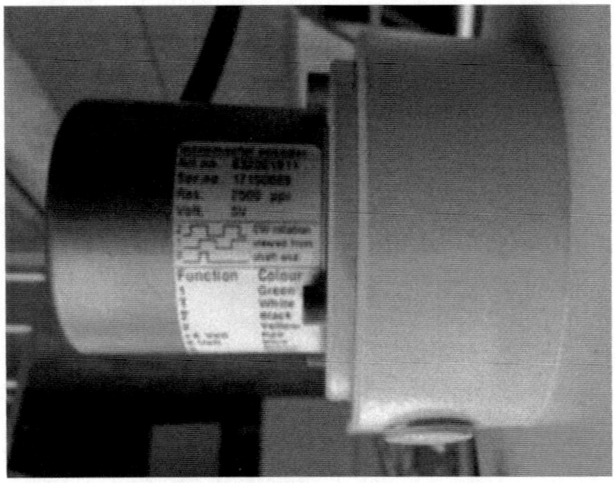

Figure 12. Digital encoder

These requirements lead to the depicted functional block diagram:

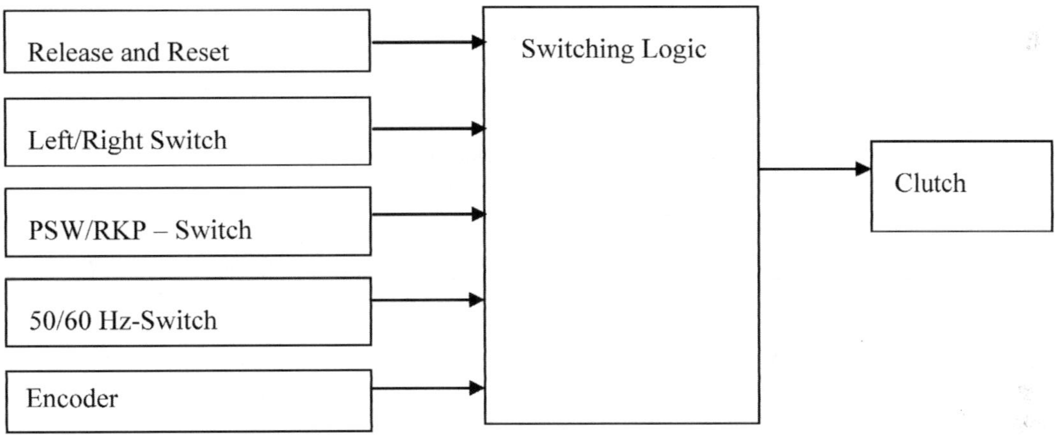

Figure 13. Block Diagram

6. Input data and calculations

The table below shows the key data of the device.

Table 1. Key Data

Maximal Engine Runtime tmax	160	s
Minimum Engine Runtime tmin	16	s
Number of Encoder Pulses	10000	Pro 360°
Clock Speed	48	MHz
Pendulul Length	0,7	m
Vmax Measured Maximal Speed	5	m/s

6.1. Clock Time:T

T = 1/f; f = frequency

T = 1/48 MHz → T = 20,83 ns

Clock = 20,83 ns at 48 Mhz

6.2. Pendulum Track: U

U is the maximum distance the pendulum can theoretically run during a full rotation.

U = circumference

U = 2π*0,7m=4,4m

Shortest pulse time tmin

tmin = U : Vmax Vmax = 5m/s

tmin = 4,4 m : 5 m/s = 0,88 s

By 5 m/s and 48 MHz are needed 0,88 s : 20 ns = 42227556 pulses.

42227556 pulses : 10000 pulses/360° = 4222 clock/pulse

If the EPLD frequency is reduced from 48 MHz to 3 MHz

48 MHz : 16 = 3 MHz

can be measured

4222 clock/pulse : 16 = 263 clock/pulse.

This clock frequency is thus sufficient for Vmax . Analog clocks per pulse can be calculated for other swing speeds.

Table 2. Cycles per pulse at different orbital periods

Rotation Time	Clock/Pulse	Clock/Pulse
S	f = 48 MHz	f = 3 MHz
0,88	4223	264
0,90	4321	270
1,00	4801	300
1,10	5281	330
1,20	5761	360
1,30	6241	390
1,40	6721	420
1,50	7201	450
1,60	7681	480
1,70	8161	510
1,80	8641	540
1,90	9121	570
2,00	9602	600
4,00	19203	1200
6,00	28805	1800
8,00	38406	2400
10,00	48008	3000
60,00	288046	18003
120,00	576092	36006
160,00	768123	48008

6.3. Counter bit width

$2x = 48008$ mit x = bit width

$x = \ln 48008 : \ln 2$

$x = 15{,}55 \sim 16$ bit

It is therefore a bit width of 16 required. With this bit width both can be detected: very fast and very slow engine rotation speed.

6.4. Logic Unit

The digital logic must include the following function blocks due to the established requirements and the calculation data supplied.

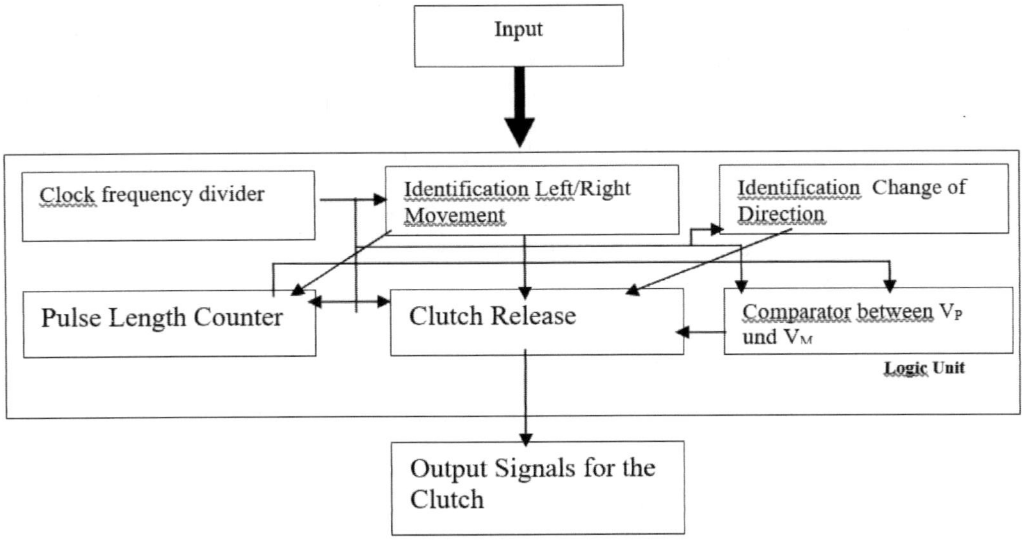

Figure 14. Logic diagram

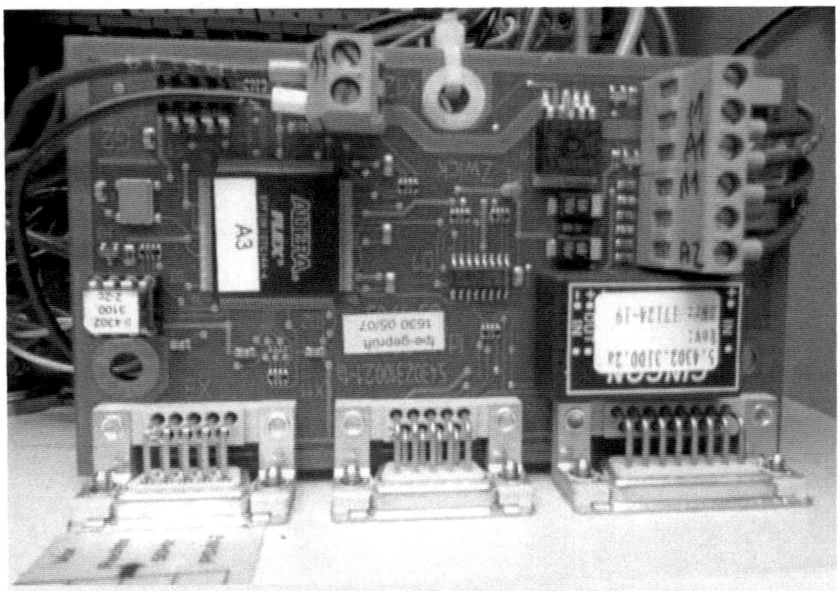

Figure 15. Logic Unit

IOP Conf. Series: Materials Science and Engineering **163** (2017) 012050 doi:10.1088/1757-899X/163/1/012050

7. Conclusion

After the circuit met all the requirements in the experimental stage, it was released for serial production. At this juncture the control unit is installed on both: On new equipment as well as an improvement to earlier devices. It contributes to the increases in sales and to security of employment.

References

[1] *Standards*: DIN EN ISO 148-1:2011-01 (Ersatz für DIN EN 10045-1), (Kerbschlagbiegeversuch); DIN 53 453, DIN 53 753 und DIN 53 448; BS 2782, P.3 Methode 350 und 359; ASTM D 256, Methode A und B; AFNOR NF T51-111;BS 7413; ISO 8256 (Bestimmung der Schlagzugzähigkeit Schlagzugversuch); ASTM D 4812, https://de.wikipedia.org/wiki/Pendelschlagwerk Zwick GmbH & Co. KG, August-Nagel-Str. 11, 89079 Ulm, Germany, https://www.zwick.de/de/aktuelles/news-detail/article/zwick-baut-das-groesste-pendelschlagwerk-der-welt.html

[2] Siebel E 1940 *Handbuch der Werkstoffprüfung*, Erster Band, Springer Verlag Berlin Heidelberg GmbH

[3] Macherauch E and Zoch H W 2014 *Praktikum in Werkstoffkunde*, 12, Auflage, Springer Fachmedien, Wiesbaden

[4] ****Hochschule Ulm*, Prittwitzstraße 10, 89075 Ulm, Germany

[5] ****Altera Corporation*, 101 Innovation Drive, San Jose, CA 95134, USA, https://www.altera.com

The curving calculation of a mechanical device attached to a multi-storey car park

C G Muscalagiu[1], I Muscalagiu[2] and D M Muscalagiu[1]

[1]University of Craiova, Faculty of Mechanical Engineering, Craiova, Romania
[2]Politehnica University of Timisoara, Department of Electrical Engineering and Industrial Informatics, 5 Revolution Street, Hunedoara, 331128, Romania

E-mail: cosmin_muscalagiu@yahoo.com

Abstract. Study bunk storage systems for motor vehicles developed much lately due to high demand for parking in congested city centers. In this paper we propose to study mechanism drive bunk platforms for dynamic request. This paper aims to improve the response mechanism on a platform behavior self during operation of the system and identify hot spots. In this paper we propose to analyze the deformations of the superposed platform in the points of application of the exterior forces produced by the weight of the vehicle in a dynamic way. This paper aims to automate the necessary computation for the analysis of the deformations of the superposed platform using Netlogo language.

1. Introduction

In previous studies [1], [2] were presented mechanical and kinematic schemes of different types of existing storey car parks:

- bunk parking system with mobile boxes, in this case the platform is supported by a cable wound on a drum;
- car parks system, in this case the platform is supported by a cable or chain driven by an electric motor;
- bunk parking system with fixed boxes, car platform is a movement in the vertical and horizontal movement, it slides in the fixed structure of the drive shaft of the electric motor via a cable;
- second places parking system, in this case the car platform is driven by a hydraulic cylinder.

These systems have a number of disadvantages to each type of structure established in part, however they are the best solution to achieve in crowded parking lots in the center of large cities where the ground surface does not allow the construction of a parking ground. It is necessary to periodically check the cable (chain), roller bearings (gear) and electric motor combined welds. There are two types of car parking systems: traditional and automated long-term automatic car parking systems are more efficient compared to traditional garages. Multi-storey car systems are less expensive automated parking place because they require a large amount of construction and a small area on the ground [1], [2] .

This paper aims to automate the necessary computation for the analysis of the deformations of the superposed platform using NetLogo language [3].

2. Patterns of kinematic scheme for a parking structure bunk

In Figures 1 and 2 are shown kinematic diagrams of two-storey parking systems with two storage locations of vehicles, upper end platform is driven by a hydraulic cylinder (Figure 1) and an electric motor (Figure 2). In both variants of unlocking the vehicle parked on the upper platform to be released lower platform which is a major disadvantage. The space required for placing the system of Figure 2 must be greater than the length of a car as car parking upper deck flips it to the lower deck. Both systems have low height, increasing the number of parking spaces for these systems can be multiplied in number to the basement blocks, shops, etc.

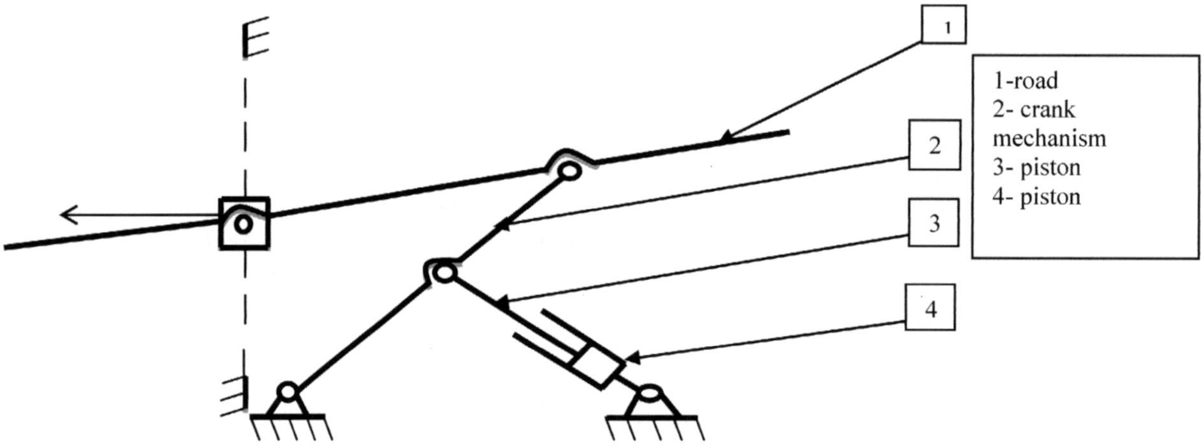

Figure 1. Hydraulic mechanism kinematic scheme

The system of Figure 1 is the simplest parking system and its installation does not require training the surrounding area, parking space height can be adjusted according to the height of the car parked. Both systems require not authorized personnel for handling; they can be operated even by the head auto. Platform system of Figure 2 can be driven by a cable wound on a drum driven by an electric motor. In the event of a power failure the system can be equipped with a mechanism for manual operation or with an electric generator.

Both systems require a maintenance contract to ensure safe operation. There are a number of areas where hydraulic systems offer many advantages that cannot be covered by electrical transmissions because hydraulics are usually characterized by a force much greater torque and power. Platforms powered or hydraulic structures are designed for a particular application payload is clearly defined, where the payload increases appreciably approximately 20% electrical system does not operate normally due to overload.

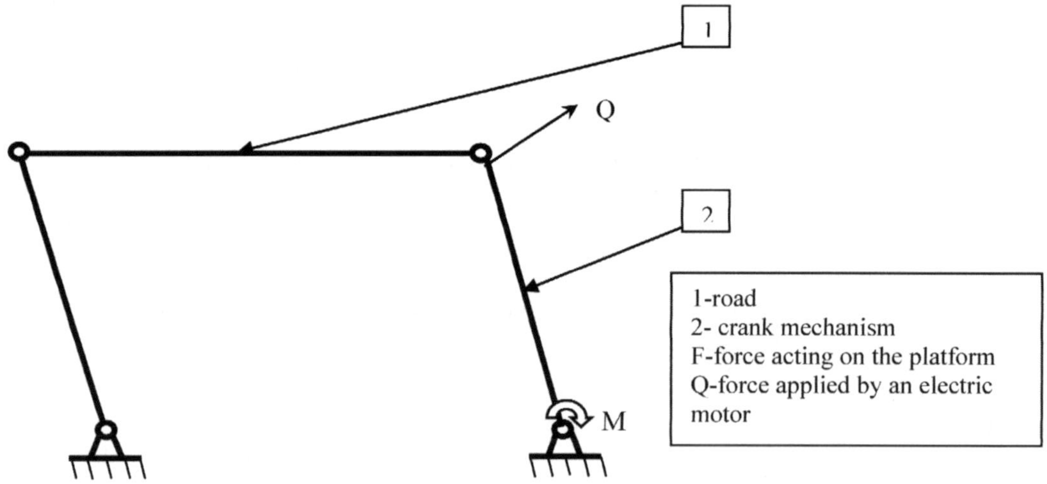

Figure 2. Scheme kinematic mechanism driven by an electric motor

IOP Conf. Series: Materials Science and Engineering **163** (2017) 012051 doi:10.1088/1757-899X/163/1/012051

On the other hand, a hydraulic system can accommodate a much wider range of operating parameters, usually a properly designed hydraulic system can be overloaded by up to 50% more efficient operation and features [4-7].

3. Curving calculation in point C

The curving application of force F can be determined with the aid of Mohr-MAXWELL [8-11]. To determine F_1 and F_2 forces the following conditions have to concure:

$F_1+F_2=F$ and $F_2(kl-a)=F_1[(1-k)l-a]$ (1)

After solving the system 1, F_1 and F_2 forces can be determined (relation 2)

$$F_1(k,a,l,F)= F\frac{k\cdot l-a}{l-2\cdot a}; \qquad F_2(k,a,l,F)= F\frac{(1-k)\cdot l-a}{l-2\cdot a} \quad (2)$$

The verifying rule for point k=0,5 $F_1=F_2$

Unknown values of the system: Q=Q(t,k,a,l,r,F); $V_1=V_1$(t,k,a,l,F); $V_2=V_2$(t,k,a,l,F); $H_1=H_1$(t,k,a,l,F); $R_{1V}=R_{1V}$(t,k,a,l,r,F); $R_{1H}=R_{1H}$(t,k,a,l,r,F); $R_{4V}=R_{4V}$(t,k,a,l,r,F) $R_{4H}=R_{4H}$(t,k,a,l,r,F)

In order to calculate reactions an equilibrium equation must be written. Is the equation of moment 1 and the equation of coordinate axes proiections (Figure 3). After solving the equations system the following unknown values can be obtained:

$$Q(t,k,a,l,r,F)=\frac{\cos\alpha\cdot[F\cdot l-a\cdot(F_1-F_2)-F_2\cdot l+r\cdot\sin\alpha\cdot F_1\cdot a+r\cdot\sin\alpha F_2\cdot l-r\cdot\sin\alpha F_{2.}\cdot a]}{\sin\alpha\cdot(\sin\alpha\cdot\cos\varepsilon+\sin\varepsilon\cdot\cos\alpha)};$$

$$V_1(t,k,a,l,F)=\frac{F\cdot l-F_1\cdot a-F_2\cdot l+F_2\cdot a}{l}; \quad V_2(t,k,a,l,F)=\frac{F_1\cdot a+F_2\cdot l-F_2\cdot a}{l}; \quad (3)$$

$$H_1(t,k,a,l,F)=\frac{\cos\alpha\cdot(F_1\cdot a+F_2\cdot l-F_2\cdot a)}{\sin\alpha\cdot l};$$

$$R_{1V}(t,k,a,l,r,F)=\frac{F\cdot l-F_1\cdot a-F_2\cdot l+F_2\cdot a}{l}-Q\sin\varepsilon;$$

$$R_{4V}(t,k,a,l,r,F)=-\frac{F_1\cdot a+F_2\cdot l-F_2\cdot a}{l};$$

$$R_{1H}(t,k,a,l,r,F)=\frac{\cos\alpha\cdot(F_1\cdot a+F_2\cdot l-F_2\cdot a)}{\sin\alpha\cdot l}-Q\cos\varepsilon;$$

$$R_{4H}(t,k,a,l,r,F)=-\frac{\cos\alpha\cdot(F_1\cdot a+F_2\cdot l-F_2\cdot a)}{\sin\alpha\cdot l}$$

We have a moment equation of real charging values as seen in Figure 3 on each of the all 5 steps:

-moment 1-2 equation:

$M_{12}(t,k,a,l,r,F,x)= -x\cdot(H_1(t,k,a,l,F)\cdot\sin\alpha+ V_1(t,k,a,l,F)\cdot\cos\alpha)+Q(t,k,a,l,F)\cdot x\cdot(\sin\varepsilon\cdot\cos\alpha+\cos\varepsilon\cdot\sin\alpha)$ (4)

-moment 3-4 equation:

$M_{34}(t,k,a,l,F,x)=x\cdot(H_1(t,k,a,l,F)\cdot\sin\alpha- V_2(t,k,a,l,F)\cdot\cos\alpha)$ (5)

-moment 6-3 equation:

$M_{25}(t,k,a,l,F,x)= V_1(t,k,a,l,F)\cdot x$

$M_{56}(t,k,a,l,F,x)= V_1(t,k,a,l,F)\cdot x-F_1(k,a,l,F)(x-a)$ (6)

$M_{63}(t,k,a,l,F,x)= V_1(t,k,a,l,F)\cdot x-F_1(k,a,l,F)\cdot(x-a)-F_2(k,a,l,F)(x-l+a)$

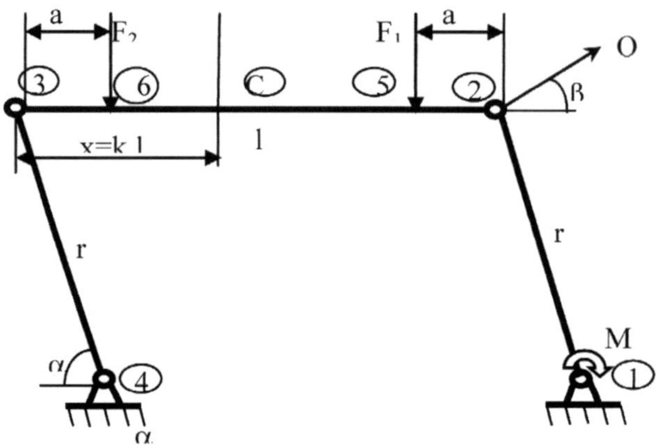

Figure 3. Charging figure

The single force F produces a moment equation presented in Figure 4.
This equation are: [3-5]

$m_{25}(t,k,a,l,x)= V_{C1}(t,k,a,l)\cdot x$ (7)
$m_{5C}(t,k,a,l,x)= V_{C1}(t,k,a,l)\cdot x$
$m_{C6}(t,k,a,l,x)= V_{C1}(t,k,a,l)\cdot x-1(x-l+kl)$
$m_{63}(t,k,a,l,x)= V_{C1}(t,k,a,l)\cdot x-1(x-l+kl)$

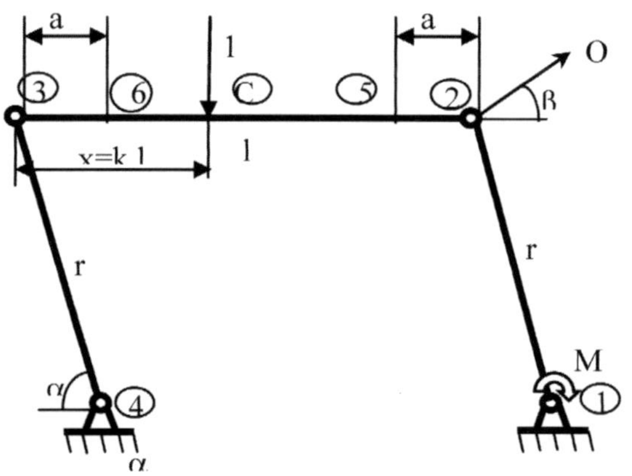

Figure 4. Single force charging

After solving the system equation the following unknown values can be found:
$V_{C1}=k;\quad V_{C2}=1-k$ (8)

Over the point F where the force is pressing, a curving point do occures. This curving can be calculated using the MOHR-MAXWELL method, using relations (4-7). Considering the type profile the deformation point is shown (relation 9)

$$\mathbf{fc}=\frac{1}{E\cdot I}\Big(\int_0^a M_{25}(t,k,a,l,r,F,x)\cdot m_{25}(t,k,a,l,r,x)dx+\int_a^{l-a} M_{56}(t,k,a,l,r,F,x)\cdot m_{5C}(t,k,a,l,r,x)dx$$
$$+\int_a^{l-a} M_{56}(t,k,a,l,r,F,x)\cdot m_{C6}(t,k,a,l,r,x)dx+\int_{l-a}^{l} M_{63}(t,k,a,l,r,F,x)\cdot m_{63}(t,k,a,l,r,x)dx\Big) \quad (9)$$

Relation 9 can be introduced in an application. An automatic calculation programme can calculate the analitical equation of the curving point where the force F is pressing on relation (10).

$$\text{fc(t,k,a,l,r,F,E,I,x)}= \frac{1}{E\cdot I}\cdot\left[\frac{l}{3}\cdot F\cdot k^2\cdot a^3-\frac{l}{6}\cdot F\cdot a\cdot k\cdot f_{c1}(t,k,a,l,r,F,E,I,x)\right]+$$
$$+\frac{1}{E\cdot I}\cdot\left[\frac{(-l)}{6}\cdot a^3\cdot F\cdot\frac{4\cdot k\cdot a+a-3\cdot k\cdot l}{(-l+2\cdot a)}\cdot k-\frac{l}{6}\cdot a\cdot F\cdot f_{c2}(t,k,a,l,r,F,E,I,x)\right]+$$
$$\frac{1}{E\cdot I}\cdot\left[\frac{(-l)}{6}\cdot a^2\cdot F\cdot f_{c3}(t,k,a,l,r,F,E,I,x)+\frac{(-l)}{6}\cdot l^2\cdot F\cdot f_{c4}(t,k,a,l,r,F,E,I,x)\right]+$$
$$+\frac{1}{E\cdot I}\cdot\left[\frac{l}{6}\cdot F\cdot f_{c5}(t,k,a,l,r,F,E,I,x)\right] \tag{10}$$

Due to the big size of the curving equation the following notations are being calculated in relations (11-15)

$\text{fc}_1\text{(t,k,a,l,r,F,E,I,x)}=$

$$\frac{(-k\ )\cdot l^3+6\cdot k\ \cdot l^2\cdot a+2\cdot l^3-9\cdot l^2\cdot a-9\cdot l\cdot a^2\cdot k+12\cdot l\cdot a^2+4\cdot a^3\cdot k-5\cdot a^3}{(-l+2\cdot a)} \tag{11}$$

$$\text{fc}_2\text{(t,k,a,l,r,F,E,I,x)}=\frac{(-k^2)\cdot l^3-3\cdot k^2\cdot l\cdot a^2+3\cdot l^2\cdot k\cdot a+6\cdot l\cdot a^2\cdot k+4\cdot k^2\cdot a^3}{(-1+2\cdot a)}+$$
$$+\frac{(-9)\cdot a^3\cdot k+l^3-3\cdot l^2\cdot a-3\cdot l\cdot a^2+5\cdot a^3}{(-1+2\cdot a)} \tag{12}$$

$$\text{fc}_3\text{(t,k,a,l,r,F,E,I,x)}=\frac{4\cdot k^2\cdot a^2-3\cdot k\cdot a^2-9\cdot a\cdot k^2\cdot l+6\cdot a\cdot k\cdot l+6\cdot k^2\cdot l^2-a^2}{(2\cdot a-l)}+$$
$$+\frac{3\cdot l\cdot a-6\cdot k\cdot l^2}{(2\cdot a-l)} \tag{13}$$

$$\text{fc}_4\text{(t,k,a,l,r,F,E,I,x)}=\frac{l^2\cdot k^2\cdot a+k^2\cdot l\cdot a^2-10\cdot l\cdot a^2\cdot k+4\cdot l^2\cdot k\cdot a-l^3\cdot k}{[(2\cdot a-l)\cdot(a-2\cdot l)]}+$$
$$+\frac{3\cdot a^3\cdot k-12\cdot a\cdot k\cdot l+12\cdot k\cdot a^2+3\cdot a\cdot k^2\cdot l-6\cdot k^2\cdot a^2+3\cdot k\cdot l^2+9\cdot l\cdot a^2-5\cdot l^2\cdot a}{[(2\cdot a-l)\cdot(a-2\cdot l)]}+$$
$$+\frac{l^3-3\cdot a^3+9\cdot l\cdot a-6\cdot a^2-3\cdot l^2}{[(2\cdot a-l)\cdot(a-2\cdot l)]} \tag{14}$$

$$\text{fc}_5\text{(t,k,a,l,r,F,E,I,x)}=\frac{(-12)\cdot a\cdot k\cdot l^3-3\cdot l^4+6\cdot a^4-12\cdot k\cdot a^4-3\cdot a^2\cdot l^2-9\cdot a^3\cdot l}{[(2\cdot a-l)\cdot(a-2\cdot l)]}+$$
$$+\frac{9\cdot l^3\cdot a+l^5+9\cdot a^2\cdot k\cdot l^2+12\cdot a^3\cdot k\cdot l+k^2\cdot l^4\cdot a+3\cdot k^2\cdot l^3\cdot a+44\cdot k\cdot l\cdot a^4}{[(2\cdot a-l)\cdot(a-2\cdot l)]}+$$
$$\frac{(-3)\cdot k^2\cdot l\cdot a^3-6\cdot k^2\cdot l^2\cdot a^2-19\cdot k^2\cdot l\cdot a^4-7\cdot k\cdot l^3\cdot a^2-23\cdot k\cdot l^2\cdot a^3+4\cdot k\cdot l^4\cdot a}{[(2\cdot a-l)\cdot(a-2\cdot l)]}+$$
$$\frac{k^2\cdot l^3\cdot a^2+9\cdot k^2\cdot l^2\cdot a^3+3\cdot l^4\cdot k-k\cdot l^5-17\cdot k\cdot a^5+6\cdot k^2\cdot a^4+8\cdot k^2\cdot a^5}{[(2\cdot a-l)\cdot(a-2\cdot l)]}+$$

$$+\frac{(-25)\cdot l\cdot a^4+6\cdot a^2\cdot l^3+14\cdot a^3\cdot l^2-5\cdot l^4\cdot a+9\cdot a^5}{[(2\cdot a-l)\cdot(a-2\cdot l)]} \tag{15}$$

Relation 10 is introduced into a programme of automatic calculation. This programme is using all the parameters shown below. At the end of the calculation the curving calcules of point F are being obtained.

$$F\left(\frac{f_c}{E\cdot I}\right)\text{(Table 1).}$$

Entrance parameters:
-force F (1000daN,1500daN,2000daN) ;
-climbing time and descending time of the platform is 10 s (2s,6s,10s) ;
-platform length l=430 cm.
-a- pressing distance of forces F_1 and F_2 $a=\lambda l$, $\lambda=0,25$;
-r- platform height r=180 cm;
-α crank angle $\alpha(t)=\omega(t)\cdot t$ $\omega(t)=0,5\frac{3,14}{t_u}$;
-t_u- necessary time for the crank to reach the optimal position t_u=10 s;
-ε-force angle Q, $\varepsilon(\alpha)=\text{arctg}\frac{1-\sin(\alpha)}{\cos(\alpha)}$;
-A area of the section A=76,2cm^2;
-W-module endurance W=254,79cm^3;
-k-constant k=(0,4-0,6); k=0,4
-r platform height r=(1,4-1,8);

Table 1. Point C curving

Ascending/ descending time of the platform	F=1000daN	F=1500daN	F=2000daN
t-2s	2168604602,678571	3252906904,017857	4337209205,357142
t-6s	2168604602,678571	3252906904,017857	4337209205,357142
t-10s	2168604602,678571	3252906904,017857	4337209205,357142

The curving reached in point 5 was calculated almost the same way. The results are presented in Table 2.

Table 2. Point 5 curving

Ascending/ descending time of the platform	F=1000daN	F=1500daN	F=2000daN
t-2s	3157821602.306547	4736732403.459821	6315643204.613095
t-6s	3157821602.306547	4736732403.459821	6315643204.613095
t-10s	3157821602.306547	4736732403.459821	6315643204.613095

4. Conclusions

A first conclusion that can be deducted from the curving of platform structure is that the value of deformation on point C is not influenced by the time of acending and descending of the platform (t).

The curve that accurs in point C is directly proportional to the growth of the force F and it doubles the multiplication of force F. This deformation produced in point C is very small comparing to the variation of time t and force F.

This study has the purpose to show the platform curving in point C and to present also its deformation in point 5. As it can be observed the maximum value was obtained in point 5 for F= 2000daN.

Similar to point C deformation on point 5 is not affected by the position of the platform.

Using dedicated software (NetLogo language) it allowed to automate the necessary computation for the analysis of the deformations of the superposed platform.

References

[1] Muscalagiu C and Muscalagiu D 2016 Experimental Analysis Of Tensions From An Working Mechanism Attached To A Multi-Storey Car Parks, *Annals of the Faculty of Engineering Hunedoara* **9**(3) 237-242

[2] Muscalagiu C, Ilincioiu D and Muscalagiu D 2016 Study of Handling Mechanism Attached to a Multistorey Car Parks, *Applied Mechanics and Materials* **823** 235-240

[3] Wilensky U *NetLogo itself: NetLogo.* Available: http://ccl.northwestern.edu/netlogo/. Center for Connected Learning and Computer-Based Modeling, Northwestern University, Evanston

[4] Curtu I and Sperperchez F L 1988 *Strength of Materials*, Vol. I and II, Printing Transilvania University of Brasov, Romania

[5] Mace Gh 1986 *Strength of Materials*, Academy Publisher House, Bucharest, Romania

[6] Zehfuss J and Hosser D 2008 *A parametric natural fire model for the structural fire design of multi storez buildings*, Berlin, Germany

[7] Deutsch J 1976 *Strength of Materials*, Editura Didactica si Pedagogica, Bucharest, Romania

[8] Ilincioiu D, Rosca V and Georgescu I 2005 *Mechanical resistance rating of the structure of fixed cranes*, International Conference of Structural Integrity of Welded Structures, Timisoara, Romania

[9] Ilincioiu D, Miriţoiu C and Padeanu A 2010 *Strength of Materials*, vol. I, University Publishing House, Craiova, Romania

[10] Ilincioiu D, Miriţoiu C and Padeanu A 2011 *Resistance Materials*, vol. II, University Publishing House, Craiova, Romania

[11] Miriţoiu C and Ilincioiu D 2012 *Strength of Materials. Applications*, Sitech Publishing House, Craiova, Romania

IOP Conf. Series: Materials Science and Engineering **163** (2017) 012052 doi:10.1088/1757-899X/163/1/012052

Using Moeller PLC in automation of an artesian fountains

C Barz[1], T Latinovic[2], S Deaconu[3], D Preradović[4], P P Pop[5] and A Pop-Vadean[5]

[1]Technical University of Cluj-Napoca, North University Centre of Baia Mare, Victor Babes str., no. 62A, 430083 Baia Mare, Romania.
[2]University of Banja Luka, Faculty of Mechanical Engineering, 73 Vojvode Stepe Stepanovića Blvd., 78000 Banja Luka, Bosnia and Hercegovina.
[3]Politehnica University of Timişoara, Faculty of Engineering of Hunedoara, Department of Electrical Engineering and Industrial Informatics, 5 Revolution Street, 331128 Hunedoara, Romania.
[4]University of Banja Luka, Faculty of Economics, 4 Majke Jugovića Street, 78000 Banja Luka, Bosnia and Hercegovina.
[5]Technical University of Cluj-Napoca, Department of Mechatronics and Machine Dynamics, Blvd. Muncii, no. 103-105, 400641 Cluj-Napoca, Romania.

E-mail: crbarz@yahoo.com

Abstract. The paper presents the use of a Moeller PLC in the automation of an artesian fountains from Baia Mare. The application is developed in Ladder Diagram and contains two modes of operation. The first mode is the Automatic Mode in which are defines certain operating times of the artesian fountains pumps. These set times make recurrence of the operation. The second mode is the Manual Mode in which each pump is operated independently. Manual mode is only active for service personal, according to the user accounts and authorization rights of its. All orders are made with a touch screen Weintek in menus that are created for application. Using the Service Menu can visualize defects occurred during the regime of operation, which is recorded.

1. Introduction

The PLC is a specialized industrial computer for real-time applications. The PLCs appeared in the late 60s and are integrated systems that usually contain a processor, a power supply, input modules, output modules and special modules dedicated (Figure 1).

Last standard (IEC 61131-3) attempted to combine the programming languages of PLCs in an international standard. As a result of these efforts, today we have PLCs which can be programmed through diagrams with block functions, instructions lists, the C language and structured text, all at once [1].

IEC 61131-3 is the third part (of 8) of the open international standard IEC 61131 for programmable logic controllers, and was first published in December 1993 by the IEC. The current (third) edition was published in February 2013.

Part 3 of IEC 61131 deals with programming languages and defines two graphical and two textual PLC programming language standards:

- Ladder diagram (LD), graphical
- Function block diagram (FBD), graphical
- Structured text (ST), textual

- Instruction list (IL), textual
- Sequential function chart (SFC), has elements to organize programs for sequential and parallel control processing.
- Continuous Function Chart (CFC). This language is an extension to the IEC 61131-3 standard, which gives free positioning of graphic elements.

The PLCs are some devices designed to replace the sequential relay circuits which are used to control various machines. They are used for the operation of machine tools, machinery, installations, etc. to perform certain tasks, and for start and stop them, however they are achieved through the program memorized by PLC. The PLC operates by inspecting its inputs and, depending on their condition and the program designed for automation, switching the outputs on or off.

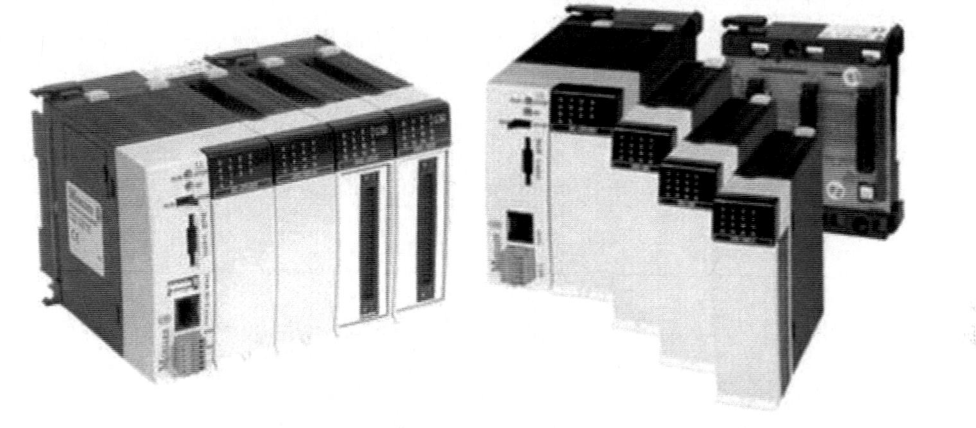

Figure 1. The Moeller PLCs

By using the PLCs, the execution times of products and services and the work performed by personnel can be reduced, improving the efficiency of industrial production.

On the PLCs market are many brands of appliances (Siemens, Moeller, Omron, Mitsubishi Electric, Schneider, Telemecanique, Omron, General Electric), every brand trying to attract with certain special specifications, but also with facile software interfaces in their programming.

2. The Moeller PLC

2.1. Moeller history

In 1899 Franz Klöckner founds a switchgear production company in Cologne, Germany and Hein Moeller joins the firm in 1911.

In 1940 Moeller starts to deliver switchgear to electrical wholesalers and undergoes a change in name to Klöckner-Moeller. Felten & Guilleaume Carlswerk is founded in Kleedorf, Lower Austria.

In 1980s, the companies that make up the Moeller Group are united into Moeller Holding Gmbh & Co. KG, achieved with the help of 7,000 employees in 75 countries. Moeller becomes one of the first suppliers of compact programmable logic controllers (PLCs), with the PS3 PLC being launched on the market to great success.

In the years 2000-2003, the economic downturn all over the world and growing competition on global markets starts to impede the company's development, which is further hampered by the owner's efforts – lasting almost two years – to sell it. Despite these difficulties, Moeller continues to invest in forward-looking production technology and the development of new product series like:

- RMQ-Titan command and signaling devices
- Easy 600/800 Control relays
- NZM 1–4/IZM circuit-breakers
- Rapid Link decentralized motor control system

- MFD-Titan multi-function relays
- LS-Titan position switches
- XC 100/600 control system
- Xpole service installation switching devices
- Xcomfort building automation system

On 2008, the US-American Eaton Corporation acquires 100 percent of the shares in the Moeller Group. Together, the two companies form a significant global player in electrical engineering [2].

2.2. The PLC design XC-CPU100

The XC-CPU100 controllers have been designed for application in machinery and plant control systems having a compact design.

These controllers are fitted with interfaces for connecting to a programming device (RS232) and for linking to decentralized CANopen expansion units, so they can form the core of a comprehensive automation system [3].

The basic unit consists of:

- Rack,
- A CPU for control or visualization, with integral power supply unit (PSU) and local inputs/outputs,
- XIOC signal modules.

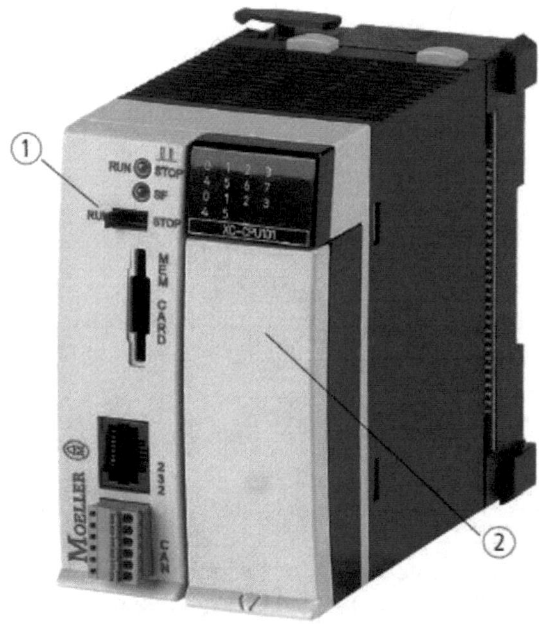

Figure 2. The PLC XC-CPU101

The CPU module of the XC100 (Figure 2) is divided into two functional units:

1. Processor unit with interfaces
2. 24 V PSU with integral digital inputs (8) and digital outputs (6).

The XC-CPU101...(-XV) types of CPU are based on a processor with an integrated CAN interface, and include battery-buffered flash and SRAM memories.

The monitoring of the system voltage ensures that the data-saving routine will be initiated if the voltage goes below a fixed preselected level. The internal real-time clock facilitates time and date dependent control functions.

The available operating and interface control devices are:

- LED display for RUN/Stop and general error
- Operating-mode selector switch RUN/Stop
- RS232 interface, e.g. for programming device interfacing
- CAN open interface as a fieldbus interface
- Interface for a multimedia memory card (MMC).

The multimedia card is used as an optional backup medium for the (boot) project and to save recipe data.

2.3. Local digital inputs/outputs

The 18-pole terminal block which has the power supply to the CPU (Figure 3), the local I/Os and the physical connection to the local inputs/outputs is located on the right half of the CPU. The 8 digital inputs (I0.0 - I0.7) and 6 semiconductor outputs (Q0.0 - Q0.5) are designed for 24 V signals and have a common 0VQ/24VQ power supply which is potentially isolated right up to the bus.

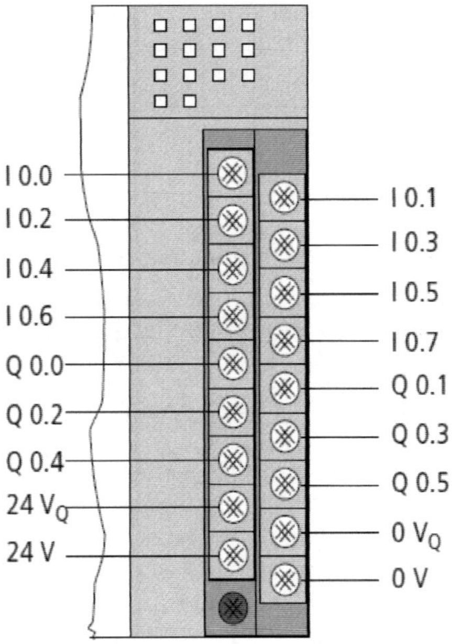

Figure 3. Connections for PSU and local I/O

2.4. The XSoft program

XSoft is a program which gives the PLC programmer an easier access to the powerful IEC language. XSoft provides us with a variety of important functions to debug, test and put quickly and efficiently our control application into operation [4], [5].

Using the editors and running functions is based on the many programs developed and demonstrated from the advanced programming languages (like Visual C ++).

XSoft contains all the programming languages for IEC 61131-3:2013. This suite consists of textual languages, Instruction List (IL) and Structured Text (ST), and graphical languages, Ladder Diagram (LD) and Function Block Diagram (FBD), Sequential Function Chart (SFC), Continuous Function Chart (CFC). The main extensions are new data types and conversion functions, references, name spaces and the object-oriented features of classes and function blocks [1].

XSoft is a system based on CoDeSys programming for industrial automation, conducted in accordance with the international standard. Developed technical features, easy operation and the widespread use of this software in the automation components used by different manufacturers guarantee the success.

It can be tested your application program without being connected to the PLC. XSoft present for this one integrated online simulator. Do not need to get out of current interface and does not change the operation mode against the online command connected.

Into a program are included the following items: POU (Program Organization Unit), characteristics information, views, resources and libraries.

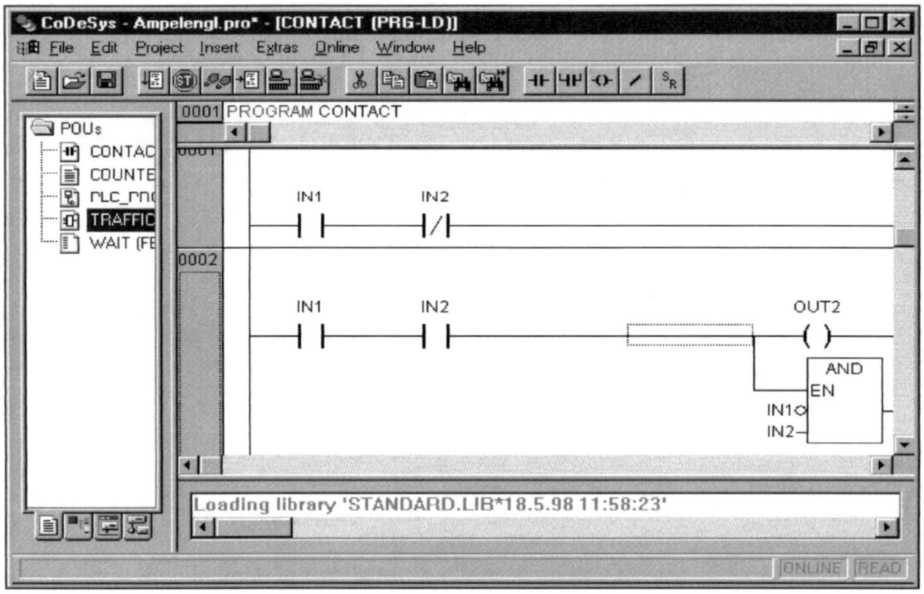

Figure 4. The Ladder Diagram (LD)

All editors of POU programs contain a declaration and a body part, the Ladder Diagram editor is a graphical editor and the most important commands are found in the content menu (Figure 4).

Each network in the Ladder Diagram contains the left side of the network of contacts (contacts are represented by two parallel lines | |) showing the condition "On" or "Off" from left to right. These conditions correspond to the boolean values TRUE and FALSE.

In the right side of the Ladder Diagram network can be multiple coils represented by parentheses ( ), but they may be connected only in parallel. A coil transmits the value of connections from left to right and copying them into a nearest boolean variable [6], [7].

3. The frequency inverter

The frequency inverters used in application are the types DF51-340-3K0 and DF51-340-4K0, from Moeller.

The DF51 frequency inverters (Figure 5) are electrical apparatus for controlling variable speed drives with three-phase motors. They are designed for installation in machines or for use in combination with other components within a machine or system.

The DF51 frequency inverters convert the voltage and frequency of an existing three-phase supply to a DC voltage and use this voltage to generate a three-phase supply with adjustable voltage and frequency. This variable three-phase supply allows stepless variability of three-phase asynchronous motors [8].

The DF51 frequency inverters are not domestic appliances. They are designed for industrial use as system components and presents different functions for control the frequency inverters.

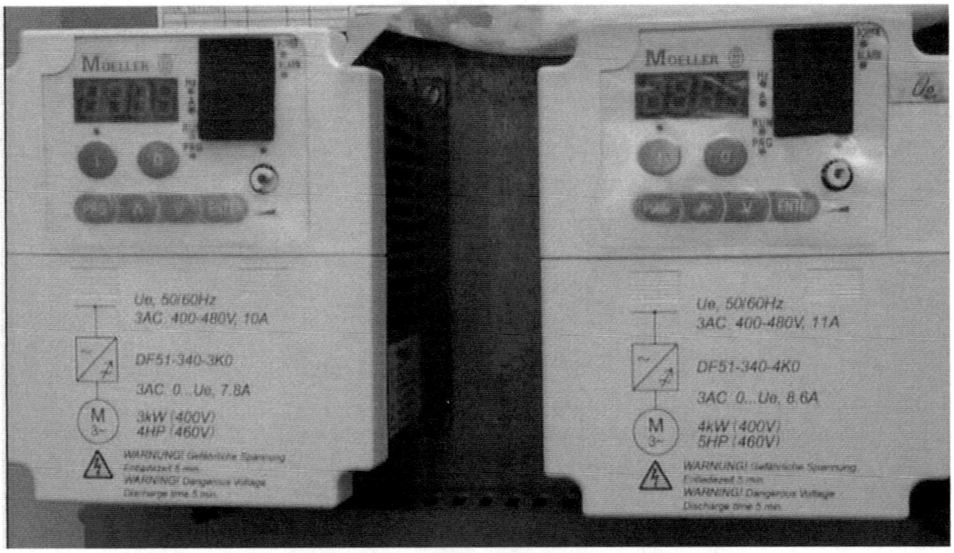

Figure 5. The DF51 frequency inverters

4. The artesian fountain application

4.1. The principle of fountain automation

The artesian fountain (Figure 6) is made with 8 circularly disposed jets, fueled into 2 groups of 4 jets, each group is powered by a 3 and 4 kW pump, controlled by two frequency inverters.

Figure 6. The artesian fountain

The inverter is used to drive the pumps for the fountain jets variation of the intensity, jets intensity being controlled by the Moeller PLC through its analog outputs and by inverters analog inputs.

The artesian fountain also presents an optic fiber lighting system for jets which function only during the evenings. The Fountain is designed for automatic mode based on a time switch and it is set to function between 07:00-23:00 daily.

The fountain has an auxiliary building such as a small waterfall which is controlled by the same automation. This consists in 3 waterfall pumps which are simultaneously operated because of the high water flow rate that need to be used for this to function. These pumps are dependent also on the level sensors which indicate presence of water in the tank.

There were established certain conditions for the operation of the fountain, which are given by the existence of recirculation water corresponding to the pumps tank. There are sensors in order to know if the minimum or maximum level of water in the tank has been reached and to signal possible alarms (lack of water, etc.).

Also, in case of unsuitable weather conditions (powerful wind, storm, ...) the fountain will be stopped. Another condition is that the fountain cannot operate during the winter when there is frost risk. For the winter period the pipes and the water pumps will be emptied.

All these commands and parameters settings of the fountain are performed in a manual operation mode only for the service personnel.

4.2. The phases of operation

In the normal regime of the fountain operation, jets are operated in 3phases, these having various values for the intensity of jets but also different function times.

In the first part of the automation ladder diagram was defined the daily operation period of the fountain, by reading the internal clock of the PLC (using the function rtc1.CDT) and comparing it with the hours' interval 07:00 to 23:00. The comparison values operate on an internal marker MX3.1 which allows or doesn't allow the jets to function (Figure 7).

Figure 7. Defined the daily operation period of the fountain

In the first phase are initialized the analog inputs values of the frequency inverters to a value of 34 Hz (corresponding to a value of 1400). If it drops below this frequency threshold for pump operation, the pumps will enter into the blank area where the pumps do not succeed to realize the pressure for jets.

Initialization at the value 1400 of the analog outputs is made on the outputs by data words type QW2 and QW4 of the PLC (Figure 8). The outputs QW2 and QW4 accept the maximum value of 2080, which corresponds to the voltage value of 10V on the analog outputs of the PLC. The voltage value is transferred to the inverter which makes the engine pump to function at 50 Hz, being the maximum operating frequency of the pump.

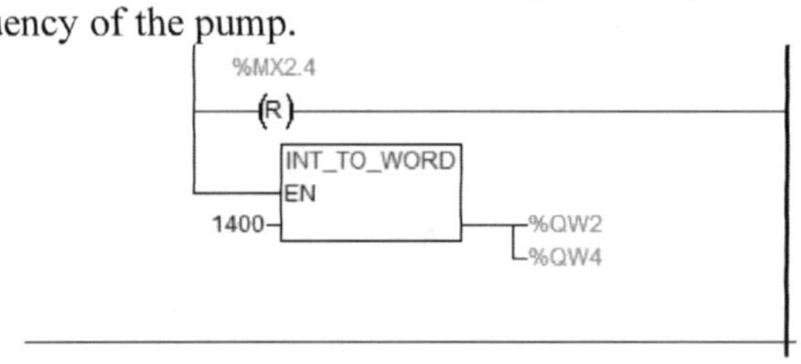

Figure 8. Initializing the analog outputs

In the first phase both pumps are turned on, their jets will ascend step by step at the same time to the maximum value established at 2080 (corresponding to 10V at the analog output), the value of a step is 30 which increments the initial values 1400 of QW2 and QW4 through the ADD block. The moment of maximum value (2080) is given by the comparison GE block which activates the internal marker MX0.7 to stop the analog outputs incrementation. From that moment begins the decrementation of analog output values with the SUB block, all with the same values of 30, like as the incrementing (Figure 9).

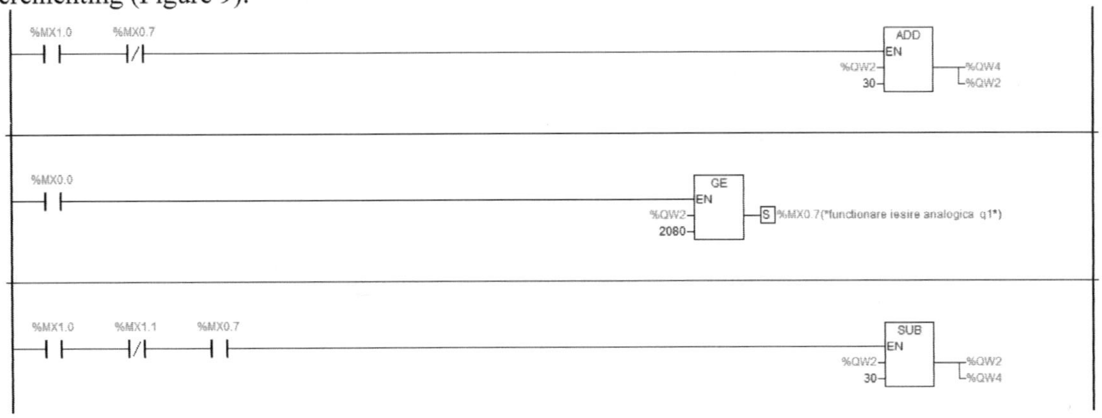

Figure 9. Increment and decrement of the analog outputs QW2 and QW4

The second phase of operation (Figure 10) begins when the jets level reach the minimum set (1400), then only the first pump starts and goes up to the maximum, activated by the analog output QW2. After reaching the maximum level (2080), the amount of analog output QW2 decreases to initial values in steps of 10secs, given by relay timer t3. From that moment the same procedure begins for pump 2.

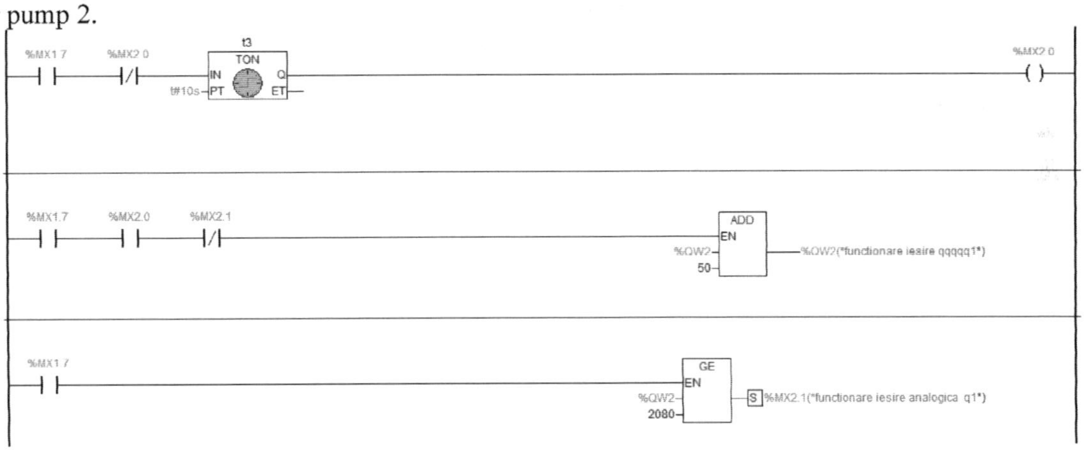

Figure 10. The incrementation of analog outputs QW2 for the second phase

In the third phase of operation the artesian fountains repeat the second phase of operation but with a higher incremented step (50), to make the jets faster.

This also applies to the decrease slope of analog outputs QW2 and QW4, after achieving the initial value for QW4 the internal marker MX2.4 (Figure 11) it's enabled to reboot the operation parameters and to restart the cycle from the beginning.

IOP Conf. Series: Materials Science and Engineering **163** (2017) 012052 doi:10.1088/1757-899X/163/1/012052

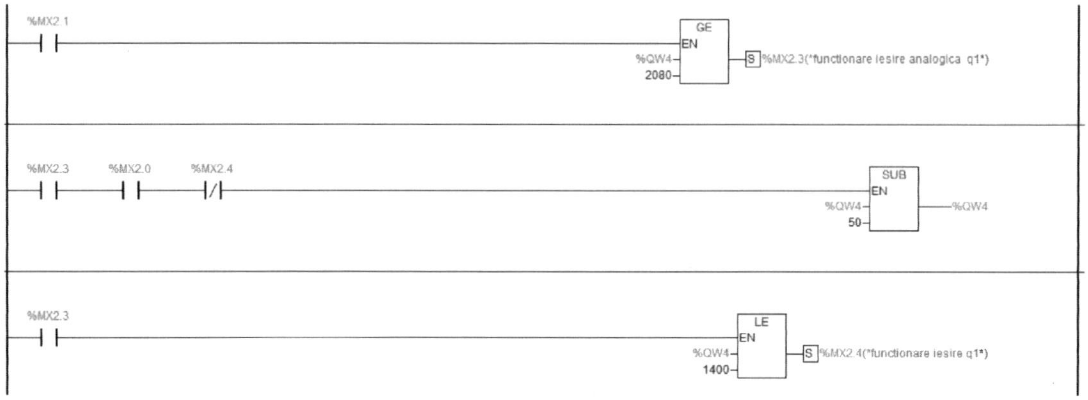

Figure 11. Decrement the analog outputs QW4 for resetting

The reset of the fountain operating parameters (Figure 12) is performed when the jets working cycle is resumed through MX2.4 marker, also after stopping one of the pumps through outputs QX0.0 and QX0.1 or through emergency command QX0.5 coming from the fountains sensors. Resetting brings the involved markers values to 0 and it also initializes the analog outputs QW2 and QW4 values to 1400 (Figure 8).

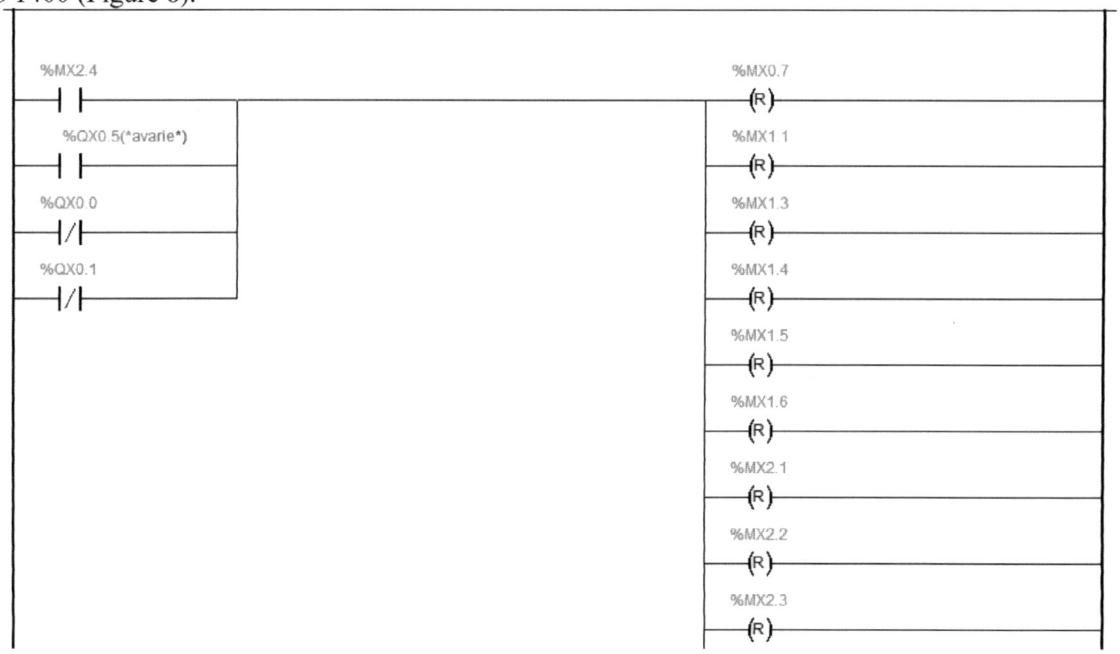

Figure 12. Resetting the artesian fountain parameters

In Figure 13 is presented the stop command of the fountain given by the low water sensor IX0.3 when tanks are not filled to full capacity, in a time of 30 minutes (timer t5). This condition is necessary so that we can control the water losses which could occur during time or the lack of water supply to the fountain.

IOP Conf. Series: Materials Science and Engineering **163** (2017) 012052 doi:10.1088/1757-899X/163/1/012052

Figure 13. Condition for lack of water

5. Graphic interface for WEINTEK panel

To view directly the programs behavior in the PLC XC-CPU101, it has been used a HMI touchscreen from Weintek series.

The application made for the WEINTEK panel (Figure 14) presents the pumps operating state, the status of the PLC inputs and outputs and it can also read and modify the values by the analog outputs of the PLC XC-CPU101. The graphics application for WEINTEK panel was made using EasyBuilder8000 [9], [10].

The main menu of the panel presents the principal information for any user (Figure 14), having also the additional menus (screens) for manual mode service and for real time behavior of the pumps operation where, from the history menu, you can see what had happened a certain time ago (Figure 15).

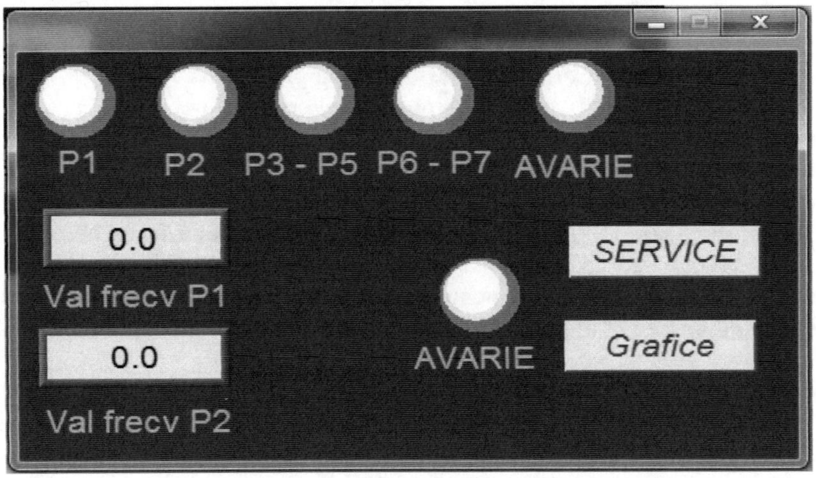

Figure 14. The main menu of interface

The P1 LED indicates the first pumps status in the automatic mode, this is turned on or off depending on the pumps switch key position. LEDs are functional only when the switches key is in automate position and the pumps are controlled by PLC XC-CPU101 and frequency inverters DF5-350-3K0 and DF5-350-4K0.

The fault LED indicates the occurrence of faults from the fountains, waterfalls, filters or frequency inverters.

The numeric indicator ***Val frecv P1*** displays the operating frequency of the pump 1 which is contained between 0 – 50 Hz, and this value is provided by frequency inverter. Similarly, in case of the pump P2.

Service button connects the main menu and the service menu, as well as ***Grafice*** button link the menu graphics.

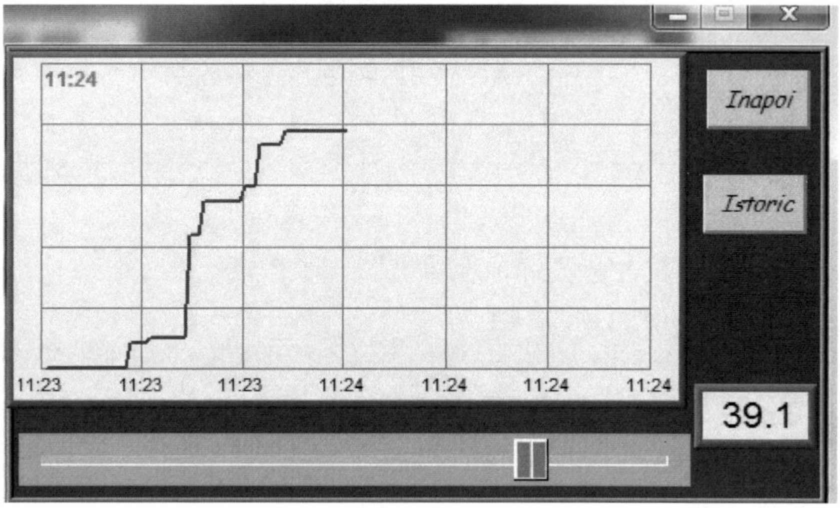

Figure 15. The Graph menu

The menu Graph gives the values plot of the frequency inverter output, to be able to follow how the program works in real time and if there is any anomaly in operation, they can be rectified.

6. Conclusions

Using the program designed in XSoft it can be built complex operation steps of the fountain jets, another main element in the automation being frequency inverters using.

By controlling them through analog inputs, it can easily juggle the motors speeds in the frequency range 0 - 50 Hz, in order to obtain maximum output to drive.

Using the HMI panels is a necessity in applications with PLCs because there is a lack of monitoring and controlling the realized program. with graphical interfaces help of the WEINTEK panel, the PLC CPU XC-101 can be controlled by operator without requiring the use of a PC.

With a minimal configuration of equipment (PLC, HMI, frequency inverters and sensors), can be built almost any automation, everything depending on the program designed in PLC.

References

[1] Chmiel M, Kulisz J, Czerwinski R et al. 2014 An IEC 61131-3-based PLC implemented by means of an FPGA, *Microprocessors And Microsystems* **44** 28-37
[2] ****Moeller History,* http://www.moeller.net/en/company/profile/history/history02.jsp
[3] ****Catalogul X-Sistem (Moeller)*
[4] ****Moeller Xsoft V 2.3 Manual*
[5] ***http://www.moeller.net/support
[6] Paez-Logreira H D, Zamora-Musa R and Bohorquez-Perez J 2015 Programming logic controllers (PLC) using ladder and structured control language (SCL) in MATLAB, *Revista Facultad De Ingenieria* **24**(39) 109-119
[7] Barz C, Oprea C, Erdei Z, Pop-Vadean A, Petrovan F 2014 *The control of an industrial process with PLC,* International Conference on Applied and Theoretical Electricity ICATE, pp 1-4
[8] ****Moeller DF51 Manual*
[9] Barz C, Latinovic T, Balan A, Pop-Vadean A and Pop P P 2015 Using HMI Weintek in

command of an industrial robot arm, *IOP Conf. Ser.: Mater. Sci. Eng.* **85** 012003
[10] Barz C, Ilia M, Ilut T, Pop-Vadean A, Pop P P, Dragan F 2016 Weintek interfaces for controlling the position of a robotic, *IOP Conf. Ser.: Mater. Sci. Eng.* **144** 012018

IOP Conf. Series: Materials Science and Engineering **163** (2017) 012053 doi:10.1088/1757-899X/163/1/012053

Frequency behavior of the residual current devices

Z Erdei[1], M Horgos[1], C Lung[1], A Pop-Vadean[2] and R Muresan[1]

[1]Technical University of Cluj-Napoca, North University Centre of Baia Mare, Department of Electrical, Electronics and Computers Engineering, Victor Babes str., no. 62A, 430083 Baia Mare, Romania.
[2]Technical University of Cluj-Napoca, Department of Mechatronics and Machine Dynamics, Blvd. Muncii, no. 103-105, 400641 Cluj-Napoca, Romania.

E-mail: erdeiz@cunbm.utcluj.ro

Abstract. This paper presents an experimental investigation into the operating characteristic of residual current devices when in presence of a residual current at a frequency of 60Hz. In order to protect persons and equipment effectively the residual current devices are made to be very sensitive to the ground fault current or the touch current. Because of their high sensitivity the residual current circuit breakers are prone to tripping under no-fault conditions.

1. Introduction

Residual current devices (RCDs) also known as residual current circuit breakers are mostly used in low voltage installations as a protective device against electric shock.

The operation of an RCD can be understood by making an analogy to the flow of water from a heating system. A leak may occur when the pipework is damaged or punctured; in the same way a "leak" of electricity may occur when the isolation of a cable in the circuit is damaged or faulty.

The residual current devices monitor the unbalanced current of live conductors from circuits of distribution and in case the breaker detects a dangerous current then that circuit is disconnected quickly. A fault causes current to flow to earth, this current, which does not flow back via the live conductors is called the earth-fault current and it represents the algebraic sum of the instantaneous values of the currents flowing in the live conductors, from where it has its name, residual current [1].

A person may be subjected to an electrical shock in two manners: direct contact – for example the person touches a damaged live conductor, or indirect contact – the person touches a metal part of an electrical machine or device with an insulation fault. The International Electrotechnical Commission elaborated a Time/Current Zones chart for the effects of AC currents on a human body Figure 1. A maximum permissible value of 30mA is considered safe given the current levels considered dangerous.

In order to protect persons and equipment effectively the residual current devices are made to be very sensitive to the ground fault current or the touch current. Because of their high sensitivity the residual current circuit breakers are prone to tripping under no-fault conditions [2]. To ensure a reliable operation of RCDs, it would be necessary to investigate the tripping characteristics of RCDs under different types of unbalanced current [3].

RCDs currently used in Europe are designed and manufactured according to IEC 61008-1 "Residual current operated circuit-breakers without integral overcurrent protection for households and similar uses (RCCBs) – Part 1: General rules" [4]. Most of the residual current devices available on the market have a rated frequency of 50Hz and in some cases 50Hz/60Hz, so the device is suitable for both frequency circuits.

Many incidents of a nuisance RCD tripping are reported every day to companies that produce electronic and electrical devices such as residual current circuit breakers, to electricity suppliers and to specialized personal from maintenance companies. For each report, there is an investigation that in majority cases reveals that there are ground faults that lead to excessive leakage currents in other healthy circuits and cause the RCD breakers to trip.

In this standard, the operating current or sensitivity of RCD breakers under the 50/60Hz current it is specified, along with the construction and operating specifications and the values of the nontripping requirements for RCDs under a surge current.

This nontripping characteristic of the RCDs allows them to be "blind" for a short specified period of time, in which the breaker does not trip. This characteristic can be found at a G-type RCD. This was brought as a defining mark for this type of breaker to avoid nuisance tripping.

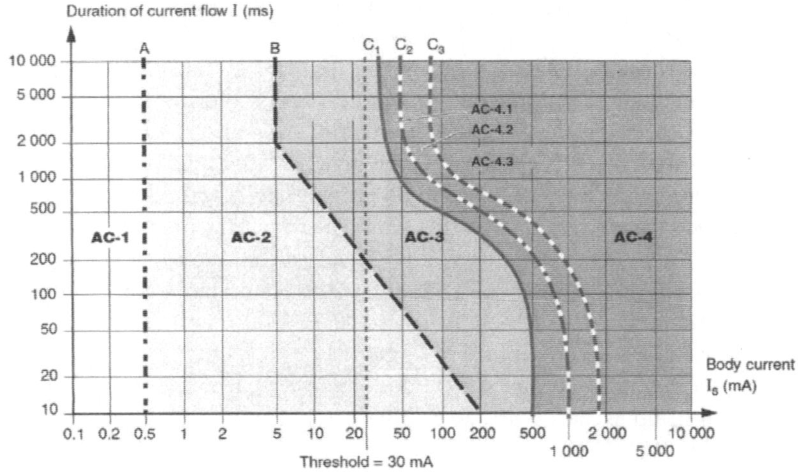

Figure 1. Time/current zones of effects of AC currents (15 Hz to 100 Hz) on persons

However, this standard does not specify the tripping characteristic of RCDs under a current with other waveforms, such as the current containing harmonics, or the surge current with nonstandard waveforms.

A further investigation into the tripping characteristic of the RCDs under a sinusoidal current at rated frequency of 50/60Hz has been carried out by the authors. The RCDs were tested with 50Hz sinusoidal current and with 60Hz sinusoidal current in the laboratory.

This paper presents the results of this experimental investigation as well as discussions. In section I there is a small description of the working principle of an RCD operating system and in section II and III there are presented the tests, the results and the discussions [5].

2. Principle of RCD operation

RCDs are designed to respond to the unbalanced current in a distribution circuit to be protected. When the load is connected to the supply through the RCD, the line and neutral conductors are connected through the primary windings on a toroid transformer.

In this arrangement, the secondary winding is used as a sensing coil and is electrically connected to a sensitive relay or solid state switching device, the operation of which triggers the tripping mechanism. In a healthy and correct circuit, the line and neutral currents are balanced and they produce equal and opposite magnetic fluxes in the toroid core resulting that there is no current generated in the sensing coil [6-8].

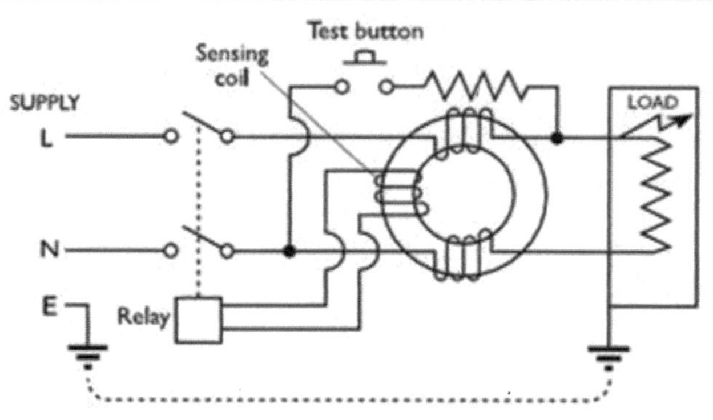

Figure 2. Principle of RCD Operation

When the line and neutral currents are not balanced they create an out-of-balance flux. This will induce a current in the secondary winding of the toroid core, which is used to operate the tripping mechanism. For each RCD breaker, the line and the neutral pass through the toroid. A common cause of nuisance tripping is failure to connect neutral through RCD. This type of breakers is usually build in compliance with single phase, three phase or three phase and neutral circuits.

Tripping current or residual current is an important parameter which characterizes the operating performance of an RCD. There are specified typical values of the rated residual operating current that includes: 6, 10, 30, 100, 300 and 500 mA. The IEC 61008-1 standard requires that the RCD should trip if the value of the residual current is higher than $I\Delta$ and – 100% and the RCD should not trip if the value of the residual current is smaller than $0.5I\,\Delta$ – 50%.

There is another specification in the standard that says that when the RCD is subject to an unbalanced surge current of up to 3kA with the 8/20μs waveform and of up to 200 or 25 ($I\delta \leq 10$ mA) with the 0.5μs/100 kHz ring waveform.

Table 1. Maximum operating times for G and S type RCDs

Type				
G	Rated Residual Current	IΔ	2IΔ	5IΔ
	Operating time[s]	0.3	0.15	0.04
S	Rated Residual Current	IΔ	2IΔ	5IΔ
	Operating time[s]	0.5	0.3	0.15

AC, type A and type B. The type A or type B RCDs are used in a power circuit with the residual current containing a pulsating or smooth dc component. RCD breakers may or may not be equipped with delay time unit – type S - Selective is a breaker with a short time delay and type G - General is a breaker that has an instantaneous tripping reaction. In this paper there is brought to discussion the general RCD breaker – type G.

The tripping unit of a residual current device contains an electromagnetic relay, a circuit board and a toroid sensing coil. The fault current, via the toroid, supplies energy to an electromagnet whose moving part is held by a permanent magnet. When the operating threshold is reached, the electromagnet counterbalances the attraction of the permanent magnet and the moving part, drawn by a spring, opens the magnetic circuit and mechanically actuates circuit-breaker opening.

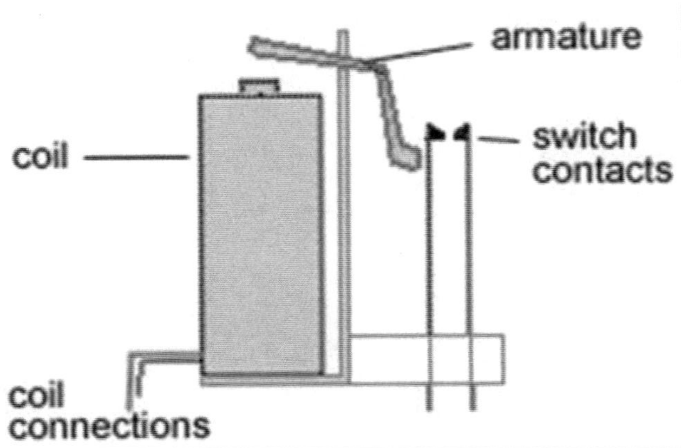

Figure 3. Electromagnetic relay

An RCD is a safety device. Whatever the technology used, it must always be equipped with a test system. Although RCDs without auxiliary sources are the most reliable, implementation of fail-safe systems on RCDs with auxiliary sources offers an enhanced degree of safety that does not, however, replace the periodical test.

For a test, a current is generated that flows in only one of the live conductors surrounded by the toroid. The resistor is sized to let through enough current to trip the RCD, taking into account any leakage currents likely to reduce the test current. The maximum permissible value is 2.5 times $I\Delta n$ (for an adjustable device, $I\Delta n$ is the lowest possible setting).

The above principle is very common because it is the means to check the entire system, i.e. toroid, relay and breaking device. It is used on earth-leakage protection socket-outlets and on residual-current circuit breakers with and without integral overcurrent protection. With respect to residual-current relays with separate toroid, the same principle is sometimes used. Certain relays, for example Merlin Gerin Vigirex relays, are equipped with a built-in "test" function and also continuously monitor the continuity of the detection circuit (toroid/relay link and toroid winding).

The electromagnetic relay consists of a multi-turn coil, with an iron core, together forming an electromagnet. If the coil is energized, by the current passing through it, then the core becomes temporarily magnetized. This magnetized core attracts the iron armature. When the coil is de-energized the armature and contacts are released [9], [10].

The sensor is a toroid transformer. It surrounds all the live conductors and is therefore excited by the magnetic field corresponding to the algebraic sum of the currents flowing in the phases and neutral. The current induced in the toroid and the electrical signal at the terminals of the secondary winding is therefore proportional to the residual current. This type of sensor can detect residual currents from a few mill amperes up to several tens of amperes.

3. Laboratory test setup

For the tests presented in this paper there were used 40 RCD breakers of one brand name. These samples are nonelectronic breakers type G/A, manufactured and assembled according to IEC 61008-1 standard.

All samples have a sensitivity of 100mA and a rated frequency of 50Hz. In this experiment, all samples were subjected to a sinusoidal current waveform at rated frequency and also at a frequency of 60Hz. There were recorded the values of the tripping time when applied a current $5I\,\Delta$ for all samples.

The laboratory setup used in order to test the breakers it is composed from a current injection system, an oscilloscope and time-measure equipment. The current injection system it is an amplifier that has the capacity to amplify the value of the rated residual current by five, resulting a residual current with a value of 500mA. The current injection system was connected to the terminals of each

RCD breaker tested. The oscilloscope used for the tests presented in this paper had a bandwidth from dc to 200MHz. The time-measure equipment was a digital device with a precision of 0.01s.

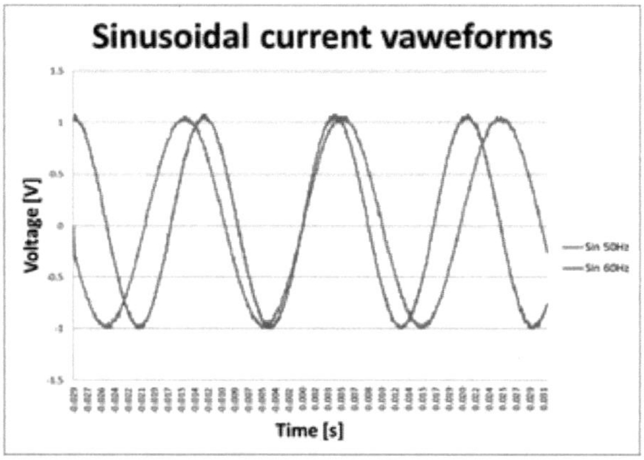

Figure 4. 50Hz and 60Hz sinusoidal current

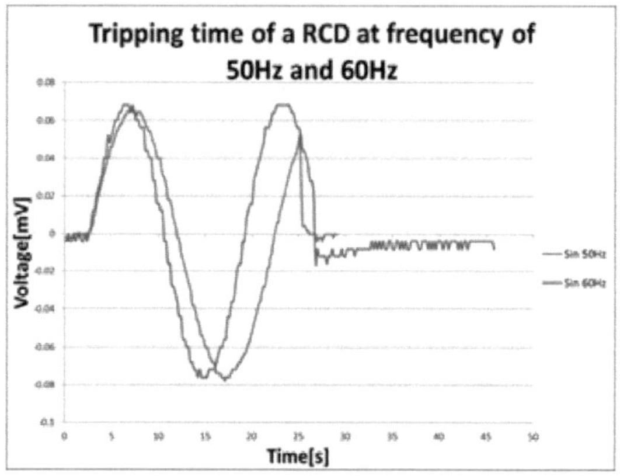

Figure 5. Tripping time of a breaker at 50Hz and 60Hz frequency

4. Steady-State current tests

The shape difference between a sinusoidal current at a frequency of 50Hz and another current sinusoidal with a frequency of 60Hz is highlighted in Figure 4. The shapes of the current were recorded with an oscilloscope. This representation is just for showing purposes and to have in mind the 2 waveforms.

An unbalanced current with a frequency of 50Hz was applied to the RCD sample under test. The test performed on the RCD breaker was a tripping time test, which means that a specific value of the current is applied to the RCD and that the breaker needs to trip in a specified range of time. In case of the G-type and S-type RCDs, the values of the tripping time are presented in Table I.

When applied a sinusoidal current of 500mA, the breaker must trip within 40ms, for 50Hz and 60Hz but not faster than 8.3ms.

In Figure 5 is presented the tripping time of a breaker for 50Hz (blue) and for 60Hz (red). The data presented in Figure 5 was recorded with an oscilloscope. As shown in the graphic from Figure 5, it can be observed that the breaker has a small delay when applied a sinusoidal current of 60Hz than when applied a sinusoidal current of 50Hz.

When applying a sinusoidal current that has a constant value of 500mA at a frequency of 50Hz, the breaker will trip after 1 full period.

In the case of a constant current at 60Hz frequency the breaker will be slightly delayed, tripping after 1 and a half periods. Although, there is a half a period difference between the two tripping time situations, the value of the tripping time is extremely close, for 50Hz frequency the breaker tripped in 25ms and for 60Hz frequency the RCD tripped in 26ms.

Taking this experiment to a larger scale, in the next paragraphs it is presented the tripping characteristic of 40 identical breakers. When testing several breakers that have the same specifications, in the same conditions and with the same equipment the results can highlight the similar behavior of the samples during the tests.

The test can be explained in easy steps:

- A constant unbalanced current with a sinusoidal shape is applied to the terminals;
- At the terminals it is measured the tripping time of the breaker in presence of a residual current with a value of 5xIΔn;
- The current shape is monitored on the oscilloscope;

- The value of the tripping time it is measured with a high sensitive chronometer that has a tolerance of 0.01ms;

In Figure 6 there is a graphic of the breaking times of 40ps. RCD breakers G-type when applying a constant sinusoidal current with a value of 500mA, at a frequency of 50 Hz. This test was performed in the same conditions as the first two tests presented in the paper. The average value of tripping time for 40 breakers is 24ms. In Figure 7 is presented the graphic of tripping time for the same 40 RCD breakers G-type when applied a constant sinusoidal current with a value of 500mA at frequency 60Hz.

The values of the tripping time are close with the ones from tripping time 50Hz current. The average value of the tripping time at 60Hz is 24.7ms.

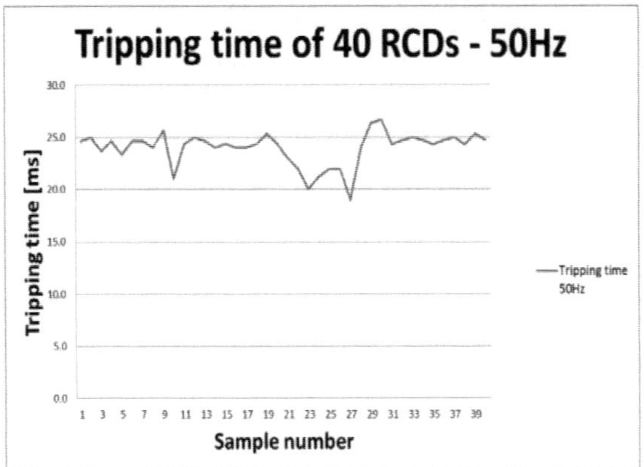

Figure 6. Breaking times at 50Hz for 40 RCDs

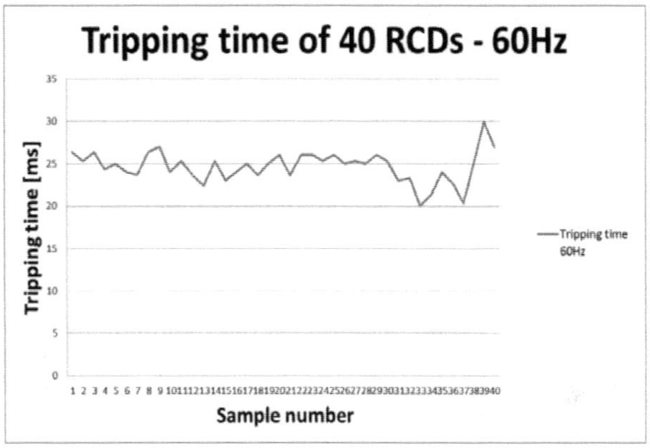

Figure 7. Breaking times at 60Hz for 40 RCDs

5. Conclusions

In conclusion, when applying a constant sinusoidal current constant at a frequency of 60Hz for a G-type breaker, this breaker will be delayed within the limit of a 1ms.

The influence of the toroid transformer, the electromagnetic relay and the electronics board are crucial in operating a residual current device.

References

[1] Boyd R R 2002 *Tolerance analysis of Electronic Circuits using Matlab*, CRC Press, Washington, USA

[2] Dumitriu L and Iordache M 1998 *Teoria moderna a circuitelor electrice - Vol. I – Fundamentare teoretica, Aplicatii, Algoritmi si Programe de calcul,* Editura All Educational S.A., Bucuresti, Romania

[3] Iordache M and Dumitriu L 2000 *Teoria moderna a circuitelor electrice - Vol. II – Fundamentare teoretica, Aplicatii, Algoritmi si Programe de calcul,* Editura All Educational S.A., Bucuresti, Romania

[4] Iordache M and Dumitriu L 2002 *Simularea asistata de calculator a circuitelor analogice*, Editura Politehnica Press, Bucuresti, Romania

[5] Felic G K, Ng D and Skafidas E 2013 Investigation of Frquency Dependent Effects in Inductive Coils for Implantable Electronics, *IEEE Transactions on Magnetics* **49**(4) 1353-1360

[6] Chiver O, Micu E and Barz C 2008 *Stator winding leakage inductances determination using finite elements method*, OPTIM 2008, Proceeding of the 11 th International Conference on Optimization of Electrical and Electronic Equipment, vol. III – Industrial Automation and Control, Edited by The Transilvania University of Brasov, Faculty of Electrical Engineering and Computer Science, Braşov, Romania, May 22-23, pp 69-74

IOP Conf. Series: Materials Science and Engineering **163** (2017) 012053 doi:10.1088/1757-899X/163/1/012053

[7] Chiver O, Neamt L, Barz C and Costea C 2014 *Frequency domain numerical analysis of rotor cage induction motor*, Proceedings of the International Conference and Exposition on Electrical and Power Engineering, Iasi, Romania, October 16-18, pp 327-331

[8] Neamt L, Chiver O, Barz C, Costea C and Erdei Z 2014 *Considerations about power system grounding for different soil structure*, Proceedings of the International Conference and Exposition on Electrical and Power Engineering, Iasi, Romania, October 16-18, pp 1034-1038

[9] Oprea C and Barz C 2011 Contribution to automate regulation after the speed of actions with asynchronized synchronous motor, *Annals of Faculty Engineeering Hunedoara-International Journal of Engineering* **IX** 111-122

[10] Barz C and Oprea C 2011 Study of electromagnetic field in claw-poles alternator, *Annals of Faculty Engineeering Hunedoara-International Journal of Engineering* **IX** 359-362

Institute of Physics
Dirac House, Temple Back
Bristol BS1 6BE UK

ISSN: 1757-8981
ISBN 978-1-5108-3526-9

CURRAN ASSOCIATES INC.
proceedings.com